Retrovirus-Cell Interactions

Retrovirus-Cell Interactions

Edited by

Leslie J. Parent
Penn State College of Medicine
Hershey, PA, United States

ELSEVIER

ACADEMIC PRESS
An imprint of Elsevier

Academic Press is an imprint of Elsevier
125 London Wall, London EC2Y 5AS, United Kingdom
525 B Street, Suite 1650, San Diego, CA 92101, United States
50 Hampshire Street, 5th Floor, Cambridge, MA 02139, United States
The Boulevard, Langford Lane, Kidlington, Oxford OX5 1GB, United Kingdom

Library of Congress Cataloging-in-Publication Data
A catalog record for this book is available from the Library of Congress

British Library Cataloguing-in-Publication Data
A catalogue record for this book is available from the British Library

ISBN: 978-0-12-811185-7

Publisher: Andre Gerhard Wolff
Acquisition Editor: Linda Versteeg-Buschman
Editorial Project Manager: Pat Gonzalez
Production Project Manager: Mohanapriyan Rajendran
Cover Designer: Christian J. Bilbow
Conceptualization of Cover Image: Stephen P. Goff

Typeset by TNQ Technologies

This book is dedicated to those people and their families who have been affected by retrovirus infections; scientists who have worked throughout the years to study retroviruses and develop preventive strategies and treatments; and students who will follow in their footsteps to find a cure.

Contents

List of Contributors xiii
Author Biographies xv
Preface xxvii
Introduction xxix

1. Retrovirus Receptor Interactions and Entry

Lorraine M. Albritton

Envelope Glycoproteins: Domain Structure 2
Retrovirus Entry Receptors 4
Virus Attachment 10
The Basics of Membrane Fusion 13
Do Receptor Interactions Contribute to the Envelope Proteins Functions That Drive Entry? 16
Do Env–Receptor Interactions Contribute to Pathogenesis? 23
Host Defenses That Inhibit Retroviral Entry Also Drive Envelope Protein Variation 26
Coevolution of Virus and Receptor 28
Captured Retroviral Envelope Proteins in the Development of Mammalian Placenta and Male Muscle Mass 30
References 31

2. Cellular Factors That Regulate Retrovirus Uncoating and Reverse Transcription

Stephen P. Goff

A Brief Description of Early Events of Infection 53
Monitoring the Course of Infection in Early Stages 56
Host Factors Promoting Early Events of Infection 62
Host Factors Restricting Infection in Early Stages of Infection 71
Sensing and Responding to Infection: Innate Immunity 85
Applications for Therapy 88
Conclusions: Parting Words 89
References 90

3. Nucleoporins in Retroviral Replication: What's Nup Got to Do with It?

Anne Monette and Andrew J. Mouland

Nuclear Pore Complexes, Nucleoporins, and Nucleocytoplasmic Transport 114
Nucleoporins Implicated in Retroviral Replication 115
Caution: Depletion of Any Single Multifunctional Nup Can Modify Overall NPC Architecture 138
The Importance of FG-Repeat Nups to Disease 139
Viruses Commonly Cause the Displacement of FG–Nups to Remodel NPCs 142
Acknowledgments 143
References 143

4. Virus–Host Interactions in Retrovirus Integration

Alan N. Engelman and Goedele N. Maertens

Introduction 163
Integrase-Interacting Proteins 167
Lentiviruses 167
Lens Epithelium–Derived Growth Factor/p75 167
Hepatoma-derived Growth Factor Like 2 (HDGFL2) 170
Gammaretroviruses 170
Bromodomain and Extra-Terminal Domain Proteins 171
Deltaretroviruses 172
B′-Protein Phosphatase 2A (PP2A) 173
Gag-Interacting Proteins 177
HIV-1 CA–Interacting Proteins 177
Foamy Virus 179
Allosteric IN Inhibitors That Target the LEDGF/p75–IN Interaction 180
Conclusions 182
Acknowledgments 183
References 183

5. Transcriptional Control and Latency of Retroviruses

Bryan C. Nikolai and Andrew P. Rice

Introduction 200
Retroviral Integration Occurs at Transcriptionally Active Sites in Host Genome 200
Transcriptional Regulation by Endogenous Retroviruses 201
Retroviral Genome Complexity Confers a Benefit for Transcriptional Regulation 203
Transcriptional Control of HIV-1 209
HIV-1 Latent Infection 214

Future Research of Retroviral Transcription 217
Acknowledgments 218
References 219

6. Teetering on the Edge: The Critical Role of RNA Processing Control During HIV-1 Replication

Ahalya Balachandran, Liang Ming and Alan Cochrane

Introduction 229
Role of hnRNPs in the Regulation of HIV-1 RNA Processing 231
Role of SR Proteins in the Regulation of HIV-1 RNA Processing 236
Manipulation of HIV-1 RNA Processing With Small Molecules 240
Acknowledgments 245
References 245

7. Cellular RNA Helicases Support Early and Late Events in Retroviral Replication

Gatikrushna Singh, Xiao Heng and Kathleen Boris-Lawrie

RNA Helicase and Retroviruses in the Advent of "Omics Technology" 258
Early Events: Reverse Transcription and Integration 259
DHX9/RNA Helicase A Activity in the Genomic RNP 259
MOV10 Activity in Virions Remains Undefined 262
Late Events: Provirus Transcription, Primary RNA Processing, Export, Translation, Formation of Genomic RNP 262
Nuclear Cap–Binding Proteins and RNA Helicase: Translation Evading Nonsense RNA–Mediated Decay 265
Steady-State Translation: Switching 5′Cap–Binding Proteins to Gain eIF4E 266
Therapeutic Targeting at the Interface of RNA Helicase and Cognate Retroviral RNA 267
References 268

8. Role of Host Factors in the Subcellular Trafficking of Gag Proteins and Genomic RNA Leading to Virion Assembly

Eunice C. Chen and Leslie J. Parent

Introduction 274
Nuclear Export of Viral RNAs 275
Retroviral Genome Trafficking in the Cytoplasm 279
Retroviral Gag Protein Trafficking 281
Factors That Interact With Gag or Viral RNA to Restrict Virus Replication 296
Concluding Remarks 298
Acknowledgments 299
References 299

9. Tumor Suppressor Gene 101: A Virus' Multifunctional Conduit to the ESCRT Trafficking Machinery

Carol A. Carter

Introduction 318
Tumor Suppressor Gene 101 Structure 323
Tumor Suppressor Gene 101 Participation 328
Cellular Factors That Work With Tumor Suppressor Gene 101 330
Tumor Suppressor Gene 101 Role in Budding of Other Viruses 335
Questions Remaining 339
Acknowledgments 340
References 340

10. The Role of Lipids in Retroviral Replication

Abdul A. Waheed and Eric O. Freed

Introduction 353
Lipids in Retroviral Entry 360
Retroviral Assembly 363
Lipids in Retroviral Cell–Cell Transfer 375
Inhibition of Retroviral Replication With Lipid-Modifying Agents 379
Concluding Remarks 381
References 382

11. Cellular Immune Responses to Retroviruses

Susan R. Ross

Introduction 401
Steps of the Retrovirus Infection Pathway Targeted by the Host Intrinsic/Innate Response 402
Host Pathways Implicated in Cellular Control of Retrovirus Infection 406
Subversion of Cellular Immune Responses by Retroviruses 412
Conclusions 414
References 414

12. Noncoding RNAs in Retrovirus Replication

Alice A. Duchon and Karin Musier-Forsyth

Introduction 421
Host RNAs Packaged Into Virions 423
Long Noncoding RNA and Endogenous Retroviruses 434
Functional Transactivating Response Element RNAs 439
RNA Interference and miRNAs 440
Therapeutics 450
Concluding Remarks 453
Acknowledgments 453
References 453

13. Cellular Control of Endogenous Retroviruses and Retroelements

Suzanne Sandmeyer

Introduction	480
Background	480
Activity	486
Control of Retroelement Expression	488
Regulation of Retroelements by the Innate Immune System	502
Perspectives	507
Acknowledgments	509
References	510

14. Strategies to Discover Novel Cellular Factors Involved in Retrovirus Replication

Rebecca J. Kaddis Maldonado, Eunice C. Chen and Leslie J. Parent

Introduction	528
Identifying Protein–Protein Interactions in Retroviruses Using Two-Hybrid Screens	528
Mass Spectrometry Approaches to Discovery of Retrovirus–Cell Protein–Protein Interactions	536
Mass Spectrometry Using Viral DNA as Bait	541
Mass Spectrometry Using Viral RNA as Bait	542
RNA Interference and Genome-Wide Screens to Assess the Contribution of Host Factors to Virus Replication	544
Host Genome Editing Using CRISPR/Cas-9 to Find Retrovirus Dependency Genes	549
Gain-of-Function Genetic Approaches Using cDNA Overexpression Screens	550
Identification of Cellular RNAs That Interact With Retroviruses	550
Concluding Remarks	554
Acknowledgments	560
References	560

Index 569

List of Contributors

Lorraine M. Albritton, The University of Tennessee Health Science Center, Memphis, TN, United States

Ahalya Balachandran, University of Toronto, Toronto, ON, Canada

Kathleen Boris-Lawrie, University of Minnesota, Saint Paul, MN, United States

Carol A. Carter, Stony Brook University, Stony Brook, NY, United States

Eunice C. Chen, Penn State College of Medicine, Hershey, PA, United States

Alan Cochrane, University of Toronto, Toronto, ON, Canada

Alice A. Duchon, The Ohio State University, Columbus, OH, United States

Alan N. Engelman, Dana-Farber Cancer Institute, Boston, MA, United States; Harvard Medical School, Boston, MA, United States

Eric O. Freed, National Cancer Institute, Frederick, MD, United States

Stephen P. Goff, HHMI, Columbia University, New York, NY, United States

Xiao Heng, University of Missouri, Columbia, MO, United States

Rebecca J. Kaddis Maldonado, Penn State College of Medicine, Hershey, PA, United States

Goedele N. Maertens, Imperial College London, London, United Kingdom

Liang Ming, University of Toronto, Toronto, ON, Canada

Anne Monette, Lady Davis Institute at the Jewish General Hospital and McGill University, Montréal, QC, Canada

Andrew J. Mouland, Lady Davis Institute at the Jewish General Hospital and McGill University, Montréal, QC, Canada

Karin Musier-Forsyth, The Ohio State University, Columbus, OH, United States

Bryan C. Nikolai, Baylor College of Medicine, Houston, TX, United States

Leslie J. Parent, Penn State College of Medicine, Hershey, PA, United States

Andrew P. Rice, Baylor College of Medicine, Houston, TX, United States

Susan R. Ross, UIC College of Medicine, Chicago, IL, United States

Suzanne Sandmeyer, University of California, Irvine, CA, United States

Gatikrushna Singh, University of Minnesota, Saint Paul, MN, United States

Abdul A. Waheed, National Cancer Institute, Frederick, MD, United States

Author Biographies

CHAPTER 1

Lorraine M. Albritton, PhD, is Professor of Microbiology, Immunology, and Biochemistry in the College of Medicine at the University of Tennessee Health Sciences Center. She earned a Bachelor of Science degree in Physics magna cum laude at Northeast Louisiana University and then attended graduate school at The University of Tennessee Oak Ridge National Laboratories. Earning her doctoral degree as a visiting graduate student with David Housman, PhD, in the Biology Department of Massachusetts Institute of Technology, she completed her postdoctoral research fellowship in molecular retrovirology with James Cunningham, MD, at Harvard Medical School, during which she isolated and characterized the host entry receptor for the ecotropic murine leukemia viruses. She has been honored with the Excellence in Teaching Award from the Student Government Association Executive Council and is Codirector of Peer-Coaching for Educators. She is a member of the NIH Director's Recombinant DNA Advisory Committee and the Editorial Board of the *Journal of Virology*, and has served as a member of NIH review committees. Dr. Albritton's research, funded by the National Institutes of Health, thc Department of Defense and the West Cancer Research Foundation, focuses on understanding the molecular mechanisms of envelope glycoprotein function and retrovirus–host cell interactions during entry, and developing methods to restrict retroviral and lentiviral vector entry to cell-type specific receptors for targeted gene delivery.

CHAPTER 2

Stephen P. Goff, PhD is Higgins Professor of Biochemistry at the College of Physicians and Surgeons of Columbia University and an Investigator of the Howard Hughes Medical Institute. He received the AB degree in Biophysics Summa Cum Laude from Amherst College in 1973. His graduate work with Dr. Paul Berg at Stanford University focused on the genetic analysis of the replication of simian virus 40 (SV40) and its use as a viral vector for the expression of foreign DNAs in mammalian cells. He did postdoctoral work with Dr. David Baltimore at MIT on the replication of the murine leukemia viruses as a Jane Coffin Childs fellow, and joined the Columbia faculty in 1981. Goff's current work is centered on the study of the retrovirus life cycle and the host restriction systems that inhibit virus replication. His lab has identified and characterized a novel host protein, termed ZAP for zinc finger antiviral protein, which blocks gene expression of murine leukemia viruses, Ebola, Sindbis, and HIV-1, by degrading viral mRNAs and inhibiting their translation. The lab has also characterized a protein complex responsible for the silencing of retroviral DNAs in embryonic stem (ES) cells, and identified a zinc finger protein, ZFP809, as an ES-cell specific recognition molecule that binds the proviral DNA and brings TRIM28 to locally modify chromatin. Goff was a Searle Scholar and has received two MERIT awards from the National Institutes of Health. He was elected to membership in the National Academy of Science, the Institute of Medicine, the American Academy of Arts and Sciences, and the American Academy of Microbiology, and he is a fellow of the American Association for the Advancement of Science. He received an honorary Doctor of Science degree from Amherst College in 1997 and was the inaugural recipient of the Retrovirus Prize.

CHAPTER 3

Andrew J. Mouland, PhD, is a Professor of Medicine at McGill University in Montréal, Canada. He earned his undergraduate degree at McGill University in nutrition and then a PhD in endocrinology with a later focus on mRNA translation and metabolism. His postdoctoral training included the study of translational control in *Drosophila melanogaster* in the Department of Genetics at Case Western University in Cleveland. Dr. Mouland began work on the molecular virology of HIV-1 as a postdoctoral fellow at the Université de Montréal and then as an independent researcher in 1998. Dr. Mouland joined McGill University in late 1999 and has made key contributions to the field of HIV-1 RNA fate and metabolism and provoked studies on how HIV-1 commandeers host cell machineries and counters antiviral stress responses. His work has been funded continuously by the Canadian Institutes of Health Research, Canadian Foundation for HIV-1/AIDS Research, and the Fonds pour la recherche en santé du Québec.

Anne Monette, MSc, PhD, earned her BSc (2004) and MSc degrees (2006) at Concordia University in cell and molecular biology with focus on bacterial genetics. As a Canadian Institutes of Health Research (CIHR) scholar, Dr. Monette then earned her PhD in molecular virology at McGill University, with a focus on biochemistry and imaging techniques for the discovery of novel host nucleocytoplasmic shuttling proteins and nucleoporins required for HIV-1 replication (2011). Dr. Monette was supported by Mitacs during her pursuant postdoctoral fellowship (PDF) training in cancer immunology at the Université de Montréal (2017) toward vaccine development, and was funded by CIHR for the development of biomaterials for immunotherapies. During her PDF, Dr. Monette coupled numerous advanced imaging, biochemical, transcriptomic, and bioinformatic techniques, toward discovery of immune checkpoint mechanisms and liquid biopsy-based immune biomarkers across various cancers. Currently, Dr. Monette aims to understand mechanisms of circulating and tissue infiltrating prognostic immune programs common to differing pathophysiologies and is a member of two Society for Immunotherapy of Cancer (SITC) initiatives pursuing biomarkers of cancer immunotherapy responsiveness.

CHAPTER 4

Alan N. Engelman, PhD has studied retroviruses since his graduate school training with Naomi Rosenberg at Tufts University School of Medicine, where he identified temperature-sensitive mutants of the Abelson murine leukemia virus protein tyrosine kinase. Dr. Engelman began work on retroviral integration while serving as a postdoctoral fellow with Drs. Robert Craigie and Kiyoshi Mizuuchi at the US National Institutes of Health, and has continued to focus on integration since establishing his own lab at the Dana-Farber Cancer Institute/Harvard Medical School in 1995. Key contributions of the Engelman lab include the structural basis for retroviral DNA integration, the roles played by virus-interacting host factors LEDGF/p75 and CPSF6 in HIV integration targeting, and the mechanism of action of the allosteric class of HIV-1 integrase inhibitors. Dr. Engelman is the recipient of a MERIT award from the National Institutes of Health, and is an elected Fellow of the American Association for the Advancement of Science and of the American Academy of Microbiology.

Goedele N. Maertens, PhD started her scientific career by measuring intracellular diffusion of fluorescently tagged proteins inside living cells under the supervision of Professors Yves Engelborghs and Zeger Debyser at the Katholieke Universiteit Leuven (Belgium). Her subject under investigation was HIV-1 integrase and its host factor LEDGF/p75. Two years into her PhD she joined the laboratory of Prof. Alan Engelman at the Dana-Farber Cancer Institute/Harvard Medical School (USA) where she further investigated the role of LEDGF/p75 in the cellular localization of HIV-1 integrase and identified a novel binding partner of LEDGF/p75. In 2005, she became a postdoctoral fellow at Cancer Research UK in the laboratory of Dr. Gordon Peters where she investigated the modulation of Polycomb Group proteins that repress the tumor suppressor locus *INK4A*. In 2009, she joined the lab of Prof. Peter Cherepanov at Imperial College London (UK) where she solved the X-ray structures of the foamy virus target capture and strand transfer complexes, and followed the integration process *in crystallo*. She also characterized binding partners of human TNPO3 and characterized the interaction interface and role of TNPO3 in nuclear import of its substrates. Since establishing her lab at Imperial College London in 2011, she identified a novel host factor for delta-retroviral integration. Dr. Maertens is currently a Senior Lecturer at Imperial College London and is a Wellcome Trust Investigator.

CHAPTER 5

Andrew P. Rice, PhD received a BA in Biology from the University of California at Santa Cruz and a PhD in Biology from Brandeis University. Following postdoctoral studies at the University of Cambridge and the Imperial Cancer Research Fund (London), Dr. Rice was a Staff Scientist at Cold Spring Harbor Laboratories. He joined the faculty at Baylor College of Medicine in 1990 and is currently the Nancy Chang Professor in the Department of Molecular Virology and Microbiology. Dr. Rice's research interest, from his PhD studies to the present, has been the study of viral–host interactions.

Bryan C. Nikolai, PhD received his Bachelor of Science in Cell Biology and Biochemistry and a Master of Biological Science graduate degree from University of Minnesota. Dr. Nikolai attended Baylor College of Medicine for his PhD and is currently a Postdoctoral Fellow of the Brockman Medical Research Foundation in the Department of Molecular and Cellular Biology at Baylor. The primary interest of Dr. Nikolai's research is eukaryotic gene transcription and host cell coregulators involved in viral latency and immune evasion.

CHAPTER 6

Alan Cochrane, PhD initiated his studies in HIV-1 during his postdoctoral training with Dr. Craig Rosen at the Roche Institute of Molecular Biology. During that time, his work was focused on understanding the mechanism of action of Rev and characterization of its interaction with the Rev Responsive Element (RRE) within viral RNA. Since establishing his own laboratory, first at McGill University then the University of Toronto, Dr. Cochrane's work has expanded to include the analysis of HIV-1 RNA splicing, transport, and translation and the role that host factors play in regulating these processes. More recent studies have focused on the identification of small molecule inhibitors of HIV-1 replication that function by modulating the posttranscriptional regulation of viral RNA. He has been the recipient of multiple awards including FRSQ Chercheur Boursier, MRC Scholar, and Ontario HIV Treatment Network Scientist.

Ahalya Balachandran, MSc, ventured into the field of virology with an interest in exploring retrovirus manipulation of host cell machinery and new therapeutic strategies to prevent viral infection. She completed her graduate training with Dr. Alan Cochrane at the University of Toronto, where her thesis project focused on the characterization of small molecules that alter RNA processing and examination of their effect on HIV gene expression and replication. She has a keen interest in engaging community members in scientific topics through outreach groups such as Let's Talk Science and Science Rendezvous and as a Science Experience Host at the Ontario Science Centre. Ahalya Balachandran is a recipient of a Kalpesh Oza New Investigator Award from the Canadian Association for HIV Research.

Liang Ming, Ph.D. received his undergraduate degree from Zhejiang University, MSc from Beijing University, China, and completed his PhD training under the supervision of Dr. Howard Lipshitz at the University of Toronto doing work that is focused on understanding the mechanisms by which the transcription factor Hindsight controls *Drosophila* development. On joining the Cochrane lab, his work centered on identifying human host factors crucial for regulating viral gene expression, as well as testing the effects of small molecules in inhibiting HIV-1 replication in humanized mouse model to develop novel antiretroviral therapies. He was the member of the Botanical Society of China, Genetics Society of America, and Canadian Society of Microbiologists.

CHAPTER 7

Kathleen Boris-Lawrie, PhD, an elected fellow of the American Academy of Microbiology and the American Association for the Advancement of Science, is an appointed member of the NIH Director's Recombinant DNA Advisory Committee and frequently serves the NIH Center for Scientific Review. She received her PhD in Molecular Genetics at the George Washington University School of Medicine while holding the Hoffman La-Roche Predoctoral Fellowship, the National Cancer Institute Research Training Predoctoral Fellowship, and appointment of Visiting Researcher in the Laboratory of Molecular Virology at the NIH in Bethesda, MD. As an NCI-supported postdoctoral fellow, she worked closely with Howard M. Temin at the University of Wisconsin McArdle Lab for Cancer Research and developed patent-protected retroviral structural gene vectors and novel prophylactic vaccines in animal models, which opened new lines of investigation of posttranscriptional control between mammalian hosts and viruses. As a faculty of The Ohio State University, she held the David White Professorship and was the founding Executive Director of the Life Sciences Network. As an active member of the Centers for Retrovirus Research and RNA Biology at Ohio State, Professor Boris-Lawrie and her trainees discovered a unique class of RNA structural element in proto-oncogenes and infectious cancer viruses, characterized secondary structural basis for interaction with RNA helicase A and mechanism for regulating translation of growth control genes of viruses and cancer cells. Presently at the University of Minnesota, Professor Boris-Lawrie is Chairman of the Department of Veterinary and Biomedical Sciences and her research, funded by the National Institutes of Health, focuses on decoding conserved RNA signatures and partner RNA-binding proteins that dysregulate growth of neoplastic cells and promote viruses infecting animals and people.

Gatikrushna Singh, PhD earned his doctoral degree at the International Center for Genetic Engineering and Biotechnology of India, and presently is a Research Associate at the University of Minnesota, Twin Cities, where he is focused on molecular basis for host–pathogen interactions and application of mechanistic findings to next-generation antiviral therapeutics. During his graduate and postdoctoral studies, he investigated viral suppression of host RNA interference, identified antiviral drug targets and developed small molecule for attenuating HIV infection. As an expert in RNA biology, Dr. Singh has developed several methods that isolated ribonucleoprotein complexes commandeered by viruses, leading to their characterization and strategies for translating this information into RNA-based therapies that prevent microbial pathogenesis.

Xiao Heng, PhD earned her doctoral degree and received postdoctoral training at the University of Maryland, Baltimore County. As a structural biologist, she investigated the structure and protein interactive properties of the 5′-Untranslated region of the HIV-1 genome, and how it regulates viral genome packaging. She helped develop modern NMR techniques that enabled direct detection of structural elements within large RNAs. As an Assistant Professor at the University of Missouri, Dr. Heng continues to work on the viral RNA structures and begin to decipher the viral replication mechanism. Her current research involves virus:host interactions that facilitates HIV replication. By using a variety of biophysical approaches, Dr. Heng's laboratory recently reported the structural basis of DHX9/RNA helicase A recruitment during HIV-1 virion assembly.

CHAPTER 8

Leslie J. Parent, MD is Vice Dean for Research and Graduate Studies at the College of Medicine and the Associate Vice President for Health Sciences Research at Penn State University. She earned her undergraduate degree at Dartmouth College magna cum laude and graduated from Duke University Medical School where she was a member of the Alpha Omega Alpha honor society. She was in the inaugural class of Howard Hughes Medical Institute–National Institutes of Health Research Scholars in the laboratory of Dinah Singer, PhD in the National Cancer Institute. She performed her residency in Internal Medicine at Duke University Medical Center in the Clinical Investigator Pathway in the laboratory of Warner Greene, MD, PhD. She completed her fellowship in Infectious Diseases at Penn State Milton Hershey Medical Center and a postdoctoral fellowship in retrovirology with John Wills, PhD. Dr. Parent joined the faculty at Penn State College of Medicine in 1995 and is currently Professor of Medicine and Microbiology & Immunology. She has been honored with the Dean's Award for Excellence in Teaching and the Excellence in Mentoring Award from the Department of Medicine. She is an elected fellow in the American Association for the Advancement of Science, American Academy of Microbiology, American College of Physicians, and the Infectious Disease Society of America. Dr. Parent's laboratory research, funded by the National Institutes of Health, focuses on understanding how retroviruses hijack host cell machinery to assemble new virus particles.

Eunice C. Chen, MS, is an MD/PhD candidate at the Penn State College of Medicine in Hershey, PA. She earned her undergraduate degree in Molecular Biology and her Master of Science degree in Neuroscience at the University of California, San Diego. Her master's thesis research was conducted in the laboratory of Dr. David Kang, PhD, studying protein interactions in the development of Alzheimer's disease. After graduating, she worked in the laboratories of Dr. Joseph DeRisi, PhD, and Dr. Charles Chiu, MD, PhD, at the University of California San Francisco using microarray and next-generation sequencing techniques to identify novel pathogens in patient samples. Her work has resulted in multiple publications and the development of a patent. Currently she is performing her dissertation research in the laboratory of Dr. Leslie Parent, MD, studying mechanisms of retrovirus genome dimerization using the oncogenic retrovirus Rous sarcoma virus as a model. Her work is funded by an F30 NRSA fellowship from the National Institutes of Health.

CHAPTER 9

Carol A. Carter, PhD, a Professor in the Department of Molecular Genetics and Microbiology at Stony Brook University in New York, received a Bachelor of Science degree in Biology and Chemistry from the City College of New York, and a Master of Philosophy degree and a Doctor of Philosophy degree in Microbiology from Yale University. She received postdoctoral training in protein biochemistry and virology at the Roche Institute of Molecular Biology. Dr. Carter's research on the human immunodeficiency virus (HIV) focuses on identifying events in HIV replication that might serve as targets for antiviral drug development. Her career progression has been featured in *The Scientist*, a professional magazine intended for Life Scientists. Her research has been supported by public and private funding, and she has been recognized for contributions to STEM teaching, mentorship, and community service.

CHAPTER 10

Eric O. Freed, PhD received his doctoral degree in 1990 in the laboratories of Drs. Rex Risser and Howard Temin at the University of Wisconsin–Madison and did postdoctoral work with Dr. Temin at UW-Madison in 1991. His work in Madison focused on the function of the murine leukemia virus and HIV envelope glycoproteins in membrane fusion and virus entry. He joined the Laboratory of Molecular Microbiology at the National Institute of Allergy and Infectious Diseases (LMM/NIAID) in 1992, where he worked with Dr. Malcolm Martin on HIV assembly and entry/postentry events in the HIV replication cycle. In 1997 he was appointed as a Tenure-Track Investigator in LMM/NIAID, and he was promoted to a tenured Senior Investigator position in 2002. In 2003 Dr. Freed joined the HIV Drug Resistance Program. In 2009, he was appointed as the first Editor-in-Chief of *Viruses*; he also currently serves on several other editorial boards. He was selected as an NCI Mentor of Merit for excellence in mentoring and guiding the careers of trainees in cancer research. In 2011, he was appointed to the NCI Senior Biomedical Research Service. In 2014, Dr. Freed was appointed as the Deputy Director of the HIV DRP and received the 2014 Outstanding Science Alumni Award from Penn State University. He became the Director of the HIV DRP in 2015. In 2016, Dr. Freed received the NCI Research Highlights Award.

Abdul A. Waheed, PhD, is a Staff Scientist in Dr. Eric Freed's lab in the Virus–Cell Interaction Section, HIV Dynamics and Replication Program, NCI-Frederick. He received his PhD in Membrane Biochemistry from the University of the Ryukyus, Okinawa, Japan in 1999. During his postdoctoral training at Tokyo Metropolitan Institute of Gerontology, Tokyo, Dr. Waheed showed the specific binding of theta-toxin (perfringolysin O) to cholesterol in lipid rafts. He joined the National Institute of Diabetes and Digestive and Kidney Diseases in 2001, where he worked on the role of lipid rafts in G-protein signaling. Since 2003 he has been working on host-factors involved in HIV-1 replication and small-molecule inhibitors of HIV-1 replication. Dr. Waheed was appointed as Lead Guest Editor in 2012 for a special issue of *Molecular Biology International* on Host–Pathogen Interactions of Retroviruses, and in 2016 for a special issue of *Current Topics in Medicinal Chemistry* on Current and Emerging Drug Targets for Human Immunodeficiency Virus. He also serves as an Editorial Board member of *Journal of Venereology* and *The Scientific World Journal*.

CHAPTER 11

Susan R. Ross, PhD is Sweeney Basic Sciences Professor of and Head of the Department of Microbiology and Immunology at the UIC College of Medicine. Dr. Ross's research interests are in the genetics of host–virus interactions, particularly retroviruses and new world arenaviruses. Dr. Ross was on the faculty in Biochemistry at UIC from 1983–94. In 1994, she moved to the Microbiology Department at the University of Pennsylvania, where she also served as Associate Dean for Biomedical Graduate Studies from 2002–12. In 2015, she moved to UIC to assume the Head position. Dr. Ross has served on numerous review panels and editorial boards, including the Recombinant DNA Advisory Committee of the NIH, Senior Editor for the Journal of Virology, Section Editor of PLOS Pathogens, and on the Editorial Committee of the Annual Review of Virology. She is the Councilor for Animal Virology of the American Society for Virology and Council Delegate, Section on Biological Sciences, for the AAAS. Dr. Ross has received several awards for teaching and research, including the ASM Wellcome Visiting Professorship, the ASM International Professorship, and the Center for Retrovirus Research Distinguished Research Career Award (Ohio State University). Dr. Ross was elected a Fellow of the American Academy of Microbiology in 2002 and an AAAS Fellow in 2009.

CHAPTER 12

Karin Musier-Forsyth, PhD, earned her undergraduate degree at Eckerd College and her PhD in Chemistry from Cornell University under the direction of

Gordon G. Hammes. She was an American Cancer Society Postdoctoral Fellow in the laboratory of Paul Schimmel at Massachusetts Institute of Technology and joined the faculty at the University of Minnesota as an Assistant Professor of Chemistry in 1992. She was named Merck Professor of Chemistry in 2003 and Distinguished McKnight University Professor in 2006. In 2007, she moved to her current position at The Ohio State University, where she is the Ohio Eminent Scholar in Biological Macromolecular Structure and Professor of Chemistry and Biochemistry. She was awarded the Camille Dreyfus Teacher-Scholar Award, the Pfizer Award in Enzyme Chemistry from the Biological Division of the American Chemical Society, and is an elected Fellow of the American Association for the Advancement of Science. She currently serves as an Associate Editor of the *Journal of Biological Chemistry.* Her National Institutes of Health–funded research focuses on investigations of quality control mechanisms in protein translation, the role of host translation factors in retroviral replication, and retroviral genomic RNA packaging.

Alice A. Duchon, PhD, earned her undergraduate degree in Biochemistry and Molecular Biology from Dickinson College as a member of the Phi Beta Kappa honors society, and then attended The Ohio State University where she was a member of the Sigma Xi research honor society. As a graduate student she studied the noncanonical function of cellular translational machinery in HIV-1 assembly and packaging in the laboratory of Karin Musier-Forsyth. Alice Duchon has continued her study of HIV-1 assembly at the National Institutes of Health as a postdoctoral researcher in the laboratory of Wei-Shau Hu.

CHAPTER 13

Suzanne Sandmeyer, PhD, is Vice Dean for Research at the University of California Irvine School of Medicine, Professor and Grace Beekhuis Bell Chair of Biological Chemistry, with joint appointments in the departments of Microbiology and Molecular Genetics in the School of Medicine, and Chemical Engineering and Materials Science in the Samueli School of Engineering. Dr. Sandmeyer earned her undergraduate degree from Carleton College and received her PhD in biochemistry from the University of Washington where she was a Damon Runyon postdoctoral fellow in Genetics. She joined the UC Irvine Department of Microbiology and Molecular Genetics and was previously Chair of the Department of Biological Chemistry. Dr. Sandmeyer has served as chair of the Senior Editors of the Genetics journal and has served on numerous national advisory panels including the NCI Basic Sciences Board of Scientific Counselors, Genetics Study Section, Shared Instrument Study Section, NIH Director's New Innovator Award Editorial Panel, AAAS Electorate Nominating Committee, and California Cancer Research Coordinating Committee. At UCI, She founded and currently directs the UCI Genomics High Throughput Facility platform for emerging genomics technologies and serves as Associate Director of the Institute for Genomics and Bioinformatics. Dr. Sandmeyer is a recipient of the School of Medicine Athalie Clarke Research Associates Award. She is an elected fellow of American Academy of Microbiology and the American Association for the Advancement of Science. Dr. Sandmeyer's research focuses on the molecular genetics and biochemistry of retrovirus-like elements. In particular, her laboratory studies Ty3 as a model for understanding both retrotransposons and retroviruses. Ty3 studies have elucidated the roles of virus structural protein in capsid assembly, roles of nucleoporins in nuclear entry, roles of transcription factors in integration specificity, and how RNA processing proteins help to localize and package Ty3 genomic RNA.

CHAPTER 14

Leslie J. Parent, MD, is Vice Dean for Research and Graduate Studies at the College of Medicine and the Associate Vice President for Health Sciences Research at Penn State University. She earned her undergraduate degree at Dartmouth College magna cum laude and graduated from Duke University Medical School where she was a member of the Alpha Omega Alpha honor society. She was in the inaugural class of Howard Hughes Medical Institute–National Institutes of Health Research Scholars in the laboratory of Dinah Singer, PhD, in the National Cancer Institute. She performed her residency in Internal Medicine at Duke University Medical Center in the Clinical Investigator Pathway in the laboratory of Warner Greene, MD, PhD. She completed her fellowship in Infectious Diseases at Penn State Milton Hershey Medical Center and a postdoctoral fellowship in retrovirology with John Wills, PhD. Dr. Parent joined the faculty at Penn State College of Medicine in 1995 and is currently Professor of Medicine and Microbiology & Immunology. She has been honored with the Dean's Award for Excellence in Teaching and the Excellence in Mentoring Award from the Department of Medicine. She is an elected fellow in the American Association for the Advancement of Science, American Academy of Microbiology, American College of Physicians, and the Infectious Disease Society of America. Dr. Parent's laboratory research, funded by the National Institutes of Health, focuses on understanding how retroviruses hijack host cell machinery to assemble new virus particles.

Rebecca J. Kaddis Maldonado, PhD, is a postdoctoral scholar in the laboratory of Dr. Leslie Parent at Penn State College of Medicine. She received her Bachelor of Science degree in Biology from the University of Scranton in 2009 and completed graduate school with a degree in Cell and Molecular Biology at Penn State. Her dissertation research, which was funded by an F31 NRSA predoctoral fellowship from the National Institutes of Health, focused on nuclear trafficking of the retroviral Gag protein.

Eunice C. Chen, MS, is an MD/PhD candidate at the Penn State College of Medicine in Hershey, PA. She earned her undergraduate degree in Molecular Biology and her Master of Science degree in Neuroscience at the University of California San Diego. Her master's thesis research was conducted in the laboratory of Dr. David Kang, PhD, studying protein interactions in the development of Alzheimer's disease. After graduating, she worked in the laboratories of Dr. Joseph DeRisi, PhD, and Dr. Charles Chiu, MD, PhD, at the University of California San Francisco using microarray and next-generation sequencing techniques to identify novel pathogens in patient samples. Her work has resulted in multiple publications and the development of a patent. Currently she is performing her dissertation research in the laboratory of Dr. Leslie Parent, MD studying mechanisms of retrovirus genome dimerization using the oncogenic retrovirus Rous sarcoma virus as a model. Her work is funded by an F30 NRSA fellowship from the National Institutes of Health.

Preface

The genesis of this book came from Jill Leonard, who was formerly the books editor for Molecular Biology, Biochemistry, and Virology at Elsevier Academic Press. Jill approached me after learning about my work investigating retrovirus–cell interactions and encouraged me to submit a book proposal to Elsevier. After considering her offer, I realized that a book on this topic had not been released recently, outside of conferences summaries and special issues in journals. I drafted a proposal, which was reviewed by several experts in the field. These scientists made insightful comments and suggestions for chapter topics, and overall they were quite supportive about developing the topic into a book. I was especially grateful for the constructive input from Stephen Goff, PhD, and Alan Engelman, PhD, and both of them later agreed to provide chapters. Fortunately, additional preeminent leaders in retrovirology joined the project as chapter authors, culminating in the collection of works in the book. These colleagues wrote outstanding current reviews of topics in their areas of expertise, and working with them was a pleasure. Their passion for their research is evident, and the book only became a reality because of their dedication. After Jill Leonard left Elsevier, Linda Versteeg, Senior Acquisitions Editor, took over the project leadership, and Pat Gonzales, the Editorial Project Manager, guided me through the publishing process. I am appreciative of Pat for her patience and guidance and also Mohanapriyan Rajendran, the Senior Project Manager, who pulled everything together at the end. I would like to acknowledge many mentors and colleagues who shaped my scientific development, and my students and postdoctoral fellows who developed the projects in our lab exploring how retroviruses co-opt host cell machinery during the replication cycle. I am grateful to the National Cancer Institute at the National Institutes of Health, which has supported our research for over 20 years. We also appreciate the research funding received from the Penn State College of Medicine and the Pennsylvania Department of Health. I am especially thankful for encouragement throughout the project from my husband and daughters, who have always supported my career as a physician and a scientist—they continuously inspire me with their insight, humor, and compassion.

Introduction

The history of retroviruses is quite remarkable, fortuitously driven by the curiosity of scientists investigating diseases they observed in animals in the early part of the 20th century. These pioneers designed simple, yet elegant, experiments to understand illnesses such as "swamp fever" in horses (Vallée and Carré, 1904), sarcomas in chickens (Rous, 1910), and mammary tumors transmitted to offspring in mice (Lathrop and Loeb, 1915; Bittner, 1942), never imagining their studies would ultimately lay the groundwork for understanding a worldwide epidemic involving tens of millions of people that would appear decades into the future. If these scientists had not wondered how and why diseases were transmitted between animals over a century ago, the acquired immunodeficiency syndrome (AIDS) epidemic might have been even more devastating, because effective drug treatments could not have been developed nearly as rapidly as they were.

The earliest account of a disease ultimately found to be caused by a retrovirus was described in 1904, when French veterinarians Vallée and Carré (1904), professors of contagious diseases at the National Veterinary School in Alfort, France, published their investigation of swamp fever, an acquired infectious anemia of horses. Swamp fever had been reported as early as 1843, but the etiology remained unknown (Ligné, 1843). To investigate how the disease was transmitted, Vallée and Carré passed horse serum through a filter and found that cell-free filtrates could transit acute hemolytic anemia to healthy horses. This transmissible agent was aptly named equine infectious anemia virus (EIAV). The genetic material of the virus was later shown to be RNA (Nakajima et al., 1970), and this class of viruses was referred to as RNA tumor viruses, or oncornaviruses. EIAV and the rest of the RNA tumor viruses were subsequently renamed retraviruses (re-, reverse and tra-, transcriptase) (Dalton et al., 1974), later modified to retroviruses; once EIAV was sequenced, it was classified as a member of the *Lentivirus* genus, to which the causative agent of AIDS, the human immunodeficiency virus (HIV), also belongs Stephens et al., 1986.

Only a few years later, at the University of Copenhagen, a physician named Vilhelm Ellermann teamed up with veterinarian Olaf Bang and found that leukemia in chickens was caused by an agent that could be passed through a filter (Ellermann and Bang, 1908). This agent was the first tumor-causing virus to be discovered, although at the time leukemia was not considered to be a malignant process; therefore, the magnitude of this discovery was not immediately appreciated. The virus was later named avian leukosis virus, becoming the first member of the *Alpharetrovirus* genus of retroviruses.

In 1910, a woman in New York noticed that one of her chickens developed a large tumor protruding from its chest wall. Her concern led her to take the Plymouth Rock hen to the Rockefeller Institute for Medical Research, where she met a young medical pathologist named Francis Peyton Rous, recently put in charge of the laboratory for cancer research. He took the time to talk with the woman, and finding the phenomenon to be quite interesting, he agreed to take the chicken for further study to investigate why this hen had developed a large tumor.

Rous (1910) injected tumor cells from the original hen into healthy chickens, and they developed similar spindle cell tumors a short time later. He performed experiments along the same lines as those of Vallée and Carré. After grinding up the tumor and passing it through a filter, he injected the filtrate into a new set of chickens and again, tumors developed (Rous, 1911). He published his results in a series of publications, but again the significance was not appreciated at the time. Even Rous himself could not predict the impact his discovery would have years later as the first virus discovered to cause solid tumors. The virus was named Rous sarcoma virus (RSV) and joined the group of avian leukosis sarcoma viruses as a member of the *Alpharetrovirus* genus. Remarkably, it took 55 years for the groundbreaking work of Peyton Rous to be fully recognized, when he was awarded the Novel Prize for Physiology or Medicine in 1966.

RSV was not yet finished contributing to medical history. Two more Nobel prizes would be awarded to scientists studying this chicken virus. It was quite surprising when Howard Temin (1964) at the University of Wisconsin Madison reported that RSV-infected cells contained DNA that showed homology to RSV RNA. He proposed the viral DNA was generated from the viral RNA template, and he began speaking at meetings and writing about the "provirus hypothesis" (Temin, 1976). He was convinced that the viral RNA was somehow converted to a DNA intermediate that then became joined to the DNA of the cell. Other scientists were skeptical—the central dogma of molecular biology stated that DNA was transcribed into RNA, which was translated into protein; there was no known mechanism for converting RNA into DNA. Temin was convinced that this was the most logical explanation, so he persisted, continuing to perform experiments to gather more evidence in support of the proviral hypothesis. Finally, he was able to isolate an enzyme from RSV that converted RNA into DNA (Temin and Mizutani, 1970) (Chapter 2). He had proven that retroviral RNA is reverse transcribed into DNA, which then undergoes integration into cellular DNA to form the provirus (Chapter 4).

In a series of independent experiments conducted at the Massachusetts Institute of Technology during the same time, David Baltimore isolated an RNA-dependent DNA polymerase from virions derived from a different RNA tumor virus, Rauscher mouse leukemia virus (Baltimore, 1970). He too had discovered an enzyme that converted viral RNA to DNA, which was called "reverse transcriptase" because it worked backward compared to the traditional transcription process, as described in a *Nature* editorial accompanying the back-to-back Baltimore and Temin manuscripts (Anonymous, 1970). Temin and Baltimore shared the Nobel Prize for Physiology or Medicine in 1975 for their discovery of the mechanism by which RNA tumor viruses changed their genetic material into DNA. RNA tumor viruses were thereafter renamed retroviruses, as mentioned previously, in homage to the reverse transcriptase enzyme. In the ensuing years, retroviral reverse transcriptase developed a life of its own, as scientists understood that it could be used to transform messenger RNA (mRNA) isolated from cells into DNA to produce complementary DNA (cDNA) (Spiegelman et al., 1971). The generation of cDNAs using reverse transcriptase revolutionized molecular biology, and cloning genes became not only possible but straightforward and easy. More recently, reverse transcriptase has been combined with the polymerase chain reaction (RT-PCR), allowing extremely sensitive and accurate measurement of RNA from viruses and even from single cells. Today, all types of RNA species from cells (mRNA, ribosomal RNA, microRNA, long noncoding RNA) can be isolated, reverse transcribed, and sequenced to generate the RNA profile of a cell.

The third Nobel Prize for the study of RSV was awarded to Harold Varmus and J. Michael Bishop for their study of the mechanism of RSV transformation at the University of California San Francisco. Varmus joined Bishop's lab as a postdoctoral fellow, ultimately becoming an independent faculty member and continuing to collaborate with Bishop. Their studies focused on the RSV gene called *src*, named because it caused sarcomas to form after infection with RSV. In a seminal experiment, they looked for the presence of *src* DNA in infected cells, including uninfected cells as a control, expecting to find *src* only in the RSV-infected cells (Stehelin et al., 1976). Much to their surprise, uninfected control cells contained *src* DNA, leading the investigative team to show that *src* was actually of cellular origin. Through a series of elegant studies that followed, they figured out that the cellular *src* (*c-src*) gene had been captured by RSV and integrated into its proviral sequence during an integration event, through a process known as transduction (Swanstrom et al., 1983; Varmus, 1990). Furthermore, the virus did not need *src* to replicate and in fact, expression of the *src* gene alone was sufficient to induce tumors in chicken and other avian cells (Martin, 1970). Thus, the c-*src* gene was the original "oncogene," a cellular gene that regulates cell growth and, if unregulated, can lead to uncontrolled cellular growth, or transformation. This concept remains the basis for understanding the molecular basis of cancer, and dozens of other cellular and viral oncogenes have been discovered and characterized. Src and other oncoproteins are tyrosine kinases, and drugs targeting tyrosine kinases are currently used in clinical practice to effectively treat certain types of cancer.

As the methods used to study retroviruses became more sophisticated, several key observations were made and progress was rapid. Using sedimentation analysis, studies of RSV demonstrated that the RNA extracted from virions, the genetic material contained within RSV, could be reduced to a fraction half its size, indicating that there were two copies of the viral genomic RNA in virus particles (Duesberg, 1968) in the form of a dimer. Other smaller RNA species isolated from virions were later identified as host RNAs, which were also incorporated into virus particles (Chapter 12 and references therein). Biochemical and structural properties of retroviruses and the enzymes they encoded became a primary focus. Electron microscopy and biochemical techniques were used to show that retrovirus particles assembled as immature particles that are released from the cell by budding from the plasma membrane (Chapters 8, 9, and 10) (Hall et al., 1967). The major structural protein of the virus, called Gag for *g*roup-specific *a*nti*g*en, was shown to orchestrate the assembly and release of virus particles, and expression of Gag alone resulted in production of immature virus-like particles (Dickson et al., 1984). Scientists determined that immature virus particles quickly underwent a maturation process mediated by a protease encoded by the virus (von der Helm, 1977; Yoshinaka and Luftig, 1977). In addition, receptors for retroviruses were discovered and factors controlling cellular tropism were characterized (Chapter 1 and references therein). The virally encoded enzyme that joins the proviral DNA to host DNA was discovered (Grandgenett et al., 1978) and named integrase (Chapter 3).

In the early 1980s, AIDS, a uniformly fatal disease was sweeping through major US cities including San Francisco, Miami, and New York. As I was beginning medical school in 1983, Luc Montagnier and Françoise Barré-Sinoussi in the Virology Department at the Institut Pasteur discovered that AIDS was caused by a retrovirus they named lymphadenopathy-associated virus, or LAV (Barré-Sinoussi et al., 1983). Robert Gallo and colleagues at the National Institutes of Health reported the isolation of a human retrovirus he called human lymphotropic virus type III (HTLV-III) (Popovic et al., 1984). Jay Levy et al. (1984) at the University of California San Francisco isolated a virus from infected patients with AIDS and called it the AIDS-associated retrovirus (ARV). Later, LAV, HTLV-III, and ARV were found to be the same virus and renamed HIV-1. Montagnier and Barré-Sinoussi were awarded the Nobel Prize for Physiology or Medicine in 2008 for the discovery of HIV.

Due to the preceding decades of work on animal retroviruses, reverse transcriptase, protease, and integrase were known to be vital to virus replication, so they quickly became the focus of efforts to develop antiretroviral therapies. Scientists in academia, research institutes, and pharmaceutical companies across the globe worked rapidly to develop inhibitors of HIV enzymes that crippled the virus. By combining drugs that inhibited reverse transcriptase and protease to produce highly active antiretroviral therapy, virus replication was controlled and there was a dramatic improvement in survival. With the addition of integrase inhibitors and long acting formulations, drug treatment could be simplified to a single pill once a day. Death rates dropped profoundly, and those HIV-1 infected people who begin antiretroviral treatment early and achieve an undetectable viral load can now expect to live a nearly normal lifespan. This remarkable progress occurred over a relatively short period, made possible by the pioneering study of animal retroviruses beginning over a century ago.

Retroviruses are stealth invaders, entering the host by binding to specific proteins studding the cell surface (Chapter 1), traveling into the nucleus (Chapter 2), and then knitting the proviral DNA into the genetic material of each cell (Chapter 3). How devious for the virus to become a permanent part of the cell; in this way, the virus irreversibly recruits the cell to serve as a virus-making machine by hijacking RNA polymerase II and transcription machinery (Chapter 5) to synthesize viral genes, host splicing machinery to create a panoply of viral proteins from a limited viral genome (Chapter 6), promote translation of viral proteins by co-opting cellular helicases (Chapter 7), and direct the viral genome and structural proteins (Chapter 8) to the plasma membrane, where they are released by budding to spread infection to new cells (Chapter 9). Each step of virus infection requires the recruitment of a myriad of host factors.

The next frontier in retrovirology, which is the focus of this book, is understanding the mechanisms by which the virus enlists cellular proteins, nucleic acids (Chapter 12), and lipids (Chapter 10) to promote virus replication. Engaged in a battle for survival, the cell has developed defensive maneuvers in an attempt to block the invading virus; of course, the virus has responded with desperation, generating countermeasures to "one-up" the cell (Chapter 11). This back-and-forth struggle for dominance leads to conscription of existing cellular factors to adopt new antiviral roles followed by the evolution of novel viral factors that arise by mutation and are refined by selective pressure.

The vestigial existence of ancient predecessors of retroviruses, called retrotransposons and retroelements (Chapter 13), have preconditioned the cell to tolerate retrovirus infection because they have coexisted for millions of years. Owing to this primordial cohabitation of retroelements with eukaryotic cells, host organisms have wisely co-opted endogenous retroviral proteins for their own purposes—fascinating examples include placental development in humans (Frendo et al., 2003), recruitment of extra copies of amylase in saliva and the pancreas to digest starch (Samuelson et al., 1990), and the use of retroelements to develop plasticity in the hippocampus to allow memory and learning to flourish (Bachiller et al., 2017) (Chapter 13).

This intimate interaction of retroviruses with their cellular hosts makes for a life-and-death epic story of invaders and defenders, vicious competitors that take advantage of unguarded portals of entry, and thieves that steal precious cargo meant for cell survival. As scientists unravel the intertwined relationships between retroviruses and the cell (Chapter 14), we hope that novel therapies will be developed as the next generation of antiviral defense, ultimately leading to a cure for retroviral diseases.

REFERENCES

Anonymous. 1970. News and Views: Central Dogma Reversed. Nature 226, 1198–1199.

Bachiller, S., Del-Pozo-Martín, Y., Carrión, Á.M., August 2017. L1 retrotransposition alters the hippocampal genomic landscape enabling memory formation. Brain Behav. Immun. 64, 65–70.

Baltimore, D., 1970. RNA-dependent DNA polymerase in virions of RNA tumor viruses. Nature 226, 1209–1211.

Barré-Sinoussi, F., Chermann, J.C., Rey, F., Nugeyre, M.T., Chamaret, S., Gruest, J., Dauguet, C., Axler-Blin, C., Vézinet-Brun, F., Rouzioux, C., Rozenbaum, W., Montagnier, L., 1983. Isolation of a T-lymphotropic retrovirus from a patient at risk for acquired immune deficiency syndrome (AIDS). Science 220, 868–871.

Bittner, J.J., 1942. The milk-influence of breast tumors in mice. Science 95, 462–463.

Dalton, A.J., Melnick, J.L., Bauer, H., Beaudreau, G., Bentvelzen, P., Bolognesi, D., Gallo, R., Grafi, A., Haguenau, F., Heston, W., Huebner, R., Todaro, G., Heine, U.I., 1974. The case for a family of reverse transcriptase viruses: Retraviridae. Intervirology 4, 201–206.

Dickson, C., Eisenman, R., Fan, H., Hunter, E., Teich, N., 1984. Protein biosynthesis and assembly. In: Weiss, R., Teich, N., Varmus, H., Coffin, J. (Eds.), RNA Tumor Viruses, vol. 1. Cold Spring Harbor Laboratory, Cold Spring Harbor, NY, pp. 513–648.

Duesberg, P.H., 1968. Physical properties of Rous sarcoma virus RNA. Proc. Natl. Acad. Sci. U.S.A. 60, 1511–1518.

Ellermann, V., Bang, O., 1908. Experimentelle Leukämie bei Hühnern. Zentralbl. Bakteriol. Parasitenkd. Infectionskr. Hyg. Abt. Orig. 46, 595–609.

Frendo, J.L., Olivier, D., Cheynet, V., Blond, J.L., Bouton, O., Vidaud, M., Rabreau, M., Evain-Brion, D., Mallet, F., 2003. Direct involvement of HERV-W Env glycoprotein in human trophoblast cell fusion and differentiation. Mol. Cell Biol. 10, 3566–3574.

Grandgenett, D.P., Vora, A.C., Schiff, R.D., 1978. A 32,000-dalton nucleic acid-binding protein from avian retravirus cores possesses DNA endonuclease activity. Virology 89, 119–132.

Hall, W.T., Andresen, W.F., Evans, V.J., December 1967. In vitro and in vivo observations on a murine C-type virus. J. Virol. 1, 1238–1254.

Lathrop, A.E., Loeb, L., 1915. Further investigations on the origins of tumors in mice: I. Tumor incidence and tumor age in various strains of mice. J. Exp. Med. 22, 646–673.

Levy, J.A., Hoffman, A.D., Kramer, S.M., Landis, J.A., Shimabukuro, J.M., Oshiro, L.S., August 24, 1984. Isolation of lymphocytopathic retroviruses from San Francisco patients with AIDS. Science 225, 840–842.

Ligné, M., 1843. Mémoire et observations sur une maladie de sang, connue sous le nom d'anhémie hydrohémie, cachexie acquise du cheval. Rec. Med. Vet. Ec. Alfort 20, 30–44.

Martin, G.S., 1970. Rous sarcoma virus: a function required for the maintenance of the transformed state. Nature 227, 1021–1023.

Nakajima, H., Tanaka, S., Ushimi, C., 1970. Physico-chemical studies of equine infectious anemia virus. IV. Determination of the nucleic acid type in the virus. Arch. Gesamte Virusforsch. 31, 273–280.

Popovic, M., Sarngadharan, M.G., Read, E., Gallo, R.C., 1984. Detection, isolation, and continuous production of cytopathic retroviruses (HTLV-III) from patients with AIDS and pre-AIDS. Science 224, 497–500.

Rous, P.A., 1910. Transmissible avian neoplasm. (Sarcoma of the common fowl). J. Exp. Med. 12, 696–705.

Rous, P., 1911. A sarcoma of the fowl transmissible by an agent separable from the tumor cells. J. Exp. Med. 13, 397–411.

Samuelson, L.C., Wiebauer, K., Snow, C.M., Meisler, M.H., 1990. Retroviral and pseudogene insertion sites reveal the lineage of human salivary and pancreatic amylase genes from a single gene during primate evolution. Mol. Cell Biol. 10, 2513–2520.

Spiegelman, S., Watson, K.F., Kacian, D.L., 1971. Synthesis of DNA complements of natural RNAs: a general approach. Proc. Natl. Acad. Sci. U.S.A. 68, 2843–2845.

Stehelin, D., Varmus, H.E., Bishop, J.M., Vogt, P.K., 1976. DNA related to the transforming gene(s) of avian sarcoma viruses is present in normal avian DNA. Nature 260, 170–173.

Stephens, R.M., Casey, J.W., Rice, N.R., 1986. Equine infectious anemia virus gag and pol genes: relatedness to Visna and AIDS virus. Science 231, 589–594.

Swanstrom, R., Parker, R.C., Varmus, H.E., Bishop, J.M., 1983. Transduction of a cellular oncogene: the genesis of Rous sarcoma virus. Proc. Natl. Acad. Sci. U.S.A. 80, 2519–2523.

Temin, H.M., 1964. Homology between RNA from Rous sarcoma virus and DNA from Rous sarcoma virus-infected cells. Proc. Natl. Acad. Sci. U.S.A. 52, 323–329.

Temin, H.M., 1976. The DNA provirus hypothesis. Science 192, 1075–1080.

Temin, H., Mizutani, S., 1970. RNA-dependent DNA polymerase in virions of Rous sarcoma virus. Nature 226, 1211–1213.

Vallée, H., Carré, H., 1904. Sur la nature infectieuse de l'anémie du cheval. C.R. Hebd. Seances Acad. Sci. Ser. D Sci. Nat. 139, 331–333.

Varmus, H.E., 1990. Nobel lecture. Retroviruses and oncogenes I. Biosci. Rep. 10, 413–430.

von der Helm, K., 1977. Cleavage of Rous sarcoma viral polyprotein precursor into internal structural proteins in vitro involves viral protein p15. Proc. Natl. Acad. Sci. U.S.A. 74, 911–915.

Yoshinaka, Y., Luftig, R.B., 1977. Murine leukemia virus morphogenesis: cleavage of P70 in vitro can be accompanied by a shift from a concentrically coiled internal strand ("immature") to a collapsed ("mature") form of the virus core. Proc. Natl. Acad. Sci. U.S.A. 74, 3446–3450.

Chapter 1

Retrovirus Receptor Interactions and Entry

Lorraine M. Albritton
The University of Tennessee Health Science Center, Memphis, TN, United States

Chapter Outline

Envelope Glycoproteins: Domain Structure 2
Retrovirus Entry Receptors 4
Virus Attachment 10
The Basics of Membrane Fusion 13
Do Receptor Interactions Contribute to the Envelope Proteins Functions That Drive Entry? 16
The Triggers 16
Conserved Motifs in Envelope Proteins Are Critical to Regulation of Fusion 17
Regulation by Control of the Conformation of Surface Subunit 18
Regulation by Control of Disulfide Bond Isomerization and Surface Subunit Conformation 18
Lentiviruses: Receptor-Triggered Conformational Changes in the Surface Subunit 19
Gammaretroviruses: Receptor- and Cellular Protease–Driven Disulfide Bond Isomerization 20
Alpharetroviruses: Receptor- and Low pH-Driven Disulfide Bond Isomerization 22
Do Env–Receptor Interactions Contribute to Pathogenesis? 23
CD4+ T-Cell Depletion and AIDS 23
Neurological Damage and HAM/TSP 24
Mutations That Adapt Virus to Low Receptor Levels Can Also Increase Pathogenicity 24
Envelope Protein–Driven Neoplasia in Betaretrovirus Infection 25
Host Defenses That Inhibit Retroviral Entry Also Drive Envelope Protein Variation 26
Challenges in Vaccine Development 26
Interferon-Induced Transmembrane Proteins and Envelope Proteins Variation 28
Coevolution of Virus and Receptor 28
Captured Retroviral Envelope Proteins in the Development of Mammalian Placenta and Male Muscle Mass 30
References 31

Retrovirus-Cell Interactions. https://doi.org/10.1016/B978-0-12-811185-7.00001-7

Interactions that occur during entry into naïve host cells are basic determinants of the host range and cell tropism of each retrovirus. The principal players are the retroviral envelope proteins (Env) and their host cell receptors. This chapter briefly describes the functional domains of Env, attributes of their receptors, and the basics of membrane fusion. It then uses these fundamental concepts to consider two questions. What role do Env and host cell receptor interactions play in entry? Do these interactions contribute to viral pathogenesis? The final sections discuss challenges in creating vaccines, coevolution of retroviruses and their entry receptors, and the impact of captured retroviral Env on mammalian evolution and development.

ENVELOPE GLYCOPROTEINS: DOMAIN STRUCTURE

Retroviral genomes contain a single *env* gene that is expressed from a spliced mRNA. In the host cell endoplasmic reticulum, newly synthesized Env associate with two additional molecules to generate a trimer. A host protease in the Golgi apparatus cleaves each Env protomer into two subunits, the surface subunit (SU) and the transmembrane subunit (TM), and they mature into native Env trimers that are incorporated into the membrane of new virus particles (Fig. 1.1A). Env trimers provide the only productive interactions with a potential host cell–specific virus attachment to entry receptors and fusion of viral and cellular membranes. The binding sites for the entry receptors are in SU and the membrane fusion activities belong to TM.

The domain structure of all but the epsilonretrovirus and spumavirus Env occurs in two arrangements and each correlates with a basic entry mechanism. The hallmarks of the first mechanism are a covalent bond between a central CXXC motif in SU and a $C(X)_6CCF$ motif in TM, an N-terminal histidine motif in SU, and a conserved HR1 trimerization domain and relatively short HR2 in their TM. Type D betaretroviruses and alpha-, delta-, and gammaretroviruses have these features (Fig. 1.1B). Lentiviruses and the betaretroviruses, including mouse mammary tumor virus (MMTV), jaagsiekte sheep retrovirus (JSRV), and enzootic nasal tumor virus (ENTV), use the second mechanism. Its hallmarks are a lack of covalent bond between SU and TM and a $C(X)_{5\text{-}7}C$ motif with long HR2 in TM (Fig. 1.1C).

Gamma-, beta-, and deltaretroviruses maintain receptor-binding sequences within an N-terminal receptor-binding domain (RBD), which is followed by a short flexible region that is proline-rich and a C-terminal domain containing the CXXC motif that bonds with TM. Soluble forms of these RBD are active and useful in identifying entry receptors and receptor-binding sites (Manel et al., 2003a; Kim et al., 2004; Kinet et al., 2007). In contrast, the receptor-binding sequences of SU from lentiviruses, alpha-, epsilonretroviruses, and spumaviruses are distributed across their length; there are no discrete N- and C-terminal domains within SU (Sun et al., 2008; Melder et al., 2015). These genera also vary in the size and function of the cytoplasmic tail domains on TM.

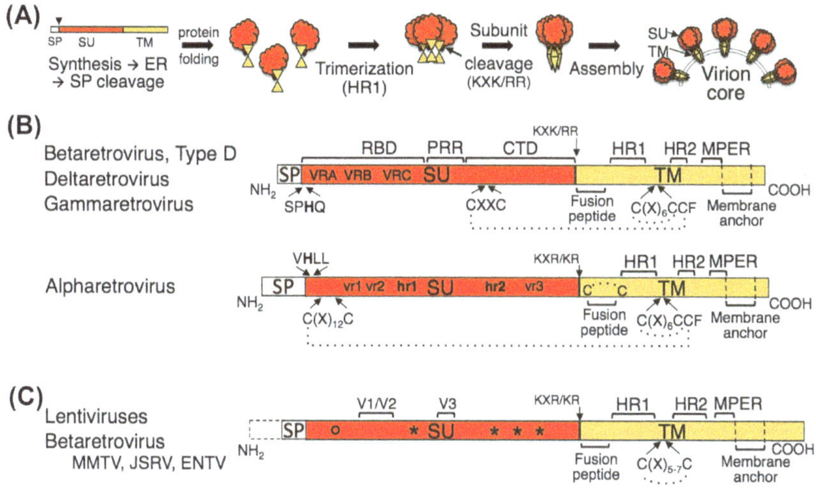

FIGURE 1.1 (A) Sequential synthesis, trimerization of HR1, and host protease processing of envelope proteins (Env) at the KXK/RR site create fusion-competent trimers for assembly into virions. The domain structure of Env occurs in two functional forms that differ primarily in the presence (B) or absence (C) of a covalent bond between SU and TM. *Dotted connectors* indicate disulfide bonds. KXK/RR, SPHQ, CXXC, VHLL, $C(X)_{12}C$, $C(X)_6CCF$, and $C(X)_{5\text{-}7}C$ denote functional motifs using the single-letter amino acid code. Variable regions are denoted by VR, vr, or V followed by a letter or number. The *open circle* and *asterisks* denote segments of HIV-1 SU that interact with CD4. hr1 and hr2 denote the host range sequences of alpharetrovirus Env. *CTD*, C-terminal domain; *ER*, endoplasmic reticulum.; *HR1 and 2*, heptad repeat 1 and 2; *MPER*, membrane proximal external region; *PRR*, proline-rich region; *RBD*, receptor-binding domain; *SP*, signal peptide; *SU*, surface subunit; *TM*, transmembrane subunit.

SUs contain regions that vary considerably between strains and isolate even within the receptor subgroups. These variable regions are typically bounded by cysteine residues engaged in disulfide bonds and are aptly called variable loops. The SU of human immunodeficiency virus type 1 (HIV-1) has many variable loops but the largest three, V1, V2, and V3, are the most important for receptor interactions. The V1 and V2 loops are bounded by a shared disulfide bond and referred to as the V1/V2 loop. The gammaretroviral SUs have three variable regions, which are lettered as VRA, VRB, and VRC. A different designation is used for the variable receptor-binding segments of the avian sarcoma and leukosis virus (ASLV) subgroups; they are called host range 1 and 2 (hr1 and hr2). Where it is known, the receptor-binding sites are within the variable regions of SU.

There is less variation in the domain structure of the TM. The fusion peptide is at the new N-terminus created in the Golgi by cleavage into SU and TM. It is followed by two 4-3 heptad repeats, HR1 and HR2, which are separated by a short segment containing a conserved intramolecular disulfide bond. The membrane proximal region (MPER), membrane-spanning anchor, and cytoplasmic tail complete the domain map (Fig. 1.1). The alpharetrovirus TMs differ somewhat; their fusion peptides are 12 residues from the N-terminus of TM and restrained by a disulfide bond between flanking cysteine residues (Pike et al., 2011).

All Env are glycosylated at multiple sites. Most SUs have relatively modest and strategically placed glycosylation that assists in evading host antibody responses. The lentiviral SUs have more extensive glycosylation. In primary HIV-1 isolates, the carbohydrates typically comprise almost half of the gp120 mass and consist of modification at more than 30 N-linked glycosylation sites and numerous O-linked sites (Leonard et al., 1990; Wei et al., 2003; Lee et al., 2016). Their branched carbohydrate chains pack into networks creating a "glycan shield" over the top and sides of native trimers that is extremely difficult for host germline rearranged antibodies to penetrate (Wei et al., 2003; Stewart-Jones et al., 2016; Lee et al., 2016). Additional immune shielding is provided by glycosylation of the TM of all genera except the gammaretroviruses.

A unique domain with two separate functions occurs in the Env of primate, feline, bovine, and equine spumaviruses (commonly called foamy viruses). Their exceptionally long signal peptides of >130 residues are cleaved from SU by the same furin-like protease that separates SU and TM (Lindemann and Goepfert, 2003; Duda et al., 2004; Tobaly-Tapiero et al., 2000). The signal peptide contains a membrane anchor and remains associated with the mature SU–TM heterotrimers (Lindemann et al., 2001; Geiselhart et al., 2004; Materniak et al., 2006; Duda et al., 2004). Its retention in native Env trimers on virions was recently confirmed by cryo-electron microscopy (cryo-EM) of intact virions and is consistent with reports that it provides a critical interaction with capsid protein inside virions (Lindemann et al., 2001; Lindemann and Goepfert, 2003; Wilk et al., 2001; Effantin et al., 2016). Interestingly, the presence of the leader peptide in native trimers may also be involved in the ordered interactions among trimer ectodomains, which appear as interconnected networks of hexamers on virions (Effantin et al., 2016).

Dual function, long signal peptides also occur in some betaretroviruses. At 98 residues in length, the signal peptide of the MMTV Env is roughly three times the length of the leader in membrane proteins in general (Hiss and Schneider, 2009). Unlike shorter signal peptides, it is not degraded but accumulates and serves a dual function as signal peptide for the Env precursor and separately, by virtue of a nuclear localization signal (NLS) and a CRM1-dependent nuclear export signal (NES), to enhance nuclear export of unspliced genomic mRNAs for assembly into virions and translation of the Gag and enzymatic protein precursors (Indik et al., 2005; Mertz et al., 2005). This protein was named Rem after its function in nuclear export of unspliced MMTV RNA. Two other betaretroviruses, JSRV and ENTV, and the alpharetroviruses have similarly long signal peptides with potential NLSs, but these sequences have not yet been examined for Rem-like activity.

RETROVIRUS ENTRY RECEPTORS

Assignment to a retroviral genus is based on genome structure, and the viruses within each genus are assigned to subgroups based on the entry receptor(s) that they use. The host cell entry receptors provide attachment and the majority play a role in membrane fusion. Also referred to as cell surface receptors, they come from several classes of membrane proteins. Table 1.1 lists the known entry receptors organized by type of membrane protein: type I proteins spanning the membrane a single time with an extracellular N-terminus, glycosylphosphatidylinositol (GPI)-anchored proteins, and multiple transmembrane-spanning proteins.

The tissue and cell type distribution of the virus receptors together with the availability of downstream host replication factors determines the sites of robust virus replication. MMTV and gammaretroviruses use ubiquitously expressed membrane proteins as entry receptors but replicate preferentially in activated B cells due to the presence of host transcription factors that increase the viral promoter activity. GLUT-1, the primate deltaretrovirus receptor, is broadly expressed but under inducible control that correlates more closely with disease pathogenesis and its transmission by sexual contact (Manel et al., 2003b; Jones et al., 2005a; Kinet et al., 2007). Similarly, the range of host cells infected by the primate or feline lentiviruses is limited by expression of their CD4 or feCD134 attachment receptors rather than more widely expressed fusion coreceptors.

Other than their location and abundance in the host cell plasma membrane, are there other attributes important for a retrovirus entry receptor? One universal characteristic is an association with cholesterol- and sphingolipid-rich, liquid-ordered lipid domains called lipid rafts. GPI-anchored proteins, CCR5 and other G-protein coupled receptors, many multiple membrane-spanning proteins, palmitoylated proteins, and some Type I membrane proteins including CD4 localize to rafts (Dietrich et al., 2001; Diaz-Rohrer et al., 2014; Levental et al., 2010; Popik et al., 2002). While all CXCR4 do not localize to lipid rafts, gp120 binding to CD4 recruits the lipid raft–associated forms that mediate HIV-1 entry (Lapham et al., 1996; Popik et al., 2002; Kamiyama et al., 2009). The ecotropic murine leukemia virus (MLV) receptor mCAT-1/SLC7A1 and the amphotropic MLV receptor PiT-2/SLC20A2 are located in cholesterol-rich caveolae (Lu and Silver, 2000; Beer et al., 2005; Beer and Pedersen, 2006). This association is critical to viral-host membrane fusion; depletion of cholesterol from the host cell membrane disrupts lipid rafts and inhibits retrovirus infection (Wielgosz et al., 2005; Rawat et al., 2003; Liao et al., 2001).

Is the normal cellular function relevant to its usefulness or selection as an entry receptor? While it is controversial for a few viruses, most retroviruses are endocytosed after attachment and membrane fusion occurs in the endocytic compartment (Andersen and Nexo, 1983; Andersen, 1983; McClure et al., 1990; Risco et al., 1995; Miyauchi et al., 2009; Diaz-Griffero et al., 2002). The mouse transferrin receptor-1 (mTfR1) is an excellent choice for a retrovirus with a low pH-dependent Env. Binding of its natural ligand, iron-loaded transferrin, to TfR1 stimulates rapid endocytosis into the low pH compartments where acidic conditions change the transferrin structure and release iron (El Hage Chahine et al., 2012; Luck and Mason, 2012). MMTV Env mimics this sequence of events and uses low pH to initiate structural changes in Env during

TABLE 1.1 Retrovirus Entry Receptors

Protein Type	Viruses	Receptor(s)	Cellular Function	Distribution	Fusion Triggers	References
Type I and GPI-anchored forms	ASLV-A	Tva	Unknown member of LDLR family	Broad	Receptor + low pH (Melikyan et al., 2004; Matsuyama et al., 2004; Jha et al., 2011; Padilla-Parra et al., 2012)	Young et al. (1993), Bates et al. (1993) and Melder et al. (2015)
Type I	ASLV-B, -D, -E	Tvb	? Avian TRAIL receptors ?	Broad	Receptor + low pH (Melikyan et al., 2004)	Brojatsch et al. (1996) and Adkins et al. (1997, 2000)
	ASLV-C	Tvc	Avian butyrophilin	Broad	Receptor + low pH (by inference)	Elleder et al. (2005)
	MMTV	mTfR1	Transferrin receptor	Broad	Low pH (Redmond et al., 1984; Ross et al., 2002)	Ross et al. (2002), Wang et al. (2006) and Garcia-Paredes et al. (1991)
	EIAV	eqELR1	Unknown member of TNFR family	Not well characterized	Unknown	Zhang et al. (2005, 2008) and Du et al. (2015)
	Visna virus	? Ovine MR ?	Mannose receptor	Ubiquitous	Unknown	Crespo et al. (2011, 2012)

GPI-anchored	JSRV ENTV	Hyal2	Hyaluronidase	Broad	Receptor + low pH (Cote et al., 2008, 2009, 2011)	Csoka et al. (1999) and Dirks et al. (2002)
Type I and multiple transmembrane-spanning	HIV-1 HIV-2 SIVmne SIVcpz	CD4 (attachment) + CCR5, CXCR4, CCR3, CCR7, and others (fusion)	T-cell activation and costimulation; Chemokine receptors	T cells, monocytes and macrophage; Immune cells	Receptors	Deen et al. (1988), Maddon et al. (1986), Feng et al. (1996), Dragic et al. (1996), Alkhatib et al. (1996), Deng et al. (1996), He et al. (1997), and Xiao et al. (1998)
	FIV	feCD134/OX40 (attachment) + CXCR4 and CCR3 or CCR5 (fusion)	T-cell activation and costimulation; Chemokine receptors	T cells, monocytes and macrophage; Immune cells	? Receptors ?	Poeschla and Looney (1998), Richardson et al. (1999), Johnston and Power (2002), Shimojima et al. (2004), and de Parseval et al. (2005)
	HTLV-1 HTLV-2 STLV-1	GLUT-1/SLC2A1 or NRP-1 (attach)	Glucose transporter VEGF165 receptor	Broad but not on quiescent T cells	Receptors (Gross and Thoma-Kress, 2016)	Manel et al. (2003a), Kinet et al. (2007), Jones et al. (2005a), and Ghez et al. (2006)

Continued

TABLE 1.1 Retrovirus Entry Receptors —cont'd

Protein Type	Viruses	Receptor(s)	Cellular Function	Distribution	Fusion Triggers	References
Multiple transmembrane-spanning	MoMLV Fr57 MLV	mCAT-1/SLC7A1	Cationic amino acid transporter	Ubiquitous	Receptor + host cathepsin (Kumar et al., 2007; Yoshii et al., 2009)	Albritton et al. (1989), Kim et al. (1991), and Wang et al. (1991)
	Ampho4070A dualtropic 10A1	PiT-2/SLC20A2	Inorganic phosphate transporter	Ubiquitous	Receptor + host cathepsin (Kamiyama et al., 2011)	Takeuchi et al. (1992), Kavanaugh et al. (1994), Miller et al. (1994), van Zeijl et al. (1994), and Wilson et al. (1995)
	GALV FeLV-B KoRV-A	PiT-1/SLC20A1	Inorganic phosphate transporter	Ubiquitous	Receptor + ? host cathepsin ?	O'Hara et al. (1990), Takeuchi et al. (1992), Wilson et al. (1995), and Kavanaugh et al. (1994)
	FeLV-A KoRV-B	TC1/THTR1/ SLC19A2	Thiamine transporter	Limited	Receptor + ? host cathepsin ?	Mendoza et al. (2006) and Xu et al. (2013)
	Xenotropic and polytropic MLV	Xpr-1	Phosphate exporter	Broad	Receptor + host cathepsin (Kamiyama et al., 2011; Cote et al., 2012)	Tailor et al. (1999a), Battini et al. (1999), and Yang et al. (1999a)

FeLV-C	FLVCR1 and 2/ SLC49A1 and A2	Heme transporters		Unknown	Tailor et al. (1999c) and Quigley et al. (2000)
PERV-A	PAR2/SLC52A1	Riboflavin transporter	Broad	Unknown	Ericsson et al. (2003)
SRV-1 SRV-2 MPMV/SRV-3 BaboonEV HERV-W FeEV RD114 SNV REV	ASCT2/SLC1A5	Neutral amino acid transporter	Ubiquitous	Receptor + ? host cathepsin ?	Rasko et al. (1999), Tailor et al. (1999b), Marin et al. (2000), and Rasko et al. (1999)
ASLV-J	NHE1/SLC9A1	Na^+/H^+ exchanger	Broad	Unknown	Chai and Bates (2006)

ASLV, avian sarcoma and leukosis virus; *ENTV*, enzootic nasal tumor virus; *GPI*, glycosylphosphatidylinositol; *HTLV*, human T-lymphotropic virus; *JSRV*, jaagsiekte sheep retrovirus; *LDLR*, low-density lipoprotein receptor; *MMTV*, mouse mammary tumor virus; *MoMLV*, Moloney murine leukemia virus; *NRP-1*, neuropilin type 1. Flanking "?" indicate a suspected trigger that has not yet established.

entry, although it binds to a different location on mTfR1 than that of transferrin (Ross et al., 2002; Wang et al., 2006; Garcia-Paredes et al., 1991).

The remaining receptors do not seem to fit the bill as well; binding of their natural ligands does not stimulate endocytosis. Internalization of virus–receptor complexes appears to represent extraordinary events. It may represent subversion of normal membrane protein recycling that occurs as part of quality control inspection for misfolded or damaged proteins and to maintain proper membrane localization. Lipid rafts are one of the major sites for recycling membrane proteins (Diaz-Rohrer et al., 2014), which is, by coincidence or evolutionary design, the normal location of many of the entry receptors.

VIRUS ATTACHMENT

Specific virus attachment to entry receptors leads to productive routes into host cells. The best characterized attachment sites are on gammaretroviruses and lentiviruses. The RBD of the gammaretroviruses resides on the N-terminal half of their SU. The crystal structures of two gammaretroviral RBDs, ecotropic Friend clone 57 MLV and feline leukemia virus subgroup B (FeLV-B), are highly similar (Fig. 1.2A). The conserved residues fold into beta strands that form a skewed immunoglobulin-like core similar in both RBDs (Heard and Danos, 1991; Fass et al., 1997; Barnett et al., 2003). The highly variable regions

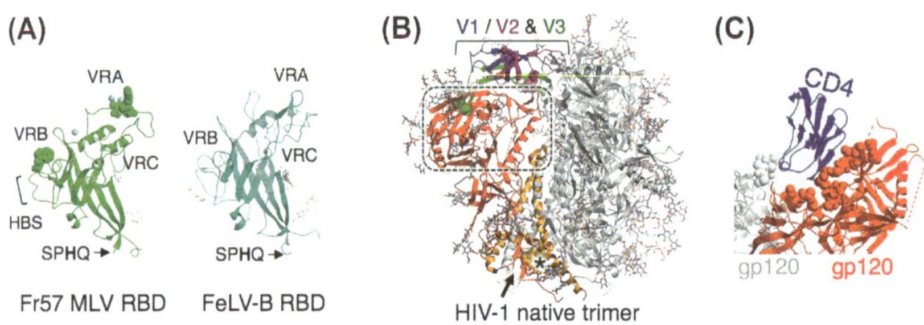

FIGURE 1.2 The structure of envelope proteins and receptor-binding sites. (A) The receptor-binding domain (RBD) of gammaretroviruses Friend clone 57 murine leukemia virus (Fr57 MLV; PDB 1AOL) and feline leukemia virus subgroup B (FeLV-B; PDB 1LCS) are highly similar. The variable regions sit atop a skewed beta-barrel core structure with the heparin-binding site nearby. Aspartate and serine residues in VRA and tryptophan in VRB in the Fr57 MLV RBD are shown space filled; these conserved residues interact with the shared ecotropic MLV entry receptor. The SPHQ motif is located at the base of the barrel adjacent to the beta strand that leads into the proline-rich region. (B) Native HIV-1 trimer ectodomain structure (PDB 3J5M). The coreceptor-binding sites in the V1/V2 and V3 loops lay folded together at the apex of HIV-1 native Env trimers. The CD4-binding site is indicated by a *dotted line*. For simplicity only one of the three protomers is labeled. gp120 (surface subunit) is shown in red except for the V1 (blue), V2 (purple), and V3 (green) loops. gp41 (transmembrane subunit) is shown in gold. The *arrow points* to the two beta strands of gp120 around which gp41 wraps. The *asterisk* indicates the location of the conserved disulfide bond in the $C(X)_5C$ motif of gp41. (C) The CD4-binding site is formed by the residues shown in space-filled form in the cleft between two adjacent gp120 protomers (PDB 5U1F).

VRA, VRB, and VRC form structural elements that present the receptor-binding site residues identified in genetic analyses (Morgan et al., 1993; MacKrell et al., 1996; Zavorotinskaya and Albritton, 1999a; Qian et al., 2004). The attachment sites for the ecotropic MLV and feline retrovirus are located in VRA and VRB (Morgan et al., 1993; MacKrell et al. 1996; Zavorotinskaya and Albritton, 1999a; Qian et al., 2004; Tailor and Kabat, 1997; Tailor et al., 2000).

The structures of the other gammaretroviral RBDs are evidently similar. Chimeric SUs made by swapping RBDs are functional and use the entry receptor of the gammaretrovirus from which the RBD was derived (Heard and Danos, 1991; Battini et al., 1992, 1995; Morgan et al., 1993; Linder et al., 1994). The wild-type RBDs from different gammaretrovirus receptor subgroups complement in *trans* and in *cis* RBD mutants of other receptor subgroups (Lavillette et al., 2000; Anderson et al., 2000; Lauring et al., 2001; Barnett et al., 2001, 2003). The FeLIX protein is a naturally occurring instance of this type of *cis* complementation of the entry defects in the T-cell tropic FeLV (Anderson et al., 2000; Lauring et al., 2001; Barnett et al., 2003).

Lentiviral SUs have two binding sites, one for their primary attachment receptor CD4 and the other for their coreceptors. The initial SU structures solved were of monomeric core gp120, a version containing several mutations required for crystallization and in which the V1/V2 and V3 loops were deleted. This gp120 core bound to soluble CD4 but not to coreceptors, supporting genetic data identifying the deleted variable loops as involved in coreceptor binding (Kwong et al., 1998, 2000; Myszka et al., 2000; Davenport et al., 2015). Core gp120 also bound neutralizing antibody (nAb) 17b known to recognize CD4-induced but not native gp120 (Thali et al., 1993). The structure of the CD4-core gp120-nAb complex showed that the nAb recognized an epitope near the deleted V3 loop (Kwong et al., 1998, 2000) and confirmed genetic and biochemical evidence that the variable loops, particularly V3, are coreceptor-binding sites (Wyatt et al., 1995; Cocchi et al., 1996; Scarlatti et al., 1997; Maeda et al., 2000).

Advances in crystallography and low-resolution cryo-EM made it possible to resolve virtually all of the trimer structure including the receptor-binding sites (Lyumkis et al., 2013; Lee et al., 2016; Stewart-Jones et al., 2016; Liu et al., 2017). The V1/V2 and V3 loops harboring the coreceptor-binding sites lay folded together at the apex of native trimers (Fig. 1.2B). The CD4-binding site is formed by five segments (Fig. 1.2C). Four segments scattered along the C-terminal half of SU comprise the largest area of contact with CD4 (Liu et al., 2017). The fifth segment is near the N-terminus of SU and provides a smaller contact area contributed by the adjacent SU in the trimer (Liu et al., 2017). No structure is yet available for bound coreceptors.

The strength of SU-receptor binding is in the low to middle nM range and comparable to that of monoclonal antibodies. The equilibrium dissociation constant, Kd, for gp120 binding to the HIV-1 attachment receptor CD4 was ~5 nM measured using CD4+ T cells (Schnittman et al., 1988) or purified soluble CD4 (Myszka et al., 2000). The Kd for purified SU from ASLV subgroup A binding to purified soluble Tva receptor was 16.4 nM (Yu et al., 2003), and that of the RBD of Fr57 MLV to its mCAT-1 receptor was 55 nM (Davey et al., 1997).

While SUs have a high affinity for their receptors, many potential host cell types express relatively low levels of the receptors. Retroviruses have evolved alternatives and receptor binding is not necessarily their first interaction with cells. Attachment to nonreceptor, cell surface molecules provides important opportunities to increase encounters with entry receptors or a way to "hang out" avoiding inactivation while waiting to encounter a cell with entry receptors. The first biologically relevant nonreceptor interaction discovered was binding to heparan sulfate proteoglycans on host cells. Some SUs contain consensus heparin-binding sites (HBSs) whose variation increases, and mutation reduces virion binding and infection (Zhang et al., 2003; Vives et al., 2005; Crublet et al., 2008; Hu et al., 2014; Jinno-Oue et al., 2001; Masuda et al., 1996). Soluble heparin inhibits HIV-1, human T-lymphotropic virus type 1 (HTLV-1), MMTV, gammaretroviruses, feline immunodeficiency virus (FIV), and human and feline foamy virus infection (Mondor et al., 1998; Patel et al., 1993; Jones et al., 2005b; Saphire et al., 2001; Okuma et al., 2003; Zacharopoulos and Phillips, 1997; Pinon et al., 2003; Hu et al., 2014; Walker et al., 2002; Plochmann et al., 2012). The SUs of HTLV-1 and -2 also have a binding site for neuropilin type 1 (NRP-1), an alternative VEGF receptor, which appears to provide additional opportunities for attachment particularly to cells expressing very low levels of the entry receptor GLUT-1 (Ghez et al., 2006; Lambert et al., 2009). NRP-1 may be more than an attachment enhancer. It may be an alternative entry receptor for HTLV-1 (Lambert et al., 2009).

Other nonreceptor interactions involve components of the viral envelope. In a nonproductive interaction, lectin receptors on the host cell bind carbohydrate moieties on the viral membrane, and virions are endocytosed without encountering entry receptors. The internalized viruses remain infectious for long periods, effectively converting the cells into Trojan horses. Subsequent contact with entry receptor–positive cells initiates a process called transinfection in which virus is exocytosed and transferred to the susceptible cell (Dutartre et al., 2016). HIV and HTLV-1 particles bind to the lectin receptors dendritic cell–specific intercellular adhesion molecule-3-grabbing nonintegrin (DC-SIGN) on dendritic cells and Siglec-1/CD169 or Siglec-7 expressed on macrophages (Turville et al., 2002; Jain et al., 2009; Izquierdo-Useros et al., 2012, 2014; Varchetta et al., 2013). Originally thought to be the primary cell protein involved in transinfection of HIV-1, DC-SIGN plays a minor role, and the major transinfection pathway uses Siglec-1 on macrophages (Izquierdo-Useros et al., 2014). In contrast, DC-SIGN remains important in HTLV-1 transmission to T cells (Jain et al., 2009). Ecotropic MLV particles also bind Siglec-1 to initiate nonproductive endocytosis in mouse macrophages and these cells subsequently transfer virus for active infection of proliferating lymphocytes, particularly B cells (Erikson et al., 2015). After injection into mouse footpads, MLV and HIV-1 virions rapidly appear in the draining lymph nodes bound to Siglec-1+ CD11b+ macrophage, which subsequently transferred virus particles to B cells (Sewald et al., 2015).

Phosphatidylserine exposed on retroviral membranes provides another opportunity for enhancing infection through a process called apoptotic mimicry. Viruses often become enveloped by membrane in which phosphatidylserine flipped from the inner to the outer leaflet has accumulated (Zwaal et al., 2005). Cellular phosphatidylserine receptors and scavenger systems, including Gas6–Axl and T cell immunoglobulin and mucin domain-1 and -4 (TIM-1 and TIM-4), enhance HIV-1 and gammaretrovirus entry (Li et al., 2014). Phosphatidylserine in the HIV-1 envelope is particularly important to infection of primary human macrophages (Callahan et al., 2003). These same cellular systems also inhibit egress of new HIV-1 particles, albeit the entry enhancing effects of TIM-1 were greater than its egress-inhibiting activity (Li et al., 2014). In addition, small deletions in TIM-1 associate with a lower risk of HIV-1 transmission, higher CD4 counts once infected and longer survival times in antiretroviral-naïve cohorts from Thailand, Italy, Peru, and Columbia (Wichukchinda et al., 2010; Biasin et al., 2017), suggesting that in the balance of things, the entry enhancement afforded by apoptotic mimicry outweighs the inhibitory effects on virion release.

THE BASICS OF MEMBRANE FUSION

Retrovirus particles are covered by a membrane obtained from the host cell during virus budding. This lipid bilayer entraps the viral enzymes and assists virus in evading host immune recognition. It also complicates entry by requiring special arrangements for the merger of the virus membrane with the membranes of a new host cell. That task falls to the retroviral Env proteins. Their domain and tertiary structure identify them as class I membrane fusion proteins.

Membrane fusion occurs in discrete well-characterized steps (reviewed in Melikyan, 2008). First the membranes are brought into close juxtaposition. At the point of closest contact, a hemifusion intermediate forms. It is a small stalk-like structure in which the outer leaflet lipids mix but the inner leaflets remain intact and separate. In the absence of a catalyst, this hemifusion intermediate is unstable. It resolves by either mixing of inner leaflet lipids to form a small pore or by collapsing. If a fusion pore forms and expands, the two bilayers mix freely and fusion is completed. If the pore does not form or expand, fusion fails.

Why is membrane fusion dependent on the TM? The net negative charge on the outside of biological membranes is a barrier to their close contact. The repulsive force increases as the two membranes approach closer to each other (Rand and Parsegian, 1984; Cowley et al., 1978; Marra and Israelachvili, 1985). When they are within 30 Å, the hydration layer surrounding the polar portions of the membrane lipids generates a repulsive force too large to be overcome without a catalyst of some kind (Rand and Parsegian, 1984; Marra and Israelachvili, 1985). Enter the TM proteins; one of their primary functions is to displace the hydration layer and eliminate the repulsion. Once the hydration layer separating the membranes is gone, outer leaflet lipids mix and fusion begins.

How does TM displace the hydration layers and catalyze formation of the hemifusion intermediate? The process begins with extension of the fusion peptide from the apex of Env trimers and its insertion into the opposing membrane. While there is no consensus sequence for retroviral fusion peptides, they are characteristically 26–30 residues in length, form an alpha helix in lipid bilayers, and are rich in the hydrophobic amino acids alanine, leucine, and glycine interspersed with a few polar threonine or serine residues. As they move toward the host membrane, the hydrophobic fusion peptides drive out water of hydration, insert, and pull the membranes together. Outer leaflet lipids begin to flow between the membranes and the stalk structure of the hemifusion intermediate forms. At this point the alpha helix of the fusion peptide aligns with the end of the HR1 helix to span completely across the host membrane (Fig. 1.3A).

How does TM catalyze formation of the fusion pore? There is a turn in the peptide chain immediately after HR1, which is maintained in all TMs by a conserved disulfide bond. Once the hemifusion intermediate is formed, the extended fusion peptide–HR1 helix pivots at this enforced turn to snap back, associates with the HR2 helices, and forms a six-helix bundle. The six-helix bundle is the postfusion conformation and the pivoting step is often referred to as chain reversal (Fig. 1.3B). While there is as yet no experimental evidence that supports or refutes the concept, it is thought that during this step the fusion peptide catalyzes formation of a fusion pore, perhaps by dragging lipids from the host membrane along with it as it snaps back to lay along side the trimerization domain.

In these steps, TM is aided by the membrane composition of the host cell. Boundaries of cholesterol-rich lipid domains are energetically favorable environments into which fusion proteins can insert (Schroeder, 2010; Yang et al., 2015, 2016). Cholesterol depletion disrupts rafts, disperses the entry receptors, and reduces infection (Liao et al., 2001, 2003; Wielgosz et al., 2005; Schroeder, 2010). Mechanistically cholesterol depletion of host membranes inhibits resolution of the hemifusion intermediate into a fusion pore and disrupts MPER function (Biswas et al., 2008; Schroeder, 2010; San-Juan-Vergara et al., 2012). The favorable energetics of lipid rafts for virus fusion explains a benefit of using entry receptors characteristically located in cholesterol-rich lipid rafts. This benefit is apparently so critical that HIV-1 can escape the small molecule CCR5 inhibitors maraviroc and vicriviroc by mutating the fusion peptide and V3 loop to preferentially use CCR5 clusters in cholesterol-rich lipid rafts (Berro et al., 2011).

The cholesterol found on lentiviral and gammaretroviral membranes also favors fusion. Cholesterol depletion of virus particles inactivates them (Liao et al., 2003; Ablan et al., 2006). While depletion using methyl-beta-cyclodextrin permeabilizes virions causing loss of the replication enzymes reverse transcriptase and integrase, its depletion using 1-phenyl-2-hexadecanoylamino-3-morpholino-1-propanol removes cholesterol without permeabilizing virions and

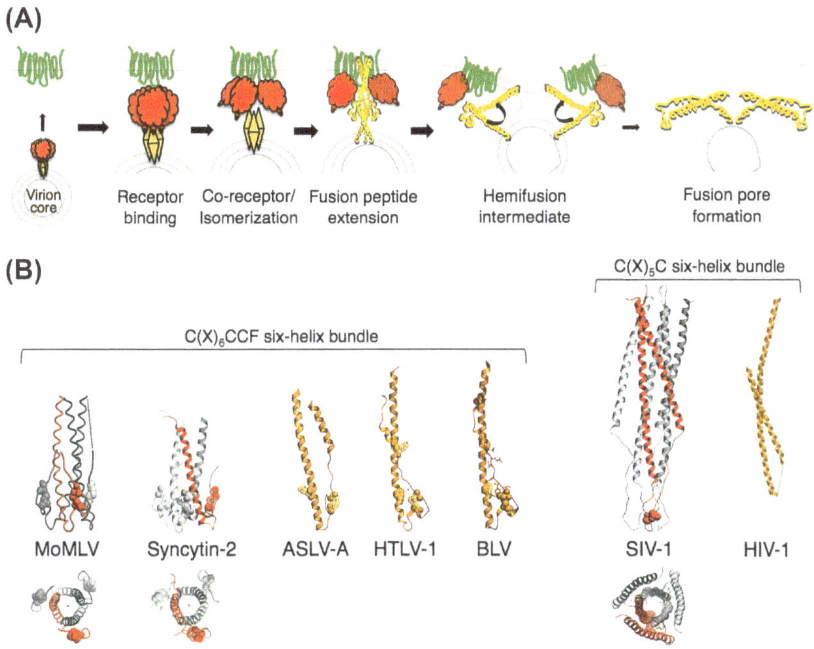

FIGURE 1.3 Membrane fusion and the fusion machinery. (A) Diagram of membrane fusion depicted from the plane parallel to the host cell membrane. Surface subunits (SUs) are in red. Transmembrane subunits (TMs) are gold. The prototype entry receptor is green. (B) The postfusion structure of TM. The six-helix bundles are shorter and more compact in viruses with a covalent disulfide bond between SU and TM than in viruses that lack an intermolecular bond between subunits, yet the core coiled-coil is of similar cross-sectional dimension. Trimers are shown where trimer coordinates are available in public databases. Top, view parallel to the plane of the membrane; disulfide-bonded cysteines in the chain reversal region are shown space filled. Bottom, perpendicular view looking down the trimer axis from the membrane. A chloride ion sits in the center of the trimer axis view of MoMLV and syncytin-2. *ASLV-A*, avian leukosis and sarcoma virus subgroup A (PDB 5H9C); *BLV*, bovine leukemia virus (PDB 2XZ3); *HIV-1*, human immunodeficiency virus 1 (PDB 1ENV); *HTLV-1*, human T-lymphotropic virus type 1 (PDB 1MG1); *MoMLV*, Moloney murine leukemia virus trimer (PDB 1MOF); *SIV-1*, simian immunodeficiency virus 1 trimer (PDB 2EZO); *Syncytin-2*, ERV-FRD (PDB 1Y4M).

yet still inactivates them (Ablan et al., 2006). The contribution of viral membrane rafts may simply be the ease of mixing lipids of similar composition. Env trimers localize within lipid raft–like domains on virions by virtue of palmitoylation on their membrane-spanning anchor and the cholesterol-binding motif in MPER (Hensel et al., 1995; Yang and Compans, 1996; Ochsenbauer-Jambor et al., 2001; Yang et al., 1995; Gonzalez et al., 2012; Li et al., 2002; Rousso et al., 2000; Li and Papadopoulos, 1998; Epand et al., 2003, 2006; Schroeder, 2010). The site in the HIV-1 MPER is so proficient at cholesterol binding that it became a standard for studying cholesterol binding (Epand et al., 2003, 2006; Fantini and Barrantes, 2013; Apellaniz et al., 2014).

How many trimers give maximal fusion? The density of Env on HIV particles is low with an average of 8–10 trimers per virion suggesting that maximal fusion efficiency does not require large numbers of trimers (Zhu et al., 2003). A single trimer of gp160 is sufficient to support in vitro HIV entry but maximal merger of viral and cell membranes requires interaction of 2–3 trimers with receptors (Yang et al., 2005; Magnus et al., 2009; Brandenberg et al., 2015). Strain differences in the minimal number of trimers are also seen and these correlate with infectivity and the rate of entry in vitro (Brandenberg et al., 2015).

DO RECEPTOR INTERACTIONS CONTRIBUTE TO THE ENVELOPE PROTEINS FUNCTIONS THAT DRIVE ENTRY?

To answer this question, consider first the mechanisms by which entry functions are controlled, and then consider how these constraints are relaxed when a host cell is encountered. Similar to all class I fusion proteins, retroviral Env are synthesized as a single protein chain, which folds into *its* lowest energy conformation, then it is cleaved into two subunits and associates with two other heterodimers to form trimers. The cleavage into two subunits creates two characteristics of class I fusion proteins—a new N-terminus positioned before a hydrophobic fusion peptide and an alternative Env conformation with lower free energy than the native one. Tight but flexible control of the binding and fusion activities of Env is important in focusing viruses on receptive host cells.

The native trimer structure is a closed one held together by a coiled-coil formed by the three HR1 alpha helices and by associations of their membrane anchors within the viral membrane. The fusion peptides are not extended; they are trapped beneath SU and within the trimer core. To initiate membrane fusion, a stimulus or trigger must move SU out of the way and open up a passage through which fusion peptides can extend.

Membrane fusion is controlled by the triggers that induce SU to rotate out of the trimer center and away from the fusion peptides. These changes give an open conformation through which the fusion peptide can move toward the host membrane. They are essential preludes to formation of the hemifusion intermediate, which is near the energy barrier between the native and the lower energy alternative conformation. Pivoting of TM around its conserved intramolecular disulfide bond in the chain reversal region and the descent of TM into the lowest energy conformation, e.g., the six-helix bundle, completes membrane fusion.

The Triggers

Three triggers are known for retrovirus entry: (1) receptor binding; (2) low pH; and (3) host protease cleavage. The triggers vary among and within the seven genera (Table 1.1). The lentiviruses use a combination of two receptors to trigger fusion peptide extension and six-helix bundle formation (reviewed in Wilen et al., 2012). The gammaretroviruses use receptor binding and host cathepsin cleavage as triggers (Wallin et al., 2004; Barnett et al., 2001; Barnett and Cunningham, 2001; Kumar et al., 2007; Yoshii et al., 2009; Kamiyama et al., 2011). The alpharetroviruses and the betaretroviruses, JSRV and ENTV, use receptor binding combined with low pH (reviewed in Barnard and Young, 2003). Low pH is an essential trigger for MMTV and spumaviruses, but it is not clear if receptor binding acts as the other trigger or if acidic conditions alone induce both conformational changes (Redmond et al., 1984; Ross et al., 2002; Picard-Maureau et al., 2003). The triggers for the deltaretroviruses, and the type D simian retroviruses of the betaretrovirus genus, and baboon and feline endogenous virus members of the gammaviruses have not yet been determined. However, highly conserved motifs within their Env suggest that these retroviruses use mechanisms similar to the gammaretroviruses.

Conserved Motifs in Envelope Proteins Are Critical to Regulation of Fusion

Four motifs involved in entry regulation are shared by all retroviruses. First, a conserved motif provides a furin-like, host protease cleavage site Arg-X-Lys/Arg-Arg (RXK/RR) between SU and TM (Hosaka et al., 1991). After Env precursors fold into the metastable native conformation, then this motif provides a mechanism to lower the energy barrier between the native and postfusion conformations, e.g., break a peptide bond and separate the two subunits. Second, the C-terminal portion of HR1 serves as the trimerization domain for the native conformation and as the core on which the lowest energy postfusion conformation forms. Its sequence is a 4-3 repeat of hydrophobic residues analogous to the heptad repeat in the influenza A hemagglutinin fusion subunit (Fass and Kim, 1995; Fass et al., 1996; Kobe et al., 1999; Lamb et al., 2011; Aydin et al., 2014). The third conserved motif consists of the segment containing the two disulfide-bonded cysteine residues located between the HR1 and HR2 that enforce the turn around which TM pivots as it collapses into the six-helix bundle. Often called the chain reversal region, this conserved disulfide bond is essential to the transition from extended fusion peptide to six-helix bundle in native trimers but not for the formation of the six-helix bundle in isolated TM ectodomains. An HIV-1-derived gp41 core in which the chain reversal region was replaced by a flexible linker and a simian immunodeficiency virus type 1 (SIV-1) gp41 core in which the cysteines were mutated to alanines both folded spontaneously into six-helix bundles (Fig. 1.3B) (Weissenhorn et al., 1997; Yang et al., 1999b). Lastly, the MPERs contain cholesterol-binding motifs thought to aid in lipid mixing by lifting cholesterol to dimple or perturb the viral membrane as the fusion peptide extends during hemifusion intermediate formation (Schroeder, 2010; Apellaniz et al., 2014).

Regulation by Control of the Conformation of Surface Subunit

Primary isolates and laboratory strains contain unique additions of other motifs and subtle variations in some of the conserved motifs. These fall into two groups that correlate with two mechanisms for controlling retrovirus entry. One group of motifs correlates with regulation primarily by control of the conformation of SU. This group is present in all lentiviruses and the betaretroviruses MMTV, JSRV, and ENTV. Viruses having this group of motifs use type I membrane receptors alone or in combination with multiple membrane-spanning coreceptors.

The SUs of these viruses are not covalently bound to their TM; the heterodimers are held together by noncovalent interactions between the subunits. The consensus for the disulfide motif in the chain reversal region of TM is **C**$(X)_{5\text{-}7}$**C**, where C is cysteine, X is any amino acid and the number of residues between the two cysteines varies from five to seven (Fig. 1.1C). In native trimers the fusion peptide and N-terminal half of HR1 fold underneath and next to segments of SU. Rearrangement of their SU disrupts these interactions and opens up an internal pathway to the apex of the trimer through which the fusion peptide can extend. Consequently, the entry triggers focus on inducing conformational changes in SU as the major factor controlling membrane fusion and entry.

Regulation by Control of Disulfide Bond Isomerization and Surface Subunit Conformation

The second group of motifs is consistent with regulation of fusion by control of disulfide bond isomerization and SU conformation. These motifs are found in the Env of gamma-, delta-, and alpharetroviruses and the type D betaretroviruses. Their consensus chain reversal motif is **C**$(X)_6$**CC**F, where F is a phenylalanine residue. The first two cysteine residues are covalently bonded and form the pivotal turn in TM. The third cysteine is covalently bonded to a cysteine residue in SU. This striking difference introduces an additional layer of control: the disulfide bond between SU and TM inhibits resolution of the hemifusion intermediate and fusion pore formation. Disruption of this bond by isomerization breaks the final restraint on six-helix bundle formation (Li et al., 2007, 2008).

How is this isomerization controlled? The SUs have two unique motifs: a **CXXC** motif located downstream of the RBD and an SP**HQ** motif located at the N-terminus of SU, where S is serine, P is proline, H is histidine, and Q is glutamine. One cysteine in the CXXC motif is unpaired and the other is covalently bonded to the third cysteine in the **C**$(X)_6$**CC**F of TM (Li et al., 2008) (Fig. 1.1B). Receptor binding triggers conformation changes that rotate SU and presumably bring the SP**HQ** motif close to this critical disulfide bond where it mediates isomerization (Li et al., 2007). After isomerization, the third cysteine of the **C**$(X)_6$**CC**F in TM becomes unpaired, and the disulfide bond is swapped into an intramolecular one between the two cysteines in the **CXXC** motif in SU.

The alpharetroviruses use a variation of this isomerization. A V**H**LL motif at the N-terminus of SU likely provides the histidine and their **CXXC**-like motif is located in the first 45 residues of mature SU instead of at its approximate center as in gamma- and deltaretroviruses.

Other features that bear on control of entry are the relatively short trimerization and HR2 domains and lack of a clustered glycosylation domain between their chain reversal motif and HR2. The trimerization domain is also uniquely conserved with a consensus sequence of $(Hy)_3$**QNR**RGLDLLF, where Hy is any hydrophobic residue although valine, leucine, and alanine prevail, N is asparagine, R is arginine, G is glycine, and L is leucine. The asparagine residue in **QNR** provides a chloride ion–binding site that stabilizes the coiled-coil structure (Fass et al., 1996; Renard et al., 2005; Kobe et al., 1999; Lamb et al., 2011; Aydin et al., 2014). As a result of these unique features, the distance the fusion peptide can extend is reduced and the six-helix bundle is substantially shorter (Fig. 1.3B). It has been proposed that multiple membrane-spanning proteins are preferred receptors for MLV because they position virions close to the host membrane (Katane et al., 2002), a characteristic that would accommodate the shorter "reach" and more compact size of their Env.

Lentiviruses: Receptor-Triggered Conformational Changes in the Surface Subunit

Our clearest understanding of Env–receptor interactions and the nature of the fusion triggers are those of HIV-1 Env, its attachment receptor CD4 and fusion coreceptors CCR5, CXCR4, and other chemokine receptors. The restraints preventing fusion peptide extension prior to receptor binding are evident in the structures of native HIV-1 trimers (Lee et al., 2016; Stewart-Jones et al., 2016; Lyumkis et al., 2013; Julien et al., 2013). The N-terminal 52 residues of gp41, including the fusion peptide and the N-terminal half of HR1, are held within native trimers by interactions with gp120 residues that are highly conserved in primary isolates. While some gp120 residues fold on top of part of gp41, the most constraining interactions involve a two-stranded β-sheet formed by the first and last β-strands of gp120 and around which the fusion peptide and N-terminal HR1 residues wrap (Fig. 1.2B) (Lee et al., 2016; Stewart-Jones et al., 2016).

Comparisons of the structure of native Env trimers to that of the CD4-bound trimers suggest that CD4 binding triggers the following changes (Kaplan et al., 2016; Guttman et al., 2014; Lee et al., 2016; Stewart-Jones et al., 2016; Lyumkis et al., 2013). The N-terminal IgG-like domain of CD4 binds to a pocket in between gp120 monomers where there is a gap in the glycan shield. Binding in this pocket disrupts interactions between two β-strands in gp120, which together had been holding the V1/V2 loops within the structure. This change releases the V1/V2 loops and they flip outward opening up the interior of the trimer. CD4-bound HIV-2 trimers show similar changes (Davenport et al., 2015). The apex of the trimer is still blocked by the V3 loops after CD4 binding. Trimers bound to soluble CD4 (sCD4) and nAbs (used as surrogates for coreceptor binding) have open structures consistent with V3 loop extension and gp120 rotation after coreceptor binding (Liu et al., 2017). In agreement with this interpretation, native trimers of the CD4-independent strain CP-MAC of SIV are in a similar open conformation (White et al., 2010).

Strikingly, there is almost no change in the conformation of gp41 even in the open trimers of the sCD4- and antibody-bound structures (Liu et al., 2017). The fusion peptide and N-terminal half of HR1 remain wrapped around the N-terminal and C-terminal beta strands of gp120 as well as under SU and maintain similar interactions to those seen in native and sCD4-bound trimers. Pancera and coworkers noted that transition from the CD4-bound structure would likely require that the two terminal beta strands of gp120 be pulled away from its wrapping of gp41 helices (Pancera et al., 2014). For example, coreceptor binding likely induces more extensive conformational changes than the nAbs and might induce the terminal beta strands of gp120 to flip upward and join the half barrel formed by the residues of gp120 that lay over the fusion peptide (Fig. 1.4B). This change would relieve the strongest conformational constraints and free the fusion peptide and N-terminus of HR1 to extend upward and insert into the host membrane. Incubation of sCD4 and HIV-1 virions results in dissociation of sCD4–gp120 complexes (Moore et al., 1990; Hart et al., 1991; Sattentau and Moore, 1991; Willey et al., 1994). A similar dissociation of gp120–CD4–coreceptor complexes might be the final trigger that induces the extended fusion peptide-HR1 helices to snap back into a six-helix bundle (Fig. 1.4D).

Gammaretroviruses: Receptor- and Cellular Protease–Driven Disulfide Bond Isomerization

Env–receptor interactions are also well characterized for gammaretroviruses. Here receptor binding induces conformational changes, which are enhanced by host cell cathepsin cleavage of SU during endocytosis. The net effect of receptor and protease action is to induce rotation of SU and open a pathway for fusion peptide extension and hemifusion intermediate formation and then induce dissociation of SU by isomerization of the covalent SU–TM bond. Cryo-EM tomograms of isolated ecotropic MLV trimer ectodomains and native trimers on virions are high resolution and do not show atomic details but indicate that the native trimer apex is closed or partially closed (Forster et al., 2005; Wu et al., 2008; Riedel et al., 2017). In activated but isomerization-arrested Moloney murine leukemia virus (MoMLV) trimers, the protomers have rotated and opened the apex while the base of the trimer remains unchanged (Wu et al., 2008), similar to the changes seen in sCD4- and antibody-bound HIV-1 trimers.

The covalent bond between SU and TM is labile and involves the **CXXC** motif, which is **CWLC**, and resembles the active site in thiol-exchange disulfide isomerases (Pinter et al., 1978, 1997). In native trimers, the third cysteine in the

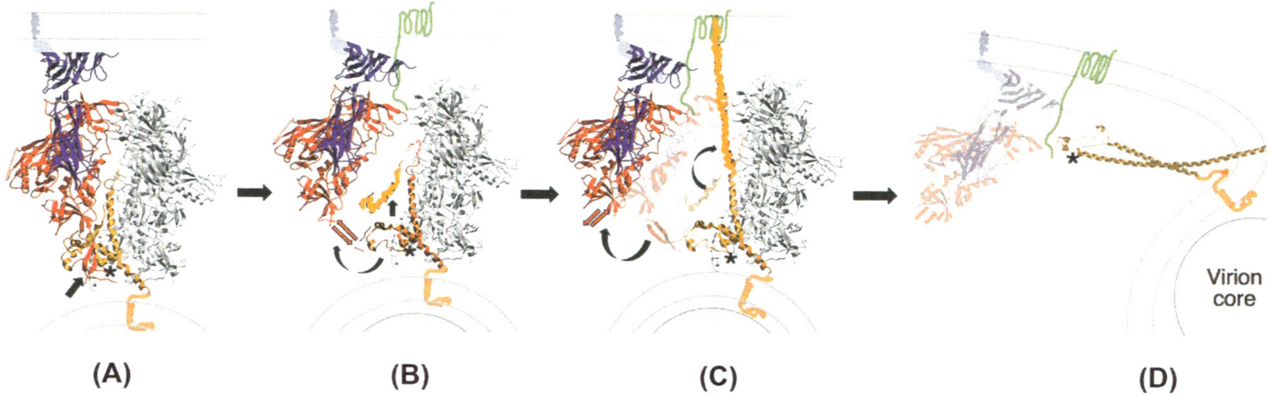

FIGURE 1.4 Model of fusion triggering mechanisms in HIV-1 entry. For simplicity, structural changes are modeled in one SU–TM heterodimer (red and orange) and one CD4 (blue). CD4 binding opens the apex of the trimer, rotates gp120, and induces the V1/V2 and V3 loops to flip upward where they become accessible to coreceptors. gp41 remains wrapped around the N- and C-terminal beta strands of gp120 (*arrow*) constraining its structural rearrangement (PDB 5U1F). Coreceptor (green) binding is predicted to trigger long range changes that pull the gp120 terminal beta strands out of gp41 and open up the central axis of the trimer releasing the fusion peptide to move toward the host membrane. The short alpha helices of the fusion peptide join end to end with those of HR1 and insert in the target membrane. The terminal beta strands of gp120 collapse into a half barrel at the base of SU and the six-helix bundle forms as a fusion pore is opened. The hemifusion intermediate is not shown. *Orange lines* and *red arrows* indicate hypothetical TM and SU structure. An *asterisk* indicates the intramolecular disulfide bond in the chain reversal region. (A) CD4 binding. (B) Coreceptor binding. (C) Fusion peptide extension. (D) Fusion pore formation. *SU*, surface subunit; *TM*, transmembrane subunits.

C(X)$_6$**CC**F motif of TM forms a disulfide bond with a cysteine in the **C**WL**C** motif on SU. This SU–TM disulfide bond is the final barrier to membrane fusion. Mutation of the first cysteine residue of this motif inhibits infection and blocks isomerization, both of which were rescued by the reducing agent dithiothreitol (Li et al., 2008).

Receptor binding primes the isomerization by triggering a change in the location of the SU–TM bond within trimers (Wallin et al., 2004; Barnett et al., 2001; Barnett and Cunningham, 2001), evidently placing it close to the SP**H**Q motif, which initiates isomerization by ionizing the labile bond with **C**WL**C** (Li et al., 2007). The histidine residue is critical for membrane fusion; its mutation blocks virus-cell fusion at the hemifusion intermediate (Bae et al., 1997; Zavorotinskaya and Albritton, 1999b; Zavorotinskaya et al., 2004; Qian and Albritton, 2004). Gammaretroviruses have a genus-specific PXXIGPNPVLADQ motif at the end of their RBD that apparently interacts with the SP**H**Q motif and influences fusion. Natural variations in this motif suppress the fusion activity of Env (Lavillette et al., 1998), whereas other changes in this motif can suppress the fusion defects resulting from mutation of the active histidine in the SP**H**Q motif (Zavorotinskaya and Albritton, 1999b).

Host cathepsins encountered in the endosomal compartment also play an important role. Early observations suggested that a host protease likely a cathepsin was involved in regulating MLV fusion (McClure et al., 1990; Andersen, 1983). Cathepsin B inhibitors reduce ecotropic MoMLV infection and embryonic fibroblasts from cathepsin B knockout mice are resistant to MoMLV (Kumar et al., 2007). The protease acts by enhancing the membrane fusion activity of Env. A short preincubation of virions with purified cathepsin B increases MoMLV infection of NIH3T3 and cathepsin B−/− fibroblasts, and addition of exogenous cathepsin B increases Env-mediated cell–cell fusion (Kumar et al., 2007). Cathepsins B and L are similarly involved in potentiating fusion during amphotropic, xenotropic, and polytropic MLV infection (Yoshii et al., 2009; Kamiyama et al., 2011). Cathepsins do not degrade TM or SU instead they cleave SU into precise fragments (Kumar et al., 2007). These findings suggest that the enhancing effect relates to cleaved SU fragments being more responsive to receptor-induced conformational changes.

Alpharetroviruses: Receptor- and Low pH-Driven Disulfide Bond Isomerization

While the TMs of alpharetroviruses contain the **C**(X)$_6$**CC**F motif and the third cysteine residue covalently bonds to a cysteine in SU, their **C**XX**C**-like motif is located in the first 45 residues of mature SU instead of at its approximate center as in gamma- and deltaretroviruses. Its consensus is a **C**LSTQSATSPFQ/RT**C** where the flanking cysteines are the first and second ones in SU. The intersubunit bond of native trimers forms between the third cysteine of the **C**(X)$_6$**CC**F motif and the first cysteine residue in their N-terminal **C**LSTQSATSPFQ/RT**C** motif (Pike et al., 2011). The second cysteine in SU is unpaired in native trimers (Pike et al., 2011).

Here the isomerization is triggered by low pH (Earp et al., 2003; Barnard et al., 2004; Jha et al., 2011). It is thought that after isomerization a new intramolecular disulfide bond forms between the cysteine residues flanking the N-terminal **C**LSTQSATSPFQ/RT**C** motif. The fate of the third cysteine in the C(X)$_6$**C**CF motif is unknown and it may not be unpaired. Perhaps it swaps partners and disrupts the native disulfide bond holding the fusion peptide in a loop. If it paired with the second cysteine at the C-terminus of the fusion peptide, the N-terminal cysteine would be unpaired and the fusion peptide could extend its full length. While there is no SPHQ motif, the sequence V**H**LLQXPG near the N-terminus of their SU could provide a histidine residue to mediate isomerization. It has not yet been examined for this function.

DO ENV–RECEPTOR INTERACTIONS CONTRIBUTE TO PATHOGENESIS?

CD4+ T-Cell Depletion and AIDS

In the absence of treatment, the numbers of circulating CD4+ T cells decline during the acute stage of HIV-1 infection, rebound after seroconversion albeit seldom to normal levels, and decline again as the latent stage ends. While CD4+ T cells are lysed within a day of infection, their depletion outpaces the level of infection suggesting that nearby uninfected cells are also killed (reviewed in Garg and Blumenthal, 2008). Bystander killing might then contribute to development of AIDS. What is the mechanism for bystander killing and does it occur in HIV-infected patients? In vitro studies show that HIV-1 Env interaction with CCR5+ cells correlates with bystander CD4+ T-cell depletion and it is now thought to be a major mechanism for in vivo T-cell depletion (Joshi et al., 2011, 2016; Tsao et al., 2016). Another lentivirus, FIV, causes an AIDS-like immune deficiency in cats. FIV infection mediates CD4+ T cell loss through a similar mechanism, in this case involving Env-CXCR4 coreceptor interactions (Garg et al., 2004).

Env–receptor interactions also contribute to pathogenesis through enhancement of virus spread. The virological synapse is a stable junction for efficient transmission of HIV-1 sequestered from immune surveillance (Jolly et al., 2004; Jolly and Sattentau, 2004; Durham et al., 2012; Deng et al., 2016). In addition to binding sites for CD4, the coreceptors, and the attachment enhancers heparan sulfate and TIM-1, gp120 has a binding site for integrin α4β7, the intestinal mucosal homing receptor on gut T cells. This interaction occurs after CD4 binding and does not involve enhanced attachment. Instead gp120 binding to integrin α4β7 on T cells activates the T-cell membrane protein lymphocyte function–associated antigen 1 and stimulates virological synapse formation for highly efficient cell-to-cell spread (Fortin et al., 1998; Arthos et al., 2008).

Through this mechanism, gp120 interactions are believed to be responsible for the rapid depletion of the gut CD4+ T cells (Arthos et al., 2008).

Neurological Damage and HAM/TSP

In HTLV-1-infected individuals, the downregulation of the virus entry receptor GLUT-1 in chronically infected cells not only establishes resistance to superinfection but also correlates with increased pathogenesis. While the majority of HTLV-1-infected individuals are assymptomatic carriers with very low proviral loads, about 2%–3% experience more virulent infections that progress over decades to HTLV-1-associated myelopathy or tropical spastic paraparesis (HAM/TSP), a neurodegenerative disease (Oh and Jacobson, 2008). Discovery of the HTLV-1 entry receptor GLUT-1 led to new insights into the rather mysterious neuropathology of HAM/TSP. Interaction of GLUT-1 with newly synthesized HTLV-1 Env results in the downregulation of the receptor (Ghez et al., 2010). Using fluorine-18 fluorodeoxyglucose (18)F-FDG positron emission tomography to measure glucose uptake in the brains of HTLV-1-infected individuals, two studies found that glucose transport function was lower in the brains of HTLV-1-infected individuals than in the brains of uninfected controls (Taniguchi et al., 2015; Schutze et al., 2017). Neurons are particularly sensitive to the loss of full glucose transport activity and these authors speculated that at least part of HTLV-1-associated neuropathy is due to the mechanism of resistance to superinfection.

Mutations That Adapt Virus to Low Receptor Levels Can Also Increase Pathogenicity

The ability to enter efficiently into cells with low receptor numbers can expand the cell tropism or accelerate the rate of virus spread. One adaptation to use of low receptor levels is through increased receptor affinity. Primary isolates from a cohort of cats naturally infected by a subgroup A feline gammaretrovirus caused multicentric lymphoma of non–T-cell origin instead of the typical T-cell lymphoma associated with FeLV-A infection (Chandhasin et al., 2005). This atypical and more aggressive pathogenesis mapped to a single residue substitution within SU that conferred a two-fold increase in binding affinity for feTHTR1, the FeLV-A entry receptor (Bolin et al., 2011, 2013).

Neurotropic variants of MLV can also map to the *env* gene. Neurovirulence typically relates to expanded tropism resulting from mutations that increase affinity for heparin sulfate proteoglycans on cell types with sparse receptors. For example, the neurotropic PVC-211 variant of ecotropic MLV harbors two point mutations that enlarge the HBS on SU and render it high affinity (Masuda et al., 1993; Jinno-Oue et al., 2001). The variant is better able to attach to the brain capillary endothelial cells, which make up the blood–brain barrier (Jinno-Oue et al., 2001, 2003). The *env* gene is also the source of neuropathy in ecotropic

Cas-Br-E MLV-infected mice; transgenic *env* mice develop neuropathy in the absence of viral replication and other viral genes (Kay et al., 1993). This virus first gains access to the central nervous system through increased ability to infect brain capillary endothelial cells (Gravel et al., 1993). Polytropic FR98 MLV is neurovirulent by virtue of expansion of its tropism to microglial cells in which nonneurotropic variants of polytropic MLV do not infect (Robertson et al., 1997). Alternatively, neurotropism can result from subtle alteration of the fusion triggers. The neurovirulence of the TR1.3 variant of ecotropic FB29 MLV also maps to its *env* gene and specifically to a single point mutation in its RBD. This mutation increases the propensity of SU to respond to receptor binding and move out of the way of the fusion peptide (Park et al., 1994; Chung et al., 1999; Landers et al., 2004). Interestingly, the increased responsiveness to fusion triggers came at the price of a 10-fold reduction in receptor-binding affinity (Murphy et al., 2006), suggesting that enhanced triggering was more important to brain capillary endothelial cell infection than receptor binding per se.

Envelope Protein–Driven Neoplasia in Betaretrovirus Infection

The ovine betaretroviruses, JSRV, and ENTV are the causative agents of ovine pulmonary adenocarcinoma (York et al., 1992; Palmarini et al., 1999; DeMartini et al., 2001) and enzootic nasal adenocarcinoma (Walsh et al., 2013), respectively. These neoplasia result from Env-induced cellular transformation (Allen et al., 2002; Chow et al., 2003; Hull and Fan, 2006; Dirks et al., 2002; Alberti et al., 2002). The transformation process is best understood for JSRV. Env interactions with HyaL2, the shared receptor for JSRV and ENTV, have no apparent role in this neoplasia (Chow et al., 2003). Instead it results from interactions of TM and specifically of its cytoplasmic tail with nonreceptor cell proteins.

Studies using the yeast two-hybrid system identified two cellular proteins that coimmunoprecipitate with JSRV Env. The cellular RalA-binding protein-1 (RALBP1), a transcription suppressor in Ras family signaling was identified using the CT domain of Env as bait and a human HeLa cell library as prey (Monot et al., 2015). Independently the cellular Kruppel family zinc finger protein 111 (Zfp111) was identified in a screen using full-length Env as bait and mouse liver cDNAs as prey (Hsu et al., 2015). RALBP1 expression is repressed in JSRV-induced lung tumor cells but in untransformed MDCK cells RALBP1 colocalized with Env in the ER and at the plasma membrane and RALBP1–Env complexes coimmunoprecipitated (Monot et al., 2015). In contrast, the Env bound to Zfp11 was not glycosylated and the complexes localized to the nucleus (Monot et al., 2015). It is possible that both interactions occur in infected lung cells and may cooperate in transformation with reduced RALBP1 expression providing opportunities for Zfp111– glycosylation-negative Env complexes to form early in Env synthesis, which could divert some of the Env mRNAs from normal route of membrane translation in the ER to wholly cytoplasmic translation.

HOST DEFENSES THAT INHIBIT RETROVIRAL ENTRY ALSO DRIVE ENVELOPE PROTEIN VARIATION

Challenges in Vaccine Development

In untreated HIV-1-infected individuals, anti-gp41 and anti-gp120 IgG and IgM appear by 4 weeks after virus is detectable in blood, however, these early antibodies are not neutralizing because they recognize epitopes not displayed on native Env trimers (Tomaras et al., 2008). The mucosal anti-gp41 IgA that is elicited in genital fluids within weeks of transmission is also not neutralizing and persists for only a few days (Yates et al., 2013). nAbs take months to evolve and for many years remain specific only to the transmitted founder virus and its quasispecies (Richman et al., 2003; Wei et al., 2003). These early autologous nAbs continually evolve to keep up with the genetic drift of Env as it races to escape each version of nAb (Richman et al., 2003; Wei et al., 2003; Murphy et al., 2013). The low density of 8–10 Env trimers on HIV-1 virions has been proposed to be in part responsible for the slow rate of nAb maturation and have evolved as an immune evasion mechanism (Schiller and Chackerian, 2014). The sparse trimers preclude bivalent antibody binding, thereby reducing opportunities to mature antibodies of high affinity (Schiller and Chackerian, 2014). They also leave room for extraordinary levels of membrane proteins from the host to be incorporated, and these dense self-antigens might push antibody mutation toward tolerization (Schiller and Chackerian, 2014).

While nAbs develop within months of infection, broadly neutralizing antibodies (bnAbs), i.e., antibodies that neutralize most of the HIV-1 subtypes, do not develop for years and most often do not become immunodominant in treatment naïve individuals (Burton and Mascola, 2015). Among the well-characterized bnAb, most recognize highly conserved epitopes present in difficult to reach regions of native trimers protected by the glycan shield or within narrow clefts of the trimers, including the CD4-binding site, the exposed stem of the V2 loop at the top of gp120, unusual mannose epitopes on the glycan shield, the MPER on gp41 or the small cleft where the fusion peptide lies (Burton and Mascola, 2015). The bnAbs have unusual characteristics particularly unusually long heavy chain complementarity-determining region (CDR H3) that can insert into Env structure and between glycans to reach their epitopes (Pancera et al., 2010; Pejchal et al., 2010; Spurrier et al., 2011), reviewed in McCoy and Burton (2017) and Kwong and Mascola (2012). Many of these are among of the longest CDR H3 loop protrusions known in human antibodies. One such bnAb is PGT151 whose CDR H3 loop forms a slender extension that slips under a carbohydrate in the glycan shield and into a small cavity to cradle the N-terminus of the fusion peptide as it lies nestled between gp41 monomers (Lee et al., 2016) (Fig. 1.5A). Another is bnAb PGT145 whose heavy chain loop extends even further and inserts deep within the apex of Env trimers where it blocks the V3 loop from flipping outward and the fusion peptide exit path (Liu et al., 2017) (Fig. 1.5B).

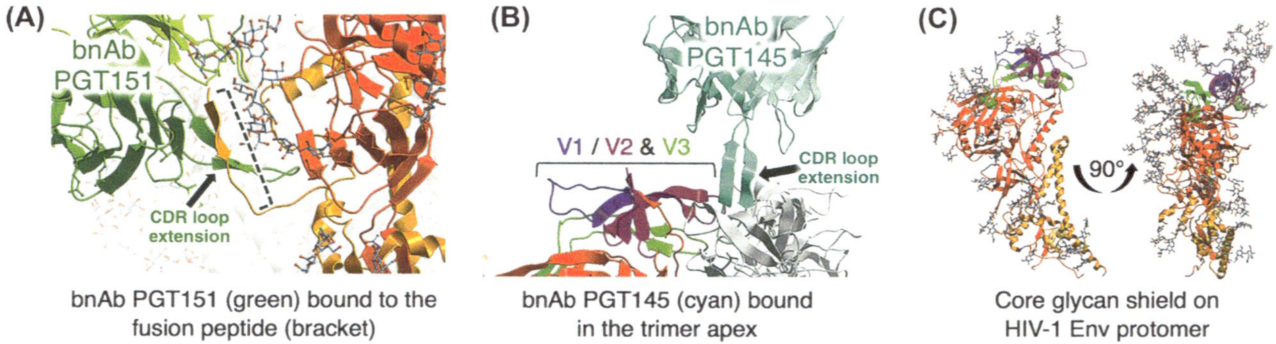

FIGURE 1.5 Extended complementarity-determining region (CDR) loops in broadly neutralizing anti-HIV-1 antibodies typically reach in between protomers and around the glycan shield. (A) A long CDR loop of broadly neutralizing antibody (bnAb) PGT151 (green) cradles the N-terminus of the fusion peptide (gold) and reaches deep within the structure to interact with gp120 residues (PDB 5FUU). (B) The extended CDR loop of bnAb PGT145 inserts within the trimer apex and interacts with the trimer interface formed by V1/V2 and V3 loops (PDB 5U1F). (C) The dense core glycans in the shield shown around one protomer of a native trimer (PDB 5FYJ).

What will it take to create an effective HIV vaccine? Mounting evidence points toward a major challenge being the need to devise a vaccine that rapidly and reproducibly elicits the editing of germline antibody genes required to generate antibody with the unusual characteristics seen in bnAbs from patients. Germline rearrangement of immunoglobulin genes produces initial antibodies that lack these features and are not able to reach the poorly accessible bnAb epitopes. They must undergo extensive hypermutation, insertions, and deletions to attain the ability to neutralize broadly. Vaccine strategies that elicit this type of bnAb should give a potent response against many strains and subtypes.

Interferon-Induced Transmembrane Proteins and Envelope Proteins Variation

In addition to nAbs, inhibition of the fusion step of entry by the interferon-induced transmembrane proteins IFITM1, IFITM2, and IFITM3 of the human innate immune response also drives Env variation. IFITMs act in part by inhibiting the membrane fusion step of entry of a broad range of viral pathogens (Brass et al., 2009; Perreira et al., 2013; Bailey et al., 2014). They block entry of the betaretrovirus JSRV and human influenza A virus at a step after receptor binding and before formation of the hemifusion intermediate (Li et al., 2013). Human IFITMs do not affect amphotropic MLV infection, but mutation of a C-terminal sorting signal in human IFITM1 results in a gain of ability to inhibit amphotropic MLV fusion and entry and increased restriction of JSRV entry (Brass et al., 2009; Li et al., 2015). The increased inhibition correlated with greater access to internalized virus; mutant IFITM1 shifted into the compartment where receptor-internalized JSRV particles were located (Li et al., 2015). Interestingly, mouse IFITM1 lacks this sorting signal raising the possibility that mouse IFITM1 may be effective against MLV.

A study of transmitted founder viruses and primary HIV-1 isolates showed the relevance of IFITM1 restriction to control of natural infection. R5-tropic-transmitted founder viruses were more sensitive to IFITM1, whereas the R4-tropic variants that developed in these patients 6 months later were more sensitive to IFITM2 and IFITM3 (Foster et al., 2016). In Jurkat T cells, IFITM1 was located in the plasma membrane where R5-tropic founder viruses would be expected to fuse, whereas IFITM2 and IFITM3 were located in internal compartments where the 6-month R4-tropic variants would internalize before fusing (Foster et al., 2016). Later studies showed that the block to HIV-1 entry occurs after binding of the V3 loop to a coreceptor and variants resistant to IFITM inhibition typically have V3 loop mutations and increased CD4 affinity (Wang et al., 2017).

COEVOLUTION OF VIRUS AND RECEPTOR

The history of retroviruses is written in the "fossil record" left in the genomes of host organisms after infection of germ cells and endogenization of the retroviral sequences. It tells us that retroviruses are very ancient and that vertebrates are favorable hosts in which these viruses flourish and are able to invade the germline. The emergence of vertebrate species marked an unprecedented burst of retrovirus endogenization and all major vertebrate lineages examined contain endogenized retroviral sequences, whereas none have been found in species lower than vertebrates (Escalera-Zamudio and Greenwood, 2016; Johnson, 2015).

As vertebrate species expanded and diverged, vertebrate retroviruses evolved to use the new host populations. Members of beta- and gammaretrovirus-like ancestral "genera" were the most adept at transitioning to new host species and endogenizing. Whole genome analysis of endogenized sequences in 65 diverse vertebrate species showed that greater than 80% of the endogenized sequences occurred in mammalian species and were from beta- and gamma-like retroviruses (Hayward et al., 2015). Alpha-like retroviral sequences predominated in birds, whereas fish species contained only epsilon-like sequences but together these represented only a small percentage of the total endogenized retroviruses found (Hayward et al., 2015). Very few spumavirus- and no lentivirus- or delta-like retroviral sequences were found in that study.

The cross-species aptitude of ancestral beta- and gamma-like viruses may reflect the ease with which a jump to a closely related species can occur. Transitions to a new host species within a given mammalian class could occur through use of an ortholog of the original entry receptor with few if any changes in the Env, although the host restriction factors must have played critical roles too. Clustering of related endogenized beta- and gamma-like retroviral sequences was consistent with cross-species transmission within but not between mammalian classes (Hayward et al., 2015). Another study of the fossil record for cross-species transmission of an endogenous bat retrovirus in 100 mammalian species also found clustering among but not between classes, with the ancient bat virus preferentially expanding among feline species (Zhuo and Feschotte, 2015).

Once introduced into a new host species, retroviral Env rapidly evolve to adapt to the species-specific version of the receptor and evolve to evade immune responses. While lentivirus endogenization is rare, cross-species transmission occurs. The zoonotic transmission of SIV-1 from chimpanzees to humans and its rapid evolution to efficient use of human CD4 and chemokine coreceptors to generate HIV-1 is a modern example. KoRV-A is another modern example of this type of coevolution. The Indonesian rodent *Melomys burtoni* migrated from Papua New Guinea to Australia carrying with it a GALV-related gammaretrovirus that the rodent transmitted to Koalas (Hanger et al., 2000; Oliveira et al., 2006). The newly speciated KoRV-A has endogenized in Australian Koala populations and is now vertically transmitted (Tarlinton et al., 2006). Separate relatively recent transmissions of two other gammaretroviruses to the Koala gave rise to KoRV-B and KoRV-J, which have not yet been endogenized (Xu et al., 2013; Shojima et al., 2013).

Transmission between vertebrate classes is a rarer event in the "fossil record" and occurred by swapping envelope genes rather than by evolution to use of orthologs of entry receptors. The first evidence that *env* swapping was involved came from recovery of an endogenous avian retrovirus. Its *gag* and *pol* genes were alpha-like but its *env* was gamma-like (Henzy et al., 2014). Analysis of sequences reconstructed from ancestral endogenous retrovirus-Fc showed extensive global spread during which the virus was transmitted to 11 of the 50 mammalian species examined (Diehl et al., 2016). An estimated 20 cross-species jumps occurred with many of them representing transmission between species from different classes by viruses that had recombined to acquire gamma-like *env* genes (Diehl et al., 2016). Specific information about the receptors used in the transmissions is not available because these studies used only the TM-encoding portions of *env* to avoid the high level of variation introduced by erosion of SU-encoding sequences as this subunit drifts to evade immune responses.

CAPTURED RETROVIRAL ENVELOPE PROTEINS IN THE DEVELOPMENT OF MAMMALIAN PLACENTA AND MALE MUSCLE MASS

No other example of the theft of a gene from a parasite has had such a profound and striking influence on a host species as a particular set of captured retroviral glycoproteins have had on mammals. These endogenized Env mediate highly efficient formation of the boundary layer between fetal and maternal tissue in the placenta. The stable and extensive boundary layer for nutrient and waste exchange is thought to be a major evolutionary development enabling more rapid propagation of mammalian species.

Discovered serendipitously, tissue analysis showed that a newly isolated cDNA was highly expressed in the human placenta, specifically in the syncytiotrophoblasts that form the permeable boundary between fetal and maternal tissue (Mi et al., 2000). The syncytiotrophoblasts were known to originate by fusion of fetal cytotrophoblast cells but the mechanism by which this crucial cell–cell fusion occurred was not known. When antiserum to the protein blocked human cytotrophoblast fusion in vitro and its ectopic expression induced cell–cell fusion in many cell types, Sha Mi and coworkers named the gene syncytin (Mi et al., 2000).

Their database search revealed that the syncytin clone was a cDNA of the *env* mRNA from human endogenous retrovirus W (HERV-W), which is closely related to the simian, baboon, and feline RD114 betaretrovirus *env* genes. Subsequent studies showed that syncytin dysregulation was associated with placental failure in preeclampsia and hemolysis, elevated liver enzymes, and low platelet count syndrome (Knerr et al., 2002; Lee et al., 2001). The fusion receptor used by syncytin was identified as human ASCT2/SLC1A5, the neutral amino acid transporter also used as an entry receptor by the betaretroviruses to which syncytin was closely related (Rasko et al., 1999; Tailor et al., 1999b; Marin et al., 2000).

Because it was known that HERV-W had endogenized only into Old World primates, it was doubtful that syncytin was a universal determinant in placenta development. The primate placental fusion mechanism might be an exception and another more general mechanism might exist in other mammals. Early comments urged researchers to determine if other endogenized *env* genes were "put to work" in mediating formation of the syncytiotrophoblast boundary layer (Stoye and Coffin, 2000). Discovery of two captured retroviral Env, named syncytin-A and syncytin-B, in mice and their characterization using transgenic knockout mice confirmed that *env* genes were critical to boundary layer formation in another mammal (Dupressoir et al., 2005, 2011; Koornstra and Exalto, 1991). Additional captured retroviral *env* genes with syncytin function have been reported since the discovery of the human and mouse syncytins. Genetic evidence indicates that each underwent purifying selection over the past 25–30 million years. Carnivora to which cats and dogs belong, ruminants, including cows, sheep, and goats, and the squirrel-related clade of Rodentia all express endogenized Env that mediate fusion of their fetal–maternal placental boundary (Cornelis et al., 2012, 2013; Redelsperger et al., 2014). Even marsupials use a captured retroviral Env to induce cell fusion and a boundary layer for nutrient and waste exchange in the early fetal stages prior to birth and maturation in the marsupial pouch (Cornelis et al., 2015).

The contributions of these captured Env may not be limited to placental development. They may also be important in gender-specific differences in muscle mass seen in mammals. Male, but not female, mice transgenic for knockout of mouse syncytin-A and -B showed a reduction in skeletal muscle that resulted in muscle mass similar to that of females (Redelsperger et al., 2016). Introduction of species-specific antisyncytin siRNAs in primary myoblasts from mice, sheep, dogs, and humans reduced myoblast fusion in vitro consistent with their role in muscle fiber formation (Redelsperger et al., 2016). Increasingly, these insights into host–pathogen interactions indicate that man and other mammals coevolved with the retroviruses that provided the fusion activities critical to mammalian expansion, an arrangement beneficial to both host and virus.

REFERENCES

Ablan, S., Rawat, S.S., Viard, M., Wang, J.M., Puri, A., Blumenthal, R., 2006. The role of cholesterol and sphingolipids in chemokine receptor function and HIV-1 envelope glycoprotein-mediated fusion. Virol. J. 3, 104.

Adkins, H.B., Brojatsch, J., Naughton, J., Rolls, M.M., Pesola, J.M., Young, J.A., 1997. Identification of a cellular receptor for subgroup E avian leukosis virus. Proc. Natl. Acad. Sci. U.S.A. 94, 11617–11622.

Adkins, H.B., Brojatsch, J., Young, J.A., 2000. Identification and characterization of a shared TNFR-related receptor for subgroup B, D, and E avian leukosis viruses reveal cysteine residues required specifically for subgroup E viral entry. J. Virol. 74, 3572–3578.

Alberti, A., Murgia, C., Liu, S.L., Mura, M., Cousens, C., Sharp, M., Miller, A.D., Palmarini, M., 2002. Envelope-induced cell transformation by ovine betaretroviruses. J. Virol. 76, 5387–5394.

Albritton, L.M., Tseng, L., Scadden, D., Cunningham, J.M., 1989. A putative murine ecotropic retrovirus receptor gene encodes a multiple membrane-spanning protein and confers susceptibility to virus infection. Cell 57, 659–666.

Alkhatib, G., Combadiere, C., Broder, C.C., Feng, Y., Kennedy, P.E., Murphy, P.M., Berger, E.A., 1996. CC CKR5: a RANTES, MIP-1alpha, MIP-1beta receptor as a fusion cofactor for macrophage-tropic HIV-1. Science 272, 1955–1958.

Allen, T.E., Sherrill, K.J., Crispell, S.M., Perrott, M.R., Carlson, J.O., DeMartini, J.C., 2002. The jaagsiekte sheep retrovirus envelope gene induces transformation of the avian fibroblast cell line DF-1 but does not require a conserved SH2 binding domain. J. Gen. Virol. 83, 2733–2742.

Andersen, K.B., Nexo, B.A., 1983. Entry of murine retrovirus into mouse fibroblasts. Virology 125, 85–98.

Andersen, K.B., 1983. Leupeptin inhibits retrovirus infection in mouse fibroblasts. J. Virol. 48, 765–769.

Anderson, M.M., Lauring, A.S., Burns, C.C., Overbaugh, J., 2000. Identification of a cellular cofactor required for infection by feline leukemia virus. Science 287, 1828–1830.

Apellaniz, B., Rujas, E., Carravilla, P., Requejo-Isidro, J., Huarte, N., Domene, C., Nieva, J.L., 2014. Cholesterol-dependent membrane fusion induced by the gp41 membrane-proximal external region-transmembrane domain connection suggests a mechanism for broad HIV-1 neutralization. J. Virol. 88, 13367–13377.

Arthos, J., Cicala, C., Martinelli, E., Macleod, K., Van Ryk, D., Wei, D., Xiao, Z., Veenstra, T.D., Conrad, T.P., Lempicki, R.A., McLaughlin, S., Pascuccio, M., Gopaul, R., McNally, J., Cruz, C.C., Censoplano, N., Chung, E., Reitano, K.N., Kottilil, S., Goode, D.J., Fauci, A.S., 2008. HIV-1 envelope protein binds to and signals through integrin alpha4beta7, the gut mucosal homing receptor for peripheral T cells. Nat. Immunol. 9, 301–309.

Aydin, H., Cook, J.D., Lee, J.E., 2014. Crystal structures of beta- and gammaretrovirus fusion proteins reveal a role for electrostatic stapling in viral entry. J. Virol. 88, 143–153.

Bae, Y., Kingsman, S.M., Kingsman, A.J., 1997. Functional dissection of the Moloney murine leukemia virus envelope protein gp70. J. Virol. 71, 2092–2099.

Bailey, C.C., Zhong, G., Huang, I.C., Farzan, M., 2014. IFITM-family proteins: the cell's first line of antiviral defense. Annu. Rev. Virol. 1, 261–283.

Barnard, R.J., Young, J.A., 2003. Alpharetrovirus envelope-receptor interactions. Curr. Top. Microbiol. Immunol. 281, 107–136.

Barnard, R.J., Narayan, S., Dornadula, G., Miller, M.D., Young, J.A., 2004. Low pH is required for avian sarcoma and leukosis virus Env-dependent viral penetration into the cytosol and not for viral uncoating. J. Virol. 78, 10433–10441.

Barnett, A.L., Cunningham, J.M., 2001. Receptor binding transforms the surface subunit of the mammalian C-type retrovirus envelope protein from an inhibitor to an activator of fusion. J. Virol. 75, 9096–9105.

Barnett, A.L., Davey, R.A., Cunningham, J.M., 2001. Modular organization of the Friend murine leukemia virus envelope protein underlies the mechanism of infection. Proc. Natl. Acad. Sci. U.S.A. 98, 4113–4118.

Barnett, A.L., Wensel, D.L., Li, W., Fass, D., Cunningham, J.M., 2003. Structure and mechanism of a coreceptor for infection by a pathogenic feline retrovirus. J. Virol. 77, 2717–2729.

Bates, P., Young, J.A., Varmus, H.E., 1993. A receptor for subgroup A Rous sarcoma virus is related to the low density lipoprotein receptor. Cell 74, 1043–1051.

Battini, J.L., Heard, J.M., Danos, O., 1992. Receptor choice determinants in the envelope glycoproteins of amphotropic, xenotropic, and polytropic murine leukemia viruses. J. Virol. 66, 1468–1475.

Battini, J.L., Danos, O., Heard, J.M., 1995. Receptor-binding domain of murine leukemia virus envelope glycoproteins. J. Virol. 69, 713–719.

Battini, J.L., Rasko, J.E., Miller, A.D., 1999. A human cell-surface receptor for xenotropic and polytropic murine leukemia viruses: possible role in G protein-coupled signal transduction. Proc. Natl. Acad. Sci. U.S.A. 96, 1385–1390.

Beer, C., Pedersen, L., 2006. Amphotropic murine leukemia virus is preferentially attached to cholesterol-rich microdomains after binding to mouse fibroblasts. Virol. J. 3, 21.

Beer, C., Andersen, D.S., Rojek, A., Pedersen, L., 2005. Caveola-dependent endocytic entry of amphotropic murine leukemia virus. J. Virol. 79, 10776–10787.

Berro, R., Klasse, P.J., Lascano, D., Flegler, A., Nagashima, K.A., Sanders, R.W., Sakmar, T.P., Hope, T.J., Moore, J.P., 2011. Multiple CCR5 conformations on the cell surface are used differentially by human immunodeficiency viruses resistant or sensitive to CCR5 inhibitors. J. Virol. 85, 8227–8240.

Biasin, M., Sironi, M., Saulle, I., Pontremoli, C., Garziano, M., Cagliani, R., Trabattoni, D., Lo Caputo, S., Vichi, F., Mazzotta, F., Forni, D., Riva, S., Aguilar-Jimenez, W., Cedeno, S., Sanchez, J., Brander, C., Zapata, W., Rugeles, M.T., Clerici, M., 2017. A 6-amino acid insertion/deletion polymorphism in the mucin domain of TIM-1 confers protections against HIV-1 infection. Microbes Infect. 19, 69–74.

Biswas, S., Yin, S.R., Blank, P.S., Zimmerberg, J., 2008. Cholesterol promotes hemifusion and pore widening in membrane fusion induced by influenza hemagglutinin. J. Gen. Physiol. 131, 503–513.

Bolin, L.L., Chandhasin, C., Lobelle-Rich, P.A., Albritton, L.M., Levy, L.S., 2011. Distinctive receptor binding properties of the surface glycoprotein of a natural feline leukemia virus isolate with unusual disease spectrum. Retrovirology 8, 35.

Bolin, L.L., Ahmad, S., Lobelle-Rich, P.A., Ooms, T.G., Alvarez-Hernandez, X., Didier, P.J., Levy, L.S., 2013. The surface glycoprotein of feline leukemia virus isolate FeLV-945 is a determinant of altered pathogenesis in the presence or absence of the unique viral long terminal repeat. J. Virol. 87, 10874–10883.

Brandenberg, O.F., Magnus, C., Rusert, P., Regoes, R.R., Trkola, A., 2015. Different infectivity of HIV-1 strains is linked to number of envelope trimers required for entry. PLoS Pathog. 11, e1004595.

Brass, A.L., Huang, I.C., Benita, Y., John, S.P., Krishnan, M.N., Feeley, E.M., Ryan, B.J., Weyer, J.L., van der Weyden, L., Fikrig, E., Adams, D.J., Xavier, R.J., Farzan, M., Elledge, S.J., 2009. The IFITM proteins mediate cellular resistance to influenza A H1N1 virus, West Nile virus, and dengue virus. Cell 139, 1243–1254.

Brojatsch, J., Naughton, J., Rolls, M.M., Zingler, K., Young, J.A., 1996. CAR1, a TNFR-related protein, is a cellular receptor for cytopathic avian leukosis-sarcoma viruses and mediates apoptosis. Cell 87, 845–855.

Burton, D.R., Mascola, J.R., 2015. Antibody responses to envelope glycoproteins in HIV-1 infection. Nat. Immunol. 16, 571–576.

Callahan, M.K., Popernack, P.M., Tsutsui, S., Truong, L., Schlegel, R.A., Henderson, A.J., 2003. Phosphatidylserine on HIV envelope is a cofactor for infection of monocytic cells. J. Immunol. 170, 4840–4845.

Chai, N., Bates, P., 2006. Na^+/H^+ exchanger type 1 is a receptor for pathogenic subgroup J avian leukosis virus. Proc. Natl. Acad. Sci. U.S.A. 103, 5531–5536.

Chandhasin, C., Coan, P.N., Levy, L.S., 2005. Subtle mutational changes in the SU protein of a natural feline leukemia virus subgroup A isolate alter disease spectrum. J. Virol. 79, 1351–1360.

Chow, Y.H., Alberti, A., Mura, M., Pretto, C., Murcia, P., Albritton, L.M., Palmarini, M., 2003. Transformation of rodent fibroblasts by the jaagsiekte sheep retrovirus envelope is receptor independent and does not require the surface domain. J. Virol. 77, 6341–6350.

Chung, M., Kizhatil, K., Albritton, L.M., Gaulton, G.N., 1999. Induction of syncytia by neuropathogenic murine leukemia viruses depends on receptor density, host cell determinants, and the intrinsic fusion potential of envelope protein. J. Virol. 73, 9377–9385.

Cocchi, F., DeVico, A.L., Garzino-Demo, A., Cara, A., Gallo, R.C., Lusso, P., 1996. The V3 domain of the HIV-1 gp120 envelope glycoprotein is critical for chemokine-mediated blockade of infection. Nat. Med. 2, 1244–1247.

Cornelis, G., Heidmann, O., Bernard-Stoecklin, S., Reynaud, K., Veron, G., Mulot, B., Dupressoir, A., Heidmann, T., 2012. Ancestral capture of syncytin-Car1, a fusogenic endogenous retroviral envelope gene involved in placentation and conserved in Carnivora. Proc. Natl. Acad. Sci. U.S.A. 109, E432–E441.

Cornelis, G., Heidmann, O., Degrelle, S.A., Vernochet, C., Lavialle, C., Letzelter, C., Bernard-Stoecklin, S., Hassanin, A., Mulot, B., Guillomot, M., Hue, I., Heidmann, T., Dupressoir, A., 2013. Captured retroviral envelope syncytin gene associated with the unique placental structure of higher ruminants. Proc. Natl. Acad. Sci. U.S.A. 110, E828–E837.

Cornelis, G., Vernochet, C., Carradec, Q., Souquere, S., Mulot, B., Catzeflis, F., Nilsson, M.A., Menzies, B.R., Renfree, M.B., Pierron, G., Zeller, U., Heidmann, O., Dupressoir, A., Heidmann, T., 2015. Retroviral envelope gene captures and syncytin exaptation for placentation in marsupials. Proc. Natl. Acad. Sci. U.S.A. 112, E487–E496.

Cote, M., Zheng, Y.M., Albritton, L.M., Liu, S.L., 2008. Fusogenicity of Jaagsiekte sheep retrovirus envelope protein is dependent on low pH and is enhanced by cytoplasmic tail truncations. J. Virol. 82, 2543–2554.

Cote, M., Zheng, Y.M., Liu, S.L., 2009. Receptor binding and low pH coactivate oncogenic retrovirus envelope-mediated fusion. J. Virol. 83, 11447–11455.

Cote, M., Zheng, Y.M., Albritton, L.M., Liu, S.L., 2011. Single residues in the surface subunits of oncogenic sheep retrovirus envelopes distinguish receptor-mediated triggering for fusion at low pH and infection. Virology 421, 173–183.

Cote, M., Zheng, Y.M., Liu, S.L., 2012. Membrane fusion and cell entry of XMRV are pH-independent and modulated by the envelope glycoprotein's cytoplasmic tail. PLoS One 7, e33734.

Cowley, A.C., Fuller, N.L., Rand, R.P., Parsegian, V.A., 1978. Measurement of repulsive forces between charged phospholipid bilayers. Biochemistry 17, 3163–3168.

Crespo, H., Reina, R., Glaria, I., Ramirez, H., de Andres, X., Jauregui, P., Lujan, L., Martinez-Pomares, L., Amorena, B., de Andres, D.F., 2011. Identification of the ovine mannose receptor and its possible role in Visna/Maedi virus infection. Vet. Res. 42, 28.

Crespo, H., Jauregui, P., Glaria, I., Sanjose, L., Polledo, L., Garcia-Marin, J.F., Lujan, L., de Andres, D., Amorena, B., Reina, R., 2012. Mannose receptor may be involved in small ruminant lentivirus pathogenesis. Vet. Res. 43, 43.

Crublet, E., Andrieu, J.P., Vives, R.R., Lortat-Jacob, H., 2008. The HIV-1 envelope glycoprotein gp120 features four heparan sulfate binding domains, including the co-receptor binding site. J. Biol. Chem. 283, 15193–15200.

Csoka, A.B., Scherer, S.W., Stern, R., 1999. Expression analysis of six paralogous human hyaluronidase genes clustered on chromosomes 3p21 and 7q31. Genomics 60, 356–361.

Davenport, Y.W., West Jr., A.P., Bjorkman, P.J., 2015. Structure of an HIV-2 gp120 in complex with CD4. J. Virol. 90, 2112–2118.

Davey, R.A., Hamson, C.A., Healey, J.J., Cunningham, J.M., 1997. In vitro binding of purified murine ecotropic retrovirus envelope surface protein to its receptor, MCAT-1. J. Virol. 71, 8096–8102.

de Parseval, A., Chatterji, U., Morris, G., Sun, P., Olson, A.J., Elder, J.H., 2005. Structural mapping of CD134 residues critical for interaction with feline immunodeficiency virus. Nat. Struct. Mol. Biol. 12, 60–66.

Deen, K.C., McDougal, J.S., Inacker, R., Folena-Wasserman, G., Arthos, J., Rosenberg, J., Maddon, P.J., Axel, R., Sweet, R.W., 1988. A soluble form of CD4 (T4) protein inhibits AIDS virus infection. Nature 331, 82–84.

DeMartini, J.C., Bishop, J.V., Allen, T.E., Jassim, F.A., Sharp, J.M., de las Heras, M., Voelker, D.R., Carlson, J.O., 2001. Jaagsiekte sheep retrovirus proviral clone JSRV(JS7), derived from the JS7 lung tumor cell line, induces ovine pulmonary carcinoma and is integrated into the surfactant protein A gene. J. Virol. 75, 4239–4246.

Deng, H., Liu, R., Ellmeier, W., Choe, S., Unutmaz, D., Burkhart, M., Di Marzio, P., Marmon, S., Sutton, R.E., Hill, C.M., Davis, C.B., Peiper, S.C., Schall, T.J., Littman, D.R., Landau, N.R., 1996. Identification of a major co-receptor for primary isolates of HIV-1. Nature 381, 661–666.

Deng, J., Mitsuki, Y.Y., Shen, G., Ray, J.C., Cicala, C., Arthos, J., Dustin, M.L., Hioe, C.E., 2016. HIV envelope gp120 alters T cell receptor mobilization in the immunological synapse of uninfected CD4 T cells and augments T cell activation. J. Virol. 90, 10513–10526.

Diaz-Griffero, F., Hoschander, S.A., Brojatsch, J., 2002. Endocytosis is a critical step in entry of subgroup B avian leukosis viruses. J. Virol. 76, 12866–12876.

Diaz-Rohrer, B.B., Levental, K.R., Simons, K., Levental, I., 2014. Membrane raft association is a determinant of plasma membrane localization. Proc. Natl. Acad. Sci. U.S.A. 111, 8500–8505.

Diehl, W.E., Patel, N., Halm, K., Johnson, W.E., 2016. Tracking interspecies transmission and long-term evolution of an ancient retrovirus using the genomes of modern mammals. Elife 5, e12704.

Dietrich, C., Volovyk, Z.N., Levi, M., Thompson, N.L., Jacobson, K., 2001. Partitioning of Thy-1, GM1, and cross-linked phospholipid analogs into lipid rafts reconstituted in supported model membrane monolayers. Proc. Natl. Acad. Sci. U.S.A. 98, 10642–10647.

Dirks, C., Duh, F.M., Rai, S.K., Lerman, M.I., Miller, A.D., 2002. Mechanism of cell entry and transformation by enzootic nasal tumor virus. J. Virol. 76, 2141–2149.

Dragic, T., Litwin, V., Allaway, G.P., Martin, S.R., Huang, Y., Nagashima, K.A., Cayanan, C., Maddon, P.J., Koup, R.A., Moore, J.P., Paxton, W.A., 1996. HIV-1 entry into CD4+ cells is mediated by the chemokine receptor CC-CKR-5. Nature 381, 667–673.

Du, C., Ma, J., Liu, Q., Li, Y.F., He, X.J., Lin, Y.Z., Wang, X.F., Meng, Q.W., Wang, X., Zhou, J.H., 2015. Mice transgenic for equine cyclin T1 and ELR1 are susceptible to equine infectious anemia virus infection. Retrovirology 12, 36.

Duda, A., Stange, A., Luftenegger, D., Stanke, N., Westphal, D., Pietschmann, T., Eastman, S.W., Linial, M.L., Rethwilm, A., Lindemann, D., 2004. Prototype foamy virus envelope glycoprotein leader peptide processing is mediated by a furin-like cellular protease, but cleavage is not essential for viral infectivity. J. Virol. 78, 13865–13870.

Dupressoir, A., Marceau, G., Vernochet, C., Benit, L., Kanellopoulos, C., Sapin, V., Heidmann, T., 2005. Syncytin-A and syncytin-B, two fusogenic placenta-specific murine envelope genes of retroviral origin conserved in Muridae. Proc. Natl. Acad. Sci. U.S.A. 102, 725–730.

Dupressoir, A., Vernochet, C., Harper, F., Guegan, J., Dessen, P., Pierron, G., Heidmann, T., 2011. A pair of co-opted retroviral envelope syncytin genes is required for formation of the two-layered murine placental syncytiotrophoblast. Proc. Natl. Acad. Sci. U.S.A. 108, E1164–E1173.

Durham, N.D., Yewdall, A.W., Chen, P., Lee, R., Zony, C., Robinson, J.E., Chen, B.K., 2012. Neutralization resistance of virological synapse-mediated HIV-1 Infection is regulated by the gp41 cytoplasmic tail. J. Virol. 86, 7484–7495.

Dutartre, H., Claviere, M., Journo, C., Mahieux, R., 2016. Cell-free versus cell-to-cell infection by human immunodeficiency virus type 1 and human T-lymphotropic virus type 1: exploring the link among viral source, viral trafficking, and viral replication. J. Virol. 90, 7607–7617.

Earp, L.J., Delos, S.E., Netter, R.C., Bates, P., White, J.M., 2003. The avian retrovirus avian sarcoma/leukosis virus subtype A reaches the lipid mixing stage of fusion at neutral pH. J. Virol. 77, 3058–3066.

Effantin, G., Estrozi, L.F., Aschman, N., Renesto, P., Stanke, N., Lindemann, D., Schoehn, G., Weissenhorn, W., 2016. Cryo-electron microscopy structure of the native prototype foamy virus glycoprotein and virus architecture. PLoS Pathog. 12, e1005721.

El Hage Chahine, J.M., Hemadi, M., Ha-Duong, N.T., 2012. Uptake and release of metal ions by transferrin and interaction with receptor 1. Biochim. Biophys. Acta 1820, 334–347.

Elleder, D., Stepanets, V., Melder, D.C., Senigl, F., Geryk, J., Pajer, P., Plachy, J., Hejnar, J., Svoboda, J., Federspiel, M.J., 2005. The receptor for the subgroup C avian sarcoma and leukosis viruses, Tvc, is related to mammalian butyrophilins, members of the immunoglobulin superfamily. J. Virol. 79, 10408–10419.

Epand, R.M., Sayer, B.G., Epand, R.F., 2003. Peptide-induced formation of cholesterol-rich domains. Biochemistry 42, 14677–14689.

Epand, R.F., Thomas, A., Brasseur, R., Vishwanathan, S.A., Hunter, E., Epand, R.M., 2006. Juxtamembrane protein segments that contribute to recruitment of cholesterol into domains. Biochemistry 45, 6105–6114.

Ericsson, T.A., Takeuchi, Y., Templin, C., Quinn, G., Farhadian, S.F., Wood, J.C., Oldmixon, B.A., Suling, K.M., Ishii, J.K., Kitagawa, Y., Miyazawa, T., Salomon, D.R., Weiss, R.A., Patience, C., 2003. Identification of receptors for pig endogenous retrovirus. Proc. Natl. Acad. Sci. U.S.A. 100, 6759–6764.

Erikson, E., Wratil, P.R., Frank, M., Ambiel, I., Pahnke, K., Pino, M., Azadi, P., Izquierdo-Useros, N., Martinez-Picado, J., Meier, C., Schnaar, R.L., Crocker, P.R., Reutter, W., Keppler, O.T., 2015. Mouse Siglec-1 mediates trans-infection of surface-bound murine leukemia virus in a sialic acid *N*-acyl side chain-dependent manner. J. Biol. Chem. 290, 27345–27359.

Escalera-Zamudio, M., Greenwood, A.D., 2016. On the classification and evolution of endogenous retrovirus: human endogenous retroviruses may not be 'human' after all. APMIS 124, 44–51.

Fantini, J., Barrantes, F.J., 2013. How cholesterol interacts with membrane proteins: an exploration of cholesterol-binding sites including CRAC, CARC, and tilted domains. Front. Physiol. 4, 31.

Fass, D., Kim, P.S., 1995. Dissection of a retrovirus envelope protein reveals structural similarity to influenza hemagglutinin. Curr. Biol. 5, 1377–1383.

Fass, D., Harrison, S.C., Kim, P.S., 1996. Retrovirus envelope domain at 1.7 angstrom resolution. Nat. Struct. Biol. 3, 465–469.

Fass, D., Davey, R.A., Hamson, C.A., Kim, P.S., Cunningham, J.M., Berger, J.M., 1997. Structure of a murine leukemia virus receptor-binding glycoprotein at 2.0 angstrom resolution. Science 277, 1662–1666.

Feng, Y., Broder, C.C., Kennedy, P.E., Berger, E.A., 1996. HIV-1 entry cofactor: functional cDNA cloning of a seven-transmembrane, G protein-coupled receptor. Science 272, 872–877.

Forster, F., Medalia, O., Zauberman, N., Baumeister, W., Fass, D., 2005. Retrovirus envelope protein complex structure in situ studied by cryo-electron tomography. Proc. Natl. Acad. Sci. U.S.A. 102, 4729–4734.

Fortin, J.F., Cantin, R., Tremblay, M.J., 1998. T cells expressing activated LFA-1 are more susceptible to infection with human immunodeficiency virus type 1 particles bearing host-encoded ICAM-1. J. Virol. 72, 2105–2112.

Foster, T.L., Wilson, H., Iyer, S.S., Coss, K., Doores, K., Smith, S., Kellam, P., Finzi, A., Borrow, P., Hahn, B.H., Neil, S.J., 2016. Resistance of transmitted founder HIV-1 to IFITM-mediated restriction. Cell Host Microbe 20, 429–442.

Garcia-Paredes, J., Diaz Rubio, M., Llenas, F., Taxonera, C., Pardo, L., 1991. Comparison of sucralfate and ranitidine in the treatment of duodenal ulcers. Am. J. Med. 91, 64S–67S.

Garg, H., Blumenthal, R., 2008. Role of HIV Gp41 mediated fusion/hemifusion in bystander apoptosis. Cell. Mol. Life Sci. 65, 3134–3144.

Garg, H., Joshi, A., Tompkins, W.A., 2004. Feline immunodeficiency virus envelope glycoprotein mediates apoptosis in activated PBMC by a mechanism dependent on gp41 function. Virology 330, 424–436.

Geiselhart, V., Bastone, P., Kempf, T., Schnolzer, M., Lochelt, M., 2004. Furin-mediated cleavage of the feline foamy virus Env leader protein. J. Virol. 78, 13573–13581.

Ghez, D., Lepelletier, Y., Lambert, S., Fourneau, J.M., Blot, V., Janvier, S., Arnulf, B., van Endert, P.M., Heveker, N., Pique, C., Hermine, O., 2006. Neuropilin-1 is involved in human T-cell lymphotropic virus type 1 entry. J. Virol. 80, 6844–6854.

Ghez, D., Lepelletier, Y., Jones, K.S., Pique, C., Hermine, O., 2010. Current concepts regarding the HTLV-1 receptor complex. Retrovirology 7, 99.

Gonzalez, S.A., Paladino, M.G., Affranchino, J.L., 2012. Palmitoylation of the feline immunodeficiency virus envelope glycoprotein and its effect on fusion activity and envelope incorporation into virions. Virology 428, 1–10.

Gravel, C., Kay, D.G., Jolicoeur, P., 1993. Identification of the infected target cell type in spongiform myeloencephalopathy induced by the neurotropic Cas-Br-E murine leukemia virus. J. Virol. 67, 6648–6658.

Gross, C., Thoma-Kress, A.K., 2016. Molecular mechanisms of HTLV-1 cell-to-cell transmission. Viruses 8, 74.

Guttman, M., Garcia, N.K., Cupo, A., Matsui, T., Julien, J.P., Sanders, R.W., Wilson, I.A., Moore, J.P., Lee, K.K., 2014. CD4-induced activation in a soluble HIV-1 Env trimer. Structure 22, 974–984.

Hanger, J.J., Bromham, L.D., McKee, J.J., O'Brien, T.M., Robinson, W.F., 2000. The nucleotide sequence of koala (*Phascolarctos cinereus*) retrovirus: a novel type C endogenous virus related to Gibbon ape leukemia virus. J. Virol. 74, 4264–4272.

Hart, T.K., Kirsh, R., Ellens, H., Sweet, R.W., Lambert, D.M., Petteway Jr., S.R., Leary, J., Bugelski, P.J., 1991. Binding of soluble CD4 proteins to human immunodeficiency virus type 1 and infected cells induces release of envelope glycoprotein gp120. Proc. Natl. Acad. Sci. U.S.A. 88, 2189–2193.

Hayward, A., Cornwallis, C.K., Jern, P., 2015. Pan-vertebrate comparative genomics unmasks retrovirus macroevolution. Proc. Natl. Acad. Sci. U.S.A. 112, 464–469.

He, J., Chen, Y., Farzan, M., Choe, H., Ohagen, A., Gartner, S., Busciglio, J., Yang, X., Hofmann, W., Newman, W., Mackay, C.R., Sodroski, J., Gabuzda, D., 1997. CCR3 and CCR5 are coreceptors for HIV-1 infection of microglia. Nature 385, 645–649.

Heard, J.M., Danos, O., 1991. An amino-terminal fragment of the Friend murine leukemia virus envelope glycoprotein binds the ecotropic receptor. J. Virol. 65, 4026–4032.

Hensel, J., Hintz, M., Karas, M., Linder, D., Stahl, B., Geyer, R., 1995. Localization of the palmitoylation site in the transmembrane protein p12E of Friend murine leukaemia virus. Eur. J. Biochem. 232, 373–380.

Henzy, J.E., Gifford, R.J., Johnson, W.E., Coffin, J.M., 2014. A novel recombinant retrovirus in the genomes of modern birds combines features of avian and mammalian retroviruses. J. Virol. 88, 2398–2405.

Hiss, J.A., Schneider, G., 2009. Architecture, function and prediction of long signal peptides. Brief. Bioinform. 10, 569–578.

Hosaka, M., Nagahama, M., Kim, W.S., Watanabe, T., Hatsuzawa, K., Ikemizu, J., Murakami, K., Nakayama, K., 1991. Arg-X-Lys/Arg-Arg motif as a signal for precursor cleavage catalyzed by furin within the constitutive secretory pathway. J. Biol. Chem. 266, 12127–12130.

Hsu, T., Phung, A., Choe, K., Kim, J.W., Fan, H., 2015. Role for a zinc finger protein (Zfp111) in transformation of 208F rat fibroblasts by jaagsiekte sheep retrovirus envelope protein. J. Virol. 89, 10453–10466.

Hu, Q.Y., Fink, E., Grant, C.K., Elder, J.H., 2014. Selective interaction of heparin with the variable region 3 within surface glycoprotein of laboratory-adapted feline immunodeficiency virus. PLoS One 9, e115252.

Hull, S., Fan, H., 2006. Mutational analysis of the cytoplasmic tail of jaagsiekte sheep retrovirus envelope protein. J. Virol. 80, 8069–8080.

Indik, S., Gunzburg, W.H., Salmons, B., Rouault, F., 2005. A novel, mouse mammary tumor virus encoded protein with Rev-like properties. Virology 337, 1–6.

Izquierdo-Useros, N., Lorizate, M., Puertas, M.C., Rodriguez-Plata, M.T., Zangger, N., Erikson, E., Pino, M., Erkizia, I., Glass, B., Clotet, B., Keppler, O.T., Telenti, A., Krausslich, H.G., Martinez-Picado, J., 2012. Siglec-1 is a novel dendritic cell receptor that mediates HIV-1 trans-infection through recognition of viral membrane gangliosides. PLoS Biol. 10, e1001448.

Izquierdo-Useros, N., Lorizate, M., McLaren, P.J., Telenti, A., Krausslich, H.G., Martinez-Picado, J., 2014. HIV-1 capture and transmission by dendritic cells: the role of viral glycolipids and the cellular receptor Siglec-1. PLoS Pathog. 10, e1004146.

Jain, P., Manuel, S.L., Khan, Z.K., Ahuja, J., Quann, K., Wigdahl, B., 2009. DC-SIGN mediates cell-free infection and transmission of human T-cell lymphotropic virus type 1 by dendritic cells. J. Virol. 83, 10908–10921.

Jha, N.K., Latinovic, O., Martin, E., Novitskiy, G., Marin, M., Miyauchi, K., Naughton, J., Young, J.A., Melikyan, G.B., 2011. Imaging single retrovirus entry through alternative receptor isoforms and intermediates of virus-endosome fusion. PLoS Pathog. 7, e1001260.

Jinno-Oue, A., Oue, M., Ruscetti, S.K., 2001. A unique heparin-binding domain in the envelope protein of the neuropathogenic PVC-211 murine leukemia virus may contribute to its brain capillary endothelial cell tropism. J. Virol. 75, 12439–12445.

Jinno-Oue, A., Wilt, S.G., Hanson, C., Dugger, N.V., Hoffman, P.M., Masuda, M., Ruscetti, S.K., 2003. Expression of inducible nitric oxide synthase and elevation of tyrosine nitration of a 32-kilodalton cellular protein in brain capillary endothelial cells from rats infected with a neuropathogenic murine leukemia virus. J. Virol. 77, 5145–5151.

Johnson, W.E., 2015. Endogenous retroviruses in the genomics era. Annu. Rev. Virol. 2, 135–159.

Johnston, J.B., Power, C., 2002. Feline immunodeficiency virus xenoinfection: the role of chemokine receptors and envelope diversity. J. Virol. 76, 3626–3636.

Jolly, C., Sattentau, Q.J., 2004. Retroviral spread by induction of virological synapses. Traffic 5, 643–650.

Jolly, C., Kashefi, K., Hollinshead, M., Sattentau, Q.J., 2004. HIV-1 cell to cell transfer across an Env-induced, actin-dependent synapse. J. Exp. Med. 199, 283–293.

Jones, K.S., Akel, S., Petrow-Sadowski, C., Huang, Y., Bertolette, D.C., Ruscetti, F.W., 2005a. Induction of human T cell leukemia virus type I receptors on quiescent naive T lymphocytes by TGF-beta. J. Immunol. 174, 4262–4270.

Jones, K.S., Petrow-Sadowski, C., Bertolette, D.C., Huang, Y., Ruscetti, F.W., 2005b. Heparan sulfate proteoglycans mediate attachment and entry of human T-cell leukemia virus type 1 virions into CD4+ T cells. J. Virol. 79, 12692–12702.

Joshi, A., Nyakeriga, A.M., Ravi, R., Garg, H., 2011. HIV ENV glycoprotein-mediated bystander apoptosis depends on expression of the CCR5 co-receptor at the cell surface and ENV fusogenic activity. J. Biol. Chem. 286, 36404–36413.

Joshi, A., Sedano, M., Beauchamp, B., Punke, E.B., Mulla, Z.D., Meza, A., Alozie, O.K., Mukherjee, D., Garg, H., 2016. HIV-1 Env glycoprotein phenotype along with immune activation determines CD4 T cell loss in HIV patients. J. Immunol. 196, 1768–1779.

Julien, J.P., Cupo, A., Sok, D., Stanfield, R.L., Lyumkis, D., Deller, M.C., Klasse, P.J., Burton, D.R., Sanders, R.W., Moore, J.P., Ward, A.B., Wilson, I.A., 2013. Crystal structure of a soluble cleaved HIV-1 envelope trimer. Science 342, 1477–1483.

Kamiyama, H., Yoshii, H., Tanaka, Y., Sato, H., Yamamoto, N., Kubo, Y., 2009. Raft localization of CXCR4 is primarily required for X4-tropic human immunodeficiency virus type 1 infection. Virology 386, 23–31.

Kamiyama, H., Kakoki, K., Yoshii, H., Iwao, M., Igawa, T., Sakai, H., Hayashi, H., Matsuyama, T., Yamamoto, N., Kubo, Y., 2011. Infection of XC cells by MLVs and Ebola virus is endosome-dependent but acidification-independent. PLoS One 6, e26180.

Kaplan, G., Roitburd-Berman, A., Lewis, G.K., Gershoni, J.M., 2016. Range of CD4-bound conformations of HIV-1 gp120, as defined using conditional CD4-induced antibodies. J. Virol. 90, 4481–4493.

Katane, M., Takao, E., Kubo, Y., Fujita, R., Amanuma, H., 2002. Factors affecting the direct targeting of murine leukemia virus vectors containing peptide ligands in the envelope protein. EMBO Rep. 3, 899–904.

Kavanaugh, M.P., Miller, D.G., Zhang, W., Law, W., Kozak, S.L., Kabat, D., Miller, A.D., 1994. Cell-surface receptors for gibbon ape leukemia virus and amphotropic murine retrovirus are inducible sodium-dependent phosphate symporters. Proc. Natl. Acad. Sci. U.S.A. 91, 7071–7075.

Kay, D.G., Gravel, C., Pothier, F., Laperriere, A., Robitaille, Y., Jolicoeur, P., 1993. Neurological disease induced in transgenic mice expressing the env gene of the Cas-Br-E murine retrovirus. Proc. Natl. Acad. Sci. U.S.A. 90, 4538–4542.

Kim, J.W., Closs, E.I., Albritton, L.M., Cunningham, J.M., 1991. Transport of cationic amino acids by the mouse ecotropic retrovirus receptor. Nature 352, 725–728.

Kim, F.J., Manel, N., Garrido, E.N., Valle, C., Sitbon, M., Battini, J.L., 2004. HTLV-1 and -2 envelope SU subdomains and critical determinants in receptor binding. Retrovirology 1, 41.

Kinet, S., Swainson, L., Lavanya, M., Mongellaz, C., Montel-Hagen, A., Craveiro, M., Manel, N., Battini, J.L., Sitbon, M., Taylor, N., 2007. Isolated receptor binding domains of HTLV-1 and HTLV-2 envelopes bind Glut-1 on activated CD4+ and CD8+ T cells. Retrovirology 4, 31.

Knerr, I., Beinder, E., Rascher, W., 2002. Syncytin, a novel human endogenous retroviral gene in human placenta: evidence for its dysregulation in preeclampsia and HELLP syndrome. Am. J. Obstet. Gynecol. 186, 210–213.

Kobe, B., Center, R.J., Kemp, B.E., Poumbourios, P., 1999. Crystal structure of human T cell leukemia virus type 1 gp21 ectodomain crystallized as a maltose-binding protein chimera reveals structural evolution of retroviral transmembrane proteins. Proc. Natl. Acad. Sci. U.S.A. 96, 4319–4324.

Koornstra, G., Exalto, N., 1991. [Echography in the first pregnancy trimester has prognostic value]. Ned. Tijdschr. Geneeskd. 135, 2231–2235.

Kumar, P., Nachagari, D., Fields, C., Franks, J., Albritton, L.M., 2007. Host cell cathepsins potentiate Moloney murine leukemia virus infection. J. Virol. 81, 10506–10514.

Kwong, P.D., Mascola, J.R., 2012. Human antibodies that neutralize HIV-1: identification, structures, and B cell ontogenies. Immunity 37, 412–425.

Kwong, P.D., Wyatt, R., Robinson, J., Sweet, R.W., Sodroski, J., Hendrickson, W.A., 1998. Structure of an HIV gp120 envelope glycoprotein in complex with the CD4 receptor and a neutralizing human antibody. Nature 393, 648–659.

Kwong, P.D., Wyatt, R., Majeed, S., Robinson, J., Sweet, R.W., Sodroski, J., Hendrickson, W.A., 2000. Structures of HIV-1 gp120 envelope glycoproteins from laboratory-adapted and primary isolates. Structure 8, 1329–1339.

Lamb, D., Schuttelkopf, A.W., van Aalten, D.M., Brighty, D.W., 2011. Charge-surrounded pockets and electrostatic interactions with small ions modulate the activity of retroviral fusion proteins. PLoS Pathog. 7, e1001268.

Lambert, S., Bouttier, M., Vassy, R., Seigneuret, M., Petrow-Sadowski, C., Janvier, S., Heveker, N., Ruscetti, F.W., Perret, G., Jones, K.S., Pique, C., 2009. HTLV-1 uses HSPG and neuropilin-1 for entry by molecular mimicry of VEGF165. Blood 113, 5176–5185.

Landers, M.C., Dugger, N., Quadros, M., Hoffman, P.M., Gaulton, G.N., 2004. Neuropathogenic murine leukemia virus TR1.3 induces selective syncytia formation of brain capillary endothelium. Virology 321, 57–64.

Lapham, C.K., Ouyang, J., Chandrasekhar, B., Nguyen, N.Y., Dimitrov, D.S., Golding, H., 1996. Evidence for cell-surface association between fusin and the CD4-gp120 complex in human cell lines. Science 274, 602–605.

Lauring, A.S., Anderson, M.M., Overbaugh, J., 2001. Specificity in receptor usage by T-cell-tropic feline leukemia viruses: implications for the in vivo tropism of immunodeficiency-inducing variants. J. Virol. 75, 8888–8898.

Lavillette, D., Maurice, M., Roche, C., Russell, S.J., Sitbon, M., Cosset, F.L., 1998. A proline-rich motif downstream of the receptor binding domain modulates conformation and fusogenicity of murine retroviral envelopes. J. Virol. 72, 9955–9965.

Lavillette, D., Ruggieri, A., Russell, S.J., Cosset, F.L., 2000. Activation of a cell entry pathway common to type C mammalian retroviruses by soluble envelope fragments. J. Virol. 74, 295–304.

Lee, X., Keith Jr., J.C., Stumm, N., Moutsatsos, I., McCoy, J.M., Crum, C.P., Genest, D., Chin, D., Ehrenfels, C., Pijnenborg, R., van Assche, F.A., Mi, S., 2001. Downregulation of placental syncytin expression and abnormal protein localization in pre-eclampsia. Placenta 22, 808–812.

Lee, J.H., Ozorowski, G., Ward, A.B., 2016. Cryo-EM structure of a native, fully glycosylated, cleaved HIV-1 envelope trimer. Science 351, 1043–1048.

Leonard, C.K., Spellman, M.W., Riddle, L., Harris, R.J., Thomas, J.N., Gregory, T.J., 1990. Assignment of intrachain disulfide bonds and characterization of potential glycosylation sites of the type 1 recombinant human immunodeficiency virus envelope glycoprotein (gp120) expressed in Chinese hamster ovary cells. J. Biol. Chem. 265, 10373–10382.

Levental, I., Lingwood, D., Grzybek, M., Coskun, U., Simons, K., 2010. Palmitoylation regulates raft affinity for the majority of integral raft proteins. Proc. Natl. Acad. Sci. U.S.A. 107, 22050–22054.

Li, H., Papadopoulos, V., 1998. Peripheral-type benzodiazepine receptor function in cholesterol transport. Identification of a putative cholesterol recognition/interaction amino acid sequence and consensus pattern. Endocrinology 139, 4991–4997.

Li, M., Yang, C., Tong, S., Weidmann, A., Compans, R.W., 2002. Palmitoylation of the murine leukemia virus envelope protein is critical for lipid raft association and surface expression. J. Virol. 76, 11845–11852.

Li, K., Zhang, S., Kronqvist, M., Ekstrom, M., Wallin, M., Garoff, H., 2007. The conserved His8 of the Moloney murine leukemia virus Env SU subunit directs the activity of the SU-TM disulphide bond isomerase. Virology 361, 149–160.

Li, K., Zhang, S., Kronqvist, M., Wallin, M., Ekstrom, M., Derse, D., Garoff, H., 2008. Intersubunit disulfide isomerization controls membrane fusion of human T-cell leukemia virus Env. J. Virol. 82, 7135–7143.

Li, K., Markosyan, R.M., Zheng, Y.M., Golfetto, O., Bungart, B., Li, M., Ding, S., He, Y., Liang, C., Lee, J.C., Gratton, E., Cohen, F.S., Liu, S.L., 2013. IFITM proteins restrict viral membrane hemifusion. PLoS Pathog. 9, e1003124.

Li, M., Ablan, S.D., Miao, C., Zheng, Y.M., Fuller, M.S., Rennert, P.D., Maury, W., Johnson, M.C., Freed, E.O., Liu, S.L., 2014. TIM-family proteins inhibit HIV-1 release. Proc. Natl. Acad. Sci. U.S.A. 111, E3699–E3707.

Li, K., Jia, R., Li, M., Zheng, Y.M., Miao, C., Yao, Y., Ji, H.L., Geng, Y., Qiao, W., Albritton, L.M., Liang, C., Liu, S.L., 2015. A sorting signal suppresses IFITM1 restriction of viral entry. J. Biol. Chem. 290, 4248–4259.

Liao, Z., Cimakasky, L.M., Hampton, R., Nguyen, D.H., Hildreth, J.E., 2001. Lipid rafts and HIV pathogenesis: host membrane cholesterol is required for infection by HIV type 1. AIDS Res. Hum. Retrovir. 17, 1009–1019.

Liao, Z., Graham, D.R., Hildreth, J.E., 2003. Lipid rafts and HIV pathogenesis: virion-associated cholesterol is required for fusion and infection of susceptible cells. AIDS Res. Hum. Retrovir. 19, 675–687.

Lindemann, D., Goepfert, P.A., 2003. The foamy virus envelope glycoproteins. Curr. Top. Microbiol. Immunol. 277, 111–129.

Lindemann, D., Pietschmann, T., Picard-Maureau, M., Berg, A., Heinkelein, M., Thurow, J., Knaus, P., Zentgraf, H., Rethwilm, A., 2001. A particle-associated glycoprotein signal peptide essential for virus maturation and infectivity. J. Virol. 75, 5762–5771.

Linder, M., Wenzel, V., Linder, D., Stirm, S., 1994. Structural elements in glycoprotein 70 from polytropic Friend mink cell focus-inducing virus and glycoprotein 71 from ecotropic Friend murine leukemia virus, as defined by disulfide-bonding pattern and limited proteolysis. J. Virol. 68, 5133–5141.

Liu, Q., Acharya, P., Dolan, M.A., Zhang, P., Guzzo, C., Lu, J., Kwon, A., Gururani, D., Miao, H., Bylund, T., Chuang, G.Y., Druz, A., Zhou, T., Rice, W.J., Wigge, C., Carragher, B., Potter, C.S., Kwong, P.D., Lusso, P., 2017. Quaternary contact in the initial interaction of CD4 with the HIV-1 envelope trimer. Nat. Struct. Mol. Biol. 24, 370–378.

Lu, X., Silver, J., 2000. Ecotropic murine leukemia virus receptor is physically associated with caveolin and membrane rafts. Virology 276, 251–258.

Luck, A.N., Mason, A.B., 2012. Transferrin-mediated cellular iron delivery. Curr. Top. Membr. 69, 3–35.

Lyumkis, D., Julien, J.P., de Val, N., Cupo, A., Potter, C.S., Klasse, P.J., Burton, D.R., Sanders, R.W., Moore, J.P., Carragher, B., Wilson, I.A., Ward, A.B., 2013. Cryo-EM structure of a fully glycosylated soluble cleaved HIV-1 envelope trimer. Science 342, 1484–1490.

MacKrell, A.J., Soong, N.W., Curtis, C.M., Anderson, W.F., 1996. Identification of a subdomain in the Moloney murine leukemia virus envelope protein involved in receptor binding. J. Virol. 70, 1768–1774.

Maddon, P.J., Dalgleish, A.G., McDougal, J.S., Clapham, P.R., Weiss, R.A., Axel, R., 1986. The T4 gene encodes the AIDS virus receptor and is expressed in the immune system and the brain. Cell 47, 333–348.

Maeda, Y., Foda, M., Matsushita, S., Harada, S., 2000. Involvement of both the V2 and V3 regions of the CCR5-tropic human immunodeficiency virus type 1 envelope in reduced sensitivity to macrophage inflammatory protein 1alpha. J. Virol. 74, 1787–1793.

Magnus, C., Rusert, P., Bonhoeffer, S., Trkola, A., Regoes, R.R., 2009. Estimating the stoichiometry of human immunodeficiency virus entry. J. Virol. 83, 1523–1531.

Manel, N., Kim, F.J., Kinet, S., Taylor, N., Sitbon, M., Battini, J.L., 2003a. The ubiquitous glucose transporter GLUT-1 is a receptor for HTLV. Cell 115, 449–459.

Manel, N., Kinet, S., Battini, J.L., Kim, F.J., Taylor, N., Sitbon, M., 2003b. The HTLV receptor is an early T-cell activation marker whose expression requires de novo protein synthesis. Blood 101, 1913–1918.

Marin, M., Tailor, C.S., Nouri, A., Kabat, D., 2000. Sodium-dependent neutral amino acid transporter type 1 is an auxiliary receptor for baboon endogenous retrovirus. J. Virol. 74, 8085–8093.

Marra, J., Israelachvili, J., 1985. Direct measurements of forces between phosphatidylcholine and phosphatidylethanolamine bilayers in aqueous electrolyte solutions. Biochemistry 24, 4608–4618.

Masuda, M., Hoffman, P.M., Ruscetti, S.K., 1993. Viral determinants that control the neuropathogenicity of PVC-211 murine leukemia virus in vivo determine brain capillary endothelial cell tropism of the virus in vitro. J. Virol. 67, 4580–4587.

Masuda, M., Hanson, C.A., Alvord, W.G., Hoffman, P.M., Ruscetti, S.K., Masuda, M., 1996. Effects of subtle changes in the SU protein of ecotropic murine leukemia virus on its brain capillary endothelial cell tropism and interference properties. Virology 215, 142–151.

Materniak, M., Bicka, L., Kuzmak, J., 2006. Isolation and partial characterization of bovine foamy virus from Polish cattle. Pol. J. Vet. Sci. 9, 207–211.

Matsuyama, S., Delos, S.E., White, J.M., 2004. Sequential roles of receptor binding and low pH in forming prehairpin and hairpin conformations of a retroviral envelope glycoprotein. J. Virol. 78, 8201–8209.

McClure, M.O., Sommerfelt, M.A., Marsh, M., Weiss, R.A., 1990. The pH independence of mammalian retrovirus infection. J. Gen. Virol. 71 (Pt 4), 767–773.

McCoy, L.E., Burton, D.R., 2017. Identification and specificity of broadly neutralizing antibodies against HIV. Immunol. Rev. 275, 11–20.

Melder, D.C., Pike, G.M., VanBrocklin, M.W., Federspiel, M.J., 2015. Model of the TVA receptor determinants required for efficient infection by subgroup A avian sarcoma and leukosis viruses. J. Virol. 89, 2136–2148.

Melikyan, G.B., Barnard, R.J., Markosyan, R.M., Young, J.A., Cohen, F.S., 2004. Low pH is required for avian sarcoma and leukosis virus Env-induced hemifusion and fusion pore formation but not for pore growth. J. Virol. 78, 3753–3762.

Melikyan, G.B., 2008. Common principles and intermediates of viral protein-mediated fusion: the HIV-1 paradigm. Retrovirology 5, 111.

Mendoza, R., Anderson, M.M., Overbaugh, J., 2006. A putative thiamine transport protein is a receptor for feline leukemia virus subgroup A. J. Virol. 80, 3378–3385.

Mertz, J.A., Simper, M.S., Lozano, M.M., Payne, S.M., Dudley, J.P., 2005. Mouse mammary tumor virus encodes a self-regulatory RNA export protein and is a complex retrovirus. J. Virol. 79, 14737–14747.

Mi, S., Lee, X., Li, X., Veldman, G.M., Finnerty, H., Racie, L., LaVallie, E., Tang, X.Y., Edouard, P., Howes, S., Keith Jr., J.C., McCoy, J.M., 2000. Syncytin is a captive retroviral envelope protein involved in human placental morphogenesis. Nature 403, 785–789.

Miller, D.G., Edwards, R.H., Miller, A.D., 1994. Cloning of the cellular receptor for amphotropic murine retroviruses reveals homology to that for gibbon ape leukemia virus. Proc. Natl. Acad. Sci. U.S.A. 91, 78–82.

Miyauchi, K., Kim, Y., Latinovic, O., Morozov, V., Melikyan, G.B., 2009. HIV enters cells via endocytosis and dynamin-dependent fusion with endosomes. Cell 137, 433–444.

Mondor, I., Ugolini, S., Sattentau, Q.J., 1998. Human immunodeficiency virus type 1 attachment to HeLa CD4 cells is CD4 independent and gp120 dependent and requires cell surface heparans. J. Virol. 72, 3623–3634.

Monot, M., Erny, A., Gineys, B., Desloire, S., Dolmazon, C., Aublin-Gex, A., Lotteau, V., Archer, F., Leroux, C., 2015. Early steps of jaagsiekte sheep retrovirus-mediated cell transformation involve the interaction between Env and the RALBP1 cellular protein. J. Virol. 89, 8462–8473.

Moore, J.P., McKeating, J.A., Weiss, R.A., Sattentau, Q.J., 1990. Dissociation of gp120 from HIV-1 virions induced by soluble CD4. Science 250, 1139–1142.

Morgan, R.A., Nussbaum, O., Muenchau, D.D., Shu, L., Couture, L., Anderson, W.F., 1993. Analysis of the functional and host range-determining regions of the murine ectropic and amphotropic retrovirus envelope proteins. J. Virol. 67, 4712–4721.

Murphy, S.L., Chung-Landers, M., Honczarenko, M., Gaulton, G.N., 2006. Linkage of reduced receptor affinity and superinfection to pathogenesis of TR1.3 murine leukemia virus. J. Virol. 80, 4601–4609.

Murphy, M.K., Yue, L., Pan, R., Boliar, S., Sethi, A., Tian, J., Pfafferot, K., Karita, E., Allen, S.A., Cormier, E., Goepfert, P.A., Borrow, P., Robinson, J.E., Gnanakaran, S., Hunter, E., Kong, X.P., Derdeyn, C.A., 2013. Viral escape from neutralizing antibodies in early subtype A HIV-1 infection drives an increase in autologous neutralization breadth. PLoS Pathog. 9, e1003173.

Myszka, D.G., Sweet, R.W., Hensley, P., Brigham-Burke, M., Kwong, P.D., Hendrickson, W.A., Wyatt, R., Sodroski, J., Doyle, M.L., 2000. Energetics of the HIV gp120-CD4 binding reaction. Proc. Natl. Acad. Sci. U.S.A. 97, 9026–9031.

Ochsenbauer-Jambor, C., Miller, D.C., Roberts, C.R., Rhee, S.S., Hunter, E., 2001. Palmitoylation of the Rous sarcoma virus transmembrane glycoprotein is required for protein stability and virus infectivity. J. Virol. 75, 11544–11554.

Oh, U., Jacobson, S., 2008. Treatment of HTLV-I-associated myelopathy/tropical spastic paraparesis: toward rational targeted therapy. Neurol. Clin. 26, 781–797 ix–x.

O'Hara, B., Johann, S.V., Klinger, H.P., Blair, D.G., Rubinson, H., Dunn, K.J., Sass, P., Vitek, S.M., Robins, T., 1990. Characterization of a human gene conferring sensitivity to infection by gibbon ape leukemia virus. Cell Growth Differ. 1, 119–127.

Okuma, K., Dalton, K.P., Buonocore, L., Ramsburg, E., Rose, J.K., 2003. Development of a novel surrogate virus for human T-cell leukemia virus type 1: inhibition of infection by osteoprotegerin. J. Virol. 77, 8562–8569.

Oliveira, N.M., Farrell, K.B., Eiden, M.V., 2006. In vitro characterization of a koala retrovirus. J. Virol. 80, 3104–3107.

Padilla-Parra, S., Matos, P.M., Kondo, N., Marin, M., Santos, N.C., Melikyan, G.B., 2012. Quantitative imaging of endosome acidification and single retrovirus fusion with distinct pools of early endosomes. Proc. Natl. Acad. Sci. U.S.A. 109, 17627–17632.

Palmarini, M., Sharp, J.M., de las Heras, M., Fan, H., 1999. Jaagsiekte sheep retrovirus is necessary and sufficient to induce a contagious lung cancer in sheep. J. Virol. 73, 6964–6972.

Pancera, M., McLellan, J.S., Wu, X., Zhu, J., Changela, A., Schmidt, S.D., Yang, Y., Zhou, T., Phogat, S., Mascola, J.R., Kwong, P.D., 2010. Crystal structure of PG16 and chimeric dissection with somatically related PG9: structure-function analysis of two quaternary-specific antibodies that effectively neutralize HIV-1. J. Virol. 84, 8098–8110.

Pancera, M., Zhou, T., Druz, A., Georgiev, I.S., Soto, C., Gorman, J., Huang, J., Acharya, P., Chuang, G.Y., Ofek, G., Stewart-Jones, G.B., Stuckey, J., Bailer, R.T., Joyce, M.G., Louder, M.K., Tumba, N., Yang, Y., Zhang, B., Cohen, M.S., Haynes, B.F., Mascola, J.R., Morris, L., Munro, J.B., Blanchard, S.C., Mothes, W., Connors, M., Kwong, P.D., 2014. Structure and immune recognition of trimeric pre-fusion HIV-1 Env. Nature 514, 455–461.

Park, B.H., Matuschke, B., Lavi, E., Gaulton, G.N., 1994. A point mutation in the env gene of a murine leukemia virus induces syncytium formation and neurologic disease. J. Virol. 68, 7516–7524.

Patel, M., Yanagishita, M., Roderiquez, G., Bou-Habib, D.C., Oravecz, T., Hascall, V.C., Norcross, M.A., 1993. Cell-surface heparan sulfate proteoglycan mediates HIV-1 infection of T-cell lines. AIDS Res. Hum. Retrovir. 9, 167–174.

Pejchal, R., Walker, L.M., Stanfield, R.L., Phogat, S.K., Koff, W.C., Poignard, P., Burton, D.R., Wilson, I.A., 2010. Structure and function of broadly reactive antibody PG16 reveal an H3 subdomain that mediates potent neutralization of HIV-1. Proc. Natl. Acad. Sci. U.S.A. 107, 11483–11488.

Perreira, J.M., Chin, C.R., Feeley, E.M., Brass, A.L., 2013. IFITMs restrict the replication of multiple pathogenic viruses. J. Mol. Biol. 425, 4937–4955.

Picard-Maureau, M., Jarmy, G., Berg, A., Rethwilm, A., Lindemann, D., 2003. Foamy virus envelope glycoprotein-mediated entry involves a pH-dependent fusion process. J. Virol. 77, 4722–4730.

Pike, G.M., Madden, B.J., Melder, D.C., Charlesworth, M.C., Federspiel, M.J., 2011. Simple, automated, high resolution mass spectrometry method to determine the disulfide bond and glycosylation patterns of a complex protein: subgroup A avian sarcoma and leukosis virus envelope glycoprotein. J. Biol. Chem. 286, 17954–17967.

Pinon, J.D., Klasse, P.J., Jassal, S.R., Welson, S., Weber, J., Brighty, D.W., Sattentau, Q.J., 2003. Human T-cell leukemia virus type 1 envelope glycoprotein gp46 interacts with cell surface heparan sulfate proteoglycans. J. Virol. 77, 9922–9930.

Pinter, A., Lieman-Hurwitz, J., Fleissner, E., 1978. The nature of the association between the murine leukemia virus envelope proteins. Virology 91, 345–351.

Pinter, A., Kopelman, R., Li, Z., Kayman, S.C., Sanders, D.A., 1997. Localization of the labile disulfide bond between SU and TM of the murine leukemia virus envelope protein complex to a highly conserved CWLC motif in SU that resembles the active-site sequence of thiol-disulfide exchange enzymes. J. Virol. 71, 8073–8077.

Plochmann, K., Horn, A., Gschmack, E., Armbruster, N., Krieg, J., Wiktorowicz, T., Weber, C., Stirnnagel, K., Lindemann, D., Rethwilm, A., Scheller, C., 2012. Heparan sulfate is an attachment factor for foamy virus entry. J. Virol. 86, 10028–10035.

Poeschla, E.M., Looney, D.J., 1998. CXCR4 is required by a nonprimate lentivirus: heterologous expression of feline immunodeficiency virus in human, rodent, and feline cells. J. Virol. 72, 6858–6866.

Popik, W., Alce, T.M., Au, W.C., 2002. Human immunodeficiency virus type 1 uses lipid raft-colocalized CD4 and chemokine receptors for productive entry into CD4(+) T cells. J. Virol. 76, 4709–4722.

Qian, Z., Albritton, L.M., 2004. An aromatic side chain is required at residue 8 of SU for fusion of ecotropic murine leukemia virus. J. Virol. 78, 508–512.

Qian, Z., Wang, H., Empig, C., Anderson, W.F., Albritton, L.M., 2004. Complementation of a binding-defective retrovirus by a host cell receptor mutant. J. Virol. 78, 5766–5772.

Quigley, J.G., Burns, C.C., Anderson, M.M., Lynch, E.D., Sabo, K.M., Overbaugh, J., Abkowitz, J.L., 2000. Cloning of the cellular receptor for feline leukemia virus subgroup C (FeLV-C), a retrovirus that induces red cell aplasia. Blood 95, 1093–1099.

Rand, R.P., Parsegian, V.A., 1984. Physical force considerations in model and biological membranes. Can. J. Biochem. Cell Biol. 62, 752–759.

Rasko, J.E., Battini, J.L., Gottschalk, R.J., Mazo, I., Miller, A.D., 1999. The RD114/simian type D retrovirus receptor is a neutral amino acid transporter. Proc. Natl. Acad. Sci. U.S.A. 96, 2129–2134.

Rawat, S.S., Viard, M., Gallo, S.A., Rein, A., Blumenthal, R., Puri, A., 2003. Modulation of entry of enveloped viruses by cholesterol and sphingolipids (Review). Mol. Membr. Biol. 20, 243–254.

Redelsperger, F., Cornelis, G., Vernochet, C., Tennant, B.C., Catzeflis, F., Mulot, B., Heidmann, O., Heidmann, T., Dupressoir, A., 2014. Capture of syncytin-Mar1, a fusogenic endogenous retroviral envelope gene involved in placentation in the Rodentia squirrel-related clade. J. Virol. 88, 7915–7928.

Redelsperger, F., Raddi, N., Bacquin, A., Vernochet, C., Mariot, V., Gache, V., Blanchard-Gutton, N., Charrin, S., Tiret, L., Dumonceaux, J., Dupressoir, A., Heidmann, T., 2016. Genetic evidence that captured retroviral envelope syncytins contribute to myoblast fusion and muscle sexual dimorphism in mice. PLoS Genet. 12, e1006289.

Redmond, S., Peters, G., Dickson, C., 1984. Mouse mammary tumor virus can mediate cell fusion at reduced pH. Virology 133, 393–402.

Renard, M., Varela, P.F., Letzelter, C., Duquerroy, S., Rey, F.A., Heidmann, T., 2005. Crystal structure of a pivotal domain of human syncytin-2, a 40 million years old endogenous retrovirus fusogenic envelope gene captured by primates. J. Mol. Biol. 352, 1029–1034.

Richardson, J., Pancino, G., Merat, R., Leste-Lasserre, T., Moraillon, A., Schneider-Mergener, J., Alizon, M., Sonigo, P., Heveker, N., 1999. Shared usage of the chemokine receptor CXCR4 by primary and laboratory-adapted strains of feline immunodeficiency virus. J. Virol. 73, 3661–3671.

Richman, D.D., Wrin, T., Little, S.J., Petropoulos, C.J., 2003. Rapid evolution of the neutralizing antibody response to HIV type 1 infection. Proc. Natl. Acad. Sci. U.S.A. 100, 4144–4149.

Riedel, C., Vasishtan, D., Siebert, C.A., Whittle, C., Lehmann, M.J., Mothes, W., Grunewald, K., 2017. Native structure of a retroviral envelope protein and its conformational change upon interaction with the target cell. J. Struct. Biol. 197, 172–180.

Risco, C., Menendez-Arias, L., Copeland, T.D., Pinto da Silva, P., Oroszlan, S., 1995. Intracellular transport of the murine leukemia virus during acute infection of NIH 3T3 cells: nuclear import of nucleocapsid protein and integrase. J. Cell Sci. 108 (Pt 9), 3039–3050.

Robertson, S.J., Hasenkrug, K.J., Chesebro, B., Portis, J.L., 1997. Neurologic disease induced by polytropic murine retroviruses: neurovirulence determined by efficiency of spread to microglial cells. J. Virol. 71, 5287–5294.

Ross, S.R., Schofield, J.J., Farr, C.J., Bucan, M., 2002. Mouse transferrin receptor 1 is the cell entry receptor for mouse mammary tumor virus. Proc. Natl. Acad. Sci. U.S.A. 99, 12386–12390.

Rousso, I., Mixon, M.B., Chen, B.K., Kim, P.S., 2000. Palmitoylation of the HIV-1 envelope glycoprotein is critical for viral infectivity. Proc. Natl. Acad. Sci. U.S.A. 97, 13523–13525.

San-Juan-Vergara, H., Sampayo-Escobar, V., Reyes, N., Cha, B., Pacheco-Lugo, L., Wong, T., Peeples, M.E., Collins, P.L., Castano, M.E., Mohapatra, S.S., 2012. Cholesterol-rich microdomains as docking platforms for respiratory syncytial virus in normal human bronchial epithelial cells. J. Virol. 86, 1832–1843.

Saphire, A.C., Bobardt, M.D., Zhang, Z., David, G., Gallay, P.A., 2001. Syndecans serve as attachment receptors for human immunodeficiency virus type 1 on macrophages. J. Virol. 75, 9187–9200.

Sattentau, Q.J., Moore, J.P., 1991. Conformational changes induced in the human immunodeficiency virus envelope glycoprotein by soluble CD4 binding. J. Exp. Med. 174, 407–415.

Scarlatti, G., Tresoldi, E., Bjorndal, A., Fredriksson, R., Colognesi, C., Deng, H.K., Malnati, M.S., Plebani, A., Siccardi, A.G., Littman, D.R., Fenyo, E.M., Lusso, P., 1997. In vivo evolution of HIV-1 co-receptor usage and sensitivity to chemokine-mediated suppression. Nat. Med. 3, 1259–1265.

Schiller, J., Chackerian, B., 2014. Why HIV virions have low numbers of envelope spikes: implications for vaccine development. PLoS Pathog. 10, e1004254.

Schnittman, S.M., Lane, H.C., Roth, J., Burrows, A., Folks, T.M., Kehrl, J.H., Koenig, S., Berman, P., Fauci, A.S., 1988. Characterization of GP120 binding to CD4 and an assay that measures ability of sera to inhibit this binding. J. Immunol. 141, 4181–4186.

Schroeder, C., 2010. Cholesterol-binding viral proteins in virus entry and morphogenesis. Subcell. Biochem. 51, 77–108.

Schutze, M., Romanelli, L.C., Rosa, D.V., Carneiro-Proietti, A.B., Nicolato, R., Romano-Silva, M.A., Brammer, M., de Miranda, D.M., 2017. Brain metabolism changes in patients infected with HTLV-1. Front. Mol. Neurosci. 10, 52.

Sewald, X., Ladinsky, M.S., Uchil, P.D., Beloor, J., Pi, R., Herrmann, C., Motamedi, N., Murooka, T.T., Brehm, M.A., Greiner, D.L., Shultz, L.D., Mempel, T.R., Bjorkman, P.J., Kumar, P., Mothes, W., 2015. Retroviruses use CD169-mediated trans-infection of permissive lymphocytes to establish infection. Science 350, 563–567.

Shimojima, M., Miyazawa, T., Ikeda, Y., McMonagle, E.L., Haining, H., Akashi, H., Takeuchi, Y., Hosie, M.J., Willett, B.J., 2004. Use of CD134 as a primary receptor by the feline immunodeficiency virus. Science 303, 1192–1195.

Shojima, T., Yoshikawa, R., Hoshino, S., Shimode, S., Nakagawa, S., Ohata, T., Nakaoka, R., Miyazawa, T., 2013. Identification of a novel subgroup of Koala retrovirus from Koalas in Japanese zoos. J. Virol. 87, 9943–9948.

Spurrier, B., Sampson, J.M., Totrov, M., Li, H., O'Neal, T., Williams, C., Robinson, J., Gorny, M.K., Zolla-Pazner, S., Kong, X.P., 2011. Structural analysis of human and macaque mAbs 2909 and 2.5B: implications for the configuration of the quaternary neutralizing epitope of HIV-1 gp120. Structure 19, 691–699.

Stewart-Jones, G.B., Soto, C., Lemmin, T., Chuang, G.Y., Druz, A., Kong, R., Thomas, P.V., Wagh, K., Zhou, T., Behrens, A.J., Bylund, T., Choi, C.W., Davison, J.R., Georgiev, I.S., Joyce, M.G., Kwon, Y.D., Pancera, M., Taft, J., Yang, Y., Zhang, B., Shivatare, S.S., Shivatare, V.S., Lee, C.C., Wu, C.Y., Bewley, C.A., Burton, D.R., Koff, W.C., Connors, M., Crispin, M., Baxa, U., Korber, B.T., Wong, C.H., Mascola, J.R., Kwong, P.D., 2016. Trimeric HIV-1-Env structures define glycan shields from clades A, B, and G. Cell 165, 813–826.

Stoye, J.P., Coffin, J.M., 2000. A provirus put to work. Nature 403, 715–717.

Sun, C., Zhang, B., Jin, J., Montelaro, R.C., 2008. Binding of equine infectious anemia virus to the equine lentivirus receptor-1 is mediated by complex discontinuous sequences in the viral envelope gp90 protein. J. Gen. Virol. 89, 2011–2019.

Tailor, C.S., Kabat, D., 1997. Variable regions A and B in the envelope glycoproteins of feline leukemia virus subgroup B and amphotropic murine leukemia virus interact with discrete receptor domains. J. Virol. 71, 9383–9391.

Tailor, C.S., Nouri, A., Kabat, D., 2000. A comprehensive approach to mapping the interacting surfaces of murine amphotropic and feline subgroup B leukemia viruses with their cell surface receptors. J. Virol. 74, 237–244.

Tailor, C.S., Nouri, A., Lee, C.G., Kozak, C., Kabat, D., 1999a. Cloning and characterization of a cell surface receptor for xenotropic and polytropic murine leukemia viruses. Proc. Natl. Acad. Sci. U.S.A. 96, 927–932.

Tailor, C.S., Nouri, A., Zhao, Y., Takeuchi, Y., Kabat, D., 1999b. A sodium-dependent neutral-amino-acid transporter mediates infections of feline and baboon endogenous retroviruses and simian type D retroviruses. J. Virol. 73, 4470–4474.

Tailor, C.S., Willett, B.J., Kabat, D., 1999c. A putative cell surface receptor for anemia-inducing feline leukemia virus subgroup C is a member of a transporter superfamily. J. Virol. 73, 6500–6505.

Takeuchi, Y., Vile, R.G., Simpson, G., O'Hara, B., Collins, M.K., Weiss, R.A., 1992. Feline leukemia virus subgroup B uses the same cell surface receptor as gibbon ape leukemia virus. J. Virol. 66, 1219–1222.

Taniguchi, A., Mochizuki, H., Nagamachi, S., Ebihara, Y., Ishii, N., Shiomi, K., Nakazato, M., 2015. Hypometabolism of watershed areas of the brain in HTLV-1-associated myelopathy/tropical spastic paraparesis. Neurol. Sci. 36, 2117–2120.

Tarlinton, R.E., Meers, J., Young, P.R., 2006. Retroviral invasion of the koala genome. Nature 442, 79–81.

Thali, M., Moore, J.P., Furman, C., Charles, M., Ho, D.D., Robinson, J., Sodroski, J., 1993. Characterization of conserved human immunodeficiency virus type 1 gp120 neutralization epitopes exposed upon gp120-CD4 binding. J. Virol. 67, 3978–3988.

Tobaly-Tapiero, J., Bittoun, P., Neves, M., Guillemin, M.C., Lecellier, C.H., Puvion-Dutilleul, F., Gicquel, B., Zientara, S., Giron, M.L., de The, H., Saib, A., 2000. Isolation and characterization of an equine foamy virus. J. Virol. 74, 4064–4073.

Tomaras, G.D., Yates, N.L., Liu, P., Qin, L., Fouda, G.G., Chavez, L.L., Decamp, A.C., Parks, R.J., Ashley, V.C., Lucas, J.T., Cohen, M., Eron, J., Hicks, C.B., Liao, H.X., Self, S.G., Landucci, G., Forthal, D.N., Weinhold, K.J., Keele, B.F., Hahn, B.H., Greenberg, M.L., Morris, L., Karim, S.S., Blattner, W.A., Montefiori, D.C., Shaw, G.M., Perelson, A.S., Haynes, B.F., 2008. Initial B-cell responses to transmitted human immunodeficiency virus type 1: virion-binding immunoglobulin M (IgM) and IgG antibodies followed by plasma anti-gp41 antibodies with ineffective control of initial viremia. J. Virol. 82, 12449–12463.

Tsao, L.C., Guo, H., Jeffrey, J., Hoxie, J.A., Su, L., 2016. CCR5 interaction with HIV-1 Env contributes to Env-induced depletion of CD4 T cells in vitro and in vivo. Retrovirology 13, 22.

Turville, S.G., Cameron, P.U., Handley, A., Lin, G., Pohlmann, S., Doms, R.W., Cunningham, A.L., 2002. Diversity of receptors binding HIV on dendritic cell subsets. Nat. Immunol. 3, 975–983.

van Zeijl, M., Johann, S.V., Closs, E., Cunningham, J., Eddy, R., Shows, T.B., O'Hara, B., 1994. A human amphotropic retrovirus receptor is a second member of the gibbon ape leukemia virus receptor family. Proc. Natl. Acad. Sci. U.S.A. 91, 1168–1172.

Varchetta, S., Lusso, P., Hudspeth, K., Mikulak, J., Mele, D., Paolucci, S., Cimbro, R., Malnati, M., Riva, A., Maserati, R., Mondelli, M.U., Mavilio, D., 2013. Sialic acid-binding Ig-like lectin-7 interacts with HIV-1 gp120 and facilitates infection of CD4pos T cells and macrophages. Retrovirology 10, 154.

Vives, R.R., Imberty, A., Sattentau, Q.J., Lortat-Jacob, H., 2005. Heparan sulfate targets the HIV-1 envelope glycoprotein gp120 coreceptor binding site. J. Biol. Chem. 280, 21353–21357.

Walker, S.J., Pizzato, M., Takeuchi, Y., Devereux, S., 2002. Heparin binds to murine leukemia virus and inhibits Env-independent attachment and infection. J. Virol. 76, 6909–6918.

Wallin, M., Ekstrom, M., Garoff, H., 2004. Isomerization of the intersubunit disulphide-bond in Env controls retrovirus fusion. EMBO J. 23, 54–65.

Walsh, S.R., Linnerth-Petrik, N.M., Yu, D.L., Foster, R.A., Menzies, P.I., Diaz-Mendez, A., Chalmers, H.J., Wootton, S.K., 2013. Experimental transmission of enzootic nasal adenocarcinoma in sheep. Vet. Res. 44, 66.

Wang, H., Kavanaugh, M.P., North, R.A., Kabat, D., 1991. Cell-surface receptor for ecotropic murine retroviruses is a basic amino-acid transporter. Nature 352, 729–731.

Wang, E., Albritton, L., Ross, S.R., 2006. Identification of the segments of the mouse transferrin receptor 1 required for mouse mammary tumor virus infection. J. Biol. Chem. 281, 10243–10249.

Wang, Y., Pan, Q., Ding, S., Wang, Z., Yu, J., Finzi, A., Liu, S.L., Liang, C., 2017. The V3 loop of HIV-1 Env determines viral susceptibility to IFITM3 impairment of viral infectivity. J. Virol. 91.

Wei, X., Decker, J.M., Wang, S., Hui, H., Kappes, J.C., Wu, X., Salazar-Gonzalez, J.F., Salazar, M.G., Kilby, J.M., Saag, M.S., Komarova, N.L., Nowak, M.A., Hahn, B.H., Kwong, P.D., Shaw, G.M., 2003. Antibody neutralization and escape by HIV-1. Nature 422, 307–312.

Weissenhorn, W., Dessen, A., Harrison, S.C., Skehel, J.J., Wiley, D.C., 1997. Atomic structure of the ectodomain from HIV-1 gp41. Nature 387, 426–430.

White, T.A., Bartesaghi, A., Borgnia, M.J., Meyerson, J.R., de la Cruz, M.J., Bess, J.W., Nandwani, R., Hoxie, J.A., Lifson, J.D., Milne, J.L., Subramaniam, S., 2010. Molecular architectures of trimeric SIV and HIV-1 envelope glycoproteins on intact viruses: strain-dependent variation in quaternary structure. PLoS Pathog. 6, e1001249.

Wichukchinda, N., Nakajima, T., Saipradit, N., Nakayama, E.E., Ohtani, H., Rojanawiwat, A., Pathipvanich, P., Ariyoshi, K., Sawanpanyalert, P., Shioda, T., Kimura, A., 2010. TIM1 haplotype may control the disease progression to AIDS in a HIV-1-infected female cohort in Thailand. AIDS 24, 1625–1631.

Wielgosz, M.M., Rauch, D.A., Jones, K.S., Ruscetti, F.W., Ratner, L., 2005. Cholesterol dependence of HTLV-I infection. AIDS Res. Hum. Retrovir. 21, 43–50.

Wilen, C.B., Tilton, J.C., Doms, R.W., 2012. HIV: cell binding and entry. Cold Spring Harb. Perspect. Med. 2.

Wilk, T., Geiselhart, V., Frech, M., Fuller, S.D., Flugel, R.M., Lochelt, M., 2001. Specific interaction of a novel foamy virus Env leader protein with the N-terminal Gag domain. J. Virol. 75, 7995–8007.

Willey, R.L., Martin, M.A., Peden, K.W., 1994. Increase in soluble CD4 binding to and CD4-induced dissociation of gp120 from virions correlates with infectivity of human immunodeficiency virus type 1. J. Virol. 68, 1029–1039.

Wilson, C.A., Eiden, M.V., Anderson, W.B., Lehel, C., Olah, Z., 1995. The dual-function hamster receptor for amphotropic murine leukemia virus (MuLV), 10A1 MuLV, and gibbon ape leukemia virus is a phosphate symporter. J. Virol. 69, 534–537.

Wu, S.R., Sjoberg, M., Wallin, M., Lindqvist, B., Ekstrom, M., Hebert, H., Koeck, P.J., Garoff, H., 2008. Turning of the receptor-binding domains opens up the murine leukaemia virus Env for membrane fusion. EMBO J. 27, 2799–2808.

Wyatt, R., Moore, J., Accola, M., Desjardin, E., Robinson, J., Sodroski, J., 1995. Involvement of the V1/V2 variable loop structure in the exposure of human immunodeficiency virus type 1 gp120 epitopes induced by receptor binding. J. Virol. 69, 5723–5733.

Xiao, L., Rudolph, D.L., Owen, S.M., Spira, T.J., Lal, R.B., 1998. Adaptation to promiscuous usage of CC and CXC-chemokine coreceptors in vivo correlates with HIV-1 disease progression. AIDS 12, F137–F143.

Xu, W., Stadler, C.K., Gorman, K., Jensen, N., Kim, D., Zheng, H., Tang, S., Switzer, W.M., Pye, G.W., Eiden, M.V., 2013. An exogenous retrovirus isolated from koalas with malignant neoplasias in a US zoo. Proc. Natl. Acad. Sci. U.S.A. 110, 11547–11552.

Yang, C., Compans, R.W., 1996. Palmitoylation of the murine leukemia virus envelope glycoprotein transmembrane subunits. Virology 221, 87–97.

Yang, C., Spies, C.P., Compans, R.W., 1995. The human and simian immunodeficiency virus envelope glycoprotein transmembrane subunits are palmitoylated. Proc. Natl. Acad. Sci. U.S.A. 92, 9871–9875.

Yang, Y.L., Guo, L., Xu, S., Holland, C.A., Kitamura, T., Hunter, K., Cunningham, J.M., 1999a. Receptors for polytropic and xenotropic mouse leukaemia viruses encoded by a single gene at Rmc1. Nat. Genet. 21, 216–219.

Yang, Z.N., Mueser, T.C., Kaufman, J., Stahl, S.J., Wingfield, P.T., Hyde, C.C., 1999b. The crystal structure of the SIV gp41 ectodomain at 1.47 A resolution. J. Struct. Biol. 126, 131–144.

Yang, X., Kurteva, S., Ren, X., Lee, S., Sodroski, J., 2005. Stoichiometry of envelope glycoprotein trimers in the entry of human immunodeficiency virus type 1. J. Virol. 79, 12132–12147.

Yang, S.T., Kiessling, V., Simmons, J.A., White, J.M., Tamm, L.K., 2015. HIV gp41-mediated membrane fusion occurs at edges of cholesterol-rich lipid domains. Nat. Chem. Biol. 11, 424–431.

Yang, S.T., Kiessling, V., Tamm, L.K., 2016. Line tension at lipid phase boundaries as driving force for HIV fusion peptide-mediated fusion. Nat. Commun. 7, 11401.

Yates, N.L., Stacey, A.R., Nolen, T.L., Vandergrift, N.A., Moody, M.A., Montefiori, D.C., Weinhold, K.J., Blattner, W.A., Borrow, P., Shattock, R., Cohen, M.S., Haynes, B.F., Tomaras, G.D., 2013. HIV-1 gp41 envelope IgA is frequently elicited after transmission but has an initial short response half-life. Mucosal Immunol. 6, 692–703.

York, D.F., Vigne, R., Verwoerd, D.W., Querat, G., 1992. Nucleotide sequence of the jaagsiekte retrovirus, an exogenous and endogenous type D and B retrovirus of sheep and goats. J. Virol. 66, 4930–4939.

Yoshii, H., Kamiyama, H., Minematsu, K., Goto, K., Mizota, T., Oishi, K., Katunuma, N., Yamamoto, N., Kubo, Y., 2009. Cathepsin L is required for ecotropic murine leukemia virus infection in NIH3T3 cells. Virology 394, 227–234.

Young, J.A., Bates, P., Varmus, H.E., 1993. Isolation of a chicken gene that confers susceptibility to infection by subgroup A avian leukosis and sarcoma viruses. J. Virol. 67, 1811–1816.

Yu, X., Wang, Q.Y., Guo, Y., Dolmer, K., Young, J.A., Gettins, P.G., Rong, L., 2003. Kinetic analysis of binding interaction between the subgroup A Rous sarcoma virus glycoprotein SU and its cognate receptor Tva: calcium is not required for ligand binding. J. Virol. 77, 7517–7526.

Zacharopoulos, V.R., Phillips, D.M., 1997. Cell-mediated HTLV-I infection of a cervix-derived epithelial cell line. Microb. Pathog. 23, 225–233.

Zavorotinskaya, T., Albritton, L.M., 1999a. A hydrophobic patch in ecotropic murine leukemia virus envelope protein is the putative binding site for a critical tyrosine residue on the cellular receptor. J. Virol. 73, 10164–10172.

Zavorotinskaya, T., Albritton, L.M., 1999b. Suppression of a fusion defect by second site mutations in the ecotropic murine leukemia virus surface protein. J. Virol. 73, 5034–5042.

Zavorotinskaya, T., Qian, Z., Franks, J., Albritton, L.M., 2004. A point mutation in the binding subunit of a retroviral envelope protein arrests virus entry at hemifusion. J. Virol. 78, 473–481.

Zhang, Y., Rassa, J.C., deObaldia, M.E., Albritton, L.M., Ross, S.R., 2003. Identification of the receptor binding domain of the mouse mammary tumor virus envelope protein. J. Virol. 77, 10468–10478.

Zhang, B., Jin, S., Jin, J., Li, F., Montelaro, R.C., 2005. A tumor necrosis factor receptor family protein serves as a cellular receptor for the macrophage-tropic equine lentivirus. Proc. Natl. Acad. Sci. U.S.A. 102, 9918–9923.

Zhang, B., Sun, C., Jin, S., Cascio, M., Montelaro, R.C., 2008. Mapping of equine lentivirus receptor 1 residues critical for equine infectious anemia virus envelope binding. J. Virol. 82, 1204–1213.

Zhu, P., Chertova, E., Bess Jr., J., Lifson, J.D., Arthur, L.O., Liu, J., Taylor, K.A., Roux, K.H., 2003. Electron tomography analysis of envelope glycoprotein trimers on HIV and simian immunodeficiency virus virions. Proc. Natl. Acad. Sci. U.S.A. 100, 15812–15817.

Zhuo, X., Feschotte, C., 2015. Cross-species transmission and differential fate of an endogenous retrovirus in three mammal lineages. PLoS Pathog. 11, e1005279.

Zwaal, R.F., Comfurius, P., Bevers, E.M., 2005. Surface exposure of phosphatidylserine in pathological cells. Cell. Mol. Life Sci. 62, 971–988.

Chapter 2

Cellular Factors That Regulate Retrovirus Uncoating and Reverse Transcription

Stephen P. Goff
HHMI, Columbia University, New York, NY, United States

Chapter Outline

A Brief Description of Early Events of Infection 53
Monitoring the Course of Infection in Early Stages 56
- Monitoring Viral Nucleic Acids 56
- Monitoring Incoming Viral Proteins 59
- Imaging 61

Host Factors Promoting Early Events of Infection 62
- Breaking and Entering: Subcortical Actin 62
- Factors Affecting Reverse Transcription 64
- Cytoskeleton and Motors: Trafficking 65
- Cyclophilin A, a Host Factor Binding Capsid 68
- Modifications of Viral Proteins During Infection 70

Host Factors Restricting Infection in Early Stages of Infection 71
- Fv1, a Gene for Resistance to the Friend Leukemia Virus 72
- TRIM5α, a Major Postentry Block 74
- TRIM-Cyp: Evolution of a Restriction Factor by Gene Fusion 76
- Mov10 77
- Lv2: An Entry-Specific Block 77
- APOBECs, the Cytidine Deaminases 78
- SAMHD1, a Nucleotidase 81
- MX2, Another Factor Targeting Capsid 82
- SUN2 84
- More Restriction Factors? 84

Sensing and Responding to Infection: Innate Immunity 85
- TRIM5α as Sensor 85
- RIG-I and MDA5 as Retroviral Sensors 86
- cGAS Involvement in Sensing Infection 86
- IFI16, a DNA Sensor 87
- TREX1, a Nuclease 88

Applications for Therapy 88
Conclusions: Parting Words 89
References 90

Retrovirus-Cell Interactions. https://doi.org/10.1016/B978-0-12-811185-7.00002-9

This chapter will present our current understanding of the events occurring soon after delivery of the virion core into the cytoplasm—the earliest intracellular steps of the early phase of the virus life cycle. We will discuss what is known about the incoming virion, where it goes, what transformations it undergoes, and how it is thought to find its way to the nucleus. We thus cover the events occurring after fusion of the virion membrane with the host membranes (as discussed in Chapter 1), up to the events involved in nuclear entry (as discussed in Chapter 3). The material here will serve as an update to earlier reviews of this portion of the life cycle (e.g., Goff, 2001; Nisole and Saib, 2004; Dvorin and Malim, 2003). We will then discuss what is known about the identity and functions of the host proteins that impact these events, including proteins that facilitate infection and those that inhibit infection.

The most prominent feature of this part of the life cycle—the defining hallmark of the retroviruses—is the reverse transcription of the incoming viral genomic RNA to form the linear double-stranded DNA copy of the genome. Much work examining early steps of infection has always focused on reverse transcription in part because the synthesis of the viral DNA is clearly an essential step in the life cycle, firmly relevant to successful infection, and also because we have highly sensitive tools to monitor the levels and structures of the viral nucleic acids during infection. In contrast, we have less understanding of the behavior of the viral proteins that take part in these early events in part because the tools for their detection are somewhat less sensitive. The viral proteins are also less attractive for study because we cannot be sure that those we can detect are all necessarily involved in infection. Some are likely associated with defective viruses, and some may have fallen away from the genome and are no longer significant. Nevertheless their behavior and their interactions with host proteins are central to our understanding of infection, and much effort is being invested in their characterization.

The virus undergoes dramatic changes over the course of the early phase of the life cycle. We can describe these events very crudely as a course of "uncoating," although the term itself is much derided and is certainly vague and ill-defined. There are many discrete steps in the process (for discussion, see Greber et al., 1994). We will discuss the various assays that provide some limited conceptions of what occurs during these early steps of infection. We do not know the details of how the viral proteins change their conformation or the exact timing of their dissociation from the nucleic acids, but these issues are the focus of much of the ongoing work in the field.

We will especially concentrate in this chapter on the interactions of various host proteins with the virus during these early stages (for an earlier review of this topic, see Aiken, 2006). A large number of host proteins of many diverse functions have been identified over the last decade, mainly as a result of genetic screens. These can be divided broadly into two categories: so-called "restriction factors" that act to limit or inhibit the course of infection; and "dependency factors" on which the virus is dependent for infection. In some cases, we have some information about how these factors work—what viral proteins they interact with and what they do. In other cases, we have almost no clue how they act and can only infer their role through monitoring the effects of manipulating their levels—either depleting them from the cell by knockdown or knockout methods or overexpressing them by transfection of gene constructs. Uncovering their various mechanisms of action is an area of very active study. We will report what is known about many such factors.

A BRIEF DESCRIPTION OF EARLY EVENTS OF INFECTION

The early steps of infection under consideration here begin with the fusion of the virion membrane with a host cell membrane and the delivery of the virion core into the cytoplasm. The fusion event separates the virion membrane and the envelope glycoprotein proteins from the core, releases the MA matrix protein lying under the membrane, and exposes the naked core into the cytoplasm. This core contains two copies of genomic RNA, held together into a dimer by noncovalent base-pairing interactions, and condensed by association with the small zinc finger nucleocapsid protein (NC). The RNA complex is enclosed in a protein shell, typically in the size range of 50–80 μm, composed of several hundred copies of the capsid protein (CA). This shell is built as an array of about 100 hexamers of CA curved into a closed shell by the specific insertion of 12 pentamers at strategic positions (Li et al., 2000). The shell can exhibit one of the variety of shapes, ranging from spherical (in the case of the murine leukemia viruses (MLVs), feline leukemia viruses, avian leukosis viruses, and others) to cylindrical (in the case of the human T-cell leukemia viruses), to conical (in the case of HIV-1), determined by the positioning of the pentamers. The structure of the CA hexamers is known in atomic detail by high-resolution crystallography, and the structure of the assembled core has been deduced nearly as well by a variety of techniques (cryo-electron microscopy, nuclear magnetic resonance, and others), though our knowledge here is ultimately limited by the inherent heterogeneity of the cores. The capsid shell is thin and appears to be somewhat porous to small molecules but is substantial enough to be isolated after gentle cell lysis and to withstand some steps of purification (e.g., sedimentation by centrifugation). The core contains, in addition to the virion RNA and NC, multiple copies of the reverse transcriptase (RT), integrase (IN), and (probably) protease (PR) enzymes.

Different retroviruses enter the cell and traffic within the cell by different pathways (for reviews, see Marsh and Helenius, 2006; Smith and Helenius, 2004; Marx and Alian, 2015b). For some viruses, the fusion event can occur at the cell surface plasma membrane, mediated by a viral envelope protein (Env) that functions at neutral pH, triggered by conformational rearrangements of Env after its interaction with receptor and sometimes coreceptors. In these cases, the virion core may encounter a meshwork of actin under the plasma membrane that must be traversed for infection to proceed. The HIV-1 Nef protein can act to surmount this barrier (Luo et al., 1998; Campbell et al., 2004). How Nef does this is unclear, but it may induce alterations in virions at the time of their assembly that enhance infectivity (Aiken and Trono, 1995). Other viruses enter the cell by endocytosis into endosomes, and here the fusion event occurs deeper in the cytoplasm and often only after acidification of the endosome triggers the Env rearrangements to allow fusion. This route of entry may allow evasion of the actin meshwork. HIV-1 seems to be able to use either of these pathways (Schaeffer et al., 2001). The preferred pathway used by a given virus can be different in different cell types (Kizhatil and Albritton, 1997). These two routes of entry may set the core on different pathways toward the nucleus and likely determine different interactions with host proteins and different sensitivities to virus restriction factors. Being able to choose alternative pathways may give viruses flexibility and allow for escape from particular restrictions (Marx and Alian, 2015a,b).

Soon after entry into the cytoplasm, the virion core is somehow triggered to begin the process of reverse transcription. The events that activate reverse transcription are quite obscure. RT is thought to be bound to the RNA template and the annealed tRNA primer, poised to begin DNA synthesis, even in the extracellular virion particle. RT in a portion of the virions may have already begun to extend the tRNA primer by the synthesis of a few nucleotides of DNA, presumably in the producer cell that released it (Trono, 1992; Zhu and Cunningham, 1993). One school of thought holds that it is simply the lack of deoxynucleotide precursors in the virion that prevent elongation, and in this model the introduction of the core into a cytoplasm with an adequate concentration of deoxynucleotides is the triggering event for DNA synthesis. This line of thinking is encouraged by the observation that when purified virion particles, after disruption of the membrane by detergent, are incubated with deoxynucleotides in vitro, the cores can initiate and even complete reverse transcription (the so-called "endogenous reverse transcription" reaction) (Rothenberg and Baltimore, 1975; Rothenberg et al., 1977; Gilboa et al., 1979a). New structural studies of the HIV-1 capsid raise the intriguing idea that conformational changes may open pores in the core that permit entry of deoxynucleotides inside to allow DNA synthesis (Jacques et al., 2016).

Host factors may be required to initiate DNA synthesis in vivo. Some biochemistry supports this: the reactions carried out in vitro are not highly efficient, and in some hands the addition of cell lysates can stimulate their efficiency, suggestive of host cofactors. This thinking is supported by the existence of host cell mutants that allow virus entry but do not support efficient initiation of DNA synthesis (Gao and Goff, 1999). Reverse transcription can also be blocked by engineered depletion of a number of host proteins (see below). These observations collectively suggest that the core may need to be activated by one or more host proteins. This step is also a target for negative regulation by many host factors, as indicated by its inhibition by overexpression of selected host proteins (Naghavi et al., 2007), or by the action of several restriction factors (e.g., Fv1 Stoye, 1998 and TRIM5α Stremlau et al., 2004; see below).

The viral nucleic acids during and after the process of reverse transcription are in the form of a very large protein complex, closely approximating the size and perhaps the organization of the intact virion core. We refer to this structure in the process of DNA synthesis as the "Reverse Transcriptase Complex" or RTC; when DNA synthesis is complete, we refer to the structure as the "Preintegration Complex" or PIC. It is not likely that the RTCs or PICs are very discrete or homogeneous entities. Lysates of recently infected cells can be fractionated on sucrose gradients, and the viral RNA is found in a broad peak, with a rough sedimentation value of ~500 S, cosedimenting with a portion of the CA protein, and with the RT enzymatic activity. The efficient completion of DNA synthesis is thought to require this large complex. This thinking comes in part from studies of the endogenous RT reactions, in which disruption of the cores blocks the formation of longer DNA products (Rothenberg and Baltimore, 1975). The stoichiometry of the CA protein and the viral RNA is not well established but the levels of CA probably drop over time. At early times, there may be several hundred CA monomers per RTC, and at later times there may be many fewer, reflecting a gradual loss of CA. For discussion of assays for in vitro uncoating, see Aiken (2009).

The details of the process we loosely term "uncoating" remain quite murky. Much evidence suggests that the CA protein is the main player in the process and is critical for controlling the timing and course of movement of the core toward the nucleus and promoting reverse transcription (see Campbell and Hope, 2015; Fassati, 2012 for reviews). The best evidence for these claims is the existence of many CA mutants that are specifically defective in these events (Hulme et al., 2015b). Some CA mutants do not allow for reverse transcription, suggesting that these mutant cores have not maintained a functional environment for the complex gymnastics that need to be carried out by RT (Alin and Goff, 1996; Tang et al., 2001; Auerbach et al., 2003). Additional evidence for the importance of CA comes from the identification of small molecule inhibitors of the early stages of infection that bind CA (see below). Much as uncoating can control the initiation of reverse transcription, it is conversely true that reverse transcription can affect uncoating. The timing of the two events is certainly at least overlapping: as reverse transcription proceeds, CA is lost from the RTC. However, there is also good evidence of a causal link because inhibitors of RT delay uncoating (Hulme et al., 2011; Yang et al., 2013).

The capsid protein is also implicated in the process of nuclear entry (Ambrose and Aiken, 2014). Thus, several CA mutants have been found that allow for reverse transcription but then fail to enter the nucleus and generate the circular and integrated proviruses (Dismuke and Aiken, 2006). CA interacts with the cyclophilin-like domain of a nuclear pore component (Nup358) (Schaller et al., 2011) and may undergo conformational changes that could affect or perhaps facilitate the entry process (Bichel et al., 2013). While most of the CA is lost in the first hours after entry into the cell (Miller et al., 1997), at least some CA remains with the viral DNA even in the nucleus (Hulme et al., 2015a;

Chin et al., 2015; Chen et al., 2016). Genetics provides strong evidence that some CA remains on the PIC in the nucleus: CA can affect the course and site preferences of DNA integration (Schaller et al., 2011; Koh et al., 2013), drugs targeting CA such as coumermycin A1 can affect integration (Chen et al., 2016), and the PIC may be directed to preferential sites of DNA integration through the interaction of CA with CPSF6 (Rasheedi et al., 2016).

MONITORING THE COURSE OF INFECTION IN EARLY STAGES

A variety of methods have been applied to the study of the early events of the retrovirus life cycle. The most direct approach is the monitoring of viral nucleic acids, viral proteins, and various markers that enter the cell in the infecting virion, using imaging or biochemical detection assays. A major problem afflicting all biochemical study of the early events is the unavoidable fact that the vast majority of the incoming particles do not successfully complete infection and do not succeed in generating a provirus. This is in part because the viruses are inherently inefficient at carrying out each of the many steps of infection, and the consequence is that there is a very large "particle-to-pfu" ratio—the number of particles that are needed to establish a successfully integrated provirus. The number is typically in the range of many tens or hundreds or even thousands to one. It is also in part because many of the incoming virus particles carry defective genomes, with lethal point mutations or deletions. All biochemical assays of incoming viral components detect both on-course and off-course virions, and must be interpreted cautiously.

Monitoring Viral Nucleic Acids

Incoming viral RNA can be detected with low sensitivity by direct assays such as Northern blots or RNase protection of annealed DNA probes but can be more readily and more sensitively monitored by reverse transcriptase polymerase chain reaction (RT-PCR). The levels of RNA can be quantified by real-time RT-PCR. The location of the RNA can be scored by fractionation of lysates before assay. A more elaborate assay for the incoming viral RNA has been developed that monitors the accessibility of 5-ethynyl uridine–tagged viral RNA to staining by dyes, detected by confocal microscopy (Xu et al., 2013). These assays have documented the role of CA in determining the time of uncoating.

The defining event early in infection is the reverse transcription of the viral RNA into DNA. The process consists of the synthesis of a long, linear double-stranded DNA molecule from the viral genomic RNA. The steps of the process are shown in Fig. 2.1. The template for the reaction is the incoming single-stranded, plus or "sense" strand of genomic RNA, bearing a 5′ cap and a 3′ poly(A) tail. The course of reverse transcription carried out by the RTC, and the intermediate structures that are generated during the synthesis of the complete DNA, are understood in great detail (originally outlined in Gilboa et al., 1979b;

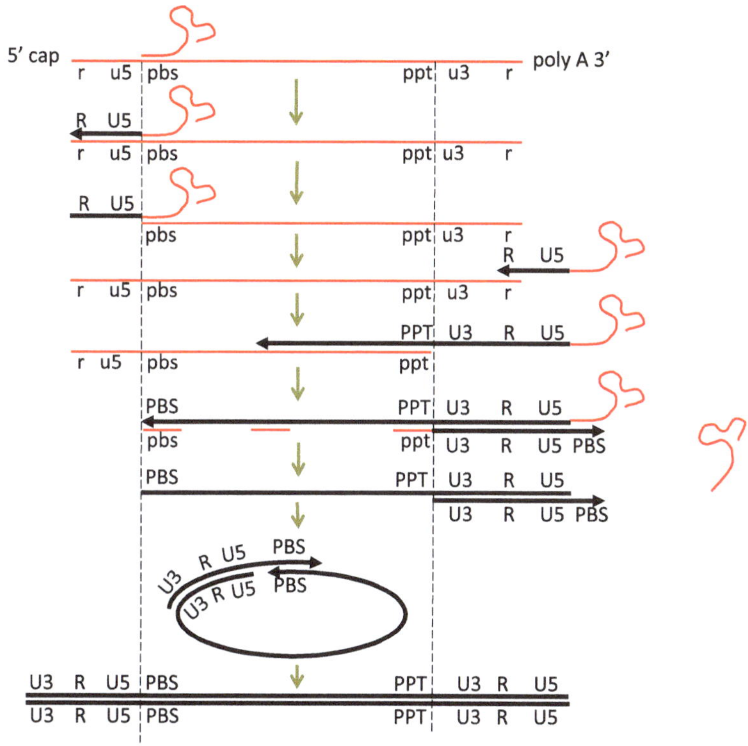

FIGURE 2.1 Reverse transcription of retroviral genomes. The course of reverse transcription of the single-stranded retroviral genomic RNA (top) to form the double-stranded DNA copy of the genome suitable for integration into the host genome (bottom). RNAs are shown in *thin red lines*, and DNAs are shown in *thick black lines*. The primer tRNA is indicated by the thin red floral structure. Sequence blocks involved in reverse transcription are shown in lower case when present in RNA and in upper case when present in DNA. See text for description of the synthesis of each of the intermediates.

for review, see Telesnitsky and Goff, 1997). Briefly, synthesis begins with the elongation of a tRNA primer annealed at a specific region of complementarity, the so-called primer-binding site (PBS), forming a short DNA product called minus-strand strong stop DNA. The 3′ portion of this DNA is then exposed by the action of RNase H, an activity inherent in the RT enzyme, which degrades the 5′ portion of the RNA genome that is in RNA/DNA hybrid form. The strong stop DNA then "jumps" or translocates to the 3′ end of the genome by annealing to a complementary sequence just upstream of the poly(A) tail, and is elongated up the genome to make a long minus-strand DNA. The RNA genome is digested as DNA synthesis proceeds, again by the action of RNase H, and among the products is a polypurine tract (PPT), a relatively resistant oligonucleotide that remains annealed to the DNA. The PPT serves as the primer for the initiation of the plus strand of DNA, and is extended to form the plus-strand strong stop DNA.

Synthesis pauses at a modified base in the tRNA, which is still present at the 5′ end of the minus strand. The 3′ end of this DNA is exposed by RNase H removal of the tRNA, and is then annealed to the 3′ end of the minus-strand DNA to form a circular intermediate. Continued elongation of the paired 3′ end of the minus strand, with displacement of the annealed DNA ahead of synthesis, opens the circle to a linear form. Finally, extension of both strands to the ends of their templates yields a fully double-stranded linear DNA. This event marks the conversion of an RTC into a PIC. All these steps occur in the cytoplasm.

The complex series of "jumps" during reverse transcription generates a large block of duplicated sequences, called the long terminal repeat (LTR), located at both ends of the linear DNA. The initial double-stranded DNA is largely blunt-ended, to the extent that RT fully completes extension of the 3′ termini. The viral DNA undergoes several modifications that can be detected in the course of establishment of the integrated provirus. Over the next few hours, the viral IN enzyme processes the 3′ ends, releasing one or two terminal bases by endonucleolytic cleavage at a conserved CA dinucleotide. As the PICs enter the nucleus, additional DNA forms appear under the direction of the host machinery of the nonhomologous end joining pathway or the homologous recombination pathway. A portion of the linear DNA is cyclized by ligation of the termini to form the so-called "2-LTR" circles. The joints formed are somewhat imperfect or "ragged" in that some bases are often lost before the ends are joined. Another portion of the linear DNA is cyclized by homologous recombination between the two LTRs, generating "one-LTR" circles. Both circular forms are dead-end or off-pathway products and do not proceed to integrate into the host genome. They can persist in the infected cell for a brief period of time (hours or days), but they do not replicate, and they decay and disappear relatively quickly in most settings. Finally, a portion of the processed linear DNAs is integrated into the host genome to form the integrated provirus. The proviral DNA is a permanent addition to the host genome, copied during host DNA replication, and will be retained in the genomes of the daughter cells and their descendants.

The DNA products of reverse transcription can be detected by Southern blots, which reveal the structures of the unintegrated DNA forms: the linear double-stranded DNA containing the LTRs, and the two circular forms containing one or two tandem LTRs. They are much more commonly and sensitively detected by PCR. A crucial control in PCR assays of incoming viral nucleic acids is testing for contaminating nucleic acids present in the virion preparations, which need to be removed by nuclease pretreatment. A typical source of such contaminating DNAs is the high concentrations of plasmid DNA that are in transfections of cells used to generate the virus stocks. The presence of contaminants can be monitored by infection with heat-inactivated virus, or virus lacking the Env, or in the presence of an RT inhibitor. The PCR primers used in the detection of viral DNA can be chosen to reveal either the very first product of reverse transcription, the minus-strand strong stop DNA; or the later extended products; or the second-strand portion that is formed only after both of two strong stop "jumps" have occurred; or the LTR–LTR junction of the circular DNAs that is formed by ligation of the termini of the linear DNA. The integrated provirus DNA can be monitored by so-called Alu-LTR PCR, in which DNA products are amplified using one primer in the highly repeated Alu sequences in the human genome and one in the viral LTR. Similar assays can be performed in mouse cells by using one primer in the B2 repeat sequences and one in the viral LTR.

Studies of the kinetics of viral DNA synthesis have revealed that reverse transcription is initiated very soon after infection (Hulme et al., 2011). The time of initiation may vary among viruses and cells but is typically in the range of only one to two or so hours. The time to completion of DNA synthesis is also variable but is often in the range of 6–12 h. This timing can vary widely depending on the cell type and cell physiology. HIV-1 infection of nonproliferating T cells can result in the initiation of DNA synthesis but not complete synthesis of the full-length genome, and this partial genomic DNA can persist for some time, though it is labile (Zack et al., 1990). Entry into the nucleus, as marked by the appearance of circular DNA forms and the integrated provirus, typically occurs between 10 and 24 h after infection.

Monitoring Incoming Viral Proteins

Viral proteins can be detected in the early stages by direct assays such as Western blots, though as noted above, much of the protein seen here is not on a productive track. The viral proteins CA, NC, RT, and IN are present in very large, high-molecular weight complexes, and the functional subset is in association with the viral RNA and DNA. The size and structure of these complexes can be evaluated by sedimentation analysis after gentle lysis at various times after infection, with the proteins detected by Western blot (Farnet and Haseltine, 1991; Bukrinsky et al., 1993; Miller et al., 1997; Fassati and Goff, 1999, 2001). Depending on the virus and conditions of isolation, matrix, RT, and IN can be shown to be associated with viral nucleic acids in these complexes. A substantial portion of the viral CA protein is also found in rapidly sedimenting forms, and as infection proceeds, more and more CA is stripped away into low-molecular weight, slowly sedimenting forms. The size of the RTC thus gradually shifts in its sedimentation properties from as large as 560 S to as small as 100 S (Fassati and Goff, 1999, 2001). Thus, the process of uncoating is marked by the gradual loss of CA protein from the core, and the timing of that loss is orchestrated to promote efficient infection. Many CA mutants affect the efficiency and timing of core disassembly (Hulme et al., 2015b).

Changes in the state of the virion core provide the conceptual basis for the so-called "fate of capsid" assay, in which the disassembly of the virus is monitored by whether or not the capsid and genomes in a lysate are pelletable through a layer of a sucrose solution (Yang et al., 2014b). Various CA mutants show distinctive behavior in these assays, revealing the fact that some mutants are slow to uncoat or dissociate from the complex ("hyperstable" mutants), whereas others uncoat prematurely ("unstable" mutants) (Yufenyuy and Aiken, 2013). The hyperstable mutants exhibit a stiffer coat as measured by atomic force microscopy (Ramalho et al., 2016). The correct time of uncoating is crucial for efficient infection, and mutants that are too early or late in uncoating are poorly infectious (Forshey et al., 2002). The events that are occurring during uncoating are probably more complex than the simple "fate of capsid" assay might suggest. Suppressor mutations that restore normal infection to CA mutants do not always correct the capsid stability measured biochemically (Yang et al., 2012b). While CA is clearly the key determinant of uncoating, it has been suggested that the viral virion infectivity factor (Vif) protein can affect the core stability (Ohagen and Gabuzda, 2000).

Some host proteins have been implicated in either stabilizing or destabilizing the cores and so can impact the course of infection. One such protein is cyclophilin A, which is thought to stabilize the capsid and so promote infection especially in settings and genetic backgrounds where premature uncoating occurs. Another such protein is PDZD8, which binds to the HIV-1 capsid (Henning et al., 2010). Knockdown of PDZD8 destabilizes capsids and inhibits reverse transcription (Guth and Sodroski, 2014). Its mode of action likely involves the cytoskeleton. Other such host proteins include CPSF6 and TRIM5α (see below).

Many CA mutants affect the efficiency of reverse transcription, suggesting that a proper core particle is needed for this process. These mutants support early observations of the so-called "endogenous reverse transcription reactions," carried out with purified virions, which demonstrate a requirement for intact particles for synthesis of complete viral DNA (Rothenberg and Baltimore, 1975, 1977; Rothenberg et al., 1977). Conversely, reverse transcription affects the course of uncoating (Hulme et al., 2011; Yang et al., 2013). Studies of the RTC during the course of reverse transcription carried out in the endogenous RT reactions have shown changes in the particle morphology as DNA synthesis proceeds (Zhang et al., 2000). These observations suggest that the complex events of reverse transcription may rearrange many of the proteins inside the core, and the DNA product itself may induce stresses on the core that promote its disassembly.

The course of uncoating as monitored by the state of the CA protein has been linked to the course of nuclear entry (Ambrose and Aiken, 2014). While much of the CA protein is lost in the cytoplasm, these studies suggest that there may be more extensive removal of CA on nuclear entry. CA is likely involved in the entry process itself. Several CA mutants carry out normal reverse transcription but are defective in the later steps of nuclear entry and DNA integration (Dismuke and Aiken, 2006). CA can interact directly with the NUP358 component of the nuclear pore (Dharan et al., 2016), and structural studies raise the intriguing notion that its proline isomerase activity may induce conformational changes in CA that could be involved in nuclear import of the PIC (Bichel et al., 2013).

The multimeric state of CA protein in the core is probably essential for its interaction with many host factors and small molecule inhibitors. Large surfaces of CA that span more than one monomer, as are present in the CA hexamer, provide the targets for the interactions with such proteins as Fv1, TRIM5α, CPSF6 and the nuclear pore protein NUP153, and the antiviral compounds PF74 and BI-2 (Price et al., 2014; Saito et al., 2016). The binding sites for the proteins and these drugs overlap, and CA mutants often coordinately affect binding and sensitivity to both (Zhou et al., 2015; Bhattacharya et al., 2014).

Imaging

Much of our understanding of the course of infection has come from direct observation of infected cells. Incoming virion particles can be seen most simply by immunofluorescent staining of fixed cells with antibodies directed against CA or pol proteins. Viral proteins can also be monitored by immunolabeling using gold probes (Risco et al., 1995). Following the localization of both CA and viral nucleic acids by versions of FISH allow more detailed monitoring of the RTCs and PICs (Chin et al., 2015). But the trafficking of incoming virions can be monitored in real time by live imaging of tagged virus moving in acutely infected cells (McDonald et al., 2002; for reviews, see Campbell and Hope, 2008; Hulme and Hope, 2014b). One method to label virion cores of HIV-1 particles is to heavily load the particles with GFP in the form of Vpr–GFP fusion proteins expressed in virus producer cells. Vpr–GFP is incorporated into virions at high stoichiometry, perhaps only a few fold less than Gag itself, resulting in hundreds of copies per virion. These virions can be isolated and applied to cells, such that the course of infection can be followed by fluorescence microscopy. These experiments show that virus particles move along curvilinear paths in the cytoplasm and accumulate near the nucleus, often near the microtubule-organizing center (MTOC). The particles colocalize with particles imaged by staining with antibodies to CA, and surprisingly, also to MA. The majority of the particles are found on or close to microtubules and can often be seen moving along microtubules (McDonald et al., 2002). Studies of the related human foamy viruses have shown similar movement of the incoming viral cores along microtubules toward the centrosome (Saib et al., 1997).

An issue with Vpr-based tagging is that on dissolution of the core, the Vpr–GFP is rather quickly lost from the virion and travels on its own to the nucleus (Desai et al., 2015). An alternative method to label virions is to make use of fluorescently tagged cyclophilin A (CypA-DsRed), which is incorporated into assembling virions by the interaction of the CypA protein with CA (Francis et al., 2016). While much of this protein is lost from the cores soon after entry into the cell, a portion remains with the cores and can be followed during later steps of infection. The rate of this later loss is accelerated by the process of reverse transcription.

It is problematic that many of the particles that are imaged are not on the infectious pathway but are in endosomes or otherwise destined for degradation. One approach to distinguish particles that were delivered into the cell by fusion of the viral membrane to the cell membrane from those that were non-specifically endocytosed is to double label the virions with both Vpr–GFP and an N-terminal peptide of c-Src fused to another reporter (S15-mCherry), which labels the virion membrane (Campbell et al., 2007). Those particles that have lost the mCherry tag are more likely on the infectious pathway. At least a subset of the particles can also be stained by injection of the cells with a fluorescently labeled deoxynucleotide (Alexa-594–dUTP) before infection, indicating that they are actively carrying out reverse transcription. These particles too are presumably more likely on a productive pathway.

More sophisticated monitoring of appropriately labeled virion cores has allowed for the detection of two stages of uncoating. Here virions are prepared with GFP that is not attached to Gag proteins and is localized both within the core and outside the core but within the membrane (a "fluid-phase" marker). The biphasic uncoating is manifest as two steps of loss of the intravirion GFP—first the loss of GFP from the volume between the core and the membrane, and then loss of GFP from the core (Xu et al., 2013; Mamede and Hope, 2016).

Another approach taken to follow HIV-1 cores has been to tag another protein, the IN, with a small tetracysteine tag, allowing for specific labeling with a bis-arsenical fluorescein derivative (the so-called FlAsH reagent) (Arhel et al., 2006). These experiments revealed the kinetics of movement of the virion core in three dimensions, both in the cytoplasm and in the nucleus, over the fourth dimension of time.

HOST FACTORS PROMOTING EARLY EVENTS OF INFECTION

We now will discuss the various host proteins known or thought to interact with the viral core during the early phase of the life cycle and regulate infection. Some have supportive roles in infection; others act as restriction factors and inhibit infection (see Colomer-Lluch et al., 2016 for recent review). A graphic summary of the supportive factors is presented in Fig. 2.2.

Breaking and Entering: Subcortical Actin

The actin cytoskeleton may play either positive or negative roles in infection. Actin may promote the multimerization and colocalization of the CD4 receptor and CXCR4 coreceptor on the cell surface, facilitating the fusion of the viral and host membrane mediated by the HIV-1 Env (Iyengar et al., 1998). CD4+ memory T cells, which exhibit a higher density of cortical actin, are more susceptible to HIV-1 infection than that of naïve T cells with a lower cortical actin density (Permanyer et al., 2013). Here the actin is suggested to enhance virus internalization. Another report has suggested that the HIV-1 gp120 envelope activates the LIM domain kinase to promote filamentous actin polymerization and enhance infection (Vorster et al., 2011).

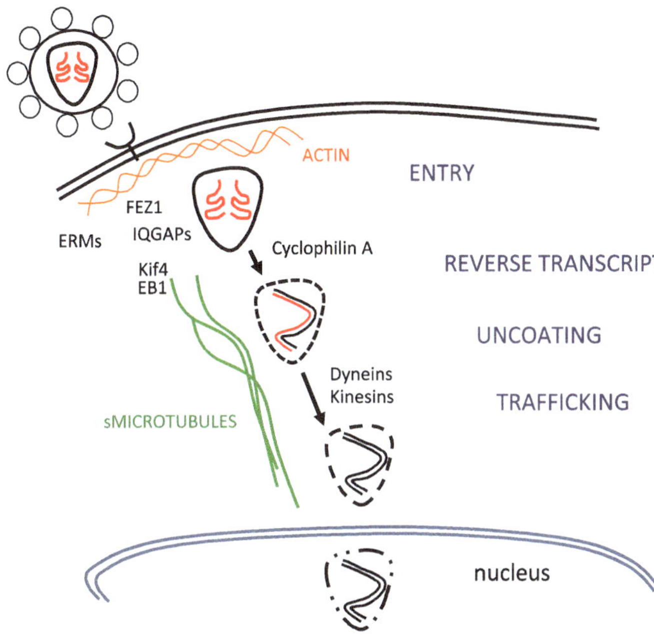

FIGURE 2.2 Host factors promoting early steps of retrovirus infection. A schematic of the early events of infection, from entry of the virion core into the cytoplasm (top) to delivery into the nucleus (bottom), are indicated by labels in blue on the right. Viral RNAs are indicated by *red lines* and DNAs by *black lines*. The cortical actin meshwork is indicated in orange, and microtubules are indicated by *wavy green lines*. The major host factors that facilitate or promote infection are indicated in black, placed at the approximate location of their activities.

When retroviruses enter the cell by fusion of the virion membrane with the plasma membrane at the cell surface, the core particle is immediately confronted in many cells with a barrier of a meshwork of subcortical actin. For these viruses the actin will inhibit, or at least delay, the course of infection. HIV-1 envelope does mediate fusion at the cell surface and so HIV-1 cores are subject to this restriction. The HIV-1 envelope interaction with the coreceptors CXCR4 has been shown to activate cofilin, an actin-depolymerizing factor, providing one means to overcome this block (Yoder et al., 2008). The HIV-1 Nef protein may also facilitate entry by disrupting the actin cytoskeleton (Schaeffer et al., 2001). Viral mutants lacking the Nef protein are significantly impaired at the step of entry into the cytoplasm, and the infectivity of these mutants can be enhanced by artificially disrupting the cytoskeleton (Campbell et al., 2004). This enhancement is only observed for viruses entering the cell via envelopes that mediate fusion at the cell surface, including the HIV-1 envelope and the murine amphotropic envelope, and is not seen for viruses entering by endocytosis, which can bypass the actin barrier.

Nef has been shown to mediate profound rearrangements of the actin cytoskeleton in some cell types (Lu et al., 2008). Nef interacts with diaphanous interacting protein (DIP), a regulator of Rho and Rac signaling to the cytoskeleton, as well as Vav2. Nef utilizes the Src kinase to induce phosphorylation of DIP and p190RhoGAP, thereby inhibiting RhoA. These events are likely aspects of the ability of Nef to enhance infectivity. Nef has been also shown to interact with the cellular kinase PAK2 (Stolp et al., 2010) and thereby block chemokine-mediated actin remodeling required for cell motility. This interaction is widely conserved across many HIV-1 and simian immunodeficiency virus (SIV) isolates. The control of actin by Nef may also impact virus production and cell-to-cell transmission (Fackler and Krausslich, 2006; Stolp and Fackler, 2011; Pearce-Pratt et al., 1994).

Another player acting at this stage of infection is gelsolin, a protein that restructures cortical F-actin and thereby controls the localization and clustering of surface molecules such as CD4 (Garcia-Exposito et al., 2013). The overexpression or knockdown of gelsolin affects the distribution of actin and impairs virus entry.

Factors Affecting Reverse Transcription

Among the many host proteins that are incorporated into virions during assembly, only a few have been demonstrated to have any major impact on virus infectivity (Warren et al., 2009). One example of such a protein is eukaryotic elongation factor 1, which directly interacts with RT and stimulates DNA synthesis carried out in isolated virion cores, promoting first strand strong stop DNA transfer and synthesis of late DNA products (Li et al., 2015; Warren et al., 2012). Another is Upf1, best known as a member of the SF1 family of helicases and a component of the nonsense-mediated decay pathway controlling the destruction of defective mRNAs. HIV-1 virions have been shown to contain significant levels of Upf1, and its presence and activity are critical for infectivity. Cells depleted of endogenous Upf1 (but not of another NMD core component) produce HIV-1 virions of substantially lower specific infectivity. Furthermore, expressing a mutant version of Upf1 lacking helicase activity potently inhibits HIV-1 virion infectivity at the level of reverse transcription (Serquina et al., 2013). Two other proteins have been shown to bind directly to RT, specifically to the RNase H domain, and modulate the course of DNA synthesis: the RNA-binding protein HuR, whose function may involve an interaction with APOBEC3G; and the PKA regulator AKAP149 (Lemay et al., 2008). The host DNA topoisomerase I has been shown to modestly enhance reverse transcription (Takahashi et al., 1995; Shoya et al., 2003). Another candidate modulator of RT is the complex of Sin3a, a scaffold protein, and its associated proteins SAP18, SAP30, and histone deacetylase 1 (Sorin et al., 2009). Yet one more is Gemin2, first implicated through an interaction with the HIV-1 IN protein (Hamamoto et al., 2006). The mechanism of action of all these host proteins is unknown.

Cytoskeleton and Motors: Trafficking

A large body of work over many years, using diverse technologies and methods, has shown that retrovirus cores—the RTCs and PICs—are transported from a periplasmic location to the nucleus along microtubules constituting the cytoplasmic cytoskeleton. The movement toward the nucleus is apparent by direct visualization of tagged cores (McDonald et al., 2002) and can also be documented by scoring the effects of manipulating the cytoskeleton on the efficiency of infection (Bukrinskaya et al., 1998). A number of reviews summarize our understanding of the mechanisms and players mediating this transport for retroviruses and other viruses (e.g., Sodeik, 2000, 2002; Campbell and Hope, 2003; Fackler and Krausslich, 2006; Dohner and Sodeik, 2005; Radtke et al., 2006; Naghavi and Goff, 2007; Stolp et al., 2010; Gaudin et al., 2013; Arriagada, 2017).

The inward movement of many viruses is mediated by dyneins, involving direct interaction with dynein light chain 8 (Petit et al., 2003; Su et al., 2010) and the kinesin motor proteins, most notably, kinesin-1 (Dohner et al., 2005; Dodding and Way, 2011) and others (Martinez et al., 2008). Peptides from dynein light chain 1 have been shown to be able to interfere with infection (Fontenot et al., 2007), and direct interactions of HIV-1 IN with dynein light chain 1 have been reported (Jayappa et al., 2015). Some components of this machinery are important for many retroviruses and not others: for example, a specific light chain of the cytoplasmic dynein complex, DYNLRB2, seems to be important for all the MLVs tested, though not HIV-1 (Opazo et al., 2017). The involvement of some of the motor proteins in infection by a given virus seems to depend on the route of entry into the cell. Thus, infection by the ecotropic MLVs, which enter the cell via the mCAT receptor, is profoundly affected by depleting either of two regulators of the dynein complex, dynamitin (dynactin subunit 2), and NudEL (Valle-Tenney et al., 2016). Viruses utilizing this pathway enter the cell via endosomes and require acidification for release into the cytoplasm. But infection by the amphotropic MLVs, utilizing a distinct receptor and entry pathway, is not affected by depleting these regulators. Viruses using this pathway enter the cell via caveola-mediated coated vesicles and macropinocytosis and do not require acidification for release. These finding suggest that the receptor choice can determine the route of passage of the virus and commit the virus to distinctive cytoskeletal motor requirements.

Though the simplest model is that dyneins mediate movement inward (toward the MTOC) and kinesins mediate outward movement, the true situation is more complex. Achieving net inward motion most likely requires both machines and results from a coordinated or combined action of both working in tension to move their various cargoes. The opposing directions of their activities may be regulated by the members of the JIP (for JNK-interacting protein) family (Fu and Holzbaur, 2013).

Most microtubules are dynamic and grow and retract continuously, but a small subset is stable over much longer time spans (Gundersen, 2002). These stable

MTs (sMTs) are stabilized by capping with a protein complex that prevents the normal growing and shrinking of conventional dynamic MTs. These sMTs can be most easily detected and quantified by staining cells for the levels of specific modifications of the MTs (acetylation and detyrosination—the removal of the C-terminal tyrosine) that accumulate on sMTs but are erased in dynamic MTs. A complex signal pathway made up of numerous regulatory proteins controls the formation of the sMTs. Retrovirus infection proceeds much more efficiently in cells with high levels of sMTs and much less efficiently in cells with no or low levels of sMTs (Naghavi et al., 2007). Remarkably, retroviruses not only make use of existing microtubule networks as they find them in the infected cell but also manipulate them to their advantage. In particular, very soon after entry into the cytoplasm, HIV-1 actually induces the formation of the so-called sMTs to enhance its own infection (Sabo et al., 2013). This induction is likely mediated by the interaction of the incoming MA protein with Kif4, a regulator of sMTs (Kim et al., 1998; Tang et al., 1999).

Many of the known regulators of sMTs affect virus infection, consistent with the importance of sMTs for the viruses. These include the formins, such as mDias 1, 2, and 3; and so-called capping proteins such as CLIP-170, EB1, APC, and Kif4 (Bartolini and Gundersen, 2010; Wen et al., 2004; Martinez et al., 2008). These proteins either regulate the activity of capping molecules or directly bind to the plus end tips of microtubules and prevent or slow the dissociation of tubulin. Knockdown of proteins such as EB1 can have profound effects on infection by retroviruses and herpes viruses (Sabo et al., 2013; Jovasevic et al., 2015).

The efficiency of transport and infection is also affected by many other proteins that interact with the microtubules, such as MAP1 (Fernandez et al., 2015). Yeast two-hybrid screens for proteins interacting with HIV-1 CA identified MAP1A and MAP1S, related proteins that stabilize microtubules and are cleaved to heavy and light chains; CKAP1, which binds tubulin and promotes polymerization; and WIRE, which regulates actin filament polymerization. Knockdown of MAP1A or MAP1S reduced trafficking of virus to the nucleus and susceptibility to infection without major effects on cytoskeletal structures. This work suggests a direct binding of the p24 CA protein to bridging proteins that link the PIC to microtubules. Knockdown of MAP4 and DNAL1 also showed modest effects on HIV-1, acting early in the course of infection (Gallo and Hope, 2012).

A large number of cytoskeletal-related proteins have been implicated in virus trafficking, most often acting through regulation of the structure or function of the microtubules. One example is FEZ1 (fasciculation and elongation protein zeta-1) thought to serve as an adaptor protein for Kinesin-1 (Butkevich et al., 2016). FEZ1 function is regulated at the level of phosphorylation by the microtubule affinity–regulating kinases (MARK/PAR-1). Either overexpression (Naghavi et al., 2005) or RNAi-mediated knockdown (Malikov et al., 2015) of FEZ1 can profoundly block HIV-1 infection, after viral DNA synthesis in the cytoplasm but before appearance of the circular DNAs in the nucleus. This work suggests that FEZ1, acting as a bridge from the PIC to dynein and kinesin motors, has a significant role in HIV-1 trafficking.

Another family of regulators of microtubules with major effects on retrovirus infection is the IQGAP family. The IQGAPs are scaffolding proteins involved in cytokinesis, intracellular signaling, vesicle trafficking and cytoskeletal dynamics; they regulate cell migration and adhesion, through their ability to promote microtubule meshwork and to crosslink microtubules to actin filaments (Briggs and Sacks, 2003). These proteins were identified in a yeast two-hybrid screen as interacting with the MA portion of the MLV Gag protein, and noninteracting Gag mutants were found to be poorly infectious with defects in both the early and late parts of the life cycle (Leung et al., 2006). Revertant viruses selected for replication invariably recovered the interaction with IQGAPs. Overexpression of a C-terminal fragment of IQGAP1 was found to exhibit a potent dominant negative block to MLV replication. IQGAP1 was also found to interact with the core of classical swine fever virus (Gladue et al., 2011). But the role of IQGAPs in the early events of retrovirus infection may not be universal: tests on HIV-1 reveal very little effects of IQGAP knockdown or overexpression on the early part of the life cycle.

Studies of inhibition of measles virus by selected monoclonal antibodies resulted in the identification of moesin, a member of the ERM family (for ezrin–radixin–moesin), as a protein required for efficient infection (Dunster et al., 1994). Moesin acts as a crosslinker between plasma membrane and actin filaments, as well as a signal transducer in responses involving cytoskeletal remodeling. Screening overexpression libraries for proteins able to block MLV infection resulted in the identification of moesin (Naghavi et al., 2007). Moesin downregulates the levels of stable microtubules, and this may be an important mechanism of its action. The close family member ezrin has similar activities (Haedicke et al., 2008), and indeed all three members affect HIV-1 infection (Kubo et al., 2008).

The interaction of virus cores with the cytoskeleton is not only involved in the trafficking of virus within the cell but also impacts the course of uncoating. The cytoplasmic dyneins have significant effects on the rate of uncoating (Lukic et al., 2014); blocking the dyneins results in accumulation of cores that remain intact, as scored by their pelletability after cell lysis (Pawlica and Berthoux, 2014; Pawlica et al., 2014). Blocking dynein or the kinesin-1 heavy chain KIF5B delays uncoating, again suggesting that motors affect not only trafficking but also uncoating (Lukic et al., 2014). The interaction with the cytoskeleton through dynein also affects the susceptibility of virus to the TRIM5α and TRIM-Cyp restriction factors. Blocking transport with drugs or depleting cells of dynein, led to decreased restriction, likely by antagonizing the destructive effects of premature uncoating (Pawlica et al., 2015).

The kinesins are well established as important for inward movement of many virus genomes, but they may also help in delivery to and even into the nucleus. The heavy chain of Kinesin-1, Kif5c, has been found to interact with several components of the nuclear pore, notably the nucleoporin Nup358 (Dharan et al., 2016), and can help disrupt adenovirus capsids and the nuclear pore itself, allowing nuclear access (Strunze et al., 2011). Other kinesins may be variously involved in trafficking particular viruses into the nucleus. The process of nuclear entry will be discussed in detail in another chapter.

Cyclophilin A, a Host Factor Binding Capsid

One of the first cellular proteins found to interact with the HIV-1 CA was cyclophilin A (CypA). The cyclophilins are ubiquitous cytoplasmic peptidyl–proline isomerases best known for their role in the immune response (Wang and Heitman, 2005). Inhibition of the cyclophilins by the small molecule cyclosporine potently blocks T-cell signaling and suppresses immunity. Cyclophilin was first implicated as a host factor for retroviruses in a yeast two-hybrid screen for cellular proteins interacting with HIV-1 Gag (Luban et al., 1993). CypA binding was found to depend critically on a proline residue (P90) in a loop exposed on the surface of HIV-1 CA. CypA did not bind the CA of HIV-2, SIV, or the MLVs. The structure of CypA in association with CA is known in great detail (Gamble et al., 1996; Vajdos et al., 1997). Dimerization of HIV-1 CA was required for the strongest interaction (Colgan et al., 1996), and crystallography much later demonstrated that CypA bridged two CA molecules present on adjacent hexamers in the virion core (Liu et al., 2016). CypA was found to be incorporated at high levels into virion particles (Franke et al., 1994; Thali et al., 1994), but the CypA already resident in the cytoplasm of the infected cell is suggested to be the determinant of virus replication (Sokolskaja et al., 2004).

A number of studies showed that CypA usually promotes infection by HIV-1. Disrupting the interaction with cyclosporine, or by mutation of CA, reduced the efficiency of infection, with a major block occurring at the time of initiation of reverse transcription (Braaten et al., 1996b). The efficacy of cyclosporine and analogues in inhibiting virus correlated well with their ability to disrupt the CypA–CA interaction (Franke and Luban, 1996). Tests of the various clades of HIV-1 revealed that clades A, B, and D of the group M isolates all required the interaction for efficient replication but that the outlier group O isolates did not (Braaten et al., 1996c). Knockout of the CypA gene in a human T-cell line impaired virus replication, and reexpression of CypA restored virus replication (Braaten and Luban, 2001). These experiments all suggested that CypA was providing a useful function to HIV-1 Gag that promoted early steps of infection. Measurements of cores carried out in vitro suggest that CypA acts to stabilize the virion core and so prevent premature uncoating (Shah et al., 2013). Tests of the stability of various CA mutants and the effects of cyclosporine on the timing of uncoating of these mutants suggested that CypA could act to either stabilize or destabilize the PIC and control the efficiency of infection (Li et al., 2009). The effects of CypA in different settings could go either way—the simplest statement that one can make is that CypA binding can modulate capsid stability through subtle changes in CA conformation, with consequences that vary from virus to virus.

Additional experiments have revealed that the levels of CypA in the virion core, or available to the PIC in the infected cell, may be an important modulator of virus infectivity. Various T-cell lines, and presumably various T cells in vivo, turn out to express distinctive levels of CypA, and this difference leads to distinctive effects on replication of wild-type and mutant viruses. Jurkat cells have a "normal" level of CypA, and the addition of cyclosporine impairs replication of wild-type virus in these cells. A particular CA mutant (A224E) that was selected for resistance to cyclosporine grows well in these cells, with or without the drug. The H9 cell line has high levels of CypA and can support the replication of wild-type virus, but the line cannot support replication of the A224E mutant—the virus packages too much CypA. Reducing the packaged CypA by the addition of cyclosporine rescues the mutant virus in these cells—thus, the mutant is cyclosporine-dependent in these cells. The conclusion is that the virus needs a narrow range of CypA levels for efficient replication, and this range can be altered by CA mutations (Yin et al., 1998).

A number of other mutations affecting the CA of HIV-1, HIV-2, and SIVmac have been found to sensitize the virus to restriction by CypA (Lahaye et al., 2016). The mutants are blocked at the time of nuclear entry. The restriction required the host factor SUN2; and further tests of wild-type virus showed that the normal positive role of CypA on wild-type virus also required SUN2. These studies provide further evidence of a complex interaction of CypA with CA with varying outcomes depending on CA sequence and structure.

Studies selecting for mutant viruses that could replicate in the face of cyclosporine painted a still more complex picture. These viruses acquired mutations in the proline-rich loop of CA, but surprisingly these mutations did not affect CypA binding nor did they change the ability of cyclosporine to block CypA binding to CA. The mutant viruses simply seem to have acquired the ability to replicate without the need for CypA (Braaten et al., 1996a). These and other results led to the notion that perhaps CypA is promoting virus replication not by a direct activity but rather by blocking or modulating the binding of a restriction factor—such as TRIM5α or another factor that acts by binding CA (Towers et al., 2003; Shah and Aiken, 2014) (for summaries of this model, see Luban, 2007; Sokolskaja and Luban, 2006). The effects of CypA on the function of the restriction factor could go in either direction in different settings. Thus, in African green monkeys (AGM) or rhesus macaques, CypA binding to CA would alter the CA conformation such that it is readily recognized by TRIM5α or similar restriction factors, sensitizing the virus to the block (Keckesova et al., 2006). Addition of cyclosporin to these cells would relieve the block and allow for infection. But in human cells, CypA binding to CA would block or reduce the binding of the restriction factor and so promote infection; addition of cyclosporine here would remove CypA and allow reestablishment of the block.

Searches for any novel restriction factor that would act in concert with CypA to block infection by wild-type virus have not yet succeeded (Shah and Aiken, 2014) (though see comments on SUN2 above).

Most studies of the time of action of CypA indicated that it affected reverse transcription or early steps of uncoating. Recent work, however, suggests that the critical point of the restriction controlled by CypA is not at reverse transcription but at the time of nuclear entry and formation of the 2-LTR circles (De Iaco and Luban, 2014). As with Fv1, it may that several aspects of infection are impacted by CypA during its interaction with the PIC. Another set of CA mutants (A92E, T54A, and R132K) exhibited a phenotype that implicated CypA binding to CA in still later stages of infection—even after nuclear entry of the PIC. These mutants were impaired for infection of nondividing HeLa cells, though generating normal levels of 2-LTR circles, and this block was relieved by cyclosporine (Qi et al., 2008). Thus, CypA can apparently remain bound to CA and impact viral DNA integration in some cells.

Modifications of Viral Proteins During Infection

The retroviral proteins in the virion core that are delivered into the cell are generally thought of as fully ready to perform the tasks assigned to them to mediate infection. The Gag and Pol precursors are already processed to their mature cleavage products, and the fact that virions can be induced to carry out reverse transcription in vitro (in the so-called endogenous reaction) would suggest that no major modifications by the host are required. Examination of the incoming Gag proteins of the MLVs, however, has revealed that some covalent modifications do occur and seems to play significant roles.

One modification occurring early in infection is SUMOylation. A yeast two-hybrid screen for host proteins that interact with the CA protein of the Moloney MLV resulted in the identification of Ubc9 and PIASy, the E2 and E3 small ubiquitin-like modifier (SUMO)-conjugating enzymes (Yueh et al., 2006). Mapping of the binding sites for these two hits on CA yielded interesting sequences. The binding site for Ubc9 mapped to a near-perfect consensus sequence for SUMOylation (ΨKxE; at residues 202–220); and the binding site for PIASy mapped to residues 114–176, directly centered on the so-called major homology region (MHR), a highly conserved sequence found on nearly all retroviruses and retroelements. These sites were recognized and mediated SUMOylation of CA. Expression of CA and a tagged SUMO-1 protein resulted in covalent transfer of SUMO-1 to CA in vivo. Mutations of lysine residues to arginines near the Ubc9-binding site and mutations at the PIASy-binding site reduced or eliminated CA SUMOylation. These sequences were also important in the context of a replication-competent genome. Introduction of these mutations into the complete viral genome blocked virus replication. The mutants exhibited no defects in the late stages of viral gene expression or virion assembly. On infection, the mutant viruses were able to carry out reverse transcription to synthesize normal levels of linear viral DNA but were unable to produce the circular viral DNAs or integrated provirus normally found in the nucleus. The results suggest that the SUMOylation of CA, mediated by an interaction with Ubc9 and PIASy, is required for early events of infection, after reverse transcription and before nuclear entry and viral DNA integration. Presumably the SUMOylated CA is part of the signal for nuclear import or retention of the PIC. The identity of the proteins that might recognize the SUMOylated CA and control its movement is unknown.

The CA protein of HIV-1 has been shown to be modified by phosphorylation at multiple serine residues, and mutations affecting any of these three resulted in delays in uncoating and reverse transcription (Cartier et al., 1999). It is also known that another Gag protein, MLV p12, is modified by phosphorylation at serine residues. Mutant viruses with changes at S61 and S78 were severely impaired for replication, and S61 was shown to be required for normal levels of phosphorylation of p12 in vivo. These defective mutant viruses could produce normal virions, but acutely infected cells failed to form circular viral DNAs, suggesting a block at the stage of entry of the PIC into the nucleus or retention of the PIC in the nucleus (Yueh and Goff, 2003). Insertion into Gag of sequences from other viruses that are phosphorylated could rescue the p12 mutants (Brzezinski et al., 2016). The kinase involved is not known. Revertant viruses selected for replication, however, acquired suppressor mutations that did not always restore normal levels of phosphorylation, indicating that phosphorylation per se is not absolutely required for function. The most important role of p12 is now established as tethering the PIC to chromatin to promote integration (Elis et al., 2012; Schneider et al., 2013). Thus, perhaps the mechanism by which p12 mutants prevent formation of the circular DNAs is by preventing retention of the PIC in the nucleus when it reforms after mitosis.

In addition, modifications of INs have been noted. The avian IN is a phosphoprotein (denoted pp32), and the site of phosphorylation has been localized to a serine five residues from the C-terminus (Horton et al., 1991). Mutations in this region that prevent phosphorylation show impaired replication. The kinase responsible is unknown. Furthermore, the HIV-1 IN is modified by acetylation of three lysine residues near the C-terminus, perhaps by the cellular p300 acetyltransferase. Mutations that prevent the acetylation abolish replication (Cereseto et al., 2005). It has been suggested that another acetyltransferase, GCN5, may be involved (Terreni et al., 2010). The HIV-1 IN has been shown to be modified by the addition of SUMO; but the importance of the SUMOylation of IN has been questioned (Topper et al., 2007).

HOST FACTORS RESTRICTING INFECTION IN EARLY STAGES OF INFECTION

We now discuss the many host factors that act to inhibit or restrict virus replication at the early stage of the life cycle. A graphic depiction of these factors is presented in Fig. 2.3.

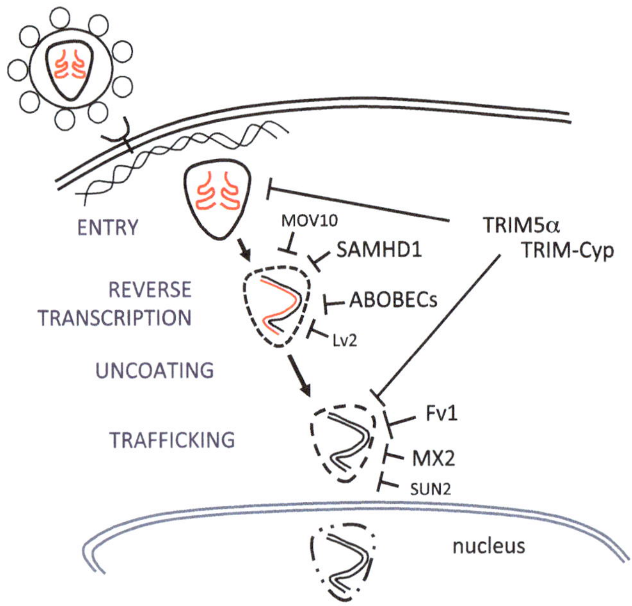

FIGURE 2.3 Host factors restricting early steps of retrovirus infection. A schematic of the early events of infection, from entry of the virion core into the cytoplasm (top) to delivery into the nucleus (bottom), are indicated by labels in blue on the left. Viral RNAs are indicated by *red lines* and DNAs by *black lines*. The major restriction factors, and their approximate time of action, are indicated on the right.

Fv1, a Gene for Resistance to the Friend Leukemia Virus

The oldest known "restriction factor"—a host protein that restricts virus replication—is the product of the mouse Fv1 gene, first identified in the 1970s by Frank Lilly and colleagues as a locus controlling infection by the murine leukemia viruses (Rowe and Hartley, 1972; Rowe et al., 1973; Pincus et al., 1975; for reviews, see Jolicoeur, 1979; Stoye, 1998). Early work revealed two naturally occurring Fv1 alleles in various mouse strains: an $Fv1^n$ allele, found in NIH swiss mice, allowing replication of the so-called N-tropic strains of virus and blocking B-tropic viruses; and an $Fv1^b$ allele in Balb/c mice, allowing replication of B-tropic viruses and blocking N-tropics (Rommelaere et al., 1979). The resistance was dominant over susceptibility, and $Fv1^{n/b}$ heterozygous animals were resistant to both N- and B-tropic viruses. Many strains lacked Fv1 restriction activity and were designated null ($Fv1^0$) (Kozak, 1985). Recently alleles with more complex patterns of restriction of various viruses have been identified in other murine species (Yap et al., 2014). These alleles strongly suggest that Fv1 evolved under pressure from a variety of viruses. The block to virus infection occurred early in infection, specifically after reverse transcription and before nuclear entry and formation of the circular viral DNAs and the integrated proviruses (Jolicoeur and Baltimore, 1976; Jolicoeur and Rassart,

1980, 1981; Yang et al., 1980; Chinsky and Soeiro, 1981). Different strains of the murine leukemia viruses were typed as either N-tropic (able to infect Fv1n/n mice), or B-tropic (able to infect Fv1b/b mice) or NB-tropic (able to infect mice of any Fv1 genotype). The critical determinant of the susceptibility to restriction was mapped to the CA domain of the Gag protein (Hopkins et al., 1977; Ou et al., 1983; DesGroseillers and Jolicoeur, 1983). A single residue of CA (amino acid #110) determined the tropism of some virus strains, and changes of other nearby CA residues also affected virus tropism (Kozak and Chakraborti, 1996; Stevens et al., 2004). The simplest model that explained these observations was that the Fv1 protein recognized the CA protein of the incoming viral PIC and somehow blocked nuclear entry of the susceptible strains. Viruses that were generated with a mixture of N-tropic and B-tropic CA proteins were susceptible to restriction by both $Fv1^b$ and $Fv1^n$ alleles (Rein et al., 1976), consistent with the idea of an attack on the PIC by Fv1.

The Fv1 block was found to be overcome by infection with very high levels of virus, and exposure of the cells to inactivated virion particles could "abrogate" the block and allow infection by subsequent challenge with a virus that would normally be restricted. These observations suggested that high levels of incoming CA proteins could titer out the Fv1 protein and allow any remaining virus to escape restriction. The abrogation did not occur if the cells were exposed to an unrestricted virus, suggesting that the unrestricted virus could not saturate, or presumably bind, to the restriction factor. Abrogation could be achieved by defective viruses without RT, and without a genome (Boone et al., 1990), and by viruses that enters the cell using many different viral envelopes. However, immature virus made up of uncleaved Gag precursors were not effective at abrogation, and expression of the Gag protein within a cell did not abrogate or otherwise affect Fv1 restriction of incoming virus (Bassin et al., 1980; Duran-Troise et al., 1981). Abrogation only occurred on infection by incoming virions with the appropriate CA protein in a mature conformation.

The Fv1 gene itself was isolated in a heroic effort by Stoye and colleagues using a positional cloning approach (Best et al., 1996). Remarkably, the gene had sequence similarity to the *gag* gene of the ERV-L family of endogenous retroviral DNAs in mouse and human genomes. There were no sequences related to *pol* or *env* genes nearby. This suggests that the gene arose from an ancient defective provirus and was retained or fixed in the genome over evolutionary times under the selective pressure of frequent infection by murine leukemia viruses (Yan et al., 2009) or other retroviruses (Yap et al., 2014). The N and B alleles encoded very similar proteins, differing at two residues near the center of the protein (#358, 399) and also in the very C-terminus. Expression of the cloned DNAs blocked infection and recreated the appropriate specificity for N- and B-tropic viruses. Mutational studies defined portions of the molecule important for its activity, including a so-called MHR found on all retroviral CA proteins (Bishop et al., 2001). Though the genetics strongly suggests that Fv1 interacts directly with the incoming CA protein, the proteins bind very weakly in vitro. A likely explanation is that the proteins normally interact as multimers, with Fv1 making multiple contacts to a large multimeric array of CA proteins on the surface of the PIC (Hilditch et al., 2011). Fv1 indeed contains a dimerization domain that is important for activity (Bishop et al., 2006b). The mechanism of action is not yet very clear; the interaction somehow leads to a destabilization of the PIC and prevents its proper movement into the nucleus. The interaction may occur in the trans-Golgi network (Yap and Stoye, 2003).

TRIM5α, a Major Postentry Block

One of the most important antiviral genes is TRIM5α (for reviews, see Towers, 2007; Grutter and Luban, 2012; Lukic and Campbell, 2012; Luban, 2016). Early tests of cells from a variety of mammals, including both humans and nonhuman primates, revealed the presence of a restriction factor with many features similar to those of the murine Fv1 factor (Towers et al., 2000; Besnier et al., 2002, 2003; Cowan et al., 2002; Hatziioannou et al., 2004a). Infection of restricting cells by N- and B-tropic murine leukemia viruses, pseudotyped with the VSV-G envelope to allow entry into the cells, demonstrated that N-tropic viruses were blocked and B-tropics were not. Many primate cells also blocked infection by HIV-1 and certain other lentiviruses but were permissive for the simian immunodeficiency virus SIVmac (Munk et al., 2002; Hatziioannou et al., 2003). As with Fv1, restriction was targeted to the CA protein, and susceptibility was even affected by alterations of the same CA residues (Hatziioannou et al., 2004b). Infection was again blocked at an early stage, though in this case the block occurred before reverse transcription, an earlier stage than in the case of Fv1 (Jolicoeur and Baltimore, 1976; Jolicoeur and Rassart, 1980). As with Fv1, the block could be saturated by the addition of noninfectious virus-like particles (Towers et al., 2002; Shi and Aiken, 2006).

In a high-profile competition, the gene responsible for this restriction in rhesus macaques was cloned from a cDNA expression library by selecting for resistance to infection by HIV-1-based vectors (Stremlau et al., 2004). The activity proved to be attributable to the TRIM5α protein encoded by a splice variant of a member of the large TRIM (for tripartite motif-containing) gene family. TRIM5α contained the three canonical domains—the RING, B-box 2, and coiled-coil domains of all the TRIMs—and also contained a so-called B30.2 or PRY/SPRY domain at its C-terminus. TRIM5α was quickly demonstrated to exhibit all the properties of the previously defined restriction activities (Keckesova et al., 2004; Hatziioannou et al., 2004c; Perron et al., 2004; Yap et al., 2004). The rhesus TRIM5α restricted N-tropic and not B-tropic murine leukemia viruses; it also potently restricted HIV-1, while the human version had only very weak activity. The TRIM5α genes of the primate lineage show rapid evolution consistent with its mutation in response to changes in the virus capsids it targets (Sawyer et al., 2005). Other TRIM family members may also bind CA and may also inhibit some retroviruses (Uchil et al., 2008; Yuan et al., 2016).

The detailed mechanism of action of TRIM5α, as with Fv1, is still unclear. TRIM5a is oligomeric (Mische et al., 2005), and its multimerization is important for its antiviral activity. All the genetics suggest that TRIM5α should interact with CA proteins in an allele-specific way, and detection of weak interactions with the appropriate specificity has been reported (Sebastian and Luban, 2005). Much insight has come from detailed studies of the structure of the PIC and TRIM5α. We now know that the PIC is made up of an array of CA protein hexamers, arranged in a fullerene conical structure by virtue of judicious placement of 12 pentamers (Ganser et al., 1999). TRIM5α also forms a hexameric array but in a much larger grid that can overlay onto the smaller CA grid (Ganser-Pornillos et al., 2011). The long arms of the TRIM5α grid consist of antiparallel dimers of the coil–coil region (Sanchez et al., 2014), connected at the intersections with threefold symmetry by the RING-B-box 2 domains (Li and Sodroski, 2008; Wagner et al., 2016). Oligomerization of TRIM5α is required for antiviral activity (Javanbakht et al., 2006), and intermolecular contacts between various domains of TRIM5α are required for proper assembly of the antiviral structure (Sastri et al., 2014). The TRIM5α array then contacts CA by the SPRY domain positioned in the middle of the long arms (Yang et al., 2012a, 2014a; Biris et al., 2012, 2013; Zhao et al., 2011). The structure allows TRIM5α to form a net around the PIC and grants it flexibility to bind many different viral cores with a variety of overall shapes and sizes (Li et al., 2016). These complexes could be visualized by light microscopy as cytoplasmic bodies forming around the virus (Campbell et al., 2008). The interaction likely depends on the presentation of the large array of CA proteins of the PIC and on a certain stability of that array. One other finding connects TRIM5α function to capsid stability: the abrogation of restriction by VLPs depends on the VLP stability (Shi and Aiken, 2006).

What ensues after TRIM5α binding to CA to mediate restriction is not clear, but TRIM5α most likely destabilizes or accelerates the uncoating of the PIC (Stremlau et al., 2006; Perron et al., 2007; Diaz-Griffero et al., 2007; Kutluay et al., 2013). Some aspects of the process can be visualized in infected cells (Takeda et al., 2015). The disruption of CA structures can also be induced by TRIM5α in vitro (Black and Aiken, 2010; Zhao et al., 2011), providing something of a window into the process. Mixed virions containing both sensitive and insensitive CA proteins show that retention of a modest portion of sensitive CAs is sufficient to confer overall sensitivity to TRIM5α (Shi et al., 2013), analogous to the inhibition of mixed virions by Fv1. Microtubules and dynein motors may also be required for restriction, either acting directly with TRIM5α or perhaps in delivering TRIM5α to appropriate intracellular locations (Pawlica et al., 2014). The localization of TRIM5α to structures known as cytoplasmic bodies may be (Sastri et al., 2010) or may not be (Song et al., 2005; Diaz-Griffero et al., 2009) important for its activity. The RING domain seems to be important for the first steps of inhibition of RT and capsid destabilization (Roa et al., 2012). Dimerization of the RING domains of TRIM5α, as would occur on assembly onto the PIC, results in activation of the ubiquitin ligase activity, which generates

K63-linked polyubiquitin chains (Yudina et al., 2015). It is not known what might be the relevant substrate for ubiquitinylation. One possibility is that autoubiquitinylation of TRIM5α, with the help of the E2 ligase Ube2W, is the key event; this seems to be important in restricting reverse transcription (Fletcher et al., 2015). Another puzzling piece of the story is the finding that the ubiquitin ligase activity of TRIM5α was required for the restriction of some viruses, but not others (Lienlaf et al., 2011; Kim et al., 2011; Campbell et al., 2016). It should be noted that the TRIMs often exhibit both ubiquitin and SUMO transferase activity, and TRIM5α may bind SUMO and induce SUMOylation of unknown substrates (Arriagada et al., 2011; Lukic et al., 2013; Nepveu-Traversy et al., 2016). There are hints that Fv1 restriction may also involve SUMO transfers.

Studies suggested that there are two blocks mediated by TRIM5α, or two steps in the block, as revealed by treating cells with a proteasome inhibitor at the time of infection. While untreated cells blocked both total viral DNA synthesis and circular viral DNA formation, cells treated with the MG132 inhibitor lost the initial block to viral DNA synthesis and only retained the block to circle formation (Wu et al., 2006; Anderson et al., 2006). The association of TRIM5α with the proteasome can be observed directly in infected cells (Danielson et al., 2012). Thus, one mechanism of TRIM5α restriction might be the targeting of the proteasome to the PIC to mediate some aspect of its destruction.

TRIM5α expression is induced by the interferons (Asaoka et al., 2005), and the gene can thus be added to the long list of interferon-stimulated genes (ISGs) contributing to the antiviral state induced by IFN (Carthagena et al., 2008). Treatment of cells with alpha IFN indeed increases TRIM5α restriction activity (Sakuma et al., 2007).

TRIM-Cyp: Evolution of a Restriction Factor by Gene Fusion

A bizarre merger of two stories of CA-targeted host factors occurred with the near-simultaneous discoveries of a remarkable restriction factor in the genome of the owl monkey *Aotus trivirgatus*. These monkeys were unusual among New World primates in efficiently restricting HIV-1, and their restriction was unusual in being released by cyclosporine. RNAi-mediated knockdown experiments targeting CypA expression also released the block, but reexpression of CypA did not restore it, suggesting the existence of a novel CypA-related block. Searching for CypA-related genes yielded TRIMCyp, producing an RNAi-responsive messenger RNA encoding a TRIM5-CypA fusion protein (Sayah et al., 2004). Cloning of TRIM5α sequences from owl monkeys, using sequence similarity to other TRIM5α genes, resulted in isolation of the same TRIM-Cyp gene (Nisole et al., 2004). Examination of the structure of the gene revealed that it arose by the insertion of a DNA copy of CypA mRNA, a pseudogene, into an intron of the canonical TRIM5 gene. The C-terminal PRY/SPRY domain of TRIM5α was thus replaced by the CypA sequences. The resulting gene fusion has potent antiviral activity, targeting the CA protein using the CypA domain rather than the PRY/SPRY domain of the parental TRIM5α. The formation of pseudogenes is thought to be mediated by the agency of the LINE-1 retrotransposons acting on cellular mRNAs in trans. The TRIM-Cyp fusion would have been fixed in the owl monkey genome by the selective pressure for protection from infection by an as-yet-unknown virus. Remarkably, this was not a unique event; similar but independent TRIM-Cyp fusions with antiviral activity have now been found in the genomes of rhesus (Wilson et al., 2008) and pig-tailed macaques (Liao et al., 2007; Virgen et al., 2008; Newman et al., 2008), and the tree shrew *Tupaia belangeri* (Mu et al., 2014). Paleogenomic searches in several simian genomes have revealed inactive TRIM-Cyp fusions that were likely once active but have decayed over evolutionary times (Brennan et al., 2008; Malfavon-Borja et al., 2013). The TRIM-Cyp fusions, and indeed the TRIM5 genes as well, show evidence of rapid evolution in the face of viral challenges. The discovery of these genes has changed our views on the significance of retroviruses as pathogens, on the importance of antiviral genes in primates, and on the frequency of pseudogene formation as an aspect of gene evolution.

The discovery of the TRIM-CypA restriction factors, and the availability of cyclosporine as a small molecule inhibitor of this restriction activity, made possible "washout" assays that monitored the time of acquisition of resistance of the restriction to the addition of inhibitor (Hulme and Hope, 2014a). This assay, now widely used, allowed for measurement of the progression of infection through and beyond the point of restriction (Perez-Caballero et al., 2005).

Mov10

Searches for additional restriction factors have revealed that overexpression of Mov10, a putative RNA helicase belonging to the Upf1-like helicase family, localized to mRNA processing P-bodies, could block replication of HIV-1 and other retroviruses (Burdick et al., 2010; Furtak et al., 2010; Wang et al., 2010). Mov10 is packaged into virions by virtue of contacts with NC (Abudu et al., 2012). Mov10-overexpressing cells modestly inhibited virion production, and the resulting virions were impaired for infectivity, with effects on reverse transcription. The effects of overexpression seem clear, but knockdown had little or no effect in increasing production of infectious viruses, suggesting that the natural levels of Mov10 do not play a significant role in restricting exogenous viruses (Arjan-Odedra et al., 2012). Knockdown did increase the replication of both LTR and non-LTR endogenous retroelements; and overexpression of Mov10 also impaired the retrotransposition of endogenous retroelements such as the intracisternal A-type particle genomes (Lu et al., 2012). The mechanism of its action is obscure.

Lv2: An Entry-Specific Block

Another block to early events, dubbed Lv2, has been identified and found to exhibit a curious set of characteristics (Schmitz et al., 2004). Particular human cell lines were found to have the ability to restrict HIV-2 after reverse transcription but before nuclear entry. The block only occurred when the virus entered the cell via the HIV-2 envelope; virus pseudotypes entering via the VSV-G envelope or the HIV-1 envelope were not restricted. Mapping the portions of the viral genome that provided sensitivity to the restriction defined two features: a specific sequence of CA and the use of specific HIV-2 envelopes. The HIV-1 core was found to also be sensitive to restriction if pseudotyped by the HIV-2 envelope. A single residue in the HIV-2 envelope provided sensitivity to restriction, associated with delayed delivery of the viral core into the cytoplasm (Reuter et al., 2005). A number of compounds that affect endocytic pathways abrogated the block, and use of a tailless CD4 receptor also bypassed the block. These observations suggest that the route of entry determined exposure to the restriction (Marchant et al., 2005), providing a striking example of how alternate routes of trafficking can have effects far downstream in the course of infection. Specifically, the data suggested that the sensitive envelope delivered the virus to a cellular compartment where the Lv2 factor was active in restricting CA, whereas a resistant envelope delivered the core to a compartment where Lv2 was not active. HeLa/CD4 cell lines similarly restricted virus with MLV cores and MLV CAs when pseudotyped with HIV-2 envelopes (Oliveira et al., 2010). This restriction occurred with RD114 envelopes, but not GALV, HIV-1, or amphotropic MLV envelopes. Subsequent tests provided some definition of the determinants of the restrictive entry pathway: endocytic routes of entry using calthrin/AP2-mediated uptake were sensitive to Lv2 restriction, whereas nonclathrin routes were not sensitive (Harrison and McKnight, 2011). A whole-genome siRNA screen has perhaps identified the gene, or at least a key component required for its activity, named RNA-associated early-stage antiviral factor (REAF) (Marno et al., 2017). The expression profile of REAF across cells, and the selectivity of its antiviral activity, was all consistent with REAF being the factor responsible for Lv2 activity. REAF may associate with incoming viral nucleic acids, in a CA stability-dependent manner, and block replication during reverse transcription. Its activity is strongest for viruses entering by endocytic pathways.

APOBECs, the Cytidine Deaminases

A remarkable detective story, played out over years, resulted in the identification of the APOBEC3G gene as a major restriction factor blocking HIV-1 replication (for reviews, see Cullen, 2006; Holmes et al., 2007; Goila-Gaur and Strebel, 2008; Harris and Dudley, 2015). The hunt for this restriction factor began with the search for the mechanism of action of the Vif protein encoded by HIV-1. Viruses lacking Vif could not replicate in primary human T cells and some T-cell lines, such as CEM cells, but replicated well in other lines such as 293T and the CEM-SS subclone (Simon et al., 1998). Vif was found to be important in producer cells for the generation of infectious virion particles; expression of

Vif-minus virus in restrictive cell types resulted in the production of virions that seemed normal by most indications but nevertheless were uninfectious (von Schwedler et al., 1993). Cell–cell fusion experiments showed that the restriction phenotype was dominant over permissivity, suggesting that the restrictive cells produced a factor that was inactivated by Vif. To find the gene responsible, cDNA clones from restrictive cells were used as probes to examine RNAs from permissive and restrictive cells, ultimately identifying a gene (CEM15) specific to restrictive lines with antiviral activity (Sheehy et al., 2002).

The encoded protein, dubbed APOBEC3G, was found to be a member of the APOBEC family of deaminases (named for the founding member, APOBEC1, apolipoprotein B mRNA-editing enzyme-catalytic polypeptide, an RNA deaminase modifying a single base of a single mRNA). The gene product was found to be a deaminase that triggers massive deamination of deoxycytidine to deoxyuridine on the minus-strand DNA formed during reverse transcription, effectively destroying the incoming genome (Harris et al., 2003; Mangeat et al., 2003). The deoxyuridines in the viral DNA can be targets of the uracil N-glycosidases and the resulting abasic sites can then be attacked by nucleases that degrade the DNA. Any rare surviving genomes are heavily mutated. APOBEC3G is specific for single-stranded DNA, inducing C-to-U changes in the minus strand and thus G-to-A mutations in the plus strand, throughout the genome. The mutations are often clustered—a phenomenon termed hypermutation–suggestive of a concerted attack on an exposed patch of DNA. There are also C-to-T changes induced in the 5′ U3 region and the PBS, which are transiently single-stranded during reverse transcription (Yu et al., 2004). It is possible that APOBECs mutate viral RNA (Bishop et al., 2004b; Sharma et al., 2016), though the viral DNA is the major target. The mutations induced by the APOBECs likely represent a major mechanism for rapid genome variation during virus evolution (Zhang et al., 2003). The APOBECs can inhibit other viruses, including the foamy viruses (Delebecque et al., 2006), endogenous retroviruses (Esnault et al., 2006), hepatitis B (Turelli et al., 2004), as well as LTR-retrotransposons (Bogerd et al., 2006a) and non-LTR retrotransposons (Bogerd et al., 2006b; Hulme et al., 2007).

HIV-1 makes use of the virus-encoded Vif protein to overcome restriction by the APOBEC3s. The APOBEC3s are incorporated into virions through interaction with the RNA genome, perhaps with NC and the 5′ UTR of the viral RNA genome (Khan et al., 2005). The Vif protein protects virus by preventing APOBEC3G incorporation into the nascent virions. The major mechanism of action is the induction of the polyubiquitinylation of APOBEC3G, triggering its degradation by the proteasome (Sheehy et al., 2003; Marin et al., 2003). Vif recruits the Cullin CUL5, elongins B and C, and Rbx1 to form a ubiquitin ligase complex, similar to Skp1–Cullin–Fbox complexes, which targets APOBEC3G and mediates its ubiquitinylation and degradation (Yu et al., 2003). The binding and targeting of APOBEC3G by Vif is very specific: for example, the HIV-1 Vif can target the human APOBEC3G but not most simian orthologues, and

SIVagm Vif can target the AGM APOBEC3S but not the human (Mariani et al., 2003). APOBEC3G sensitivity to Vif can be altered by the change of a single-amino acid residue (Bogerd et al., 2004; Mangeat et al., 2004). There can also be proteasome-independent methods of APOBEC degradation (Kao et al., 2003b). Vif may lower the levels of APOBEC3G by inhibiting its translation (Stopak et al., 2003). In addition, Vif may prevent encapsidation of APOBEC3G and APOBEC3F without inducing their degradation. Finally, Vif may inactivate APOBEC3G even when it cannot exclude it from virions (Takeuchi et al., 2005). It has been noted that Vif is present in the virions, at levels that vary widely depending on the cell (Kao et al., 2003a), but the significance of this is unclear.

Evasion of the APOBECs is not unique to the lentiviruses. Even the simple retroviruses, which do not contain genes for accessory proteins, can block the action of the APOBECs. The MLVs utilize a glycosylated form of the Gag protein, termed glyco-Gag, for this purpose (Stavrou et al., 2013). Glyco-Gag is synthesized by translation of the unspliced viral mRNA from an alternative initiation codon (CUG) present upstream of the major AUG start codon and in the same reading frame. Translation from this codon generates a longer protein with a leader sequence that directs glyco-Gag to the ER and through the Golgi apparatus to the cell surface. The virion cores of MLVs lacking glyco-Gag were found to be less stable than wild type and less infectious in the presence of APOBEC3. Mutant viruses lacking glyco-Gag quickly reverted to reexpress glyco-Gag, but did so only in wild-type mice and not in APOBEC3 knockout mice, confirming that the major need for glyco-Gag is the evasion of APOBEC3. The mechanism of glyco-Gag action is totally distinct from that of Vif, apparently the stabilization of the virion core to protect the viral DNA from attack by the APOBEC3 present in the cytoplasm of the infected cell.

APOBEC3G proved to be only one member of a large family of cytidine deaminases in human and other mammals. The proteins are Zn-containing enzymes with a highly conserved folded structure. These genes are rapidly evolving, with different numbers of family members in human, nonhuman primates, and other mammals. Several of the genes encode tandem copies of the so-called Z domains that make up the conserved structure, with an inactive N-terminal copy and a catalytically active C-terminal copy. While APOBEC3G is the most abundant and effective member of the human family in many cell types, other members of the family are also important (Bishop et al., 2004a). Thus, the various isoforms APOBEC3A, B, C, DE, F, and H of human and other primates have all been shown to demonstrate antiviral activity in various settings (Liddament et al., 2004; Wiegand et al., 2004; Zheng et al., 2004; Doehle et al., 2005; Rose et al., 2005; Dang et al., 2006). Many of the APOBECs are induced by interferons (i.e., they are ISGs; Tanaka et al., 2006) and thus specific members of the family may be recruited to block virus only after induction of an antiviral state. It is possible that some of the family members can exert their activity when expressed in the cytoplasm of the infected cell, and thus for these it may not be necessary that they be incorporated into the virion particle during virion assembly, though this functionality may only be induced with high overexpression. Each of the APOBECs has a limited, but characteristic and distinctive preference for the sequences surrounding the target cytidine. The preferences of the enzymes in vitro correlate well with the spectra of mutations that are found in the viruses in vivo, suggesting that the APOBECs represent a major source of the mutations arising in replication (Beale et al., 2004). It has been noted that the potency of the APOBECs does not correlate well with their deamination activity, indicating that there are other limiting determinants of function or other mechanisms of action (Bishop et al., 2006a). Indeed, catalytically inactive mutants still retain antiviral activity, perhaps by interfering with the progress of the RT enzyme (Newman et al., 2005).

SAMHD1, a Nucleotidase

SAMHD1 (sterile α motif and histidine-aspartate domain–containing protein 1) is a restriction factor that was first detected as blocking HIV-1 infection of macrophages and myeloid and dendritic cells at the level of synthesis of the viral DNA (for reviews, see St Gelais and Wu, 2011; Ayinde et al., 2012; Pan et al., 2013). The protein was found to be a deoxynucleoside triphosphate triphosphohydrolase, removing the triphosphate from dNTPs and releasing the free nucleoside, thereby depleting the cell of the intracellular deoxynucleotide triphosphates required for reverse transcription (Goldstone et al., 2011; Powell et al., 2011; Lahouassa et al., 2012). The restriction is most significant in those cell types that are quiescent or otherwise have very low basal levels of the triphosphates, including resting CD4+ T cells (Baldauf et al., 2012). Very low levels of SAMHD1 are sufficient to induce a block in some cells such as monocyte-derived macrophages (Hollenbaugh et al., 2014). The block can be partially relieved by the addition of extracellular deoxynucleosides, which are converted to dNTPs by the salvage pathway (Lahouassa et al., 2012), in strong support for its function as a dNTPase. SAMHD1 is counteracted by the Vpx proteins of SIV and HIV-2 and the Vpr protein of SIVagm (Laguette et al., 2011; Hrecka et al., 2011; Descours et al., 2012), which induce the proteasomal degradation of SAMHD1 and allow for increase in the levels of dNTPs. Mutations in the gene are also known to be responsible for a familial disease in humans, Aicardi-Goutieres syndrome.

The enzymatic activity of SAMHD1 is subject to complex regulation. The activity is strongly stimulated by dGTP—an aspect of regulation that is difficult to study because dGTP is also a substrate of the enzyme. The activity is also stimulated by other triphosphates, including dATP and even GTP (Amie et al., 2013). The mechanism of activation is the induced formation of tetramers, the active form of the enzyme (Ji et al., 2013; Yan et al., 2013); nucleotide binding stabilizes two inactive dimers to yield the activated tetramer. Nucleotides bind at two distinct allosteric sites; one site binds dGTP or GTP, whereas the other site binds any dNTP (Amie et al., 2013; Ji et al., 2014). The tetramerization induces conformational changes at the active site that increase the binding of substrate (Zhu et al., 2013; Ji et al., 2013; Yan et al., 2013). This regulatory system suggests that the enzyme is self-regulated in a negative feedback loop—as the dNTP levels are lowered by SAMHD1, its activity is shut off, resulting in a homoeostatic control of the dNTP levels in the cell. There may be multiple conformations of tetramers—some short-lived and some long-lived in vivo—with different levels of antiviral activity.

SAMHD1 activity has been found to be negatively regulated by phosphorylation, most likely by CDK1/cyclin A2 and CDK2 modification of threonine 592 (Cribier et al., 2013). Phosphorylation was associated with reduced formation of tetramers and reduced dNTP turnover (Arnold et al., 2015). A phosphomimetic mutation (T592E) strongly impaired the antiviral activity (Welbourn et al., 2013). Exposure to type I interferon resulted in reduced phosphorylation and increased activation of SAMHD1, as well as increased antiviral activity. SAMHD1 function is apparently also regulated during cellular development; actively dividing cells had high levels of phosphorylated SAMHD1, and differentiated cells that were postmitotic showed higher levels of dephosphorylated SAMHD1, in the form of a highly active stable tetramer. These observations together suggest that SAMHD1 is regulated by the cell as a mechanism to control the levels of dNTPs as needed for cellular DNA synthesis. Reducing these levels in resting cells, or in response to IFN, serves to protect these cells from infection.

Similar regulation also was observed for the murine SAMHD1. The murine homolog of the human gene product is active against both HIV-1 and MLV (Wittmann et al., 2015). The murine SAMHD1 is also phosphorylated, and its phosphorylation also inhibits its block to HIV-1 infection (though curiously not to MLV infection) (Wang et al., 2016). Interestingly, one isoform of the mouse enzyme is not phosphorylatable, is constitutively tetrameric, and active even without GTP, essentially seeming to be an unregulated variant (Bloch et al., 2017).

SAMHD1 has also been reported to exhibit nuclease activity, and it has been suggested that degradation of the viral RNA is responsible for its antiviral function (Ryoo et al., 2014; Choi et al., 2015). The specific activity is extremely low, however (barely detectable above backgrounds), and does not correlate with antiviral activity (Antonucci et al., 2016). This mechanism would not be consistent with the rescue by addition of dNTPs. On several grounds, the triphosphatase activity is most likely the key enzymatic function mediating the antiviral effects.

MX2, Another Factor Targeting Capsid

Recently a new relative of a very old antiviral gene has been discovered to be active against retroviruses. This gene is MX2 (also MxB), related to the heavily studied MX1 (also MxA), one of the first antiviral genes ever to be discovered. The MX1 gene was long known as a potent inhibitor of influenza, measles, parainfluenza, reovirus, pox virus, and myxoviruses; it is the major restriction activity against influenza induced by the type I and III interferons (for reviews, see Haller and Kochs, 2002; Haller et al., 2015). MX2 is a related gene that previously had no known activity but was conserved in a variety of mammals (though not in rodents; the mouse gene named MX2 is actually a variant of mouse MX1 and not closely related to the human MX2 gene Haller, 2013). MX2 became a subject of much interest when it was discovered to block HIV-1 replication.

MX2 was found through expression profiling as a gene selectively expressed in monocytic cell lines showing IFN-inducible resistance to HIV-1 and not expressed in cells that did not respond to IFN (Kane et al., 2013). It had also been noted in genome-wide overexpression screens for virus resistance genes induced by type I IFN (Schoggins et al., 2011). Ectopic expression of human MX2 blocked infection by HIV-1, HIV-2, and several SIVs but not gammaretroviruses such as the MLVs (Goujon et al., 2013; Liu et al., 2013). Several primate MX2 also had antiviral activity against a range of lentiviruses. HIV-1 reverse transcription occurred normally in MX2-expressing cells, but the levels of the 2-LTR circular and integrated proviral DNAs were reduced, suggesting a block to nuclear import (Liu et al., 2013). Several HIV-1 mutants with alterations in the CA protein were found to be resistant to the MX2 block (Matreyek et al., 2014), indicating that MX2 might somehow target wild-type CA and prevent the PIC from entering the nucleus. MX2 does bind to the HIV-1 core, and the N-terminal region of MX2 contains the binding site (Schulte et al., 2015). A small molecule known to bind CA could block MX2 binding (Fricke et al., 2014b). HIV-1 CA mutant N57S, which does not infect nondividing cells and is thought to be defective for one route of nuclear entry, was resistant to MX2; and mutant G89V, which eliminates CypA binding and the requirement for Nup358 in import, was similarly resistant to MX2. Several so-called transmitted founder isolates were found to be resistant to MX2 by virtue of changes in residues 86, 87, 88, 92, and 208 of CA, all reducing the interaction with CypA (Liu et al., 2015). Selecting for viral variants that could escape the block resulted in the recovery of mutants with similar alterations in the CA protein, specifically with mutations of alanine 88 in the CyPA-binding loop. This suggested that MX2 might work together with CypA. Indeed MX2 was found to bind to CypA, and either knockdown of CypA or inhibition of the CypA–Gag interaction with cyclosporine led to the loss of the MX2 restriction activity (Liu et al., 2013). One model would be that MX2 forms a complex with CypA and binds CA to trap the PIC, perhaps preventing the normal course of uncoating (Buffone et al., 2015). Studies of additional escape mutants revealed that mutations affecting a number of surfaces of CA could also protect the virus from MX2 (Busnadiego et al., 2014), suggesting that MX2 might make more extended interactions with CA.

MX2 (and MX1) belong to a family of dynamin-like GTPases, raising the possibility that they function in trafficking on the cytoskeleton. They contain an N-terminal GTPase domain, a central spacer domain, and a C-terminal GTPase effector domain similar to dynamin. Mutations predicted to eliminate GTP binding or hydrolysis did not affect the antiviral activity of MX2, unlike MX1

(Matreyek et al., 2014). The C-terminal domains together form an extended stalk of uncertain function. MX2 forms an extended antiparallel dimer, and the dimerization is required for function—monomers are not active (Fribourgh et al., 2014; Dicks et al., 2015). MX1 dimers can then further polymerize to form extended oligomers or possibly hexameric rings. The higher oligomerization of MX2 may promote higher affinity binding to the viral core but is not required for antiviral activity (Fribourgh et al., 2014; Dicks et al., 2015). MX2 is localized mainly in the nucleus, or to the cytoplasmic side of the nuclear pore, by virtue of a nuclear localization signal near the aminoterminus. This NLS and sequences immediately downstream of the NLS, including a triple-arginine motif, were required for binding to CA and for antiviral activity (Kane et al., 2013; Fricke et al., 2014b; Fribourgh et al., 2014; Matreyek et al., 2014; Goujon et al., 2015). The aminoterminal region also contained the determinants of the spectrum of viruses blocked by the various primate MX2s, presumably reflecting the binding to CA (Busnadiego et al., 2014). Finally, there are hints that the MX proteins might be involved with the SUMOylation machinery and become SUMOylated (Brantis-de-Carvalho et al., 2015), but the significance of this for their functionality is not yet clear. At this time we only know that MX2, like Fv1, TRIM5α, and CypA, interacts with CA to somehow modulate PIC trafficking or uncoating. We will need more information to understand the details of the restriction.

SUN2

SUN2 was identified in large-scale screens for genes that block HIV-1 replication (Schoggins et al., 2011). Overexpression was found to restrict many HIV-1 strains, but not MLV or other viruses, at the time of nuclear entry (Donahue et al., 2016). SUN2 protein is localized to the inner nuclear membrane. Overexpression caused dramatic change in the cell: formation of a multilobular flower-like nuclear shape. The restriction activity, and the change in nuclear morphology, both required the SUN2 domain that interacts with the nuclear lamina. The block required the presence of CypA, suggesting a cooperative interaction of CypA and SUN2 with CA. Selection for escape mutants of HIV-1 that replicated in resistant cells resulted in mutants with substitution mutations in CA. The knockdown of SUN2 expression, however, did not promote HIV-1 replication, suggesting that the endogenous levels of SUN2 were not sufficient to offer significant restriction. Rather, knockdown in T cells reduced virus production, likely through inhibition of T-cell activation (Donahue et al., 2017). Our limited understanding of the mode of action of this protein suggests that it acts through blocking passage of the PIC through the nuclear membrane.

More Restriction Factors?

It seems certain that we have not yet uncovered all the restriction factors that can impact the early events of infection. Indeed there are hints of still more activities and new times of their function. For example, monocytes treated to activate the TLR7/8 receptors show resistance to HIV-1 and HIV-2 that is not accounted for by any of the known factors, and involving destabilization of the viral RNA, unlike known restrictions (Hofmann et al., 2016). The many screens for antiviral factors already completed have produced numerous candidates that are as yet unexplored, and several more targeted screens that will undoubtedly add to the lists are underway.

SENSING AND RESPONDING TO INFECTION: INNATE IMMUNITY

We have discussed many host genes that encode restriction factors acting early in infection, but in addition, there are proteins that do not directly block infection but serve as sensors of virus, and that sound the alarm to induce the expression or function of effectors. These proteins—pattern recognition receptors or PRRs—lead to rapid sensing of many pathogenic viruses and strong induction of IFN. Retroviruses are known traditionally neither to activate a strong innate immune response nor to induce IFN production. But in some settings and cell types, retroviruses are sensed by various PRRs and do induce IFNs. We will here report what is known about a few of these sensor proteins.

TRIM5α as Sensor

While TRIM5α clearly has antiviral activity, recent work has revealed that it also serves as a signaling molecule, and this may play a major role in its antiviral functions. The notion is that TRIM5α recognition of the incoming viral capsid can initiate production of IFN and establish an antiviral state—serving as a PRR. A key aspect of TRIM5α activation is the strong requirement that the CA be in a highly polymeric form, like the hexameric lattice present in the incoming PIC. How might TRIM5α signal? TRIM5α overexpression was found to activate reporter genes that respond to the AP-1 transcription factor or NFkB (Pertel et al., 2011), and knockdown of TRIM5α reduced the expression of a handful of genes known to be responsive to AP-1 or NFkB. The related TRIM5Cyp fusion protein had similar properties. To search for downstream mediators of these effects, proteomic screens of factors that bind to TRIM5 were carried out, leading to the identification of TAK1, TAB2, and TAB3, the subunits of the TAK1 kinase complex that is a key component of the lipopolysaccharide response pathway. Both TAK1 and an associated heterodimeric E2 ubiquitin ligase UBC13/UEV1A were found to be required for TRIM5α signaling. The product of the ligase activity may be a protein target but certainly included free, unlinked K63 polyubiquitin chains, and these may be the more important reaction product. These chains result in the autophosphorylation of TAK1 and the downstream phosphorylation of IRF3, ultimately activating IFN expression. The final model is thus that TRIM5α is a PRR that senses incoming capsids and triggers the formation of the antiviral state. Various tests of this hypothesis confirmed many predictions of the model. Viruses with different capsids demonstrated good correlation of binding avidity for TRIM5α with the resulting strength of the induction of ISGs. Similarly, tests of variety of mammalian TRIM5 orthologues showed good correlation of AP-1 activation with restriction activity (sometimes tested in the context of TRIM-Cyp fusions (Lascano et al., 2015).

One might expect that TRIM5α restriction and TRIM5α sensing of virus would work together to limit infection, but they may be competing or even antagonistic functions in some cells. In dendritic cells, TRIM5α intracellular localization is regulated by SUMOylation. TRIM5α SUMOylation promotes its movement to the nucleus, resulting in loss of restriction activity but strong induction of IFN during infection (Portilho et al., 2016). Blocking SUMOylation relocates TRIM5α to the cytoplasm, restoring restriction activity and blocking IFN induction. The utility to the cell of this trade-off between these functions is unclear.

Because TRIM5α acts as a PRR by virtue of its interaction with CA at the CypA interaction loop, CypA can modulate HIV-1 virion core detection. CypA can block TRIM5α access or modulate TRIM5α recognition likely by inducing subtle shifts in CA conformation.

RIG-I and MDA5 as Retroviral Sensors

RIG-I (a.k.a. DDX58) and MDA5 (a.k.a. IFIH1) are major sensors of RNA viral infection, recognizing double-stranded RNA, 5′-triphosphorylated RNAs, and other specific nucleic acid structures as pathogen-associated molecular patterns (PAMPs) that signal the presence of virus. They act in concert with several accessory cofactors, including the cytoplasmic helicase DDX60 (Miyashita et al., 2011). These PRRs are generally poorly able to sense retrovirus infection because the viruses produce few of the structures generated by other RNA viruses. However, retroviral infection does induce the production of IFN with at least some efficiency. The viral RNAs transcribed from the provirus are single-stranded and capped, but secondary structures do activate RIG-I (Berg et al., 2012). The RNA–DNA hybrids that are formed during reverse transcription may also be PAMPs that are recognized by these or other PRRs. This activation of RIG-I is inhibited by the viral PR, which can bind RIG-I and sequester it into lysosomes, blocking IRF-3 phosphorylation (Solis et al., 2011). Thus, these PPRs likely play at least some role in limiting HIV-1 infection.

cGAS Involvement in Sensing Infection

Cyclic guanosine monophosphate–adenosine monophosphate (cGAMP) synthase (cGAS) is a cytosolic sensor of infection by many pathogens (Sun et al., 2013). Activation of cGAS results in formation of the signaling molecule cGAMP, which binds to the adaptor protein STING to induce the production of type I IFNs and other cytokines (Wu et al., 2013). HIV-1 infection can activate cGAS and IFN production, and this is blocked by RT inhibitors but not IN inhibitors, suggesting that viral DNA (or other RT intermediates) may be the trigger for this activation (Gao et al., 2013). Various sequence-specific DNA structures formed during reverse transcription may be particularly effective in activating cGAS (Herzner et al., 2015). Insofar as the restriction factor SAMHD1 can block reverse transcription, it also can limit the activation of the immune response to HIV-1 infection (Maelfait et al., 2016). In a surprising extension of this system, it has been reported that cGAMP can be transferred by virion particles between cells, from virus-producing cells to neighboring cells, to activate IFN production (Bridgeman et al., 2015; Gentili et al., 2015). cGAMP can even spread between cells through viral envelope–generated pores (Xu et al., 2016).

There are good indications that the capsid plays a role in controlling the sensing of virus by cGAS, perhaps by affecting the accessibility or timing of exposure of the encapsidated nucleic acids. These events are likely very cell type-dependent. Thus, in dendritic cells, infection by viruses with HIV-1 capsids is poorly sensed, whereas infection by those with HIV-2 capsids is strongly sensed (Lahaye et al., 2013). CA mutations that affect the interaction with CypA also affect virus sensing, consistent with CypA's roles in affecting CA stability and CA interactions with TRIM5α. Indeed, CA structures are likely to affect interactions with many host proteins—CypA, CPSF6, and TRIM5α—that can indirectly affect activation of cGAS. We note that IFN induction may also be regulated by the HIV-1 accessory proteins. In activated CD4+ T cells, infection by HIV-1 can induce IFN in a cGAS-dependent manner that is enhanced by Vpr and suppressed by Vpu (Vermeire et al., 2016). The details of the mechanisms of actions of these proteins remain to be determined.

IFI16, a DNA Sensor

Interferon gamma-inducible protein 16 (IFI16) is a PRR that, like cGAS, can sense DNAs and inhibit viral replication (Thompson et al., 2014). Tests of various HIV-1 DNA sequences and structures revealed that single-stranded DNAs could induce expression of IFNs in primary human macrophages, and that this induction depended on IFI16 (Jakobsen et al., 2013). IFI16 bound these immunostimulatory DNAs directly, activating STING and leading to the downstream phosphorylation of IRF3. Knockdown of IFI16 in macrophages allowed for enhanced HIV-1 replication, highlighting the significance of its restriction activity in this setting.

The sensing of viral DNAs by IFI16 in some cell types and settings may also lead to cell death by pyroptosis, a form of programmed cell death associated with the inflammatory response (Doitsh et al., 2014). This can result in death of productively infected cells and may even represent a major route of CD4+ T-cell depletion in AIDS patients (Galloway et al., 2015).

TREX1, a Nuclease

TREX1 is a ubiquitous cellular 3′-5′ exonuclease that degrades DNA fragments in the cytoplasm. One of the consequences of its activity is to eliminate viral DNAs before their detection by cGAS or other PRRs, thereby preventing IFN induction and inhibiting the innate immune response (Yan et al., 2010). When TREX1 is depleted by knockdown or knockout, infection by HIV-1 induces type I IFN expression, limiting subsequent virus replication (Wheeler et al., 2016). Complicating experiments in whole animals is the fact that TREX1 depletion and IFN induction can also lead to inflammatory responses and increased T-cell activation, potentially providing increased numbers of susceptible cells and better cell environments for virus replication.

APPLICATIONS FOR THERAPY

The early steps of retroviral infection described here provide a number of potential opportunities for intervention by antiviral drugs. The uncoating of virion cores has been studied as a particularly attractive target (for recent review, see Thenin-Houssier and Valente, 2016). Molecules that bind CA to block uncoating or conversely enhance premature uncoating could in principle strongly inhibit virus infection and most attractively would block infection before formation of the integrated provirus. Such compounds do indeed exist, including PF74 (Shi et al., 2011), BI-1 and BI-2 (Lamorte et al., 2013; Price et al., 2014; Fricke et al., 2014a), I-XW-053 (Kortagere et al., 2012), Ebselen (Thenin-Houssier et al., 2016), and PF-1385801 (Blair et al., 2010); some have been entered into clinical trials.

PF74 is the most extensively studied. PF74 binds in a pocket formed by the CA NTD and CTD interface, within the binding site for CPSF6 (Bhattacharya et al., 2014). The binding is thought to destabilize the capsid (Shi et al., 2011). HIV-1 escape from inhibition by PF74 requires multiple mutations in the binding pocket of CA that collectively reduce drug binding (Shi et al., 2015). These substitution mutations also reduced binding of the host protein CPSF6 and enhanced replication in the presence of the dominant-acting inhibitory protein fragment CPSF6-358. Many of the effects of PF47 on virus replication are plausibly mediated by its effects on host factor binding (Saito et al., 2016). Some of the mutations are associated with considerable loss in fitness, but others compensate for that fitness cost. The selection for PF74 resistance in some cases was found to result in the appearance of mutants that were not merely resistant but dependent on the drug for replication (Zhou et al., 2015). These mutants also showed altered dependence on the host factors CPSF6, the nucleoporin Nup153, and the importin TNP03.

Other compounds under investigation, BI-1 and -2, also bind CA in the context of incoming capsids and block events postreverse transcription; escape mutations also arise in CA (Lamorte et al., 2013; Price et al., 2014; Fricke et al., 2014a). The binding sites here also lie within the binding site for CPSF6 and Nup153, and the compounds induce conformational changes that prevent host factor binding.

These results are encouraging in the sense that escape from the inhibitors will be constrained by effects on host factor functions. While viral escape will surely arise in vivo, these compounds may ultimately have a significant role in combination therapies. Ebselen is particularly interesting and unusual in forming a covalent disulfide bond with CA (Thenin-Houssier et al., 2016).

The many host factors discussed here all raise the potential for antiviral therapies that act through modulation of the interaction between host and viral proteins or of the specific activity of the host factor. Targeting host factors in this way has always drawn much attention, as a possible way to avoid the viral escape from inhibitors through rapid mutation. This is not likely to be a perfect solution to the problem—viruses can escape even those inhibitors that target a host protein (such as the CCR5 inhibitor maraviroc) by finding alternate routes of infection—but at a minimum these efforts will provide novel mechanisms to block virus replication.

Inhibitors of several host factors have been considered or tested. Microtubule modulators such as taxol, vinblastine, colchicine, and nocodazole have been examined, but their antiviral effects are limited by issues of redundant pathways for virus trafficking (Yoder et al., 2011), and cell toxicity. A number of drugs targeting the cyclophilins show some promise as antivirals. Cyclosporin, the prototypical inhibitor of CypA, can meaningfully inhibit HIV-1 in some settings, but its inhibition of T-cell signaling and the immune response is a dangerous unwanted consequence. There exist nonimmunosuppressive derivatives, most notably SDZ NIM811, which binds CypA and prevent its interaction with HIV-1 Gag but does not block T-cell signaling (Rosenwirth et al., 1994; Billich et al., 1995; Steinkasserer et al., 1995). Their effect is to block events after reverse transcription, either PIC uncoating or movement of the PIC to the nucleus. Studies suggest that these compounds could be promising antiretroviral drugs. Lastly we note that any molecules that activate the innate immune response and IFN production have the potential to serve as antivirals. Indeed IFN itself is one of the oldest antivirals to have been tested and is used to treat other viral infections. Generally this approach is fraught with toxic side effects due to the induction of dozens, if not hundreds, of ISGs with unwanted consequences. Molecules that would activate expression of only select ISGs, rather than all of them, could be more successful antivirals. Thus, inducers of only TRIM5α, or MX2, or the APOBECs, should have more limited deleterious effects on the host. Even here, however, there may be unacceptable toxicities. The APOBECs, for example, are potent mutagens, with some limited activity even on double-stranded DNA, and the potential for oncogenesis is probably too high to seriously contemplate their intentional induction as an antiviral therapy.

CONCLUSIONS: PARTING WORDS

Our understanding of the early events of retroviral infection has undergone a huge expansion in the past decade, with the definition of multiple discrete steps in the process and the revelation of the involvement of multiple host factors.

We have come to appreciate the involvement of the cytoskeleton as a mediator and regulator of virus core movement; to uncover the identity of numerous proteins that interact with the incoming capsid and affect uncoating and reverse transcription; and to see how restriction factors inhibit infection and in turn how the virus evades them. Most important in these advances has been the development of new tools and assays that allow probing of these events, in the face of the difficulties inherent in studying inefficient processes. Imaging, biochemistry, cell biology, and especially genetics have all played important roles in building the picture and model that we currently have for this most mysterious of all the parts of the life cycle. We need to emphasize that genetics has always been a critically important tool to validate the biochemical analyses of viral and host protein functions.

Many questions remain about the early events in retroviral infection. The variety of routes that different viral cores take within the cell are not all understood, and in particular the way different envelopes and receptors specify the choice of trafficking route is not clear. The roles of SUMOylation, ubiquitinylation, and other modifications of capsid are not fully understood. The link between uncoating and reverse transcription is unclear, and new proximal triggers for activation of reverse transcription may be found. We can be confident that more host factors that act to promote or to inhibit virus infection are yet to be discovered. Indeed many such factors have already been "found" in the sense that they appear on the long lists of host proteins that have been generated through genome-wide proteomic and functional screens of overexpression, knockout, and knockdown libraries. They lie hiding in plain site in such lists, diamonds in the rough waiting to be uncovered and confirmed as playing significant roles in infection. The subsequent determination of their various mechanisms of action will be the subject for many future investigations and will surely reveal entirely new aspects of virus replication.

REFERENCES

Abudu, A., Wang, X., Dang, Y., Zhou, T., Xiang, S.H., Zheng, Y.H., 2012. Identification of molecular determinants from Moloney leukemia virus 10 homolog (MOV10) protein for virion packaging and anti-HIV-1 activity. J. Biol. Chem. 287, 1220–1228.

Aiken, C., 2006. Viral and cellular factors that regulate HIV-1 uncoating. Curr. Opin. HIV AIDS 1, 194–199.

Aiken, C., 2009. Cell-free assays for HIV-1 uncoating. Methods Mol. Biol. 485, 41–53.

Aiken, C., Trono, D., 1995. Nef stimulates human immunodeficiency virus type 1 proviral DNA synthesis. J. Virol. 69, 5048–5056.

Alin, K., Goff, S.P., 1996. Amino acid substitutions in the CA protein of Moloney murine leukemia virus that block early events in infection. Virology 222, 339–351.

Ambrose, Z., Aiken, C., 2014. HIV-1 uncoating: connection to nuclear entry and regulation by host proteins. Virology 454–455, 371–379.

Amie, S.M., Bambara, R.A., Kim, B., 2013. GTP is the primary activator of the anti-HIV restriction factor SAMHD1. J. Biol. Chem. 288, 25001–25006.

Anderson, J.L., Campbell, E.M., Wu, X., Vandegraaff, N., Engelman, A., Hope, T.J., 2006. Proteasome inhibition reveals that a functional preintegration complex intermediate can be generated during restriction by diverse TRIM5 proteins. J. Virol. 80, 9754–9760.

Antonucci, J.M., St Gelais, C., de Silva, S., Yount, J.S., Tang, C., Ji, X., Shepard, C., Xiong, Y., Kim, B., Wu, L., 2016. SAMHD1-mediated HIV-1 restriction in cells does not involve ribonuclease activity. Nat. Med. 22, 1072–1074.

Arhel, N., Genovesio, A., Kim, K.A., Miko, S., Perret, E., Olivo-Marin, J.C., Shorte, S., Charneau, P., 2006. Quantitative four-dimensional tracking of cytoplasmic and nuclear HIV-1 complexes. Nat. Methods 3, 817–824.

Arjan-Odedra, S., Swanson, C.M., Sherer, N.M., Wolinsky, S.M., Malim, M.H., 2012. Endogenous MOV10 inhibits the retrotransposition of endogenous retroelements but not the replication of exogenous retroviruses. Retrovirology 9, 53.

Arnold, L.H., Groom, H.C., Kunzelmann, S., Schwefel, D., Caswell, S.J., Ordonez, P., Mann, M.C., Rueschenbaum, S., Goldstone, D.C., Pennell, S., Howell, S.A., Stoye, J.P., Webb, M., Taylor, I.A., Bishop, K.N., 2015. Phospho-dependent regulation of SAMHD1 oligomerisation couples catalysis and restriction. PLoS Pathog. 11, e1005194.

Arriagada, G., 2017. Retroviruses and microtubule associated motor proteins. Cell. Microbiol. 19, e12759.

Arriagada, G., Muntean, L.N., Goff, S.P., 2011. SUMO-interacting motifs of human TRIM5alpha are important for antiviral activity. PLoS Pathog. 7, e1002019.

Asaoka, K., Ikeda, K., Hishinuma, T., Horie-Inoue, K., Takeda, S., Inoue, S., 2005. A retrovirus restriction factor TRIM5alpha is transcriptionally regulated by interferons. Biochem. Biophys. Res. Commun. 338, 1950–1956.

Auerbach, M.R., Shu, C., Kaplan, A., Singh, I.R., 2003. Functional characterization of a portion of the Moloney murine leukemia virus gag gene by genetic footprinting. Proc. Natl. Acad. Sci. U.S.A. 100, 11678–11683.

Ayinde, D., Casartelli, N., Schwartz, O., 2012. Restricting HIV the SAMHD1 way: through nucleotide starvation. Nat. Rev. Microbiol. 10, 675–680.

Baldauf, H.M., Pan, X., Erikson, E., Schmidt, S., Daddacha, W., Burggraf, M., Schenkova, K., Ambiel, I., Wabnitz, G., Gramberg, T., Panitz, S., Flory, E., Landau, N.R., Sertel, S., Rutsch, F., Lasitschka, F., Kim, B., Konig, R., Fackler, O.T., Keppler, O.T., 2012. SAMHD1 restricts HIV-1 infection in resting CD4(+) T cells. Nat. Med. 18, 1682–1687.

Bartolini, F., Gundersen, G.G., 2010. Formins and microtubules. Biochim. Biophys. Acta 1803, 164–173.

Bassin, R.H., Gerwin, B.I., Levin, J.G., Duran-Troise, G., Benjers, B.M., Rein, A., 1980. Macromolecular requirements for abrogation of Fv-1 restriction by murine leukemia viruses. J. Virol. 35, 287–297.

Beale, R.C., Petersen-Mahrt, S.K., Watt, I.N., Harris, R.S., Rada, C., Neuberger, M.S., 2004. Comparison of the differential context-dependence of DNA deamination by APOBEC enzymes: correlation with mutation spectra in vivo. J. Mol. Biol. 337, 585–596.

Berg, R.K., Melchjorsen, J., Rintahaka, J., Diget, E., Soby, S., Horan, K.A., Gorelick, R.J., Matikainen, S., Larsen, C.S., Ostergaard, L., Paludan, S.R., Mogensen, T.H., 2012. Genomic HIV RNA induces innate immune responses through RIG-I-dependent sensing of secondary-structured RNA. PLoS One 7, e29291.

Besnier, C., Takeuchi, Y., Towers, G., 2002. Restriction of lentivirus in monkeys. Proc. Natl. Acad. Sci. U.S.A. 99, 11920–11925.

Besnier, C., Ylinen, L., Strange, B., Lister, A., Takeuchi, Y., Goff, S.P., Towers, G.J., 2003. Characterization of murine leukemia virus restriction in mammals. J. Virol. 77, 13403–13406.

Best, S., Le Tissier, P., Towers, G., Stoye, J.P., 1996. Positional cloning of the mouse retrovirus restriction gene Fv1. Nature 382, 826–829.

Bhattacharya, A., Alam, S.L., Fricke, T., Zadrozny, K., Sedzicki, J., Taylor, A.B., Demeler, B., Pornillos, O., Ganser-Pornillos, B.K., Diaz-Griffero, F., Ivanov, D.N., Yeager, M., 2014. Structural basis of HIV-1 capsid recognition by PF74 and CPSF6. Proc. Natl. Acad. Sci. U.S.A. 111, 18625–18630.

Bichel, K., Price, A.J., Schaller, T., Towers, G.J., Freund, S.M., James, L.C., 2013. HIV-1 capsid undergoes coupled binding and isomerization by the nuclear pore protein NUP358. Retrovirology 10, 81.

Billich, A., Hammerschmid, F., Peichl, P., Wenger, R., Zenke, G., Quesniaux, V., Rosenwirth, B., 1995. Mode of action of SDZ NIM 811, a nonimmunosuppressive cyclosporin A analog with activity against human immunodeficiency virus (HIV) type 1: interference with HIV protein-cyclophilin A interactions. J. Virol. 69, 2451–2461.

Biris, N., Tomashevski, A., Bhattacharya, A., Diaz-Griffero, F., Ivanov, D.N., 2013. Rhesus monkey TRIM5alpha SPRY domain recognizes multiple epitopes that span several capsid monomers on the surface of the HIV-1 mature viral core. J. Mol. Biol. 425, 5032–5044.

Biris, N., Yang, Y., Taylor, A.B., Tomashevski, A., Guo, M., Hart, P.J., Diaz-Griffero, F., Ivanov, D.N., 2012. Structure of the rhesus monkey TRIM5alpha PRYSPRY domain, the HIV capsid recognition module. Proc. Natl. Acad. Sci. U.S.A. 109, 13278–13283.

Bishop, K.N., Bock, M., Towers, G., Stoye, J.P., 2001. Identification of the regions of Fv1 necessary for murine leukemia virus restriction. J. Virol. 75, 5182–5188.

Bishop, K.N., Holmes, R.K., Malim, M.H., 2006a. Antiviral potency of APOBEC proteins does not correlate with cytidine deamination. J. Virol. 80, 8450–8458.

Bishop, K.N., Holmes, R.K., Sheehy, A.M., Davidson, N.O., Cho, S.J., Malim, M.H., 2004a. Cytidine deamination of retroviral DNA by diverse APOBEC proteins. Curr. Biol. 14, 1392–1396.

Bishop, K.N., Holmes, R.K., Sheehy, A.M., Malim, M.H., 2004b. APOBEC-mediated editing of viral RNA. Science 305, 645.

Bishop, K.N., Mortuza, G.B., Howell, S., Yap, M.W., Stoye, J.P., Taylor, I.A., 2006b. Characterization of an amino-terminal dimerization domain from retroviral restriction factor Fv1. J. Virol. 80, 8225–8235.

Black, L.R., Aiken, C., 2010. TRIM5alpha disrupts the structure of assembled HIV-1 capsid complexes in vitro. J. Virol. 84, 6564–6569.

Blair, W.S., Pickford, C., Irving, S.L., Brown, D.G., Anderson, M., Bazin, R., Cao, J., Ciaramella, G., Isaacson, J., Jackson, L., Hunt, R., Kjerrstrom, A., Nieman, J.A., Patick, A.K., Perros, M., Scott, A.D., Whitby, K., Wu, H., Butler, S.L., 2010. HIV capsid is a tractable target for small molecule therapeutic intervention. PLoS Pathog. 6, e1001220.

Bloch, N., Glasker, S., Sitaram, P., Hofmann, H., Shepard, C.N., Schultz, M.L., Kim, B., Landau, N.R., 2017. A highly active isoform of lentivirus restriction factor SAMHD1 in mouse. J. Biol. Chem. 292, 1068–1080.

Bogerd, H.P., Doehle, B.P., Wiegand, H.L., Cullen, B.R., 2004. A single amino acid difference in the host APOBEC3G protein controls the primate species specificity of HIV type 1 virion infectivity factor. Proc. Natl. Acad. Sci. U.S.A. 101, 3770–3774.

Bogerd, H.P., Wiegand, H.L., Doehle, B.P., Lueders, K.K., Cullen, B.R., 2006a. APOBEC3A and APOBEC3B are potent inhibitors of LTR-retrotransposon function in human cells. Nucleic Acids Res. 34, 89–95.

Bogerd, H.P., Wiegand, H.L., Hulme, A.E., Garcia-Perez, J.L., O'shea, K.S., Moran, J.V., Cullen, B.R., 2006b. Cellular inhibitors of long interspersed element 1 and Alu retrotransposition. Proc. Natl. Acad. Sci. U.S.A. 103, 8780–8785.

Boone, L.R., Innes, C.L., Heitman, C.K., 1990. Abrogation of Fv-1 restriction by genome-deficient virions produced by a retrovirus packaging cell line. J. Virol. 64, 3376–3381.

Braaten, D., Aberham, C., Franke, E.K., Yin, L., Phares, W., Luban, J., 1996a. Cyclosporine A-resistant human immunodeficiency virus type 1 mutants demonstrate that Gag encodes the functional target of cyclophilin A. J. Virol. 70, 5170–5176.

Braaten, D., Franke, E.K., Luban, J., 1996b. Cyclophilin A is required for an early step in the life cycle of human immunodeficiency virus type 1 before the initiation of reverse transcription. J. Virol. 70, 3551–3560.

Braaten, D., Franke, E.K., Luban, J., 1996c. Cyclophilin A is required for the replication of group M human immunodeficiency virus type 1 (HIV-1) and simian immunodeficiency virus SIV(CPZ)GAB but not group O HIV-1 or other primate immunodeficiency viruses. J. Virol. 70, 4220–4227.

Braaten, D., Luban, J., 2001. Cyclophilin A regulates HIV-1 infectivity, as demonstrated by gene targeting in human T cells. EMBO J. 20, 1300–1309.

Brantis-de-Carvalho, C.E., Maarifi, G., Goncalves Boldrin, P.E., Zanelli, C.F., Nisole, S., Chelbi-Alix, M.K., Valentini, S.R., 2015. MxA interacts with and is modified by the SUMOylation machinery. Exp. Cell Res. 330, 151–163.

Brennan, G., Kozyrev, Y., Hu, S.L., 2008. TRIMCyp expression in Old World primates *Macaca nemestrina* and *Macaca fascicularis*. Proc. Natl. Acad. Sci. U.S.A. 105, 3569–3574.

Bridgeman, A., Maelfait, J., Davenne, T., Partridge, T., Peng, Y., Mayer, A., Dong, T., Kaever, V., Borrow, P., Rehwinkel, J., 2015. Viruses transfer the antiviral second messenger cGAMP between cells. Science 349, 1228–1232.

Briggs, M.W., Sacks, D.B., 2003. IQGAP proteins are integral components of cytoskeletal regulation. EMBO Rep. 4, 571–574.

Brzezinski, J.D., Felkner, R., Modi, A., Liu, M., Roth, M.J., 2016. Phosphorylation requirement of murine leukemia virus p12. J. Virol. 90, 11208–11219.

Buffone, C., Schulte, B., Opp, S., Diaz-Griffero, F., 2015. Contribution of MxB oligomerization to HIV-1 capsid binding and restriction. J. Virol. 89, 3285–3294.

Bukrinskaya, A., Brichacek, B., Mann, A., Stevenson, M., 1998. Establishment of a functional human immunodeficiency virus type 1 (HIV-1) reverse transcription complex involves the cytoskeleton. J. Exp. Med. 188, 2113–2125.

Bukrinsky, M.I., Sharova, N., McDonald, T.L., Pushkarskaya, T., Tarpley, W.G., Stevenson, M., 1993. Association of integrase, matrix, and reverse transcriptase antigens of human immunodeficiency virus type 1 with viral nucleic acids following acute infection. Proc. Natl. Acad. Sci. U.S.A. 90, 6125–6129.

Burdick, R., Smith, J.L., Chaipan, C., Friew, Y., Chen, J., Venkatachari, N.J., Delviks-Frankenberry, K.A., Hu, W.S., Pathak, V.K., 2010. P body-associated protein Mov10 inhibits HIV-1 replication at multiple stages. J. Virol. 84, 10241–10253.

Busnadiego, I., Kane, M., Rihn, S.J., Preugschas, H.F., Hughes, J., Blanco-Melo, D., Strouvelle, V.P., Zang, T.M., Willett, B.J., Boutell, C., Bieniasz, P.D., Wilson, S.J., 2014. Host and viral determinants of Mx2 antiretroviral activity. J. Virol. 88, 7738–7752.

Butkevich, E., Hartig, W., Nikolov, M., Erck, C., Grosche, J., Urlaub, H., Schmidt, C.F., Klopfenstein, D.R., Chua, J.J., 2016. Phosphorylation of FEZ1 by microtubule affinity regulating kinases regulates its function in presynaptic protein trafficking. Sci. Rep. 6, 26965.

Campbell, E.M., Hope, T.J., 2003. Role of the cytoskeleton in nuclear import. Adv. Drug Deliv. Rev. 55, 761–771.

Campbell, E.M., Hope, T.J., 2008. Live cell imaging of the HIV-1 life cycle. Trends Microbiol. 16, 580–587.

Campbell, E.M., Hope, T.J., 2015. HIV-1 capsid: the multifaceted key player in HIV-1 infection. Nat. Rev. Microbiol. 13, 471–483.

Campbell, E.M., Nunez, R., Hope, T.J., 2004. Disruption of the actin cytoskeleton can complement the ability of Nef to enhance human immunodeficiency virus type 1 infectivity. J. Virol. 78, 5745–5755.

Campbell, E.M., Perez, O., Anderson, J.L., Hope, T.J., 2008. Visualization of a proteasome-independent intermediate during restriction of HIV-1 by rhesus TRIM5alpha. J. Cell Biol. 180, 549–561.

Campbell, E.M., Perez, O., Melar, M., Hope, T.J., 2007. Labeling HIV-1 virions with two fluorescent proteins allows identification of virions that have productively entered the target cell. Virology 360, 286–293.

Campbell, E.M., Weingart, J., Sette, P., Opp, S., Sastri, J., O'connor, S.K., Talley, S., Diaz-Griffero, F., Hirsch, V., Bouamr, F., 2016. TRIM5alpha-mediated ubiquitin chain conjugation is required for inhibition of HIV-1 reverse transcription and capsid destabilization. J. Virol. 90, 1849–1857.

Carthagena, L., Parise, M.C., Ringeard, M., Chelbi-Alix, M.K., Hazan, U., Nisole, S., 2008. Implication of TRIM alpha and TRIMCyp in interferon-induced anti-retroviral restriction activities. Retrovirology 5, 59.

Cartier, C., Sivard, P., Tranchat, C., Decimo, D., Desgranges, C., Boyer, V., 1999. Identification of three major phosphorylation sites within HIV-1 capsid. Role of phosphorylation during the early steps of infection. J. Biol. Chem. 274, 19434–19440.

Cereseto, A., Manganaro, L., Gutierrez, M.I., Terreni, M., Fittipaldi, A., Lusic, M., Marcello, A., Giacca, M., 2005. Acetylation of HIV-1 integrase by p300 regulates viral integration. EMBO J. 24, 3070–3081.

Chen, N.Y., Zhou, L., Gane, P.J., Opp, S., Ball, N.J., Nicastro, G., Zufferey, M., Buffone, C., Luban, J., Selwood, D., Diaz-Griffero, F., Taylor, I., Fassati, A., 2016. HIV-1 capsid is involved in post-nuclear entry steps. Retrovirology 13, 28.

Chin, C.R., Perreira, J.M., Savidis, G., Portmann, J.M., Aker, A.M., Feeley, E.M., Smith, M.C., Brass, A.L., 2015. Direct visualization of HIV-1 replication intermediates shows that capsid and CPSF6 modulate HIV-1 intra-nuclear invasion and integration. Cell Rep. 13, 1717–1731.

Chinsky, J., Soeiro, R., 1981. Fv-1 host restriction of Friend leukemia virus: analysis of unintegrated proviral DNA. J. Virol. 40, 45–55.

Choi, J., Ryoo, J., Oh, C., Hwang, S., Ahn, K., 2015. SAMHD1 specifically restricts retroviruses through its RNase activity. Retrovirology 12, 46.

Colgan, J., Yuan, H.E., Franke, E.K., Luban, J., 1996. Binding of the human immunodeficiency virus type 1 Gag polyprotein to cyclophilin A is mediated by the central region of capsid and requires Gag dimerization. J. Virol. 70, 4299–4310.

Colomer-Lluch, M., Gollahon, L.S., Serra-Moreno, R., 2016. Anti-HIV factors: targeting each step of HIV's replication cycle. Curr. HIV Res. 14, 175–182.

Cowan, S., Hatziioannou, T., Cunningham, T., Muesing, M.A., Gottlinger, H.G., Bieniasz, P.D., 2002. Cellular inhibitors with Fv1-like activity restrict human and simian immunodeficiency virus tropism. Proc. Natl. Acad. Sci. U.S.A. 99, 11914–11919.

Cribier, A., Descours, B., Valadao, A.L., Laguette, N., Benkirane, M., 2013. Phosphorylation of SAMHD1 by cyclin A2/CDK1 regulates its restriction activity toward HIV-1. Cell Rep. 3, 1036–1043.

Cullen, B.R., 2006. Role and mechanism of action of the APOBEC3 family of antiretroviral resistance factors. J. Virol. 80, 1067–1076.

Dang, Y., Wang, X., Esselman, W.J., Zheng, Y.H., 2006. Identification of APOBEC3DE as another antiretroviral factor from the human APOBEC family. J. Virol. 80, 10522–10533.

Danielson, C.M., Cianci, G.C., Hope, T.J., 2012. Recruitment and dynamics of proteasome association with rhTRIM5alpha cytoplasmic complexes during HIV-1 infection. Traffic 13, 1206–1217.

De Iaco, A., Luban, J., 2014. Cyclophilin A promotes HIV-1 reverse transcription but its effect on transduction correlates best with its effect on nuclear entry of viral cDNA. Retrovirology 11, 11.

Delebecque, F., Suspene, R., Calattini, S., Casartelli, N., Saib, A., Froment, A., Wain-Hobson, S., Gessain, A., Vartanian, J.P., Schwartz, O., 2006. Restriction of foamy viruses by APOBEC cytidine deaminases. J. Virol. 80, 605–614.

Desai, T.M., Marin, M., Sood, C., Shi, J., Nawaz, F., Aiken, C., Melikyan, G.B., 2015. Fluorescent protein-tagged Vpr dissociates from HIV-1 core after viral fusion and rapidly enters the cell nucleus. Retrovirology 12, 88.

Descours, B., Cribier, A., Chable-Bessia, C., Ayinde, D., Rice, G., Crow, Y., Yatim, A., Schwartz, O., Laguette, N., Benkirane, M., 2012. SAMHD1 restricts HIV-1 reverse transcription in quiescent CD4(+) T-cells. Retrovirology 9, 87.

DesGroseillers, L., Jolicoeur, P., 1983. Physical mapping of the Fv-1 tropism host range determinant of BALB/c murine leukemia viruses. J. Virol. 48, 685–696.

Dharan, A., Talley, S., Tripathi, A., Mamede, J.I., Majetschak, M., Hope, T.J., Campbell, E.M., 2016. KIF5B and Nup358 cooperatively mediate the nuclear import of HIV-1 during infection. PLoS Pathog. 12, e1005700.

Diaz-Griffero, F., Kar, A., Perron, M., Xiang, S.H., Javanbakht, H., Li, X., Sodroski, J., 2007. Modulation of retroviral restriction and proteasome inhibitor-resistant turnover by changes in the TRIM5alpha B-box 2 domain. J. Virol. 81, 10362–10378.

Diaz-Griffero, F., Qin, X.R., Hayashi, F., Kigawa, T., Finzi, A., Sarnak, Z., Lienlaf, M., Yokoyama, S., Sodroski, J., 2009. A B-box 2 surface patch important for TRIM5alpha self-association, capsid binding avidity, and retrovirus restriction. J. Virol. 83, 10737–10751.

Dicks, M.D., Goujon, C., Pollpeter, D., Betancor, G., Apolonia, L., Bergeron, J.R., Malim, M.H., 2015. Oligomerization requirements for MX2-mediated suppression of HIV-1 infection. J. Virol. 90, 22–32.

Dismuke, D.J., Aiken, C., 2006. Evidence for a functional link between uncoating of the human immunodeficiency virus type 1 core and nuclear import of the viral preintegration complex. J. Virol. 80, 3712–3720.

Dodding, M.P., Way, M., 2011. Coupling viruses to dynein and kinesin-1. EMBO J. 30, 3527–3539.

Doehle, B.P., Schafer, A., Cullen, B.R., 2005. Human APOBEC3B is a potent inhibitor of HIV-1 infectivity and is resistant to HIV-1 Vif. Virology 339, 281–288.

Dohner, K., Nagel, C.H., Sodeik, B., 2005. Viral stop-and-go along microtubules: taking a ride with dynein and kinesins. Trends Microbiol. 13, 320–327.

Dohner, K., Sodeik, B., 2005. The role of the cytoskeleton during viral infection. Curr. Top. Microbiol. Immunol. 285, 67–108.

Doitsh, G., Galloway, N.L., Geng, X., Yang, Z., Monroe, K.M., Zepeda, O., Hunt, P.W., Hatano, H., Sowinski, S., Munoz-Arias, I., Greene, W.C., 2014. Cell death by pyroptosis drives CD4 T-cell depletion in HIV-1 infection. Nature 505, 509–514.

Donahue, D.A., Amraoui, S., Di Nunzio, F., Kieffer, C., Porrot, F., Opp, S., Diaz-Griffero, F., Casartelli, N., Schwartz, O., 2016. SUN2 overexpression deforms nuclear shape and inhibits HIV. J. Virol. 90, 4199–4214.

Donahue, D.A., Porrot, F., Couespel, N., Schwartz, O., 2017. SUN2 silencing impairs CD4 T cell proliferation and alters sensitivity to HIV-1 infection independently of Cyclophilin A. J. Virol. 91.

Dunster, L.M., Schneider-Schaulies, J., Loffler, S., Lankes, W., Schwartz-Albiez, R., Lottspeich, F., Ter Meulen, V., 1994. Moesin: a cell membrane protein linked with susceptibility to measles virus infection. Virology 198, 265–274.

Duran-Troise, G., Bassin, R.H., Wallace, B.F., Rein, A., 1981. Balb/3T3 cells chronically infected with N-tropic murine leukemia virus continue to express Fv-1b restriction. Virology 112, 795–799.

Dvorin, J.D., Malim, M.H., 2003. Intracellular trafficking of HIV-1 cores: journey to the center of the cell. Curr. Top. Microbiol. Immunol. 281, 179–208.

Elis, E., Ehrlich, M., Prizan-Ravid, A., Laham-Karam, N., Bacharach, E., 2012. p12 tethers the murine leukemia virus pre-integration complex to mitotic chromosomes. PLoS Pathog. 8, e1003103.

Esnault, C., Millet, J., Schwartz, O., Heidmann, T., 2006. Dual inhibitory effects of APOBEC family proteins on retrotransposition of mammalian endogenous retroviruses. Nucleic Acids Res. 34, 1522–1531.

Fackler, O.T., Krausslich, H.G., 2006. Interactions of human retroviruses with the host cell cytoskeleton. Curr. Opin. Microbiol. 9, 409–415.

Farnet, C.M., Haseltine, W.A., 1991. Determination of viral proteins present in the human immunodeficiency virus type 1 preintegration complex. J. Virol. 65, 1910–1915.

Fassati, A., 2012. Multiple roles of the capsid protein in the early steps of HIV-1 infection. Virus Res. 170, 15–24.

Fassati, A., Goff, S.P., 1999. Characterization of intracellular reverse transcription complexes of Moloney murine leukemia virus. J. Virol. 73, 8919–8925.

Fassati, A., Goff, S.P., 2001. Characterization of intracellular reverse transcription complexes of human immunodeficiency virus type 1. J. Virol. 75, 3626–3635.

Fernandez, J., Portilho, D.M., Danckaert, A., Munier, S., Becker, A., Roux, P., Zambo, A., Shorte, S., Jacob, Y., Vidalain, P.O., Charneau, P., Clavel, F., Arhel, N.J., 2015. Microtubule-associated proteins 1 (MAP1) promote human immunodeficiency virus type I (HIV-1) intracytoplasmic routing to the nucleus. J. Biol. Chem. 290, 4631–4646.

Fletcher, A.J., Christensen, D.E., Nelson, C., Tan, C.P., Schaller, T., Lehner, P.J., Sundquist, W.I., Towers, G.J., 2015. TRIM5alpha requires Ube2W to anchor Lys63-linked ubiquitin chains and restrict reverse transcription. EMBO J. 34, 2078–2095.

Fontenot, D.R., den Hollander, P., Vela, E.M., Newman, R., Sastry, J.K., Kumar, R., 2007. Dynein light chain 1 peptide inhibits human immunodeficiency virus infection in eukaryotic cells. Biochem. Biophys. Res. Commun. 363, 901–907.

Forshey, B.M., von Schwedler, U., Sundquist, W.I., Aiken, C., 2002. Formation of a human immunodeficiency virus type 1 core of optimal stability is crucial for viral replication. J. Virol. 76, 5667–5677.

Francis, A.C., Marin, M., Shi, J., Aiken, C., Melikyan, G.B., 2016. Time-resolved imaging of single HIV-1 uncoating in vitro and in living cells. PLoS Pathog. 12, e1005709.

Franke, E.K., Luban, J., 1996. Inhibition of HIV-1 replication by cyclosporine A or related compounds correlates with the ability to disrupt the Gag-cyclophilin A interaction. Virology 222, 279–282.

Franke, E.K., Yuan, H.E., Luban, J., 1994. Specific incorporation of cyclophilin A into HIV-1 virions. Nature 372, 359–362.

Fribourgh, J.L., Nguyen, H.C., Matreyek, K.A., Alvarez, F.J., Summers, B.J., Dewdney, T.G., Aiken, C., Zhang, P., Engelman, A., Xiong, Y., 2014. Structural insight into HIV-1 restriction by MxB. Cell Host Microbe 16, 627–638.

Fricke, T., Buffone, C., Opp, S., Valle-Casuso, J., Diaz-Griffero, F., 2014a. BI-2 destabilizes HIV-1 cores during infection and Prevents Binding of CPSF6 to the HIV-1 Capsid. Retrovirology 11, 120.

Fricke, T., White, T.E., Schulte, B., de Souza Aranha Vieira, D.A., Dharan, A., Campbell, E.M., Brandariz-Nunez, A., Diaz-Griffero, F., 2014b. MxB binds to the HIV-1 core and prevents the uncoating process of HIV-1. Retrovirology 11, 68.

Fu, M.M., Holzbaur, E.L., 2013. JIP1 regulates the directionality of APP axonal transport by coordinating kinesin and dynein motors. J. Cell Biol. 202, 495–508.

Furtak, V., Mulky, A., Rawlings, S.A., Kozhaya, L., Lee, K., Kewalramani, V.N., Unutmaz, D., 2010. Perturbation of the P-body component Mov10 inhibits HIV-1 infectivity. PLoS One 5, e9081.

Gallo, D.E., Hope, T.J., 2012. Knockdown of MAP4 and DNAL1 produces a post-fusion and pre-nuclear translocation impairment in HIV-1 replication. Virology 422, 13–21.

Galloway, N.L., Doitsh, G., Monroe, K.M., Yang, Z., Munoz-Arias, I., Levy, D.N., Greene, W.C., 2015. Cell-to-Cell transmission of HIV-1 is required to trigger pyroptotic death of lymphoid-tissue-derived CD4 T cells. Cell Rep. 12, 1555–1563.

Gamble, T.R., Vajdos, F.F., Yoo, S., Worthylake, D.K., Houseweart, M., Sundquist, W.I., Hill, C.P., 1996. Crystal structure of human cyclophilin A bound to the amino-terminal domain of HIV-1 capsid. Cell 87, 1285–1294.

Ganser, B.K., Li, S., Klishko, V.Y., Finch, J.T., Sundquist, W.I., 1999. Assembly and analysis of conical models for the HIV-1 core. Science 283, 80–83.

Ganser-Pornillos, B.K., Chandrasekaran, V., Pornillos, O., Sodroski, J.G., Sundquist, W.I., Yeager, M., 2011. Hexagonal assembly of a restricting TRIM5alpha protein. Proc. Natl. Acad. Sci. U.S.A. 108, 534–539.

Gao, D., Wu, J., Wu, Y.T., Du, F., Aroh, C., Yan, N., Sun, L., Chen, Z.J., 2013. Cyclic GMP-AMP synthase is an innate immune sensor of HIV and other retroviruses. Science 341, 903–906.

Gao, G., Goff, S.P., 1999. Somatic cell mutants resistant to retrovirus replication: intracellular blocks during the early stages of infection. Mol. Biol. Cell 10, 1705–1717.

Garcia-Exposito, L., Ziglio, S., Barroso-Gonzalez, J., de Armas-Rillo, L., Valera, M.S., Zipeto, D., Machado, J.D., Valenzuela-Fernandez, A., 2013. Gelsolin activity controls efficient early HIV-1 infection. Retrovirology 10, 39.

Gaudin, R., de Alencar, B.C., Arhel, N., Benaroch, P., 2013. HIV trafficking in host cells: motors wanted! Trends Cell Biol. 23, 652–662.

Gentili, M., Kowal, J., Tkach, M., Satoh, T., Lahaye, X., Conrad, C., Boyron, M., Lombard, B., Durand, S., Kroemer, G., Lowe, D., Dalod, M., Thery, C., Manel, N., 2015. Transmission of innate immune signaling by packaging of cGAMP in viral particles. Science 349, 1232–1236.

Gilboa, E., Goff, S., Shields, A., Yoshimura, F., Mitra, S., Baltimore, D., 1979a. In vitro synthesis of a 9 kbp terminally redundant DNA carrying the infectivity of Moloney murine leukemia virus. Cell 16, 863–874.

Gilboa, E., Mitra, S.W., Goff, S., Baltimore, D., 1979b. A detailed model of reverse transcription and tests of crucial aspects. Cell 18, 93–100.

Gladue, D.P., Holinka, L.G., Fernandez-Sainz, I.J., Prarat, M.V., O'donnell, V., Vepkhvadze, N.G., Lu, Z., Risatti, G.R., Borca, M.V., 2011. Interaction between Core protein of classical swine fever virus with cellular IQGAP1 protein appears essential for virulence in swine. Virology 412, 68–74.

Goff, S.P., 2001. Intracellular trafficking of retroviral genomes during the early phase of infection: viral exploitation of cellular pathways. J. Gene Med. 3, 517–528.

Goila-Gaur, R., Strebel, K., 2008. HIV-1 Vif, APOBEC, and intrinsic immunity. Retrovirology 5, 51.

Goldstone, D.C., Ennis-Adeniran, V., Hedden, J.J., Groom, H.C., Rice, G.I., Christodoulou, E., Walker, P.A., Kelly, G., Haire, L.F., Yap, M.W., de Carvalho, L.P., Stoye, J.P., Crow, Y.J., Taylor, I.A., Webb, M., 2011. HIV-1 restriction factor SAMHD1 is a deoxynucleoside triphosphate triphosphohydrolase. Nature 480, 379–382.

Goujon, C., Greenbury, R.A., Papaioannou, S., Doyle, T., Malim, M.H., 2015. A triple-arginine motif in the amino-terminal domain and oligomerization are required for HIV-1 inhibition by human MX2. J. Virol. 89, 4676–4680.

Goujon, C., Moncorge, O., Bauby, H., Doyle, T., Ward, C.C., Schaller, T., Hue, S., Barclay, W.S., Schulz, R., Malim, M.H., 2013. Human MX2 is an interferon-induced post-entry inhibitor of HIV-1 infection. Nature 502, 559–562.

Greber, U.F., Singh, I., Helenius, A., 1994. Mechanisms of virus uncoating. Trends Microbiol. 2, 52–56.

Grutter, M.G., Luban, J., 2012. TRIM5 structure, HIV-1 capsid recognition, and innate immune signaling. Curr. Opin. Virol. 2, 142–150.

Gundersen, G.G., 2002. Microtubule capture: IQGAP and CLIP-170 expand the repertoire. Curr. Biol. 12, R645–R647.

Guth, C.A., Sodroski, J., 2014. Contribution of PDZD8 to stabilization of the human immunodeficiency virus type 1 capsid. J. Virol. 88, 4612–4623.

Haedicke, J., de Los Santos, K., Goff, S.P., Naghavi, M.H., 2008. The Ezrin-radixin-moesin family member ezrin regulates stable microtubule formation and retroviral infection. J. Virol. 82, 4665–4670.

Haller, O., 2013. Dynamins are forever: MxB inhibits HIV-1. Cell Host Microbe 14, 371–373.

Haller, O., Kochs, G., 2002. Interferon-induced mx proteins: dynamin-like GTPases with antiviral activity. Traffic 3, 710–717.

Haller, O., Staeheli, P., Schwemmle, M., Kochs, G., 2015. Mx GTPases: dynamin-like antiviral machines of innate immunity. Trends Microbiol. 23, 154–163.

Hamamoto, S., Nishitsuji, H., Amagasa, T., Kannagi, M., Masuda, T., 2006. Identification of a novel human immunodeficiency virus type 1 integrase interactor, Gemin2, that facilitates efficient viral cDNA synthesis in vivo. J. Virol. 80, 5670–5677.

Harris, R.S., Bishop, K.N., Sheehy, A.M., Craig, H.M., Petersen-Mahrt, S.K., Watt, I.N., Neuberger, M.S., Malim, M.H., 2003. DNA deamination mediates innate immunity to retroviral infection. Cell 113, 803–809.

Harris, R.S., Dudley, J.P., 2015. APOBECs and virus restriction. Virology 479–480, 131–145.

Harrison, I.P., McKnight, A., 2011. Cellular entry via an actin and clathrin-dependent route is required for Lv2 restriction of HIV-2. Virology 415, 47–55.

Hatziioannou, T., Cowan, S., Bieniasz, P.D., 2004a. Capsid-dependent and -independent postentry restriction of primate lentivirus tropism in rodent cells. J. Virol. 78, 1006–1011.

Hatziioannou, T., Cowan, S., Goff, S.P., Bieniasz, P.D., Towers, G.J., 2003. Restriction of multiple divergent retroviruses by Lv1 and Ref1. EMBO J. 22, 385–394.

Hatziioannou, T., Cowan, S., von Schwedler, U.K., Sundquist, W.I., Bieniasz, P.D., 2004b. Species-specific tropism determinants in the human immunodeficiency virus type 1 capsid. J. Virol. 78, 6005–6012.

Hatziioannou, T., Perez-Caballero, D., Yang, A., Cowan, S., Bieniasz, P.D., 2004c. Retrovirus resistance factors Ref1 and Lv1 are species-specific variants of TRIM5alpha. Proc. Natl. Acad. Sci. U.S.A. 101, 10774–10779.

Henning, M.S., Morham, S.G., Goff, S.P., Naghavi, M.H., 2010. PDZD8 is a novel Gag-interacting factor that promotes retroviral infection. J. Virol. 84, 8990–8995.

Herzner, A.M., Hagmann, C.A., Goldeck, M., Wolter, S., Kubler, K., Wittmann, S., Gramberg, T., Andreeva, L., Hopfner, K.P., Mertens, C., Zillinger, T., Jin, T., Xiao, T.S., Bartok, E., Coch, C., Ackermann, D., Hornung, V., Ludwig, J., Barchet, W., Hartmann, G., Schlee, M., 2015. Sequence-specific activation of the DNA sensor cGAS by Y-form DNA structures as found in primary HIV-1 cDNA. Nat. Immunol. 16, 1025–1033.

Hilditch, L., Matadeen, R., Goldstone, D.C., Rosenthal, P.B., Taylor, I.A., Stoye, J.P., 2011. Ordered assembly of murine leukemia virus capsid protein on lipid nanotubes directs specific binding by the restriction factor, Fv1. Proc. Natl. Acad. Sci. U.S.A. 108, 5771–5776.

Hofmann, H., Vanwalscappel, B., Bloch, N., Landau, N.R., 2016. TLR7/8 agonist induces a post-entry SAMHD1-independent block to HIV-1 infection of monocytes. Retrovirology 13, 83.

Hollenbaugh, J.A., Tao, S., Lenzi, G.M., Ryu, S., Kim, D.H., Diaz-Griffero, F., Schinazi, R.F., Kim, B., 2014. dNTP pool modulation dynamics by SAMHD1 protein in monocyte-derived macrophages. Retrovirology 11, 63.

Holmes, R.K., Malim, M.H., Bishop, K.N., 2007. APOBEC-mediated viral restriction: not simply editing? Trends Biochem. Sci. 32, 118–128.

Hopkins, N., Schindler, J., Hynes, R., 1977. Six NB-tropic murine leukemia viruses derived from a B-tropic virus of BALB/c have altered p30. J. Virol. 21, 309–318.

Horton, R., Mumm, S.R., Grandgenett, D.P., 1991. Phosphorylation of the avian retrovirus integration protein and proteolytic processing of its carboxyl terminus. J. Virol. 65, 1141–1148.

Hrecka, K., Hao, C., Gierszewska, M., Swanson, S.K., Kesik-Brodacka, M., Srivastava, S., Florens, L., Washburn, M.P., Skowronski, J., 2011. Vpx relieves inhibition of HIV-1 infection of macrophages mediated by the SAMHD1 protein. Nature 474, 658–661.

Hulme, A.E., Bogerd, H.P., Cullen, B.R., Moran, J.V., 2007. Selective inhibition of Alu retrotransposition by APOBEC3G. Gene 390, 199–205.

Hulme, A.E., Hope, T.J., 2014a. The cyclosporin A washout assay to detect HIV-1 uncoating in infected cells. Methods Mol. Biol. 1087, 37–46.

Hulme, A.E., Hope, T.J., 2014b. Live cell imaging of retroviral entry. Annu. Rev. Virol. 1, 501–515.

Hulme, A.E., Kelley, Z., Foley, D., Hope, T.J., 2015a. Complementary assays reveal a low level of CA associated with viral complexes in the nuclei of HIV-1-infected cells. J. Virol. 89, 5350–5361.

Hulme, A.E., Kelley, Z., Okocha, E.A., Hope, T.J., 2015b. Identification of capsid mutations that alter the rate of HIV-1 uncoating in infected cells. J. Virol. 89, 643–651.

Hulme, A.E., Perez, O., Hope, T.J., 2011. Complementary assays reveal a relationship between HIV-1 uncoating and reverse transcription. Proc. Natl. Acad. Sci. U.S.A. 108, 9975–9980.

Iyengar, S., Hildreth, J.E., Schwartz, D.H., 1998. Actin-dependent receptor colocalization required for human immunodeficiency virus entry into host cells. J. Virol. 72, 5251–5255.

Jacques, D.A., Mcewan, W.A., Hilditch, L., Price, A.J., Towers, G.J., James, L.C., 2016. HIV-1 uses dynamic capsid pores to import nucleotides and fuel encapsidated DNA synthesis. Nature 536, 349–353.

Jakobsen, M.R., Bak, R.O., Andersen, A., Berg, R.K., Jensen, S.B., Tengchuan, J., Laustsen, A., Hansen, K., Ostergaard, L., Fitzgerald, K.A., Xiao, T.S., Mikkelsen, J.G., Mogensen, T.H., Paludan, S.R., 2013. IFI16 senses DNA forms of the lentiviral replication cycle and controls HIV-1 replication. Proc. Natl. Acad. Sci. U.S.A. 110, E4571–E4580.

Javanbakht, H., Yuan, W., Yeung, D.F., Song, B., Diaz-Griffero, F., Li, Y., Li, X., Stremlau, M., Sodroski, J., 2006. Characterization of TRIM5alpha trimerization and its contribution to human immunodeficiency virus capsid binding. Virology 353, 234–246.

Jayappa, K.D., Ao, Z., Wang, X., Mouland, A.J., Shekhar, S., Yang, X., Yao, X., 2015. Human immunodeficiency virus type 1 employs the cellular dynein light chain 1 protein for reverse transcription through interaction with its integrase protein. J. Virol. 89, 3497–3511.

Ji, X., Tang, C., Zhao, Q., Wang, W., Xiong, Y., 2014. Structural basis of cellular dNTP regulation by SAMHD1. Proc. Natl. Acad. Sci. U.S.A. 111, E4305–E4314.

Ji, X., Wu, Y., Yan, J., Mehrens, J., Yang, H., Delucia, M., Hao, C., Gronenborn, A.M., Skowronski, J., Ahn, J., Xiong, Y., 2013. Mechanism of allosteric activation of SAMHD1 by dGTP. Nat. Struct. Mol. Biol. 20, 1304–1309.

Jolicoeur, P., 1979. The Fv-1 gene of the mouse and its control of murine leukemia virus replication. Curr. Top. Microbiol. Immunol. 86, 67–122.

Jolicoeur, P., Baltimore, D., 1976. Effect of Fv-1 gene product on proviral DNA formation and integration in cells infected with murine leukemia viruses. Proc. Natl. Acad. Sci. U.S.A. 73, 2236–2240.

Jolicoeur, P., Rassart, E., 1980. Effect of Fv-1 gene product on synthesis of linear and supercoiled viral DNA in cells infected with murine leukemia virus. J. Virol. 33, 183–195.

Jolicoeur, P., Rassart, E., 1981. Fate of unintegrated viral DNA in Fv-1 permissive and resistant mouse cells infected with murine leukemia virus. J. Virol. 37, 609–619.

Jovasevic, V., Naghavi, M.H., Walsh, D., 2015. Microtubule plus end-associated CLIP-170 initiates HSV-1 retrograde transport in primary human cells. J. Cell Biol. 211, 323–337.

Kane, M., Yadav, S.S., Bitzegeio, J., Kutluay, S.B., Zang, T., Wilson, S.J., Schoggins, J.W., Rice, C.M., Yamashita, M., Hatziioannou, T., Bieniasz, P.D., 2013. MX2 is an interferon-induced inhibitor of HIV-1 infection. Nature 502, 563–566.

Kao, S., Akari, H., Khan, M.A., Dettenhofer, M., Yu, X.F., Strebel, K., 2003a. Human immunodeficiency virus type 1 Vif is efficiently packaged into virions during productive but not chronic infection. J. Virol. 77, 1131–1140.

Kao, S., Khan, M.A., Miyagi, E., Plishka, R., Buckler-White, A., Strebel, K., 2003b. The human immunodeficiency virus type 1 Vif protein reduces intracellular expression and inhibits packaging of APOBEC3G (CEM15), a cellular inhibitor of virus infectivity. J. Virol. 77, 11398–11407.

Keckesova, Z., Ylinen, L.M., Towers, G.J., 2004. The human and African green monkey TRIM5alpha genes encode Ref1 and Lv1 retroviral restriction factor activities. Proc. Natl. Acad. Sci. U.S.A. 101, 10780–10785.

Keckesova, Z., Ylinen, L.M., Towers, G.J., 2006. Cyclophilin A renders human immunodeficiency virus type 1 sensitive to Old World monkey but not human TRIM5 alpha antiviral activity. J. Virol. 80, 4683–4690.

Khan, M.A., Kao, S., Miyagi, E., Takeuchi, H., Goila-Gaur, R., Opi, S., Gipson, C.L., Parslow, T.G., Ly, H., Strebel, K., 2005. Viral RNA is required for the association of APOBEC3G with human immunodeficiency virus type 1 nucleoprotein complexes. J. Virol. 79, 5870–5874.

Kim, J., Tipper, C., Sodroski, J., 2011. Role of TRIM5alpha RING domain E3 ubiquitin ligase activity in capsid disassembly, reverse transcription blockade, and restriction of simian immunodeficiency virus. J. Virol. 85, 8116–8132.

Kim, W., Tang, Y., Okada, Y., Torrey, T.A., Chattopadhyay, S.K., Pfleiderer, M., Falkner, F.G., Dorner, F., Choi, W., Hirokawa, N., Morse, H.C., 1998. Binding of murine leukemia virus Gag polyproteins to KIF4, a microtubule-based motor protein. J. Virol. 72, 6898–6901.

Kizhatil, K., Albritton, L.M., 1997. Requirements for different components of the host cell cytoskeleton distinguish ecotropic murine leukemia virus entry via endocytosis from entry via surface fusion. J. Virol. 71, 7145–7156.

Koh, Y., Wu, X., Ferris, A.L., Matreyek, K.A., Smith, S.J., Lee, K., Kewalramani, V.N., Hughes, S.H., Engelman, A., 2013. Differential effects of human immunodeficiency virus type 1 capsid and cellular factors nucleoporin 153 and LEDGF/p75 on the efficiency and specificity of viral DNA integration. J. Virol. 87, 648–658.

Kortagere, S., Madani, N., Mankowski, M.K., Schon, A., Zentner, I., Swaminathan, G., Princiotto, A., Anthony, K., Oza, A., Sierra, L.J., Passic, S.R., Wang, X., Jones, D.M., Stavale, E., Krebs, F.C., Martin-Garcia, J., Freire, E., Ptak, R.G., Sodroski, J., Cocklin, S., Smith, A.B., 2012. Inhibiting early-stage events in HIV-1 replication by small-molecule targeting of the HIV-1 capsid. J. Virol. 86, 8472–8481.

Kozak, C., 1985. Analysis of wild-derived mice for Fv-1 and Fv-2 murine leukemia virus restriction loci: a novel wild mouse Fv-1 allele responsible for lack of host range restriction. J. Virol. 55, 281–285.

Kozak, C.A., Chakraborti, A., 1996. Single amino acid changes in the murine leukemia virus capsid protein gene define the target of Fv1 resistance. Virology 225, 300–305.

Kubo, Y., Yoshii, H., Kamiyama, H., Tominaga, C., Tanaka, Y., Sato, H., Yamamoto, N., 2008. Ezrin, Radixin, and Moesin (ERM) proteins function as pleiotropic regulators of human immunodeficiency virus type 1 infection. Virology 375, 130–140.

Kutluay, S.B., Perez-Caballero, D., Bieniasz, P.D., 2013. Fates of retroviral core components during unrestricted and TRIM5-restricted infection. PLoS Pathog. 9, e1003214.

Laguette, N., Sobhian, B., Casartelli, N., Ringeard, M., Chable-Bessia, C., Segeral, E., Yatim, A., Emiliani, S., Schwartz, O., Benkirane, M., 2011. SAMHD1 is the dendritic- and myeloid-cell-specific HIV-1 restriction factor counteracted by Vpx. Nature 474, 654–657.

Lahaye, X., Satoh, T., Gentili, M., Cerboni, S., Conrad, C., Hurbain, I., El Marjou, A., Lacabaratz, C., Lelievre, J.D., Manel, N., 2013. The capsids of HIV-1 and HIV-2 determine immune detection of the viral cDNA by the innate sensor cGAS in dendritic cells. Immunity 39, 1132–1142.

Lahaye, X., Satoh, T., Gentili, M., Cerboni, S., Silvin, A., Conrad, C., Ahmed-Belkacem, A., Rodriguez, E.C., Guichou, J.F., Bosquet, N., Piel, M., Le Grand, R., King, M.C., Pawlotsky, J.M., Manel, N., Nuclear envelope protein SUN2 promotes cyclophilin-a-dependent steps of HIV replication. Cell Rep. 15, 879–892.

Lahouassa, H., Daddacha, W., Hofmann, H., Ayinde, D., Logue, E.C., Dragin, L., Bloch, N., Maudet, C., Bertrand, M., Gramberg, T., Pancino, G., Priet, S., Canard, B., Laguette, N., Benkirane, M., Transy, C., Landau, N.R., Kim, B., Margottin-Goguet, F., 2012. SAMHD1 restricts the replication of human immunodeficiency virus type 1 by depleting the intracellular pool of deoxynucleoside triphosphates. Nat. Immunol. 13, 223–228.

Lamorte, L., Titolo, S., Lemke, C.T., Goudreau, N., Mercier, J.F., Wardrop, E., Shah, V.B., von Schwedler, U.K., Langelier, C., Banik, S.S., Aiken, C., Sundquist, W.I., Mason, S.W., 2013. Discovery of novel small-molecule HIV-1 replication inhibitors that stabilize capsid complexes. Antimicrob. Agents Chemother. 57, 4622–4631.

Lascano, J., Uchil, P.D., Mothes, W., Luban, J., 2015. TRIM5 retroviral restriction activity correlates with the ability to induce innate immune signaling. J. Virol. 90, 308–316.

Lemay, J., Maidou-Peindara, P., Bader, T., Ennifar, E., Rain, J.C., Benarous, R., Liu, L.X., 2008. HuR interacts with human immunodeficiency virus type 1 reverse transcriptase, and modulates reverse transcription in infected cells. Retrovirology 5, 47.

Leung, J., Yueh, A., Appah Jr., F.S., Yuan, B., de Los Santos, K., Goff, S.P., 2006. Interaction of Moloney murine leukemia virus matrix protein with IQGAP. EMBO J. 25, 2155–2166.

Li, D., Wei, T., Rawle, D.J., Qin, F., Wang, R., Soares, D.C., Jin, H., Sivakumaran, H., Lin, M.H., Spann, K., Abbott, C.M., Harrich, D., 2015. Specific interaction between eEF1A and HIV RT is critical for HIV-1 reverse transcription and a potential anti-HIV target. PLoS Pathog. 11, e1005289.

Li, S., Hill, C.P., Sundquist, W.I., Finch, J.T., 2000. Image reconstructions of helical assemblies of the HIV-1 CA protein. Nature 407, 409–413.

Li, X., Sodroski, J., 2008. The TRIM5alpha B-box 2 domain promotes cooperative binding to the retroviral capsid by mediating higher-order self-association. J. Virol. 82, 11495–11502.

Li, Y., Kar, A.K., Sodroski, J., 2009. Target cell type-dependent modulation of human immunodeficiency virus type 1 capsid disassembly by cyclophilin A. J. Virol. 83, 10951–10962.

Li, Y.L., Chandrasekaran, V., Carter, S.D., Woodward, C.L., Christensen, D.E., Dryden, K.A., Pornillos, O., Yeager, M., Ganser-Pornillos, B.K., Jensen, G.J., Sundquist, W.I., 2016. Primate TRIM5 proteins form hexagonal nets on HIV-1 capsids. Elife 5.

Liao, C.H., Kuang, Y.Q., Liu, H.L., Zheng, Y.T., Su, B., 2007. A novel fusion gene, TRIM5-Cyclophilin A in the pig-tailed macaque determines its susceptibility to HIV-1 infection. AIDS 21 (Suppl. 8), S19–S26.

Liddament, M.T., Brown, W.L., Schumacher, A.J., Harris, R.S., 2004. APOBEC3F properties and hypermutation preferences indicate activity against HIV-1 in vivo. Curr. Biol. 14, 1385–1391.

Lienlaf, M., Hayashi, F., Di Nunzio, F., Tochio, N., Kigawa, T., Yokoyama, S., Diaz-Griffero, F., 2011. Contribution of E3-ubiquitin ligase activity to HIV-1 restriction by TRIM5alpha(rh): structure of the RING domain of TRIM5alpha. J. Virol. 85, 8725–8737.

Liu, C., Perilla, J.R., Ning, J., Lu, M., Hou, G., Ramalho, R., Himes, B.A., Zhao, G., Bedwell, G.J., Byeon, I.J., Ahn, J., Gronenborn, A.M., Prevelige, P.E., Rousso, I., Aiken, C., Polenova, T., Schulten, K., Zhang, P., 2016. Cyclophilin A stabilizes the HIV-1 capsid through a novel non-canonical binding site. Nat. Commun. 7, 10714.

Liu, Z., Pan, Q., Ding, S., Qian, J., Xu, F., Zhou, J., Cen, S., Guo, F., Liang, C., 2013. The interferon-inducible MxB protein inhibits HIV-1 infection. Cell Host Microbe 14, 398–410.

Liu, Z., Pan, Q., Liang, Z., Qiao, W., Cen, S., Liang, C., 2015. The highly polymorphic cyclophilin A-binding loop in HIV-1 capsid modulates viral resistance to MxB. Retrovirology 12, 1.

Lu, C., Luo, Z., Jager, S., Krogan, N.J., Peterlin, B.M., 2012. Moloney leukemia virus type 10 inhibits reverse transcription and retrotransposition of intracisternal a particles. J. Virol. 86, 10517–10523.

Lu, T.C., He, J.C., Wang, Z.H., Feng, X., Fukumi-Tominaga, T., Chen, N., Xu, J., Iyengar, R., Klotman, P.E., 2008. HIV-1 Nef disrupts the podocyte actin cytoskeleton by interacting with diaphanous interacting protein. J. Biol. Chem. 283, 8173–8182.

Luban, J., 2007. Cyclophilin A, TRIM5, and resistance to human immunodeficiency virus type 1 infection. J. Virol. 81, 1054–1061.

Luban, J., 2016. Closing the net on retroviruses. Elife 5, e18243.

Luban, J., Bossolt, K.L., Franke, E.K., Kalpana, G.V., Goff, S.P., 1993. Human immunodeficiency virus type 1 Gag protein binds to cyclophilins A and B. Cell 73, 1067–1078.

Lukic, Z., Campbell, E.M., 2012. The cell biology of TRIM5alpha. Curr. HIV AIDS Rep. 9, 73–80.

Lukic, Z., Dharan, A., Fricke, T., Diaz-Griffero, F., Campbell, E.M., 2014. HIV-1 uncoating is facilitated by dynein and kinesin 1. J. Virol. 88, 13613–13625.

Lukic, Z., Goff, S.P., Campbell, E.M., Arriagada, G., 2013. Role of SUMO-1 and SUMO interacting motifs in rhesus TRIM5alpha-mediated restriction. Retrovirology 10, 10.

Luo, T., Douglas, J.L., Livingston, R.L., Garcia, J.V., 1998. Infectivity enhancement by HIV-1 Nef is dependent on the pathway of virus entry: implications for HIV-based gene transfer systems. Virology 241, 224–233.

Maelfait, J., Bridgeman, A., Benlahrech, A., Cursi, C., Rehwinkel, J., 2016. Restriction by SAMHD1 limits cGAS/STING-dependent innate and adaptive immune responses to HIV-1. Cell Rep. 16, 1492–1501.

Malfavon-Borja, R., Wu, L.I., Emerman, M., Malik, H.S., 2013. Birth, decay, and reconstruction of an ancient TRIMCyp gene fusion in primate genomes. Proc. Natl. Acad. Sci. U.S.A. 110, E583–E592.

Malikov, V., da Silva, E.S., Jovasevic, V., Bennett, G., de Souza Aranha Vieira, D.A., Schulte, B., Diaz-Griffero, F., Walsh, D., Naghavi, M.H., 2015. HIV-1 capsids bind and exploit the kinesin-1 adaptor FEZ1 for inward movement to the nucleus. Nat. Commun. 6, 6660.

Mamede, J.I., Hope, T.J., 2016. Detection and tracking of dual-labeled HIV particles using wide-field live cell imaging to follow viral core integrity. Methods Mol. Biol. 1354, 49–59.

Mangeat, B., Turelli, P., Caron, G., Friedli, M., Perrin, L., Trono, D., 2003. Broad antiretroviral defence by human APOBEC3G through lethal editing of nascent reverse transcripts. Nature 424, 99–103.

Mangeat, B., Turelli, P., Liao, S., Trono, D., 2004. A single amino acid determinant governs the species-specific sensitivity of APOBEC3G to Vif action. J. Biol. Chem. 279, 14481–14483.

Marchant, D., Neil, S.J., Aubin, K., Schmitz, C., Mcknight, A., 2005. An envelope-determined, pH-independent endocytic route of viral entry determines the susceptibility of human immunodeficiency virus type 1 (HIV-1) and HIV-2 to Lv2 restriction. J. Virol. 79, 9410–9418.

Mariani, R., Chen, D., Schrofelbauer, B., Navarro, F., Koenig, R., Bollman, B., Muenk, C., Nymark-Mcmahon, H., Landau, N.R., 2003. Species-specific exclusion of AOPBEC3G from HIV-1 virions by Vif. Cell 114, 21–31.

Marin, M., Rose, K.M., Kozak, S.L., Kabat, D., 2003. HIV-1 Vif protein binds the editing enzyme APOBEC3G and induces its degradation. Nat. Med. 9, 1398–1403.

Marno, K.M., O'Sullivan, E., Jones, C.E., Diaz-Delfin, J., Pardieu, C., Sloan, R.D., Mcknight, A., 2017. RNA-associated early-stage antiviral factor is a major component of Lv2 restriction. J. Virol. 91.

Marsh, M., Helenius, A., 2006. Virus entry: open sesame. Cell 124, 729–740.

Martinez, N.W., Xue, X., Berro, R.G., Kreitzer, G., Resh, M.D., 2008. Kinesin KIF4 regulates intracellular trafficking and stability of the human immunodeficiency virus type 1 Gag polyprotein. J. Virol. 82, 9937–9950.

Marx, A., Alian, A., 2015a. Rerouting resistance: escaping restriction using alternative cellular pathways. Trends Microbiol. 23, 595–597.

Marx, A., Alian, A., 2015b. The road less traveled: HIV's use of alternative routes through cellular pathways. J. Virol. 89, 5204–5212.

Matreyek, K.A., Wang, W., Serrao, E., Singh, P.K., Levin, H.L., Engelman, A., 2014. Host and viral determinants for MxB restriction of HIV-1 infection. Retrovirology 11, 90.

McDonald, D., Vodicka, M.A., Lucero, G., Svitkina, T.M., Borisy, G.G., Emerman, M., Hope, T.J., 2002. Visualization of the intracellular behavior of HIV in living cells. J. Cell Biol. 159, 441–452.

Miller, M.D., Farnet, C.M., Bushman, F.D., 1997. Human immunodeficiency virus type 1 preintegration complexes: studies of organization and composition. J. Virol. 71, 5382–5390.

Mische, C.C., Javanbakht, H., Song, B., Diaz-Griffero, F., Stremlau, M., Strack, B., Si, Z., Sodroski, J., 2005. Retroviral restriction factor TRIM5alpha is a trimer. J. Virol. 79, 14446–14450.

Miyashita, M., Oshiumi, H., Matsumoto, M., Seya, T., 2011. DDX60, a DEXD/H box helicase, is a novel antiviral factor promoting RIG-I-like receptor-mediated signaling. Mol. Cell. Biol. 31, 3802–3819.

Mu, D., Yang, H., Zhu, J.W., Liu, F.L., Tian, R.R., Zheng, H.Y., Han, J.B., Shi, P., Zheng, Y.T., 2014. Independent birth of a novel TRIMCyp in *Tupaia belangeri* with a divergent function from its paralog TRIM5. Mol. Biol. Evol. 31, 2985–2997.

Munk, C., Brandt, S.M., Lucero, G., Landau, N.R., 2002. A dominant block to HIV-1 replication at reverse transcription in simian cells. Proc. Natl. Acad. Sci. U.S.A. 99, 13843–13848.

Naghavi, M.H., Goff, S.P., 2007. Retroviral proteins that interact with the host cell cytoskeleton. Curr. Opin. Immunol. 19, 402–407.

Naghavi, M.H., Hatziioannou, T., Gao, G., Goff, S.P., 2005. Overexpression of fasciculation and elongation protein zeta-1 (FEZ1) induces a post-entry block to retroviruses in cultured cells. Genes Dev. 19, 1105–1115.

Naghavi, M.H., Valente, S., Hatziioannou, T., de Los Santos, K., Wen, Y., Mott, C., Gundersen, G.G., Goff, S.P., 2007. Moesin regulates stable microtubule formation and limits retroviral infection in cultured cells. EMBO J. 26, 41–52.

Nepveu-Traversy, M.E., Demogines, A., Fricke, T., Plourde, M.B., Riopel, K., Veillette, M., Diaz-Griffero, F., Sawyer, S.L., Berthoux, L., 2016. A putative SUMO interacting motif in the B30.2/SPRY domain of rhesus macaque TRIM5alpha important for NF-kappaB/AP-1 signaling and HIV-1 restriction. Heliyon 2, e00056.

Newman, E.N., Holmes, R.K., Craig, H.M., Klein, K.C., Lingappa, J.R., Malim, M.H., Sheehy, A.M., 2005. Antiviral function of APOBEC3G can be dissociated from cytidine deaminase activity. Curr. Biol. 15, 166–170.

Newman, R.M., Hall, L., Kirmaier, A., Pozzi, L.A., Pery, E., Farzan, M., O'Neil, S.P., Johnson, W., 2008. Evolution of a TRIM5-CypA splice isoform in old world monkeys. PLoS Pathog. 4, e1000003.

Nisole, S., Lynch, C., Stoye, J.P., Yap, M.W., 2004. A Trim5-cyclophilin A fusion protein found in owl monkey kidney cells can restrict HIV-1. Proc. Natl. Acad. Sci. U.S.A. 101, 13324–13328.

Nisole, S., Saib, A., 2004. Early steps of retrovirus replicative cycle. Retrovirology 1, 9.

Ohagen, A., Gabuzda, D., 2000. Role of Vif in stability of the human immunodeficiency virus type 1 core. J. Virol. 74, 11055–11066.

Oliveira, N.M., Trikha, R., Mcknight, A., 2010. A novel envelope mediated post entry restriction of murine leukaemia virus in human cells is Ref1/TRIM5alpha independent. Retrovirology 7, 81.

Opazo, T., Garces, A., Tapia, D., Barraza, F., Bravo, A., Schwenke, T., Cancino, J., Arriagada, G., 2017. Functional evidence of the involvement of the dynein light chain DYNLRB2 in murine leukemia virus infection. J. Virol. 91, e00129–17

Ou, C.-Y., Boone, L.R., Koh, C.K., Tennant, R.W., Yang, W.K., 1983. Nucleotide sequences of gag-pol regions that determine the Fv-1 host range property of BALB/c N-tropic and B-tropic murine leukemia viruses. J. Virol. 48, 779–784.

Pan, X., Baldauf, H.M., Keppler, O.T., Fackler, O.T., 2013. Restrictions to HIV-1 replication in resting CD4+ T lymphocytes. Cell Res. 23, 876–885.

Pawlica, P., Berthoux, L., 2014. Cytoplasmic dynein promotes HIV-1 uncoating. Viruses 6, 4195–4211.

Pawlica, P., Dufour, C., Berthoux, L., 2015. Inhibition of microtubules and dynein rescues human immunodeficiency virus type 1 from owl monkey TRIMCyp-mediated restriction in a cellular context-specific fashion. J. Gen. Virol. 96, 874–886.

Pawlica, P., Le Sage, V., Poccardi, N., Tremblay, M.J., Mouland, A.J., Berthoux, L., 2014. Functional evidence for the involvement of microtubules and dynein motor complexes in TRIM5alpha-mediated restriction of retroviruses. J. Virol. 88, 5661–5676.

Pearce-Pratt, R., Malamud, D., Phillips, D.M., 1994. Role of the cytoskeleton in cell-to-cell transmission of human immunodeficiency virus. J. Virol. 68, 2898–2905.

Perez-Caballero, D., Hatziioannou, T., Zhang, F., Cowan, S., Bieniasz, P.D., 2005. Restriction of human immunodeficiency virus type 1 by TRIM-CypA occurs with rapid kinetics and independently of cytoplasmic bodies, ubiquitin, and proteasome activity. J. Virol. 79, 15567–15572.

Permanyer, M., Pauls, E., Badia, R., Este, J.A., Ballana, E., 2013. The cortical actin determines different susceptibility of naive and memory CD4+ T cells to HIV-1 cell-to-cell transmission and infection. PLoS One 8, e79221.

Perron, M.J., Stremlau, M., Lee, M., Javanbakht, H., Song, B., Sodroski, J., 2007. The human TRIM5alpha restriction factor mediates accelerated uncoating of the N-tropic murine leukemia virus capsid. J. Virol. 81, 2138–2148.

Perron, M.J., Stremlau, M., Song, B., Ulm, W., Mulligan, R.C., Sodroski, J., 2004. TRIM5alpha mediates the postentry block to N-tropic murine leukemia viruses in human cells. Proc. Natl. Acad. Sci. U.S.A. 101, 11827–11832.

Pertel, T., Hausmann, S., Morger, D., Zuger, S., Guerra, J., Lascano, J., Reinhard, C., Santoni, F.A., Uchil, P.D., Chatel, L., Bisiaux, A., Albert, M.L., Strambio-De-Castillia, C., Mothes, W., Pizzato, M., Grutter, M.G., Luban, J., 2011. TRIM5 is an innate immune sensor for the retrovirus capsid lattice. Nature 472, 361–365.

Petit, C., Giron, M.L., Tobaly-Tapiero, J., Bittoun, P., Real, E., Jacob, Y., Tordo, N., De The, H., Saib, A., 2003. Targeting of incoming retroviral Gag to the centrosome involves a direct interaction with the dynein light chain 8. J. Cell Sci. 116, 3433–3442.

Pincus, T., Hartley, J.W., Rowe, W.P., 1975. A major genetic locus affecting resistance to infection with murine leukemia viruses. IV. Dose-response relationships in *Fv-1* sensitive and resistant cell cultures. Virology 65, 333–342.

Portilho, D.M., Fernandez, J., Ringeard, M., Machado, A.K., Boulay, A., Mayer, M., Muller-Trutwin, M., Beignon, A.S., Kirchhoff, F., Nisole, S., Arhel, N.J., 2016. Endogenous TRIM5alpha function is regulated by SUMOylation and nuclear sequestration for efficient innate sensing in dendritic cells. Cell Rep. 14, 355–369.

Powell, R.D., Holland, P.J., Hollis, T., Perrino, F.W., 2011. Aicardi-Goutieres syndrome gene and HIV-1 restriction factor SAMHD1 is a dGTP-regulated deoxynucleotide triphosphohydrolase. J. Biol. Chem. 286, 43596–43600.

Price, A.J., Jacques, D.A., Mcewan, W.A., Fletcher, A.J., Essig, S., Chin, J.W., Halambage, U.D., Aiken, C., James, L.C., 2014. Host cofactors and pharmacologic ligands share an essential interface in HIV-1 capsid that is lost upon disassembly. PLoS Pathog. 10, e1004459.

Qi, M., Yang, R., Aiken, C., 2008. Cyclophilin A-dependent restriction of human immunodeficiency virus type 1 capsid mutants for infection of nondividing cells. J. Virol. 82, 12001–12008.

Radtke, K., Dohner, K., Sodeik, B., 2006. Viral interactions with the cytoskeleton: a hitchhiker's guide to the cell. Cell. Microbiol. 8, 387–400.

Ramalho, R., Rankovic, S., Zhou, J., Aiken, C., Rousso, I., 2016. Analysis of the mechanical properties of wild type and hyperstable mutants of the HIV-1 capsid. Retrovirology 13, 17.

Rasheedi, S., Shun, M.C., Serrao, E., Sowd, G.A., Qian, J., Hao, C., Dasgupta, T., Engelman, A.N., Skowronski, J., 2016. The cleavage and polyadenylation specificity factor 6 (CPSF6) subunit of the capsid-recruited pre-messenger RNA cleavage factor I (CFIm) complex mediates HIV-1 integration into genes. J. Biol. Chem. 291, 11809–11819.

Rein, A., Kashmiri, S.V.S., Bassin, R.H., Gerwin, B.I., Duran-Troise, G., 1976. Phenotypic mixing between N- and B-tropic murine leukemia viruses: infectious particles with dual sensitivity to Fv-1 restriction. Cell 7, 373–379.

Reuter, S., Kaumanns, P., Buschhorn, S.B., Dittmar, M.T., 2005. Role of HIV-2 envelope in Lv2-mediated restriction. Virology 332, 347–358.

Risco, C., Menendez-Arias, L., Copeland, T.D., Pinto da Silva, P., Oroszlan, S., 1995. Intracellular transport of the murine leukemia virus during acute infection of NIH 3T3 cells: nuclear import of nucleocapsid protein and integrase. J. Cell Sci. 108, 3039–3050.

Roa, A., Hayashi, F., Yang, Y., Lienlaf, M., Zhou, J., Shi, J., Watanabe, S., Kigawa, T., Yokoyama, S., Aiken, C., Diaz-Griffero, F., 2012. RING domain mutations uncouple TRIM5alpha restriction of HIV-1 from inhibition of reverse transcription and acceleration of uncoating. J. Virol. 86, 1717–1727.

Rommelaere, J., Donis-Keller, H., Hopkins, N., 1979. RNA sequencing provides evidence for allelism of determinants of the N-, B- or NB-tropism of murine leukemia viruses. Cell 16, 43–50.

Rose, K.M., Marin, M., Kozak, S.L., Kabat, D., 2005. Regulated production and anti-HIV type 1 activities of cytidine deaminases APOBEC3B, 3F, and 3G. AIDS Res. Hum. Retroviruses 21, 611–619.

Rosenwirth, B., Billich, A., Datema, R., Donatsch, P., Hammerschmid, F., Harrison, R., Hiestand, P., Jaksche, H., Mayer, P., Peichl, P., et al., 1994. Inhibition of human immunodeficiency virus type 1 replication by SDZ NIM 811, a nonimmunosuppressive cyclosporine analog. Antimicrob. Agents Chemother. 38, 1763–1772.

Rothenberg, E., Baltimore, D., 1975. Synthesis of long, representative DNA copies of the murine RNA tumor virus genome. J. Virol. 17, 168–174.

Rothenberg, E., Baltimore, D., 1977. Increased length of DNA made by virions of murine leukemia virus at limiting magnesium ion concentration. J. Virol. 21, 168–178.

Rothenberg, E., Smotkin, D., Baltimore, D., Weinberg, R.A., 1977. In vitro synthesis of infectious DNA of murine leukaemia virus. Nature 269, 122–126.

Rowe, W.P., Hartley, J.W., 1972. Studies of genetic transmission of murine leukemia virus by AKR mice. II. Crosses with Fv-1 b strains of mice. J. Exp. Med. 136, 1286–1301.

Rowe, W.P., Humphrey, J.B., Lilly, F., 1973. A major genetic locus affecting resistance to infection with murine leukemia viruses. 3. Assignment of the Fv-1 locus to linkage group 8 of the mouse. J. Exp. Med. 137, 850–853.

Ryoo, J., Choi, J., Oh, C., Kim, S., Seo, M., Kim, S.Y., Seo, D., Kim, J., White, T.E., Brandariz-Nunez, A., Diaz-Griffero, F., Yun, C.H., Hollenbaugh, J.A., Kim, B., Baek, D., Ahn, K., 2014. The ribonuclease activity of SAMHD1 is required for HIV-1 restriction. Nat. Med. 20, 936–941.

Sabo, Y., Walsh, D., Barry, D.S., Tinaztepe, S., de Los Santos, K., Goff, S.P., Gundersen, G.G., Naghavi, M.H., 2013. HIV-1 induces the formation of stable microtubules to enhance early infection. Cell Host Microbe 14, 535–546.

Saib, A., Puvion-Dutilleul, F., Schmid, M., Peries, J., De The, H., 1997. Nuclear targeting of incoming human foamy virus Gag proteins involves a centriolar step. J. Virol. 71, 1155–1161.

Saito, A., Ferhadian, D., Sowd, G.A., Serrao, E., Shi, J., Halambage, U.D., Teng, S., Soto, J., Siddiqui, M.A., Engelman, A.N., Aiken, C., Yamashita, M., 2016. Roles of capsid-interacting host factors in multimodal inhibition of HIV-1 by PF74. J. Virol. 90, 5808–5823.

Sakuma, R., Mael, A.A., Ikeda, Y., 2007. Alpha interferon enhances TRIM5alpha-mediated antiviral activities in human and rhesus monkey cells. J. Virol. 81, 10201–10206.

Sanchez, J.G., Okreglicka, K., Chandrasekaran, V., Welker, J.M., Sundquist, W.I., Pornillos, O., 2014. The tripartite motif coiled-coil is an elongated antiparallel hairpin dimer. Proc. Natl. Acad. Sci. U.S.A. 111, 2494–2499.

Sastri, J., Johnsen, L., Smolin, N., Imam, S., Mukherjee, S., Lukic, Z., Brandariz-Nunez, A., Robia, S.L., Diaz-Griffero, F., Wiethoff, C., Campbell, E.M., 2014. Restriction of HIV-1 by rhesus TRIM5alpha is governed by alpha helices in the Linker2 region. J. Virol. 88, 8911–8923.

Sastri, J., O'Connor, C., Danielson, C.M., Mcraven, M., Perez, P., Diaz-Griffero, F., Campbell, E.M., 2010. Identification of residues within the L2 region of rhesus TRIM5alpha that are required for retroviral restriction and cytoplasmic body localization. Virology 405, 259–266.

Sawyer, S.L., Wu, L.I., Emerman, M., Malik, H.S., 2005. Positive selection of primate TRIM5{alpha} identifies a critical species-specific retroviral restriction domain. Proc. Natl. Acad. Sci. U.S.A 102, 2832–2837

Sayah, D.M., Sokolskaja, E., Berthoux, L., Luban, J., 2004. Cyclophilin A retrotransposition into TRIM5 explains owl monkey resistance to HIV-1. Nature 430, 569–573.

Schaeffer, E., Geleziunas, R., Greene, W.C., 2001. Human immunodeficiency virus type 1 Nef functions at the level of virus entry by enhancing cytoplasmic delivery of virions. J. Virol. 75, 2993–3000.

Schaller, T., Ocwieja, K.E., Rasaiyaah, J., Price, A.J., Brady, T.L., Roth, S.L., Hue, S., Fletcher, A.J., Lee, K., Kewalramani, V.N., Noursadeghi, M., Jenner, R.G., James, L.C., Bushman, F.D., Towers, G.J., 2011. HIV-1 capsid-cyclophilin interactions determine nuclear import pathway, integration targeting and replication efficiency. PLoS Pathog. 7, e1002439.

Schmitz, C., Marchant, D., Neil, S.J., Aubin, K., Reuter, S., Dittmar, M.T., Mcknight, A., 2004. Lv2, a novel postentry restriction, is mediated by both capsid and envelope. J. Virol. 78, 2006–2016.

Schneider, W.M., Brzezinski, J.D., Aiyer, S., Malani, N., Gyuricza, M., Bushman, F.D., Roth, M.J., 2013. Viral DNA tethering domains complement replication-defective mutations in the p12 protein of MuLV Gag. Proc. Natl. Acad. Sci. U.S.A. 110, 9487–9492.

Schoggins, J.W., Wilson, S.J., Panis, M., Murphy, M.Y., Jones, C.T., Bieniasz, P., Rice, C.M., 2011. A diverse range of gene products are effectors of the type I interferon antiviral response. Nature 472, 481–485.

Schulte, B., Buffone, C., Opp, S., Di Nunzio, F., de Souza Aranha Vieira, D.A., Brandariz-Nunez, A., Diaz-Griffero, F., 2015. Restriction of HIV-1 requires the N-Terminal region of MxB as a capsid-binding motif but not as a nuclear localization signal. J. Virol. 89, 8599–8610.

Sebastian, S., Luban, J., 2005. TRIM5alpha selectively binds a restriction-sensitive retroviral capsid. Retrovirology 2, 40.

Serquina, A.K., Das, S.R., Popova, E., Ojelabi, O.A., Roy, C.K., Gottlinger, H.G., 2013. UPF1 is crucial for the infectivity of human immunodeficiency virus type 1 progeny virions. J. Virol. 87, 8853–8861.

Shah, V.B., Aiken, C., 2014. Gene expression analysis of a panel of cell lines that differentially restrict HIV-1 CA mutants infection in a cyclophilin a-dependent manner. PLoS One 9, e92724.

Shah, V.B., Shi, J., Hout, D.R., Oztop, I., Krishnan, L., Ahn, J., Shotwell, M.S., Engelman, A., Aiken, C., 2013. The host proteins transportin SR2/TNPO3 and cyclophilin A exert opposing effects on HIV-1 uncoating. J. Virol. 87, 422–432.

Sharma, S., Patnaik, S.K., Taggart, R.T., Baysal, B.E., 2016. The double-domain cytidine deaminase APOBEC3G is a cellular site-specific RNA editing enzyme. Sci. Rep. 6, 39100.

Sheehy, A.M., Gaddis, N.C., Choi, J.D., Malim, M.H., 2002. Isolation of a human gene that inhibits HIV-1 infection and is suppressed by the viral Vif protein. Nature 418, 646–650.

Sheehy, A.M., Gaddis, N.C., Malim, M.H., 2003. The antiretroviral enzyme APOBEC3G is degraded by the proteasome in response to HIV-1 Vif. Nat. Med. 9, 1404–1407.

Shi, J., Aiken, C., 2006. Saturation of TRIM5 alpha-mediated restriction of HIV-1 infection depends on the stability of the incoming viral capsid. Virology 350, 493–500.

Shi, J., Friedman, D.B., Aiken, C., 2013. Retrovirus restriction by TRIM5 proteins requires recognition of only a small fraction of viral capsid subunits. J. Virol. 87, 9271–9278.

Shi, J., Zhou, J., Halambage, U.D., Shah, V.B., Burse, M.J., Wu, H., Blair, W.S., Butler, S.L., Aiken, C., 2015. Compensatory substitutions in the HIV-1 capsid reduce the fitness cost associated with resistance to a capsid-targeting small-molecule inhibitor. J. Virol. 89, 208–219.

Shi, J., Zhou, J., Shah, V.B., Aiken, C., Whitby, K., 2011. Small-molecule inhibition of human immunodeficiency virus type 1 infection by virus capsid destabilization. J. Virol. 85, 542–549.

Shoya, Y., Tokunaga, K., Sawa, H., Maeda, M., Ueno, T., Yoshikawa, T., Hasegawa, H., Sata, T., Kurata, T., Hall, W.W., Cullen, B.R., Takahashi, H., 2003. Human topoisomerase I promotes HIV-1 proviral DNA synthesis: implications for the species specificity and cellular tropism of HIV-1 infection. Proc. Natl. Acad. Sci. U.S.A. 100, 8442–8447.

Simon, J.H., Gaddis, N.C., Fouchier, R.A., Malim, M.H., 1998. Evidence for a newly discovered cellular anti-HIV-1 phenotype. Nat. Med. 4, 1397–1400.

Smith, A.E., Helenius, A., 2004. How viruses enter animal cells. Science 304, 237–242.

Sodeik, B., 2000. Mechanisms of viral transport in the cytoplasm. Trends Microbiol. 8, 465–472.

Sodeik, B., 2002. Unchain my heart, baby let me go–the entry and intracellular transport of HIV. J. Cell Biol. 159, 393–395.

Sokolskaja, E., Luban, J., 2006. Cyclophilin, TRIM5, and innate immunity to HIV-1. Curr. Opin. Microbiol. 9, 404–408.

Sokolskaja, E., Sayah, D.M., Luban, J., 2004. Target cell cyclophilin A modulates human immunodeficiency virus type 1 infectivity. J. Virol. 78, 12800–12808.

Solis, M., Nakhaei, P., Jalalirad, M., Lacoste, J., Douville, R., Arguello, M., Zhao, T., Laughrea, M., Wainberg, M.A., Hiscott, J., 2011. RIG-I-mediated antiviral signaling is inhibited in HIV-1 infection by a protease-mediated sequestration of RIG-I. J. Virol. 85, 1224–1236.

Song, B., Diaz-Griffero, F., Park, D.H., Rogers, T., Stremlau, M., Sodroski, J., 2005. TRIM5alpha association with cytoplasmic bodies is not required for antiretroviral activity. Virology 343, 201–211.

Sorin, M., Cano, J., Das, S., Mathew, S., Wu, X., Davies, K.P., Shi, X., Cheng, S.W., Ott, D., Kalpana, G.V., 2009. Recruitment of a SAP18-HDAC1 complex into HIV-1 virions and its requirement for viral replication. PLoS Pathog. 5, e1000463.

St Gelais, C., Wu, L., 2011. SAMHD1: a new insight into HIV-1 restriction in myeloid cells. Retrovirology 8, 55.

Stavrou, S., Nitta, T., Kotla, S., Ha, D., Nagashima, K., Rein, A.R., Fan, H., Ross, S.R., 2013. Murine leukemia virus glycosylated Gag blocks apolipoprotein B editing complex 3 and cytosolic sensor access to the reverse transcription complex. Proc. Natl. Acad. Sci. U.S.A. 110, 9078–9083.

Steinkasserer, A., Harrison, R., Billich, A., Hammerschmid, F., Werner, G., Wolff, B., Peichl, P., Palfi, G., Schnitzel, W., Mlynar, E., 1995. Mode of action of SDZ NIM 811, a nonimmunosuppressive cyclosporin A analog with activity against human immunodeficiency virus type 1 (HIV-1): interference with early and late events in HIV-1 replication. J. Virol. 69, 814–824.

Stevens, A., Bock, M., Ellis, S., Letissier, P., Bishop, K.N., Yap, M.W., Taylor, W., Stoye, J.P., 2004. Retroviral capsid determinants of Fv1 NB and NR tropism. J. Virol. 78, 9592–9598.

Stolp, B., Abraham, L., Rudolph, J.M., Fackler, O.T., 2010. Lentiviral Nef proteins utilize PAK2-mediated deregulation of cofilin as a general strategy to interfere with actin remodeling. J. Virol. 84, 3935–3948.

Stolp, B., Fackler, O.T., 2011. How HIV takes advantage of the cytoskeleton in entry and replication. Viruses 3, 293–311.

Stopak, K., de Noronha, C., Yonemoto, W., Greene, W.C., 2003. HIV-1 Vif blocks the antiviral activity of APOBEC3G by impairing both its translation and intracellular stability. Mol. Cell 12, 591–601.

Stoye, J.P., 1998. Fv1, the mouse retrovirus resistance gene. Rev. Sci. Tech. 17, 269–277.

Stremlau, M., Owens, C.M., Perron, M.J., Kiessling, M., Autissier, P., Sodroski, J., 2004. The cytoplasmic body component TRIM5alpha restricts HIV-1 infection in Old World monkeys. Nature 427, 848–853.

Stremlau, M., Perron, M., Lee, M., Li, Y., Song, B., Javanbakht, H., Diaz-Griffero, F., Anderson, D.J., Sundquist, W.I., Sodroski, J., 2006. Specific recognition and accelerated uncoating of retroviral capsids by the TRIM5alpha restriction factor. Proc. Natl. Acad. Sci. U.S.A. 103, 5514–5519.

Strunze, S., Engelke, M.F., Wang, I.H., Puntener, D., Boucke, K., Schleich, S., Way, M., Schoenenberger, P., Burckhardt, C.J., Greber, U.F., 2011. Kinesin-1-mediated capsid disassembly and disruption of the nuclear pore complex promote virus infection. Cell Host Microbe 10, 210–223.

Su, Y., Qiao, W., Guo, T., Tan, J., Li, Z., Chen, Y., Li, X., Li, Y., Zhou, J., Chen, Q., 2010. Microtubule-dependent retrograde transport of bovine immunodeficiency virus. Cell. Microbiol. 12, 1098–1107.

Sun, L., Wu, J., Du, F., Chen, X., Chen, Z.J., 2013. Cyclic GMP-AMP synthase is a cytosolic DNA sensor that activates the type I interferon pathway. Science 339, 786–791.

Takahashi, H., Matsuda, M., Kojima, A., Sata, T., Andoh, T., Kurata, T., Nagashima, K., Hall, W.W., 1995. Human immunodeficiency virus type 1 reverse transcriptase: enhancement of activity by interaction with cellular topoisomerase I. Proc. Natl. Acad. Sci. U.S.A. 92, 5694–5698.

Takeda, E., Kono, K., Hulme, A.E., Hope, T.J., Nakayama, E.E., Shioda, T., 2015. Fluorescent image analysis of HIV-1 and HIV-2 uncoating kinetics in the presence of old world monkey TRIM5alpha. PLoS One 10, e0121199.

Takeuchi, H., Kao, S., Miyagi, E., Khan, M.A., Buckler-White, A., Plishka, R., Strebel, K., 2005. Production of infectious SIVagm from human cells requires functional inactivation but not viral exclusion of human APOBEC3G. J. Biol. Chem. 280, 375–382.

Tanaka, Y., Marusawa, H., Seno, H., Matsumoto, Y., Ueda, Y., Kodama, Y., Endo, Y., Yamauchi, J., Matsumoto, T., Takaori-Kondo, A., Ikai, I., Chiba, T., 2006. Anti-viral protein APOBEC3G is induced by interferon-alpha stimulation in human hepatocytes. Biochem. Biophys. Res. Commun. 341, 314–319.

Tang, Y., Winkler, U., Freed, E.O., Torrey, T.A., Li, H., Goff, S.P., Morse, H.C., III., 1999. Cellular motor protein KIF-4 associates with retroviral Gag. J. Virol. 73, 10508–10513.

Tang, S., Murakami, T., Agresta, B.E., Campbell, S., Freed, E.O., Levin, J.G., 2001. Human immunodeficiency virus type 1 N-terminal capsid mutants that exhibit aberrant core morphology and are blocked in initiation of reverse transcription in infected cells. J. Virol. 75, 9357–9366.

Telesnitsky, A., Goff, S.P., 1997. Reverse transcriptase and the generation of retroviral DNA. In: Coffin, J.M., Hughes, S.H., Varmus, H.E. (Eds.), Retroviruses. Cold Spring Harbor Press, Cold Spring Harbor, NY.

Terreni, M., Valentini, P., Liverani, V., Gutierrez, M.I., Di Primio, C., Di Fenza, A., Tozzini, V., Allouch, A., Albanese, A., Giacca, M., Cereseto, A., 2010. GCN5-dependent acetylation of HIV-1 integrase enhances viral integration. Retrovirology 7, 18.

Thali, M., Bukovsky, A., Kondo, E., Rosenwirth, B., Walsh, C.T., Sodroski, J., Gottlinger, H.G., 1994. Functional association of cyclophilin A with HIV-1 virions. Nature 372, 363–365.

Thenin-Houssier, S., de Vera, I.M., Pedro-Rosa, L., Brady, A., Richard, A., Konnick, B., Opp, S., Buffone, C., Fuhrmann, J., Kota, S., Billack, B., Pietka-Ottlik, M., Tellinghuisen, T., Choe, H., Spicer, T., Scampavia, L., Diaz-Griffero, F., Kojetin, D.J., Valente, S.T., 2016. Ebselen, a small-molecule capsid inhibitor of HIV-1 replication. Antimicrob. Agents Chemother. 60, 2195–2208.

Thenin-Houssier, S., Valente, S.T., 2016. HIV-1 capsid inhibitors as antiretroviral agents. Curr. HIV Res. 14, 270–282.

Thompson, M.R., Sharma, S., Atianand, M., Jensen, S.B., Carpenter, S., Knipe, D.M., Fitzgerald, K.A., Kurt-Jones, E.A., 2014. Interferon gamma-inducible protein (IFI) 16 transcriptionally regulates type i interferons and other interferon-stimulated genes and controls the interferon response to both DNA and RNA viruses. J. Biol. Chem. 289, 23568–23581.

Topper, M., Luo, Y., Zhadina, M., Mohammed, K., Smith, L., Muesing, M.A., 2007. Posttranslational acetylation of the human immunodeficiency virus type 1 integrase carboxyl-terminal domain is dispensable for viral replication. J. Virol. 81, 3012–3017.

Towers, G., Bock, M., Martin, S., Takeuchi, Y., Stoye, J.P., Danos, O., 2000. A conserved mechanism of retrovirus restriction in mammals. Proc. Natl. Acad. Sci. U.S.A. 97, 12295–12299.

Towers, G., Collins, M., Takeuchi, Y., 2002. Abrogation of Ref1 retrovirus restriction in human cells. J. Virol. 76, 2548–2550.

Towers, G.J., 2007. The control of viral infection by tripartite motif proteins and cyclophilin A. Retrovirology 4, 40.

Towers, G.J., Hatziioannou, T., Cowan, S., Goff, S.P., Luban, J., Bieniasz, P.D., 2003. Cyclophilin A modulates the sensitivity of HIV-1 to host restriction factors. Nat. Med. 9, 1138–1143.

Trono, D., 1992. Partial reverse transcripts in virions from human immunodeficiency and murine leukemia viruses. J. Virol. 66, 4893–4900.

Turelli, P., Mangeat, B., Jost, S., Vianin, S., Trono, D., 2004. Inhibition of hepatitis B virus replication by APOBEC3G. Science 303, 1829.

Uchil, P.D., Quinlan, B.D., Chan, W.T., Luna, J.M., Mothes, W., 2008. TRIM E3 ligases interfere with early and late stages of the retroviral life cycle. PLoS Pathog. 4, e16.

Vajdos, F.F., Yoo, S., Houseweart, M., Sundquist, W.I., Hill, C.P., 1997. Crystal structure of cyclophilin A complexed with a binding site peptide from the HIV-1 capsid protein. Protein Sci. 6, 2297–2307.

Valle-Tenney, R., Opazo, T., Cancino, J., Goff, S.P., Arriagada, G., 2016. Dynein regulators are important for ecotropic murine leukemia virus infection. J. Virol. 90, 6896–6905.

Vermeire, J., Roesch, F., Sauter, D., Rua, R., Hotter, D., Van Nuffel, A., Vanderstraeten, H., Naessens, E., Iannucci, V., Landi, A., Witkowski, W., Baeyens, A., Kirchhoff, F., Verhasselt, B., 2016. HIV triggers a cGAS-dependent, Vpu- and Vpr-regulated type I interferon response in CD4+ T cells. Cell Rep. 17, 413–424.

Virgen, C.A., Kratovac, Z., Bieniasz, P.D., Hatziioannou, T., 2008. Independent genesis of chimeric TRIM5-cyclophilin proteins in two primate species. Proc. Natl. Acad. Sci. U.S.A. 105, 3563–3568.

von Schwedler, U., Song, J., Aiken, C., Trono, D., 1993. Vif is crucial for human immunodeficiency virus type 1 proviral DNA synthesis in infected cells. J. Virol. 67, 4945–4955.

Vorster, P.J., Guo, J., Yoder, A., Wang, W., Zheng, Y., Xu, X., Yu, D., Spear, M., Wu, Y., 2011. LIM kinase 1 modulates cortical actin and CXCR4 cycling and is activated by HIV-1 to initiate viral infection. J. Biol. Chem. 286, 12554–12564.

Wagner, J.M., Roganowicz, M.D., Skorupka, K., Alam, S.L., Christensen, D., Doss, G., Wan, Y., Frank, G.A., Ganser-Pornillos, B.K., Sundquist, W.I., Pornillos, O., 2016. Mechanism of B-box 2 domain-mediated higher-order assembly of the retroviral restriction factor TRIM5alpha. Elife 5.

Wang, F., St Gelais, C., de Silva, S., Zhang, H., Geng, Y., Shepard, C., Kim, B., Yount, J.S., Wu, L., 2016. Phosphorylation of mouse SAMHD1 regulates its restriction of human immunodeficiency virus type 1 infection, but not murine leukemia virus infection. Virology 487, 273–284.

Wang, P., Heitman, J., 2005. The cyclophilins. Genome Biol. 6, 226.

Wang, X., Han, Y., Dang, Y., Fu, W., Zhou, T., Ptak, R.G., Zheng, Y.H., 2010. Moloney leukemia virus 10 (MOV10) protein inhibits retrovirus replication. J. Biol. Chem. 285, 14346–14355.

Warren, K., Warrilow, D., Meredith, L., Harrich, D., 2009. Reverse transcriptase and cellular factors: regulators of HIV-1 reverse transcription. Viruses 1, 873–894.

Warren, K., Wei, T., Li, D., Qin, F., Warrilow, D., Lin, M.H., Sivakumaran, H., Apolloni, A., Abbott, C.M., Jones, A., Anderson, J.L., Harrich, D., 2012. Eukaryotic elongation factor 1 complex subunits are critical HIV-1 reverse transcription cofactors. Proc. Natl. Acad. Sci. U.S.A. 109, 9587–9592.

Welbourn, S., Dutta, S.M., Semmes, O.J., Strebel, K., 2013. Restriction of virus infection but not catalytic dNTPase activity is regulated by phosphorylation of SAMHD1. J. Virol. 87, 11516–11524.

Wen, Y., Eng, C.H., Schmoranzer, J., Cabrera-Poch, N., Morris, E.J., Chen, M., Wallar, B.J., Alberts, A.S., Gundersen, G.G., 2004. EB1 and APC bind to mDia to stabilize microtubules downstream of Rho and promote cell migration. Nat. Cell Biol. 6, 820–830.

Wheeler, L.A., Trifonova, R.T., Vrbanac, V., Barteneva, N.S., Liu, X., Bollman, B., Onofrey, L., Mulik, S., Ranjbar, S., Luster, A.D., Tager, A.M., Lieberman, J., 2016. TREX1 knockdown induces an interferon response to HIV that delays viral infection in humanized mice. Cell Rep. 15, 1715–1727.

Wiegand, H.L., Doehle, B.P., Bogerd, H.P., Cullen, B.R., 2004. A second human antiretroviral factor, APOBEC3F, is suppressed by the HIV-1 and HIV-2 Vif proteins. EMBO J. 23, 2451–2458.

Wilson, S.J., Webb, B.L., Ylinen, L.M., Verschoor, E., Heeney, J.L., Towers, G.J., 2008. Independent evolution of an antiviral TRIMCyp in rhesus macaques. Proc. Natl. Acad. Sci. U.S.A. 105, 3557–3562.

Wittmann, S., Behrendt, R., Eissmann, K., Volkmann, B., Thomas, D., Ebert, T., Cribier, A., Benkirane, M., Hornung, V., Bouzas, N.F., Gramberg, T., 2015. Phosphorylation of murine SAMHD1 regulates its antiretroviral activity. Retrovirology 12, 103.

Wu, J., Sun, L., Chen, X., Du, F., Shi, H., Chen, C., Chen, Z.J., 2013. Cyclic GMP-AMP is an endogenous second messenger in innate immune signaling by cytosolic DNA. Science 339, 826–830.

Wu, X., Anderson, J.L., Campbell, E.M., Joseph, A.M., Hope, T.J., 2006. Proteasome inhibitors uncouple rhesus TRIM5alpha restriction of HIV-1 reverse transcription and infection. Proc. Natl. Acad. Sci. U.S.A. 103, 7465–7470.

Xu, H., Franks, T., Gibson, G., Huber, K., Rahm, N., Strambio De Castillia, C., Luban, J., Aiken, C., Watkins, S., Sluis-Cremer, N., Ambrose, Z., 2013. Evidence for biphasic uncoating during HIV-1 infection from a novel imaging assay. Retrovirology 10, 70.

Xu, S., Ducroux, A., Ponnurangam, A., Vieyres, G., Franz, S., Musken, M., Zillinger, T., Malassa, A., Ewald, E., Hornung, V., Barchet, W., Haussler, S., Pietschmann, T., Goffinet, C., 2016. cGAS-mediated innate immunity spreads intercellularly through HIV-1 env-induced membrane fusion sites. Cell Host Microbe 20, 443–457.

Yan, J., Kaur, S., Delucia, M., Hao, C., Mehrens, J., Wang, C., Golczak, M., Palczewski, K., Gronenborn, A.M., Ahn, J., Skowronski, J., 2013. Tetramerization of SAMHD1 is required for biological activity and inhibition of HIV infection. J. Biol. Chem. 288, 10406–10417.

Yan, N., Regalado-Magdos, A.D., Stiggelbout, B., Lee-Kirsch, M.A., Lieberman, J., 2010. The cytosolic exonuclease TREX1 inhibits the innate immune response to human immunodeficiency virus type 1. Nat. Immunol. 11, 1005–1013.

Yan, Y., Buckler-White, A., Wollenberg, K., Kozak, C.A., 2009. Origin, antiviral function and evidence for positive selection of the gammaretrovirus restriction gene Fv1 in the genus Mus. Proc. Natl. Acad. Sci. U.S.A. 106, 3259–3263.

Yang, H., Ji, X., Zhao, G., Ning, J., Zhao, Q., Aiken, C., Gronenborn, A.M., Zhang, P., Xiong, Y., 2012a. Structural insight into HIV-1 capsid recognition by rhesus TRIM5alpha. Proc. Natl. Acad. Sci. U.S.A. 109, 18372–18377.

Yang, R., Shi, J., Byeon, I.J., Ahn, J., Sheehan, J.H., Meiler, J., Gronenborn, A.M., Aiken, C., 2012b. Second-site suppressors of HIV-1 capsid mutations: restoration of intracellular activities without correction of intrinsic capsid stability defects. Retrovirology 9, 30.

Yang, W.K., Yang, D.-M., Kiggans, J.O.J., 1980. Synthesis and circularization of N- and B-tropic retroviral DNA in Fv-1 permissive and restrictive mouse cells. Proc. Natl. Acad. Sci. U.S.A. 77, 2994–2998.

Yang, Y., Brandariz-Nunez, A., Fricke, T., Ivanov, D.N., Sarnak, Z., Diaz-Griffero, F., 2014a. Binding of the rhesus TRIM5alpha PRYSPRY domain to capsid is necessary but not sufficient for HIV-1 restriction. Virology 448, 217–228.

Yang, Y., Fricke, T., Diaz-Griffero, F., 2013. Inhibition of reverse transcriptase activity increases stability of the HIV-1 core. J. Virol. 87, 683–687.

Yang, Y., Luban, J., Diaz-Griffero, F., 2014b. The fate of HIV-1 capsid: a biochemical assay for HIV-1 uncoating. Methods Mol. Biol. 1087, 29–36.

Yap, M.W., Colbeck, E., Ellis, S.A., Stoye, J.P., 2014. Evolution of the retroviral restriction gene Fv1: inhibition of non-MLV retroviruses. PLoS Pathog. 10, e1003968.

Yap, M.W., Nisole, S., Lynch, C., Stoye, J.P., 2004. Trim5alpha protein restricts both HIV-1 and murine leukemia virus. Proc. Natl. Acad. Sci. U.S.A. 101, 10786–10791.

Yap, M.W., Stoye, J.P., 2003. Intracellular localisation of Fv1. Virology 307, 76–89.

Yin, L., Braaten, D., Luban, J., 1998. Human immunodeficiency virus type 1 replication is modulated by host cyclophilin A expression levels. J. Virol. 72, 6430–6436.

Yoder, A., Guo, J., Yu, D., Cui, Z., Zhang, X.E., Wu, Y., 2011. Effects of microtubule modulators on HIV-1 infection of transformed and resting CD4 T cells. J. Virol. 85, 3020–3024.

Yoder, A., Yu, D., Dong, L., Iyer, S.R., Xu, X., Kelly, J., Liu, J., Wang, W., Vorster, P.J., Agulto, L., Stephany, D.A., Cooper, J.N., Marsh, J.W., Wu, Y., 2008. HIV envelope-CXCR4 signaling activates cofilin to overcome cortical actin restriction in resting CD4 T cells. Cell 134, 782–792.

Yu, Q., Konig, R., Pillai, S., Chiles, K., Kearney, M., Palmer, S., Richman, D., Coffin, J.M., Landau, N.R., 2004. Single-strand specificity of APOBEC3G accounts for minus-strand deamination of the HIV genome. Nat. Struct. Mol. Biol. 11, 435–442.

Yu, X., Yu, Y., Liu, B., Luo, K., Kong, W., Mao, P., Yu, X.F., 2003. Induction of APOBEC3G ubiquitination and degradation by an HIV-1 Vif-Cul5-SCF complex. Science 302, 1056–1060.

Yuan, T., Yao, W., Tokunaga, K., Yang, R., Sun, B., 2016. An HIV-1 capsid binding protein TRIM11 accelerates viral uncoating. Retrovirology 13, 72.

Yudina, Z., Roa, A., Johnson, R., Biris, N., de Souza Aranha Vieira, D.A., Tsiperson, V., Reszka, N., Taylor, A.B., Hart, P.J., Demeler, B., Diaz-Griffero, F., Ivanov, D.N., 2015. RING dimerization links higher-order assembly of TRIM5alpha to synthesis of K63-Linked polyubiquitin. Cell Rep. 12, 788–797.

Yueh, A., Goff, S.P., 2003. Phosphorylated serine residues and an arginine-rich domain of the moloney murine leukemia virus p12 protein are required for early events of viral infection. J. Virol. 77, 1820–1829.

Yueh, A., Leung, J., Bhattacharyya, S., Perrone, L.A., de Los Santos, K., Pu, S.Y., Goff, S.P., 2006. Interaction of moloney murine leukemia virus capsid with Ubc9 and PIASy mediates SUMO-1 addition required early in infection. J. Virol. 80, 342–352.

Yufenyuy, E.L., Aiken, C., 2013. The NTD-CTD intersubunit interface plays a critical role in assembly and stabilization of the HIV-1 capsid. Retrovirology 10, 29.

Zack, J.A., Arrigo, S.J., Weitsman, S.R., Go, A.S., Haislip, A., Chen, I.S., 1990. HIV-1 entry into quiescent primary lymphocytes: molecular analysis reveals a labile, latent viral structure. Cell 61, 213–222.

Zhang, H., Dornadula, G., Orenstein, J., Pomerantz, R.J., 2000. Morphologic changes in human immunodeficiency virus type 1 virions secondary to intravirion reverse transcription: evidence indicating that reverse transcription may not take place within the intact viral core. J. Hum. Virol. 3, 165–172.

Zhang, H., Yang, B., Pomerantz, R.J., Zhang, C., Arunachalam, S.C., Gao, L., 2003. The cytidine deaminase CEM15 induces hypermutation in newly synthesized HIV-1 DNA. Nature 424, 94–98.

Zhao, G., Ke, D., Vu, T., Ahn, J., Shah, V.B., Yang, R., Aiken, C., Charlton, L.M., Gronenborn, A.M., Zhang, P., 2011. Rhesus TRIM5alpha disrupts the HIV-1 capsid at the inter-hexamer interfaces. PLoS Pathog. 7, e1002009.

Zheng, Y.H., Irwin, D., Kurosu, T., Tokunaga, K., Sata, T., Peterlin, B.M., 2004. Human APOBEC3F is another host factor that blocks human immunodeficiency virus type 1 replication. J. Virol. 78, 6073–6076.

Zhou, J., Price, A.J., Halambage, U.D., James, L.C., Aiken, C., 2015. HIV-1 resistance to the capsid-targeting inhibitor PF74 results in altered dependence on host factors required for virus nuclear entry. J. Virol. 89, 9068–9079.

Zhu, C., Gao, W., Zhao, K., Qin, X., Zhang, Y., Peng, X., Zhang, L., Dong, Y., Zhang, W., Li, P., Wei, W., Gong, Y., Yu, X.F., 2013. Structural insight into dGTP-dependent activation of tetrameric SAMHD1 deoxynucleoside triphosphate triphosphohydrolase. Nat. Commun. 4, 2722.

Zhu, J., Cunningham, J.M., 1993. Minus-strand DNA is present within murine type C ecotropic retroviruses prior to infection. J. Virol. 67, 2385–2388.

Chapter 3

Nucleoporins in Retroviral Replication: What's Nup Got to Do with It?

Anne Monette, Andrew J. Mouland
Lady Davis Institute at the Jewish General Hospital and McGill University, Montréal, QC, Canada

Chapter Outline

Nuclear Pore Complexes, Nucleoporins, and Nucleocytoplasmic Transport 114
Nucleoporins Implicated in Retroviral Replication 115
Normal Cellular Functions of the Nup358/RanBP2 Nuclear Pore Complex Protein 124
Functions of Nu358/RanBP2 in Retroviral Replication 125
Normal Cellular Functions of the Nup214/CAN Nuclear Pore Complex Protein 126
Functions of Nup214/CAN in Retroviral Replication 126
Normal Cellular Functions of the Nup98 Nuclear Pore Complex Protein 127
Functions of Nup98 in Retroviral Replication 128
Normal Cellular Functions of the Nup85 and Nup160 Components of the Nup107–160 Nuclear Pore Complex 129
Functions of the Nup107–160 Complex (Y-Complex) during Retroviral Replication 131
Normal Cellular Functions of the Nup62 Nuclear Pore Complex Protein 132
Functions of Nup62 in Retroviral Replication 133
Normal Cellular Functions of the Nup155 Nuclear Pore Complex Protein 134
Functions of Nup155 in Retroviral Replication 134
Normal Cellular Functions of the Nup153 Nuclear Pore Complex Protein 134
Functions of Nup153 in Retroviral Replication 135
Normal Cellular Functions of the Tetratricopeptide Repeat Nuclear Pore Complex Protein 137
Functions of Tetratricopeptide Repeat in Retroviral Replication 137

Retrovirus-Cell Interactions. https://doi.org/10.1016/B978-0-12-811185-7.00003-0

Caution: Depletion of Any Single Multifunctional Nup Can Modify Overall NPC Architecture 138
The Importance of FG-Repeat Nups to Disease 139
Viruses Commonly Cause the Displacement of FG–Nups to Remodel NPCs 142
Acknowledgments 143
References 143

NUCLEAR PORE COMPLEXES, NUCLEOPORINS, AND NUCLEOCYTOPLASMIC TRANSPORT

Every aspect of eukaryotic cellular metabolism, from gene expression to cell growth and division, is under control of continuous nucleocytoplasmic transport of micro- and macromolecules between the nucleus and cytoplasm and is made possible by nuclear pore complexes (NPCs) studding the nuclear envelope (NE) separating these two main cellular environments (Beck and Hurt, 2017). NPCs are large macromolecular complexes spanning NEs by fusing with the outer and inner nuclear membranes (INMs). These are well conserved throughout eukaryotic evolution (DeGrasse et al., 2009; Cronshaw et al., 2002; Rout et al., 2000) and represent refined, flexible, gate keepers that form a selective channel for molecular transport (Shahin, 2016).

Human NPCs are among the largest and longest lived protein complex in the cell (Savas et al., 2012; Kabachinski and Schwartz, 2015), consisting of a molecular mass of 125 MDa, and are 120 nm in spoke diameter by 70 nm in overall height (Reichelt et al., 1990). NPCs are composed of approximately 30 unique proteins termed nucleoporins (Nups) (Knockenhauer and Schwartz, 2016), whose nomenclature generally follow their molecular masses (i.e., Nup62 is a Nup of 62 kDa) (Hoelz et al., 2011), and which are present in at multiple copies of 8–16 (Mi et al., 2015), for a total estimation of 500–1000 Nups per NPC (Peters, 2005; Sorokin et al., 2007). Multiple copies of Nups assemble into approximately 10 stable subcomplexes to form NPCs (Stoffler et al., 1999; Suntharalingam and Wente, 2003), providing them an eightfold rotational symmetry and separating them into their principal structures, including (from cytoplasm to nucleus): the cytoplasmic filaments, which are peripheral structures deriving from the cytoplasmic and nuclear rings; the cytoplasmic and nuclear rings, which are anchored by the inner pore ring; the inner pore ring, which is fused to the outer and INMs; and the nuclear basket, which is anchored to the inner pore ring (Beck and Hurt, 2017). Indeed, certain Nups, such as Nup50, Nup62, Nup98, and Nup153, do not reside constitutively within the NPC but instead shuttle through the nuclear pore in both directions, dynamically associating with other nuclear and cytoplasmic structures (Nofrini et al., 2016). Interactions among most members of Nup subcomplexes are believed to be preserved during mitosis (Belgareh et al., 2001; Matsuoka et al., 1999; Dultz et al., 2008) and initiate NPC formation thereafter. A key feature of Nups that directly interact with cargo transiting through NPCs is their possessing of numerous (i.e., 5–50) flexible phenylalanine–glycine (FG)-repeat motifs that are thought to populate the central channel of NPCs and which are responsible for the maintenance of its selective permeability barrier (Wente and Rout, 2010). While there is no selection for the nucleocytoplasmic transport through NPCs for molecules smaller than 10 nm in diameter (Paine et al., 1975; Keminer and Peters, 1999), larger macromolecules, measuring up to 39 nm (Pante and Kann, 2002), are required to contain an FG-repeat (i.e., F, phenylalanine; G, glycine) interacting passport for their selective transport, termed a nuclear localization signal (NLS) or nuclear export signals (NES), usually found on specific carrier molecules, collectively known as transport receptors or karyopherins, and which can be further classified as importins, transportins, or exportins (Chook and Suel, 2011; Strambio-De-Castillia et al., 2010; Wente and Rout, 2010).

Aside from its role in nucleocytoplasmic transport, the NPC has functional importance to the cytoskeleton and during mitosis, and has also been documented to serve several other dynamic cellular functions (Chatel and Fahrenkrog, 2012), including spatial organization of molecular complexes at the nuclear periphery, chromatin organization (Light and Brickner, 2013), and transcriptional regulation (Ptak et al., 2014; Pascual-Garcia and Capelson, 2014). The NPC and its constituent Nups become polyfunctional as a result of both their spatial localizations and their posttranslational modifications, including phosphorylation, O- and N-glycosylation, ubiquitylation, acetylation, and sumoylation (Beck and Glavy, 2014).

From their essential and multifunctional nature, it is no surprise that the interaction with Nups has also been well demonstrated to be a requirement toward the successful replication of numerous viruses, including retroviruses, at numerous steps of viral replication, including docking, import, integration site selection and integration, transcription, and nuclear export (Table 3.1) (for reviews on Nups and viruses see Lusic and Siliciano, 2017; Campbell and Hope, 2015; Labokha and Fassati, 2013; Flather and Semler, 2015; Gustin, 2003; Walker and Ghildyal, 2017; Flatt and Greber, 2015; Fontoura et al., 2005; Ding et al., 2010; Cohen et al., 2011; Woodward and Chow, 2010; Le Sage and Mouland, 2013; Di Nunzio, 2013; Matreyek and Engelman, 2013; Yarbrough et al., 2014; Fay and Pante, 2015a,b; Bin Hamid et al., 2016; Yamauchi and Greber, 2016). In this chapter, we will summarize the natural functions and known structures of Nups associated with retroviruses, detail their mechanism of function during key stages of retroviral replication, and highlight the functional similarities of Nups in the replication of other viruses (Table 3.2).

NUCLEOPORINS IMPLICATED IN RETROVIRAL REPLICATION

The key steps in retroviral replication can be categorized into the main processes: binding, trafficking to nucleus, docking, import, integration, transcription, export, translation, cytoplasmic egress, encapsidation, morphogenesis, budding, and release. Because viruses are small packages of genetic material

TABLE 3.1 Nucleoporin Function Classified According to Human Immunodeficiency Virus Type 1 Replication

Stage	Retrovirus	Nucleoporin*	References
Docking	HIV-1	Nup358/RanBP2	Di Nunzio et al. (2012)
		CG1	Le Rouzic et al. (2002)
		Nup107	Brass et al. (2008)
		Nup133	Brass et al. (2008)
		Nup153	Bhattacharya et al. (2014), Brass et al. (2008), Burdick et al. (2013), Chen et al. (2016), De Iaco and Luban (2011), Di Nunzio et al. (2012), Di Nunzio et al. (2013), Konig et al. (2008), Lee et al. (2010), Matreyek and Engelman (2011), Matreyek et al. (2013), Price et al. (2014), Saito et al. (2016), Swaminathan et al. (2012), Zhou et al. (2015) and Lelek et al. (2015)
Import	HIV-1	Nup155	Brass et al. (2008), Lee et al. (2010) and Zhou et al. (2008)
		Nup160	Brass et al. (2008) and Lee et al. (2010)
		Nup214/CAN	Konig et al. (2008)
		Nup358/RanBP2	Bichel et al. (2013), Brass et al. (2008), Dharan et al. (2016), Konig et al. (2008), Lin et al. (2013), Meehan et al. (2014), Price et al. (2012), Rasaiyaah et al. (2013), Schaller et al. (2011), Walde et al. (2012), Zhang et al. (2010) and Zhou et al. (2015)
		Nup85	Brass et al. (2008) and Lee et al. (2010)
		Pom121	Saito et al. (2017)
		Sec13	Zhu et al. (2014)
	EIAV	Nup153	Matreyek et al. (2013)
	SIV	Nup153	Matreyek et al. (2013)
	FIV	Nup155	Lee et al. (2010)
		Nup160	Lee et al. (2010)

TABLE 3.1 Nucleoporin Function Classified According to Human Immunodeficiency Virus Type 1 Replication—cont'd

Stage	Retrovirus	Nucleoporin*	References
Integration	HIV-1	Nup153	Di Nunzio et al. (2013), Koh et al. (2013), Marini et al. (2015), Schaller et al. (2011) and Woodward et al. (2009)
		Nup358/RanBP2	Ocwieja et al. (2011)
		Nup62	Ao et al. (2012)
		Nup98	Konig et al. (2008) and Di Nunzio et al. (2012, 2013)
	HTLV	Pom121	Saito et al. (2017)
		Nup62	Tsuji et al. (2007)
Transcription	HIV-1	Nup153	Marini et al. (2015)
		Nup62	Marini et al. (2015)
		Nup98	Marini et al. (2015)
		Tpr	Lelek et al. (2015) and Marini et al. (2015)
Export	HIV-1	Nup153	Hofmann et al. (2001) and Di Nunzio et al. (2012)
		Nup214/CAN	Bogerd et al. (1998), Di Nunzio et al. (2012), Hofmann et al. (2001), Hutten and Kehlenbach (2006), Mertz et al. (2009) and Zolotukhin and Felber (1999)
		Nup62	Ajamian et al. (2015), Hofmann et al. (2001) and Monette et al. (2011)
		Nup98	Hofmann et al. (2001), Takeda et al. (2010) and Zolotukhin and Felber (1999)
		Tpr	Rajanala and Nandicoori (2012)
	HTLV	Nup214/CAN	Bogerd et al. (1998) and Mertz et al. (2009)
	RSV	Nup214/CAN	Scheifele et al. (2005)
		Nup98	Scheifele et al. (2005)

* Nup highlighted colors in table correspond to Nup subcomplex colours presented in Figure 3.1.

and a minimal set of proteins, they have evolved so that all steps of viral replication largely depend on their hijacking of host protein functions. Following infection, retroviruses must penetrate nuclei, integrate into host chromosomes, be transcribed, exported back to the cytoplasm, and translated prior to being repackaged into new infectious virions for release. Due to their dependency on a nuclear existence, reports have identified numerous Nups assisting retroviral replication processes.

The main Nups that have been repeatedly identified as being required toward HIV-1 and retroviral replication are Nup358/RanBP2, Nup214/CAN,

TABLE 3.2 Nucleoporins and Other Viruses

Function	Species	Abbreviations	Nucleoporin	References
Docking	Herpes simplex virus	HSV	Nup214/CAN	Copeland et al. (2009) and Pasdeloup et al. (2009)
			Nup358/RanBP2	Copeland et al. (2009)
Import	Epstein–Barr virus	EBV	Nup107	Li et al. (2015)
			Nup133	
			Nup188	
			Nup210	
			Nup35	
			Nup50	
			Nup98	
			Pom121	
			Tpr	
	Hepatitis B virus	HBV	Nup153	Chang et al. (2012, 2015), Li et al. (2015), Fang et al. (2012), Gallucci and Kann (2017), Schmitz et al. (2010), and Onder and Moroianu (2014)
	Hepatitis D virus	HDV		
	Human papillomavirus type 8	HPV8		
	Human papillomavirus type 11	HPV11	Nup62	Onder and Moroianu (2014), McKee et al. (2013), Eberhard et al. (2013), Mire et al. (2010), and Chang et al. (2012, 2015)
	Human papillomavirus type 16	HPV16		
	Vesicular stomatitis virus	VSIV		
	Epstein–Barr virus	EBV		

	Adenovirus type 5	Ad5	Nup214/CAN	Li et al. (2015), Cassany et al. (2015), Strunze et al. (2011), and Trotman et al. (2001)
	Adenovirus type 2	Ad2		
	Epstein–Barr virus	EBV	Nup358/RanBP2	Strunze et al. (2011) and Li et al. (2015)
Transcription	Adenovirus type 5	Ad5	Nup205	Lu et al. (2014)
	Sindbis virus	SINV	Nup98	Panda et al. (2015), Enninga et al. (2002), and Rajani et al. (2012)
	Vesicular stomatitis virus	VSIV		
Export	Herpes simplex virus	HSV	Nup62	Malik et al. (2012) and Morita et al. (2013)
	Influenza A virus	IVA	Nup98	Chen et al. (2010)
	Vesicular stomatitis virus	VSIV	Nup98	Quan et al. (2014)
Replication and assembly	Hepatitis C virus	HCV	Nup107	Neufeldt et al. (2013)
			Nup153	
			Nup155	
			Nup88 Neufeldt et al. (2013)	
	Enterovirus 71	EV71	Nup214/CAN	Wang et al. (2013)
	Hepatitis A virus	HAV	Nup98	Neufeldt et al. (2013)
	Dengue virus	DENV		
	Hepatitis C virus	HCV		
	Vaccinia virus	VACV	Nup358/RanBP2	Neufeldt et al. (2013) and Khuperkar et al. (2017)
	Influenza A virus	IVA	Nup62	Khuperkar et al. (2017), Morita et al. (2013), and Munier et al. (2013)

Continued

TABLE 3.2 **Nucleoporins and Other Viruses—cont'd**

Function	Species	Abbreviations	Nucleoporin	References
Morphogenesis and budding	Hendra virus	HeV	Nup107	Pentecost et al. (2015)
	Newcastle disease virus	NDV		
	Nipah virus	NiV		
	Sendai virus	SeV		
	Hendra virus	HeV	Nup133	Pentecost et al. (2015)
	Newcastle disease virus	NDV		
	Nipah virus	NiV		
	Sendai virus	SeV		
	Hendra virus	HeV	Nup155	Pentecost et al. (2015)
	Newcastle disease virus	NDV		
	Nipah virus	NiV		
	Hendra virus	HeV	Nup160	Pentecost et al. (2015)
	Newcastle disease virus	NDV		
	Nipah virus	NiV		
	Sendai virus	SeV		
	Newcastle disease virus	NDV	Nup188	Pentecost et al. (2015)
	Newcastle disease virus	NDV	Nup205	Pentecost et al. (2015)
	Nipah virus	NiV		

	Newcastle disease virus	NDV	Nup358/RanBP2	Pentecost et al. (2015)
	Hendra virus	HeV	Nup37	Pentecost et al. (2015)
	Newcastle disease virus	NDV		
	Nipah virus	NiV		
	Hendra virus	HeV	Nup93	Pentecost et al. (2015)
	Newcastle disease virus	NDV		
	Nipah virus	NiV		
	Sendai virus	SeV		
	Newcastle disease virus	NDV	Nup62	Pentecost et al. (2015) and Sivan et al. (2013)
	Vaccinia virus	VACV		
	Hepatitis C virus	HCV	Nup98	Lussignol et al. (2016)
Host defense evasion via phosphorylation	Encephalomyocarditis virus	ECMV	Nup153	Porter and Palmenberg (2009)
			Nup62	Bardina et al. (2009), Basta et al. (2014), Ciomperlik et al. (2016), and Porter and Palmenberg (2009)
			Nup98	Porter and Palmenberg (2009)

Continued

TABLE 3.2 Nucleoporins and Other Viruses—cont'd

Function	Species	Abbreviations	Nucleoporin	References
Host defense evasion via proteolysis	Rhinovirus	RHV	Nup214/CAN	Ghildyal et al. (2009)
			Nup358/RanBP2	
	Poliovirus	PV	Nup153	Fitzgerald et al. (2013), Gustin and Sarnow (2001), Ghildyal et al. (2009), Walker et al. (2013, 2015), Watters and Palmenberg (2011), and Lidsky et al. (2006)
	Rhinovirus	RHV		
	Coxsackievirus B3	CB3		
	Poliovirus	PV	Nup62	Lidsky et al. (2006), Fitzgerald et al. (2013), Gustin and Sarnow (2001), Park et al. (2010), Walker et al. (2013, 2015), and Watters and Palmenberg (2011)
	Rhinovirus	RHV		
	Poliovirus	PV	Nup98	Walker et al. (2013, 2015), Watters and Palmenberg (2011), Fitzgerald et al. (2013), and Park et al. (2015)
	Enterovirus	EV		
Downregulated	Herpes simplex virus 1	HSV-1	Nup153	Leuzinger et al. (2005)
Relocalized	Herpes simplex virus 1	HSV-1	Nup153	Leuzinger et al. (2005)
Infection	Herpes simplex virus 1	HSV-1	Nup358/RanBP2	Hofemeister and O'Hare (2008)
Modified	Japanese encephalitis virus	JEV	Nup358/RanBP2	Zhang et al. (2013)
Unknown	Respiratory syncytial virus	RSV	Nup96	Munday et al. (2010)
Unknown	Respiratory syncytial virus	RSV	Nup98	Munday et al. (2010)
Unknown	Avian Reovirus	ARV	Tpr	Huang et al. (2015)

Tpr, tetratricopeptide repeat.

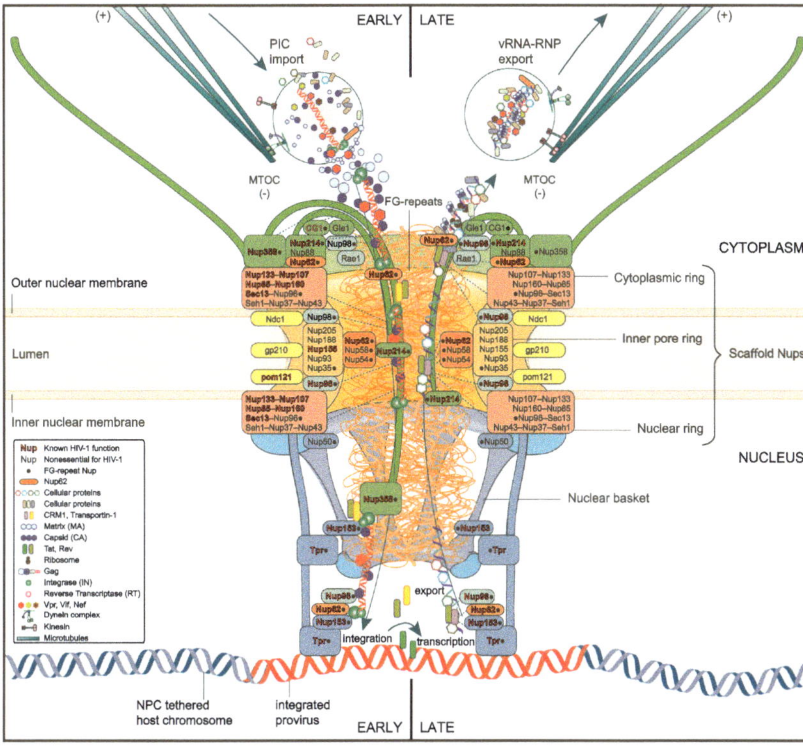

FIGURE 3.1 HIV-1 interaction with the nuclear pore complex (NPC) toward its early and late replication stages. Two-dimensional mirrored location of nucleoporins (Nups) within the overall architecture of the human NPC to illustrate Nups involved in either early (left) or late (right) stages of HIV-1 replication. Nups are organized into subcomplexes displayed as *colored boxes* superimposed onto the general NPC structure. Y-complexes are shown in rose; the inner ring complex in beige; the transmembrane NUPs in yellow; the NUP62 complex in tangerine; cytoplasmic complexes in green; and nuclear basket complexes in blue. Phenylalanine–glycine repeats of Nups filling the central channel are displayed as spaghetti-like filaments and are colored beige and orange. The ability to reach through the pore to the opposite sides from which they are anchored has been graphically displayed for Nup358 and Nup214 to illustrate how they may be involved in both cytoplasmic and nuclear stages of HIV-1 replication. Nup62, Nup98, and Nup153 have been graphically displaced from mapped positions to illustrate where they are acting in the HIV-1 replication cycle. Nup62 color is kept tangerine orange despite it also being a part of the Nup214/Nup88 subcomplex. Names of HIV-1 dependency factors identified from screens are red in early stages of HIV-1 replication because NPCs lacking these would have modified architecture and would also have fewer pores in general, thereby potentially crippling the nuclear import of HIV-1. Dashed lines identify known NPC–HIV-1 and NPC–karyopherin interactions. See Table 3.1 for supporting references. *GC1*, candidate gene 1; *gp210*, glycoprotein 210; myosin-like protein; *Ndc1*, nuclear division cycle 1; *pom*, pore membrane protein; *Tpr*, translocated promoter region. Additional interactions between Nups, karyopherins, and HIV-1 proteins are illustrated by close contact or by *black dotted lines* and referenced in Yu et al. (2012), Neville et al. (1997), Di Nunzio (2013), Woodward et al. (2009), Kinoshita et al. (2012), Schaller et al. (2011) and Matreyek and Engelman (2011). *(Image modified from Beck, M., Hurt, E., 2017. The nuclear pore complex: understanding its function through structural insight. Nat. Rev. Mol. Cell Biol. 18, 73–89 and Mouland, A.J., Milev, M.P., 2012. Role of dynein in viral pathogenesis. In: King, S.M. (Ed.), Dyneins: Structure, Biology and Disease. Elsevier.)*

Nup98, Nup62, Nup153, and tetratricopeptide repeat (Tpr) (Fig. 3.1). Other Nups that have been identified through in HIV-1 host dependency factor screens, or where little is known mechanistically, include Nup133, Nup107, Nup155, Nup160, Nup85, Pom121, and Sec13. The main Nups repeatedly identified for HIV-1 and other retroviruses are the same as those that have been linked to tumorigenesis (Simon and Rout, 2014), implicating that these are those Nups most studied or that these may represent core Nups implicated in a broader set of human pathologies than is currently appreciated (i.e., immune diseases, neurological and neurodegenerative diseases, cardiovascular disorders, inherited diseases, dementias, and cancers) (Nofrini et al., 2016; Cronshaw and Matunis, 2004; Sheffield et al., 2006; Sakuma and D'Angelo, 2017; Chou et al., 2018). Examination of the crystal structures and detailed functional interactions of these important Nups may lead to the development of precision medicines that can target these Nups to alleviate various diseases including HIV-1 infection. The following sections examine the individual functions of Nups in normal, nondisease settings and then provide an overview on how these are co-opted by retroviral replication.

Normal Cellular Functions of the Nup358/RanBP2 Nuclear Pore Complex Protein

The giant Nup, Nup358 (also known as Ran-binding protein 2 or RanBP2), is the largest NUP (358 KDa) at 36 nm in length (Delphin et al., 1997; Yokoyama et al., 1995) and is a major component of the eight rod-shaped filaments emanating from the cytoplasmic ring of the NPC (Wu et al., 1995; Goldberg and Allen, 1996). Nup358 proteins possess several distinct protein domains (Wu et al., 1995) required for its numerous roles: an N-terminal Tpr domain, FG-repeats, an alpha-helical region, Ran-binding domains, zinc fingers, a SUMO E3 ligase domain, and a C-terminal domain (CTD) sharing sequence homology to cyclophilins (Lin et al., 2013; Hutten et al., 2008).

Localization of Nup385 to the outermost region of the NPC depends on its N-terminal domain (NTD) anchoring to the Nup133 and Nup96 components of the Y-complex (Hamada et al., 2011), constituting the scaffolding rings of the NPC core (Kosinski et al., 2016), and on its association with Nup214 and Nup88 (Bernad et al., 2004). As the first and last nucleocytoplasmic cargo interactor, Nup358 is the preliminary gatekeeper of the nucleus, and many of its other dynamic and diverse functions include mediation with microtubules, nuclear import interaction with transcription factors, nuclear export, and RNA processing, mitosis, and SUMOylation (Pichler et al., 2002; Chatel and Fahrenkrog, 2012; Goldberg, 2017; Xylourgidis and Fornerod, 2009).

Nup358 is important for the translocation of molecules through the NPC (Hutten et al., 2008; Singh et al., 1999; Walde et al., 2012; Forler et al., 2004; Mahadevan et al., 2013). Nup358 mediates import and export across the NPC by interacting with karyopherins via its FG-repeats (Tran and Wente, 2006), which are features of many of the core Nups permitting selective translocation of nuclear or cytoplasmic cargo (Weis, 2007), and which are required for mRNA export (Terry and Wente, 2007). Nup358 forms a complex with RanGAP1, the GTPase-activating protein for Ran, to facilitate the hydrolysis of GTP by Ran, resulting in export complex disassembly and the establishment of a RanGTP gradient across the NE required for the continuation of the majority of nucleocytoplasmic transport pathways including those that are CRMI-mediated (Hutten et al., 2008; Bernad et al., 2004). Crystal structures of Nup358 have been resolved, including its CTD (Lin et al., 2013), uncomplexed second Ran-binding domain (Geyer et al., 2005), NTD (Kassube et al., 2012), and cyclophilin domain in complex with the N-terminal capsid (CA) of HIV-1 (Bichel et al., 2013).

Functions of Nu358/RanBP2 in Retroviral Replication

In 2008, Nup358 was one of the proteins found from siRNA screens identifying host factors essential for HIV-1 replication (Brass et al., 2008; Konig et al., 2008). Functional studies demonstrated that HIV-1 relied on Nup358 for nuclear import, with the HIV-1 capsid (CA) binding directly to the cyclophilin domain of Nup358 (Lee et al., 2010; Schaller et al., 2011; Zhang et al., 2010; Rasaiyaah et al., 2013). It was later determined that the more precise mechanism was Nup358-mediated docking of HIV-1 cores to NPCs (Di Nunzio et al., 2012), which turns out to be a concerted process involving other host nuclear import factors RanBP2, Nup153, TNPO3, and CPSF6. This process could be abrogated by a spontaneously arising HIV-1 CA mutation N74D to result in a loss of viral dependence on these host cofactors (Price et al., 2012).

Nup358's role in HIV-1 replication has been hotly debated, as it has been suggested by further comprehensive functional studies that the important role of Nup358 in HIV-1's replication cycle may depend on its FG-repeats or on its critical housekeeping role of maintaining the Ran-GTP cycle, NLS-mediated cargo import, and host cell physiology (Meehan et al., 2014). Nup358 has also been suggested to mediate viral uncoating prior to nuclear import of HIV-1 preintegration complexes (PICs) (Bichel et al., 2013), and it has also been implicated in chromosomal integration site selection of gene dense regions favoring HIV-1 replication (Ocwieja et al., 2011) (Fig. 3.1). From later studies, it appears that Nup358 is displaced from NPCs during HIV-1 infection, where it associates with viral cores during their anterograde trafficking to the nucleus by kinesin motor proteins (Dharan et al., 2016). The NTD of Nup358 has also been demonstrated to be responsible for the transportin-dependent import of Rev (Hutten et al., 2009; Walde et al., 2012). Therefore, Nup358 plays essential roles in both early and late stages of HIV-1 replication.

Normal Cellular Functions of the Nup214/CAN Nuclear Pore Complex Protein

Nup214 is another peripheral, fibrillous, FG-repeat Nup that is localized to the cytoplasmic face of the NPC but has been proposed to form shorter, more central fibers, and be more closely associated with the NPC cytoplasmic ring than is Nup358 (Kraemer et al., 1994; Xu and Powers, 2009). Nup214 is anchored to the cytoplasmic face ring structure of the NPC by its N-terminus, whereas its C-terminal repeat domain has been observed to reach through the center of the pore, and possibly as far as the nuclear basket (Paulillo et al., 2005). This extended reach may be one possible explanation why Nup214 is found in complex with Nup62 (Macaulay et al., 1995; Stochaj et al., 2006), which could also be due to cell type–associated differences in NPC composition (Kinoshita et al., 2012).

Nup214 is predominantly found in a subcomplex with Nup88 (Bastos et al., 1997; Fornerod et al., 1997), and this Nup214/Nup88 subcomplex has been found to interact with Nup358, Nup98, and Nup62 (Xu and Powers, 2009). Persistence of this subcomplex at NPCs is Nup214-dependent, as its depletion results in a codepletion of Nup88 (Hutten and Kehlenbach, 2006; Bernad et al., 2004). The anchoring N-terminus of Nup214 has been predicted to form a β-propeller structure, similar to its yeast ortholog, Nup159 (Weirich et al., 2004). This structure is followed by a stretchrich in proline and serine residues, two predicted coiled-coil structures, and a structurally uncharacterized region (Xu and Powers, 2009). Nup214 contains C-terminus FG- and FxFG-repeats having high affinity for the nuclear protein export receptor, CRM1/exportin-1 (Fornerod et al., 1997; Roth et al., 2003), and together with Nup88, is required for CRM1-mediated export (Bernad et al., 2006; Hutten and Kehlenbach, 2006), where their function has been proposed to be in the attenuation of nuclear export rather than in its enhancement (Xylourgidis et al., 2006). However, the absence of Nup214/Nup88 and Nup358 in NPCs results in defective nuclear export (Hutten and Kehlenbach, 2006) while maintaining fully functional nuclear import.

Crystal structures of Nup214 have been resolved, including its NTD (Napetschnig et al., 2007), the NTD in complex with Ddx19 in its adenosine diphosphate–bound state (Napetschnig et al., 2009), the NTD in complex with Dbp5 bound to RNA and AMP-PNP (von Moeller et al., 2009), and the FG-repeat-containing fragment in its quaternary nuclear export complex with CRM1, snurportin 1, and RanGTP (Port et al., 2015; Monecke et al., 2015).

Functions of Nup214/CAN in Retroviral Replication

One of the HIV-1 screens in 2008 also uncovered Nup214 as an HIV-1 essential host factor, where it was suspected to act to accelerate PIC translocation (Konig et al., 2008), but with the caveat that RNAi depletion could alter the localization of other proteins required for HIV-1 nuclear import. Accordingly, all other reports of Nup214 and HIV-1 have validated its functional role toward HIV-1 export, starting as early as 1998, where Nup214 mutants were shown to block nuclear export of the HIV-1 Rev-CRM1 complex, as well as the analogous human T-lymphotropic virus type I (HTLV-1) Rex-CRM1 complex (Bogerd et al., 1998); in addition, this mechanism was also demonstrated to involve Nup98 (Zolotukhin and Felber, 1999) (Fig. 3.1). This nuclear export complex consisting of Nups 214 and 98 was expanded to contain additional NUPs and other cellular factors, including Nup153, Nup62, eIF-5A, and actin, but not Nup358, by use of mab414 (Hofmann et al., 2001). The requirement for Nup214 in CRM1-mediated nuclear protein export was later confirmed in vivo (Hutten and Kehlenbach, 2006), where Nup214 was also demonstrated to serve as a terminal-binding site for nuclear export complexes. The requirement of Nup214 for export was to be essential for both Rem and Rex, viral homologs of Rev from mouse mammary tumor virus and HTLV, respectively (Bogerd et al., 1998; Mertz et al., 2009). Nup214 has furthermore been implicated as an important cofactor for Rous sarcoma virus (RSV) Gag protein nuclear export (Scheifele et al., 2005). Finally, as laboratory techniques became more refined, in a study comparing many Nups and various HIV-1 replication stages, Nup214 was again validated to exert its functions in HIV-1 mRNA nuclear export (Di Nunzio et al., 2012). Identification of Nup214 as an essential factor by an HIV-1 screen designed to reveal early cellular factors required for HIV-1 replication (Konig et al., 2008) serves as an instructive example for (1) how the careful design of such screens is critical to avoid the identification of false positives and negatives that can only be validated through functional assays, and (2) how caution should also be used in the interpretation of the effects from depletion of any single Nup, many of which are responsible for the overall stability of NPCs.

Normal Cellular Functions of the Nup98 Nuclear Pore Complex Protein

Nup98 is another peripheral FG-repeat-containing Nup that localizes to both cytoplasmic and nuclear faces of the central channel of the NPC and also within the nucleus itself (Rout et al., 2000; Griffis et al., 2002, 2003), where its localization has been linked to active transcription (Griffis et al., 2004). Nup98 is translated as a single precursor, the Nup98–Nup96 polypeptide, which is then autocleaved to generate the two proteins Nup98 and Nup96, and the interaction of C-terminal of Nup98 and N-terminal of Nup96 is maintained thereafter (Hodel et al., 2002). Even after assembly into the NPC, on the nuclear face of the pore, Nup98 continues to bind directly to Nup96, which is a component of the Nup133 subcomplex (Fontoura et al., 1999; Vasu et al., 2001). On the cytoplasmic side of NPCs, Nup98 binds to Nup88 of the NUP214/Nup88/Nup62 subcomplex only found on the cytoplasmic face of pores (Bastos et al., 1997; Fornerod et al., 1997; Griffis et al., 2003).

Nup98 is distinct from the other eukaryotic FG Nups as it bears a unique FG-repeat signature of Gly-Leu-Phe-Gly (GLFG) (Powers et al., 1995; Radu et al., 1995b), which has been determined to be substantially more cohesive and resistant to chemicals than regular FG-repeats (Xu and Powers, 2009). Similarly, GLFG domains are possessed by yeast Nups whose distributions are not restricted to a single face of the NPC (Rout et al., 2000). The GLFG-repeat region of Nup98 is essential for its participation in nuclear transport (Radu et al., 1995b; Powers et al., 1997), as it is the site where import cargo carried by transportin/karyopherin β2 is released from the transport complex (Fontoura et al., 2000). The FG-repeat of Nup98 has also been demonstrated to be the region bound by several other importin-β family proteins, as reviewed in Iwamoto et al. (2010), strongly implicating Nup98 in nuclear import. In addition to import, several studies have also identified a role for Nup98 in mRNA nuclear export, where its GLFG-repeat region harbors regions implicated in direct binding to mRNA transport factors Rae1 (Brown et al., 1995; Bharathi et al., 1997; Kraemer and Blobel, 1997) and TAP/Mex67 (Strawn et al., 2001; Blevins et al., 2003). Nup98 has roles in the regulation of gene expression (Wang et al., 2009; Xu and Powers, 2009; Kalverda et al., 2010; Capelson et al., 2010), which are also powered by its GLFG-repeat, creating nuclear Nup98 "GLFG bodies" (Griffis et al., 2002) associated with transcriptional coactivators (Kasper et al., 1999). Finally, over 20 genes, many of which are DNA-binding proteins, have been reported as fusion partners of the GLFG-repeat of Nup98 in leukemic patients, further implicating this important region in the regulation of transcriptional activation (Nakamura, 2005; Xu and Powers, 2009).

Crystal structures of Nup98 have been resolved, including its C-terminal autoproteolytic domain (Sun and Guo, 2008), its C-terminal autoproteolytic domain in complex with the NTD of Nup82 and the C-terminal tail fragment of Nup159 (Stuwe et al., 2012), the Gle2-binding sequence motif in complex with Rae1 (Ren et al., 2010) and in complex with Rae1 and the vesicular stomatitis virus matrix protein (Quan et al., 2014).

Functions of Nup98 in Retroviral Replication

One of the HIV-1 screens in 2008 also uncovered Nup98 as an HIV-1 essential host factor, but in contrast to other Nups identified, e.g., Nup358 and Nup153, Nup98 was concluded to be required for viral integration rather than PIC translocation (Konig et al., 2008). Indeed, the earlier studies of Nup98 function for HIV-1 had found it to directly participate in Rev-CRM1-mediated nuclear export (Zolotukhin and Felber, 1999), in complex with other factors Nup153, Nup62, eIF-5A, and actin (Hofmann et al., 2001). Nup98 fusions were found to dysregulate transcription and inhibition of all CRM1-mediated nuclear export with aberrant nuclear retention of transcriptional regulators (Takeda et al., 2010). Nup98 has also been implicated as an important cofactor for retrovirus RSV Gag protein nuclear export (Scheifele et al., 2005). Nup98 depletion or inhibition has provided evidence that it may also specifically participate in efficiency of nuclear entry of HIV-1 cDNA (Ebina et al., 2004). In other studies comparing many Nups and their roles in HIV-1 replication, however, Nup98 was not shown to play a role in import, but rather one in proviral DNA integration into the host chromosomal genome (Di Nunzio et al., 2012), despite the observation that Nup98 binds to HIV-1 CA–NC complexes (Di Nunzio et al., 2013). Nup98 has also recently been observed to be immunoprecipitated with transcriptionally active HIV-1 proviral DNA in primary CD4$^+$ T cells (Marini et al., 2015) (Fig. 3.1). With its ability to position itself at both cytoplasmic and nuclear faces of the central channel of the NPC, though debated, it is not surprising that Nup98 could be engaged in numerous stages of HIV-1 replication.

Normal Cellular Functions of the Nup85 and Nup160 Components of the Nup107–160 Nuclear Pore Complex

The conserved Nup107–160 subcomplex is located on both sides of NPCs, forming its cytoplasmic and nucleoplasmic rings (Walther et al., 2003a; Bui et al., 2013) and is referred to as the Y-complex because of its Y-shaped structure (Lutzmann et al., 2002). It is related to vesicle coats and is critical for NPC assembly (Walther et al., 2003a), where it is believed to stabilize NPC membrane curvature (Brohawn et al., 2008). Two distinct NPC assembly pathways during mitosis and interphase both require the Y-complex as an essential structural component of NPCs (D'Angelo et al., 2006; Dultz and Ellenberg, 2010; Doucet et al., 2010; Harel et al., 2003), and Nup153 is required for its assembly during interphase (Vollmer et al., 2015).

The Nup107 subcomplex is composed of nine Nups: Nup133, Nup107, Nup96, Nup160, Nup85, Nup43, Nup37, Sec13, and Seh1. This NPC subcomplex has been implicated in various cellular functions, including NPC biogenesis and organization, mRNA export, gene activation, transcriptional elongation, membrane stabilization, kinetochore function, spindle assembly, MAD1 localization, centrosome attachment, telomere anchoring and silencing, DNA repair, immunity, neural differentiation, hybrid inviability, and cold stress tolerance, as reviewed in Gonzalez-Aguilera and Askjaer (2012). Numerous members of the Y-complex have been studied crystallographically, as reviewed in Kelley et al. (2015) and Gonzalez-Aguilera and Askjaer (2012). Importantly, for the purposes of this chapter, none of the Nup107 subcomplex Nups contain FG-repeats (Table 3.1) (Denning et al., 2003). Only the Nup107 subcomplex Nups that have been identified to be required for retroviral replication (along with Nup96) will be reviewed in more detail below.

Nup133 consists of an N-terminal β-propeller and C-terminal α-solenoid-like folds (Devos et al., 2006) and is located at the end of the stalk of the Nup107 subcomplex, and by interaction with Nup107, it connects to the NE (Fernandez-Martinez et al., 2012). Nup133 is highly conserved and plays roles in both interphase and postmitotic NPC biogenesis (Doucet et al., 2010; Walther et al., 2003a), along with anchoring of the dynein/dynactin complex for centrosomes tethering to NEs at prophase (Bolhy et al., 2011). A loop in its N-terminal β-propeller contains an ArfGAP1 lipid packing sensor motif, an NE membrane curvature sensor required for proper NPC biogenesis during interphase by permitting Nup133–membrane interactions (Doucet et al., 2010; Drin et al., 2007). Mutations in Nup133 cause NPC clustering to discrete regions of NE (Doye et al., 1994), where Nup133 has been found to contribute to overall stabilization of NE membrane curvature (Fernandez-Martinez et al., 2012), presumably through a membrane-coating mechanism (Kim et al., 2014b).

Nup107 contains a leucine zipper motif in its C-terminal region and numerous kinase consensus sites (Radu et al., 1995b). Nup107 forms the center of the subcomplex stalk, linking Nup96 and Nup133; (Fig. 3.1) (Gonzalez-Aguilera and Askjaer, 2012), where Nup107 is the critical anchor for Nup133 positioning at the periphery of the NPC (Boehmer et al., 2008). Nup107 and Nup133 remain associated during mitosis, they are among the first Nups recruited during reformation of the NEs, and they are tightly attached to NPCs during interphase, only exchanged once every cell cycle (Seo et al., 2009). The depletion of Nup107 results in failure of Nup assembly at NPCs, leading to both their degradation and defective nuclear export of mRNAs (Boehmer et al., 2003), and ultimately, to cellular apoptosis (Banerjee et al., 2010).

Nup96 is centrally positioned at crossroads of the stalk and "Y" of the Nup107/Y-complex, at the intersection of Nup160, Nup85, and Sec13 (Gonzalez-Aguilera and Askjaer, 2012). It is synthesized as a 186kDa precursor, which is proteolytically cleaved to yield Nup96 and Nup98 (Fontoura et al., 1999). Making use of RNAi strategies to define its contribution to cell function have yieded unclear results, and a monoclonal antibody specific to this Nup has only recently been produced (Mizuguchi et al., 2010). Nup96 interacts with Sec13 (Enninga et al., 2003; Lutzmann et al., 2002), an NPC protein and component of ER COPII–coated vesicles (Bickford et al., 2004). Nup96 functions in mRNA export, chromatin metabolism, and defense against pathogen invasion via its requirement for proper expression of interferon-regulated responses (Dockendorff et al., 1997; Emtage et al., 1997; Galy et al., 2000; Teixeira et al., 1997; Zhang and Li, 2005; Faria et al., 2006).

Nup160 is a branch point of the Y-complex and has been found to interact with Nup85, Sec13, Nup98, and Nup153, where the abilities of Nup98 and Nup153 in tethering to nuclei or targeting pores depends on this interaction (Vasu et al., 2001). Nup160 has not been thoroughly studied in isolation from the Nup107 complex, although initial results have shown that it plays an important role in mRNA export (Vasu et al., 2001; Wiermer et al., 2012; Siniossoglou et al., 1996), but not in general protein export or import. Its depletion results in defects in NPC assembly and in SMAD proteins that are signal transducers and transcriptional modulators mediating multiple signaling pathways (Chen and Xu, 2010; Siniossoglou et al., 1996). Nup160 has also been shown to be involved in the polymerization of microtubules (Mishra et al., 2010) and disease resistance (Wiermer et al., 2012).

Nup85 is the other branch point of the Y-complex and also has not been extensively studies in isolation from the Nup107 complex. Nup85 seems to be required for NPC assembly and maintenance, and its depletion has also led to defective immune responses and reduced response to osmotic stress (Zhu et al., 2017). Disruption of Nup85 causes defects in NE and NPC organization, as well as in mRNA transport (Siniossoglou et al., 1996).

Sec13 is a member of the WD-repeat protein family (Niu et al., 2014) and is part of the Nup107 complex, where it stably interacts with Nup96 of the Nup107 complex (Lutzmann et al., 2002) and has been demonstrated to have essential functions related to NPC biogenesis (Siniossoglou et al., 1996). Sec13 also functions as a core component of cytoplasmic COPII complexes (Stagg et al., 2006), responsible for mediating protein trafficking from the ER (Sato and Nakano, 2007). Sec13 is exchanged between the intranuclear and cytoplasmic pools and is actively imported into the nucleus via its NLS (Enninga et al., 2003).

Functions of the Nup107–160 Complex (Y-Complex) during Retroviral Replication

The Nups, Nup133, Nup107, Nup160, Nup85, and Sec13 of the Nup107/Y-complex have all been implicated in the replication cycle of retroviruses. Depletions of Nup107 or Nup133 have been demonstrated to destabilize the Y-complex (Boehmer et al., 2008; Walther et al., 2003a), where importantly, in a domino effect, the depletion of keystone Nup107 prevents the assembly of a subset of Nup proteins at NPCs but has no effect on their mRNA levels (Boehmer et al., 2003). Accordingly, knockdown of Nup107 affects the assembly of Nup133, Nup358, Nup214, Nup153, and Tpr, whereas Nup96, Sec13, and Nup62 are entirely unaffected. This again indicates that caution should be used when interpreting results from the depletion of any single Nup.

In focused functional HIV-1 studies examining Nups previously identified by HIV-1 dependency factor screens that included Nup85, Nup107, Nup133, Nup153, Nup155, and Nup160 (Brass et al., 2008), Nups were depleted to examine their effects on HIV-1 resistance to CPSF6-358 through the N74D mutation of the CA altering its nuclear entry pathway (Lee et al., 2010) (Fig. 3.1). Here, neither Nup107 nor Nup133 depletions in viable cells impacted virus infection (Lee et al., 2010). By contrast, in another study Nup107 knockdown was observed to reduce p24 CA levels, where it also diminished virus production, slightly increased virus infectivity, and appeared to cause alterations to NPC localization at NEs (Monette et al., 2011). Wild-type HIV-1, the HIV-1 N74D mutant, and FIV all shared reproducible reduction of infection resulting from Nup160 depletion, whereas Nup85 depletion also diminished N74D HIV-1 infection, but to a lesser extent (Lee et al., 2010). In addition, Pom121 anchors NPCs to NEs via interaction with Nup160 and has also been shown to be required for HIV-1 replication (Saito et al., 2017).

Other Nups of the Y-complex that have been implicated in HIV-1 replication are Sec13 and Nup96, and depletion of Sec13 in a more recent HIV-1 screen was found to inhibit HIV-1 infection in primary human $CD4^+$ T cells (Zhu et al., 2014). These experiments suggested that this factor is necessary after PIC nuclear entry and prior to viral DNA integration. Nup96 has not been directly identified as an HIV-1 dependency host factor, but is another cautionary tale in the interpretation of Nup depletion studies, where effects of Nup98 depletion on HIV-1 may be due to the concomitant depletion of Nup96 because Nup98 and Nup96 are translated as a polyprotein from the same mRNA (Fontoura et al., 1999; Di Nunzio et al., 2012).

Normal Cellular Functions of the Nup62 Nuclear Pore Complex Protein

Nup62 belongs to the most central, pore facing subcomplex of the NPC, and assembled from O-glycosylated proteins Nup62, Nup58, Nup58 splice-variant Nup45, and Nup54 (Guan et al., 1995), which are connected to each other through conserved coiled-coil domains flanked by FG-repeats (Hu et al., 1996), to the NPC scaffold by direct coiled-coil interaction with inner ring Nup93 (Nic96) (Bailer et al., 2001; Grandi et al., 1995b; Schrader et al., 2008b), and also predicted to Nup53 of the inner ring (Kim et al., 2014a). Each Nup62 subcomplex member also contains a less conserved FG-repeat domain, forming filaments filling the central plug of the NPC pore (>150 FG–Nups per pore) that are essential to nuclear transport (Finlay et al., 1991; Ulrich et al., 2014) by acting as a selective barrier to passive transport and by orchestrating nucleocytoplasmic transport of larger cargo across NPCs (Mohr et al., 2009).

Nup62 is composed of three domains: an N-terminal FG-repeat, a central threonine/alanine-rich linker, and a C-terminal α-helical coiled-coil (Hu et al., 1996; Guan et al., 1995; Hurt, 1988). The FG-repeat of Nup62 serves as a docking site for nuclear transport factor 2 (NTF2), facilitating the nuclear import of the Ran GTPase (Paschal and Gerace, 1995; Clarkson et al., 1996) and providing a gradient for directionally to nucleocytoplamic transport (Ribbeck et al., 1998). The C-terminus of Nup62 is predicted to adopt a coiled-coil structure that anchors it to the NPC (Devos et al., 2006; Alber et al., 2007) and interacts with transport receptors (Percipalle et al., 1997) to mediate interactions with other members of the Nup62 complex. Nup62 has been demonstrated to bind directly to the mucin 1 C-terminal subunit (Leng et al., 2007), and to the glucocorticoid receptor–hsp90 complex (Echeverria et al., 2009) to mediate their nuclear import. Nup62 has been found to serve other functions outside of NPCs, including cell migration via interaction with cytoplasmic exocyst complex component Exo70 (Hubert et al., 2009), altering dendritic architecture in chronic stress (Kinoshita et al., 2014), transcription and chromatin organization (Liang and Hetzer, 2011; Capelson et al., 2010; Kalverda et al., 2010), and centrosome homeostasis (Hashizume et al., 2013).

Nup62 is also a part of the Nup82 complex located exclusively at the NPC cytoplasmic face, which is predominantly involved in nuclear mRNA export (Grandi et al., 1995a; Belgareh et al., 1998), and which is composed of C-terminal coiled coil–binding subunits Nup214 and Nup88 (Gaik et al., 2015), Nup98 (Bailer et al., 2000), Dbp5 and Gle1 (Noble et al., 2011), and Dyn2 (Stelter et al., 2007). Aside from transport, the Nup82 complex has also been determined to be required for inheritance of functional NPCs by daughter cells (Makio et al., 2013). Crystal structures of Nup62 have been resolved, including its coiled-coil motif (Dewangan et al., 2017), and Nup62 complex (Nup54, Nup58, and Nup62) (Chug et al., 2015; Solmaz et al., 2011; Sharma et al., 2015).

Functions of Nup62 in Retroviral Replication

Various functions of Nup62 have been assigned to retroviral replication, including integration (Ao et al., 2012), transcription (Marini et al., 2015), viral protein import (Tsuji et al., 2007), viral protein nuclear export, and vRNA export and encapsidation (Ajamian et al., 2015; Monette et al., 2011). Finding these pleomorphic effects is not unexpected due to the many functions of Nup62 and the fact that it is found in both the Nup62 and Nup88 NPC subcomplexes occupying the central plug and cytoplasmic face, respectively (Beck and Hurt, 2017) (Fig. 3.1).

For HIV-1, Nup62 has predominantly been observed to directly participate in Rev-CRM1-mediated export (Hofmann et al., 2001). Nup62 was shown to be critical for Rev-mediated unspiced viral RNA (vRNA) export, as its depletion causes nuclear retention of vRNA (Ajamian et al., 2015; Monette et al., 2011). Nup62 has also been demonstrated to interact with the HIV-1 Rev RNP complex consisting of Rev, CRM1, DDX3, and UPF1 (Ajamian et al., 2015; Hofmann et al., 2001; Monette et al., 2011). Nup62 depletion decreases virus production and infectivity, and HIV-1 infection causes Nup62 to be displaced from NPCs and packaged into new HIV-1 virions (Monette et al., 2011). In this proteomics-based study, NEs were deficient in the entire Nup62 complex (Nup62, Nup58/Nup45, and Nup54), along with the Nup62 complex NPC-anchoring protein Nup93 (Grandi et al., 1995a; Bailer et al., 2001; Monette et al., 2011). The observed encapsidation of Nup62 into budding HIV-1 virions was speculated to facilitate integrase (IN)-associated viral PIC translocation into the nucleus and/or the targeting of the PIC to chromatin (Ao et al., 2012). Nup62 was shown to be bound by chromatin and specifically interact with HIV-1 IN in the nucleus, whereby its depletion affected IN association with host chromatin and significantly impaired both viral integration and replication (Ao et al., 2012). Nup62 has since been validated to immunoprecipitate with proviral HIV-1 DNA in primary $CD4^+$ T cells (Marini et al., 2015). Finally, Nup62 has also been demonstrated to be responsible for the nuclear import of the HTLV-1 Tax protein, which directly interacts with the FG-repeat region of Nup62 (Tsuji et al., 2007).

Normal Cellular Functions of the Nup155 Nuclear Pore Complex Protein

Nup155, a member the of the inner ring subcomplex of the NPC termed the Nup93/Nic96 complex, consists of Nup205, Nup188, Nup93, and Nup53 (Vollmer and Antonin, 2014), and recruits and establishes a continuous contact with the Nup62 complex (Beck and Hurt, 2017; Chug et al., 2015; Sachdev et al., 2012; Breuer and Ohkura, 2015). Nup155 possesses an N-terminal β-propeller region (Devos et al., 2004; Lin et al., 2016; Mans et al., 2004; Seo et al., 2013) and C-terminal α-helical domain (Flemming et al., 2009; Whittle and Schwartz, 2009; Lin et al., 2016). Nup155 plays critical functions in both NE membrane fusion and NPC assembly via its interactions with membranes (Franz et al., 2005; De Magistris et al., 2018), and interacts with transmembrane Nups Ndc1 and Pom121, and with Nup160 for NPC anchoring to the pore membrane (Mitchell et al., 2010; Yavuz et al., 2010). Nup155 depletion causes alterations in NE structure, loss of targeting of membrane proteins to INMs, resulting in dramatic decreases in NPC numbers (Mitchell et al., 2010). Nup155 interacts with Nup53 and the mRNA export factor Gle1 on both cytoplasmic and nuclear sides of the NPC and is proposed to facilitate mRNA export (Rayala et al., 2004; Hawryluk-Gara et al., 2005).

Functions of Nup155 in Retroviral Replication

Two HIV-1 screens have identified Nup155 as an essential HIV-1 host factor (Brass et al., 2008; Zhou et al., 2008), and functional studies have suggested that Nup155 directly interacts with PICs derived from the N74D mutant of HIV-1 and with wild-type FIV PICs (Lee et al., 2010). However, from its normal assignment in NE biogenesis and NPC assembly, caution should be used in the interpretation of results from its depletion. In addition, Pom121, also shown to be required for HIV-1 replication (Saito et al., 2017), anchors NPCs to NEs via an interaction with Nup155 (Fig. 3.1).

Normal Cellular Functions of the Nup153 Nuclear Pore Complex Protein

Eight Nup153 proteins are estimated to exist per NPC (Cronshaw et al., 2002), localizing primarily to proximal and distal regions of the nuclear pore basket (Sukegawa and Blobel, 1993; Fahrenkrog et al., 2002). Other detailed studies demonstrated that Nup153 localizes to many parts of NPCs (Fahrenkrog et al., 2002; Krull et al., 2004; Pante et al., 1994, 2000), and its positioning appears to be dependent on active transcription, as it has been demonstrated for Nup98 (Griffis et al., 2004). Nup153 is a serine-/threonine-enriched glycoprotein, having a C-terminal region containing approximately 30 copies of degenerate FG-repeats (Denning et al., 2003; Ball and Ullman, 2005). This ~200 nm long region is expected to reach all the way through to the cytoplasmic side of the

NPC (Fahrenkrog et al., 2002), shifting in spatial distribution as a function of NPC transport dynamics (Paulillo et al., 2005; Cardarelli et al., 2012) and attributing a crucial role to Nup153 in nucleocytoplasmic transport (Bastos et al., 1996; Shah and Forbes, 1998; Ullman et al., 1999). Nup153 can be phosphorylated at interphase and hyperphosphorylated during mitosis (Bodoor et al., 1999; Favreau et al., 1996; Walther et al., 2003b) or modified with O-linked N-acetylglucosamine (Meier et al., 1995; Radu et al., 1995a), another modification proposed to counteract its phosphorylation (Hart, 1997). Nup153 harbors four C_2–C_2 zinc finger motifs in its central region, responsible for its association to the distal ring of the nuclear basket (Fahrenkrog et al., 2002). These zinc finger motifs are unique in their class, sharing highest similarity with the zinc fingers of Nup358 (Higa et al., 2007), the other Nup implicated in NE breakdown (Prunuske et al., 2006). The N-terminal region of Nup153 is unique and contains overlapping domains for its targeting and anchoring to the nuclear rim of the central NPC framework, in addition to its binding RNA (Bastos et al., 1996; Enarson et al., 1998). The N-terminal region of Nup153 is also critical for the localization of Nup50 at nuclear pores (Makise et al., 2012). Nup153 is known to be important for nuclear protein import, as well as for most RNA export pathways. This central role in nuclear transport stems from its ability to interact with virtually every karyopherin, including importin-β and -α2, Ran, transportin-1, exportin-1 and -5 (CRM1), and Kapβ3 (RanBP5); NTF2 protein transport machinery; NXF1/TAP and eIF5A of the RNA transport system; the Nup160 complex and Tpr in the NPC; and both RNA and DNA. Thus, Nup153 appears to have functional roles in protein and RNA transport, NPC/lamina interactions, membrane remodeling, transcription and translation regulation, cell cycle control, and apoptosis (Denning et al., 2003; Prunuske et al., 2006), as reviewed in Ball and Ullman (2005).

Crystal structures of the Nup153 zinc fingers in complex with Ran have been solved (Partridge and Schwartz, 2009; Schrader et al., 2008a), as their interaction may be important for both nucleocytoplasmic transport and mitosis. In addition, in a screen searching for an inhibitor of angiogenesis, a small, naturally occurring molecule named R-(-)-β-O-methylsynephrine was identified to bind directly to Nup153 to competitively inhibit mRNA binding to its RNA-binding domain (Kim et al., 2015). microRNA-155 (miR-155) has been found to exert an anti-HIV-1 effect by targeting several HIV-1 dependency factors, including Nup153 (Swaminathan et al., 2012). PF-3450074 (PF74) is a drug that binds to and inhibits Nup153 and polyadenylation specific factor 6 (CPSF6) at overlapping binding sites, and also binds to HIV-1 CA to inhibit viral replication (Bhattacharya et al., 2014; Price et al., 2014).

Functions of Nup153 in Retroviral Replication

Nup153 was the only Nup to be identified in as an HIV-1 dependency factor by all four genome-wide screens (Brass et al., 2008; Konig et al., 2008; Zhou et al., 2008; Yeung et al., 2009) and has since been the most widely studied Nup when it comes to retroviral replication. The majority of studies have centered on its role in nuclear import of viral components (Table 3.1), whereas others have also found that it has roles in integration, transcription, and nuclear export regulation.

Nup153 was first suggested to be a key factor for HIV-1 import based on an siRNA screen examining early stages of HIV-1 replication (Konig et al., 2008). Importantly, Nup153 had been previously implicated as part of the Rev-vRNA RNP nuclear export machinery (Hofmann et al., 2001). The effects of Nup153 depletion on HIV-1 infectivity were confirmed by functional studies, where it was suggested that Nup153 was interacting with CA (Lee et al., 2010) (Fig. 3.1). During this time, HIV-1 IN was also found to interact directly with Nup153, which was then suggested to permit nuclear import of the HIV-1 PIC (Woodward et al., 2009). The requirement of Nup153 for nuclear import of the PIC was then mapped to HIV-1 CA (Matreyek and Engelman, 2011) and subsequently was more precisely mapped to a direct interaction between the CA NTD and the FG-repeat enriched C-terminal domain of Nup153 (Matreyek et al., 2013). These findings were extrapolated to several additional lentiviruses including simian immunodeficiency viruses, HIV-2, and equine infectious anemia virus. In addition, Nup153 was also demonstrated as important for targeting HIV-1 DNA to preferred, transcriptionally active regions of chromosomes (Schaller et al., 2011; Koh et al., 2013). In a study comparing several Nups identified by HIV-1 screens and different stages of HIV-1 replication, Nup153 was found to be the only Nup directly participating in PIC translocation, and it was also confirmed that Nup153 is important for RNA export (Di Nunzio et al., 2012). Indeed, further functional studies confirmed that the cumulative effects of Nup153 on HIV-1 infectivity could not be explained by PIC translocation alone, and that Nup153 had to be acting on a step in replication after nuclear import (Burdick et al., 2013). Further work confirmed that Nup153 interacts with the HIV-1 core and brought additional supporting evidence that the large defect in infectivity from Nup153 depletion was due to a reduced tendency for HIV-1 to integrate in intragenic sites. The conclusion from these studies was that Nup153 was probably assisting HIV-1 exit from the nuclear basket and integration into the host chromosomal genome in concert with Nup98 (Di Nunzio et al., 2012), the other Nup found to interact with HIV-1 cores (Di Nunzio et al., 2013). Nup153 was found to participate in HIV-1 nuclear import independently of the integrity of the nuclear basket (Lelek et al., 2015). Nup153 has been found to be indispensable for integration of the provirus at the nuclear periphery in primary CD4$^+$ T cells and is required for proviral transcription (Marini et al., 2015). The finding that Nup153 appears to stabilize CA on functional PICs to protect viral genetic material during transit to the chromosome may explain its many roles at distinct stages of HIV-1 replication (Chen et al., 2016). Thus, despite being fixed to the inner ring of NPCs, this Nup can span the length of the NPC all the way to its cytoplasmic face, where its FG-repeats interact directly with HIV-1 proteins. It is therefore not a stretch that Nup153 would play many functional roles during HIV-1 replication, including import, integration, transcription, and export.

Normal Cellular Functions of the Tetratricopeptide Repeat Nuclear Pore Complex Protein

Nup Tpr (translocated promoter region) is another constituent of fibrillary NPC appendices and localizes exclusively to intranuclear filaments associated with the nucleoplasmic side of the NPC by directly binding to Nup153, and where it acts as the main architectural element of the nuclear basket (Cordes et al., 1997; Shah et al., 1998; Hase and Cordes, 2003; Krull et al., 2004). Tpr is a large, 265 kDa protein that exists as a coiled-coil-dominated homodimer of long, rodlike shape (Byrd et al., 1994; Cordes et al., 1997; Bangs et al., 1998). Its NTD forms a parallel, periodically interrupted, two-stranded coiled-coil motif required for anchoring to the NPC (Bangs et al., 1998; Cordes et al., 1998; Hase et al., 2001; Frosst et al., 2002), whereas its CTD, located deeper within the nucleus, is nonhelical, largely unfolded and flexible, and highly enriched in acidic residues (Frosst et al., 2002; Shibata et al., 2002; Hase et al., 2001; Huve et al., 2008). The C-terminal region of Tpr contains three FG motifs, which interacts with import receptor snuportin (Zhao et al., 2014), and contains many heptad repeat or leucine zipper motifs (Nakano et al., 2010), potentially for interactions with interactions with CRM1 (Frosst et al., 2002). Importantly, Tpr binds to Nup153 and is released from NPCs as a result of Nup153 depletion (Hase and Cordes, 2003).

Tpr was originally identified as an oncogenic activator through its N-terminal fusion with the met, raf, and trk proto-oncogenes (Greco et al., 1992; Park et al., 1986; Soman et al., 1991). Tpr plays functional roles in intranuclear and nucleocytoplasmic transport and also as scaffolding element in the nuclear phase of the NPC (Fontoura et al., 2001; Frosst et al., 2002; Zimowska and Paddy, 2002). Tpr has a fundamental role in chromatin organization, by excluding heterochromatin from the vicinity of NPCs and in proposed to recruit hypertranscribed genes near the NPC (Krull et al., 2010). Tpr is involved in the nuclear export of NES containing proteins (Frosst et al., 2002; Ben-Efraim et al., 2009); the regulation of export of unspliced RNA (Galy et al., 2004; Rajanala and Nandicoori, 2012); mitotic spindle checkpoint signaling (Lee et al., 2008); and the prevention of chromosome lagging during mitosis (Nakano et al., 2010).

Functions of Tetratricopeptide Repeat in Retroviral Replication

Tpr has been implicated in retroviral integration (Lelek et al., 2015), transcription (Marini et al., 2015), and nuclear export (Rajanala and Nandicoori, 2012). Tpr was first found to regulate export of intron-containing retroviral RNAs, where its depletion led to drastic enhancement of Gag expression levels (Rajanala and Nandicoori, 2012). Following this, and considering that depletion of Nup153 causes loss of

Tpr, in an elegant study elucidating the distinct and complementary involvement of Nup153 and Tpr at different steps of the HIV-1 replication cycle, the depletion of Tpr reduced HIV-1 infectivity, but not integration levels (Lelek et al., 2015). Tpr-depletion affected HIV-1's ability to become integrated in active genes, and Tpr was found to stabilize LEDGF/p75, which is responsible for promoting viral integration at active genes (Fig. 3.1). Finally, HIV-1 proviral transcription was also found to be negatively impacted by the depletion of Tpr (Marini et al., 2015).

CAUTION: DEPLETION OF ANY SINGLE MULTIFUNCTIONAL NUP CAN MODIFY OVERALL NPC ARCHITECTURE

Numerous screens and functional studies refining of the role of Nups in retroviruses have identified several Nups essential at distinct stages of their replication. What becomes evident is that several of these Nups are dynamic and multifunctional for both their normal cellular functions and their modified functions when they are hijacked during viral infection of cells.

A detailed examination of the natural function of Nups; their recruitment to sites in the NPC as well as interactions with factors in the cytoplasm and nucleus; and the manner in which they are anchored to NPCs leads to the understanding of their critical codependencies, whereby depletion of any single Nup risks modifying the overall architectural integrity of NPCs (Kim et al., 2014a). Focusing on the Nups that play essential roles in retroviral replication, such as Nup358, which is responsible for docking, import, and integration, requires Y-complex proteins Nup133 and Nup96 for its anchoring to NPCs, and its proper positioning depends on interactions with the Nup214/Nup88 subcomplex. Nup133 is located at the end of the Y-complex stalk and bridges the connection between the Y-complex via Nup107 to NPCs, so that its depletion has the potential to destabilize this entire subcomplex, and potentially the loss of NPC cytoplasmic and nucleoplasmic rings. Therefore, it makes sense that Nup133 would also be found as an HIV-1 dependency factor, as have other members of the Y-complex: Nup107, Nup160, and Nup85. Indeed, depletion of cornerstone Nup107, another HIV-1 dependency factor, has been found to prevent the assembly of Nup358, Nup214, Nup153, and Tpr. Nup96, on the other hand, has not been screened for its effects on HIV-1 replication but is also responsible for Nup358 anchoring and is a protein that originates from the same transcript as Nup98. Therefore, one can expect that targeted depletion of Nup98 causes loss of Nup96, and thus the loss in Nup358 anchoring stability, perhaps explaining how Nup98 depletion has drastic effects on retroviral integration. On the cytoplasmic side of NPCs, Nup98 binds to Nup88 of the NUP214/Nup88/Nup62 subcomplex only found on the cytoplasmic face of pores, and the stability of this subcomplex is dependent on Nup88. Nup155, another HIV-1 dependency factor, is a member of the inner ring with Nup93 and Nup53, where all three may recruit and establish continuous contact with the essential Nup62 complex locating at the central channel. Nup155 and Nup160 interactions with pore membrane protein Pom121 also found to be required for HIV-1 replication, and are crucial for NPC anchoring at NEs. Finally, Nup153 depletion causes Tpr release from NPCs, and thus retroviral functions assigned to Nup153 from depletion studies may be due to a synergism from the codepletion of these two proteins. Therefore, to avoid false positive or negative results in the assigning of HIV-1 dependency factors, such observational studies may benefit from the use of less invasive methods such as proteomics or small molecules targeting precise interaction sites between Nups and components of HIV-1.

THE IMPORTANCE OF FG-REPEAT NUPS TO DISEASE

FG-repeat regions of Nups are natively unfolded, and are highly flexible proteins that lack ordered secondary structure, as proven by use of protease sensitivity assays (Denning et al., 2003). FG-repeats on Nups are typically 200–700 amino acids in length, and are believed to facilitate the passage of cargo through NPCs by binding directly to karyopherins carrying this cargo (Rexach and Blobel, 1995; Allen et al., 2001). The main Nups that have been functionally studied for HIV-1 all contain FG-repeats motifs: Nup358, Nup214, Nup98, Nup62, Nup153, and Tpr. Other Nups that contain FG-repeats but have not as of yet been identified toward HIV-1 replication include Nup50, Nup54, Nup58, Nup96, Nup35, whereas CG1 has been found to interact with Vpr, but not via its FG-repeats (Le Rouzic et al., 2002). One important distinction in studies that have investigated Nups and HIV-1 is the common usage of mab414, which specifically recognizes FG–Nups (Davis and Blobel, 1986, 1987). It could also be possible that since these FG-repeat Nups are all positioned at the face the inner pore of the NPC, they would have the first opportunity to interact with transiting cargo bound to transport receptors. Towards HIV-1 replication, it is clear that this opportunistic virus would have more easily evolved to utilize the channel facing Nups for its replication first, rather than other scaffolding Nups that have been identified by HIV-1 dependency factor screens as essential (i.e., Nup133, Nup107, Nup155, Nup160, Nup85, Pom121, and Sec13), but where very little has been done to prove the functionality of these Nups toward retroviral replication, and where the interpretation of such studies is difficult due to their essential functions in the maintenance of the architectural integrity of NPCs.

These main FG-repeat Nups, Nup358, Nup214, Nup98, Nup62, Nup153, and Tpr, are those most studied, have been identified as important for the replication of many other viruses (see Table 3.2 and Fig. 3.2), have most frequently been identified as essential to the progression of cancer (Simon and Rout, 2014), and represent those implicated in numerous other pathologies that are independent of embryonically lethal germline mutations. Surprisingly, these FG regions are also encountered in a large numbers of oncogenic fusion proteins, with Nup358, Nup214, and Nup98 being those most prominently identified. The most promiscuous FG-repeat region is that of Nup98, found in approximately 20 different oncogenic fusion proteins (Romana et al., 2006). A second oncogenic FG-repeat is that of Nup214, which is found in two different leukemia-associated fusion proteins (von Lindern et al., 1992). The N-terminal GLFG domain of Nup98 is common to all fusion proteins and their expression is driven by the Nup98 promoter (Gough et al., 2011). Also, the FG-repeat domain of Nup98, and by extension those of Nup153 and Nup214 (Kasper et al., 1999), are firmly placed in the context of oncogenic transcription, where transactivation or transrepression coincides with the ability of FG-repeats to interact with CBP/p300 acetyltransferase (Kasper et al., 1999) or histone deacetylase HDAC1 (Bai et al., 2006); reviewed in Kalverda and Fornerod (2007). Therefore, it is feasible that FG-repeat Nups could be important for the transcription of the integrated proviral DNA. In the examination of Nups common to closely related viruses using phylograms (Fig. 3.2), it becomes obvious that many more Nups than previously imagined may be commonly employed or hijacked for the replication of both closely and distantly related viruses, possibly using similar Nup-based mechanisms. Indeed, with numerous Nup crystal structures resolved and small molecule inhibitors discovered to target Nups and their interacting partners, it may be possible to target these virus–Nups interactions in similar ways toward the development of precision medicines against numerous viral pathologies.

FIGURE 3.2 **Phylogeny mapping of nucleoporins (Nups) and virus genus or family.** Genus or family of all listed viruses that have been reported with Nups were imported into phyloT (http://phylot.biobyte.de), and ITOL (Interactive Tree of life; http://itol.embl.de) was used to add Nup name and render phylogeny trees that would classify Nups in accordance with virus genus or family. This was performed to observe known essential and functional Nups among related viruses to elucidate common replication mechanisms or reveal Nups that have not yet been observed or studied for these closely related viruses, as revealed by phylograms. Gene names appearing more than once on phylograms have been colored. From Table 3.2 (*top phylogram*), genus viruses are Alphapapillomavirus (human papillomavirus types 8, 11, and 16); Alpharetrovirus (Rous sarcoma virus); Alphavirus (Sindbis virus); Avulavirus (Newcastle disease virus); Cardiovirus (encephalomyocarditis virus); Deltaretrovirus (human T-lymphotropic virus); Deltavirus (hepatitis D virus); Enterovirus (coxsackievirus B3, enterovirus 71, poliovirus, rhinovirus); Flavivirus (dengue virus, Japanese encephalitis virus); Henipavirus (Hendra virus, Nipah virus, hepatitis C virus); Hepacivirus (Hepatitis C virus); Hepatovirus (hepatitis A virus); Influenzavirus A (influenza A virus); Lentivirus (equine infectious anemia virus [EIAV], feline immunodeficiency virus, human immunodeficiency virus type 1, simian immunodeficiency virus [SIV]); Lymphocryptovirus (Epstein–Barr virus); Mastadenovirus (adenovirus type 2 and 5); Orthohepadnavirus (hepatitis B virus); Orthopneumovirus (respiratory syncytial virus); Orthopoxvirus (vaccinia virus); Orthoreovirus (avian reovirus); Simplexvirus (herpes simplex virus); Respirovirus (Sendai virus); and Vesiculovirus (vesicular stomatitis virus). From Table 3.2 (*bottom phylogram*), family viruses are Adenoviridae (adenovirus type 2 and 5); Flaviviridae (dengue virus, hepatitis C virus, Japanese encephalitis virus); Hepadnaviridae (hepatitis B virus); Herpesviridae (Epstein–Barr virus, herpes simplex virus); Orthomyxoviridae (influenza A virus); Papillomaviridae (human papillomavirus types 8, 11, and 16); Paramyxoviridae (Hendra virus, Newcastle disease virus, Nipah virus, Sendai virus); Picornaviridae (coxsackievirus B3, encephalomyocarditis virus, enterovirus 71, hepatitis A virus, poliovirus, rhinovirus); Pneumoviridae (respiratory syncytial virus); Poxviridae (vaccinia virus); Reoviridae (avian Reovirus); Retroviridae (EIAV, Feline immunodeficiency virus, human immunodeficiency virus type 1, human T-lymphotropic virus, Rous sarcoma virus, SIV); Rhabdoviridae (Sindbis virus, vesicular stomatitis virus).

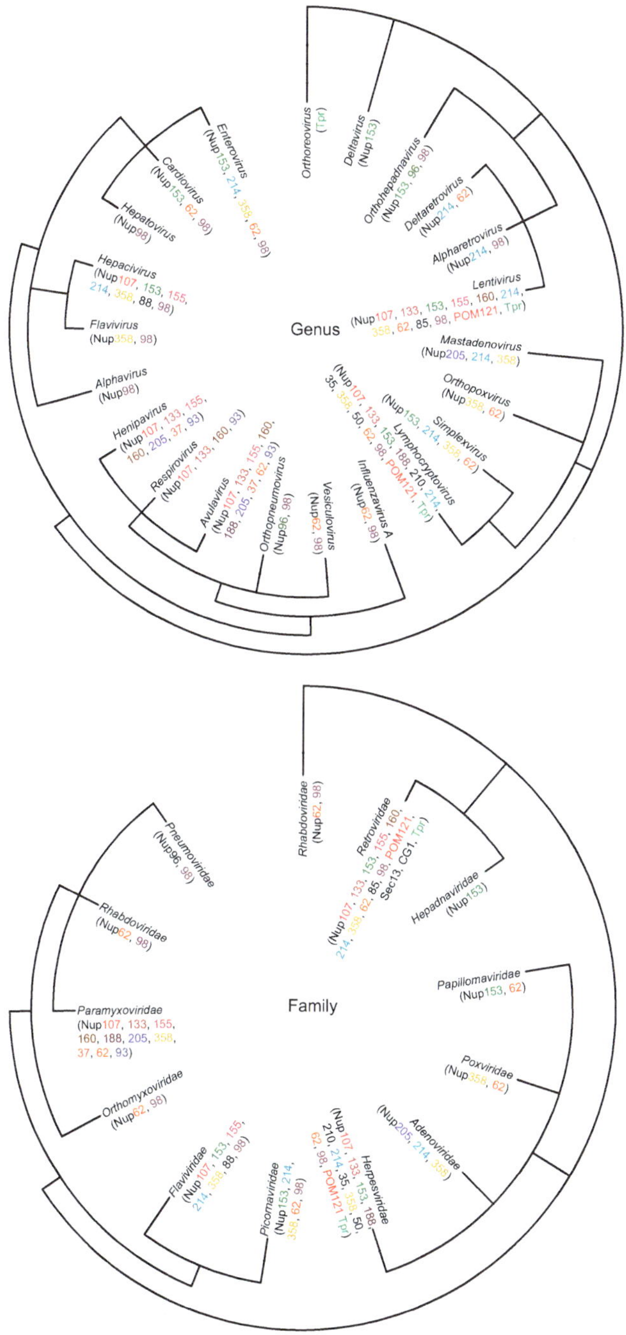
Genus
Orthoreovirus (Tpr)
Deltavirus (Nup153)
Orthohepadnavirus (Nup153, 96, 98)
Deltaretrovirus (Nup214, 62)
Alpharetrovirus (Nup214, 98)
Lentivirus (Nup107, 133, 153, 155, 160, 214, 358, 62, 85, 98, POM121, Tpr)
Mastadenovirus (Nup205, 214, 358)
Orthopoxvirus (Nup358, 62)
Simplexvirus (Nup153, 214, 358, 62)
Lymphocryptovirus (Nup107, 133, 153, 188, 210, 214, 35, 358, 50, 62, 98, POM121, Tpr)
Influenzavirus A (Nup62, 98)
Vesiculovirus (Nup62, 98)
Orthopneumovirus (Nup96, 98)
Avulavirus (Nup107, 133, 155, 160, 188, 205, 37, 62, 93)
Respirovirus (Nup107, 133, 160, 93)
Henipavirus (Nup107, 133, 155, 160, 205, 37, 93)
Alphavirus (Nup98)
Flavivirus (Nup358, 98)
Hepacivirus (Nup107, 153, 155, 214, 358, 88, 98)
Hepatovirus (Nup98)
Cardiovirus (Nup153, 62, 98)
Enterovirus (Nup153, 214, 358, 62, 98)
Family
Rhabdoviridae (Nup62, 98)
Retroviridae (Nup107, 133, 153, 155, 160, 214, 358, 62, 85, 98, POM121, Sec13, CG1, Tpr)
Hepadnaviridae (Nup153)
Papillomaviridae (Nup153, 62)
Poxviridae (Nup358, 62)
Adenoviridae (Nup205, 214, 358)
Herpesviridae (Nup107, 133, 153, 188, 210, 214, 35, 358, 50, 62, 98, POM121, Tpr)
Picornaviridae (Nup153, 214, 358, 62, 98)
Flaviviridae (Nup107, 153, 155, 214, 358, 88, 98)
Orthomyxoviridae (Nup62, 98)
Paramyxoviridae (Nup107, 133, 155, 160, 188, 205, 358, 37, 62, 93)
Rhabdoviridae (Nup62, 98)
Pneumoviridae (Nup96, 98)

In a static, two-dimensional cross-sectional picture of the NPC, Nups are placed in their predominant positions where they have most often been observed to occupy. However, from the above sections detailing their natural functions, it becomes evident that Nups are dynamic, with many of the FG-repeat Nups traveling across NPCs during their activities. Other, larger Nups are able to extend from their anchoring positions all the way across to the other side of NPC and thereby act on both the cytoplasmic and nuclear faces of the pore (Fig. 3.1). Another distinction that must be made is that FG-repeat Nups are present in multiple copies at NPCs that are continuously engaged in a multitude of cargo translocations. Therefore, it is quite possible that multiple copies of any individual Nup could simultaneously be involved in more than one stage of HIV-1 replication. As an example, Nup153 was originally debated to participate in either import, or integration, or transcription. However, independent groups have all successfully validated all three of these Nup153 functions in HIV-1 replication. The same may be true for numerous other Nups and transport molecules, whereby HIV-1 may have evolved not to necessarily depend on any one particular Nup to facilitate its replication, but rather on a common Nup protein motif such as the FG-repeat.

Therefore, Nup358 may be involved in docking the HIV-1 PIC at the NPC and then may persist for PIC import, where in cooperation with the other cytoplasmic Nups (e.g., Nup214 and Nup62 of the Nup214/Nup88 subcomplex), may facilitaten PIC transit through the pore. The duration and extent of interaction between Nup358 and the HIV-1 PIC throughout this process may be made possible due to the length and extension of Nup358 through the central channel, allowing it to remain bound to the PIC along with the original or recycled Nup62 (which interacts with IN). The complex containing the HIV-1 PIC, IN, Nup358 and Nup62 can collect Nup98 during its transit through the pore, adding Tpr on its way through the nuclear basket, finally entering the nucleus for integration site selection (Fig. 3.1). The affinity that FG–Nups possess for the PIC may then be transferred to the integrated provirus and to other host factors tethered there by these Nups during transit, since Nup153, Nup62, Nup98, and Tpr have all been implicated in proviral transcription. These four Nups are also implicated in nuclear export, in addition to Nup214 that has also been frequently observed to assist the export of the viral RNA RNP, Thus, it may very well be the same copies of these Nups that remain tethered to the viral material from HIV-1 nuclear entry through nuclear egress. Finally, HIV-1 infection has been observed to displace many Nups from NPCs and cause an increase in representation of others. Though Nup62 has been the only Nup that has throughly been examined in cytoplasmic egress of newly formed virions (Monette et al., 2011), it is also possible that other Nups displaced by the infection could persist with the vRNA–RNP during its journey to the plasma membrane and packaging into assembling virus particles.

VIRUSES COMMONLY CAUSE THE DISPLACEMENT OF FG–NUPS TO REMODEL NPCs

What happens to FG-repeat Nups during viral infection is popularly described as a "relocalization," "remodeling," "modulation," "displacement," "disruption," "redistribution," "reorganization," "mislocalisation," or "relocalization," leading to NPC "rearrangement" or "disassembly." HIV-1 infection causes the "relocalization" of FG–Nup 358 to the cytoplasm (Dharan et al., 2016), and HIV-1 infection causes an overall "remodeling" of NPCs, by displacing many Nups, including numerous FG–Nups, away from NEs and into the cytoplasm where they associate with virions (Monette et al., 2011). Herpesvirus also causes INM "remodeling" (Zeev-Ben-Mordehai et al., 2015). Cytomegalovirus pUL97 kinase induces "modulation" of FG–Nups and NPC function (Oberstein et al., 2015). Adenovirus infection causes "displacement" of FG–Nups and "disruption" NPCs and NE permeability (Strunze et al., 2011). EBV BGLF4 kinase causes the "redistribution" of FG–Nups (Chang et al., 2015) and NPC "disassembly" (Li et al., 2015). Canine parvovirus infection causes the apical distribution of FG–Nups and "reorganization" of NPCs and the lamina (Mantyla et al., 2015). Rhinovirus infection causes the "mislocalisation" of FG–Nups (Walker et al., 2013). Mengovirus infection induces the "rearrangement" of the NPCs by phosphorylating FG–Nups (Bardina et al., 2009). Hepatitis C virus infection causes the "relocalization" of numerous FG–Nups to regions of the cytoplasm populated by viral proteins (Neufeldt et al., 2013), a report in which it was suggested that viral infection may be causing this "relocalization" of FG–Nups to reposition them on other cytoplasmic structures where they may form pseudo-NPCs accelerating viral passage into other membranous structures required for viral replication and morphogenesis. Finally, there is the possibility is that the FG-repeat regions of these "displaced" Nups represent a passport commonly retained by viruses as a key host defense evasion strategy assisting their rapid and unhindered transit through the tough cellular microenvironment.

ACKNOWLEDGMENTS

We are grateful to Martin Beck, Ed Hurt and Miroslav Milev for contributing images. This work was supported by the Canadian Foundation for HIV-1/AIDS Research (CANFAR) and the Canadian Institutes of Health Research grant #MOP-56974.

REFERENCES

Ajamian, L., Abel, K., Rao, S., Vyboh, K., Garcia-de-Gracia, F., Soto-Rifo, R., Kulozik, A.E., Gehring, N.H., Mouland, A.J., 2015. HIV-1 recruits UPF1 but excludes UPF2 to promote nucleocytoplasmic export of the genomic RNA. Biomolecules 5, 2808–2839.

Alber, F., Dokudovskaya, S., Veenhoff, L.M., Zhang, W., Kipper, J., Devos, D., Suprapto, A., Karni-Schmidt, O., Williams, R., Chait, B.T., Sali, A., Rout, M.P., 2007. The molecular architecture of the nuclear pore complex. Nature 450, 695–701.

Allen, N.P., Huang, L., Burlingame, A., Rexach, M., 2001. Proteomic analysis of nucleoporin interacting proteins. J. Biol. Chem. 276, 29268–29274.

Ao, Z., Jayappa, K.D., Wang, B., Zheng, Y., Wang, X., Peng, J., Yao, X., 2012. Contribution of host nucleoporin 62 in HIV-1 integrase chromatin association and viral DNA integration. J. Biol. Chem. 287, 10544–10555.

Bai, X.T., Gu, B.W., Yin, T., Niu, C., Xi, X.D., Zhang, J., Chen, Z., Chen, S.J., 2006. Trans-repressive effect of NUP98-PMX1 on PMX1-regulated c-FOS gene through recruitment of histone deacetylase 1 by FG-repeats. Cancer Res. 66, 4584–4590.

Bailer, S.M., Balduf, C., Hurt, E., 2001. The Nsp1p carboxy-terminal domain is organized into functionally distinct coiled-coil regions required for assembly of nucleoporin subcomplexes and nucleocytoplasmic transport. Mol. Cell. Biol. 21, 7944–7955.

Bailer, S.M., Balduf, C., Katahira, J., Podtelejnikov, A., Rollenhagen, C., Mann, M., Pante, N., Hurt, E., 2000. Nup116p associates with the Nup82p-Nsp1p-Nup159p nucleoporin complex. J. Biol. Chem. 275, 23540–23548.

Ball, J.R., Ullman, K.S., 2005. Versatility at the nuclear pore complex: lessons learned from the nucleoporin Nup153. Chromosoma 114, 319–330.

Banerjee, H.N., Gibbs, J., Jordan, T., Blackshear, M., 2010. Depletion of a single nucleoporin, Nup107, induces apoptosis in eukaryotic cells. Mol. Cell. Biochem. 343, 21–25.

Bangs, P., Burke, B., Powers, C., Craig, R., Purohit, A., Doxsey, S., 1998. Functional analysis of Tpr: identification of nuclear pore complex association and nuclear localization domains and a role in mRNA export. J. Cell Biol. 143, 1801–1812.

Bardina, M.V., Lidsky, P.V., Sheval, E.V., Fominykh, K.V., van Kuppeveld, F.J., Polyakov, V.Y., Agol, V.I., 2009. Mengovirus-induced rearrangement of the nuclear pore complex: hijacking cellular phosphorylation machinery. J. Virol. 83, 3150–3161.

Basta, H.A., Bacot-Davis, V.R., Ciomperlik, J.J., Palmenberg, A.C., 2014. Encephalomyocarditis virus leader is phosphorylated by CK2 and syk as a requirement for subsequent phosphorylation of cellular nucleoporins. J. Virol. 88, 2219–2226.

Bastos, R., Lin, A., Enarson, M., Burke, B., 1996. Targeting and function in mRNA export of nuclear pore complex protein Nup153. J. Cell Biol. 134, 1141–1156.

Bastos, R., Ribas de Pouplana, L., Enarson, M., Bodoor, K., Burke, B., 1997. Nup84, a novel nucleoporin that is associated with CAN/Nup214 on the cytoplasmic face of the nuclear pore complex. J. Cell Biol. 137, 989–1000.

Beck, M., Glavy, J.S., 2014. Toward understanding the structure of the vertebrate nuclear pore complex. Nucleus 5, 119–123.

Beck, M., Hurt, E., 2017. The nuclear pore complex: understanding its function through structural insight. Nat. Rev. Mol. Cell Biol. 18, 73–89.

Belgareh, N., Rabut, G., Bai, S.W., van Overbeek, M., Beaudouin, J., Daigle, N., Zatsepina, O.V., Pasteau, F., Labas, V., Fromont-Racine, M., Ellenberg, J., Doye, V., 2001. An evolutionarily conserved NPC subcomplex, which redistributes in part to kinetochores in mammalian cells. J. Cell Biol. 154, 1147–1160.

Belgareh, N., Snay-Hodge, C., Pasteau, F., Dagher, S., Cole, C.N., Doye, V., 1998. Functional characterization of a Nup159p-containing nuclear pore subcomplex. Mol. Biol. Cell 9, 3475–3492.

Ben-Efraim, I., Frosst, P.D., Gerace, L., 2009. Karyopherin binding interactions and nuclear import mechanism of nuclear pore complex protein Tpr. BMC Cell Biol. 10, 74.

Bernad, R., Engelsma, D., Sanderson, H., Pickersgill, H., Fornerod, M., 2006. Nup214-Nup88 nucleoporin subcomplex is required for CRM1-mediated 60 S preribosomal nuclear export. J. Biol. Chem. 281, 19378–19386.

Bernad, R., van der Velde, H., Fornerod, M., Pickersgill, H., 2004. Nup358/RanBP2 attaches to the nuclear pore complex via association with Nup88 and Nup214/CAN and plays a supporting role in CRM1-mediated nuclear protein export. Mol. Cell. Biol. 24, 2373–2384.

Bharathi, A., Ghosh, A., Whalen, W.A., Yoon, J.H., Pu, R., Dasso, M., Dhar, R., 1997. The human RAE1 gene is a functional homologue of *Schizosaccharomyces pombe* rae1 gene involved in nuclear export of poly(A)+ RNA. Gene 198, 251–258.

Bhattacharya, A., Alam, S.L., Fricke, T., Zadrozny, K., Sedzicki, J., Taylor, A.B., Demeler, B., Pornillos, O., Ganser-Pornillos, B.K., Diaz-Griffero, F., Ivanov, D.N., Yeager, M., 2014. Structural basis of HIV-1 capsid recognition by PF74 and CPSF6. Proc. Natl. Acad. Sci. U.S.A. 111, 18625–18630.

Bichel, K., Price, A.J., Schaller, T., Towers, G.J., Freund, S.M., James, L.C., 2013. HIV-1 capsid undergoes coupled binding and isomerization by the nuclear pore protein NUP358. Retrovirology 10, 81.

Bickford, L.C., Mossessova, E., Goldberg, J., 2004. A structural view of the COPII vesicle coat. Curr. Opin. Struct. Biol. 14, 147–153.

Bin Hamid, F., Kim, J., Shin, C.G., 2016. Cellular and viral determinants of retroviral nuclear entry. Can. J. Microbiol. 62, 1–15.

Blevins, M.B., Smith, A.M., Phillips, E.M., Powers, M.A., 2003. Complex formation among the RNA export proteins Nup98, Rae1/Gle2, and TAP. J. Biol. Chem. 278, 20979–20988.

Bodoor, K., Shaikh, S., Salina, D., Raharjo, W.H., Bastos, R., Lohka, M., Burke, B., 1999. Sequential recruitment of NPC proteins to the nuclear periphery at the end of mitosis. J. Cell Sci. 112 (Pt 13), 2253–2264.

Boehmer, T., Enninga, J., Dales, S., Blobel, G., Zhong, H., 2003. Depletion of a single nucleoporin, Nup107, prevents the assembly of a subset of nucleoporins into the nuclear pore complex. Proc. Natl. Acad. Sci. U.S.A. 100, 981–985.

Boehmer, T., Jeudy, S., Berke, I.C., Schwartz, T.U., 2008. Structural and functional studies of Nup107/Nup133 interaction and its implications for the architecture of the nuclear pore complex. Mol. Cell 30, 721–731.

Bogerd, H.P., Echarri, A., Ross, T.M., Cullen, B.R., 1998. Inhibition of human immunodeficiency virus Rev and human T-cell leukemia virus Rex function, but not Mason-Pfizer monkey virus constitutive transport element activity, by a mutant human nucleoporin targeted to Crm1. J. Virol. 72, 8627–8635.

Bolhy, S., Bouhlel, I., Dultz, E., Nayak, T., Zuccolo, M., Gatti, X., Vallee, R., Ellenberg, J., Doye, V., 2011. A Nup133-dependent NPC-anchored network tethers centrosomes to the nuclear envelope in prophase. J. Cell Biol. 192, 855–871.

Brass, A.L., Dykxhoorn, D.M., Benita, Y., Yan, N., Engelman, A., Xavier, R.J., Lieberman, J., Elledge, S.J., 2008. Identification of host proteins required for HIV infection through a functional genomic screen. Science 319, 921–926.

Breuer, M., Ohkura, H., 2015. A negative loop within the nuclear pore complex controls global chromatin organization. Genes Dev. 29, 1789–1794.

Brohawn, S.G., Leksa, N.C., Spear, E.D., Rajashankar, K.R., Schwartz, T.U., 2008. Structural evidence for common ancestry of the nuclear pore complex and vesicle coats. Science 322, 1369–1373.

Brown, J.A., Bharathi, A., Ghosh, A., Whalen, W., Fitzgerald, E., Dhar, R., 1995. A mutation in the *Schizosaccharomyces pombe* rae1 gene causes defects in poly(A)+ RNA export and in the cytoskeleton. J. Biol. Chem. 270, 7411–7419.

Bui, K.H., von Appen, A., Diguilio, A.L., Ori, A., Sparks, L., Mackmull, M.T., Bock, T., Hagen, W., Andres-Pons, A., Glavy, J.S., Beck, M., 2013. Integrated structural analysis of the human nuclear pore complex scaffold. Cell 155, 1233–1243.

Burdick, R.C., Hu, W.S., Pathak, V.K., 2013. Nuclear import of APOBEC3F-labeled HIV-1 preintegration complexes. Proc. Natl. Acad. Sci. U.S.A. 110, E4780–E4789.

Byrd, D.A., Sweet, D.J., Pante, N., Konstantinov, K.N., Guan, T., Saphire, A.C., Mitchell, P.J., Cooper, C.S., Aebi, U., Gerace, L., 1994. Tpr, a large coiled coil protein whose amino terminus is involved in activation of oncogenic kinases, is localized to the cytoplasmic surface of the nuclear pore complex. J. Cell Biol. 127, 1515–1526.

Campbell, E.M., Hope, T.J., 2015. HIV-1 capsid: the multifaceted key player in HIV-1 infection. Nat. Rev. Microbiol. 13, 471–483.
Capelson, M., Liang, Y., Schulte, R., Mair, W., Wagner, U., Hetzer, M.W., 2010. Chromatin-bound nuclear pore components regulate gene expression in higher eukaryotes. Cell 140, 372–383.
Cardarelli, F., Lanzano, L., Gratton, E., 2012. Capturing directed molecular motion in the nuclear pore complex of live cells. Proc. Natl. Acad. Sci. U.S.A. 109, 9863–9868.
Cassany, A., Ragues, J., Guan, T., Begu, D., Wodrich, H., Kann, M., Nemerow, G.R., Gerace, L., 2015. Nuclear import of adenovirus DNA involves direct interaction of hexon with an N-terminal domain of the nucleoporin Nup214. J. Virol. 89, 1719–1730.
Chang, C.W., Lee, C.P., Huang, Y.H., Yang, P.W., Wang, J.T., Chen, M.R., 2012. Epstein-Barr virus protein kinase BGLF4 targets the nucleus through interaction with nucleoporins. J. Virol. 86, 8072–8085.
Chang, C.W., Lee, C.P., Su, M.T., Tsai, C.H., Chen, M.R., 2015. BGLF4 kinase modulates the structure and transport preference of the nuclear pore complex to facilitate nuclear import of Epstein-Barr virus lytic proteins. J. Virol. 89, 1703–1718.
Chatel, G., Fahrenkrog, B., 2012. Dynamics and diverse functions of nuclear pore complex proteins. Nucleus 3, 162–171.
Chen, J., Huang, S., Chen, Z., 2010. Human cellular protein nucleoporin hNup98 interacts with influenza A virus NS2/nuclear export protein and overexpression of its GLFG-repeat domain can inhibit virus propagation. J. Gen. Virol. 91, 2474–2484.
Chen, N.Y., Zhou, L., Gane, P.J., Opp, S., Ball, N.J., Nicastro, G., Zufferey, M., Buffone, C., Luban, J., Selwood, D., Diaz-Griffero, F., Taylor, I., Fassati, A., 2016. HIV-1 capsid is involved in post-nuclear entry steps. Retrovirology 13, 28.
Chen, X., Xu, L., 2010. Specific nucleoporin requirement for Smad nuclear translocation. Mol. Cell. Biol. 30, 4022–4034.
Chook, Y.M., Suel, K.E., 2011. Nuclear import by karyopherin-betas: recognition and inhibition. Biochim. Biophys. Acta 1813, 1593–1606.
Chou, C.C., Zhang, Y., Umoh, M.E., Vaughan, S.W., Lorenzini, I., Liu, F., Sayegh, M., Donlin-Asp, P.G., Chen, Y.H., Duong, D.M., Seyfried, N.T., Powers, M.A., Kukar, T., Hales, C.M., Gearing, M., Cairns, N.J., Boylan, K.B., Dickson, D.W., Rademakers, R., Zhang, Y.J., Petrucelli, L., Sattler, R., Zarnescu, D.C., Glass, J.D., Rossoll, W., 2018. TDP-43 pathology disrupts nuclear pore complexes and nucleocytoplasmic transport in ALS/FTD. Nat. Neurosci. 21, 228–239.
Chug, H., Trakhanov, S., Hulsmann, B.B., Pleiner, T., Gorlich, D., 2015. Crystal structure of the metazoan Nup62*Nup58*Nup54 nucleoporin complex. Science 350, 106–110.
Ciomperlik, J.J., Basta, H.A., Palmenberg, A.C., 2016. Cardiovirus leader proteins bind exportins: implications for virus replication and nucleocytoplasmic trafficking inhibition. Virology 487, 19–26.
Clarkson, W.D., Kent, H.M., Stewart, M., 1996. Separate binding sites on nuclear transport factor 2 (NTF2) for GDP-Ran and the phenylalanine-rich repeat regions of nucleoporins p62 and Nsp1p. J. Mol. Biol. 263, 517–524.
Cohen, S., Au, S., Pante, N., 2011. How viruses access the nucleus. Biochim. Biophys. Acta 1813, 1634–1645.
Copeland, A.M., Newcomb, W.W., Brown, J.C., 2009. Herpes simplex virus replication: roles of viral proteins and nucleoporins in capsid-nucleus attachment. J. Virol. 83, 1660–1668.
Cordes, V.C., Hase, M.E., Muller, L., 1998. Molecular segments of protein Tpr that confer nuclear targeting and association with the nuclear pore complex. Exp. Cell Res. 245, 43–56.
Cordes, V.C., Reidenbach, S., Rackwitz, H.R., Franke, W.W., 1997. Identification of protein p270/Tpr as a constitutive component of the nuclear pore complex-attached intranuclear filaments. J. Cell Biol. 136, 515–529.

Cronshaw, J.M., Krutchinsky, A.N., Zhang, W., Chait, B.T., Matunis, M.J., 2002. Proteomic analysis of the mammalian nuclear pore complex. J. Cell Biol. 158, 915–927.

Cronshaw, J.M., Matunis, M.J., 2004. The nuclear pore complex: disease associations and functional correlations. Trends Endocrinol. Metab. 15, 34–39.

D'Angelo, M.A., Anderson, D.J., Richard, E., Hetzer, M.W., 2006. Nuclear pores form de novo from both sides of the nuclear envelope. Science 312, 440–443.

Davis, L.I., Blobel, G., 1986. Identification and characterization of a nuclear pore complex protein. Cell 45, 699–709.

Davis, L.I., Blobel, G., 1987. Nuclear pore complex contains a family of glycoproteins that includes p62: glycosylation through a previously unidentified cellular pathway. Proc. Natl. Acad. Sci. U.S.A. 84, 7552–7556.

De Iaco, A., Luban, J., 2011. Inhibition of HIV-1 infection by TNPO3 depletion is determined by capsid and detectable after viral cDNA enters the nucleus. Retrovirology 8, 98.

De Magistris, P., Tatarek-Nossol, M., Dewor, M., Antonin, W., 2018. A self-inhibitory interaction within Nup155 and membrane binding are required for nuclear pore complex formation. J. Cell Sci. 131.

DeGrasse, J.A., Dubois, K.N., Devos, D., Siegel, T.N., Sali, A., Field, M.C., Rout, M.P., Chait, B.T., 2009. Evidence for a shared nuclear pore complex architecture that is conserved from the last common eukaryotic ancestor. Mol. Cell. Proteom. 8, 2119–2130.

Delphin, C., Guan, T., Melchior, F., Gerace, L., 1997. RanGTP targets p97 to RanBP2, a filamentous protein localized at the cytoplasmic periphery of the nuclear pore complex. Mol. Biol. Cell 8, 2379–2390.

Denning, D.P., Patel, S.S., Uversky, V., Fink, A.L., Rexach, M., 2003. Disorder in the nuclear pore complex: the FG-repeat regions of nucleoporins are natively unfolded. Proc. Natl. Acad. Sci. U.S.A. 100, 2450–2455.

Devos, D., Dokudovskaya, S., Alber, F., Williams, R., Chait, B.T., Sali, A., Rout, M.P., 2004. Components of coated vesicles and nuclear pore complexes share a common molecular architecture. PLoS Biol. 2, e380.

Devos, D., Dokudovskaya, S., Williams, R., Alber, F., Eswar, N., Chait, B.T., Rout, M.P., Sali, A., 2006. Simple fold composition and modular architecture of the nuclear pore complex. Proc. Natl. Acad. Sci. U.S.A. 103, 2172–2177.

Dewangan, P.S., Sonawane, P.J., Chouksey, A.R., Chauhan, R., 2017. The Nup62 coiled-coil motif provides plasticity for triple-helix bundle formation. Biochemistry 56, 2803–2811.

Dharan, A., Talley, S., Tripathi, A., Mamede, J.I., Majetschak, M., Hope, T.J., Campbell, E.M., 2016. KIF5B and Nup358 cooperatively mediate the nuclear import of HIV-1 during infection. PLoS Pathog. 12, e1005700.

Di Nunzio, F., 2013. New insights in the role of nucleoporins: a bridge leading to concerted steps from HIV-1 nuclear entry until integration. Virus Res. 178, 187–196.

Di Nunzio, F., Danckaert, A., Fricke, T., Perez, P., Fernandez, J., Perret, E., Roux, P., Shorte, S., Charneau, P., Diaz-Griffero, F., Arhel, N.J., 2012. Human nucleoporins promote HIV-1 docking at the nuclear pore, nuclear import and integration. PLoS One 7, e46037.

Di Nunzio, F., Fricke, T., Miccio, A., Valle-Casuso, J.C., Perez, P., Souque, P., Rizzi, E., Severgnini, M., Mavilio, F., Charneau, P., Diaz-Griffero, F., 2013. Nup153 and Nup98 bind the HIV-1 core and contribute to the early steps of HIV-1 replication. Virology 440, 8–18.

Ding, Q., Zhao, L., Guo, H., Zheng, A.C., 2010. The nucleocytoplasmic transport of viral proteins. Virol. Sin. 25, 79–85.

Dockendorff, T.C., Heath, C.V., Goldstein, A.L., Snay, C.A., Cole, C.N., 1997. C-terminal truncations of the yeast nucleoporin Nup145p produce a rapid temperature-conditional mRNA export defect and alterations to nuclear structure. Mol. Cell. Biol. 17, 906–920.

Doucet, C.M., Talamas, J.A., Hetzer, M.W., 2010. Cell cycle-dependent differences in nuclear pore complex assembly in metazoa. Cell 141, 1030–1041.

Doye, V., Wepf, R., Hurt, E.C., 1994. A novel nuclear pore protein Nup133p with distinct roles in poly(A)+ RNA transport and nuclear pore distribution. EMBO J. 13, 6062–6075.

Drin, G., Casella, J.F., Gautier, R., Boehmer, T., Schwartz, T.U., Antonny, B., 2007. A general amphipathic alpha-helical motif for sensing membrane curvature. Nat. Struct. Mol. Biol. 14, 138–146.

Dultz, E., Ellenberg, J., 2010. Live imaging of single nuclear pores reveals unique assembly kinetics and mechanism in interphase. J. Cell Biol. 191, 15–22.

Dultz, E., Zanin, E., Wurzenberger, C., Braun, M., Rabut, G., Sironi, L., Ellenberg, J., 2008. Systematic kinetic analysis of mitotic dis- and reassembly of the nuclear pore in living cells. J. Cell Biol. 180, 857–865.

Eberhard, J., Onder, Z., Moroianu, J., 2013. Nuclear import of high risk HPV16 E7 oncoprotein is mediated by its zinc-binding domain via hydrophobic interactions with Nup62. Virology 446, 334–345.

Ebina, H., Aoki, J., Hatta, S., Yoshida, T., Koyanagi, Y., 2004. Role of Nup98 in nuclear entry of human immunodeficiency virus type 1 cDNA. Microbes Infect. 6, 715–724.

Echeverria, P.C., Mazaira, G., Erlejman, A., Gomez-Sanchez, C., Piwien Pilipuk, G., Galigniana, M.D., 2009. Nuclear import of the glucocorticoid receptor-hsp90 complex through the nuclear pore complex is mediated by its interaction with Nup62 and importin beta. Mol. Cell. Biol. 29, 4788–4797.

Emtage, J.L., Bucci, M., Watkins, J.L., Wente, S.R., 1997. Defining the essential functional regions of the nucleoporin Nup145p. J. Cell Sci. 110 (Pt 7), 911–925.

Enarson, P., Enarson, M., Bastos, R., Burke, B., 1998. Amino-terminal sequences that direct nucleoporin nup153 to the inner surface of the nuclear envelope. Chromosoma 107, 228–236.

Enninga, J., Levay, A., Fontoura, B.M., 2003. Sec13 shuttles between the nucleus and the cytoplasm and stably interacts with Nup96 at the nuclear pore complex. Mol. Cell. Biol. 23, 7271–7284.

Enninga, J., Levy, D.E., Blobel, G., Fontoura, B.M., 2002. Role of nucleoporin induction in releasing an mRNA nuclear export block. Science 295, 1523–1525.

Fahrenkrog, B., Maco, B., Fager, A.M., Koser, J., Sauder, U., Ullman, K.S., Aebi, U., 2002. Domain-specific antibodies reveal multiple-site topology of Nup153 within the nuclear pore complex. J. Struct. Biol. 140, 254–267.

Fang, C., Zhao, C., Liu, X., Yang, P., Lu, H., 2012. Protein alteration of HepG2.2.15 cells induced by iron overload. Proteomics 12, 1378–1390.

Faria, A.M., Levay, A., Wang, Y., Kamphorst, A.O., Rosa, M.L., Nussenzveig, D.R., Balkan, W., Chook, Y.M., Levy, D.E., Fontoura, B.M., 2006. The nucleoporin Nup96 is required for proper expression of interferon-regulated proteins and functions. Immunity 24, 295–304.

Favreau, C., Worman, H.J., Wozniak, R.W., Frappier, T., Courvalin, J.C., 1996. Cell cycle-dependent phosphorylation of nucleoporins and nuclear pore membrane protein Gp210. Biochemistry 35, 8035–8044.

Fay, N., Pante, N., 2015a. Nuclear entry of DNA viruses. Front. Microbiol. 6, 467.

Fay, N., Pante, N., 2015b. Old foes, new understandings: nuclear entry of small non-enveloped DNA viruses. Curr. Opin. Virol. 12, 59–65.

Fernandez-Martinez, J., Phillips, J., Sekedat, M.D., Diaz-Avalos, R., Velazquez-Muriel, J., Franke, J.D., Williams, R., Stokes, D.L., Chait, B.T., Sali, A., Rout, M.P., 2012. Structure-function mapping of a heptameric module in the nuclear pore complex. J. Cell Biol. 196, 419–434.

Finlay, D.R., Meier, E., Bradley, P., Horecka, J., Forbes, D.J., 1991. A complex of nuclear pore proteins required for pore function. J. Cell Biol. 114, 169–183.

Fitzgerald, K.D., Chase, A.J., Cathcart, A.L., Tran, G.P., Semler, B.L., 2013. Viral proteinase requirements for the nucleocytoplasmic relocalization of cellular splicing factor SRp20 during picornavirus infections. J. Virol. 87, 2390–2400.

Flather, D., Semler, B.L., 2015. Picornaviruses and nuclear functions: targeting a cellular compartment distinct from the replication site of a positive-strand RNA virus. Front. Microbiol. 6, 594.

Flatt, J.W., Greber, U.F., 2015. Misdelivery at the nuclear pore complex-stopping a virus dead in its tracks. Cells 4, 277–296.

Flemming, D., Sarges, P., Stelter, P., Hellwig, A., Bottcher, B., Hurt, E., 2009. Two structurally distinct domains of the nucleoporin Nup170 cooperate to tether a subset of nucleoporins to nuclear pores. J. Cell Biol. 185, 387–395.

Fontoura, B.M., Blobel, G., Matunis, M.J., 1999. A conserved biogenesis pathway for nucleoporins: proteolytic processing of a 186-kilodalton precursor generates Nup98 and the novel nucleoporin, Nup96. J. Cell Biol. 144, 1097–1112.

Fontoura, B.M., Blobel, G., Yaseen, N.R., 2000. The nucleoporin Nup98 is a site for GDP/GTP exchange on ran and termination of karyopherin beta 2-mediated nuclear import. J. Biol. Chem. 275, 31289–31296.

Fontoura, B.M., Dales, S., Blobel, G., Zhong, H., 2001. The nucleoporin Nup98 associates with the intranuclear filamentous protein network of TPR. Proc. Natl. Acad. Sci. U.S.A. 98, 3208–3213.

Fontoura, B.M., Faria, P.A., Nussenzveig, D.R., 2005. Viral interactions with the nuclear transport machinery: discovering and disrupting pathways. IUBMB Life 57, 65–72.

Forler, D., Rabut, G., Ciccarelli, F.D., Herold, A., Kocher, T., Niggeweg, R., Bork, P., Ellenberg, J., Izaurralde, E., 2004. RanBP2/Nup358 provides a major binding site for NXF1-p15 dimers at the nuclear pore complex and functions in nuclear mRNA export. Mol. Cell. Biol. 24, 1155–1167.

Fornerod, M., van Deursen, J., van Baal, S., Reynolds, A., Davis, D., Murti, K.G., Fransen, J., Grosveld, G., 1997. The human homologue of yeast CRM1 is in a dynamic subcomplex with CAN/Nup214 and a novel nuclear pore component Nup88. EMBO J. 16, 807–816.

Franz, C., Askjaer, P., Antonin, W., Iglesias, C.L., Haselmann, U., Schelder, M., de Marco, A., Wilm, M., Antony, C., Mattaj, I.W., 2005. Nup155 regulates nuclear envelope and nuclear pore complex formation in nematodes and vertebrates. EMBO J. 24, 3519–3531.

Frosst, P., Guan, T., Subauste, C., Hahn, K., Gerace, L., 2002. Tpr is localized within the nuclear basket of the pore complex and has a role in nuclear protein export. J. Cell Biol. 156, 617–630.

Gaik, M., Flemming, D., von Appen, A., Kastritis, P., Mucke, N., Fischer, J., Stelter, P., Ori, A., Bui, K.H., Bassler, J., Barbar, E., Beck, M., Hurt, E., 2015. Structural basis for assembly and function of the Nup82 complex in the nuclear pore scaffold. J. Cell Biol. 208, 283–297.

Gallucci, L., Kann, M., 2017. Nuclear import of hepatitis B virus capsids and genome. Viruses 9.

Galy, V., Gadal, O., Fromont-Racine, M., Romano, A., Jacquier, A., Nehrbass, U., 2004. Nuclear retention of unspliced mRNAs in yeast is mediated by perinuclear Mlp1. Cell 116, 63–73.

Galy, V., Olivo-Marin, J.C., Scherthan, H., Doye, V., Rascalou, N., Nehrbass, U., 2000. Nuclear pore complexes in the organization of silent telomeric chromatin. Nature 403, 108–112.

Geyer, J.P., Doker, R., Kremer, W., Zhao, X., Kuhlmann, J., Kalbitzer, H.R., 2005. Solution structure of the Ran-binding domain 2 of RanBP2 and its interaction with the C terminus of Ran. J. Mol. Biol. 348, 711–725.

Ghildyal, R., Jordan, B., Li, D., Dagher, H., Bardin, P.G., Gern, J.E., Jans, D.A., 2009. Rhinovirus 3C protease can localize in the nucleus and alter active and passive nucleocytoplasmic transport. J. Virol. 83, 7349–7352.

Goldberg, M.W., 2017. Nuclear pore complex tethers to the cytoskeleton. Semin. Cell Dev. Biol. 68, 52–58.

Goldberg, M.W., Allen, T.D., 1996. The nuclear pore complex and lamina: three-dimensional structures and interactions determined by field emission in-lens scanning electron microscopy. J. Mol. Biol. 257, 848–865.

Gonzalez-Aguilera, C., Askjaer, P., 2012. Dissecting the NUP107 complex: multiple components and even more functions. Nucleus 3, 340–348.

Gough, S.M., Slape, C.I., Aplan, P.D., 2011. NUP98 gene fusions and hematopoietic malignancies: common themes and new biologic insights. Blood 118, 6247–6257.

Grandi, P., Emig, S., Weise, C., Hucho, F., Pohl, T., Hurt, E.C., 1995a. A novel nuclear pore protein Nup82p which specifically binds to a fraction of Nsp1p. J. Cell Biol. 130, 1263–1273.

Grandi, P., Schlaich, N., Tekotte, H., Hurt, E.C., 1995b. Functional interaction of Nic96p with a core nucleoporin complex consisting of Nsp1p, Nup49p and a novel protein Nup57p. EMBO J. 14, 76–87.

Greco, A., Pierotti, M.A., Bongarzone, I., Pagliardini, S., Lanzi, C., Della Porta, G., 1992. TRK-T1 is a novel oncogene formed by the fusion of TPR and TRK genes in human papillary thyroid carcinomas. Oncogene 7, 237–242.

Griffis, E.R., Altan, N., Lippincott-Schwartz, J., Powers, M.A., 2002. Nup98 is a mobile nucleoporin with transcription-dependent dynamics. Mol. Biol. Cell 13, 1282–1297.

Griffis, E.R., Craige, B., Dimaano, C., Ullman, K.S., Powers, M.A., 2004. Distinct functional domains within nucleoporins Nup153 and Nup98 mediate transcription-dependent mobility. Mol. Biol. Cell 15, 1991–2002.

Griffis, E.R., Xu, S., Powers, M.A., 2003. Nup98 localizes to both nuclear and cytoplasmic sides of the nuclear pore and binds to two distinct nucleoporin subcomplexes. Mol. Biol. Cell 14, 600–610.

Guan, T., Muller, S., Klier, G., Pante, N., Blevitt, J.M., Haner, M., Paschal, B., Aebi, U., Gerace, L., 1995. Structural analysis of the p62 complex, an assembly of O-linked glycoproteins that localizes near the central gated channel of the nuclear pore complex. Mol. Biol. Cell 6, 1591–1603.

Gustin, K.E., 2003. Inhibition of nucleo-cytoplasmic trafficking by RNA viruses: targeting the nuclear pore complex. Virus Res. 95, 35–44.

Gustin, K.E., Sarnow, P., 2001. Effects of poliovirus infection on nucleo-cytoplasmic trafficking and nuclear pore complex composition. EMBO J. 20, 240–249.

Hamada, M., Haeger, A., Jeganathan, K.B., van Ree, J.H., Malureanu, L., Walde, S., Joseph, J., Kehlenbach, R.H., van Deursen, J.M., 2011. Ran-dependent docking of importin-beta to RanBP2/Nup358 filaments is essential for protein import and cell viability. J. Cell Biol. 194, 597–612.

Harel, A., Orjalo, A.V., Vincent, T., Lachish-Zalait, A., Vasu, S., Shah, S., Zimmerman, E., Elbaum, M., Forbes, D.J., 2003. Removal of a single pore subcomplex results in vertebrate nuclei devoid of nuclear pores. Mol. Cell 11, 853–864.

Hart, G.W., 1997. Dynamic O-linked glycosylation of nuclear and cytoskeletal proteins. Annu. Rev. Biochem. 66, 315–335.

Hase, M.E., Cordes, V.C., 2003. Direct interaction with nup153 mediates binding of Tpr to the periphery of the nuclear pore complex. Mol. Biol. Cell 14, 1923–1940.

Hase, M.E., Kuznetsov, N.V., Cordes, V.C., 2001. Amino acid substitutions of coiled-coil protein Tpr abrogate anchorage to the nuclear pore complex but not parallel, in-register homodimerization. Mol. Biol. Cell 12, 2433–2452.

Hashizume, C., Moyori, A., Kobayashi, A., Yamakoshi, N., Endo, A., Wong, R.W., 2013. Nucleoporin Nup62 maintains centrosome homeostasis. Cell Cycle 12, 3804–3816.

Hawryluk-Gara, L.A., Shibuya, E.K., Wozniak, R.W., 2005. Vertebrate Nup53 interacts with the nuclear lamina and is required for the assembly of a Nup93-containing complex. Mol. Biol. Cell 16, 2382–2394.

Higa, M.M., Alam, S.L., Sundquist, W.I., Ullman, K.S., 2007. Molecular characterization of the Ran-binding zinc finger domain of Nup153. J. Biol. Chem. 282, 17090–17100.

Hodel, A.E., Hodel, M.R., Griffis, E.R., Hennig, K.A., Ratner, G.A., Xu, S., Powers, M.A., 2002. The three-dimensional structure of the autoproteolytic, nuclear pore-targeting domain of the human nucleoporin Nup98. Mol. Cell 10, 347–358.

Hoelz, A., Debler, E.W., Blobel, G., 2011. The structure of the nuclear pore complex. Annu. Rev. Biochem. 80, 613–643.

Hofemeister, H., O'Hare, P., 2008. Nuclear pore composition and gating in herpes simplex virus-infected cells. J. Virol. 82, 8392–8399.

Hofmann, W., Reichart, B., Ewald, A., Muller, E., Schmitt, I., Stauber, R.H., Lottspeich, F., Jockusch, B.M., Scheer, U., Hauber, J., Dabauvalle, M.C., 2001. Cofactor requirements for nuclear export of Rev response element (RRE)- and constitutive transport element (CTE)-containing retroviral RNAs. An unexpected role for actin. J. Cell Biol. 152, 895–910.

Hu, T., Guan, T., Gerace, L., 1996. Molecular and functional characterization of the p62 complex, an assembly of nuclear pore complex glycoproteins. J. Cell Biol. 134, 589–601.

Huang, W.R., Chiu, H.C., Liao, T.L., Chuang, K.P., Shih, W.L., Liu, H.J., 2015. Avian reovirus protein p17 functions as a nucleoporin Tpr suppressor leading to activation of p53, p21 and PTEN and inactivation of PI3K/AKT/mTOR and ERK signaling pathways. PLoS One 10, e0133699.

Hubert, T., Vandekerckhove, J., Gettemans, J., 2009. Exo70-mediated recruitment of nucleoporin Nup62 at the leading edge of migrating cells is required for cell migration. Traffic 10, 1257–1271.

Hurt, E.C., 1988. A novel nucleoskeletal-like protein located at the nuclear periphery is required for the life cycle of *Saccharomyces cerevisiae*. EMBO J. 7, 4323–4334.

Hutten, S., Flotho, A., Melchior, F., Kehlenbach, R.H., 2008. The Nup358-RanGAP complex is required for efficient importin alpha/beta-dependent nuclear import. Mol. Biol. Cell 19, 2300–2310.

Hutten, S., Kehlenbach, R.H., 2006. Nup214 is required for CRM1-dependent nuclear protein export in vivo. Mol. Cell. Biol. 26, 6772–6785.

Hutten, S., Walde, S., Spillner, C., Hauber, J., Kehlenbach, R.H., 2009. The nuclear pore component Nup358 promotes transportin-dependent nuclear import. J. Cell Sci. 122, 1100–1110.

Huve, J., Wesselmann, R., Kahms, M., Peters, R., 2008. 4Pi microscopy of the nuclear pore complex. Biophys. J. 95, 877–885.

Iwamoto, M., Asakawa, H., Hiraoka, Y., Haraguchi, T., 2010. Nucleoporin Nup98: a gatekeeper in the eukaryotic kingdoms. Genes Cells 15, 661–669.

Kabachinski, G., Schwartz, T.U., 2015. The nuclear pore complex–structure and function at a glance. J. Cell Sci. 128, 423–429.

Kalverda, B., Fornerod, M., 2007. The nuclear life of nucleoporins. Dev. Cell 13, 164–165.

Kalverda, B., Pickersgill, H., Shloma, V.V., Fornerod, M., 2010. Nucleoporins directly stimulate expression of developmental and cell-cycle genes inside the nucleoplasm. Cell 140, 360–371.

Kasper, L.H., Brindle, P.K., Schnabel, C.A., Pritchard, C.E., Cleary, M.L., van Deursen, J.M., 1999. CREB binding protein interacts with nucleoporin-specific FG-repeats that activate transcription and mediate NUP98-HOXA9 oncogenicity. Mol. Cell. Biol. 19, 764–776.

Kassube, S.A., Stuwe, T., Lin, D.H., Antonuk, C.D., Napetschnig, J., Blobel, G., Hoelz, A., 2012. Crystal structure of the N-terminal domain of Nup358/RanBP2. J. Mol. Biol. 423, 752–765.

Kelley, K., Knockenhauer, K.E., Kabachinski, G., Schwartz, T.U., 2015. Atomic structure of the Y complex of the nuclear pore. Nat. Struct. Mol. Biol. 22, 425–431.

Keminer, O., Peters, R., 1999. Permeability of single nuclear pores. Biophys. J. 77, 217–228.

Khuperkar, D., Kamble, A., Singh, A., Ghate, A., Nawadkar, R., Sahu, A., Joseph, J., 2017. Selective recruitment of nucleoporins on vaccinia virus factories and the role of Nup358 in viral infection. Virology 512, 151–160.

Kim, D.I., Birendra, K.C., Zhu, W., Motamedchaboki, K., Doye, V., Roux, K.J., 2014a. Probing nuclear pore complex architecture with proximity-dependent biotinylation. Proc. Natl. Acad. Sci. U.S.A. 111, E2453–E2461.

Kim, N.H., Pham, N.B., Quinn, R.J., Shim, J.S., Cho, H., Cho, S.M., Park, S.W., Kim, J.H., Seok, S.H., Oh, J.W., Kwon, H.J., 2015. The small molecule R-(-)-beta-O-methylsynephrine binds to nucleoporin 153 kDa and inhibits angiogenesis. Int. J. Biol. Sci. 11, 1088–1099.

Kim, S.J., Fernandez-Martinez, J., Sampathkumar, P., Martel, A., Matsui, T., Tsuruta, H., Weiss, T.M., Shi, Y., Markina-Inarrairaegui, A., Bonanno, J.B., Sauder, J.M., Burley, S.K., Chait, B.T., Almo, S.C., Rout, M.P., Sali, A., 2014b. Integrative structure-function mapping of the nucleoporin Nup133 suggests a conserved mechanism for membrane anchoring of the nuclear pore complex. Mol. Cell. Proteom. 13, 2911–2926.

Kinoshita, Y., Hunter, R.G., Gray, J.D., Mesias, R., Mcewen, B.S., Benson, D.L., Kohtz, D.S., 2014. Role for NUP62 depletion and PYK2 redistribution in dendritic retraction resulting from chronic stress. Proc. Natl. Acad. Sci. U.S.A. 111, 16130–16135.

Kinoshita, Y., Kalir, T., Dottino, P., Kohtz, D.S., 2012. Nuclear distributions of NUP62 and NUP214 suggest architectural diversity and spatial patterning among nuclear pore complexes. PLoS One 7, e36137.

Knockenhauer, K.E., Schwartz, T.U., 2016. The nuclear pore complex as a flexible and dynamic gate. Cell 164, 1162–1171.

Koh, Y., Wu, X., Ferris, A.L., Matreyek, K.A., Smith, S.J., Lee, K., Kewalramani, V.N., Hughes, S.H., Engelman, A., 2013. Differential effects of human immunodeficiency virus type 1 capsid and cellular factors nucleoporin 153 and LEDGF/p75 on the efficiency and specificity of viral DNA integration. J. Virol. 87, 648–658.

Konig, R., Zhou, Y., Elleder, D., Diamond, T.L., Bonamy, G.M., Irelan, J.T., Chiang, C.Y., Tu, B.P., De Jesus, P.D., Lilley, C.E., Seidel, S., Opaluch, A.M., Caldwell, J.S., Weitzman, M.D., Kuhen, K.L., Bandyopadhyay, S., Ideker, T., Orth, A.P., Miraglia, L.J., Bushman, F.D., Young, J.A., Chanda, S.K., 2008. Global analysis of host-pathogen interactions that regulate early-stage HIV-1 replication. Cell 135, 49–60.

Kosinski, J., Mosalaganti, S., von Appen, A., Teimer, R., Diguilio, A.L., Wan, W., Bui, K.H., Hagen, W.J., Briggs, J.A., Glavy, J.S., Hurt, E., Beck, M., 2016. Molecular architecture of the inner ring scaffold of the human nuclear pore complex. Science 352, 363–365.

Kraemer, D., Blobel, G., 1997. mRNA binding protein mrnp 41 localizes to both nucleus and cytoplasm. Proc. Natl. Acad. Sci. U.S.A. 94, 9119–9124.

Kraemer, D., Wozniak, R.W., Blobel, G., Radu, A., 1994. The human CAN protein, a putative oncogene product associated with myeloid leukemogenesis, is a nuclear pore complex protein that faces the cytoplasm. Proc. Natl. Acad. Sci. U.S.A. 91, 1519–1523.

Krull, S., Dorries, J., Boysen, B., Reidenbach, S., Magnius, L., Norder, H., Thyberg, J., Cordes, V.C., 2010. Protein Tpr is required for establishing nuclear pore-associated zones of heterochromatin exclusion. EMBO J. 29, 1659–1673.

Krull, S., Thyberg, J., Bjorkroth, B., Rackwitz, H.R., Cordes, V.C., 2004. Nucleoporins as components of the nuclear pore complex core structure and Tpr as the architectural element of the nuclear basket. Mol. Biol. Cell 15, 4261–4277.

Labokha, A.A., Fassati, A., 2013. Viruses challenge selectivity barrier of nuclear pores. Viruses 5, 2410–2423.

Le Rouzic, E., Mousnier, A., Rustum, C., Stutz, F., Hallberg, E., Dargemont, C., Benichou, S., 2002. Docking of HIV-1 Vpr to the nuclear envelope is mediated by the interaction with the nucleoporin hCG1. J. Biol. Chem. 277, 45091–45098.

Le Sage, V., Mouland, A.J., 2013. Viral subversion of the nuclear pore complex. Viruses 5, 2019–2042.

Lee, K., Ambrose, Z., Martin, T.D., Oztop, I., Mulky, A., Julias, J.G., Vandegraaff, N., Baumann, J.G., Wang, R., Yuen, W., Takemura, T., Shelton, K., Taniuchi, I., Li, Y., Sodroski, J., Littman, D.R., Coffin, J.M., Hughes, S.H., Unutmaz, D., Engelman, A., Kewalramani, V.N., 2010. Flexible use of nuclear import pathways by HIV-1. Cell Host Microbe 7, 221–233.

Lee, S.H., Sterling, H., Burlingame, A., Mccormick, F., 2008. Tpr directly binds to Mad1 and Mad2 and is important for the Mad1-Mad2-mediated mitotic spindle checkpoint. Genes Dev. 22, 2926–2931.

Lelek, M., Casartelli, N., Pellin, D., Rizzi, E., Souque, P., Severgnini, M., Di Serio, C., Fricke, T., Diaz-Griffero, F., Zimmer, C., Charneau, P., Di Nunzio, F., 2015. Chromatin organization at the nuclear pore favours HIV replication. Nat. Commun. 6, 6483.

Leng, Y., Cao, C., Ren, J., Huang, L., Chen, D., Ito, M., Kufe, D., 2007. Nuclear import of the MUC1-C oncoprotein is mediated by nucleoporin Nup62. J. Biol. Chem. 282, 19321–19330.

Leuzinger, H., Ziegler, U., Schraner, E.M., Fraefel, C., Glauser, D.L., Heid, I., Ackermann, M., Mueller, M., Wild, P., 2005. Herpes simplex virus 1 envelopment follows two diverse pathways. J. Virol. 79, 13047–13059.

Li, R., Liao, G., Nirujogi, R.S., Pinto, S.M., Shaw, P.G., Huang, T.C., Wan, J., Qian, J., Gowda, H., Wu, X., Lv, D.W., Zhang, K., Manda, S.S., Pandey, A., Hayward, S.D., 2015. Phosphoproteomic profiling reveals Epstein-Barr virus protein kinase integration of DNA damage response and mitotic signaling. PLoS Pathog. 11, e1005346.

Liang, Y., Hetzer, M.W., 2011. Functional interactions between nucleoporins and chromatin. Curr. Opin. Cell Biol. 23, 65–70.

Lidsky, P.V., Hato, S., Bardina, M.V., Aminev, A.G., Palmenberg, A.C., Sheval, E.V., Polyakov, V.Y., van Kuppeveld, F.J., Agol, V.I., 2006. Nucleocytoplasmic traffic disorder induced by cardioviruses. J. Virol. 80, 2705–2717.

Light, W.H., Brickner, J.H., 2013. Nuclear pore proteins regulate chromatin structure and transcriptional memory by a conserved mechanism. Nucleus 4, 357–360.

Lin, D.H., Stuwe, T., Schilbach, S., Rundlet, E.J., Perriches, T., Mobbs, G., Fan, Y., Thierbach, K., Huber, F.M., Collins, L.N., Davenport, A.M., Jeon, Y.E., Hoelz, A., 2016. Architecture of the symmetric core of the nuclear pore. Science 352, aaf1015.

Lin, D.H., Zimmermann, S., Stuwe, T., Stuwe, E., Hoelz, A., 2013. Structural and functional analysis of the C-terminal domain of Nup358/RanBP2. J. Mol. Biol. 425, 1318–1329.

Lu, Y., Kucharski, T.J., Gamache, I., Blanchette, P., Branton, P.E., Teodoro, J.G., 2014. Interaction of adenovirus type 5 E4orf4 with the nuclear pore subunit Nup205 is required for proper viral gene expression. J. Virol. 88, 13249–13259.

Lusic, M., Siliciano, R.F., 2017. Nuclear landscape of HIV-1 infection and integration. Nat. Rev. Microbiol. 15, 69–82.

Lussignol, M., Kopp, M., Molloy, K., Vizcay-Barrena, G., Fleck, R.A., Dorner, M., Bell, K.L., Chait, B.T., Rice, C.M., Catanese, M.T., 2016. Proteomics of HCV virions reveals an essential role for the nucleoporin Nup98 in virus morphogenesis. Proc. Natl. Acad. Sci. U.S.A. 113, 2484–2489.

Lutzmann, M., Kunze, R., Buerer, A., Aebi, U., Hurt, E., 2002. Modular self-assembly of a Y-shaped multiprotein complex from seven nucleoporins. EMBO J. 21, 387–397.

Macaulay, C., Meier, E., Forbes, D.J., 1995. Differential mitotic phosphorylation of proteins of the nuclear pore complex. J. Biol. Chem. 270, 254–262.

Mahadevan, K., Zhang, H., Akef, A., Cui, X.A., Gueroussov, S., Cenik, C., Roth, F.P., Palazzo, A.F., 2013. RanBP2/Nup358 potentiates the translation of a subset of mRNAs encoding secretory proteins. PLoS Biol. 11, e1001545.

Makio, T., Lapetina, D.L., Wozniak, R.W., 2013. Inheritance of yeast nuclear pore complexes requires the Nsp1p subcomplex. J. Cell Biol. 203, 187–196.

Makise, M., Mackay, D.R., Elgort, S., Shankaran, S.S., Adam, S.A., Ullman, K.S., 2012. The Nup153-Nup50 protein interface and its role in nuclear import. J. Biol. Chem. 287, 38515–38522.

Malik, P., Tabarraei, A., Kehlenbach, R.H., Korfali, N., Iwasawa, R., Graham, S.V., Schirmer, E.C., 2012. Herpes simplex virus ICP27 protein directly interacts with the nuclear pore complex through Nup62, inhibiting host nucleocytoplasmic transport pathways. J. Biol. Chem. 287, 12277–12292.

Mans, B.J., Anantharaman, V., Aravind, L., Koonin, E.V., 2004. Comparative genomics, evolution and origins of the nuclear envelope and nuclear pore complex. Cell Cycle 3, 1612–1637.

Mantyla, E., Niskanen, E.A., Ihalainen, T.O., Vihinen-Ranta, M., 2015. Reorganization of nuclear pore complexes and the lamina in late-stage parvovirus infection. J. Virol. 89, 11706–11710.

Marini, B., Kertesz-Farkas, A., Ali, H., Lucic, B., Lisek, K., Manganaro, L., Pongor, S., Luzzati, R., Recchia, A., Mavilio, F., Giacca, M., Lusic, M., 2015. Nuclear architecture dictates HIV-1 integration site selection. Nature 521, 227–231.

Matreyek, K.A., Engelman, A., 2011. The requirement for nucleoporin NUP153 during human immunodeficiency virus type 1 infection is determined by the viral capsid. J. Virol. 85, 7818–7827.

Matreyek, K.A., Engelman, A., 2013. Viral and cellular requirements for the nuclear entry of retroviral preintegration nucleoprotein complexes. Viruses 5, 2483–2511.

Matreyek, K.A., Yucel, S.S., Li, X., Engelman, A., 2013. Nucleoporin NUP153 phenylalanine-glycine motifs engage a common binding pocket within the HIV-1 capsid protein to mediate lentiviral infectivity. PLoS Pathog. 9, e1003693.

Matsuoka, Y., Takagi, M., Ban, T., Miyazaki, M., Yamamoto, T., Kondo, Y., Yoneda, Y., 1999. Identification and characterization of nuclear pore subcomplexes in mitotic extract of human somatic cells. Biochem. Biophys. Res. Commun. 254, 417–423.

McKee, C.H., Onder, Z., Ashok, A., Cardoso, R., Moroianu, J., 2013. Characterization of the transport signals that mediate the nucleocytoplasmic traffic of low risk HPV11 E7. Virology 443, 113–122.

Meehan, A.M., Saenz, D.T., Guevera, R., Morrison, J.H., Peretz, M., Fadel, H.J., Hamada, M., van Deursen, J., Poeschla, E.M., 2014. A cyclophilin homology domain-independent role for Nup358 in HIV-1 infection. PLoS Pathog. 10, e1003969.

Meier, E., Miller, B.R., Forbes, D.J., 1995. Nuclear pore complex assembly studied with a biochemical assay for annulate lamellae formation. J. Cell Biol. 129, 1459–1472.

Mertz, J.A., Lozano, M.M., Dudley, J.P., 2009. Rev and Rex proteins of human complex retroviruses function with the MMTV Rem-responsive element. Retrovirology 6, 10.

Mi, L., Goryaynov, A., Lindquist, A., Rexach, M., Yang, W., 2015. Quantifying nucleoporin stoichiometry inside single nuclear pore complexes in vivo. Sci. Rep. 5, 9372.

Mire, C.E., White, J.M., Whitt, M.A., 2010. A spatio-temporal analysis of matrix protein and nucleocapsid trafficking during vesicular stomatitis virus uncoating. PLoS Pathog. 6, e1000994.

Mishra, R.K., Chakraborty, P., Arnaoutov, A., Fontoura, B.M., Dasso, M., 2010. The Nup107-160 complex and gamma-TuRC regulate microtubule polymerization at kinetochores. Nat. Cell Biol. 12, 164–169.

Mitchell, J.M., Mansfeld, J., Capitanio, J., Kutay, U., Wozniak, R.W., 2010. Pom121 links two essential subcomplexes of the nuclear pore complex core to the membrane. J. Cell Biol. 191, 505–521.

Mizuguchi, C., Oka, M., Moriyama, T., Tachibana, T., Yoneda, Y., 2010. Specific monoclonal antibody against the nuclear pore complex protein, Nup96. Hybridoma (Larchmt) 29, 551–553.

Mohr, D., Frey, S., Fischer, T., Guttler, T., Gorlich, D., 2009. Characterisation of the passive permeability barrier of nuclear pore complexes. EMBO J. 28, 2541–2553.

Monecke, T., Dickmanns, A., Weiss, M.S., Port, S.A., Kehlenbach, R.H., Ficner, R., 2015. Combining dehydration, construct optimization and improved data collection to solve the crystal structure of a CRM1-RanGTP-SPN1-Nup214 quaternary nuclear export complex. Acta Crystallogr. F Struct. Biol. Commun. 71, 1481–1487.

Monette, A., Pante, N., Mouland, A.J., 2011. HIV-1 remodels the nuclear pore complex. J. Cell Biol. 193, 619–631.

Morita, M., Kuba, K., Ichikawa, A., Nakayama, M., Katahira, J., Iwamoto, R., Watanebe, T., Sakabe, S., Daidoji, T., Nakamura, S., Kadowaki, A., Ohto, T., Nakanishi, H., Taguchi, R., Nakaya, T., Murakami, M., Yoneda, Y., Arai, H., Kawaoka, Y., Penninger, J.M., Arita, M., Imai, Y., 2013. The lipid mediator protectin D1 inhibits influenza virus replication and improves severe influenza. Cell 153, 112–125.

Mouland, A.J., Milev, M.P., 2012. Role of dynein in viral pathogenesis. In: King, S.M. (Ed.), Dyneins: Structure, Biology and Disease. Elsevier.

Munday, D.C., Emmott, E., Surtees, R., Lardeau, C.H., Wu, W., Duprex, W.P., Dove, B.K., Barr, J.N., Hiscox, J.A., 2010. Quantitative proteomic analysis of A549 cells infected with human respiratory syncytial virus. Mol. Cell. Proteom. 9, 2438–2459.

Munier, S., Rolland, T., Diot, C., Jacob, Y., Naffakh, N., 2013. Exploration of binary virus-host interactions using an infectious protein complementation assay. Mol. Cell. Proteom. 12, 2845–2855.

Nakamura, T., 2005. NUP98 fusion in human leukemia: dysregulation of the nuclear pore and homeodomain proteins. Int. J. Hematol. 82, 21–27.

Nakano, H., Funasaka, T., Hashizume, C., Wong, R.W., 2010. Nucleoporin translocated promoter region (Tpr) associates with dynein complex, preventing chromosome lagging formation during mitosis. J. Biol. Chem. 285, 10841–10849.

Napetschnig, J., Blobel, G., Hoelz, A., 2007. Crystal structure of the N-terminal domain of the human protooncogene Nup214/CAN. Proc. Natl. Acad. Sci. U.S.A. 104, 1783–1788.

Napetschnig, J., Kassube, S.A., Debler, E.W., Wong, R.W., Blobel, G., Hoelz, A., 2009. Structural and functional analysis of the interaction between the nucleoporin Nup214 and the DEAD-box helicase Ddx19. Proc. Natl. Acad. Sci. U.S.A. 106, 3089–3094.

Neufeldt, C.J., Joyce, M.A., Levin, A., Steenbergen, R.H., Pang, D., Shields, J., Tyrrell, D.L., Wozniak, R.W., 2013. Hepatitis C virus-induced cytoplasmic organelles use the nuclear transport machinery to establish an environment conducive to virus replication. PLoS Pathog. 9, e1003744.

Neville, M., Stutz, F., Lee, L., Davis, L.I., Rosbash, M., 1997. The importin-beta family member Crm1p bridges the interaction between Rev and the nuclear pore complex during nuclear export. Curr. Biol. 7, 767–775.

Niu, X., Hong, J., Zheng, X., Melville, D.B., Knapik, E.W., Meng, A., Peng, J., 2014. The nuclear pore complex function of Sec13 protein is required for cell survival during retinal development. J. Biol. Chem. 289, 11971–11985.

Noble, K.N., Tran, E.J., Alcazar-Roman, A.R., Hodge, C.A., Cole, C.N., Wente, S.R., 2011. The Dbp5 cycle at the nuclear pore complex during mRNA export II: nucleotide cycling and mRNP remodeling by Dbp5 are controlled by Nup159 and Gle1. Genes Dev. 25, 1065–1077.

Nofrini, V., Di Giacomo, D., Mecucci, C., 2016. Nucleoporin genes in human diseases. Eur. J. Hum. Genet. 24, 1388–1395.

Oberstein, A., Perlman, D.H., Shenk, T., Terry, L.J., 2015. Human cytomegalovirus pUL97 kinase induces global changes in the infected cell phosphoproteome. Proteomics 15, 2006–2022.

Ocwieja, K.E., Brady, T.L., Ronen, K., Huegel, A., Roth, S.L., Schaller, T., James, L.C., Towers, G.J., Young, J.A., Chanda, S.K., Konig, R., Malani, N., Berry, C.C., Bushman, F.D., 2011. HIV integration targeting: a pathway involving transportin-3 and the nuclear pore protein RanBP2. PLoS Pathog. 7, e1001313.

Onder, Z., Moroianu, J., 2014. Nuclear import of cutaneous beta genus HPV8 E7 oncoprotein is mediated by hydrophobic interactions between its zinc-binding domain and FG nucleoporins. Virology 449, 150–162.

Paine, P.L., Moore, L.C., Horowitz, S.B., 1975. Nuclear envelope permeability. Nature 254, 109–114.

Panda, D., Gold, B., Tartell, M.A., Rausch, K., Casas-Tinto, S., Cherry, S., 2015. The transcription factor FoxK participates with Nup98 to regulate antiviral gene expression. mBio 6.

Pante, N., Bastos, R., Mcmorrow, I., Burke, B., Aebi, U., 1994. Interactions and three-dimensional localization of a group of nuclear pore complex proteins. J. Cell Biol. 126, 603–617.

Pante, N., Kann, M., 2002. Nuclear pore complex is able to transport macromolecules with diameters of about 39 nm. Mol. Biol. Cell 13, 425–434.

Pante, N., Thomas, F., Aebi, U., Burke, B., Bastos, R., 2000. Recombinant Nup153 incorporates in vivo into *Xenopus* oocyte nuclear pore complexes. J. Struct. Biol. 129, 306–312.

Park, M., Gonzatti-Haces, M., Dean, M., Blair, D.G., Testa, J.R., Bennett, D.D., Copeland, T., Oroszlan, S., Vande Woude, G., 1986. The met oncogene: a new member of the tyrosine kinase family and a marker for cystic fibrosis. Cold Spring Harb. Symp. Quant. Biol. 51 (Pt 2), 967–975.

Park, N., Schweers, N.J., Gustin, K.E., 2015. Selective removal of FG-repeat domains from the nuclear pore complex by enterovirus 2A(pro). J. Virol. 89, 11069–11079.

Park, N., Skern, T., Gustin, K.E., 2010. Specific cleavage of the nuclear pore complex protein Nup62 by a viral protease. J. Biol. Chem. 285, 28796–28805.

Partridge, J.R., Schwartz, T.U., 2009. Crystallographic and biochemical analysis of the Ran-binding zinc finger domain. J. Mol. Biol. 391, 375–389.

Paschal, B.M., Gerace, L., 1995. Identification of NTF2, a cytosolic factor for nuclear import that interacts with nuclear pore complex protein p62. J. Cell Biol. 129, 925–937.

Pascual-Garcia, P., Capelson, M., 2014. Nuclear pores as versatile platforms for gene regulation. Curr. Opin. Genet. Dev. 25, 110–117.

Pasdeloup, D., Blondel, D., Isidro, A.L., Rixon, F.J., 2009. Herpesvirus capsid association with the nuclear pore complex and viral DNA release involve the nucleoporin CAN/Nup214 and the capsid protein pUL25. J. Virol. 83, 6610–6623.

Paulillo, S.M., Phillips, E.M., Koser, J., Sauder, U., Ullman, K.S., Powers, M.A., Fahrenkrog, B., 2005. Nucleoporin domain topology is linked to the transport status of the nuclear pore complex. J. Mol. Biol. 351, 784–798.

Pentecost, M., Vashisht, A.A., Lester, T., Voros, T., Beaty, S.M., Park, A., Wang, Y.E., Yun, T.E., Freiberg, A.N., Wohlschlegel, J.A., Lee, B., 2015. Evidence for ubiquitin-regulated nuclear and subnuclear trafficking among Paramyxovirinae matrix proteins. PLoS Pathog. 11, e1004739.

Percipalle, P., Clarkson, W.D., Kent, H.M., Rhodes, D., Stewart, M., 1997. Molecular interactions between the importin alpha/beta heterodimer and proteins involved in vertebrate nuclear protein import. J. Mol. Biol. 266, 722–732.

Peters, R., 2005. Translocation through the nuclear pore complex: selectivity and speed by reduction-of-dimensionality. Traffic 6, 421–427.

Pichler, A., Gast, A., Seeler, J.S., Dejean, A., Melchior, F., 2002. The nucleoporin RanBP2 has SUMO1 E3 ligase activity. Cell 108, 109–120.

Port, S.A., Monecke, T., Dickmanns, A., Spillner, C., Hofele, R., Urlaub, H., Ficner, R., Kehlenbach, R.H., 2015. Structural and functional characterization of CRM1-Nup214 interactions reveals multiple FG-binding sites involved in nuclear export. Cell Rep. 13, 690–702.

Porter, F.W., Palmenberg, A.C., 2009. Leader-induced phosphorylation of nucleoporins correlates with nuclear trafficking inhibition by cardioviruses. J. Virol. 83, 1941–1951.

Powers, M.A., Forbes, D.J., Dahlberg, J.E., Lund, E., 1997. The vertebrate GLFG nucleoporin, Nup98, is an essential component of multiple RNA export pathways. J. Cell Biol. 136, 241–250.

Powers, M.A., Macaulay, C., Masiarz, F.R., Forbes, D.J., 1995. Reconstituted nuclei depleted of a vertebrate GLFG nuclear pore protein, p97, import but are defective in nuclear growth and replication. J. Cell Biol. 128, 721–736.

Price, A.J., Fletcher, A.J., Schaller, T., Elliott, T., Lee, K., Kewalramani, V.N., Chin, J.W., Towers, G.J., James, L.C., 2012. CPSF6 defines a conserved capsid interface that modulates HIV-1 replication. PLoS Pathog. 8, e1002896.

Price, A.J., Jacques, D.A., Mcewan, W.A., Fletcher, A.J., Essig, S., Chin, J.W., Halambage, U.D., Aiken, C., James, L.C., 2014. Host cofactors and pharmacologic ligands share an essential interface in HIV-1 capsid that is lost upon disassembly. PLoS Pathog. 10, e1004459.

Prunuske, A.J., Liu, J., Elgort, S., Joseph, J., Dasso, M., Ullman, K.S., 2006. Nuclear envelope breakdown is coordinated by both Nup358/RanBP2 and Nup153, two nucleoporins with zinc finger modules. Mol. Biol. Cell 17, 760–769.

Ptak, C., Aitchison, J.D., Wozniak, R.W., 2014. The multifunctional nuclear pore complex: a platform for controlling gene expression. Curr. Opin. Cell Biol. 28, 46–53.

Quan, B., Seo, H.S., Blobel, G., Ren, Y., 2014. Vesiculoviral matrix (M) protein occupies nucleic acid binding site at nucleoporin pair (Rae1 * Nup98). Proc. Natl. Acad. Sci. U.S.A. 111, 9127–9132.

Radu, A., Blobel, G., Moore, M.S., 1995a. Identification of a protein complex that is required for nuclear protein import and mediates docking of import substrate to distinct nucleoporins. Proc. Natl. Acad. Sci. U.S.A. 92, 1769–1773.

Radu, A., Moore, M.S., Blobel, G., 1995b. The peptide repeat domain of nucleoporin Nup98 functions as a docking site in transport across the nuclear pore complex. Cell 81, 215–222.

Rajanala, K., Nandicoori, V.K., 2012. Localization of nucleoporin Tpr to the nuclear pore complex is essential for Tpr mediated regulation of the export of unspliced RNA. PLoS One 7, e29921.

Rajani, K.R., Pettit Kneller, E.L., Mckenzie, M.O., Horita, D.A., Chou, J.W., Lyles, D.S., 2012. Complexes of vesicular stomatitis virus matrix protein with host Rae1 and Nup98 involved in inhibition of host transcription. PLoS Pathog. 8, e1002929.

Rasaiyaah, J., Tan, C.P., Fletcher, A.J., Price, A.J., Blondeau, C., Hilditch, L., Jacques, D.A., Selwood, D.L., James, L.C., Noursadeghi, M., Towers, G.J., 2013. HIV-1 evades innate immune recognition through specific cofactor recruitment. Nature 503, 402–405.

Rayala, H.J., Kendirgi, F., Barry, D.M., Majerus, P.W., Wente, S.R., 2004. The mRNA export factor human Gle1 interacts with the nuclear pore complex protein Nup155. Mol. Cell. Proteom. 3, 145–155.

Reichelt, R., Holzenburg, A., Buhle Jr., E.L., Jarnik, M., Engel, A., Aebi, U., 1990. Correlation between structure and mass distribution of the nuclear pore complex and of distinct pore complex components. J. Cell Biol. 110, 883–894.

Ren, Y., Seo, H.S., Blobel, G., Hoelz, A., 2010. Structural and functional analysis of the interaction between the nucleoporin Nup98 and the mRNA export factor Rae1. Proc. Natl. Acad. Sci. U.S.A. 107, 10406–10411.

Rexach, M., Blobel, G., 1995. Protein import into nuclei: association and dissociation reactions involving transport substrate, transport factors, and nucleoporins. Cell 83, 683–692.
Ribbeck, K., Lipowsky, G., Kent, H.M., Stewart, M., Gorlich, D., 1998. NTF2 mediates nuclear import of Ran. EMBO J. 17, 6587–6598.
Romana, S.P., Radford-Weiss, I., Ben Abdelali, R., Schluth, C., Petit, A., Dastugue, N., Talmant, P., Bilhou-Nabera, C., Mugneret, F., Lafage-Pochitaloff, M., Mozziconacci, M.J., Andrieu, J., Lai, J.L., Terre, C., Rack, K., Cornillet-Lefebvre, P., Luquet, I., Nadal, N., Nguyen-Khac, F., Perot, C., Van den Akker, J., Fert-Ferrer, S., Cabrol, C., Charrin, C., Tigaud, I., Poirel, H., Vekemans, M., Bernard, O.A., Berger, R., Groupe Francophone de Cytogenetique Hematologique, 2006. NUP98 rearrangements in hematopoietic malignancies: a study of the Groupe Francophone de Cytogenetique Hematologique. Leukemia 20, 696–706.
Roth, P., Xylourgidis, N., Sabri, N., Uv, A., Fornerod, M., Samakovlis, C., 2003. The Drosophila nucleoporin DNup88 localizes DNup214 and CRM1 on the nuclear envelope and attenuates NES-mediated nuclear export. J. Cell Biol. 163, 701–706.
Rout, M.P., Aitchison, J.D., Suprapto, A., Hjertaas, K., Zhao, Y., Chait, B.T., 2000. The yeast nuclear pore complex: composition, architecture, and transport mechanism. J. Cell Biol. 148, 635–651.
Sachdev, R., Sieverding, C., Flotenmeyer, M., Antonin, W., 2012. The C-terminal domain of Nup93 is essential for assembly of the structural backbone of nuclear pore complexes. Mol. Biol. Cell 23, 740–749.
Saito, A., Henning, M.S., Serrao, E., Dubose, B.N., Teng, S., Huang, J., Li, X., Saito, N., Roy, S.P., Siddiqui, M.A., Ahn, J., Tsuji, M., Hatziioannou, T., Engelman, A.N., Yamashita, M., 2016. Capsid-CPSF6 interaction is dispensable for HIV-1 replication in primary cells but is selected during virus passage in vivo. J. Virol. 90, 6918–6935.
Saito, H., Takeuchi, H., Masuda, T., Noda, T., Yamaoka, S., 2017. N-terminally truncated POM121C inhibits HIV-1 replication. PLoS One 12, e0182434.
Sakuma, S., D'Angelo, M.A., 2017. The roles of the nuclear pore complex in cellular dysfunction, aging and disease. Semin. Cell Dev. Biol. 68, 72–84.
Sato, K., Nakano, A., 2007. Mechanisms of COPII vesicle formation and protein sorting. FEBS Lett. 581, 2076–2082.
Savas, J.N., Toyama, B.H., Xu, T., Yates 3rd, J.R., Hetzer, M.W., 2012. Extremely long-lived nuclear pore proteins in the rat brain. Science 335, 942.
Schaller, T., Ocwieja, K.E., Rasaiyaah, J., Price, A.J., Brady, T.L., Roth, S.L., Hue, S., Fletcher, A.J., Lee, K., Kewalramani, V.N., Noursadeghi, M., Jenner, R.G., James, L.C., Bushman, F.D., Towers, G.J., 2011. HIV-1 capsid-cyclophilin interactions determine nuclear import pathway, integration targeting and replication efficiency. PLoS Pathog. 7, e1002439.
Scheifele, L.Z., Ryan, E.P., Parent, L.J., 2005. Detailed mapping of the nuclear export signal in the Rous sarcoma virus Gag protein. J. Virol. 79, 8732–8741.
Schmitz, A., Schwarz, A., Foss, M., Zhou, L., Rabe, B., Hoellenriegel, J., Stoeber, M., Pante, N., Kann, M., 2010. Nucleoporin 153 arrests the nuclear import of hepatitis B virus capsids in the nuclear basket. PLoS Pathog. 6, e1000741.
Schrader, N., Koerner, C., Koessmeier, K., Bangert, J.A., Wittinghofer, A., Stoll, R., Vetter, I.R., 2008a. The crystal structure of the Ran-Nup153ZnF2 complex: a general Ran docking site at the nuclear pore complex. Structure 16, 1116–1125.
Schrader, N., Stelter, P., Flemming, D., Kunze, R., Hurt, E., Vetter, I.R., 2008b. Structural basis of the nic96 subcomplex organization in the nuclear pore channel. Mol. Cell 29, 46–55.
Seo, H.S., Blus, B.J., Jankovic, N.Z., Blobel, G., 2013. Structure and nucleic acid binding activity of the nucleoporin Nup157. Proc. Natl. Acad. Sci. U.S.A. 110, 16450–16455.

Seo, H.S., Ma, Y., Debler, E.W., Wacker, D., Kutik, S., Blobel, G., Hoelz, A., 2009. Structural and functional analysis of Nup120 suggests ring formation of the Nup84 complex. Proc. Natl. Acad. Sci. U.S.A. 106, 14281–14286.

Shah, S., Forbes, D.J., 1998. Separate nuclear import pathways converge on the nucleoporin Nup153 and can be dissected with dominant-negative inhibitors. Curr. Biol. 8, 1376–1386.

Shah, S., Tugendreich, S., Forbes, D., 1998. Major binding sites for the nuclear import receptor are the internal nucleoporin Nup153 and the adjacent nuclear filament protein Tpr. J. Cell Biol. 141, 31–49.

Shahin, V., 2016. Cellular transport: gatekeepers of the nucleus. Nat. Nanotechnol. 11, 658–659.

Sharma, A., Solmaz, S.R., Blobel, G., Melcak, I., 2015. Ordered regions of channel nucleoporins Nup62, Nup54, and Nup58 form dynamic complexes in solution. J. Biol. Chem. 290, 18370–18378.

Sheffield, L.G., Miskiewicz, H.B., Tannenbaum, L.B., Mirra, S.S., 2006. Nuclear pore complex proteins in Alzheimer disease. J. Neuropathol. Exp. Neurol. 65, 45–54.

Shibata, S., Matsuoka, Y., Yoneda, Y., 2002. Nucleocytoplasmic transport of proteins and poly(A)+ RNA in reconstituted Tpr-less nuclei in living mammalian cells. Genes Cells 7, 421–434.

Simon, D.N., Rout, M.P., 2014. Cancer and the nuclear pore complex. Adv. Exp. Med. Biol. 773, 285–307.

Singh, B.B., Patel, H.H., Roepman, R., Schick, D., Ferreira, P.A., 1999. The zinc finger cluster domain of RanBP2 is a specific docking site for the nuclear export factor, exportin-1. J. Biol. Chem. 274, 37370–37378.

Siniossoglou, S., Wimmer, C., Rieger, M., Doye, V., Tekotte, H., Weise, C., Emig, S., Segref, A., Hurt, E.C., 1996. A novel complex of nucleoporins, which includes Sec13p and a Sec13p homolog, is essential for normal nuclear pores. Cell 84, 265–275.

Sivan, G., Martin, S.E., Myers, T.G., Buehler, E., Szymczyk, K.H., Ormanoglu, P., Moss, B., 2013. Human genome-wide RNAi screen reveals a role for nuclear pore proteins in poxvirus morphogenesis. Proc. Natl. Acad. Sci. U.S.A. 110, 3519–3524.

Solmaz, S.R., Chauhan, R., Blobel, G., Melcak, I., 2011. Molecular architecture of the transport channel of the nuclear pore complex. Cell 147, 590–602.

Soman, N.R., Correa, P., Ruiz, B.A., Wogan, G.N., 1991. The TPR-MET oncogenic rearrangement is present and expressed in human gastric carcinoma and precursor lesions. Proc. Natl. Acad. Sci. U.S.A. 88, 4892–4896.

Sorokin, A.V., Kim, E.R., Ovchinnikov, L.P., 2007. Nucleocytoplasmic transport of proteins. Biochemistry (Mosc.) 72, 1439–1457.

Stagg, S.M., Gurkan, C., Fowler, D.M., Lapointe, P., Foss, T.R., Potter, C.S., Carragher, B., Balch, W.E., 2006. Structure of the Sec13/31 COPII coat cage. Nature 439, 234–238.

Stelter, P., Kunze, R., Flemming, D., Hopfner, D., Diepholz, M., Philippsen, P., Bottcher, B., Hurt, E., 2007. Molecular basis for the functional interaction of dynein light chain with the nuclear-pore complex. Nat. Cell Biol. 9, 788–796.

Stochaj, U., Banski, P., Kodiha, M., Matusiewicz, N., 2006. The N-terminal domain of the mammalian nucleoporin p62 interacts with other nucleoporins of the FXFG family during interphase. Exp. Cell Res. 312, 2490–2499.

Stoffler, D., Fahrenkrog, B., Aebi, U., 1999. The nuclear pore complex: from molecular architecture to functional dynamics. Curr. Opin. Cell Biol. 11, 391–401.

Strambio-De-Castillia, C., Niepel, M., Rout, M.P., 2010. The nuclear pore complex: bridging nuclear transport and gene regulation. Nat. Rev. Mol. Cell Biol. 11, 490–501.

Strawn, L.A., Shen, T., Wente, S.R., 2001. The GLFG regions of Nup116p and Nup100p serve as binding sites for both Kap95p and Mex67p at the nuclear pore complex. J. Biol. Chem. 276, 6445–6452.

Strunze, S., Engelke, M.F., Wang, I.H., Puntener, D., Boucke, K., Schleich, S., Way, M., Schoenenberger, P., Burckhardt, C.J., Greber, U.F., 2011. Kinesin-1-mediated capsid disassembly and disruption of the nuclear pore complex promote virus infection. Cell Host Microbe 10, 210–223.

Stuwe, T., von Borzyskowski, L.S., Davenport, A.M., Hoelz, A., 2012. Molecular basis for the anchoring of proto-oncoprotein Nup98 to the cytoplasmic face of the nuclear pore complex. J. Mol. Biol. 419, 330–346.

Sukegawa, J., Blobel, G., 1993. A nuclear pore complex protein that contains zinc finger motifs, binds DNA, and faces the nucleoplasm. Cell 72, 29–38.

Sun, Y., Guo, H.C., 2008. Structural constraints on autoprocessing of the human nucleoporin Nup98. Protein Sci. 17, 494–505.

Suntharalingam, M., Wente, S.R., 2003. Peering through the pore: nuclear pore complex structure, assembly, and function. Dev. Cell 4, 775–789.

Swaminathan, G., Rossi, F., Sierra, L.J., Gupta, A., Navas-Martin, S., Martin-Garcia, J., 2012. A role for microRNA-155 modulation in the anti-HIV-1 effects of Toll-like receptor 3 stimulation in macrophages. PLoS Pathog. 8, e1002937.

Takeda, A., Sarma, N.J., Abdul-Nabi, A.M., Yaseen, N.R., 2010. Inhibition of CRM1-mediated nuclear export of transcription factors by leukemogenic NUP98 fusion proteins. J. Biol. Chem. 285, 16248–16257.

Teixeira, M.T., Siniossoglou, S., Podtelejnikov, S., Benichou, J.C., Mann, M., Dujon, B., Hurt, E., Fabre, E., 1997. Two functionally distinct domains generated by in vivo cleavage of Nup145p: a novel biogenesis pathway for nucleoporins. EMBO J. 16, 5086–5097.

Terry, L.J., Wente, S.R., 2007. Nuclear mRNA export requires specific FG nucleoporins for translocation through the nuclear pore complex. J. Cell Biol. 178, 1121–1132.

Tran, E.J., Wente, S.R., 2006. Dynamic nuclear pore complexes: life on the edge. Cell 125, 1041–1053.

Trotman, L.C., Mosberger, N., Fornerod, M., Stidwill, R.P., Greber, U.F., 2001. Import of adenovirus DNA involves the nuclear pore complex receptor CAN/Nup214 and histone H1. Nat. Cell Biol. 3, 1092–1100.

Tsuji, T., Sheehy, N., Gautier, V.W., Hayakawa, H., Sawa, H., Hall, W.W., 2007. The nuclear import of the human T lymphotropic virus type I (HTLV-1) tax protein is carrier- and energy-independent. J. Biol. Chem. 282, 13875–13883.

Ullman, K.S., Shah, S., Powers, M.A., Forbes, D.J., 1999. The nucleoporin nup153 plays a critical role in multiple types of nuclear export. Mol. Biol. Cell 10, 649–664.

Ulrich, A., Partridge, J.R., Schwartz, T.U., 2014. The stoichiometry of the nucleoporin 62 subcomplex of the nuclear pore in solution. Mol. Biol. Cell 25, 1484–1492.

Vasu, S., Shah, S., Orjalo, A., Park, M., Fischer, W.H., Forbes, D.J., 2001. Novel vertebrate nucleoporins Nup133 and Nup160 play a role in mRNA export. J. Cell Biol. 155, 339–354.

Vollmer, B., Antonin, W., 2014. The diverse roles of the Nup93/Nic96 complex proteins – structural scaffolds of the nuclear pore complex with additional cellular functions. Biol. Chem. 395, 515–528.

Vollmer, B., Lorenz, M., Moreno-Andres, D., Bodenhofer, M., De Magistris, P., Astrinidis, S.A., Schooley, A., Flotenmeyer, M., Leptihn, S., Antonin, W., 2015. Nup153 recruits the Nup107-160 complex to the inner nuclear membrane for interphasic nuclear pore complex assembly. Dev. Cell 33, 717–728.

von Lindern, M., Fornerod, M., Soekarman, N., Van Baal, S., Jaegle, M., Hagemeijer, A., Bootsma, D., Grosveld, G., 1992. Translocation t(6;9) in acute non-lymphocytic leukaemia results in the formation of a DEK-CAN fusion gene. Baillieres Clin. Haematol. 5, 857–879.

von Moeller, H., Basquin, C., Conti, E., 2009. The mRNA export protein DBP5 binds RNA and the cytoplasmic nucleoporin NUP214 in a mutually exclusive manner. Nat. Struct. Mol. Biol. 16, 247–254.

Walde, S., Thakar, K., Hutten, S., Spillner, C., Nath, A., Rothbauer, U., Wiemann, S., Kehlenbach, R.H., 2012. The nucleoporin Nup358/RanBP2 promotes nuclear import in a cargo- and transport receptor-specific manner. Traffic 13, 218–233.

Walker, E.J., Ghildyal, R., 2017. Editorial: viral interactions with the nucleus. Front. Microbiol. 8, 951.

Walker, E.J., Jensen, L.M., Croft, S., Ghildyal, R., 2015. Variation in the nuclear effects of infection by different human rhinovirus serotypes. Front. Microbiol. 6, 875.

Walker, E.J., Younessi, P., Fulcher, A.J., Mccuaig, R., Thomas, B.J., Bardin, P.G., Jans, D.A., Ghildyal, R., 2013. Rhinovirus 3C protease facilitates specific nucleoporin cleavage and mislocalisation of nuclear proteins in infected host cells. PLoS One 8, e71316.

Walther, T.C., Alves, A., Pickersgill, H., Loiodice, I., Hetzer, M., Galy, V., Hulsmann, B.B., Kocher, T., Wilm, M., Allen, T., Mattaj, I.W., Doye, V., 2003a. The conserved Nup107-160 complex is critical for nuclear pore complex assembly. Cell 113, 195–206.

Walther, T.C., Askjaer, P., Gentzel, M., Habermann, A., Griffiths, G., Wilm, M., Mattaj, I.W., Hetzer, M., 2003b. RanGTP mediates nuclear pore complex assembly. Nature 424, 689–694.

Wang, G.G., Song, J., Wang, Z., Dormann, H.L., Casadio, F., Li, H., Luo, J.L., Patel, D.J., Allis, C.D., 2009. Haematopoietic malignancies caused by dysregulation of a chromatin-binding PHD finger. Nature 459, 847–851.

Watters, K., Palmenberg, A.C., 2011. Differential processing of nuclear pore complex proteins by rhinovirus 2A proteases from different species and serotypes. J. Virol. 85, 10874–10883.

Weirich, C.S., Erzberger, J.P., Berger, J.M., Weis, K., 2004. The N-terminal domain of Nup159 forms a beta-propeller that functions in mRNA export by tethering the helicase Dbp5 to the nuclear pore. Mol. Cell 16, 749–760.

Weis, K., 2007. The nuclear pore complex: oily spaghetti or gummy bear? Cell 130, 405–407.

Wente, S.R., Rout, M.P., 2010. The nuclear pore complex and nuclear transport. Cold Spring Harb. Perspect. Biol. 2, a000562.

Whittle, J.R., Schwartz, T.U., 2009. Architectural nucleoporins Nup157/170 and Nup133 are structurally related and descend from a second ancestral element. J. Biol. Chem. 284, 28442–28452.

Wiermer, M., Cheng, Y.T., Imkampe, J., Li, M., Wang, D., Lipka, V., Li, X., 2012. Putative members of the Arabidopsis Nup107-160 nuclear pore sub-complex contribute to pathogen defense. Plant J. 70, 796–808.

Woodward, C.L., Chow, S.A., 2010. The nuclear pore complex: a new dynamic in HIV-1 replication. Nucleus 1, 18–22.

Woodward, C.L., Prakobwanakit, S., Mosessian, S., Chow, S.A., 2009. Integrase interacts with nucleoporin NUP153 to mediate the nuclear import of human immunodeficiency virus type 1. J. Virol. 83, 6522–6533.

Wu, J., Matunis, M.J., Kraemer, D., Blobel, G., Coutavas, E., 1995. Nup358, a cytoplasmically exposed nucleoporin with peptide repeats, Ran-GTP binding sites, zinc fingers, a cyclophilin A homologous domain, and a leucine-rich region. J. Biol. Chem. 270, 14209–14213.

Xu, S., Powers, M.A., 2009. Nuclear pore proteins and cancer. Semin. Cell Dev. Biol. 20, 620–630.

Xylourgidis, N., Fornerod, M., 2009. Acting out of character: regulatory roles of nuclear pore complex proteins. Dev. Cell 17, 617–625.

Xylourgidis, N., Roth, P., Sabri, N., Tsarouhas, V., Samakovlis, C., 2006. The nucleoporin Nup214 sequesters CRM1 at the nuclear rim and modulates NFkappaB activation in Drosophila. J. Cell Sci. 119, 4409–4419.

Yamauchi, Y., Greber, U.F., 2016. Principles of virus uncoating: cues and the snooker ball. Traffic 17, 569–592.

Yarbrough, M.L., Mata, M.A., Sakthivel, R., Fontoura, B.M., 2014. Viral subversion of nucleocytoplasmic trafficking. Traffic 15, 127–140.

Yavuz, S., Santarella-Mellwig, R., Koch, B., Jaedicke, A., Mattaj, I.W., Antonin, W., 2010. NLS-mediated NPC functions of the nucleoporin Pom121. FEBS Lett. 584, 3292–3298.

Yeung, M.L., Houzet, L., Yedavalli, V.S., Jeang, K.T., 2009. A genome-wide short hairpin RNA screening of jurkat T-cells for human proteins contributing to productive HIV-1 replication. J. Biol. Chem. 284, 19463–19473.

Yokoyama, N., Hayashi, N., Seki, T., Pante, N., Ohba, T., Nishii, K., Kuma, K., Hayashida, T., Miyata, T., Aebi, U., et al., 1995. A giant nucleopore protein that binds Ran/TC4. Nature 376, 184–188.

Yu, S.H., Boyce, M., Wands, A.M., Bond, M.R., Bertozzi, C.R., Kohler, J.J., 2012. Metabolic labeling enables selective photocrosslinking of O-GlcNAc-modified proteins to their binding partners. Proc. Natl. Acad. Sci. U.S.A. 109, 4834–4839.

Zeev-Ben-Mordehai, T., Weberruss, M., Lorenz, M., Cheleski, J., Hellberg, T., Whittle, C., El Omari, K., Vasishtan, D., Dent, K.C., Harlos, K., Franzke, K., Hagen, C., Klupp, B.G., Antonin, W., Mettenleiter, T.C., Grunewald, K., 2015. Crystal structure of the herpesvirus nuclear egress complex provides insights into inner nuclear membrane remodeling. Cell Rep. 13, 2645–2652.

Zhang, L.K., Chai, F., Li, H.Y., Xiao, G., Guo, L., 2013. Identification of host proteins involved in Japanese encephalitis virus infection by quantitative proteomics analysis. J. Proteome Res. 12, 2666–2678.

Zhang, R., Mehla, R., Chauhan, A., 2010. Perturbation of host nuclear membrane component RanBP2 impairs the nuclear import of human immunodeficiency virus-1 preintegration complex (DNA). PLoS One 5, e15620.

Zhang, Y., Li, X., 2005. A putative nucleoporin 96 Is required for both basal defense and constitutive resistance responses mediated by suppressor of npr1-1, constitutive 1. Plant Cell 17, 1306–1316.

Zhao, C.L., Mahboobi, S.H., Moussavi-Baygi, R., Mofrad, M.R., 2014. The interaction of CRM1 and the nuclear pore protein Tpr. PLoS One 9, e93709.

Zhou, H., Xu, M., Huang, Q., Gates, A.T., Zhang, X.D., Castle, J.C., Stec, E., Ferrer, M., Strulovici, B., Hazuda, D.J., Espeseth, A.S., 2008. Genome-scale RNAi screen for host factors required for HIV replication. Cell Host Microbe 4, 495–504.

Zhou, J., Price, A.J., Halambage, U.D., James, L.C., Aiken, C., 2015. HIV-1 resistance to the capsid-targeting inhibitor PF74 results in altered dependence on host factors required for virus nuclear entry. J. Virol. 89, 9068–9079.

Zhu, J., Davoli, T., Perriera, J.M., Chin, C.R., Gaiha, G.D., John, S.P., Sigiollot, F.D., Gao, G., Xu, Q., Qu, H., Pertel, T., Sims, J.S., Smith, J.A., Baker, R.E., Maranda, L., Ng, A., Elledge, S.J., Brass, A.L., 2014. Comprehensive identification of host modulators of HIV-1 replication using multiple orthologous RNAi reagents. Cell Rep. 9, 752–766.

Zhu, Y., Wang, B., Tang, K., Hsu, C.C., Xie, S., Du, H., Yang, Y., Tao, W.A., Zhu, J.K., 2017. An Arabidopsis Nucleoporin NUP85 modulates plant responses to ABA and salt stress. PLoS Genet. 13, e1007124.

Zimowska, G., Paddy, M.R., 2002. Structures and dynamics of Drosophila Tpr inconsistent with a static, filamentous structure. Exp. Cell Res. 276, 223–232.

Zolotukhin, A.S., Felber, B.K., 1999. Nucleoporins nup98 and nup214 participate in nuclear export of human immunodeficiency virus type 1 Rev. J. Virol. 73, 120–127.

Chapter 4

Virus–Host Interactions in Retrovirus Integration

Alan N. Engelman[1,2], Goedele N. Maertens[3]

[1]Dana-Farber Cancer Institute, Boston, MA, United States; [2]Harvard Medical School, Boston, MA, United States; [3]Imperial College London, London, United Kingdom

Chapter Outline

Introduction 163
Integrase-Interacting Proteins 167
Lentiviruses 167
Lens Epithelium–Derived Growth Factor/p75 167
Hepatoma-derived Growth Factor Like 2 (HDGFL2) 170
Gammaretroviruses 170
Bromodomain and Extra-Terminal Domain Proteins 171
Deltaretroviruses 172
B′-Protein Phosphatase 2A (PP2A) 173
Gag-Interacting Proteins 177
HIV-1 CA–Interacting Proteins 177
Foamy Virus 179
Allosteric IN Inhibitors That Target the LEDGF/p75–IN Interaction 180
Conclusions 182
Acknowledgments 183
References 183

INTRODUCTION

Retroviruses are the only animal viruses that are known to require the stable integration of their genetic information into the genomes of the host cells they infect. Because retroviruses possess RNA genomes, they undergo reverse transcription to synthesize the requisite DNA precursor molecule for integration. They additionally encode for and carry within their viral particles a specialized DNA recombinase protein known as integrase (IN). A variety of double-stranded forms of retroviral DNA, including linear with a copy of the long terminal repeat (LTR) at each end and circular forms that contain one or two copies of the LTR, are found in infected cells. The linear DNA molecule is the immediate precursor to the integrated provirus (Brown et al., 1987, 1989; Fujiwara and Mizuuchi, 1988).

Retroviral IN harbors two distinct enzymatic activities, 3′ processing and strand transfer, which are catalyzed within the context of a subviral preintegration

Retrovirus-Cell Interactions. https://doi.org/10.1016/B978-0-12-811185-7.00004-2

complex (PIC) derived from the infecting virion (Bowerman et al., 1989; Wei et al., 1997). A multimer of IN initially engages the two LTR ends of viral DNA to form the stable synaptic complex (SSC) (Hare et al., 2012; Li et al., 2006; Miller et al., 1997; Wei et al., 1997). During 3′ processing, IN hydrolyzes phosphodiester bonds adjacent to invariant CA dinucleotides, which in the majority of cases liberates a dinucleotide from each LTR end (Craigie et al., 1990; Katzman et al., 1989; Roth et al., 1989; Sherman and Fyfe, 1990). After trafficking to the nucleus and engaging chromosomal DNA, IN uses the viral DNA CA_{OH}-3′-hydroxyl groups to cut opposing target DNA strands at defined spacing, which at the same time joins the viral DNA ends to host 5′-phosphates (Engelman et al., 1991). The resulting strand transfer complex is dismantled, assumingly through the action of host proteases, to enable host-mediated repair of the hemi-integrant, yielding a short 4–6 bp duplication of target DNA flanking the provirus (reviewed in Lesbats et al., 2016).

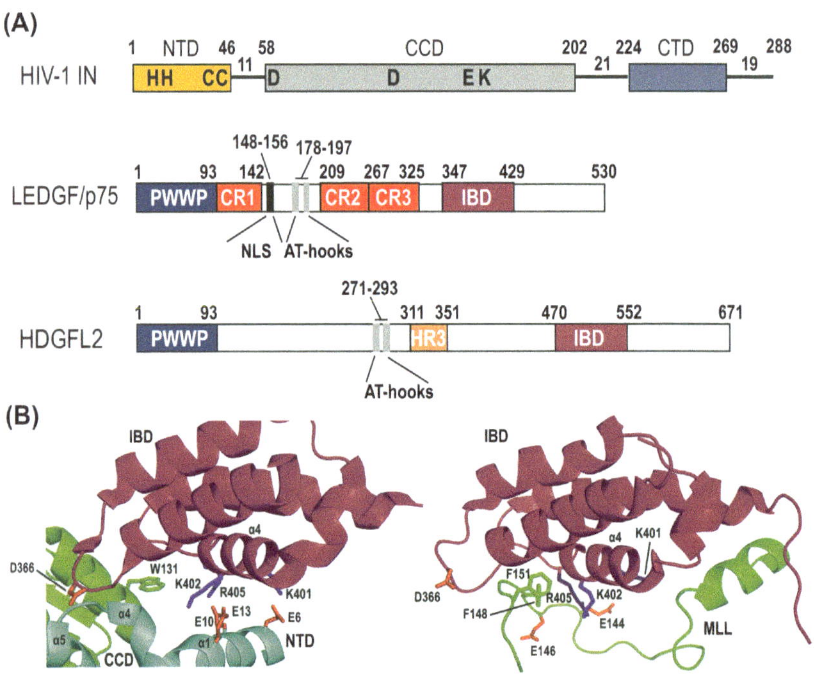

FIGURE 4.1 Proteins and complexes that mediate lentiviral integration targeting. (A) Shown are domain organizations of HIV-1 integrase (IN), LEDGF/p75, and HDGFL2, with numbers indicating domain boundaries or interdomain linker lengths. IN residues conserved throughout *Retroviridae* are also indicated. (B) X-ray crystal structure of the HIV-2 IN NTD-CCD–LEDGF/p75 IBD complex (left; PDB accession code 3F9K), highlighting IN and host factor residues that mediate the protein–protein interaction. The solution structure of an IBD–MLL fusion protein (PDB code 2MTN) was oriented similarly to highlight relative positions of IBM core residues E144, E146, F148, and F151. *CCD*, catalytic core domain; *CTD*, C-terminal domain; *HDGFL2*, hepatoma derived growth factor like 2; *HIV-1*, human immunodeficiency virus 1; *IBD*, integrase-binding domain; *IBM*, IBD-binding motif; *LEDGF*, lens epithelium–derived growth factor; *MLL*, mixed lineage leukemia; *NTD*, N-terminal domain.

Importantly, IN proteins purified from a variety of sources promote the 3′ processing and strand transfer of recombinant viral DNA substrates in vitro (Bushman and Craigie, 1991; Bushman et al., 1990; Craigie et al., 1990; Katz et al., 1990; Katzman et al., 1989; Sherman and Fyfe, 1990).

Retroviral INs harbor three conserved domains including the N-terminal domain (NTD), catalytic core domain (CCD), and C-terminal domain (CTD) (Figs. 4.1A and 4.2A). The CCD contains the enzyme active site, which is composed of Asp and Glu residues (the D,DX_{35}E motif). IN activity is Mg-dependent, and the catalytic triad coordinates two Mg ions to activate chemical nucleophiles (water for 3′ processing; viral DNA 3′-hydroxyl groups for strand transfer) and destabilize scissile phosphodiester bonds (Hare et al., 2010, 2012; Maertens et al., 2010). The NTD uses conserved His and Cys residues of the HHCC motif to bind a single Zn atom, and Zn-binding promotes IN multimerization and catalytic function (Lee and Han, 1996; Zheng et al., 1996). The CTD is the least conserved of the three canonical IN domains. The INs from spumaviruses,

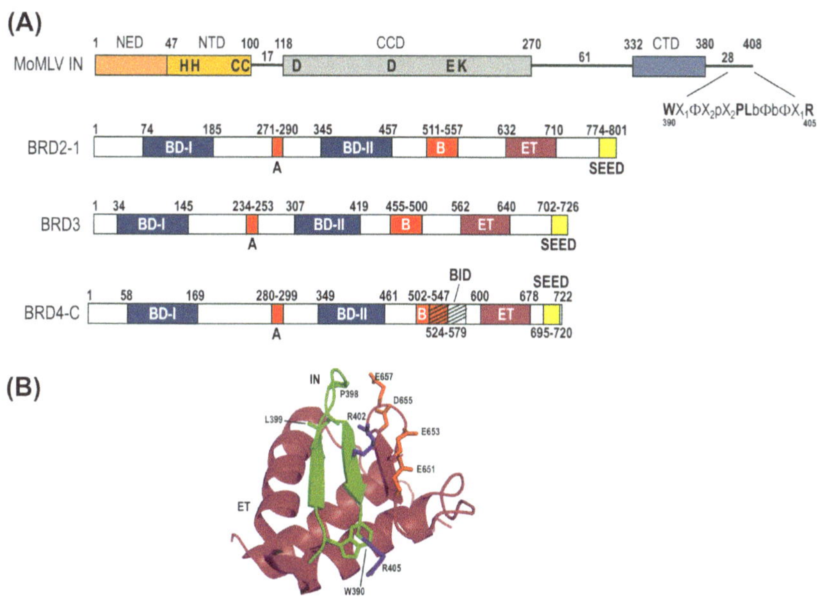

FIGURE 4.2 Key players in γ-retroviral integration. (A) The domain organizations of Moloney murine leukemia virus (MoMLV) integrase (IN) and bromodomain and extra-terminal domain proteins BRD2–4 are shown, with numbers denoting domain boundaries or interdomain linker lengths. Only the -1 and -C isoforms of BRD2 and 4, respectively, are diagrammed. The expanded amino acid sequence highlights the conserved γ-retroviral IN C-terminal tail region using the following code: *b*, basic residue; *p*, small polar residue; *X*, any residue; *Φ*, hydrophobic residue. SEED, Ser/Glu/Arg-rich regions (please refer to main text for other abbreviations). The IN sequence also shows residues conserved across *Retroviridae*. (B) Solution structure of the BRD4 extra-terminal (ET) domain in complex with a peptide from the MoMLV IN C-terminal tail region (PDB access code 2N3K), oriented to highlight the shared β sheet. Also noted by stick are conserved IN amino acid residues as well as key electronegative ET domain residues.

γ-retroviruses, and ε-retroviruses harbor an additional NTD-extension domain (Ballandras-Colas et al., 2013; Hare et al., 2010) (Fig. 4.2A). Each IN domain makes functional IN–IN and IN–DNA contacts within the intasome nucleoprotein complex (Ballandras-Colas et al., 2016, 2017; Hare et al., 2010; Passos et al., 2017; Yin et al., 2016).

Integration is nonrandom with respect to the host genome, with different aspects of nuclear architecture and chromatin structure influencing the choice of insertion site (see Craigie and Bushman, 2014; Kvaratskhelia et al., 2014; Lusic and Siliciano, 2016 for recent reviews). Global patterns of integration site preferences generally follow genera specificities. For examples, lentiviruses, such as human immunodeficiency virus 1 (HIV-1), favor the bodies of active genes and gene-rich regions of chromosomes (Schroder et al., 2002). Although the γ-retroviruses, typified by Moloney murine leukemia virus (MoMLV), also prefer active genes, promoters and enhancers are greatly preferred over interior gene regions (De Ravin et al., 2014; LaFave et al., 2014; Wu et al., 2003). Among the analyzed genera (ε-retroviral integration targeting has yet to be reported), the lenti- and γ-retroviral preferences differ most from random (Table 4.1). The molecular mechanisms of lentiviral and γ-retroviral targeting are also best understood due to the identification and characterization of host

TABLE 4.1 Representative Genomic Distributions of Retroviral Integration Sites

Virus	Genus	Intragenic[a]	TSS[b]	Gene Density[c]
Prototype foamy virus	Spuma	40.4	14.5	10.8
Avian sarcoma–leukosis virus	α	54.7	7.5	11.3
Mouse mammary tumor virus	β	40.3	3.9	8.3
HTLV-1	δ	49.8	7.4	10.7
MoMLV	γ	57.7	27.8	15.4
HIV-1	Lenti	74.6	4.1	19.9
MRC[d]	–	45.7	4.9	9.2

HIV-1, human immunodeficiency virus 1; *HTLV-1*, human T-lymphotropic virus type 1; *MoMLV*, Moloney murine leukemia virus; *TSS*, transcriptional start site.
[a]*Percent of integrations within Refseq genes.*
[b]*Percent of integrations within 5 kb (±2.5 kb) of a TSS.*
[c]*Average gene density per Mb (±0.5 Mb) surrounding integration sites.*
[d]*MRC, matched random control, the value of which can vary across studies due to bioinformatics technique and/or analyzed genome build.*
Data from Serrao, E., Ballandras-Colas, A., Cherepanov, P., Maertens, G.N., Engelman, A.N., 2015. Key determinants of target DNA recognition by retroviral intasomes. Retrovirology 12, 39.

targeting factors that directly engage each type of virus. In this chapter, we review the molecular mechanisms of retroviral integration targeting, focusing on the roles of the responsible host factors. We additionally discuss the targeting of these virus–host interactions via small molecules as novel approaches to antiretroviral inhibitor development.

INTEGRASE-INTERACTING PROTEINS

Numerous host factors have been reported to interact with retroviral INs (Jager et al., 2012; Maertens, 2016; Studamire and Goff, 2008) (reviewed in Engelman, 2007 and Debyser et al., 2015), so a key issue within the field is to separate the wheat from the chaff in terms of which proteins play physiologically relevant roles during virus replication. Numerous experimental approaches, including the use of purified proteins in binding and in in vitro integration assays, analysis of viral and/or host protein mutants in vitro and in infected cells, antagonism of host protein expression through the use of RNA interference (RNAi) and/or gene knockout, and structural characterization of interacting partners, are the types of analyses that are typically used to address physiological relevance.

LENTIVIRUSES

Mining the published literature (Allouch and Cereseto, 2011; Allouch et al., 2011; Ao et al., 2010, 2012; Christ et al., 2008; Engelman, 2007; Huang et al., 2008; Jager et al., 2012; Kobbi et al., 2011; Manganaro et al., 2010; Sorin et al., 2011; Studamire and Goff, 2008; Terreni et al., 2010; Woodward et al., 2009; Yamamoto et al., 2011; Zhang et al., 2009; Zheng et al., 2011) alongside the HIV-1 Human Interaction Database maintained by the National Center for Biotechnology Information (Ako-Adjei et al., 2015) yields a dazzling array of >200 cell factors that reportedly interact with HIV-1 IN. However, only a small fraction of these is likely to play an important role in HIV-1 replication through a specific interaction with the IN protein. Among these, lens epithelium–derived growth factor (LEDGF)/transcriptional coactivator p75 is confirmed to play a significant role in determining both the efficiency of HIV-1 integration and proviral DNA distribution.

LENS EPITHELIUM–DERIVED GROWTH FACTOR/P75

LEDGF/p75 was first identified to interact with HIV-1 IN through coimmunoprecipitation of extracts of cells that were engineered to constitutively express the viral protein (Cherepanov et al., 2003). This same study importantly showed that purified LEDGF/p75 significantly stimulated HIV-1 IN strand transfer activity in vitro. The protein interaction was soon verified by independent groups that utilized either the same basic coimmunoprecipitation approach (Turlure et al., 2004) or a yeast two-hybrid screen for HIV-1 IN interactors (Emiliani et al., 2005; Rain et al., 2009). The "LEDGF/p75" term is a misnomer; the protein at steady-state associates with chromatin (Cherepanov et al., 2003; Maertens et al., 2003; Nishizawa et al., 2001), and the mouse eye lens develops normally in animals that are knocked out for expression of the *Psip1* gene, which encodes LEDGF/p75 (Sutherland et al., 2006). Moreover, its predicted molecular mass is 60.1 kDa. Because LEDGF/p75 is highly charged (calculated isoelectric point of 9.15), it migrates more slowly through denaturing polyacrylamide gels than anticipated for a charge-neutral protein of the same mass. Considering its role in mediating DNA recombination (Daugaard et al., 2012), LEDGF/p75 might accordingly be best characterized as a high mobility group protein (Reeves, 2015).

LEDGF/p75 possesses highly conserved regions interspersed by variable sequences (Cherepanov et al., 2004) (Fig. 4.1A). Conserved elements within the N-terminal portion of the protein, including the PWWP domain, charged regions 1–3, nuclear localization signal (NLS), and a pair of AT-hooks, contribute to DNA and chromatin binding (Llano et al., 2006b; Singh et al., 2006; Turlure et al., 2006). The PWWP domain is a chromatin reader with affinity for histone H3 modified by Lys36 trimethylation (Eidahl et al., 2013; Pradeepa et al., 2012; van Nuland et al., 2013). C-terminal to these elements is the IN-binding domain (IBD), which is both necessary and sufficient to bind HIV-1 IN in vitro (Cherepanov et al., 2004). The IN CCD is the principal LEDGF/p75-binding component (Cherepanov et al., 2005; Maertens et al., 2003), whereas the NTD is required for high-affinity binding (Hare et al., 2009b; Maertens et al., 2003). LEDGF/p75 is a lentivirus-specific IN-binding protein (Busschots et al., 2005; Cherepanov, 2007; Llano et al., 2004).

Numerous cellular factors, including the transcription factor JPO2 (Bartholomeeusen et al., 2007; Maertens et al., 2006), domesticated transposon PogZ (Bartholomeeusen et al., 2009), CDC7 activator of S-phase kinase heterodimer (Hughes et al., 2010), mixed lineage leukemia (MLL)–menin complex (Čermáková et al., 2014; Murai et al., 2014; Yokoyama and Cleary, 2008), and IWS1 (Gerard et al., 2015; Tesina et al., 2015) interact with LEDGF/p75 through the IBD. These interactions are mediated via a consensus IBD-binding motif (IBM) $(E/D)X_1EX_1FX_1GF$ that is present in each of the binding partners (Tesina et al., 2015). Interestingly, lentiviral IN proteins do not possess a contiguous IBM. Instead, discontinuous sequences from the NTD and CCD dimer contribute analogous electronegative and hydrophobic residues, respectively (Hare et al., 2009b; Tesina et al., 2015) (Fig. 4.1B). Other proteins, including TOX4 and NOVA1, interact with LEDGF/p75 via its PWWP domain (Morchikh et al., 2013). LEDGF/p75 also interacts with mRNA-splicing factors (Singh et al., 2015), though which LEDGF/p75 determinant(s) mediates these interactions is unknown.

Initially it was unclear whether LEDGF/p75 played an important role in lentiviral infection. Despite efficient LEDGF/p75 depletion by RNAi, some studies failed to detect an obvious impairment in HIV-1 infection (Llano et al., 2004; Vandegraaff et al., 2006), whereas others reported approximately 2- to

5-fold defects (Vandekerckhove et al., 2006; Zielske and Stevenson, 2006). Any number of variables, including target cell type, strain of HIV-1, multiplicity of infection, or level of LEDGF/p75 knockdown, could have contributed to the differences among studies. Parallel work with short-hairpin RNA expression vectors highlighted the importance of knockdown depth, as cell clones devoid of chromatin-associated LEDGF/p75 supported significantly less HIV-1 infection than clones that retained a residual level of protein (Llano et al., 2006a). The normal cellular content of LEDGF/p75, estimated at several hundred thousand copies (Vandegraaff et al., 2006), seemingly greatly outweighs what is required for a single PIC to integrate efficiently. Knocking out the *PSIP1* gene in cell lines, which should ablate all traces of LEDGF/p75, accordingly yielded ~5- to 20-fold defects in HIV-1 infection (Fadel et al., 2014; Marshall et al., 2007; Schrijvers et al., 2012a; Shun et al., 2007; Wang et al., 2012). Knocking out the gene in primary CD4+ T cells yielded a milder ~twofold infection defect (Hultquist et al., 2016). Because the primary T cells were electroporated with purified Cas9–guide RNA complexes, residual levels of LEDGF/p75 could persist under these conditions. Although the preference for HIV-1 to integrate into active genes was reduced significantly under all conditions of LEDGF/p75 depletion, the virus at the same time retained levels of gene targeting that were significantly greater than those calculated based on random chance (Ciuffi et al., 2005; Marshall et al., 2007; Schrijvers et al., 2012b; Shun et al., 2007; Wang et al., 2012). These data hinted that other cell factors could potentially participate in gene-tropic HIV-1 integration targeting.

Both IN and chromatin binding are essential for LEDGF/p75 function during HIV-1 infection (Llano et al., 2006a; Shun et al., 2007, 2008). Thus, LEDGF/p75 functions as a molecular tether to anchor the PIC to favored genomic targets (De Rijck et al., 2010) (reviewed in Kvaratskhelia et al., 2014 and Debyser et al., 2015). Somewhat unclear is when during the infectious process the PIC encounters LEDGF/p75. The host factor can be detected at low levels in purified HIV-1 virions (Desimmie et al., 2015). However, due to the low level of incorporation—one to 2.5 copies per virion, whereas IN is present at 250–500 copies—combined with the fact that LEDGF/p75 is readily proteolyzed by the viral protease (Desimmie et al., 2015), it seems extremely unlikely that virion-incorporated protein plays any role in integration during the ensuing infection. LEDGF/p75 stimulates HIV-1 IN multimerization (Hare et al., 2009a; McKee et al., 2008), but LEDGF/p75-induced multimers are unable to form functional SSCs in vitro (Kessl et al., 2011). Thus, it may be important for IN to avoid LEDGF/p75 until after SSC formation in HIV-infected cells. Consistent with this premise, HIV-1 PICs extracted from *PSIP1* knockout cells support normal levels of strand transfer activity in vitro (Shun et al., 2007). Recently, LEDGF/p75 was shown to be an essential cofactor for maedi-visna virus intasome assembly in vitro (Ballandras-Colas et al., 2017), raising questions as to whether different lentiviruses might acquire LEDGF/p75 at different times postinfection. Our favored model is that PICs encounter LEDGF/p75 as a chromatin-bound protein, although other scenarios that posit LEDGF/p75-IN engagement during or after SSC formation, but before PIC–chromatin binding, are also plausible.

LEDGF/p75 proteins that harbor heterologous chromatin interaction modules in place of the N-terminal chromatin-binding elements support HIV-1 infection (Meehan et al., 2009) and efficiently redirect integration to novel sites (Ferris et al., 2010; Gijsbers et al., 2010; Silvers et al., 2010). Although these observations suggest approaches to potentially steer lentiviral vectors to specific genomic locations for gene therapy applications, the inability to package functional LEDGF/p75 in virions (Desimmie et al., 2015) necessitates the engineering of target cells to express the heterologous fusion proteins (Vets et al., 2013; Vranckx et al., 2016a; Wang et al., 2014), complicating the potential translational applications of such technologies.

HEPATOMA-DERIVED GROWTH FACTOR LIKE 2 (HDGFL2)

LEDGF/p75 is a member of the hepatoma-derived growth factor like (HDGFL) protein family, which consists of six members including HDGF, HDGFL1-3, LEDGF/p75, and LEDGF/p52 (Izumoto et al., 1997). The PWWP domain is the most conserved element among HDGFL family members (Izumoto et al., 1997). LEDGF/p52 is also expressed from the *PSIP1* gene, but this splice variant lacks the IBD and hence does not bind HIV-1 IN (Maertens et al., 2003). HDGFL2 is the only HDGFL family member that in addition to LEDGF/p75 possesses a functional IBD (Cherepanov et al., 2004) (Fig. 4.1A).

To address whether HDGFL2 might play a role in HIV-1 infection or integration targeting, it was depleted from cells using RNAi (Llano et al., 2006a; Schrijvers et al., 2012a; Vandegraaff et al., 2006) or gene knockout (Wang et al., 2012). In all cases, sole depletion of HDGFL2 did not adversely affect the level of HIV-1 infection or extent of integration targeting. However, HDGFL2 depletion from *PSIP1* knockout cells yielded an additional ~twofold infection defect (Schrijvers et al., 2012a; Wang et al., 2012) and reductions in gene-tropic integration targeting beyond those observed by solely depleting LEDGF/p75 (Schrijvers et al., 2012b; Wang et al., 2012). However, because these data failed to reveal a prominent role for HDGFL2 in mediating infection or integration targeting in cells that harbor LEDGF/p75, it would seem that HDGFL2 plays little if any role in determining the sites of HIV-1 integration under normal infection conditions.

GAMMARETROVIRUSES

Two observations hinted that γ-retroviral integration might also potentially rely on specific IN-binding proteins. First was the significant viral preference for particular genomic features such as gene promoters (Wu et al., 2003) (Table 4.1), and second was the prior work that established precedence for an IN-binding host targeting factor in lentiviral integration. A yeast two-hybrid screen for

MoMLV IN interacting proteins yielded 27 host cell candidates, including several proteins involved in chromatin remodeling, DNA repair, and transcription (Studamire and Goff, 2008). Included in this list was bromodomain and extra-terminal domain (BET) protein bromodomain-containing 2 (BRD2), and subsequent work has established that BET proteins BRD2, 3, and 4 play critical roles in targeting γ-retroviruses to their preferred integration sites (De Rijck et al., 2013; Gupta et al., 2013; Sharma et al., 2013).

BROMODOMAIN AND EXTRA-TERMINAL DOMAIN PROTEINS

The BET proteins encompass BRD2, 3, 4, and T, with proteins BRD1 and 7–9 included in the extended BET family. BRD2–4 are ubiquitously expressed, whereas BRD-T expression is restricted to the testis. BET proteins play key roles in several cellular processes including DNA replication, transcription, and control of the cell cycle (reviewed in Belkina and Denis, 2012 and Taniguchi, 2016).

Similar to HDGFL proteins, BET proteins consist of intermingled conserved and variable regions (Fig. 4.2A). Two N-terminal bromodomains, BD-I and BD-II, bind acetylated tales of histones H3 and H4 (Filippakopoulos et al., 2010; Morinière et al., 2009). Regions A and B are positively charged and bind DNA nonspecifically to confer high-affinity binding of BET proteins to mononucleosomes (Larue et al., 2014). The basic residue-enriched interaction domain (BID) region, specific to BRD4 isoforms, mediates binding to p53 (Wu et al., 2013). The C-terminally located extra-terminal (ET) domain binds several types of proteins, including chromatin modifying enzymes such as Jumonji C-domain-containing protein 6 (JMJD6) and histone-lysine N-methyltransferase NSDS (Rahman et al., 2011). BRD3 is expressed as a single isoform, whereas BRD2 and 4 are expressed as several isoforms. BRD4 isoform A harbors a unique C-terminal motif that interacts with positive transcription elongation factor and accordingly plays a role in the regulation of HIV-1 latency (Bisgrove et al., 2007; Boehm et al., 2013; Zhu et al., 2012).

Prior to the work with retroviruses, BET proteins were known to play important roles in the lifecycles of several types of other viruses. BET proteins regulate transcription of different herpesviruses, including Epstein–Barr virus and Kaposi's sarcoma–associated herpesvirus, as well as papillomaviruses (reviewed in Weidner-Glunde et al., 2010). BRD4 additionally plays a critical role in papillomavirus replication, physically tethering viral DNA to mitotic chromosomes, which ensures viral segregation to daughter cells following mitosis (Baxter et al., 2005; You et al., 2004).

Several lines of evidence indicate that BRD2–4 function as bimodal tethers to connect γ-retroviral PICs to favored sites of integration, akin to the role played by LEDGF/p75 in lentiviral DNA integration. First, they are γ-retroviral IN specific–binding proteins, and their addition to in vitro integration assays stimulated the strand transfer activity of γ-retroviral INs significantly (De Rijck et al., 2013; Gupta et al., 2013; Larue et al., 2014;

Serrao et al., 2015; Sharma et al., 2013). Second, the ET domains function as specific IBDs, with particular amino acid substitutions ablating the BRD–IN interaction (De Rijck et al., 2013; Gupta et al., 2013; Serrao et al., 2015; Sharma et al., 2013). Third, they colocalize with ectopically expressed IN in cells (De Rijck et al., 2013; Gupta et al., 2013). And, most importantly, their impairment led to significant shifts in γ-retroviral integration profiles from preferred integration loci such as promoter-proximal regions (De Rijck et al., 2013; Gupta et al., 2013; Sharma et al., 2013). Moreover, global BET protein–binding sites mapped by chromatin immunoprecipitation (LeRoy et al., 2012) correlated with sites of MoMLV but not HIV-1 integration (De Rijck et al., 2013; Sharma et al., 2013). Small molecule inhibitors of the BD–chromatin interaction, such as JQ-1 (Filippakopoulos et al., 2010), were particularly useful to address the role of BET proteins in γ-retroviral integration. The inhibitors specifically blocked MoMLV as compared to HIV-1 infection (De Rijck et al., 2013; Gupta et al., 2013; Sharma et al., 2013), and moreover inhibited MoMLV integration in a dose-dependent manner (Gupta et al., 2013; Sharma et al., 2013). Finally, akin to what was previously demonstrated for LEDGF/p75, artificial BET fusion proteins that replaced the BDs with a heterologous chromatin reader (residues 1-324 of LEDGF/p75) retargeted integration to novel genomic locations (De Rijck et al., 2013).

Discontinuous lentiviral IN sequences from the CCD dimerization interface and the NTD comprise the LEDGF/p75-binding site (Cherepanov et al., 2005; Hare et al., 2009b) (Fig. 4.1B). The BET proteins by contrast engage the C-terminal tail sequences of γ-retroviral INs (Aiyer et al., 2014; De Rijck et al., 2013; Larue et al., 2014). Indeed, γ-retroviral INs harbor a unique tail sequence (Aiyer et al., 2014; Kvaratskhelia et al., 2014) (Fig. 4.2A), and substitution of conserved tail residues ablated the BRD–IN interaction (Aiyer et al., 2014; De Rijck et al., 2013; Gupta et al., 2013; Larue et al., 2014). The solution structure of the BRD4 ET domain in complex with a C-terminal peptide of MoMLV IN revealed that each protein donates otherwise disordered sequences to construct a hybrid three-stranded β sheet (Crowe et al., 2016) (Fig. 4.2B). The tail region of MoMLV IN is not essential for virus replication, with deletion mutants suffering perhaps a twofold reduction in virus titer (Roth, 1991). Importantly, the preference for promoter-proximal integration was reduced significantly by IN tail region deletion, suggesting that these mutant constructs might help to reduce the genotoxicity that has been associated with the integration preferences of MoMLV-derived gene therapy vectors (Aiyer et al., 2014).

DELTARETROVIRUSES

The δ-retrovirus human T-lymphotropic virus type 1 (HTLV-1), much like HIV-1, has a complex genetic structure that encodes for accessory proteins including Tax and bZIP, which are crucial for viral persistence and are also implicated in viral-mediated oncogenesis (Matsuoka and Jeang, 2011). Expression of both of these proteins drives mitotic proliferation to create clones of daughter cells sharing the same proviral integration site, with infected individuals estimated to harbor anywhere between 10^4 and 10^5 unique clones (Laydon et al., 2014; Niederer et al., 2014). In vivo, this phenomenon results in a wide range of clones with unique integration sites (UIS) that can range in number from singletons to thousands of cells.

HTLV-1 (Derse et al., 2007; Doi et al., 2005; Gillet et al., 2011; Meekings et al., 2008; Melamed et al., 2013) as well as related δ-retroviruses HTLV-2 (Melamed et al., 2014) and bovine leukemia virus (BLV) (Gillet et al., 2013) favor integration into CpG islands, enhancers, transcriptional start sites (TSSs), genes, and regions of active transcription (Table 4.1). However, although both gene- and TSS-tropic targeting are each enriched compared with random integration, these preferences fall short of the strong preferences displayed by HIV-1 and MoMLV for these respective metrics (Table 4.1). Underscoring a balance between viral expression and immune surveillance in vivo, integration sites in de novo-infected T cells differ significantly from those recovered from chronically infected patients. However, integration site distributions in expanded clones from patients with HTLV-1-associated disease do not differ significantly from in vitro infections (Gillet et al., 2011). This finding suggests that the site of HTLV-1 integration is intimately linked to clonal expansion.

By contrast, the integration site distribution in HTLV-1 asymptomatic carriers is significantly different from de novo infection and favors transcriptionally silent regions of the genome, with the clonal landscape consisting of fewer and less expanded clones. Cook et al. (2014) reported that integration sites located on the q-arms of the acrocentric chromosomes (13, 14, 15, 21, and 22) are associated with persistence in vivo. Because acrocentric chromosomes associate with the nucleolus periphery, a transcriptionally silent region (reviewed in Prieto and McStay, 2008), absence of expression would allow escape from immune pressure and thereby help to support clonal expansion. The presence of a chronically active anti-HTLV-1 immune response in these patients could come either from expression of other clones or from transient activation of the provirus integrated in the acrocentric chromosomes when the nucleolus dissociates during cell division (Cook et al., 2014). Taken together, these observations suggest that integration site selection by δ-retroviruses is regulated differently from that of lenti- or γ-retroviruses. First, the lack of strong genomic signatures suggests integration site targeting factor(s) with a broad range of chromatin interactions. Second, integration would be expected to allow for the establishment of a wide variety of clones with different levels of proviral expression.

B′-PROTEIN PHOSPHATASE 2A (PP2A)

A proteomic screen for interaction partners using BLV and HTLV-1 INs as bait identified a range of chromatin-associated proteins including members of the DNA damage repair machinery, cohesin, condensin, and transcription factors such as p53. One protein complex in particular, PP2A, coprecipitated with the INs in near stoichiometric amounts (Maertens, 2016). PP2A is a ubiquitous Ser/Thr phosphatase which, together with protein phosphatase 1 (PP1), is responsible for 90% of the dephosphorylation events that take place in a cell. PP2A is a heterotrimer (Fig. 4.3) composed of a core dimer made up of a structural (A) and catalytic (C) subunit joined by any one of the four families of B-type regulatory subunits. The C subunit is highly similar in sequence and structure to the C subunits of related Ser/Thr protein phosphatases such as PP4, PP6, and PP1 (reviewed in Shi, 2009). For PP2A, as is the case for other phosphoprotein phosphatases, divalent metal ions play a catalytic role through the activation of a water molecule in the active site that is necessary for dephosphorylation. The catalytic metal ions are coordinated by six invariant residues: three His (H59, H167, and H241), two Asp (D57 and D85), and a single Asn (N117) (Fig. 4.3) (reviewed in Shi, 2009). The C subunit

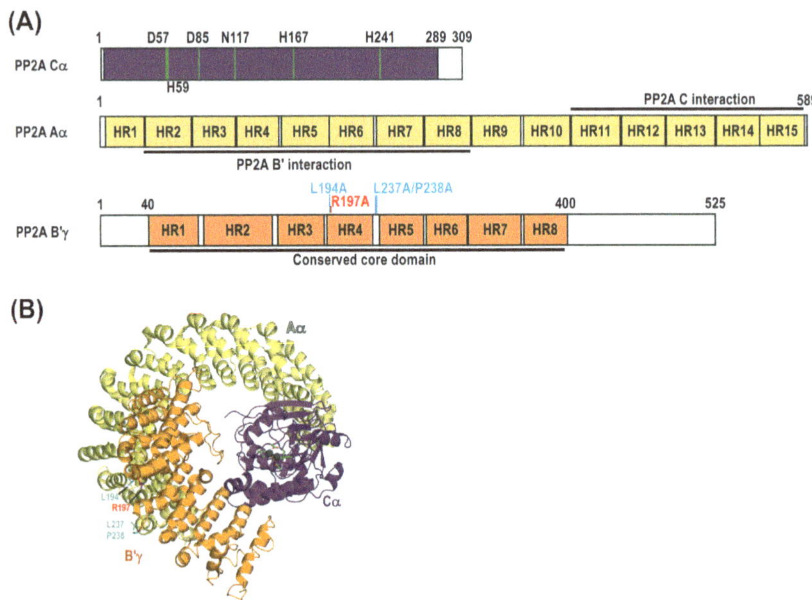

FIGURE 4.3 B′-PP2A holoenzyme, a functional δ-retroviral integrase-binding partner. (A) The domain organization of the Cα, Aα and B′γ proteins that make up the B′-PP2A holoenzyme, with numbers denoting domain boundaries. The *blue box* in Cα indicates the conserved catalytic domain, the metal chelating residues in Cα are indicated in green. Aα comprises 15 HEAT repeats (HR, yellow). HR 2–8 are involved in the interaction with B′, while HR 11–15 engage with the C subunit. B′γ is a pseudo-HEAT repeat protein; all B′ proteins comprise a highly conserved core domain (composed of 8 HR, orange), with divergent N- and C-terminal extensions likely contributing to substrate recognition. B′γ mutant R197A (red) is unable to bind δ-retroviral INs, whereas B′γ L194A or L237A/P238A (cyan) loose affinity for δ-retroviral INs (Maertens, 2016). (B) Cartoon representation of the X-ray crystallography structure of the B′-PP2A holoenzyme (PDB ID: 2IAE; Cho and Xu, 2007). Cα is shown in blue, Aα in yellow, and B′γ in orange. Cα residues D57, H59, D85, N117, H167, and H241 that chelate the active site metal ions (Mn^{2+}, *gray spheres*) are shown as *green sticks*, B′γ residues that mediate the interaction with δ-retroviral INs are shown in red (R197) and cyan (L194, L237/P238) sticks. Pymol was used to generate the B panels of Figs. 4.1–4.3.

is expressed from two alternative genes giving rise to Cα and Cβ isoforms, which share 97% sequence identity (Arino et al., 1988; da Cruz e Silva and Cohen, 1987; Stone et al., 1987). The A subunit comprises 15 HEAT (*H*untingtin–*e*longation factor 3–protein phosphatase 2*A*–yeast PI3-kinase *T*OR1) repeats, which form an elongated, horseshoe-shaped structure (Fig. 4.3) that drastically changes shape on formation of the holoenzyme (Cho and Xu, 2007; Xu et al., 2006). There are two isoforms of the A subunit, Aα and Aβ, which share 87% sequence identity (Hemmings et al., 1990). Each A isoform is expressed throughout the cell cycle, although the α isoform is expressed at 10 times greater levels than Aβ (Zhou et al., 2003), and they appear to display differential affinity for the C- and B-type subunits (Ruediger et al., 2001; Zhou et al., 2003). Distinct tumor-specific mutations in the Aα or Aβ isoform have been identified that result in the lack of holoenzyme formation with certain types of regulatory subunits (reviewed in Eichhorn et al., 2009). PP2A is aberrantly expressed in cancers (reviewed in Seshacharyulu et al., 2013), and viral proteins such as simian virus 40 small t-antigen and polyoma middle t-antigen disrupt the holoenzyme to replace the B-type regulatory subunit, thereby inducing cellular transformation (reviewed in Arroyo and Hahn, 2005). These data, together with the observation that treatment of mice with okadaic acid (a PP2A inhibitor) leads to tumorigenesis, supports the notion that PP2A acts as a tumor suppressor.

PP2A is highly regulated through phosphorylation and methylation and by the association of PP2A-specific inhibitors (reviewed in Haesen et al., 2014; Janssens and Goris, 2001; Seshacharyulu et al., 2013). Although the core dimer is stable, its association with a B-type regulatory subunit determines its intracellular localization and substrate specificity (reviewed in Shi, 2009). The four families of B subunits, B (B55), B′ (B56), B″, and B‴, are structurally and functionally distinct (reviewed in Shi, 2009), and each possesses two to five isoforms (reviewed in Eichhorn et al., 2009; Seshacharyulu et al., 2013). The δ-retroviral INs displayed affinity for PP2A holoenzymes that specifically contain one of the B′ regulatory subunits (Maertens, 2016). B′ has five different isoforms (α, β, γ, δ, and ε), each of which is expressed as a number of different splice and/or translational isoforms (reviewed in Yang and Phiel, 2010). B′s are pseudo-HEAT repeat proteins (Magnusdottir et al., 2009) and have a highly conserved core with distinct N- and C-terminal extensions (Sommer et al., 2015) (Fig. 4.3A). Expression of the B′ isoforms is both temporally and spatially regulated. The β isoform is specifically expressed in the brain, and whereas the α, β, and ε isoforms are cytoplasmic, the γ and δ isoforms harbor NLSs (Flegg et al., 2010; McCright et al., 1996). An elegant system designed by Strack et al. (2004) illustrated that the PP2A holoenzyme is critical for cell survival. Cells failed to survive when endogenous Aα was downregulated and mutant Aα that is unable to engage any member of B-type regulatory subunit was expressed. Likewise, reexpression of an Aα isoform that could bind to all regulatory subunits except for B′ was detrimental to cell survival.

B′-PP2A displays key characteristics of a retroviral IN targeting cofactor. B′–IN binding is specific for the δ-retroviruses, HTLV-1 IN and B′-PP2A colocalize in cell nuclei, and purified B′ protein significantly stimulates the concerted integration activity of HTLV-1 and -2 IN in vitro (Maertens, 2016). Each of these features is shared with the lentiviral and γ-retroviral IN targeting factors LEDGF/p75 and BRD2–4, respectively. On the flip side, whereas LEDGF/p75 and BRD2–4 proteins harbor chromatin binding domains (respectively PWWP and BD) and protein interaction domains (IBD and ET), B′-PP2A does not display these obvious features. However, PP2A is involved in gene bookmarking (Xing et al., 2008), B′-PP2A dephosphorylates certain chromatin modifiers and transcription factors such as p53 (Shouse et al., 2008, 2010, 2011) and signal transducer and activator of transcription 1 (Shanker et al., 2013), and it is recruited to the centromere (Foley et al., 2011) by shugoshin 2 (Kitajima et al., 2006), thereby regulating the phosphorylation status of cohesin to ensure timely separation of daughter chromatids. We speculate that B′-PP2A interacts with IN via the substrate binding site created at the B′:C interface (Cho and Xu, 2007; Xu et al., 2006). Maertens et al. characterized some residues on B′γ (Maertens, 2016) (Fig. 4.3) and on HTLV-1 and -2 IN (Maertens, G.N., unpublished results) that are critical for the interaction. Of interest, this interface was recently shown to be a conserved recognition motif for B′ substrates (Hertz et al., 2016; Wang et al., 2016a,b), suggesting that δ-retroviral INs mimic and compete with B′-PP2A substrate complexes in the cell. Although HTLV-1 or -2 IN do not inhibit B′-PP2A activity using a phospho-specific peptide as a substrate, it is possible that binding of HTLV IN or the HTLV PIC to B′-PP2A influences its enzymatic activity or its ability to associate with a certain substrate(s) (Maertens, 2016). However, it is worth mentioning that some of the chromatin-associated proteins that coprecipitated with HTLV-1 IN are known B′-PP2A substrates. The existence of alternative interaction sites with B′-PP2A substrates has been described (Xu et al., 2009), which could explain a lack of competition between HTLV-1 IN with these substrates. Alternatively, HTLV-1 IN could directly associate with these proteins in a complex independent of B′-PP2A. Further investigations are needed to determine the specificity and functionality of these interactions.

The verdict is still out on whether B′-PP2A plays an actual role in δ-retroviral integration targeting, although preliminary data strongly support this hypothesis (McCallin, A., Maertens, G.N., and Bangham, C.R., manuscript in preparation). A remaining question is why δ-retroviral INs would use B′-PP2A as a chromatin targeting factor. We postulate that interacting with a chromatin-associated protein that has a myriad of substrates gives the virus the advantage to establish a broad range of integration sites and hence infected clones with a plethora of UIS in different chromatin milieus. Given that high proviral expressers will be eliminated once an effective cytotoxic T-cell response is established, having a battery of proviral clones that allow for high, medium, or low expression ensures the survival of HTLV-1 DNA by succession in cells that effectively escape the immune response. Thus, the use of the ubiquitously expressed, multifaceted PP2A could be the reason that δ-retroviruses have been so successful in infecting and establishing latency in its host.

GAG-INTERACTING PROTEINS

Whereas HIV-1 efficiently infects growth-arrested and terminally differentiated cells, other types of retroviruses, such as MoMLV, require nuclear membrane dissolution during mitosis to reach their integration targets (Lewis et al., 1992; Lewis and Emerman, 1994; Roe et al., 1993). Analysis of chimeric MoMLV–HIV-1 constructs mapped infection of growth-arrested cells to the HIV-1 capsid (CA) protein (Yamashita and Emerman, 2004; Yamashita et al., 2007), leading to the hypothesis that active transport of the HIV-1 PIC through the cellular nuclear pore complex (NPC) is mediated by CA (reviewed in Ambrose and Aiken, 2014; Campbell and Hope, 2015; Matreyek and Engelman, 2013). Indeed, numerous proteins directly involved in nuclear transport, including NPC components nucleoporin (Nup) 358 and Nup153, directly bind HIV-1 CA (Matreyek et al., 2013; Schaller et al., 2011). Herein, we will discuss the roles of such proteins in integration targeting. Please refer to Chapter 3 in this treatise by Monette and Mouland for detailed overviews of PIC nuclear import and the roles of Nups in retroviral biology.

HIV-1 CA–INTERACTING PROTEINS

The analysis of MoMLV–HIV-1 chimeric viruses first revealed a role for HIV-1 Gag in integration targeting (Lewinski et al., 2006), and subsequent work with HIV-1 CA missense mutant viruses and cells depleted for expression of particular host factors further elucidated the roles of CA-binding proteins in HIV-1 integration targeting.

CA is a two domain protein consisting of an NTD and CTD, which are separated by a flexible linker (reviewed in Sundquist and Kräusslich, 2012). Several factors implicated in HIV-1 PIC nuclear import and integration targeting, including cyclophilin A (CypA) (De Iaco and Luban, 2014; Schaller et al., 2011), Nup358 (Schaller et al., 2011; Zhang et al., 2010), Nup153 (Di Nunzio et al., 2012, 2013; Matreyek and Engelman, 2011), and cleavage and polyadenylation specificity factor 6 (CPSF6) (Lee et al., 2010; Sowd et al., 2016), engage the HIV-1 CA NTD (Bichel et al., 2013; Gamble et al., 1996; Matreyek et al., 2013; Price et al., 2012; Schaller et al., 2011). The CA missense mutation N57A impairs Nup358 (Schaller et al., 2011) and Nup153 (Matreyek et al., 2013) binding, whereas the N74D change disrupts binding to CPSF6 (Lee et al., 2010) and Nup358 (Schaller et al., 2011). Although CA N57A and N74D mutant viruses were initially described to retain the wild-type preference for integrating into genes, both viruses were subsequently shown to lose the preference to target gene dense regions of chromosomes (Schaller et al., 2011). Similarly, integration into genes was only marginally impaired by Nup358 depletion, whereas gene density per megabase (Mb) surrounding integration sites was reduced from approximately 19.0 to 11.8. Although this is a significant drop, 11.8 genes/Mb is still significantly above the reported random value of 8.3 genes/Mb. The fact that the CA N57A and N74D mutations disrupt binding to more than one host factor likely accounted for the loss of gene density targeting under these infection conditions. Interesting, disrupting the CA–CypA interaction, either through CA G89V or P90A mutation or by treating target cells with the small molecule CypA inhibitor cyclosporine A, yielded the opposite phenotype. In these cases, the targeting of gene dense regions of chromosomes actually increased, from about 15 genes/Mb to 20 (Schaller et al., 2011). Targeting of gene dense regions by contrast was unaffected by LEDGF/p75 depletion (Ocwieja et al., 2011). Thus, these pioneering studies indicated that CA-binding factors generally influenced the ability of the HIV-1 PIC to target gene dense regions, whereas LEDGF/p75 directed integration to transcription units. Moreover, disruption of CA binding led to opposing effects, which spanned from the inability to sense gene dense regions to actual increases in the targeting of these chromosomal loci, depending on which host factor interaction was perturbed.

Subsequent work that analyzed increasing numbers of integration sites and employed knockout cells confirmed these earlier observations and also clarified that the distinction between IN- and CA-binding factors in directing integration to particular genomic regions was not so cut and dried. Thus, the CA N74D mutation also significantly affected integration into genes, and *PSIP1* knockout affected HIV-1 PIC targeting of gene dense regions (Koh et al., 2013). RNAi-mediated knockdown of CPSF6 unexpectedly ablated integration into genes and gene dense regions under conditions where LEDGF/p75 knockdown only marginally impaired integration targeting (Sowd et al., 2016). As discussed above, LEDGF/p75 knockdown sometimes fails to unveil the full effect of this host factor on HIV-1 integration. To assess the roles of CPSF6 and LEDGF/p75 from an even playing field, they were accordingly knocked out, either individually or in tandem (Sowd et al., 2016). While *PSIP1* knockout reduced targeting of gene dense regions from 20.7 to 13.7 genes/Mb, *CPSF6* knockout reduced the value to 5.8, which was well below the value expected from random targeting (8.7 genes/Mb). Consistent with the knockdown results, integration into genes was reduced to a greater extent by *CPSF6* as compared with *PSIP1* knockout, to approximately 57% versus 63%, respectively, whereas genes were targeted ~83% of the time in wild-type cells. Moreover, gene-tropic integration dropped precipitously in cells knocked out for both factors to 48.3% (with random targeting at 44.7%) (Sowd et al., 2016). Thus, LEDGF/p75 and CPSF6 together seem to account for the preference for HIV-1 to integrate into genes.

HIV-1 integration greatly favors transcriptionally active genes (Schroder et al., 2002), and integration sites accordingly correlate with genomic regions associated with activating epigenetic marks (Roth et al., 2011; Wang et al., 2007). Although such preferences were maintained in *PSIP1* knockout cells, by contrast HIV-1 lost its preference for integrating near activating epigenetic marks in *CPSF6* knockout cells and moreover gained some preference for repressive regions of chromatin (Sowd et al., 2016). Taken together, these data indicate that CA–CPSF6 binding initially directs the HIV-1 PIC to active chromatin, whereas LEDGF/p75 predominantly guides integration along active gene bodies (Sowd et al., 2016). The model invokes a potential interaction between

CA and CPSF6 that is then followed by IN-LEDGF/p75, but whether these are indeed independent interactions in the context of the PIC or occur simultaneously remains unknown.

Studies have indicated that HIV-1 prefers to integrate into peripheral regions of the nucleus (Albanese et al., 2008), in association with NPCs (Lelek et al., 2015; Marini et al., 2015), and moreover implicated a role for Nup153 and LEDGF/p75 in this process (Marini et al., 2015). By contrast, peripheral nuclear integration targeting was increased by CPSF6 depletion or by infection with CA mutant viruses such as N74A that are defective for CPSF6 binding (Chin et al., 2015). It therefore seems that the inability to engage CPSF6 precludes access to interior nuclear regions and shunts the PIC to regions devoid of gene density. Additional work, for example, with *PSIP1* and *CPSF6* knockout cells, should help to clarify the roles of different integration targeting factors in guiding the HIV-1 PIC to particular nuclear regions for integration.

CPSF6 is a member of the CFIm complex that regulates sites of polyadenylation in mRNA (reviewed in Hardy and Norbury, 2016). CFIm is a heterotetramer composed of two subunits of CPSF5, which can pair with either two subunits of CPSF6 or two subunits of CPSF7 (Rüegsegger et al., 1996). The lengths of mRNA 3′ untranslated regions (UTRs) are decreased significantly by CPSF6 depletion (Gruber et al., 2012; Martin et al., 2012). Although CFIm accounts for the brunt of CPSF6 in the cell (Rasheedi et al., 2016), the role of CPSF6 in integration targeting appears independent of its role in CFIm. Integration into genes with CPSF6-dependent changes in 3′ UTR length was unaffected by *CPSF6* knockout (Sowd et al., 2016). Expressing a CPSF6 mutant that was defective for CPSF5 binding and hence CFIm incorporation moreover effectively restored integration targeting to *CPSF6* knockout cells (Rasheedi et al., 2016). CPSF6 is a nuclear protein with affinity for paraspeckle regions (Cardinale et al., 2007; Dettwiler et al., 2004). Although CPSF6 is not known to directly bind to DNA or chromatin, CPSF6-dependent integration sites tracked with genomic loci recovered via CPSF6 chromatin immunoprecipitation (Katahira et al., 2013; Sowd et al., 2016). We therefore suspect that CPSF6 engages chromatin-bound factor(s) to facilitate intranuclear HIV-1 PIC trafficking and integration targeting. The identification of such factors should help to further elucidate the fate of the HIV-1 PIC once it gains access to the nuclear environment.

FOAMY VIRUS

Several features distinguish spumaviruses, which are also known as foamy viruses, from other retroviruses. For example, reverse transcription occurs during virus production, and Gag proteins remain largely intact during the acute phase of infection (reviewed in Delelis et al., 2004). Foamy virus Gag interacts with chromatin through a C-terminal chromatin-binding sequence (CBS) that is critical to tether incoming PICs to chromosomal DNA (Tobaly-Tapiero et al., 2008). Spumavirus integration disfavors genes yet favors promoters and features associated with heterochromatin (Table 4.1) (Maskell et al., 2015; Nowrouzi et al., 2006; Trobridge et al., 2006). Mutation of the Gag CBS was shown to further shift integration away from genes and also reduce promoter-proximal integration to below random (Hocum et al., 2016). Interestingly, combining the CBS mutation with IN fused to the heterochromatin reader CBX1, which recognizes H3K9me3 marks, yielded a significant additional reduction in intragenic integration (Hocum et al., 2016). Unlike prior work with LEDGF/p75 retargeting fusion proteins that require engineering of susceptible target cells (Ferris et al., 2010; Gijsbers et al., 2010; Silvers et al., 2010), these observations importantly establish effective integration retargeting through the modification of virion-incorporated proteins.

A yeast two-hybrid screen identified polo-like kinases (PLKs) as Gag interaction partners, and an S-T/S-P consensus motif in Gag was critical for the interaction (Zurnic et al., 2016). Analysis of Gag mutant viruses revealed that all late stage viral processes—up to and including reverse transcription—were intact. However, integration was delayed and reduced by about twofold. As intragenic and heterochromatin integration targeting was somewhat marginally reduced and accentuated, respectively, by the Gag mutations, the kinases could potentially help to tether incoming PICs to chromosomal targets (Zurnic et al., 2016). However, it also seems possible that the shifts in integration profile could be a consequence of the delay in integration kinetics. Additional work with foamy virus Gag and IN proteins should help to further elucidate the molecular mechanisms that enable this virus subfamily to favor intergenic regions and heterochromatin during integration.

ALLOSTERIC IN INHIBITORS THAT TARGET THE LEDGF/P75–IN INTERACTION

Expression of LEDGF/p75 IBD fusion proteins that are unable to bind chromatin dominantly interfered with HIV-1 integration and hence virus infection (De Rijck et al., 2006; Llano et al., 2006a). Moreover, combining the expression of such fusion proteins with LEDGF/p75 knockdown could effectively kill HIV-1 infection (Meehan et al., 2011). Together, these observations suggest that small molecule inhibitors of the LEDGF/p75–IN interaction might display antiviral activity. Interestingly, lead compounds that after optimization yielded highly similar quinoline-based compounds that effectively inhibit HIV-1 IN-LEDGF/p75-binding in vitro and potently inhibit virus replication were discovered using two independent approaches. Whereas one approach employed a high-throughput screen for inhibitors of IN 3′ processing activity (Fader et al., 2014; Tsantrizos et al., 2007), the other modeled compounds in silico based on the key amino acid side chains that mediate the LEDGF/p75–IN interaction (Fig. 4.1B) (Cherepanov et al., 2005; Christ et al., 2010). Multiple acronyms are used to describe these compounds, including ALLINI for allosteric IN inhibitor (Kessl et al., 2012), LEDGIN for LEDGF/p75–IN inhibitor (Christ et al., 2010), and NCINI for noncatalytic site IN inhibitor (Balakrishnan et al., 2013).

Selection of HIV-1 resistance to ALLINIs in tissue culture invariably selects for changes in IN, identifying IN as the antiviral drug target (Christ et al., 2010; Gupta et al., 2016; Sharma et al., 2014; Tsiang et al., 2012). Accordingly, HIV-1 integration is inhibited in cells treated with ALLINIs (Christ et al., 2010; Jurado et al., 2013; Le Rouzic et al., 2013; Tsiang et al., 2012). Whereas the dose response curves of clinical IN strand transfer inhibitors (INSTIs) reveal slopes of ~1, ALLINIs display steeper slopes of >2 (Kessl et al., 2012; Laskey and Siliciano, 2016; Shen et al., 2008). INSTIs have but one molecular target per infectious cycle, the intasome nucleoprotein complex (Hare et al., 2010), which seemingly accounts for the relatively shallow slopes of their dose response curves. Compounds with steeper slopes might inhibit more than one step in the replication cycle or need to inhibit the full complement of molecular targets at a specific step to be effective (Shen et al., 2008). Several hundred IN molecules accompany the HIV-1 particle into the cell, though only a handful (data suggest ~4–16) are required for intasome complex function (Ballandras-Colas et al., 2017; Passos et al., 2017). ALLINIs aberrantly multimerize IN in vitro (Christ et al., 2012; Kessl et al., 2012; Tsiang et al., 2012), which precludes SSC formation (Kessl et al., 2012). Thus, inhibition of both SSC formation and IN–LEDGF/p75 binding could potentially account for the relatively steep slopes of these compounds.

Subsequent work pinpointed a novel mechanism of ALLINI action. The compounds invariably display greater potencies during virus production than they do when treatment is restricted to the acute phase of infection when integration occurs (Balakrishnan et al., 2013; Desimmie et al., 2013; Gupta et al., 2014; Jurado et al., 2013; Le Rouzic et al., 2013; Sharma et al., 2014). Although several aspects of the late phase of HIV-1 replication, including polyprotein processing and viral protein and genomic RNA incorporation, are unaffected, the compounds interfere with a relatively late stage of HIV-1 particle maturation. HIV-1 CA forms a conical CA shell that houses the viral ribonucleoprotein (RNP) complex composed of genomic RNA and nucleocapsid protein (see Sundquist and Kräusslich, 2012). ALLINIs decouple RNP incorporation into assembled CAs. In other words, although CA shells are generally visualized by electron microscopy, the electron-dense RNP situates outside of the CA shell, most often in association with the viral membrane (Balakrishnan et al., 2013; Desimmie et al., 2013; Jurado et al., 2013; Le Rouzic et al., 2013; Sharma et al., 2014). The phenotype is reminiscent of that seen with the majority of HIV-1 IN mutations. Whereas some mutations, for example, those that replace catalytic Asp and Glu residues with conservative amino acid substitutions, specifically block integration, most mutations confer pleiotropic defects, which can interfere with particle assembly/release, maturation, and reverse transcription (reviewed in Engelman, 1999, 2011). Accordingly, HIV-1 produced from ALLINI-treated cells is defective for reverse transcription (Balakrishnan et al., 2013; Desimmie et al., 2013; Jurado et al., 2013; Le Rouzic et al., 2013; Sharma et al., 2014). Thus, ALLINIs could display relatively steep slopes due to inhibition of multiple steps in the replication cycle including maturation, reverse transcription, and integration. However, a detailed analysis has suggested that inhibition of maturation fully accounts for the potency of these compounds (Fontana et al., 2015). This analysis aside, the compounds can obviously also inhibit integration, depending on how the experiment is staged. Nevertheless, we favor the model that aberrant multimerization of the full complement of virion IN accounts for the potency and unique pharmacology of ALLINIs. Of note, ALLINI potency during virus production is independent of cellular LEDGF/p75 content (Desimmie et al., 2015; Fadel et al., 2014; Jurado et al., 2013).

How might multimerization of virion IN decouple RNP incorporation into the CA shell? HIV-1 IN can bind RNA in vitro (Allen et al., 1995), and recent data have indicated that IN binding to specific regions of genomic RNA is critical for infection (Kessl et al., 2016). IN mutations that specifically inhibited RNA binding in vitro yielded noninfectious particles with eccentric RNPs, and ALLINI treatment inhibited IN–RNA binding in virions. Thus, IN–RNA binding is apparently critical for RNP encapsidation into the CA shell and HIV-1 infectivity. These new findings raise several questions about the role of IN–RNA binding in retroviral biology, including the biochemistry behind IN–RNA binding and RNP encapsidation, and whether IN–RNA binding might play a similar critical role in the maturation of other types of retroviruses.

HIV-1 integration is retargeted to novel sites in cells treated with ALLINIs (Feng et al., 2016; Sharma et al., 2014; Vranckx et al., 2016b). It is somewhat less clear whether ALLINI-treated virions will retarget integration during the subsequent round of infection (Sharma et al., 2014), though unpublished data from one of our laboratories (Engelman) indicate that retargeting can occur. HIV-1 gains preference for gene 5′ end regions and hence TSSs in the absence of LEDGF/p75 (Shun et al., 2007; Singh et al., 2015; Sowd et al., 2016) or in the presence of ALLINIs (Vranckx et al., 2016b). Although sites of HIV-1 integration are not operationally linked with carcinogenesis in humans, the so-called promoter insertion model of oncogenesis has long been recognized in other retroviral systems (reviewed in Makowski et al., 1984). And, as discussed above, UIS are operationally linked to clonal expansion in chronically infected HTLV-1 patients (Gillet et al., 2011). Recent studies have moreover indicated that UIS may be operationally linked to the clonal expansion of cells that contribute to the reservoir of latently infected T cells in AIDS patients (Maldarelli et al., 2014; Wagner et al., 2014). It seemingly could be unwise to accordingly direct a higher fraction of HIV-1 integrations to gene promoters/5′-end regions. Thus, a thorough investigation of the integration retargeting potential of preclinical ALLINI compounds is warranted.

CONCLUSIONS

We now have fairly clear pictures of how the lentiviruses and γ-retroviruses locate their preferred genomic loci for integration. In each case a specific bimodal molecule tethers the IN from within the invading PIC to loci determined through host factor nonspecific DNA-binding elements and chromatin readers. The contributions of Gag/CA to integration targeting are somewhat less clear. For HIV-1, CPSF6 seems to be the dominant player, though Nup358, CypA, and, to a lesser extent, Nup153, also appear to contribute. It is moreover unclear why most of these factors seem to work in one direction, while inhibition of CypA–CA binding yields the opposite phenotype. Our picture is further clouded by the fact that CPSF6 is not known to bind DNA or chromatin directly. Moreover, while primate lentiviruses generally rely on the same set of CA-binding factors for infection, other types of lentiviruses do not (Lee et al., 2010). In these cases, might LEDGF/p75 alone fully do the job? Or will yet other CA-binding proteins be at play? Additional work is needed to further clarify the role of CA-binding proteins in lentiviral integration targeting.

Recent work with CBS (Hocum et al., 2016) and PLK-binding S-T/S-P motif (Zurnic et al., 2016) mutants in foamy virus Gag has established a role for this structural protein in integration site selection, although further experimentation is needed to fully characterize the underlying mechanisms. The MoMLV Gag protein p12, similarly to spumaviral Gag, tethers PICs to mitotic chromosomes (Elis et al., 2012; Schneider et al., 2013; Wight et al., 2012), but it is unknown whether this protein plays a role in integration targeting *per se*. One reason for this uncertainty is that many p12 mutant viruses are notoriously noninfectious. Nevertheless, it is tempting to speculate that in addition to the INs, Gag-derived structural proteins might play important roles in integration targeting for many different types of retroviruses. It is thus intuitive to expect that IN/Gag-interacting host factors for the remaining uncharacterized viruses (α-, β-, and ε-retroviruses) will harbor chromatin-associated lifestyles that mesh with the respective viral integration patterns. Detailed characterization of these interactions may lead to new insights on the regulation of retroviral integration targeting, as well as possible suggestions on how to improve upon the safety of retroviral gene therapy vectors.

ACKNOWLEDGMENTS

Work in the authors' laboratories is supported by grants AI039394, AI052014, and AI070042 from the US National Institutes of Health (A.N.E.) and 107005/Z/15/Z from the Wellcome Trust (G.N.M.).

REFERENCES

Aiyer, S., Swapna, G.V., Malani, N., Aramini, J.M., Schneider, W.M., Plumb, M.R., Ghanem, M., Larue, R.C., Sharma, A., Studamire, B., Kvaratskhelia, M., Bushman, F.D., Montelione, G.T., Roth, M.J., 2014. Altering murine leukemia virus integration through disruption of the integrase and BET protein family interaction. Nucleic Acids Res. 42, 5917–5928.

Ako-Adjei, D., Fu, W., Wallin, C., Katz, K.S., Song, G., Darji, D., Brister, J.R., Ptak, R.G., Pruitt, K.D., 2015. HIV-1, human interaction database: current status and new features. Nucleic Acids Res. 43, D566–D570.

Albanese, A., Arosio, D., Terreni, M., Cereseto, A., 2008. HIV-1 pre-integration complexes selectively target decondensed chromatin in the nuclear periphery. PLoS One 3, e2413.

Allen, P., Worland, S., Gold, L., 1995. Isolation of high-affinity RNA ligands to HIV-1 integrase from a random pool. Virology 209, 327–336.

Allouch, A., Cereseto, A., 2011. Identification of cellular factors binding to acetylated HIV-1 integrase. Amino Acids 41, 1137–1145.

Allouch, A., Di Primio, C., Alpi, E., Lusic, M., Arosio, D., Giacca, M., Cereseto, A., 2011. The TRIM family protein KAP1 inhibits HIV-1 integration. Cell Host Microbe 9, 484–495.

Ambrose, Z., Aiken, C., 2014. HIV-1 uncoating: connection to nuclear entry and regulation by host proteins. Virology 454–455, 371–379.

Ao, Z., Danappa Jayappa, K., Wang, B., Zheng, Y., Kung, S., Rassart, E., Depping, R., Kohler, M., Cohen, E.A., Yao, X., 2010. Importin α3 interacts with HIV-1 integrase and contributes to HIV-1 nuclear import and replication. J. Virol. 84, 8650–8663.

Ao, Z., Jayappa, K.D., Wang, B., Zheng, Y., Wang, X., Peng, J., Yao, X., 2012. Contribution of host nucleoporin 62 in HIV-1 integrase chromatin association and viral DNA integration. J. Biol. Chem. 287, 10544–10555.

Arino, J., Woon, C.W., Brautigan, D.L., Miller Jr., T.B., Johnson, G.L., 1988. Human liver phosphatase 2A: cDNA and amino acid sequence of two catalytic subunit isotypes. Proc. Natl. Acad. Sci. U.S.A. 85, 4252–4256.

Arroyo, J.D., Hahn, W.C., 2005. Involvement of PP2A in viral and cellular transformation. Oncogene 24, 7746–7755.

Balakrishnan, M., Yant, S.R., Tsai, L., O'Sullivan, C., Bam, R.A., Tsai, A., Niedziela-Majka, A., Stray, K.M., Sakowicz, R., Cihlar, T., 2013. Non-catalytic site HIV-1 integrase inhibitors disrupt core maturation and induce a reverse transcription block in target cells. PLoS One 8, e74163.

Ballandras-Colas, A., Brown, M., Cook, N.J., Dewdney, T.G., Demeler, B., Cherepanov, P., Lyumkis, D., Engelman, A.N., 2016. Cryo-EM reveals a novel octameric integrase structure for betaretroviral intasome function. Nature 530, 358–361.

Ballandras-Colas, A., Maskell, D.P., Serrao, E., Locke, J., Swuec, P., Jonsson, S.R., Kotecha, A., Cook, N.J., Pye, V.E., Taylor, I.A., Andresdottir, V., Engelman, A.N., Costa, A., Cherepanov, P., 2017. A supramolecular assembly mediates lentiviral DNA integration. Science 355, 93–95.

Ballandras-Colas, A., Naraharisetty, H., Li, X., Serrao, E., Engelman, A., 2013. Biochemical characterization of novel retroviral integrase proteins. PLoS One 8, e76638.

Bartholomeeusen, K., Christ, F., Hendrix, J., Rain, J.-C., Emiliani, S., Benarous, R., Debyser, Z., Gijsbers, R., De Rijck, J., 2009. Lens epithelium-derived growth factor/p75 interacts with the transposase-derived DDE domain of PogZ. J. Biol. Chem. 284, 11467–11477.

Bartholomeeusen, K., De Rijck, J., Busschots, K., Desender, L., Gijsbers, R., Emiliani, S., Benarous, R., Debyser, Z., Christ, F., 2007. Differential interaction of HIV-1 integrase and JPO2 with the C terminus of LEDGF/p75. J. Mol. Biol. 372, 407–421.

Baxter, M.K., McPhillips, M.G., Ozato, K., McBride, A.A., 2005. The mitotic chromosome binding activity of the papillomavirus E2 protein correlates with interaction with the cellular chromosomal protein, Brd4. J. Virol. 79, 4806–4818.

Belkina, A.C., Denis, G.V., 2012. BET domain co-regulators in obesity, inflammation and cancer. Nat. Rev. Cancer 12, 465–477.

Bichel, K., Price, A.J., Schaller, T., Towers, G.J., Freund, S.M., James, L.C., 2013. HIV-1 capsid undergoes coupled binding and isomerization by the nuclear pore protein NUP358. Retrovirology 10, 81.

Bisgrove, D.A., Mahmoudi, T., Henklein, P., Verdin, E., 2007. Conserved P-TEFb-interacting domain of BRD4 inhibits HIV transcription. Proc. Natl. Acad. Sci. U.S.A. 104, 13690–13695.

Boehm, D., Calvanese, V., Dar, R.D., Xing, S., Schroeder, S., Martins, L., Aull, K., Li, P.C., Planelles, V., Bradner, J.E., Zhou, M.M., Siliciano, R.F., Weinberger, L., Verdin, E., Ott, M., 2013. BET bromodomain-targeting compounds reactivate HIV from latency via a Tat-independent mechanism. Cell Cycle 12, 452–462.

Bowerman, B., Brown, P.O., Bishop, J.M., Varmus, H.E., 1989. A nucleoprotein complex mediates the integration of retroviral DNA. Genes Dev. 3, 469–478.

Brown, P.O., Bowerman, B., Varmus, H.E., Bishop, J.M., 1987. Correct integration of retroviral DNA in vitro. Cell 49, 347–356.

Brown, P.O., Bowerman, B., Varmus, H.E., Bishop, J.M., 1989. Retroviral integration: structure of the initial covalent product and its precursor, and a role for the viral IN protein. Proc. Natl. Acad. Sci. U.S.A. 86, 2525–2529.

Bushman, F.D., Craigie, R., 1991. Activities of human immunodeficiency virus (HIV) integration protein in vitro: specific cleavage and integration of HIV DNA. Proc. Natl. Acad. Sci. U.S.A. 88, 1339–1343.

Bushman, F.D., Fujiwara, T., Craigie, R., 1990. Retroviral DNA integration directed by HIV integration protein in vitro. Science 249, 1555–1558.

Busschots, K., Vercammen, J., Emiliani, S., Benarous, R., Engelborghs, Y., Christ, F., Debyser, Z., 2005. The interaction of LEDGF/p75 with integrase is lentivirus-specific and promotes DNA binding. J. Biol. Chem. 280, 17841–17847.

Campbell, E.M., Hope, T.J., 2015. HIV-1 capsid: the multifaceted key player in HIV-1 infection. Nat. Rev. Microbiol. 13, 471–483.

Cardinale, S., Cisterna, B., Bonetti, P., Aringhieri, C., Biggiogera, M., Barabino, S.M.L., 2007. Subnuclear localization and dynamics of the pre-mRNA 3′ end processing factor mammalian cleavage factor I 68-kDa subunit. Mol. Biol. Cell 18, 1282–1292.

Čermáková, K., Tesina, P., Demeulemeester, J., El Ashkar, S., Méreau, H., Schwaller, J., Řezáčová, P., Veverka, V., De Rijck, J., 2014. Validation and structural characterization of the LEDGF/p75–MLL interface as a new target for the treatment of MLL-dependent leukemia. Canc. Res. 74, 5139–5151.

Cherepanov, P., 2007. LEDGF/p75 interacts with divergent lentiviral integrases and modulates their enzymatic activity in vitro. Nucleic Acids Res. 35, 113–124.

Cherepanov, P., Ambrosio, A.L., Rahman, S., Ellenberger, T., Engelman, A., 2005. Structural basis for the recognition between HIV-1 integrase and transcriptional coactivator p75. Proc. Natl. Acad. Sci. U.S.A. 102, 17308–17313.

Cherepanov, P., Devroe, E., Silver, P.A., Engelman, A., 2004. Identification of an evolutionarily conserved domain in human lens epithelium-derived growth factor/transcriptional co-activator p75 (LEDGF/p75) that binds HIV-1 integrase. J. Biol. Chem. 279, 48883–48892.

Cherepanov, P., Maertens, G., Proost, P., Devreese, B., Van Beeumen, J., Engelborghs, Y., De Clercq, E., Debyser, Z., 2003. HIV-1 integrase forms stable tetramers and associates with LEDGF/p75 protein in human cells. J. Biol. Chem. 278, 372–381.

Chin, C.R., Perreira, J.M., Savidis, G., Portmann, J.M., Aker, A.M., Feeley, E.M., Smith, M.C., Brass, A.L., 2015. Direct visualization of HIV-1 replication intermediates shows that capsid and CPSF6 modulate HIV-1 intra-nuclear invasion and integration. Cell Rep. 13, 1717–1731.

Cho, U.S., Xu, W., 2007. Crystal structure of a protein phosphatase 2A heterotrimeric holoenzyme. Nature 445, 53–57.

Christ, F., Shaw, S., Demeulemeester, J., Desimmie, B.A., Marchand, A., Butler, S., Smets, W., Chaltin, P., Westby, M., Debyser, Z., Pickford, C., 2012. Small-molecule inhibitors of the LEDGF/p75 binding site of integrase block HIV replication and modulate integrase multimerization. Antimicrob. Agents Chemother. 56, 4365–4374.

Christ, F., Thys, W., De Rijck, J., Gijsbers, R., Albanese, A., Arosio, D., Emiliani, S., Rain, J.C., Benarous, R., Cereseto, A., Debyser, Z., 2008. Transportin-SR2 imports HIV into the nucleus. Curr. Biol. 18, 1192–1202.

Christ, F., Voet, A., Marchand, A., Nicolet, S., Desimmie, B.A., Marchand, D., Bardiot, D., Van der Veken, N.J., Van Remoortel, B., Strelkov, S.V., De Maeyer, M., Chaltin, P., Debyser, Z., 2010. Rational design of small-molecule inhibitors of the LEDGF/p75-integrase interaction and HIV replication. Nat. Chem. Biol. 6, 442–448.

Ciuffi, A., Llano, M., Poeschla, E., Hoffmann, C., Leipzig, J., Shinn, P., Ecker, J.R., Bushman, F., 2005. A role for LEDGF/p75 in targeting HIV DNA integration. Nat. Med. 11, 1287–1289.

Cook, L.B., Melamed, A., Niederer, H., Valganon, M., Laydon, D., Foroni, L., Taylor, G.P., Matsuoka, M., Bangham, C.R., 2014. The role of HTLV-1 clonality, proviral structure, and genomic integration site in adult T-cell leukemia/lymphoma. Blood 123, 3925–3931.

Craigie, R., Bushman, F.D., 2014. Host factors in retroviral integration and the selection of integration target sites. Microbiol. Spectr. 2.

Craigie, R., Fujiwara, T., Bushman, F., 1990. The IN protein of Moloney murine leukemia virus processes the viral DNA ends and accomplishes their integration in vitro. Cell 62, 829–837.

Crowe, B.L., Larue, R.C., Yuan, C., Hess, S., Kvaratskhelia, M., Foster, M.P., 2016. Structure of the Brd4 ET domain bound to a C-terminal motif from γ-retroviral integrases reveals a conserved mechanism of interaction. Proc. Natl. Acad. Sci. U.S.A. 113, 2086–2091.

da Cruz e Silva, O.B., Cohen, P.T., 1987. A second catalytic subunit of type-2A protein phosphatase from rabbit skeletal muscle. FEBS Lett. 226, 176–178.

Daugaard, M., Baude, A., Fugger, K., Povlsen, L.K., Beck, H., Sorensen, C.S., Petersen, N.H., Sorensen, P.H., Lukas, C., Bartek, J., Lukas, J., Rohde, M., Jaattela, M., 2012. LEDGF (p75) promotes DNA-end resection and homologous recombination. Nat. Struct. Mol. Biol. 19, 803–810.

De Iaco, A., Luban, J., 2014. Cyclophilin A promotes HIV-1 reverse transcription but its effect on transduction correlates best with its effect on nuclear entry of viral cDNA. Retrovirology 11, 11.

De Ravin, S.S., Su, L., Theobald, N., Choi, U., Macpherson, J.L., Poidinger, M., Symonds, G., Pond, S.M., Ferris, A.L., Hughes, S.H., Malech, H.L., Wu, X., 2014. Enhancers are major targets for murine leukemia virus vector integration. J. Virol. 88, 4504–4513.

De Rijck, J., Bartholomeeusen, K., Ceulemans, H., Debyser, Z., Gijsbers, R., 2010. High-resolution profiling of the LEDGF/p75 chromatin interaction in the ENCODE region. Nucleic Acids Res. 38, 6135–6147.

De Rijck, J., de Kogel, C., Demeulemeester, J., Vets, S., El Ashkar, S., Malani, N., Bushman, F.D., Landuyt, B., Husson, S.J., Busschots, K., Gijsbers, R., Debyser, Z., 2013. The BET family of proteins targets moloney murine leukemia virus integration near transcription start sites. Cell Rep. 5, 886–894.

De Rijck, J., Vandekerckhove, L., Gijsbers, R., Hombrouck, A., Hendrix, J., Vercammen, J., Engelborghs, Y., Christ, F., Debyser, Z., 2006. Overexpression of the lens epithelium-derived growth factor/p75 integrase binding domain inhibits human immunodeficiency virus replication. J. Virol. 80, 11498–11509.

Debyser, Z., Christ, F., De Rijck, J., Gijsbers, R., 2015. Host factors for retroviral integration site selection. Trends Biochem. Sci. 40, 108–116.

Delelis, O., Lehmann-Che, J., Saïb, A., 2004. Foamy viruses — a world apart. Curr. Opin. Microbiol. 7, 400–406.

Derse, D., Crise, B., Li, Y., Princler, G., Lum, N., Stewart, C., McGrath, C.F., Hughes, S.H., Munroe, D.J., Wu, X., 2007. Human T-cell leukemia virus type 1 integration target sites in the human genome: comparison with those of other retroviruses. J. Virol. 81, 6731–6741.

Desimmie, B.A., Schrijvers, R., Demeulemeester, J., Borrenberghs, D., Weydert, C., Thys, W., Vets, S., Van Remoortel, B., Hofkens, J., De Rijck, J., Hendrix, J., Bannert, N., Gijsbers, R., Christ, F., Debyser, Z., 2013. LEDGINs inhibit late stage HIV-1 replication by modulating integrase multimerization in the virions. Retrovirology 10, 57.
Desimmie, B.A., Weydert, C., Schrijvers, R., Vets, S., Demeulemeester, J., Proost, P., Paron, I., De Rijck, J., Mast, J., Bannert, N., Gijsbers, R., Christ, F., Debyser, Z., 2015. HIV-1 IN/Pol recruits LEDGF/p75 into viral particles. Retrovirology 12, 16.
Dettwiler, S., Aringhieri, C., Cardinale, S., Keller, W., Barabino, S.M.L., 2004. Distinct sequence motifs within the 68-kDa subunit of cleavage factor Im mediate RNA binding, protein-protein interactions, and subcellular localization. J. Biol. Chem. 279, 35788–35797.
Di Nunzio, F., Danckaert, A., Fricke, T., Perez, P., Fernandez, J., Perret, E., Roux, P., Shorte, S., Charneau, P., Diaz-Griffero, F., Arhel, N.J., 2012. Human nucleoporins promote HIV-1 docking at the nuclear pore, nuclear import and integration. PLoS One 7, e46037.
Di Nunzio, F., Fricke, T., Miccio, A., Valle-Casuso, J.C., Perez, P., Souque, P., Rizzi, E., Severgnini, M., Mavilio, F., Charneau, P., Diaz-Griffero, F., 2013. Nup153 and Nup98 bind the HIV-1 core and contribute to the early steps of HIV-1 replication. Virology 440, 8–18.
Doi, K., Wu, X., Taniguchi, Y., Yasunaga, J., Satou, Y., Okayama, A., Nosaka, K., Matsuoka, M., 2005. Preferential selection of human T-cell leukemia virus type I provirus integration sites in leukemic versus carrier states. Blood 106, 1048–1053.
Eichhorn, P.J., Creyghton, M.P., Bernards, R., 2009. Protein phosphatase 2A regulatory subunits and cancer. Biochim. Biophys. Acta 1795, 1–15.
Eidahl, J.O., Crowe, B.L., North, J.A., McKee, C.J., Shkriabai, N., Feng, L., Plumb, M., Graham, R.L., Gorelick, R.J., Hess, S., Poirier, M.G., Foster, M.P., Kvaratskhelia, M., 2013. Structural basis for high-affinity binding of LEDGF PWWP to mononucleosomes. Nucleic Acids Res. 41, 3924–3936.
Elis, E., Ehrlich, M., Prizan-Ravid, A., Laham-Karam, N., Bacharach, E., 2012. p12 tethers the murine leukemia virus pre-integration complex to mitotic chromosomes. PLoS Pathog. 8, e1003103.
Emiliani, S., Mousnier, A., Busschots, K., Maroun, M., Van Maele, B., Tempe, D., Vandekerckhove, L., Moisant, F., Ben-Slama, L., Witvrouw, M., Christ, F., Rain, J.C., Dargemont, C., Debyser, Z., Benarous, R., 2005. Integrase mutants defective for interaction with LEDGF/p75 are impaired in chromosome tethering and HIV-1 replication. J. Biol. Chem. 280, 25517–25523.
Engelman, A., 1999. In vivo analysis of retroviral integrase structure and function. Adv. Virus Res. 52, 411–426.
Engelman, A., 2007. Host cell factors and HIV-1 integration. Future HIV Ther. 1, 415–426.
Engelman, A., 2011. Pleiotropic nature of HIV-1 integrase mutations. In: Neamati, N. (Ed.), HIV-1 Integrase: Mechanism and Inhibitor Design. John Wiley & Sons, Inc., Hoboken, NJ, USA, pp. 67–81.
Engelman, A., Mizuuchi, K., Craigie, R., 1991. HIV-1 DNA integration: mechanism of viral DNA cleavage and DNA strand transfer. Cell 67, 1211–1221.
Fadel, H.J., Morrison, J.H., Saenz, D.T., Fuchs, J.R., Kvaratskhelia, M., Ekker, S.C., Poeschla, E.M., 2014. TALEN knockout of the PSIP1 gene in human cells: analyses of HIV-1 replication and allosteric integrase inhibitor mechanism. J. Virol. 88, 9704–9717.
Fader, L.D., Malenfant, E., Parisien, M., Carson, R., Bilodeau, F., Landry, S., Pesant, M., Brochu, C., Morin, S., Chabot, C., Halmos, T., Bousquet, Y., Bailey, M.D., Kawai, S.H., Coulombe, R., LaPlante, S., Jakalian, A., Bhardwaj, P.K., Wernic, D., Schroeder, P., Amad, M., Edwards, P., Garneau, M., Duan, J., Cordingley, M., Bethell, R., Mason, S.W., Bös, M., Bonneau, P., Poupart, M.A., Faucher, A.M., Simoneau, B., Fenwick, C., Yoakim, C., Tsantrizos, Y., 2014. Discovery of BI 224436, a noncatalytic site integrase inhibitor (NCINI) of HIV-1. ACS Med. Chem. Lett. 5, 422–427.

Feng, L., Dharmarajan, V., Serrao, E., Hoyte, A., Larue, R.C., Slaughter, A., Sharma, A., Plumb, M.R., Kessl, J.J., Fuchs, J.R., Bushman, F.D., Engelman, A.N., Griffin, P.R., Kvaratskhelia, M., 2016. The competitive interplay between allosteric HIV-1 integrase inhibitor BI/D and LEDGF/p75 during the early stage of HIV-1 replication adversely affects inhibitor potency. ACS Chem. Biol. 11, 1313–1321.

Ferris, A.L., Wu, X., Hughes, C.M., Stewart, C., Smith, S.J., Milne, T.A., Wang, G.G., Shun, M.C., Allis, C.D., Engelman, A., Hughes, S.H., 2010. Lens epithelium-derived growth factor fusion proteins redirect HIV-1 DNA integration. Proc. Natl. Acad. Sci. U.S.A. 107, 3135–3140.

Filippakopoulos, P., Qi, J., Picaud, S., Shen, Y., Smith, W.B., Fedorov, O., Morse, E.M., Keates, T., Hickman, T.T., Felletar, I., Philpott, M., Munro, S., McKeown, M.R., Wang, Y., Christie, A.L., West, N., Cameron, M.J., Schwartz, B., Heightman, T.D., La Thangue, N., French, C.A., Wiest, O., Kung, A.L., Knapp, S., Bradner, J.E., 2010. Selective inhibition of BET bromodomains. Nature 468, 1067–1073.

Flegg, C.P., Sharma, M., Medina-Palazon, C., Jamieson, C., Galea, M., Brocardo, M.G., Mills, K., Henderson, B.R., 2010. Nuclear export and centrosome targeting of the protein phosphatase 2A subunit B56alpha: role of B56alpha in nuclear export of the catalytic subunit. J. Biol. Chem. 285, 18144–18154.

Foley, E.A., Maldonado, M., Kapoor, T.M., 2011. Formation of stable attachments between kinetochores and microtubules depends on the B56-PP2A phosphatase. Nat. Cell Biol. 13, 1265–1271.

Fontana, J., Jurado, K.A., Cheng, N., Ly, N.L., Fuchs, J.R., Gorelick, R.J., Engelman, A.N., Steven, A.C., 2015. Distribution and redistribution of HIV-1 nucleocapsid protein in immature, mature, and integrase-inhibited virions: a role for integrase in maturation. J. Virol. 89, 9765–9780.

Fujiwara, T., Mizuuchi, K., 1988. Retroviral DNA integration: structure of an integration intermediate. Cell 54, 497–504.

Gamble, T.R., Vajdos, F.F., Yoo, S., Worthylake, D.K., Houseweart, M., Sundquist, W.I., Hill, C.P., 1996. Crystal structure of human cyclophilin A bound to the amino-terminal domain of HIV-1 capsid. Cell 87, 1285–1294.

Gerard, A., Segeral, E., Naughtin, M., Abdouni, A., Charmeteau, B., Cheynier, R., Rain, J.C., Emiliani, S., 2015. The integrase cofactor LEDGF/p75 associates with Iws1 and Spt6 for postintegration silencing of HIV-1 gene expression in latently infected cells. Cell Host Microbe 17, 107–117.

Gijsbers, R., Ronen, K., Vets, S., Malani, N., De Rijck, J., McNeely, M., Bushman, F.D., Debyser, Z., 2010. LEDGF hybrids efficiently retarget lentiviral integration into heterochromatin. Mol. Ther. 18, 552–560.

Gillet, N.A., Gutierrez, G., Rodriguez, S.M., de Brogniez, A., Renotte, N., Alvarez, I., Trono, K., Willems, L., 2013. Massive depletion of bovine leukemia virus proviral clones located in genomic transcriptionally active sites during primary infection. PLoS Pathog. 9, e1003687.

Gillet, N.A., Malani, N., Melamed, A., Gormley, N., Carter, R., Bentley, D., Berry, C., Bushman, F.D., Taylor, G.P., Bangham, C.R., 2011. The host genomic environment of the provirus determines the abundance of HTLV-1-infected T-cell clones. Blood 117, 3113–3122.

Gruber, A.R., Martin, G., Keller, W., Zavolan, M., 2012. Cleavage factor Im is a key regulator of 3′ UTR length. RNA Biol. 9, 1405–1412.

Gupta, K., Brady, T., Dyer, B.M., Malani, N., Hwang, Y., Male, F., Nolte, R.T., Wang, L., Velthuisen, E., Jeffrey, J., Van Duyne, G.D., Bushman, F.D., 2014. Allosteric inhibition of human immunodeficiency virus integrase: late block during viral replication and abnormal multimerization involving specific protein domains. J. Biol. Chem. 289, 20477–20488.

Gupta, K., Turkki, V., Sherrill-Mix, S., Hwang, Y., Eilers, G., Taylor, L., McDanal, C., Wang, P., Temelkoff, D., Nolte, R.T., Velthuisen, E., Jeffrey, J., Van Duyne, G.D., Bushman, F.D., 2016. Structural basis for inhibitor-induced aggregation of HIV integrase. PLoS Biol. 14, e1002584.

Gupta, S.S., Maetzig, T., Maertens, G.N., Sharif, A., Rothe, M., Weidner-Glunde, M., Galla, M., Schambach, A., Cherepanov, P., Schulz, T.F., 2013. Bromo- and extraterminal domain chromatin regulators serve as cofactors for murine leukemia virus integration. J. Virol. 87, 12721–12736.

Haesen, D., Sents, W., Lemaire, K., Hoorne, Y., Janssens, V., 2014. The basic biology of PP2A in hematologic cells and malignancies. Front. Oncol. 4, 347.

Hardy, J.G., Norbury, C.J., 2016. Cleavage factor Im (CFIm) as a regulator of alternative polyadenylation. Biochem. Soc. Trans. 44, 1051–1057.

Hare, S., Di Nunzio, F., Labeja, A., Wang, J., Engelman, A., Cherepanov, P., 2009a. Structural basis for functional tetramerization of lentiviral integrase. PLoS Pathog. 5, e1000515.

Hare, S., Shun, M.C., Gupta, S.S., Valkov, E., Engelman, A., Cherepanov, P., 2009b. A novel co-crystal structure affords the design of gain-of-function lentiviral integrase mutants in the presence of modified PSIP1/LEDGF/p75. PLoS Pathog. 5, e1000259.

Hare, S., Gupta, S.S., Valkov, E., Engelman, A., Cherepanov, P., 2010. Retroviral intasome assembly and inhibition of DNA strand transfer. Nature 464, 232–236.

Hare, S., Maertens, G.N., Cherepanov, P., 2012. 3′-processing and strand transfer catalysed by retroviral integrase in crystallo. EMBO J. 31, 3020–3028.

Hemmings, B.A., Adams-Pearson, C., Maurer, F., Muller, P., Goris, J., Merlevede, W., Hofsteenge, J., Stone, S.R., 1990. alpha- and beta-forms of the 65-kDa subunit of protein phosphatase 2A have a similar 39 amino acid repeating structure. Biochemistry 29, 3166–3173.

Hertz, E.P., Kruse, T., Davey, N.E., Lopez-Mendez, B., Sigurethsson, J.O., Montoya, G., Olsen, J.V., Nilsson, J., 2016. A conserved motif provides binding specificity to the PP2A-B56 phosphatase. Mol. Cell 63, 686–695.

Hocum, J.D., Linde, I., Rae, D.T., Collins, C.P., Matern, L.K., Trobridge, G.D., 2016. Retargeted foamy virus vectors integrate less frequently near proto-oncogenes. Sci. Rep. 6, 36610.

Huang, L., et al., 2008. Daxx interacts with HIV-1 integrase and inhibits lentiviral gene expression. Biochem. Biophys. Res. Commun. 373, 241–245.

Hughes, S., Jenkins, V., Dar, M.J., Engelman, A., Cherepanov, P., 2010. Transcriptional co-activator LEDGF interacts with Cdc7-activator of S-phase kinase (ASK) and stimulates its enzymatic activity. J. Biol. Chem. 285, 541–554.

Hultquist, J.F., Schumann, K., Woo, J.M., Manganaro, L., McGregor, M.J., Doudna, J., Simon, V., Krogan, N.J., Marson, A., 2016. A Cas9 ribonucleoprotein platform for functional genetic studies of HIV-host interactions in primary human T cells. Cell Rep. 17, 1438–1452.

Izumoto, Y., Kuroda, T., Harada, H., Kishimoto, T., Nakamura, H., 1997. Hepatoma-derived growth factor belongs to a gene family in mice showing significant homology in the amino terminus. Biochem. Biophys. Res. Commun. 238, 26–32.

Jager, S., Cimermancic, P., Gulbahce, N., Johnson, J.R., McGovern, K.E., Clarke, S.C., Shales, M., Mercenne, G., Pache, L., Li, K., Hernandez, H., Jang, G.M., Roth, S.L., Akiva, E., Marlett, J., Stephens, M., D'Orso, I., Fernandes, J., Fahey, M., Mahon, C., O'Donoghue, A.J., Todorovic, A., Morris, J.H., Maltby, D.A., Alber, T., Cagney, G., Bushman, F.D., Young, J.A., Chanda, S.K., Sundquist, W.I., Kortemme, T., Hernandez, R.D., Craik, C.S., Burlingame, A., Sali, A., Frankel, A.D., Krogan, N.J., 2012. Global landscape of HIV-human protein complexes. Nature 481, 365–370.

Janssens, V., Goris, J., 2001. Protein phosphatase 2A: a highly regulated family of serine/threonine phosphatases implicated in cell growth and signalling. Biochem. J. 353, 417–439.

Jurado, K.A., Wang, H., Slaughter, A., Feng, L., Kessl, J.J., Koh, Y., Wang, W., Ballandras-Colas, A., Patel, P.A., Fuchs, J.R., Kvaratskhelia, M., Engelman, A., 2013. Allosteric integrase inhibitor potency is determined through the inhibition of HIV-1 particle maturation. Proc. Natl. Acad. Sci. U.S.A. 110, 8690–8695.

Katahira, J., Okuzaki, D., Inoue, H., Yoneda, Y., Maehara, K., Ohkawa, Y., 2013. Human TREX component Thoc5 affects alternative polyadenylation site choice by recruiting mammalian cleavage factor I. Nucleic Acids Res. 41, 7060–7072.

Katz, R.A., Merkel, G., Kulkosky, J., Leis, J., Skalka, A.M., 1990. The avian retroviral IN protein is both necessary and sufficient for integrative recombination in vitro. Cell 63, 87–95.

Katzman, M., Katz, R.A., Skalka, A.M., Leis, J., 1989. The avian retroviral integration protein cleaves the terminal sequences of linear viral DNA at the in vivo sites of integration. J. Virol. 63, 5319–5327.

Kessl, J.J., Jena, N., Koh, Y., Taskent-Sezgin, H., Slaughter, A., Feng, L., de Silva, S., Wu, L., Le Grice, S.F., Engelman, A., Fuchs, J.R., Kvaratskhelia, M., 2012. Multimode, cooperative mechanism of action of allosteric HIV-1 integrase inhibitors. J. Biol. Chem. 287, 16801–16811.

Kessl, J.J., Kutluay, S.B., Townsend, D., Rebensburg, S., Slaughter, A., Larue, R.C., Shkriabai, N., Bakouche, N., Fuchs, J.R., Bieniasz, P.D., Kvaratskhelia, M., 2016. HIV-1 integrase binds the viral RNA genome and is essential during virion morphogenesis. Cell 166 1257–1268.e1212.

Kessl, J.J., Li, M., Ignatov, M., Shkriabai, N., Eidahl, J.O., Feng, L., Musier-Forsyth, K., Craigie, R., Kvaratskhelia, M., 2011. FRET analysis reveals distinct conformations of IN tetramers in the presence of viral DNA or LEDGF/p75. Nucleic Acids Res. 39, 9009–9022.

Kitajima, T.S., Sakuno, T., Ishiguro, K., Iemura, S., Natsume, T., Kawashima, S.A., Watanabe, Y., 2006. Shugoshin collaborates with protein phosphatase 2A to protect cohesin. Nature 441, 46–52.

Kobbi, L., Octobre, G., Dias, J., Comisso, M., Mirande, M., 2011. Association of mitochondrial lysyl-tRNA synthetase with HIV-1 GagPol involves catalytic domain of the synthetase and transframe and integrase domains of Pol. J. Mol. Biol. 410, 875–886.

Koh, Y., Wu, X., Ferris, A.L., Matreyek, K.A., Smith, S.J., Lee, K., KewalRamani, V.N., Hughes, S.H., Engelman, A., 2013. Differential effects of human immunodeficiency virus type 1 capsid and cellular factors nucleoporin 153 and LEDGF/p75 on the efficiency and specificity of viral DNA integration. J. Virol. 87, 648–658.

Kvaratskhelia, M., Sharma, A., Larue, R.C., Serrao, E., Engelman, A., 2014. Molecular mechanisms of retroviral integration site selection. Nucleic Acids Res. 42, 10209–10225.

LaFave, M.C., Varshney, G.K., Gildea, D.E., Wolfsberg, T.G., Baxevanis, A.D., Burgess, S.M., 2014. MLV integration site selection is driven by strong enhancers and active promoters. Nucleic Acids Res. 42, 4257–4269.

Larue, R.C., Plumb, M.R., Crowe, B.L., Shkriabai, N., Sharma, A., DiFiore, J., Malani, N., Aiyer, S.S., Roth, M.J., Bushman, F.D., Foster, M.P., Kvaratskhelia, M., 2014. Bimodal high-affinity association of Brd4 with murine leukemia virus integrase and mononucleosomes. Nucleic Acids Res. 42, 4868–4881.

Laskey, S.B., Siliciano, R.F., 2016. Quantitative evaluation of the antiretroviral efficacy of dolutegravir. JCI Insight 1, e90033.

Laydon, D.J., Melamed, A., Sim, A., Gillet, N.A., Sim, K., Darko, S., Kroll, J.S., Douek, D.C., Price, D.A., Bangham, C.R., Asquith, B., 2014. Quantification of HTLV-1 clonality and TCR diversity. PLoS Comput. Biol. 10, e1003646.

Le Rouzic, E., Bonnard, D., Chasset, S., Bruneau, J.M., Chevreuil, F., Le Strat, F., Nguyen, J., Beauvoir, R., Amadori, C., Brias, J., Vomscheid, S., Eiler, S., Levy, N., Delelis, O., Deprez, E., Saib, A., Zamborlini, A., Emiliani, S., Ruff, M., Ledoussal, B., Moreau, F., Benarous, R., 2013. Dual inhibition of HIV-1 replication by integrase-LEDGF allosteric inhibitors is predominant at the post-integration stage. Retrovirology 10, 144.

Lee, K., Ambrose, Z., Martin, T.D., Oztop, I., Mulky, A., Julias, J.G., Vandegraaff, N., Baumann, J.G., Wang, R., Yuen, W., Takemura, T., Shelton, K., Taniuchi, I., Li, Y., Sodroski, J., Littman, D.R., Coffin, J.M., Hughes, S.H., Unutmaz, D., Engelman, A., KewalRamani, V.N., 2010. Flexible use of nuclear import pathways by HIV-1. Cell Host Microbe 7, 221–233.

Lee, S.P., Han, M.K., 1996. Zinc stimulates Mg^{2+}-dependent 3′-processing activity of human immunodeficiency virus type 1 integrase in vitro. Biochemistry 35, 3837–3844.

Lelek, M., Casartelli, N., Pellin, D., Rizzi, E., Souque, P., Severgnini, M., Di Serio, C., Fricke, T., Diaz-Griffero, F., Zimmer, C., Charneau, P., Di Nunzio, F., 2015. Chromatin organization at the nuclear pore favours HIV replication. Nat Commun. 6, 6483.

LeRoy, G., Chepelev, I., DiMaggio, P.A., Blanco, M.A., Zee, B.M., Zhao, K., Garcia, B.A., 2012. Proteogenomic characterization and mapping of nucleosomes decoded by Brd and HP1 proteins. Genome Biol. 13, R68.

Lesbats, P., Engelman, A.N., Cherepanov, P., 2016. Retroviral DNA integration. Chem. Rev. 116, 12730–12757.

Lewinski, M.K., Yamashita, M., Emerman, M., Ciuffi, A., Marshall, H., Crawford, G., Collins, F., Shinn, P., Leipzig, J., Hannenhalli, S., Berry, C.C., Ecker, J.R., Bushman, F.D., 2006. Retroviral DNA integration: viral and cellular determinants of target-site selection. PLoS Pathog. 2, e60.

Lewis, P., Hensel, M., Emerman, M., 1992. Human immunodeficiency virus infection of cells arrested in the cell cycle. EMBO J. 11, 3053–3058.

Lewis, P.F., Emerman, M., 1994. Passage through mitosis is required for oncoretroviruses but not for the human immunodeficiency virus. J. Virol. 68, 510–516.

Li, M., Mizuuchi, M., Burke Jr., T.R., Craigie, R., 2006. Retroviral DNA integration: reaction pathway and critical intermediates. EMBO J. 25, 1295–1304.

Llano, M., Saenz, D.T., Meehan, A., Wongthida, P., Peretz, M., Walker, W.H., Teo, W., Poeschla, E.M., 2006a. An essential role for LEDGF/p75 in HIV integration. Science 314, 461–464.

Llano, M., Vanegas, M., Fregoso, O., Saenz, D., Chung, S., Peretz, M., Poeschla, E.M., 2004. LEDGF/p75 determines cellular trafficking of diverse lentiviral but not murine oncoretroviral integrase proteins and is a component of functional lentiviral preintegration complexes. J. Virol. 78, 9524–9537.

Llano, M., Vanegas, M., Hutchins, N., Thompson, D., Delgado, S., Poeschla, E.M., 2006b. Identification and characterization of the chromatin-binding domains of the HIV-1 integrase interactor LEDGF/p75. J. Mol. Biol. 360, 760–773.

Lusic, M., Siliciano, R.F., 2017. Nuclear landscape of HIV-1 infection and integration. Nat. Rev. Microbiol. 15, 69–82.

Maertens, G., Cherepanov, P., Pluymers, W., Busschots, K., De Clercq, E., Debyser, Z., Engelborghs, Y., 2003. LEDGF/p75 is essential for nuclear and chromosomal targeting of HIV-1 integrase in human cells. J. Biol. Chem. 278, 33528–33539.

Maertens, G.N., 2016. B′-protein phosphatase 2A is a functional binding partner of delta-retroviral integrase. Nucleic Acids Res. 44, 364–376.

Maertens, G.N., Cherepanov, P., Engelman, A., 2006. Transcriptional co-activator p75 binds and tethers the Myc-interacting protein JPO2 to chromatin. J. Cell Sci. 119, 2563–2571.

Maertens, G.N., Hare, S., Cherepanov, P., 2010. The mechanism of retroviral integration from X-ray structures of its key intermediates. Nature 468, 326–329.

Magnusdottir, A., Stenmark, P., Flodin, S., Nyman, T., Kotenyova, T., Graslund, S., Ogg, D., Nordlund, P., 2009. The structure of the PP2A regulatory subunit B56 gamma: the remaining piece of the PP2A jigsaw puzzle. Proteins 74, 212–221.

Makowski, D.R., Rothberg, P.G., Astrin, S.M., 1984. The role of promoter insertion in the induction of neoplasia. Surv. Synth. Pathol. Res. 3, 342–349.

Maldarelli, F., Wu, X., Su, L., Simonetti, F.R., Shao, W., Hill, S., Spindler, J., Ferris, A.L., Mellors, J.W., Kearney, M.F., Coffin, J.M., Hughes, S.H., 2014. Specific HIV integration sites are linked to clonal expansion and persistence of infected cells. Science 345, 179–183.

Manganaro, L., Lusic, M., Gutierrez, M.I., Cereseto, A., Del Sal, G., Giacca, M., 2010. Concerted action of cellular JNK and Pin1 restricts HIV-1 genome integration to activated CD4+ T lymphocytes. Nat. Med. 16, 329–333.

Marini, B., Kertesz-Farkas, A., Ali, H., Lucic, B., Lisek, K., Manganaro, L., Pongor, S., Luzzati, R., Recchia, A., Mavilio, F., Giacca, M., Lusic, M., 2015. Nuclear architecture dictates HIV-1 integration site selection. Nature 521, 227–231.

Marshall, H.M., Ronen, K., Berry, C., Llano, M., Sutherland, H., Saenz, D., Bickmore, W., Poeschla, E., Bushman, F.D., 2007. Role of PSIP1/LEDGF/p75 in lentiviral infectivity and integration targeting. PLoS One 2, e1340.

Martin, G., Gruber, A.R., Keller, W., Zavolan, M., 2012. Genome-wide analysis of pre-mRNA 3′ end processing reveals a decisive role of human cleavage factor I in the regulation of 3′ UTR length. Cell Rep. 1, 753–763.

Maskell, D.P., Renault, L., Serrao, E., Lesbats, P., Matadeen, R., Hare, S., Lindemann, D., Engelman, A.N., Costa, A., Cherepanov, P., 2015. Structural basis for retroviral integration into nucleosomes. Nature 523, 366–369.

Matreyek, K.A., Engelman, A., 2011. The requirement for nucleoporin NUP153 during human immunodeficiency virus type 1 infection Is determined by the viral capsid. J. Virol. 85, 7818–7827.

Matreyek, K.A., Engelman, A., 2013. Viral and cellular requirements for the nuclear entry of retroviral preintegration nucleoprotein complexes. Viruses 5, 2483–2511.

Matreyek, K.A., Yucel, S.S., Li, X., Engelman, A., 2013. Nucleoporin NUP153 phenylalanine-glycine motifs engage a common binding pocket within the HIV-1 capsid protein to mediate lentiviral infectivity. PLoS Pathog. 9, e1003693.

Matsuoka, M., Jeang, K.T., 2011. Human T-cell leukemia virus type 1 (HTLV-1) and leukemic transformation: viral infectivity, tax, HBZ and therapy. Oncogene 30, 1379–1389.

McCright, B., Rivers, A.M., Audlin, S., Virshup, D.M., 1996. The B56 family of protein phosphatase 2A (PP2A) regulatory subunits encodes differentiation-induced phosphoproteins that target PP2A to both nucleus and cytoplasm. J. Biol. Chem. 271, 22081–22089.

McKee, C.J., Kessl, J.J., Shkriabai, N., Dar, M.J., Engelman, A., Kvaratskhelia, M., 2008. Dynamic modulation of HIV-1 integrase structure and function by cellular lens epithelium-derived growth factor (LEDGF) protein. J. Biol. Chem. 283, 31802–31812.

Meehan, A.M., Saenz, D.T., Morrison, J., Hu, C., Peretz, M., Poeschla, E.M., 2011. LEDGF dominant interference proteins demonstrate prenuclear exposure of HIV-1 integrase and synergize with LEDGF depletion to destroy viral infectivity. J. Virol. 85, 3570–3583.

Meehan, A.M., Saenz, D.T., Morrison, J.H., Garcia-Rivera, J.A., Peretz, M., Llano, M., Poeschla, E.M., 2009. LEDGF/p75 proteins with alternative chromatin tethers are functional HIV-1 cofactors. PLoS Pathog. 5, e1000522.

Meekings, K.N., Leipzig, J., Bushman, F.D., Taylor, G.P., Bangham, C.R., 2008. HTLV-1 integration into transcriptionally active genomic regions is associated with proviral expression and with HAM/TSP. PLoS Pathog. 4, e1000027.

Melamed, A., Laydon, D.J., Gillet, N.A., Tanaka, Y., Taylor, G.P., Bangham, C.R., 2013. Genome-wide determinants of proviral targeting, clonal abundance and expression in natural HTLV-1 infection. PLoS Pathog. 9, e1003271.

Melamed, A., Witkover, A.D., Laydon, D.J., Brown, R., Ladell, K., Miners, K., Rowan, A.G., Gormley, N., Price, D.A., Taylor, G.P., Murphy, E.L., Bangham, C.R., 2014. Clonality of HTLV-2 in natural infection. PLoS Pathog. 10, e1004006.

Miller, M.D., Farnet, C.M., Bushman, F.D., 1997. Human immunodeficiency virus type 1 preintegration complexes: studies of organization and composition. J. Virol. 71, 5382–5390.

Morchikh, M., Naughtin, M., Di Nunzio, F., Xavier, J., Charneau, P., Jacob, Y., Lavigne, M., 2013. TOX4 and NOVA1 proteins are partners of the LEDGF PWWP domain and affect HIV-1 replication. PLoS One 7, e1001280.

Morinière, J., Rousseaux, S., Steuerwald, U., Soler-López, M., Curtet, S., Vitte, A.L., Govin, J., Gaucher, J., Sadoul, K., Hart, D.J., Krijgsveld, J., Khochbin, S., Müller, C.W., Petosa, C., 2009. Cooperative binding of two acetylation marks on a histone tail by a single bromodomain. Nature 461, 664–668.

Murai, M.J., Pollock, J., He, S., Miao, H., Purohit, T., Yokom, A., Hess, J.L., Muntean, A.G., Grembecka, J., Cierpicki, T., 2014. The same site on the integrase-binding domain of lens epithelium-derived growth factor is a therapeutic target for MLL leukemia and HIV. Blood 124, 3730–3737.

Niederer, H.A., Laydon, D.J., Melamed, A., Elemans, M., Asquith, B., Matsuoka, M., Bangham, C.R., 2014. HTLV-1 proviral integration sites differ between asymptomatic carriers and patients with HAM/TSP. Virol. J. 11, 172.

Nishizawa, Y., Usukura, J., Singh, D.P., Chylack, L.T.J., Shinohara, T., 2001. Spatial and temporal dynamics of two alternatively spliced regulatory factors, lens epithelium-derived growth factor (ledgf/p75) and p52, in the nucleus. Cell Tissue Res. 305, 107–114.

Nowrouzi, A., Dittrich, M., Klanke, C., Heinkelein, M., Rammling, M., Dandekar, T., von Kalle, C., Rethwilm, A., 2006. Genome-wide mapping of foamy virus vector integrations into a human cell line. J. Gen. Virol. 87, 1339–1347.

Ocwieja, K.E., Brady, T.L., Ronen, K., Huegel, A., Roth, S.L., Schaller, T., James, L.C., Towers, G.J., Young, J.A., Chanda, S.K., Konig, R., Malani, N., Berry, C.C., Bushman, F.D., 2011. HIV integration targeting: a pathway involving Transportin-3 and the nuclear pore protein RanBP2. PLoS Pathog. 7, e1001313.

Passos, D.O., Li, M., Yang, R., Regensburg, S., Ghirlando, R., Jeon, Y., Kvaratskhelia, M., Craigie, R., Lyumkis, D., 2017. CryoEM structures and atomic model of the HIV-1 strand transfer complex intasome. Science 355, 89–92.

Pradeepa, M.M., Sutherland, H.G., Ule, J., Grimes, G.R., Bickmore, W.A., 2012. Psip1/Ledgf p52 binds methylated histone H3K36 and splicing factors and contributes to the regulation of alternative splicing. PLoS Genet. 8, e1002717.

Price, A.J., Fletcher, A.J., Schaller, T., Elliott, T., Lee, K., KewalRamani, V.N., Chin, J.W., Towers, G.J., James, L.C., 2012. CPSF6 defines a conserved capsid interface that modulates HIV-1 replication. PLoS Pathog. 8, e1002896.

Prieto, J.-L., McStay, B., 2008. Pseudo-NORs: a novel model for studying nucleoli. Biochim. Biophys. Acta Mol. Cell Res. 1783, 2116–2123.

Rahman, S., Sowa, M.E., Ottinger, M., Smith, J.A., Shi, Y., Harper, J.W., Howley, P.M., 2011. The Brd4 extraterminal domain confers transcription activation independent of pTEFb by recruiting multiple proteins, including NSD3. Mol. Cell. Biol. 31, 2641–2652.

Rain, J.-C., Cribier, A., Gérard, A., Emiliani, S., Benarous, R., 2009. Yeast two-hybrid detection of integrase–host factor interactions. Methods 47, 291–297.

Rasheedi, S., Shun, M.-C., Serrao, E., Sowd, G.A., Qian, J., Hao, C., Dasgupta, T., Engelman, A.N., Skowronski, J., 2016. The cleavage and polyadenylation specificity factor 6 (CPSF6) subunit of the capsid-recruited pre-messenger RNA cleavage factor I (CFIm) complex mediates HIV-1 integration into genes. J. Biol. Chem. 291, 11809–11819.

Reeves, R., 2015. High mobility group (HMG) proteins: modulators of chromatin structure and DNA repair in mammalian cells. DNA Repair 36, 122–136.

Roe, T., Reynolds, T.C., Yu, G., Brown, P.O., 1993. Integration of murine leukemia virus DNA depends on mitosis. EMBO J. 12, 2099–2108.

Roth, M.J., 1991. Mutational analysis of the carboxyl terminus of the Moloney murine leukemia virus integration protein. J. Virol. 65, 2141–2145.

Roth, M.J., Schwartzberg, P.L., Goff, S.P., 1989. Structure of the termini of DNA intermediates in the integration of retroviral DNA: dependence on IN function and terminal DNA sequence. Cell 58, 47–54.

Roth, S.L., Malani, N., Bushman, F.D., 2011. Gammaretroviral integration into nucleosomal target DNA in vivo. J. Virol. 85, 7393–7401.

Ruediger, R., Pham, H.T., Walter, G., 2001. Alterations in protein phosphatase 2A subunit interaction in human carcinomas of the lung and colon with mutations in the A beta subunit gene. Oncogene 20, 1892–1899.

Rüegsegger, U., Beyer, K., Keller, W., 1996. Purification and characterization of human cleavage factor I involved in the 3′ end processing of messenger RNA precursors. J. Biol. Chem. 271, 6107–6113.

Schaller, T., Ocwieja, K.E., Rasaiyaah, J., Price, A.J., Brady, T.L., Roth, S.L., Hue, S., Fletcher, A.J., Lee, K., KewalRamani, V.N., Noursadeghi, M., Jenner, R.G., James, L.C., Bushman, F.D., Towers, G.J., 2011. HIV-1 capsid-cyclophilin interactions determine nuclear import pathway, integration targeting and replication efficiency. PLoS Pathog. 7, e1002439.

Schneider, W.M., Brzezinski, J.D., Aiyer, S., Malani, N., Gyuricza, M., Bushman, F.D., Roth, M.J., 2013. Viral DNA tethering domains complement replication-defective mutations in the p12 protein of MuLV Gag. Proc. Natl. Acad. Sci. U.S.A. 110, 9487–9492.

Schrijvers, R., De Rijck, J., Demeulemeester, J., Adachi, N., Vets, S., Ronen, K., Christ, F., Bushman, F.D., Debyser, Z., Gijsbers, R., 2012a. LEDGF/p75-independent HIV-1 replication demonstrates a role for HRP-2 and remains sensitive to inhibition by LEDGINs. PLoS Pathog. 8, e1002558.

Schrijvers, R., Vets, S., De Rijck, J., Malani, N., Bushman, F.D., Debyser, Z., Gijsbers, R., 2012b. HRP-2 determines HIV-1 integration site selection in LEDGF/p75 depleted cells. Retrovirology 9, 84.

Schroder, A.R., Shinn, P., Chen, H., Berry, C., Ecker, J.R., Bushman, F., 2002. HIV-1 integration in the human genome favors active genes and local hotspots. Cell 110, 521–529.

Serrao, E., Ballandras-Colas, A., Cherepanov, P., Maertens, G.N., Engelman, A.N., 2015. Key determinants of target DNA recognition by retroviral intasomes. Retrovirology 12, 39.

Seshacharyulu, P., Pandey, P., Datta, K., Batra, S.K., 2013. Phosphatase: PP2A structural importance, regulation and its aberrant expression in cancer. Cancer Lett. 335, 9–18.

Shanker, V., Trincucci, G., Heim, H.M., Duong, H.T., 2013. Protein phosphatase 2A impairs IFNalpha-induced antiviral activity against the hepatitis C virus through the inhibition of STAT1 tyrosine phosphorylation. J. Viral Hepat. 20, 612–621.

Sharma, A., Larue, R.C., Plumb, M.R., Malani, N., Male, F., Slaughter, A., Kessl, J.J., Shkriabai, N., Coward, E., Aiyer, S.S., Green, P.L., Wu, L., Roth, M.J., Bushman, F.D., Kvaratskhelia, M., 2013. BET proteins promote efficient murine leukemia virus integration at transcription start sites. Proc. Natl. Acad. Sci. U.S.A. 110, 12036–12041.

Sharma, A., Slaughter, A., Jena, N., Feng, L., Kessl, J.J., Fadel, H.J., Malani, N., Male, F., Wu, L., Poeschla, E., Bushman, F.D., Fuchs, J.R., Kvaratskhelia, M., 2014. A new class of multimerization selective inhibitors of HIV-1 integrase. PLoS Pathog. 10, e1004171.

Shen, L., Peterson, S., Sedaghat, A.R., McMahon, M.A., Callender, M., Zhang, H., Zhou, Y., Pitt, E., Anderson, K.S., Acosta, E.P., Siliciano, R.F., 2008. Dose-response curve slope sets class-specific limits on inhibitory potential of anti-HIV drugs. Nat. Med. 14, 762–766.

Sherman, P.A., Fyfe, J.A., 1990. Human immunodeficiency virus integration protein expressed in *Escherichia coli* possesses selective DNA cleaving activity. Proc. Natl. Acad. Sci. U.S.A. 87, 5119–5123.

Shi, Y., 2009. Serine/threonine phosphatases: mechanism through structure. Cell 139, 468–484.

Shouse, G.P., Cai, X., Liu, X., 2008. Serine 15 phosphorylation of p53 directs its interaction with B56gamma and the tumor suppressor activity of B56gamma-specific protein phosphatase 2A. Mol. Cell. Biol. 28, 448–456.

Shouse, G.P., Nobumori, Y., Liu, X., 2010. A B56gamma mutation in lung cancer disrupts the p53-dependent tumor-suppressor function of protein phosphatase 2A. Oncogene 29, 3933–3941.

Shouse, G.P., Nobumori, Y., Panowicz, M.J., Liu, X., 2011. ATM-mediated phosphorylation activates the tumor-suppressive function of B56gamma-PP2A. Oncogene 30, 3755–3765.

Shun, M.-C., Botbol, Y., Li, X., Di Nunzio, F., Daigle, J.E., Yan, N., Lieberman, J., Lavigne, M., Engelman, A., 2008. Identification and characterization of PWWP domain residues critical for LEDGF/p75 chromatin binding and human immunodeficiency virus type 1 infectivity. J. Virol. 82, 11555–11567.

Shun, M.C., Raghavendra, N.K., Vandegraaff, N., Daigle, J.E., Hughes, S., Kellam, P., Cherepanov, P., Engelman, A., 2007. LEDGF/p75 functions downstream from preintegration complex formation to effect gene-specific HIV-1 integration. Genes Dev. 21, 1767–1778.

Silvers, R.M., Smith, J.A., Schowalter, M., Litwin, S., Liang, Z., Geary, K., Daniel, R., 2010. Modification of integration site preferences of an HIV-1-based vector by expression of a novel synthetic protein. Hum. Gene Ther. 21, 337–349.

Singh, D.P., Kubo, E., Takamura, Y., Shinohara, T., Kumar, A., Chylack Jr., L.T., Fatma, N., 2006. DNA binding domains and nuclear localization signal of LEDGF: contribution of two helix-turn-helix (HTH)-like domains and a stretch of 58 amino acids of the N-terminal to the trans-activation potential of LEDGF. J. Mol. Biol. 355, 379–394.

Singh, P.K., Plumb, M.R., Ferris, A.L., Iben, J.R., Wu, X., Fadel, H.J., Luke, B.T., Esnault, C., Poeschla, E.M., Hughes, S.H., Kvaratskhelia, M., Levin, H.L., 2015. LEDGF/p75 interacts with mRNA splicing factors and targets HIV-1 integration to highly spliced genes. Genes Dev. 29, 2287–2297.

Sommer, L.M., Cho, H., Choudhary, M., Seeling, J.M., 2015. Evolutionary analysis of the B56 gene family of PP2A regulatory subunits. Int. J. Mol. Sci. 16, 10134–10157.

Sorin, M., Cano, J., Das, S., Mathew, S., Wu, X., Davies, K.P., Shi, X., Cheng, S.W., Ott, D., Kalpana, G.V., 2011. Recruitment of a SAP18-HDAC1 complex into HIV-1 virions and its requirement for viral replication. PLoS Pathog. 5, e1000463.

Sowd, G.A., Serrao, E., Wang, H., Wang, W., Fadel, H.J., Poeschla, E.M., Engelman, A.N., 2016. A critical role for alternative polyadenylation factor CPSF6 in targeting HIV-1 integration to transcriptionally active chromatin. Proc. Natl. Acad. Sci. U.S.A. 113, E1054–E1063.

Stone, S.R., Hofsteenge, J., Hemmings, B.A., 1987. Molecular cloning of cDNAs encoding two isoforms of the catalytic subunit of protein phosphatase 2A. Biochemistry 26, 7215–7220.

Strack, S., Cribbs, J.T., Gomez, L., 2004. Critical role for protein phosphatase 2A heterotrimers in mammalian cell survival. J. Biol. Chem. 279, 47732–47739.

Studamire, B., Goff, S.P., 2008. Host proteins interacting with the Moloney murine leukemia virus integrase: multiple transcriptional regulators and chromatin binding factors. Retrovirology 5, 48.

Sundquist, W.I., Kräusslich, H.-G., 2012. HIV-1 assembly, budding, and maturation. Cold Spring Harb. Perspect. Med. 2.

Sutherland, H.G., Newton, K., Brownstein, D.G., Holmes, M.C., Kress, C., Semple, C.A., Bickmore, W.A., 2006. Disruption of Ledgf/Psip1 results in perinatal mortality and homeotic skeletal transformations. Mol. Cell. Biol. 26, 7201–7210.

Taniguchi, Y., 2016. The bromodomain and extra-terminal domain (BET) family: functional anatomy of BET paralogous proteins. Int. J. Mol. Sci. 17, 1849.

Terreni, M., Valentini, P., Liverani, V., Gutierrez, M.I., Di Primio, C., Di Fenza, A., Tozzini, V., Allouch, A., Albanese, A., Giacca, M., Cereseto, A., 2010. GCN5-dependent acetylation of HIV-1 integrase enhances viral integration. Retrovirology 7, 18.

Tesina, P., Čermáková, K., Hořejší, M., Procházková, K., Fábry, M., Sharma, S., Christ, F., Demeulemeester, J., Debyser, Z., Rijck, J.D., Veverka, V., Řezáčová, P., 2015. Multiple cellular proteins interact with LEDGF/p75 through a conserved unstructured consensus motif. Nat. Commun. 6, 7968.

Tobaly-Tapiero, J., Bittoun, P., Lehmann-Che, J., Delelis, O., Giron, M.L., de Thé, H., Saïb, A., 2008. Chromatin tethering of incoming foamy virus by the structural Gag protein. Traffic 9, 1717–1727.

Trobridge, G.D., Miller, D.G., Jacobs, M.A., Allen, J.M., Kiem, H.-P., Kaul, R., Russell, D.W., 2006. Foamy virus vector integration sites in normal human cells. Proc. Natl. Acad. Sci. U.S.A. 103, 1498–1503.

Tsantrizos, Y.S., Boes, M., Brochu, C., Fenwick, C., Malenfant, E., Mason, S., Pesant, M., 2007. Inhibitors of human immunodeficiency virus replication. International Patent Number WO/2007/131350.

Tsiang, M., Jones, G.S., Niedziela-Majka, A., Kan, E., Lansdon, E.B., Huang, W., Hung, M., Samuel, D., Novikov, N., Xu, Y., Mitchell, M., Guo, H., Babaoglu, K., Liu, X., Geleziunas, R., Sakowicz, R., 2012. New class of HIV-1 integrase (IN) inhibitors with a dual mode of action. J. Biol. Chem. 287, 21189–21203.

Turlure, F., Devroe, E., Silver, P.A., Engelman, A., 2004. Human cell proteins and human immunodeficiency virus DNA integration. Front. Biosci. (Landmark Ed.) 9, 3187–3208.

Turlure, F., Maertens, G., Rahman, S., Cherepanov, P., Engelman, A., 2006. A tripartite DNA-binding element, comprised of the nuclear localization signal and two AT-hook motifs, mediates the association of LEDGF/p75 with chromatin in vivo. Nucleic Acids Res. 34, 1653–1665.

van Nuland, R., van Schaik, F.M., Simonis, M., van Heesch, S., Cuppen, E., Boelens, R., Timmers, H.M., van Ingen, H., 2013. Nucleosomal DNA binding drives the recognition of H3K36-methylated nucleosomes by the PSIP1-PWWP domain. Epigenet. Chromatin 6, 12.

Vandegraaff, N., Devroe, E., Turlure, F., Silver, P.A., Engelman, A., 2006. Biochemical and genetic analyses of integrase-interacting proteins lens epithelium-derived growth factor (LEDGF)/p75 and hepatoma-derived growth factor related protein 2 (HRP2) in preintegration complex function and HIV-1 replication. Virology 346, 415–426.

Vandekerckhove, L., Christ, F., Van Maele, B., De Rijck, J., Gijsbers, R., Van den Haute, C., Witvrouw, M., Debyser, Z., 2006. Transient and stable knockdown of the integrase cofactor LEDGF/p75 reveals its role in the replication cycle of human immunodeficiency virus. J. Virol. 80, 1886–1896.

Vets, S., De Rijck, J., Brendel, C., Grez, M., Bushman, F., Debyser, Z., Gijsbers, R., 2013. Transient expression of an LEDGF/p75 chimera retargets lentivector integration and functionally rescues in a model for X-CGD. Mol. Ther. Nucleic Acids 2, e77.

Vranckx, L.S., Demeulemeester, J., Debyser, Z., Gijsbers, R., 2016a. Towards a safer, more randomized lentiviral vector integration profile exploring artificial LEDGF chimeras. PLoS One 11, e0164167.

Vranckx, L.S., Demeulemeester, J., Saleh, S., Boll, A., Vansant, G., Schrijvers, R., Weydert, C., Battivelli, E., Verdin, E., Cereseto, A., Christ, F., Gijsbers, R., Debyser, Z., 2016b. LEDGIN-mediated inhibition of integrase–LEDGF/p75 interaction reduces reactivation of residual latent HIV. EBioMedicine 8, 248–264.

Wagner, T.A., McLaughlin, S., Garg, K., Cheung, C.Y., Larsen, B.B., Styrchak, S., Huang, H.C., Edlefsen, P.T., Mullins, J.I., Frenkel, L.M., 2014. Proliferation of cells with HIV integrated into cancer genes contributes to persistent infection. Science 345, 570–573.

Wang, G.P., Ciuffi, A., Leipzig, J., Berry, C.C., Bushman, F.D., 2007. HIV integration site selection: analysis by massively parallel pyrosequencing reveals association with epigenetic modifications. Genome Res. 17, 1186–1194.

Wang, H., Jurado, K.A., Wu, X., Shun, M.C., Li, X., Ferris, A.L., Smith, S.J., Patel, P.A., Fuchs, J.R., Cherepanov, P., Kvaratskhelia, M., Hughes, S.H., Engelman, A., 2012. HRP2 determines the efficiency and specificity of HIV-1 integration in LEDGF/p75 knockout cells but does not contribute to the antiviral activity of a potent LEDGF/p75-binding site integrase inhibitor. Nucleic Acids Res. 40, 11518–11530.

Wang, H., Shun, M.C., Li, X., Di Nunzio, F., Hare, S., Cherepanov, P., Engelman, A., 2014. Efficient transduction of LEDGF/p75 mutant cells by gain-of-function HIV-1 integrase mutant viruses. Mol. Ther. Methods Clin. Dev. 1, 2.

Wang, J., Wang, Z., Yu, T., Yang, H., Virshup, D.M., Kops, G.J., Lee, S.H., Zhou, W., Li, X., Xu, W., Rao, Z., 2016a. Crystal structure of a PP2A B56-BubR1 complex and its implications for PP2A substrate recruitment and localization. Protein Cell 7, 516–526.

Wang, X., Bajaj, R., Bollen, M., Peti, W., Page, R., 2016b. Expanding the PP2A interactome by defining a B56-Specific SLiM. Structure 24, 2174–2181.

Wei, S.Q., Mizuuchi, K., Craigie, R., 1997. A large nucleoprotein assembly at the ends of the viral DNA mediates retroviral DNA integration. EMBO J. 16, 7511–7520.

Weidner-Glunde, M., Ottinger, M., Schulz, T.F., 2010. WHAT do viruses BET on? Front. Biosci. (Landmark Ed.) 15, 537–549.

Wight, D.J., Boucherit, V.C., Nader, M., Allen, D.J., Taylor, I.A., Bishop, K.N., 2012. The Gammaretroviral p12 protein has multiple domains that function during the early stages of replication. Retrovirology 9, 83.

Woodward, C.L., Prakobwanakit, S., Mosessian, S., Chow, S.A., 2009. Integrase interacts with nucleoporin NUP153 to mediate the nuclear import of human immunodeficiency virus type 1. J. Virol. 83, 6522–6533.

Wu, S.-Y., Lee, A.Y., Lai, H.-T., Zhang, H., Chiang, C.-M., 2013. Phospho switch triggers Brd4 chromatin binding and activator recruitment for gene-specific targeting. Mol. Cell 49, 843–857.

Wu, X., Li, Y., Crise, B., Burgess, S.M., 2003. Transcription start regions in the human genome are favored targets for MLV integration. Science 300, 1749–1751.

Xing, H., Vanderford, N.L., Sarge, K.D., 2008. The TBP-PP2A mitotic complex bookmarks genes by preventing condensin action. Nat. Cell Biol. 10, 1318–1323.

Xu, Y., Xing, Y., Chen, Y., Chao, Y., Lin, Z., Fan, E., Yu, J.W., Strack, S., Jeffrey, P.D., Shi, Y., 2006. Structure of the protein phosphatase 2A holoenzyme. Cell 127, 1239–1251.

Xu, Z., Cetin, B., Anger, M., Cho, U.S., Helmhart, W., Nasmyth, K., Xu, W., 2009. Structure and function of the PP2A-shugoshin interaction. Mol. Cell 35, 426–441.

Yamamoto, S.P., Okawa, K., Nakano, T., Sano, K., Ogawa, K., Masuda, T., Morikawa, Y., Koyanagi, Y., Suzuki, Y., 2011. Huwe1, a novel cellular interactor of Gag-Pol through integrase binding, negatively influences HIV-1 infectivity. Microb. Infect. 13, 339–349.

Yamashita, M., Emerman, M., 2004. Capsid is a dominant determinant of retrovirus infectivity in nondividing cells. J. Virol. 78, 5670–5678.

Yamashita, M., Perez, O., Hope, T.J., Emerman, M., 2007. Evidence for direct involvement of the capsid protein in HIV infection of nondividing cells. PLoS Pathog. 3, 1502–1510.

Yang, J., Phiel, C., 2010. Functions of B56-containing PP2As in major developmental and cancer signaling pathways. Life Sci. 87, 659–666.

Yin, Z., Shi, K., Banerjee, S., Pandey, K., Bera, S., Grandgenett, D.P., Aihara, H., 2016. Crystal structure of the Rous sarcoma virus intasome. Nature 530, 362–366.

Yokoyama, A., Cleary, M.L., 2008. Menin critically links MLL proteins with LEDGF on cancer-associated target genes. Cancer Cell 14, 36–46.

You, J., Croyle, J.L., Nishimura, A., Ozato, K., Howley, P.M., 2004. Interaction of the bovine papillomavirus E2 protein with Brd4 tethers the viral DNA to host mitotic chromosomes. Cell 117, 349–360.

Zhang, J.-Q., Wang, J.-J., Li, W.-J., Huang, L., Tian, L., Xue, J.-L., Chen, J.-Z., Jia, W., 2009. Cellular protein TTRAP interacts with HIV-1 integrase to facilitate viral integration. Biochem. Biophys. Res. Commun. 387, 256–260.

Zhang, R., Mehla, R., Chauhan, A., 2010. Perturbation of host nuclear membrane component RanBP2 impairs the nuclear import of human immunodeficiency virus-1 preintegration complex (DNA). PLoS One 5, e15620.

Zheng, R., Jenkins, T.M., Craigie, R., 1996. Zinc folds the N-terminal domain of HIV-1 integrase, promotes multimerization, and enhances catalytic activity. Proc. Natl. Acad. Sci. U.S.A. 93, 13659–13664.

Zheng, Y., Ao, Z., Wang, B., Jayappa, K.D., Yao, X., 2011. Host protein Ku70 binds and protects HIV-1 integrase from proteasomal degradation and is required for HIV replication. J. Biol. Chem. 286, 17722–17735.

Zhou, J., Pham, H.T., Ruediger, R., Walter, G., 2003. Characterization of the Aalpha and Abeta subunit isoforms of protein phosphatase 2A: differences in expression, subunit interaction, and evolution. Biochem. J. 369, 387–398.

Zhu, J., Gaiha, G.D., John, S.P., Pertel, T., Chin, C.R., Gao, G., Qu, H., Walker, B.D., Elledge, S.J., Brass, A.L., 2012. Reactivation of latent HIV-1 by inhibition of BRD4. Cell Rep. 2, 807–816.

Zielske, S.P., Stevenson, M., 2006. Modest but reproducible inhibition of human immunodeficiency virus type 1 infection in macrophages following LEDGFp75 silencing. J. Virol. 80, 7275–7280.

Zurnic, I., Hütter, S., Rzeha, U., Stanke, N., Reh, J., Müllers, E., Hamann, M.V., Kern, T., Gerresheim, G.K., Lindel, F., Serrao, E., Lesbats, P., Engelman, A.N., Cherepanov, P., Lindemann, D., 2016. Interactions of prototype foamy virus capsids with host cell polo-like kinases are important for efficient viral DNA integration. PLoS Pathog. 12, e1005860.

Chapter 5

Transcriptional Control and Latency of Retroviruses

Bryan C. Nikolai, Andrew P. Rice
Baylor College of Medicine, Houston, TX, United States

Chapter Outline

Introduction 200
Retroviral Integration Occurs at Transcriptionally Active Sites in Host Genome 200
Transcriptional Regulation by Endogenous Retroviruses 201
Retroviral Elements Are Prominent in All Jawed Vertebrate Genomes 201
Endogenous Retroviruses Can Function as Long-Range Transcriptional Enhancers 202
Endogenous Retroviruses Regulate the Potency of Stem Cells 202
Retroviral Genome Complexity Confers a Benefit for Transcriptional Regulation 203
Transcriptional Regulation in Simple Retroviruses 203
Retroviral Capture of Cellular Genes for Transcription Factors or Signaling Kinases 204
Retroviral Use of Hormone Response Elements in Long Terminal Repeat Enhancer 205
Transcriptional Silencing of Simple Retroviruses 207
Complex Retroviruses Encode Transactivator Proteins 208
Transcriptional Control of HIV-1 209
Clinical Importance of HIV-1 Proviral Transcription 209
HIV-1 5′ Long Terminal Repeat Contains the Viral Promoter and Enhancer 210
RNA Polymerase II Transcriptional Initiation of the HIV-1 Provirus 210
HIV-1 Tat Activates RNA Polymerase II Elongation 212
Tat and P-TEFb 213
CDK11 and HIV-1 3′ End Processing 214
HIV-1 Latent Infection 214
Establishment of Latent Infection 214
Maintenance of Latent Infection 215
Reactivation of Latent Virus 216

Retrovirus-Cell Interactions. https://doi.org/10.1016/B978-0-12-811185-7.00005-4

Future Research of Retroviral Transcription 217
HIV-1 Latency and the Need for a Functional Cure 217
Endogenous Retroviruses in Stem Cells and Cancer 218
Acknowledgments 218
References 219

INTRODUCTION

A defining step in the retrovirus life cycle is integration of the complementary DNA (cDNA) copy of the viral RNA genome into a host cell chromosome. The integrated virus is termed the provirus and the viral replication cycle is dependent on cellular RNA polymerase II (RNAP II) to productively transcribe the provirus. Knowledge of mechanisms whereby this is achieved by RNAP II provides insight into an essential step in the retrovirus life cycle. In turn, such insight may lead to development of therapeutic strategies to treat infections by pathogenic retroviruses. Additionally, insight into these mechanisms can allow retroviruses to be modified into effective gene therapy vectors for the treatment of a variety of human diseases.

RNAP II transcription of the retroviral provirus is the topic of this chapter. We begin with a discussion of the influence of proviral integration sites on transcription. We will then describe endogenous retroviruses (ERVs) and their impact on cellular physiology and evolution. This will be followed by a description of how retroviral genome complexity can allow additional levels of regulation of RNAP II transcription. We will then describe in detail RNAP II transcription of the HIV-1 provirus. The mechanisms involved in HIV-1 proviral transcription are known in considerable detail, and elucidation of these mechanisms has contributed to general insight into RNAP II transcription. Importantly, transcriptional regulation of the HIV-1 proviral is of clinical significance to current efforts to develop a functional cure of infection. We will end the chapter with a discussion of future needs and goals of research into retroviral transcription.

RETROVIRAL INTEGRATION OCCURS AT TRANSCRIPTIONALLY ACTIVE SITES IN HOST GENOME

The retrovirus cDNA is not integrated into random sites of the host cell genome. Instead, the various genera of retroviruses display distinct preferences for sites of integration (reviewed in Demeulemeester et al., 2015). Lentiviruses such as HIV-1 and SIV have a preference for integration into actively transcribed genes (Crise et al., 2005; Schroder et al., 2002). The deltaretrovirus human T cell leukemia virus (HTLV)-1 also has a preference for integration near or in transcriptionally active genes (Gillet et al., 2011). In contrast, the gammaretrovirus (murine leukemia virus [MLV]) has a preference for promoter–proximal regions and distal enhancers of active genes (LaFave et al., 2014; Wu et al., 2003). The alpharetrovirus avian leukosis/sarcoma virus (ASLV) appears to be an outlier and has been shown to have only a weak preference for integration into actively transcribed genes (Barr et al., 2005; Mitchell et al., 2004).

The preference for integration within or near active genes reflects the chromatin-binding properties of cellular factors that associate with retroviral integrase (IN) proteins and contribute to integration. The IN protein of HIV-1 binds to a cellular protein known as LEDGF/p75 that contains a PWWP domain that binds H3K36me3-modified nucleosomes, a mark of active transcription (Eidahl et al., 2013; van Nuland et al., 2013). Thus, the binding of LEDGF/p75 to these chromatin marks targets the HIV-1 preintegration complex (PIC) to active genes. The IN protein of MLV binds to BET protein family members that include BRD2, -3, and -4; BET proteins bind to acetylated histones H3 and H4, which are also marks of active transcription (Filippakopoulos et al., 2012). Similar to LEDGF/p75 and HIV-1 IN, the association of MLV IN with BET proteins targets the PIC to active genes. The preference for retroviruses to integrate into or near actively expressed genes greatly increases the likelihood that the proviral genome will be located in a region of the host genome that favors active RNAP II transcription.

Subnuclear localization contributes to HIV-1 integration site selection. HIV-1 integration occurs preferentially in active genes located at the outer shell of the nucleus, a region of the nucleus that is near nuclear pore structures (Marini et al., 2015). Presumably, genes in this nuclear location are encountered first by the PIC after it traffics from the cytoplasm through a nuclear pore. Integration within genes near nuclear pores is likely to provide an additional advantage in HIV-1 replication, as this may enable efficient nuclear export of processed viral transcripts. It is not yet known if retroviruses other than HIV-1 have a bias for integration at chromosomal sites near nuclear pores. For more detailed information on retroviral integration, please see Chapter 4 by Engelman and Maertens.

TRANSCRIPTIONAL REGULATION BY ENDOGENOUS RETROVIRUSES

Retroviral Elements Are Prominent in All Jawed Vertebrate Genomes

The ability of retrotransposons and endogenous retroviral elements (ERVs) to function as transcriptional enhancers and to expand genome size has connected them with evolution and speciation. The field of evolutionary development has put forth the concept that speciation and branches within evolutionary trees are necessarily dictated by *cis*-regulatory elements because the structural protein code is largely unchanged during evolution (Prud'homme et al., 2007). The evolution of jawed vertebrates has been linked to ERV elements within the genomes of early and extant cartilaginous fish. Unique ERVs resembling foamy viruses

(FVs), the so-called foamy-like ERVs or FLERVs are pervasive in lobe-finned coelacanths and amphibians (Aiewsakun and Katzourakis, n.d.). Interestingly, retroviruses and ERVs are not recovered from hagfish, and sea lampreys have fewer or very different retroviral elements in their genomes (Hayward et al., 2015; Naville and Volff, 2016). These findings suggest ancient coevolution of retroviruses and their hosts, although a precise causative role of ERVs in early evolution is yet to be determined. In mammals, syncytiotrophoblast formation during placentation and totipotency of embryonic stem cells are known to involve or even require ERVs (Mi et al., 2000), suggesting a more direct role of these elements during evolutionary processes. Please see Chapter 13 by Sandmeyer for more detailed discussion of endogenous retroviruses and retroelements.

Endogenous Retroviruses Can Function as Long-Range Transcriptional Enhancers

In humans, ERVs comprise approximately about 8% of the genome. Compared to the entire exome of translated mRNA, which is less than 1%, ERVs are a significant portion of genomic material that has been pejoratively called "junk DNA" or ambiguously labeled the "dark genome." The minority of the human genome consisting of protein-coding genes is essential, no doubt—but only as much as the regulation of those genes affords. Recently, several studies have linked ERVs to gene regulation in a variety of contexts. For instance, some ERVs are highly enriched in regulatory sequences and can function as enhancer elements for interferon-induced innate immunity (Chuong et al., 2016). Specifically, the regulatory network of interferon responsive genes that are elicited by ERV enhancers represents a coopted response to pathogen/host conflict wherein the host has ironically evolved a network using pathogen regulatory sequences to respond to viral infection. Additionally, long terminal repeat (LTR) elements present at intergenic sites have been shown to regulate the globin locus in mice through long-range looping and RNA polymerase recruitment to distant sites (Hu et al., 2017; Pi et al., 2004). Most standard methods of in silico analysis remove repetitive sequences from cistromic, transcriptomic, and other sequencing datasets. Therefore, it is only in very special cases in which these transcriptional regulatory functions of ERVs have come to light. Collectively, the literature suggests that ERVs and LTR elements from retrotransposons can serve as long-range transcriptional enhancers and are likely to play key roles in regulation of normal cellular gene expression.

Endogenous Retroviruses Regulate the Potency of Stem Cells

The advent of high-throughput sequencing has allowed unbiased investigation of transcriptional programs of pluripotent germ cells and early embryonic cells. Surprisingly, approximately one-third of RNA transcripts synthesized in these cells are from repetitive elements and many protein-coding genes initiate transcription from within *cis* regulating retrotransposons or ERVs (Faulkner et al., 2009). The human endogenous retrovirus subfamily H (HERVH) class of ERVs contains a strong embryonic stem cell enhancer element that is enriched in binding motifs for the pluripotency-coupled transcription factor OCT4 (Santoni et al., 2012). Interestingly, a long noncoding RNA (lncRNA) transcribed from HERVH can function as a scaffold to assemble transcription factor and coregulator complexes (Lu et al., 2014). Indeed, HERVH transcripts are abundant in stem cells and have been suggested to be a defining marker for pluripotency. In mice, the murine endogenous retrovirus with leucine transfer RNA primer (MERVL) is associated with stem cell totipotency (Choi et al., 2017). Approximately 300 protein-coding genes transcribed in early stage embryos contain MERVL sequences, presumably regulated by a minimal element containing GATA2 response elements. The retrogene DUX4, located in the D4Z4 repeat region of chromosome 4 in humans, and the murine homolog Dux are important for the cleavage stage of two cell embryos in men and mice (Hendrickson et al., 2017). DUX and DUX4 are homeodomain transcription factors that regulate programs of zygotic transcription, including MERVs and HERVs, respectively. Collectively, ERV-mediated mechanisms of transcriptional regulation are important for pluripotency of stem cells and additional functions during reproduction and development.

RETROVIRAL GENOME COMPLEXITY CONFERS A BENEFIT FOR TRANSCRIPTIONAL REGULATION

Transcriptional Regulation in Simple Retroviruses

Retroviral genomes can be broadly divided into one of two classes for the regulation of RNA production: simple or complex (Fig. 5.1). Simple retroviruses do not encode proteins that facilitate RNA transcription, splicing, or nuclear export and therefore rely entirely on host cellular machinery. In contrast, complex retroviral genomes express one or more genes, often from multiply spliced transcripts, which regulate RNA transcription or posttranscriptional processes. It is important to note that almost all retroviruses contain a TATA box as the core promoter and retroviral transcription is therefore subject to canonical mechanisms of eukaryotic gene transcription. The TATA binding protein (TBP) recognizes the TATAA sequence on the sense strand of proviral DNA and together with TBP-associated factors forms the multiprotein complex TFIID. Subsequent assembly of the general transcription machinery results in RNAP II recruitment, sufficient to generate short strands of nascent RNA. Transcriptional pausing occurs at most cellular genes and complex retroviruses have evolved elegant mechanisms for pause release into transcriptional elongation (e.g., HIV-1, see below), whereas simple retroviruses rely strictly on host factors. Examples of simple retroviruses include members of the alpha-, beta-, and gammaretrovirus genera. The study and use of simple retroviruses within these genera has significantly contributed to diverse areas of biology including tumorigenesis, hormone-induced gene regulation, and gene therapy. Below, we provide a brief

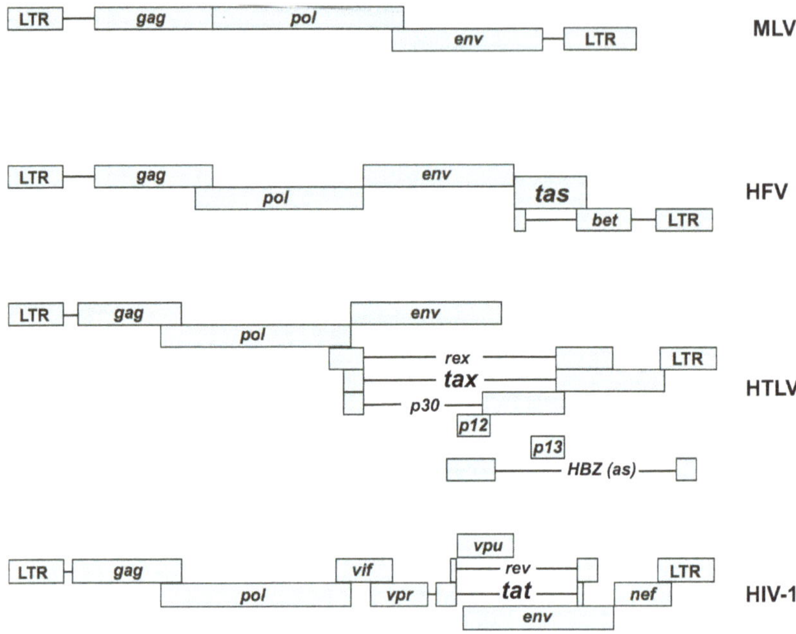

FIGURE 5.1 **Retroviral genome complexity confers transcription advantages.** The simple gammaretrovirus murine leukemia virus (MLV) has no accessory genes (top). The complex retroviruses encode accessory genes that always include a transcription factor or transcriptional coactivator. Notably, *tas*, *tax*, and *tat* all have different mechanisms of transcriptional regulation: The Tas protein is a DNA-binding transcription factor, Tax is a transcriptional coactivator, and Tat is an RNA binding protein. The most basic need of an RNA virus is transcription of its communicable genome from host DNA and complex retroviruses have evolved methods to ensure proviral transcription despite limited genome size. *HTLV*, human T cell leukemia virus; *LTR*, long terminal repeat.

overview of simple retroviruses and the impact their study has had on the understanding of host cell biology and transcriptional regulation.

Retroviral Capture of Cellular Genes for Transcription Factors or Signaling Kinases

Members of the alpharetrovirus genus are famously associated with the identification of several host cell oncogenes including *v-Src* from Rous sarcoma virus, the only retroviral oncogene (v-onc) known to coexist with intact *gag*, *pol*, and *env* sequences in a replication-competent oncoretrovirus. In contrast, v-onc genes of other transforming retroviruses replace or inactivate viral *gag*, *pol*, or *env* sequences, requiring coinfection with a complementing retrovirus. Alpharetroviruses with transforming potential include those requiring avian leucosis virus (ALV) as a helper virus for productive infection. Interestingly, there is a trend for v-onc genes to encode homologues of cellular tyrosine kinases and transcription factors. For example, the avian erythroblastosis virus encodes *v-erbA* or *v-erbB*, the cellular orthologs of which encode the thyroid hormone nuclear receptor or *HER2/ERBB2* receptor tyrosine kinase, respectively. Moreover, *v-yes*, *v-myc*, *v-myb*, and *v-jun* retroviral oncogenes were also discovered in the context of helper-dependent ALV, suggesting an evolutionary advantage in these retroviruses for constitutive tyrosine kinase signaling and expression of transcription factors. Indeed, signaling from tyrosine kinases is known to drive extensive transcriptional programs (Nikolai et al., 2016) and the anabolic output of v-onc genes may be a mechanism to ensure cellular proliferation and replication of the viral genome.

Retroviral Use of Hormone Response Elements in Long Terminal Repeat Enhancer

The mouse mammary tumor virus (MMTV), a betaretrovirus, contains canonical DNA-binding sequences for nuclear hormone receptors and is responsive to steroids including glucocorticoids, progestins, and androgens. Initially discovered by John J. Bittner, the eponymously named Bittner virus was coined the "milk factor" by its discoverer. Bittner observed that mammary tumors developed in mice due to an extrachromosomal cause, i.e., they could be transmitted to pups when fostered by mothers from a lineage that develop breast cancer (Bittner, 1936). Bittner demonstrated that the "milk factor" was filterable and could be transmitted from milk of cancer-bearing mice (Bittner, 1942). Further research would prove that MMTV was a retrovirus and in his 1989 Nobel Prize lecture, Harold Varmus acknowledged MMTV as the first-described mammalian retrovirus and a key instrument in his discovery of cellular oncogene regulation by retroviral insertion (Varmus, 1990). Additionally, studies of the MMTV promoter/enhancer region in transcription reporter systems played a major role in current understanding of steroid-induced regulation of cellular gene transcription.

The value of MMTV in studies of steroid-induced gene transcription was first reported in 1975 using dexamethasone, a synthetic corticosteroid already in clinical use by the 1950s due to its antiinflammatory properties (CECIL, 1959). Notably, the glucocorticoid receptor (GR) gene would not be cloned for another decade, but the convenience of a viral system allowed research on GR-mediated transcription to flourish. Indeed, the DNA sequences recognized and bound by GR within the MMTV enhancer (Fig. 5.2A) were reported before the cDNA for GR was cloned (Ahe von et al., 1985; Hollenberg et al., 1985; Scheidereit et al., 1983; Weinberger et al., 1985). This would lead to the discovery of mechanisms of steroid-induced retroviral transcription that peaked in the 1980s through mid-1990s (Fig. 5.2B and D). The description in 1984 of MMTV enhancer–driven *myc* tumorigenesis in mice revealed the utility of transgenic mouse models for breast cancer research (Stewart et al., 1984). These studies would have a significant impact on the field of molecular endocrinology and research of hormone-driven cancer. The tissue selective expression of MMTV in the

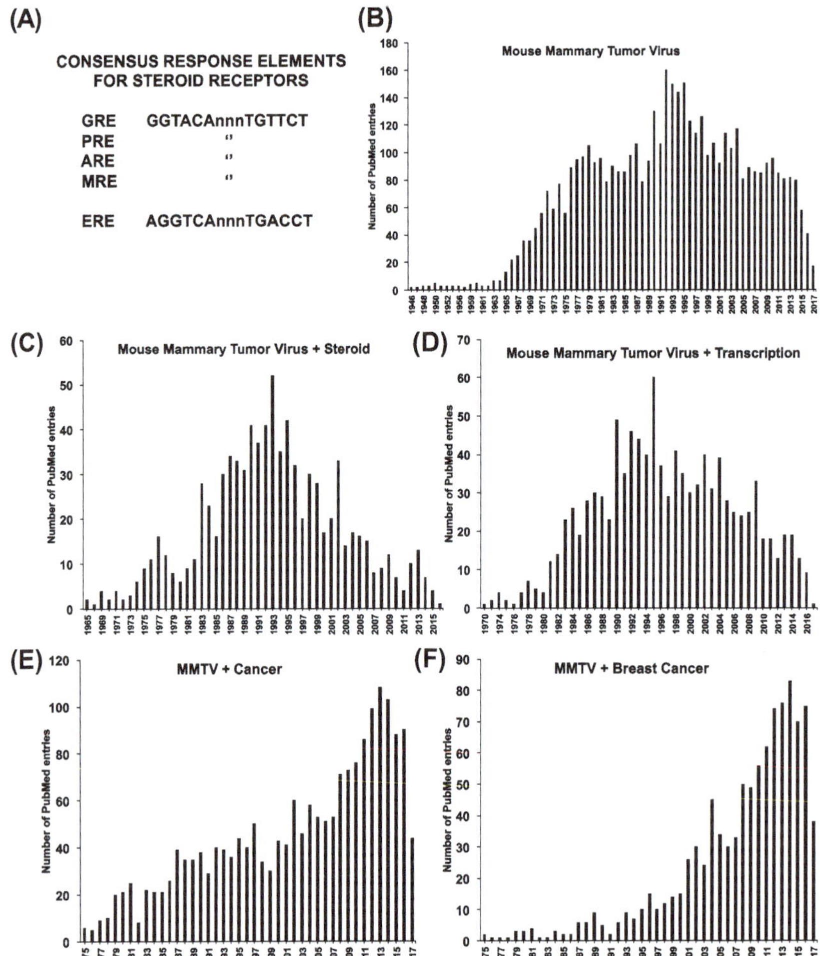

FIGURE 5.2 **Research of the mouse mammary tumor virus (MMTV) enhancer significantly contributed to molecular endocrinology and breast cancer.** (A) Early studies of MMTV helped define steroid-induced gene regulation. (B) Before the 1980s HIV/AIDS epidemic shifted focus and funding for retroviral research, MMTV was well studied for its hormone responsiveness. (C and D) Citations of MMTV peak in the mid-1990s. (E and F) MMTV use in transgenic mouse models for cancer is increasing. Research on MMTV transcription has had lasting impact on the fields of molecular endocrinology and the hormone-driven MMTV long terminal repeat has been a harbinger for transgenic mouse models of breast cancer. *((A) Adapted from Beato, M., February 10, 1989. Gene regulation by steroid hormones. Cell 56, 335–344.)*

mammary gland of mice has been exploited to ectopically express oncogenes, resulting in numerous mouse models of breast cancer. Similarly, MMTV-Cre mice are used to selectively delete tumor suppressors or otherwise target Cre recombinase activity to the mammary gland. Indeed, the use and citation of MMTV in cancer research is on the rise, most of which is in the context of breast cancer research (Fig. 5.2E and F). Collectively, the impact of hormone-driven MMTV transcription resonates beyond the field of retrovirology and has heavily influenced endocrinology and oncology research.

The replicative advantage gained by MMTV through its hormone-driven expression should not be understated. By uniting proviral transcription with essential primary and secondary reproductive functions of its host, MMTV has assured replication of its genome with minimal pathology until the time of transmission. In the absence of lactation hormones, MMTV is transcriptionally latent and avirulent, therefore ensuring host survival until transmission. On pregnancy, prolactin and progestagens stimulate productive MMTV infection in the proliferating ducts of the mammary gland. Offspring survival is dependent on nursing, and the presence of MMTV in the milk provides for vertical transmission of the virus. Many pathogens use host reproduction as a convenient route of transmission, thus forming an intimate connection between their own propagation and that of their host. This certainly appears true for some retroviruses, including MMTV and others that are known to be sexually transmitted, such as HTLV-1 and HIV-1. Moreover, emerging evidence suggests that germline infection, integration, and expansion of retroviruses during early vertebrate and mammalian evolution may have been a significant driver of evolution. Indeed, these now-ERVs fell under mechanisms for natural selection and provided an instrument for replicative genome expansion that would ultimately become intimately tied to essential reproductive functions and cellular pluripotency.

Transcriptional Silencing of Simple Retroviruses

The ability of retroviruses to transduce additional genetic cargo and their amenability to envelope pseudotyping makes them invaluable vectors for gene delivery and gene therapy. The ease of packaging and ability to establish stable cell lines expressing a gene of interest have led to the use of retroviral vectors in many applications. The MLVs, gammaretroviruses, have been extensively developed for use as vectors. Because these retroviral vectors are subject to host cell suppression, mechanisms of Moloney MLV (MoMLV) transcriptional silencing have been studied in depth. In undifferentiated cells such as embryonic stem cells and teratocarcinoma cells, expression from newly integrated MoMLV provirus is blocked by de novo DNA methylation. Despite efficient MoMLV proviral integration into the cell genome, no viral replication or virus-specific RNA transcripts can be detected in undifferentiated cells, in contrast to differentiated cell lines. Multiple histone-modifying enzymes and de novo DNA methyltransferases have been attributed to retroviral silencing, and epigenetic modification of ERVs also appears to play critical roles in embryonic stem cells (Jahner et al., 1982; Matsui et al., 2010; Okano et al., 1998; Schlesinger and Goff, 2012). Moreover, rare MLV escape mutants that can replicate in undifferentiated cells have been described, and they provide clues about mechanisms of retroviral silencing. For example, some MoMLV clones were found to have mutations in the primer binding site (PBS) and use an alternative tRNA for priming of reverse transcription (Schlesinger and Goff, 2012). The mutation in the PBS allowed expression of retroviral transcripts in embryonic stem cells and the repressive element within MoMLV PBS was attributed to binding of a host factor, "binding factor A." These observations led to the discovery of repressors of proviral transcription, most notably, tripartite motif-containing 28 (TRIM28) and ZFP809 that function together to drive latency. ZFP809 is a member of the Krüppel associated box domain family and binds to a specific sequence in the PBS of MoMLV through its zinc finger domains (Wolf and Goff, 2009). Subsequently, ZFP809 recruits TRIM28 and additional transcriptional corepressors to mediate restrictive chromatin and restrict retroviral transcription. TRIM28 can epigenetically repress transcription through heterochromatin formation, facilitated by recruitment of methyltransferases and histone deacetylases. These studies demonstrate mechanisms of retroviral silencing by the host that depend on the differentiated state of the cell (Wolf and Goff, 2009). Transcriptional silencing mechanisms may contribute to the inability of simple retroviruses to infect nondividing cells, whereas complex retroviruses have evolved mechanisms to circumvent host cell transcriptional blocks (Wolf and Goff, 2009, 2007).

Complex Retroviruses Encode Transactivator Proteins

The complex retroviruses are subdivided into three genera: spumavirus, deltaretrovirus, and lentivirus. All complex retroviruses encode accessory genes for proteins involved in various functions throughout the life cycle of the virus. Here we focus on virally encoded proteins that function as transcriptional transactivators, a feature unique to the complex retroviruses as simple retroviruses rely entirely on the transcriptional machinery of the host cell. The increase in complexity of retroviral genomes always includes a mechanism to regulate transcription (Fig. 5.1).

The deltaretrovirus genus includes the HTLVs, of which HTLV-1 is the best studied and is the etiological agent of adult T cell leukemia and additional clinical disorders. Two of the most studied HTLV-1 accessory genes are Tax and Rex, which facilitate proviral RNA transcription and processing, respectively. The Rex protein binds to a response element in the 3′ LTR to mediate nuclear export of incompletely spiced transcripts, analogous to HIV-1 Rev (Hidaka et al., 1988; Inoue et al., 1991; Rimsky et al., 1988).

Tax is primarily a nuclear protein and serves as a transactivator for proviral and cellular gene expression, predominantly through its interaction with cellular transcription factors and coregulators. The *cis*-regulatory sequences within the HTLV-1 LTR that enable Tax-driven transcription are homologous to the consensus sequence for host cell CREB, the cyclic AMP (cAMP) response element (CRE) binding protein that is characterized by a basic leucine zipper (bZIP) motif (Beimling and Moelling, 1992). Tax has been proposed to enhance the homodimerization of CREB and alter its affinity for the imperfect CRE. During this process, an N-terminal region of Tax directly contacts DNA adjacent to the CRE and the Tax/CREB complex recruits additional coactivators involved in histone acetylation and chromatin remodeling (Zhao and Giam, 1992). The histone acetyltransferases (HATs) CBP, p300, and PCAF are among the coactivators involved in Tax-mediated HTLV-1 transcription (Harrod et al., 1998; Jiang et al., 1999). In addition to CREB, another bZIP protein named ATF-4 can also serve as a host cell factor during HTLV-1 transcription, presumably by facilitating HAT-coactivator recruitment. Through these mechanisms, addition of the Tax gene to the HTLV-1 genome grants more sophisticated regulation of proviral transcription through direct contact and recruitment of host cell transcription factors and coregulators.

Other complex retroviruses have evolved similar mechanisms of proviral transcriptional regulation. The spumavirus or foamy viruses (FVs) encode a protein called Tas (transactivator of spumavirus) that is a sequence-specific DNA-binding transcription factor. Tas is necessary for efficient FV replication and recognizes a consensus sequence immediately adjacent to the core promoter element, as well as other enhancers within the proviral gene body (Campbell et al., 1996; Zou and Luciw, 1996). Tas makes contact with and is a substrate for several cellular HATs including p300 and PCAF (Bannert et al., 2004). Additionally, several lysine residues in Tas have been identified, which are critical to Tas function (Zhang et al., 2016). This is significant because post-translational modification of transcription factors and coregulators, including acetylation, is an important mechanism for transcriptional regulation. The Tas gene product of FV therefore represents another example of complex retroviruses using limited genome capacity to encode a transcriptional regulator.

TRANSCRIPTIONAL CONTROL OF HIV-1

Clinical Importance of HIV-1 Proviral Transcription

As described below, studies of HIV-1 transcriptional control have contributed to an understanding of general mechanisms involved in RNAP II transcriptional elongation. The study of HIV-1 transcriptional control is also a topic of considerable clinical importance. There are currently over 20 FDA-approved drugs that block HIV-1 replication at different stages in the replication cycle. These drugs effectively suppress viral replication when used in combination in a regime called combined antiretroviral therapy (cART) (De Clercq and Li, 2016). However, cessation of cART invariably leads to the reemergence of active HIV-1 replication from a latent reservoir. Thus, current cART must be taken for the lifetime of the patient.

The latent viral reservoir consists of memory $CD4^+$ T cells that harbor replication-competent proviruses that are transcriptionally silent (Churchill et al., 2016). Because RNAP II does not transcribe the HIV-1 provirus in this population of infected cells, viral proteins are not expressed and the cells cannot be recognized by the immune system. Insight into how latent infection is established and maintained, as well as how it can be reactivated, may allow the development of strategies to induce latent virus and reduce the size of the reservoir. With a significantly reduced latent reservoir, it is hoped that a functional cure of infection can be achieved. Such a cure will likely require enhancement of the immune system of infected individuals, perhaps by a therapeutic vaccine or a gene therapy approach to express broadly neutralizing antibodies against HIV-1. Because HIV-1 latency is controlled at the level of RNAP II transcription of the provirus, there is currently a major research effort devoted to the study of molecular mechanisms of HIV-1 proviral transcription.

HIV-1 5′ Long Terminal Repeat Contains the Viral Promoter and Enhancer

The process of reverse transcription and integration generates duplicate LTR sequences at the 5′ and 3′ ends of retroviral proviruses. The 5′ LTR directs RNAP II transcription of the provirus and it is subdivided into the U3, R3, and U5 regions (reviewed in Karn and Stoltzfus, 2012; Fig. 5.3). The core promoter of HIV-1 is composed of three tandem Sp1 binding sites (Jones et al., 1986), a TATA element (Garcia et al., 1989), and an initiator sequence. An “enhancer” element is located upstream from the core promoter and it contains tandem NF-κB sites whose number can vary between HIV-1 subtypes (van Opijnen et al., 2004). Subtype B, the predominant subtype in North America and Western Europe, contains two NF-κB sites. Subtype C, the predominant subtype in Southern Africa and Asia, contains three NF-κB sites. Although differences in the number of NF-κB sites are capable of influencing HIV-1 replication levels in vitro, their effects in vivo have been difficult to assess (van Opijnen et al., 2004). The TAR element (described below) responds to the HIV-1 Tat proteins and it is located downstream from the transcriptional initiation site, encompassing approximately nucleotides +1 to +65 (Rosen et al., 1985).

RNA Polymerase II Transcriptional Initiation of the HIV-1 Provirus

The HIV-1 core promoter has a high basal level and transcriptional initiation can be further stimulated by the binding of NF-κB or NFAT family members to the NF-κB sites in the viral enhancer (Kinoshita et al., 1998; Liu et al., 1992; Nabel and Baltimore, 1987). T cell activation leads to the release and nuclear translocation of NF-κB from sequestration in the cytoplasm by a family of inhibitors

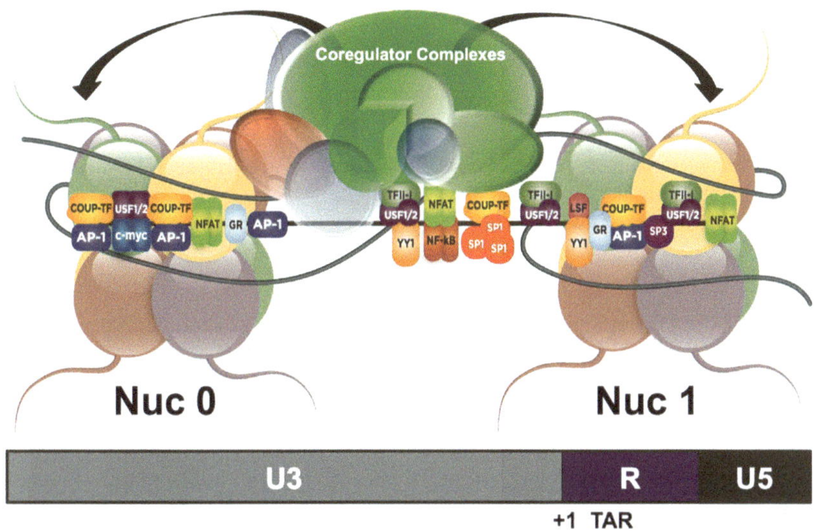

FIGURE 5.3 **The HIV-1 proviral long terminal repeat.** The HIV-1 provirus is organized into two well-positioned nucleosomes and transcription factor–binding sites are defined. However, discrete complexes of transcriptional coregulators dictate the proviral chromatin state of HIV-1 provirus. Corepressors drive latency, whereas transcriptional coactivators are responsible for reversal and transcriptional activation. The chromatin state of provirus is defined by histone modifications. *(Modified from Ahlenstiel, C., Mendez, C., Lim, S.T.H., Marks, K., Turville, S., Cooper, D.A., Kelleher, A.D., Suzuki, K., 2015. Novel RNA duplex locks HIV-1 in a latent state via chromatin-mediated transcriptional silencing. Mol. Ther. Nucleic Acids 4, e261. https://doi.org/10.1038/mtna.2015.31.)*

termed IκBs (Oeckinghaus et al., 2011). T cell activation also results in calcium signaling that activates the phosphatase calcineurin (Hogan, 2017). Calcineurin dephosphorylates NFAT proteins and this induces a conformation change that exposes an NFAT nuclear localization signal, resulting in nuclear import of NFAT. Consequently, T cell activation stimulates transcriptional initiation of the HIV-1 provirus through the induction of NF-κB and NFAT (Macian, 2005; O'Dea and Hoffmann, 2009).

Although RNAP II actively initiates transcription of the HIV-1 provirus, especially after T cell activation, transcriptional elongation is defective due to the action of two negative factors that associate with the RNAP II complex—DSIF (DRB sensitivity inducing factor) and NELF (negation elongation factor). DSIF comprises two protein subunits (Spt4 and Spt5, Wada et al., 1998); NELF comprises five protein subunits (NELF-A, NELF-B, NELF-C/D, and NELF-E, Yamaguchi et al., 1999). The pause in transcriptional elongation of the HIV-1 provirus that is specified by DSIF and NELF occurs at many, if not most, cellular genes and is thought to allow efficient modification of the 5′ end of transcript by the mRNA capping complex (Adelman and Lis, 2012; Zhou et al., 2012).

HIV-1 Tat Activates RNA Polymerase II Elongation

It was initially observed that expression of an HIV-1 LTR reporter plasmid is nearly 1000-fold higher in HIV-infected cells than in control cells (Sodroski et al., 1985a). This potent stimulatory activity of HIV-1 gene expression was ultimately found to be mediated by the activity of the HIV-1 Tat protein. Tat has been extensively investigated because it is essential for HIV-1 replication (Dayton et al., 1986) and its mechanism of action appeared to be unique when it was first discovered. The cis-regulatory sequences in the HIV-1 LTR that respond to Tat were identified to reside downstream from the site of RNAP II transcriptional initiation (Sodroski et al., 1985b), and the major function of Tat is to stimulate transcriptional elongation rather than initiation (Kao et al., 1987). Using a genetic approach, the cis-acting Tat responsive element was shown to function in vivo as a stem-loop RNA motif encompassing nucleotides +1 to +59, which was termed, TAR RNA (Feng and Holland, 1988). Biochemical experiments showed that Tat binds directly to TAR RNA in vitro (Ernberg et al., 1989), however, binding of Tat to TAR in vivo also requires a cellular TAR RNA loop-binding factor (Madore and Cullen, 1993).

A key finding in studies probing the mechanism of Tat was that recombinant Tat binds in vitro to a cellular protein kinase initially named Tat-associated kinase (TAK) (Herrmann and Rice, 1993). TAK can hyperphosphorylate Ser2 residues in the carboxyl terminal domain (CTD) of RNAP II in vitro, a pattern of phosphorylation implicated in activation of RNAP II elongation (Herrmann and Rice, 1995). An analysis of a series of wild type and mutant Tat proteins demonstrated that Tat binding to TAK in vitro correlates strictly with Tat function in vivo (Herrmann and Rice, 1995). Shortly after describing TAK were published, it was shown that TAK is identical to a general RNAP II elongation factor known as P-TEFb (Zhu et al., 1997). The catalytic subunit of TAK/P-TEFb is CDK9, a member of the cyclin-dependent protein kinase family (Yang et al., 1997; Zhu et al., 1997). The cyclin-regulatory subunit of CDK9 is cyclin T1 (Peng et al., 1998a; Wei et al., 1998), which is the previously genetically defined factor required for Tat binding to TAR RNA in vivo (Madore and Cullen, 1993). This cellular kinase composed of cyclin T1 and CDK9 that mediates Tat function is now known as P-TEFb. As well as being required for RNAP II elongation of the HIV-1 provirus, P-TEFb is a general RNAP II transcription factor that is required for elongation of most, if not all, protein-coding genes (Adelman and Lis, 2012).

These findings with Tat and P-TEFb lead to the model shown in Fig. 5.4. As illustrated in the figure, the inhibition of transcriptional elongation of the HIV-1 provirus by DSIF and NELF is overcome by Tat. A Tat/CDK9/cyclin T1 complex binds to the TAR RNA that forms at the 5′ ends of nascent HIV-1 transcripts. CDK9 is then positioned to phosphorylate specific protein subunits of both DSIF and NELF and counter their negative function. Notably, phosphorylation of the Spt5 subunit of DSIF converts the protein into a positive factor for

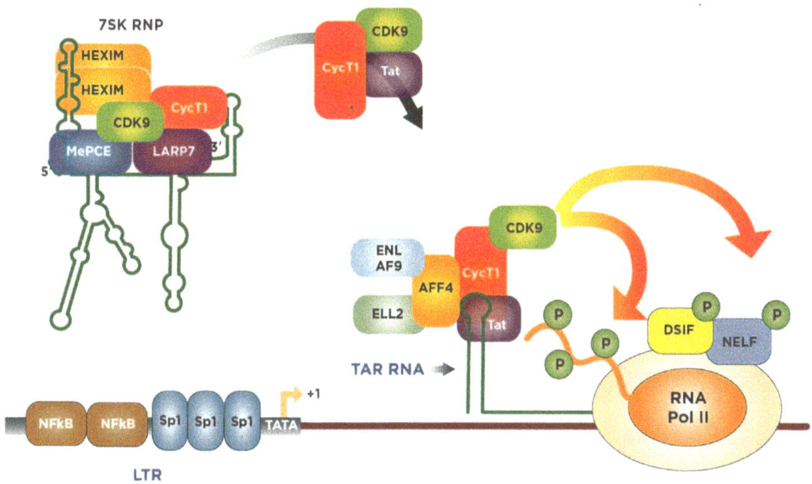

FIGURE 5.4 **Mechanism of HIV-1 Tat.** P-TEFb (CDK9/cyclin T1) is catalytically inactive in the 7SK RNP and is recruited to the TAR RNA element by the viral Tat protein (see text for details). The SEC assembles on the TAR RNA element and CDK9 phosphorylates DRB sensitivity inducing factor (DSIF), negation elongation factor (NELF), and the carboxyl terminal domain of RNAP II to potently activate transcriptional elongation.

transcriptional elongation (Bourgeois et al., 2002). CDK9 also phosphorylates Ser2 in the RNAP II CTD and this is associated with activation of elongation.

Tat and P-TEFb

P-TEFb exists in three distinct biochemical complexes. The first complex is "core P-TEFb" and it is composed of CDK9, cyclinT1, and Brd4 (Yang et al., 2005). Brd4 contains two bromodomains that bind to acetyllysine residues of histones H3 and H4 found in active chromatin. The second P-TEFb complex is the "super elongation complex" (SEC) composed of CDK9, cyclin T1, ELL1/ELL2, AFF1/AFF4, and ENL/AF9 (He et al., 2010; Sobhian et al., 2010). The third P-TEFb complex is the "7SK RNP" (ribonucleoprotein complex), composed of 7SK RNA, CDK9, cyclin T1, HEXIM1/2, LARP7, and MEPCE; CDK9 is not catalytically active in the 7SK RNP (He and Zhou, 2011; Nguyen et al., 2001; Yang et al., 2001). In cells that are rapidly dividing and metabolically active, such as transformed cell lines or activated $CD4^+$ T cells, the vast majority of P-TEFb (perhaps up to 90%) is located in the 7SK RNP (Gudipaty and D'Orso, 2016). Cyclin T2, an additional minor regulatory partner of CDK9 (Peng et al., 1998b), is encoded by a distinct gene, and its expression levels are significantly lower than cyclin T1 (Liou et al., 2006; Peng et al., 1998b). Therefore, the abundance of cyclin T2 in the various P-TEFb complexes is considerably lower than that of cyclin T1 and these particular complexes are not targeted by Tat.

Tat can utilize CDK9/cyclin T1 in each of the three P-TEFb complexes by binding directly to CDK9 and cyclin T1. Crystal structures of the Tat/CDK9/cyclin T1 complex have been determined, providing detailed information about the nature of protein–protein interactions (Tahirov et al., 2010). Tat utilizes the Core P-TEFb complex by displacing Brd4 from CDK9/cyclin T1 (Bisgrove et al., 2007). JQ1, a thienotriazolodiazepine compound, inhibits Brd4 and enhances RNAP II transcription of the HIV-1 provirus by displacing Brd4 from Core P-TEFb, presumably thereby increasing the pool of P-TEFb that can be targeted by Tat (Banerjee et al., 2012; Boehm et al., 2012; Li et al., 2012; Zhu et al., 2012). Tat can associate with the SEC and stabilize this P-TEFb complex to potently activate viral gene expression (He et al., 2010; Sobhian et al., 2010). A crystal structure of Tat and P-TEFb bound to the AFF4 subunit of the SEC has been determined (Schulze-Gahmen et al., 2014). This structure provides insight into molecular interactions whereby the SEC enhances the affinity of Tat and P-TEFb for TAR RNA. Tat can also extract active CDK9/cyclin T1 from the 7SK RNP (Barboric et al., 2007; Schulte et al., 2005; Sedore et al., 2007). Experimental evidence indicates that the 7SK RNP binds to the HIV-1 promoter at a site that is 5′ to the transcriptional start site, and following synthesis of TAR RNA by RNAP II, Tat extracts cyclin T1/CDK9 from the 7SK RNP to activate transcriptional elongation (D'Orso et al., 2012; McNamara et al., 2013). However, controversy surrounds the question of whether or not the 7SK RNP is recruited to the LTR upstream of the transcriptional start site based on chromatin-immunoprecipitation analyses indicating that CDK9/cyclin T1 is recruited to the HIV-1 LTR at the TAR RNA sequence and not to the HIV-1 promoter (Kim et al., 2011).

CDK11 and HIV-1 3′ End Processing

After RNAP II elongation is activated by Tat and P-TEFb, 3′ end processing of the viral RNA is an additional step required for production of a mature transcript. This is achieved through the action and another cyclin-dependent kinase—CDK11. CDK11 is associated with the TREX/THOC complex, a multiprotein complex that couples RNAP II transcription with mRNA processing and nuclear export (Strasser et al., 2002). When recruited by TREX/THOC to an RNAP II complex engaged in transcription of the provirus, CDK11 phosphorylates Ser2 residues in the RNAP II CTD, and these modifications recruit cleavage and polyadenylation factors that process the 3′ end of the viral transcript (Pak et al., 2015). As discussed above, HIV-1 has a preference to insert into genes near nuclear pores and this subnuclear localization is likely to facilitate TREX/THOC-directed export of mature viral transcripts to the cytoplasm.

HIV-1 LATENT INFECTION

Establishment of Latent Infection

HIV-1 infection of resting $CD4^+$ T cells is not productive (Folks et al., 1986) due to limiting levels of cellular cofactors that are essential for the viral replication cycle, as well as to the action of a restriction factor known as SAMHD1. SAMHD1 is a triphosphohydrolase that inhibits HIV-1 reverse transcription by reducing the cytoplasmic levels of deoxyribonucleoside triphosphates (Hrecka et al., 2011; Laguette et al., 2011). In activated $CD4^+$ T cells, SAMHD1 is inactivated through phosphorylation by cyclin-dependent kinases that are induced following cellular activation (Rice and Kimata, 2015). As described below, key transcription factors (NF-κB, NFAT, and P-TEFb) are downregulated in resting $CD4^+$ T cells and are therefore limited for RNAP II transcription of the provirus. Because of these blocks to HIV-1 infection in resting cells, it is thought that latency is established under the rare condition when HIV-1 infects an activated $CD4^+$ T cell that is reverting to a resting state (Han et al., 2007). This notion is supported by the finding that the majority of latent virus can be found in memory $CD4^+$ T cells (Chomont et al., 2009). It is believed that about 1 in 10^6 resting $CD4^+$ T cells harbor latent HIV-1 in patients under suppressive cART therapy. A recent study has shown that HIV-1 can infect and establish latent infection in resting $CD4^+$ T cells when SAMHD1 is inhibited (Descours et al., 2017).

Maintenance of Latent Infection

Maintenance of HIV-1 latent infection in resting $CD4^+$ T cells is the consequence of a block to RNAP II transcription of the provirus. Independent of the site of proviral integration, two precisely positioned nucleosomes assemble at the HIV-1 5′ LTR, termed Nuc-0 and Nuc-1 (Verdin et al., 1993). Nuc-1 is located around the transcriptional start site and therefore has potential to act as a potent block to RNAP II transcription. Indeed, epigenetic modification of Nuc-1 and other nucleosomes on the viral 5′ LTR have been shown in many studies to contribute to silencing proviral transcription. The nucleosomes assembled on the 5′ LTR contain deacetylated and trimethylated histones, whose modifications are found in repressive heterochromatic structures (Friedman et al., 2011; Keedy et al., 2009). Moreover, histone methyltransferases themselves have been found to be associated with the latent HIV-1 5′ LTR and these lysine methyltranseferase (KMT) enzymes are likely to contribute to maintenance of latency (Chene du et al., 2007; Friedman et al., 2011). Specifically, entry into repressive chromatin associated with latency involves H3K9 and H3K27 methylation of proviral histones by G9a, SUV39H1, and the polycomb repressive complex subunit EZH2 (Chene du et al., 2007; Marban et al., 2007; Nguyen et al., 2017). Additionally, the SET-domain KMT SMYD2 monomethylates H4K20 resulting in recruitment of the heterochromatin-forming corepressor L3MBTL1 and subsequent proviral latency (Boehm et al., 2017; Nikolai and Feng, 2017).

Key transcription factors required for proviral transcription are downregulated in resting $CD4^+$ T cells. NF-κB and NFAT are held in inhibitory states in the cytoplasm and therefore transcriptional initiation of the provirus is inhibited. Additionally, the CDK9 and cyclin T1 subunits of P-TEFb are downregulated in resting $CD4^+$ T cells. Although cyclin T1 mRNA levels are high in these cells, four miRNAs have been identified that repress translation of cyclin T1 mRNA: miR-27b, miR-29b, miR-150, and miR-223 (Chiang et al., 2012). Cyclin T2, the alternative regulatory subunit of CDK9 not targeted by Tat, is expressed at a basal level in resting CD4+ T cells and is typically not induced upon T cell activation. Presumably, cyclin T2 rather than cyclin T1 provides the basal activity of P-TEFb needed in resting cells (Herrmann et al., 1998; Liu and Herrmann, 2005). Unlike cyclin T1, CDK9 is expressed at a high basal level in resting CD4+ T cells (Haaland et al., 2003; Sung and Rice, 2006). However, phosphorylation of threonine 186 in the CDK9 T-loop, a modification required for kinase activity, is low in resting CD4+ T cells (Ramakrishnan et al., 2009). Following T cell activation, there is a rapid phosphorylation of the CDK9 T-loop by CDK7 (Larochelle et al., 2012; Ramakrishnan et al., 2009). The phosphatase PPM1A regulates T-loop dephosphorylation in resting CD4+ T cells, as siRNA depletion of PPM1A results in elevated levels of T-loop phosphorylation in resting cells (Budhiraja et al., 2012). Three additional phosphatases have been shown to dephosphorylate the CDK9 T-loop in transformed cell lines: PP2B, PP1α, and PPM1G (Chen et al., 2008; McNamara et al., 2013). The relative importance of these three phosphatases for the CDK9 T-loop in primary CD4+ T cells is not known. CDK9 is also phosphorylated on Ser175 following activation of resting CD4+ T cells, and molecular modeling suggests that phosphorylation of this residue strengthens the interaction between CDK9 and Tat (Mbonye et al., 2013). The kinase and phosphatase(s) that regulate Ser175 phosphorylation have not yet been identified.

Reactivation of Latent Virus

A variety of stimuli can activate RNAP II transcription of latent HIV-1, such as engagement of the T cell receptor or PKC agonists (Brooks et al., 2003; Kim et al., 2011). These stimuli lead to release of both NF-κB and NFAT from their inhibited states in the cytoplasm and their translocalization to the nucleus where they bind to NF-κB sites in the viral enhancer. Because NFAT and NF-κB bind in a mutually exclusive manner to overlapping sites in the proviral enhancer, they are thought to bind sequentially, with NF-κB activating initially and NFAT at latter times (Mbonye and Karn, 2014).

NF-κB recruits a coactivator complex containing HATs to the proviral enhancer, such as p300, CBP, and PCAF (Gerritsen et al., 1997; Kiernan et al., 1999; Perkins et al., 1997). Histone acetylation of Nuc-1 by these HATs allows recruitment of the chromatin remodeling complexes SWI/SNF, leading to relief of repressive epigenetic marks and an enhancement of RNAP II initiation (Rafati et al., 2011). P-TEFb is also activated by TCR engagement or PKC agonists; these stimuli result in rapid phosphorylation of the CDK9 T-loop (Ramakrishnan et al., 2009) and reduction in the relative levels of miRNAs that repress translation of cyclin T1 mRNA (Chiang et al., 2012). It is notable that agents that only minimally activate resting CD4+ T cells, such as a combination of cytokines (TNF-α, IL-2, IL-6) or the PKC agonist prostratin, can upregulate cyclin T1 and CDK9 kinase activity without inducing T cell proliferation or T cell activation markers (Ghose et al., 2001; Sung and Rice, 2006). A synthetic prostratin derivative has been developed, which is potent in reactivating latent HIV-1 in patient cells ex vivo and has only minimal effects on T cell activation markers (Beans et al., 2013). Additional PKC agonists termed ingenol esters effectively reactivate latent HIV-1 in patient cells ex vivo and also upregulate CDK9/P-TEFb (Jiang et al., 2015; Pandelo Jose et al., 2014; Spivak et al., 2015) with minimal effects on T cell activation. These findings indicate that it may be possible to selectively upregulate P-TEFb activity in resting $CD4^+$ T cells without generalized immune activation. This may be of value in efforts to develop small molecules that can reactivate latent HIV-1 in HIV-infected individuals and reduce the size of the latent reservoir.

FUTURE RESEARCH OF RETROVIRAL TRANSCRIPTION

Research into mechanisms of retroviral transcription has provided deep insight into mechanisms involved in an essential step in the viral life cycle. Additionally, this research has also provided a window into eukaryotic transcription in general. Early research of oncoretroviruses provided the first clues to the genetic nature of cancer and MMTV-driven transgenes are now standard tools for mouse models of breast cancer. Moreover, dissection of the promoter in the MMTV LTR revealed the consensus-binding sites for steroid receptors and significantly contributed to the field of molecular endocrinology. Transcriptional elongation and RNAP II pausing is best understood in the context of HIV-1 Tat/TAR RNA recruitment of P-TEFb, and current efforts to pharmacologically target HIV-1 latency have important clinical implications for a functional cure of HIV-1 infection. Transcriptional regulation by ERVs and lncRNAs is an emerging field that may have far reaching implications for basic understanding of biology.

HIV-1 Latency and the Need for a Functional Cure

Roughly 40 million people worldwide are living with HIV-1 infection, including 1 million individuals in the United States where there are an estimated 50,000 new infections each year. Living with HIV-1 infection is not without consequence. Side effects to cART and intrinsic complications of viral infection contribute to a number of significant comorbidities including metabolic syndrome, neuropathy, and nephropathy. A promising area of HIV research involves pharmacological activation of HIV-1 transcription to prevent or reverse latency. To this end, it will be important to identify novel host targets involved in HIV-1 transcriptional processes that are amendable to pharmacologic targeting with small molecule compounds capable of reaching latent reservoirs. An interesting facet of certain reservoirs, especially the gut associated lymphatic tissue, is the local immune tolerance of microbiome flora that condition a metabolic microenvironment that may be conducive to HIV-1 latency. Indeed, short-chain fatty acids are abundant in the gut and the histone deacetylase inhibitor suberoylanilide hydroxamic acid (SAHA or Vorinostat) has been shown to activate HIV-1 proviral transcription and has been pursued in clinical trials for its effectiveness in inducing HIV in cART-treated patients. However, the effectiveness of HDAC inhibitors in modulating HIV-1 transcription in patient samples has been questioned and an alternative approach to this "shock" strategy of latency reversal is to pharmacologically "lock" the virus in a dormant state by inhibiting HIV-1 transactivators or inducing proviral heterochromatin (Darcis et al., 2017). More basic science research into mechanisms of HIV-1 transcriptional regulation in the context of host cell chromatin is needed to identify new treatment strategies that may lead to a functional cure for HIV-1 infection. The fact that there is an individual previously infected with HIV-1 and living AIDS-free for over a decade without cART treatment is encouraging to the pursuit of cure-based research.

Endogenous Retroviruses in Stem Cells and Cancer

The importance of ERVs for the pluripotency of stem cells and an essential biological role for ERVs in embryogenesis, reproduction, and development suggests that ERVs may also be involved in the etiology or pathogenesis of cancer. The tumor-initiating cell or "cancer stem cell" hypothesis posits that a rare population of cells within a neoplasm is responsible for tumor maintenance, treatment resistance, and recurrence leading to metastatic disease and eventual patient mortality. It has been demonstrated that treatment with 5-azacitidine, a nucleoside analogue that promotes demethylation of DNA, can kill tumor cells through an interferon response that is mounted against aberrant ERV transcription (Chiappinelli et al., 2015; Roulois et al., 2015). However, relatively little is known about ERVs in cancer stem cells or whether ERV-mediated transcriptional events contribute to tumor cell plasticity during treatment and metastasis. Moreover, ERV function in immune cells may influence effector responses to cytokines in the tumor microenvironment. It will be important to more inclusively consider ERVs when using bioinformatics to analyze large "omics" datasets. The significant portion of the genome occupied by ERVs and their transcriptional functions during critical biological processes deserves more attention to the "ERVomics" of biological systems, including cancer and cancer stem cells.

ACKNOWLEDGMENTS

The authors wish to thank Dr. Jung Sun Kim for critical review of the manuscript. Additionally, we thank Drs. Dorothy Lewis, Jason Kimata, Edward Siwak, and Qin Feng for helpful discussion and collegial rapport. We also thank Lam Nguyen and the creative services staff at Baylor College of Medicine for assistance with figure artwork. BCN is supported by The Brockman Medical Research Foundation, and Creative and Novel Ideas in HIV Research (CNIHR) and Targeting Persistent HIV Reservoirs (TaPHIR, 1R21AI122418-01) awards to Qin Feng. APR is supported by National Institutes of Health grants AI116173.

REFERENCES

Adelman, K., Lis, J.T., 2012. Promoter-proximal pausing of RNA polymerase II: emerging roles in metazoans. Nat. Rev. Genet. 13, 720–731. https://doi.org/10.1038/nrg3293.

Ahe von der, D., Janich, S., Scheidereit, C., Renkawitz, R., Schutz, G., Beato, M., 1985. Glucocorticoid and progesterone receptors bind to the same sites in two hormonally regulated promoters. Nature 313, 706–709.

Ahlenstiel, C., Mendez, C., Lim, S.T.H., Marks, K., Turville, S., Cooper, D.A., Kelleher, A.D., Suzuki, K., 2015. Novel RNA duplex locks HIV-1 in a latent state via chromatin-mediated transcriptional silencing. Mol. Ther. Nucleic Acids 4, e261. https://doi.org/10.1038/mtna.2015.31.

Aiewsakun, P., Katzourakis, A., n.d. Marine origin of retroviruses in the early Palaeozoic Era. Nat. Commun. 8, 13954. https://doi.org/10.1038/ncomms13954.

Banerjee, C., Archin, N., Michaels, D., Belkina, A.C., Denis, G.V., Bradner, J., Sebastiani, P., Margolis, D.M., Montano, M., 2012. BET bromodomain inhibition as a novel strategy for reactivation of HIV-1. J. Leukoc. Biol. 92, 1147–1154.

Bannert, H., Muranyi, W., Ogryzko, V.V., Nakatani, Y., Flugel, R.M., 2004. Coactivators p300 and PCAF physically and functionally interact with the foamy viral trans-activator. BMC Mol. Biol. 5, 16.

Barboric, M., Yik, J.H., Czudnochowski, N., Yang, Z., Chen, R., Contreras, X., Geyer, M., Peterlin, B.M., Zhou, Q., 2007. Tat competes with HEXIM1 to increase the active pool of P-TEFb for HIV-1 transcription. Nucleic Acids Res. 35, 2003–2012.

Barr, S.D., Leipzig, J., Shinn, P., Ecker, J.R., Bushman, F.D., 2005. Integration targeting by avian sarcoma-leukosis virus and human immunodeficiency virus in the chicken genome. J. Virol. 79, 12035–12044. https://doi.org/10.1128/JVI.79.18.12035-12044.2005.

Beans, E.J., Fournogerakis, D., Gauntlett, C., Heumann, L.V., Kramer, R., Marsden, M.D., Murray, D., Chun, T.-W., Zack, J.A., Wender, P.A., 2013. Highly potent, synthetically accessible prostratin analogs induce latent HIV expression in vitro and ex vivo. Proc. Natl. Acad. Sci. U.S.A. 110, 11698–11703. https://doi.org/10.1073/pnas.1302634110.

Beato, M., 1989. Gene regulation by steroid hormones. Cell 56, 335–344.

Beimling, P., Moelling, K., 1992. Direct interaction of CREB protein with 21 bp tax-response elements of HTLV-ILTR. Oncogene 7, 257–262.

Bisgrove, D.A., Mahmoudi, T., Henklein, P., Verdin, E., 2007. Conserved P-TEFb-interacting domain of BRD4 inhibits HIV transcription. Proc. Natl. Acad. Sci. U.S.A. 104, 13690–13695.

Bittner, J.J., 1942. The milk-influence of breast tumors in mice. Science 95, 462–463. https://doi.org/10.1126/science.95.2470.462.

Bittner, J.J., 1936. Some possible effects of nursing on the mammary gland tumor incidence in mice. Science 84, 162. https://doi.org/10.1126/science.84.2172.162.

Boehm, D., Calvanese, V., Dar, R.D., Xing, S., Schroeder, S., Martins, L., Aull, K., Li, P.C., Planelles, V., Bradner, J.E., Zhoug, M.M., Siliciano, R.F., Weinberger, L., Verdin, E., Ott, M., 2012. BET bromodomain-targeting compounds reactivate HIV from latency via a Tat-independent mechanism. Cell Cycle 12.

Boehm, D., Jeng, M., Camus, G., Gramatica, A., Schwarzer, R., Johnson, J.R., Hull, P.A., Montano, M., Sakane, N., Pagans, S., Godin, R., Deeks, S.G., Krogan, N.J., Greene, W.C., Ott, M., 2017. SMYD2-mediated histone methylation contributes to HIV-1 latency. Cell Host Microbe 21, 569–579.e6. https://doi.org/10.1016/j.chom.2017.04.011.

Bourgeois, C.F., Kim, Y.K., Churcher, M.J., West, M.J., Karn, J., 2002. Spt5 cooperates with human immunodeficiency virus type 1 Tat by preventing premature RNA release at terminator sequences. Mol. Cell. Biol. 22, 1079–1093.

Brooks, D.G., Arlen, P.A., Gao, L., Kitchen, C.M., Zack, J.A., 2003. Identification of T cell-signaling pathways that stimulate latent HIV in primary cells. Proc. Natl. Acad. Sci. U.S.A. 100, 12955–12960.

Budhiraja, S., Ramakrishnan, R., Rice, A.P., 2012. Phosphatase PPM1A negatively regulates P-TEFb function in resting CD4(+) T cells and inhibits HIV-1 gene expression. Retrovirology 9, 52. https://doi.org/10.1186/1742-4690-9-52.

Campbell, M., Eng, C., Luciw, P.A., 1996. The simian foamy virus type 1 transcriptional transactivator (Tas) binds and activates an enhancer element in the gag gene. J. Virol. 70, 6847–6855.

CECIL, R.L., 1959. A decade of anti-inflammatory steroids, from cortisone to dexamethasone. Part III. Clinical studies. Introductory remarks. Ann. N.Y. Acad. Sci. 82, 886.

Chen, R., Liu, M., Li, H., Xue, Y., Ramey, W.N., He, N., Ai, N., Luo, H., Zhu, Y., Zhou, N., Zhou, Q., 2008. PP2B and PP1alpha cooperatively disrupt 7SK snRNP to release P-TEFb for transcription in response to Ca2+ signaling. Genes Dev. 22, 1356–1368. https://doi.org/10.1101/gad.1636008.

Chene du, I., Basyuk, E., Lin, Y.-L., Triboulet, R., Knezevich, A., Chable-Bessia, C., Mettling, C., Baillat, V., Reynes, J., Corbeau, P., Bertrand, E., Marcello, A., Emiliani, S., Kiernan, R., Benkirane, M., 2007. Suv39H1 and HP1gamma are responsible for chromatin-mediated HIV-1 transcriptional silencing and post-integration latency. EMBO J. 26, 424–435. https://doi.org/10.1038/sj.emboj.7601517.

Chiang, K., Sung, T.L., Rice, A.P., 2012. Regulation of cyclin T1 and HIV-1 replication by microRNAs in resting CD4+ T lymphocytes. J. Virol. 86, 3244–3252.

Chiappinelli, K.B., Strissel, P.L., Desrichard, A., Li, H., Henke, C., Akman, B., Hein, A., Rote, N.S., Cope, L.M., Snyder, A., Makarov, V., Buhu, S., Slamon, D.J., Wolchok, J.D., Pardoll, D.M., Beckmann, M.W., Zahnow, C.A., Mergoub, T., Chan, T.A., Baylin, S.B., Strick, R., 2015. Inhibiting DNA methylation causes an interferon response in cancer via dsRNA including endogenous retroviruses. Cell 162, 974–986.

Choi, Y.J., Lin, C.-P., Risso, D., Chen, S., Kim, T.A., Tan, M.H., Li, J.B., Wu, Y., Chen, C., Xuan, Z., Macfarlan, T., Peng, W., Lloyd, K.C.K., Kim, S.Y., Speed, T.P., He, L., 2017. Deficiency of microRNA miR-34a expands cell fate potential in pluripotent stem cells. Science 355. https://doi.org/10.1126/science.aag1927.

Chomont, N., Far, El, M., Ancuta, P., Trautmann, L., Procopio, F.A., Yassine-Diab, B., Boucher, G., Boulassel, M.R., Ghattas, G., Brenchley, J.M., Schacker, T.W., Hill, B.J., Douek, D.C., Routy, J.P., Haddad, E.K., Sekaly, R.P., 2009. HIV reservoir size and persistence are driven by T cell survival and homeostatic proliferation. Nat. Med. 15, 893–900.

Chuong, E.B., Elde, N.C., Feschotte, C., 2016. Regulatory evolution of innate immunity through co-option of endogenous retroviruses. Science 351, 1083–1087. https://doi.org/10.1126/science.aad5497.

Churchill, M.J., Deeks, S.G., Margolis, D.M., Siliciano, R.F., Swanstrom, R., 2016. HIV reservoirs: what, where and how to target them. Nat. Rev. Microbiol. 14, 55–60. https://doi.org/10.1038/nrmicro.2015.5.

Crise, B., Li, Y., Yuan, C., Morcock, D.R., Whitby, D., Munroe, D.J., Arthur, L.O., Wu, X., 2005. Simian immunodeficiency virus integration preference is similar to that of human immunodeficiency virus type 1. J. Virol. 79, 12199–12204. https://doi.org/10.1128/JVI.79.19.12199-12204.2005.

D'Orso, I., Jang, G.M., Pastuszak, A.W., Faust, T.B., Quezada, E., Booth, D.S., Frankel, A.D., 2012. Transition step during assembly of HIV Tat:P-TEFb transcription complexes and transfer to TAR RNA. Mol. Cell. Biol. 32, 4780–4793.

Darcis, G., Van Driessche, B., Van Lint, C., 2017. HIV latency: should we shock or lock? Trends Immunol. 38, 217–228. https://doi.org/10.1016/j.it.2016.12.003.

Dayton, A.I., Sodroski, J.G., Rosen, C.A., Goh, W.C., Haseltine, W.A., 1986. The trans-activator gene of the human T cell lymphotropic virus type III is required for replication. Cell 44, 941–947.

De Clercq, E., Li, G., 2016. Approved antiviral drugs over the past 50 years. Clin. Microbiol. Rev. 29, 695–747. https://doi.org/10.1128/CMR.00102-15.

Demeulemeester, J., De Rijck, J., Gijsbers, R., Debyser, Z., 2015. Retroviral integration: site matters. Bioessays 37, 1202–1214. https://doi.org/10.1002/bies.201500051.

Descours, B., Petitjean, G., Lopez-Zaragoza, J.-L., Bruel, T., Raffel, R., Psomas, C., Reynes, J., Lacabaratz, C., Levy, Y., Schwartz, O., Lelievre, J.D., Benkirane, M., 2017. CD32a is a marker of a CD4 T-cell HIV reservoir harbouring replication-competent proviruses. Nature 543, 564–567. https://doi.org/10.1038/nature21710.

Eidahl, J.O., Crowe, B.L., North, J.A., McKee, C.J., Shkriabai, N., Feng, L., Plumb, M., Graham, R.L., Gorelick, R.J., Hess, S., Poirier, M.G., Foster, M.P., Kvaratskhelia, M., 2013. Structural basis for high-affinity binding of LEDGF PWWP to mononucleosomes. Nucleic Acids Res. 41, 3924–3936. https://doi.org/10.1093/nar/gkt074.

Ernberg, I., Gait, M.J., Green, S.M., Heaphy, S., Karn, J., Lowe, A.D., Singh, M., Skinner, M.A., Valerio, R., 1989. Human immunodeficiency virus 1 tat protein binds trans- activation-responsive region (TAR) RNA in vitro. Proc. Natl. Acad. Sci. U.S.A. 86, 6925–6929.

Faulkner, G.J., Kimura, Y., Daub, C.O., Wani, S., Plessy, C., Irvine, K.M., Schroder, K., Cloonan, N., Steptoe, A.L., Lassmann, T., Waki, K., Hornig, N., Arakawa, T., Takahashi, H., Kawai, J., Forrest, A.R.R., Suzuki, H., Hayashizaki, Y., Hume, D.A., Orlando, V., Grimmond, S.M., Carninci, P., 2009. The regulated retrotransposon transcriptome of mammalian cells. Nat. Genet. 41, 563–571. https://doi.org/10.1038/ng.368.

Feng, S., Holland, E.C., 1988. HIV-1 tat trans-activation requires the loop sequence within tar. Nature 334, 165–167.

Filippakopoulos, P., Picaud, S., Mangos, M., Keates, T., Lambert, J.-P., Barsyte-Lovejoy, D., Felletar, I., Volkmer, R., Muller, S., Pawson, T., Gingras, A.-C., Arrowsmith, C.H., Knapp, S., 2012. Histone recognition and large-scale structural analysis of the human bromodomain family. Cell 149, 214–231. https://doi.org/10.1016/j.cell.2012.02.013.

Folks, T., Kelly, J., Benn, S., Kinter, A., Justement, J., Gold, J., Redfield, R., Sell, K.W., Fauci, A.S., 1986. Susceptibility of normal human lymphocytes to infection with HTLV-III/LAV. J. Immunol. 136, 4049–4053.

Friedman, J., Cho, W.-K., Chu, C.K., Keedy, K.S., Archin, N.M., Margolis, D.M., Karn, J., 2011. Epigenetic silencing of HIV-1 by the histone H3 lysine 27 methyltransferase enhancer of Zeste 2. J. Virol. 85, 9078–9089.

Garcia, J.A., Harrich, D., Soultanakis, E., Wu, F., Mitsuyasu, R., Gaynor, R.B., 1989. Human immunodeficiency virus type 1 LTR TATA and TAR region sequences required for transcriptional regulation. EMBO J. 8, 765–778.

Gerritsen, M.E., Williams, A.J., Neish, A.S., Moore, S., Shi, Y., Collins, T., 1997. CREB-binding protein/p300 are transcriptional coactivators of p65. Proc. Natl. Acad. Sci. U.S.A. 94, 2927–2932.

Ghose, R., Liou, L.Y., Herrmann, C.H., Rice, A.P., 2001. Induction of TAK (cyclin T1/P-TEFb) in purified resting CD4(+) T lymphocytes by combination of cytokines. J. Virol. 75, 11336–11343. https://doi.org/10.1128/JVI.75.23.11336-11343.2001.

Gillet, N.A., Malani, N., Melamed, A., Gormley, N., Carter, R., Bentley, D., Berry, C., Bushman, F.D., Taylor, G.P., Bangham, C.R.M., 2011. The host genomic environment of the provirus determines the abundance of HTLV-1-infected T-cell clones. Blood 117, 3113–3122. https://doi.org/10.1182/blood-2010-10-312926.

Gudipaty, S.A., D'Orso, I., 2016. Functional interplay between PPM1G and the transcription elongation machinery. RNA Dis. 3.

Haaland, R.E., Herrmann, C.H., Rice, A.P., 2003. Increased association of 7SK snRNA with Tat cofactor P-TEFb following activation of peripheral blood lymphocytes. AIDS 17, 2429–2436.

Han, Y., Wind-Rotolo, M., Yang, H.-C., Siliciano, J.D., Siliciano, R.F., 2007. Experimental approaches to the study of HIV-1 latency. Nat. Rev. Microbiol. 5, 95–106. https://doi.org/10.1038/nrmicro1580.

Harrod, R., Tang, Y., Nicot, C., Lu, H.S., Vassilev, A., Nakatani, Y., Giam, C.Z., 1998. An exposed KID-like domain in human T-cell lymphotropic virus type 1 Tax is responsible for the recruitment of coactivators CBP/p300. Mol. Cell. Biol. 18, 5052–5061.

Hayward, A., Cornwallis, C.K., Jern, P., 2015. Pan-vertebrate comparative genomics unmasks retrovirus macroevolution. Proc. Natl. Acad. Sci. U.S.A. 112, 464–469.

He, N., Liu, M., Hsu, J., Xue, Y., Chou, S., Burlingame, A., Krogan, N.J., Alber, T., Zhou, Q., 2010. HIV-1 Tat and host AFF4 recruit two transcription elongation factors into a bifunctional complex for coordinated activation of HIV-1 transcription. Mol. Cell 38, 428–438.

He, N., Zhou, Q., 2011. New insights into the control of HIV-1 transcription: when Tat meets the 7SK snRNP and super elongation complex (SEC). J. Neuroimmune Pharmacol. 6, 260–268. https://doi.org/10.1007/s11481-011-9267-6.

Hendrickson, P.G., Dorais, J.A., Grow, E.J., Whiddon, J.L., Lim, J.-W., Wike, C.L., Weaver, B.D., Pflueger, C., Emery, B.R., Wilcox, A.L., Nix, D.A., Peterson, C.M., Tapscott, S.J., Carrell, D.T., Cairns, B.R., 2017. Conserved roles of mouse DUX and human DUX4 in activating cleavage-stage genes and MERVL/HERVL retrotransposons. Nat. Genet. 49, 925–934. https://doi.org/10.1038/ng.3844.

Herrmann, C.H., Carroll, R.G., Wei, P., Jones, K.A., Rice, A.P., 1998. Tat-associated kinase, TAK, activity is regulated by distinct mechanisms in peripheral blood lymphocytes and promonocytic cell lines. J. Virol. 72, 9881–9888.

Herrmann, C.H., Rice, A.P., 1995. Lentivirus Tat proteins specifically associate with a cellular protein kinase, TAK, that hyperphosphorylates the carboxyl-terminal domain of the large subunit of RNA polymerase II: candidate for a Tat cofactor. J. Virol. 69, 1612–1620.

Herrmann, C.H., Rice, A.P., 1993. Specific interaction of the human immunodeficiency virus Tat proteins with a cellular protein kinase. Virology 197, 601–608. https://doi.org/10.1006/viro.1993.1634.

Hidaka, M., Inoue, J., Yoshida, M., Seiki, M., 1988. Post-transcriptional regulator (rex) of HTLV-1 initiates expression of viral structural proteins but suppresses expression of regulatory proteins. EMBO J. 7, 519–523.

Hogan, P.G., 2017. Calcium-NFAT transcriptional signalling in T cell activation and T cell exhaustion. Cell Calcium. https://doi.org/10.1016/j.ceca.2017.01.014.

Hollenberg, S.M., Weinberger, C., Ong, E.S., Cerelli, G., Oro, A., Lebo, R., Thompson, E.B., Rosenfeld, M.G., Evans, R.M., 1985. Primary structure and expression of a functional human glucocorticoid receptor cDNA. Nature 318, 635–641.

Hrecka, K., Hao, C., Gierszewska, M., Swanson, S.K., Kesik-Brodacka, M., Srivastava, S., Florens, L., Washburn, M.P., Skowronski, J., 2011. Vpx relieves inhibition of HIV-1 infection of macrophages mediated by the SAMHD1 protein. Nature 474, 658–661. https://doi.org/10.1038/nature10195.

Hu, T., Pi, W., Zhu, X., Yu, M., Ha, H., Shi, H., Choi, J.-H., Tuan, D., 2017. Long non-coding RNAs transcribed by ERV-9 LTR retrotransposon act in cis to modulate long-range LTR enhancer function. Nucleic Acids Res. 45, 4479–4492. https://doi.org/10.1093/nar/gkx055.

Inoue, J., Itoh, M., Akizawa, T., Toyoshima, H., Yoshida, M., 1991. HTLV-1 Rex protein accumulates unspliced RNA in the nucleus as well as in cytoplasm. Oncogene 6, 1753–1757.

Jahner, D., Stuhlmann, H., Stewart, C.L., Harbers, K., Lohler, J., Simon, I., Jaenisch, R., 1982. De novo methylation and expression of retroviral genomes during mouse embryogenesis. Nature 298, 623–628.

Jiang, G., Mendes, E.A., Kaiser, P., Wong, D.P., Tang, Y., Cai, I., Fenton, A., Melcher, G.P., Hildreth, J.E.K., Thompson, G.R., Wong, J.K., Dandekar, S., 2015. Synergistic reactivation of latent HIV expression by ingenol-3-angelate, PEP005, targeted NF-kB signaling in combination with JQ1 induced p-TEFb activation. PLoS Pathog. 11, e1005066.

Jiang, H., Lu, H., Schiltz, R.L., Pise-Masison, C.A., Ogryzko, V.V., Nakatani, Y., Brady, J.N., 1999. PCAF interacts with tax and stimulates tax transactivation in a histone acetyltransferase-independent manner. Mol. Cell. Biol. 19, 8136–8145.

Jones, K.A., Kadonaga, J.T., Luciw, P.A., Tjian, R., 1986. Activation of the AIDS retrovirus promoter by the cellular transcription factor, Sp1. Science 232, 755–759.

Kao, S.Y., Calman, A.F., Luciw, P.A., Peterlin, B.M., 1987. Anti-termination of transcription within the long terminal repeat of HIV-1 by tat gene product. Nature 330, 489–493.

Karn, J., Stoltzfus, C.M., 2012. Transcriptional and posttranscriptional regulation of HIV-1 gene expression. Cold Spring Harb. Perspect. Med. 2, a006916.

Keedy, K.S., Archin, N.M., Gates, A.T., Espeseth, A., Hazuda, D.J., Margolis, D.M., 2009. A limited group of class I histone deacetylases acts to repress human immunodeficiency virus type 1 expression. J. Virol. 83, 4749–4756. https://doi.org/10.1128/JVI.02585-08.

Kiernan, R.E., Vanhulle, C., Schiltz, L., Adam, E., Xiao, H., Maudoux, F., Calomme, C., Burny, A., Nakatani, Y., Jeang, K.T., Benkirane, M., Van Lint, C., 1999. HIV-1 tat transcriptional activity is regulated by acetylation. EMBO J. 18, 6106–6118.

Kim, Y.K., Mbonye, U., Hokello, J., Karn, J., 2011. T-cell receptor signaling enhances transcriptional elongation from latent HIV proviruses by activating P-TEFb through an ERK-dependent pathway. J. Mol. Biol. 410, 896–916. https://doi.org/10.1016/j.jmb.2011.03.054.

Kinoshita, S., Chen, B.K., Kaneshima, H., Nolan, G.P., 1998. Host control of HIV-1 parasitism in T cells by the nuclear factor of activated T cells. Cell 95, 595–604.

LaFave, M.C., Varshney, G.K., Gildea, D.E., Wolfsberg, T.G., Baxevanis, A.D., Burgess, S.M., 2014. MLV integration site selection is driven by strong enhancers and active promoters. Nucleic Acids Res. 42, 4257–4269. https://doi.org/10.1093/nar/gkt1399.

Laguette, N., Sobhian, B., Casartelli, N., Ringeard, M., Chable-Bessia, C., Segeral, E., Yatim, A., Emiliani, S., Schwartz, O., Benkirane, M., 2011. SAMHD1 is the dendritic- and myeloid-cell-specific HIV-1 restriction factor counteracted by Vpx. Nature 474, 654–657. https://doi.org/10.1038/nature10117.

Larochelle, S., Amat, R., Glover-Cutter, K., Sanso, M., Zhang, C., Allen, J.J., Shokat, K.M., Bentley, D.L., Fisher, R.P., 2012. Cyclin-dependent kinase control of the initiation-to-elongation switch of RNA polymerase II. Nat. Struct. Mol. Biol. 19, 1108–1115.

Li, Z., Guo, J., Wu, Y., Zhou, Q., 2012. The BET bromodomain inhibitor JQ1 activates HIV latency through antagonizing Brd4 inhibition of Tat-transactivation. Nucleic Acids Res.

Liou, L.Y., Haaland, R.E., Herrmann, C.H., Rice, A.P., 2006. Cyclin T1 but not cyclin T2a is induced by a post-transcriptional mechanism in PAMP-activated monocyte-derived macrophages. J. Leukoc. Biol. 79, 388–396.

Liu, H., Herrmann, C.H., 2005. Differential localization and expression of the Cdk9 42k and 55k isoforms. J. Cell. Physiol. 203, 251–260. https://doi.org/10.1002/jcp.20224.

Liu, J., Perkins, N.D., Schmid, R.M., Nabel, G.J., 1992. Specific NF-kappa B subunits act in concert with Tat to stimulate human immunodeficiency virus type 1 transcription. J. Virol. 66, 3883–3887.

Lu, X., Sachs, F., Ramsay, L., Jacques, P.-E., Goke, J., Bourque, G., Ng, H.-H., 2014. The retrovirus HERVH is a long noncoding RNA required for human embryonic stem cell identity. Nat. Struct. Mol. Biol. 21, 423–425. https://doi.org/10.1038/nsmb.2799.

Macian, F., 2005. NFAT proteins: key regulators of T-cell development and function. Nat. Rev. Immunol. 5, 472–484. https://doi.org/10.1038/nri1632.

Madore, S.J., Cullen, B.R., 1993. Genetic analysis of the cofactor requirement for human immunodeficiency virus type 1 Tat function. J. Virol. 67, 3703–3711.

Marban, C., Suzanne, S., Dequiedt, F., de Walque, S., Redel, L., Van Lint, C., Aunis, D., Rohr, O., 2007. Recruitment of chromatin-modifying enzymes by CTIP2 promotes HIV-1 transcriptional silencing. EMBO J. 26, 412–423. https://doi.org/10.1038/sj.emboj.7601516.

Marini, B., Kertesz-Farkas, A., Ali, H., Lucic, B., Lisek, K., Manganaro, L., Pongor, S., Luzzati, R., Recchia, A., Mavilio, F., Giacca, M., Lusic, M., 2015. Nuclear architecture dictates HIV-1 integration site selection. Nature 521, 227–231. https://doi.org/10.1038/nature14226.

Matsui, T., Leung, D., Miyashita, H., Maksakova, I.A., Miyachi, H., Kimura, H., Tachibana, M., Lorincz, M.C., Shinkai, Y., 2010. Proviral silencing in embryonic stem cells requires the histone methyltransferase ESET. Nature 464, 927–931. https://doi.org/10.1038/nature08858.

Mbonye, U., Karn, J., 2014. Transcriptional control of HIV latency: cellular signaling pathways, epigenetics, happenstance and the hope for a cure. Virology 454–455, 328–339.

Mbonye, U.R., Gokulrangan, G., Datt, M., Dobrowolski, C., Cooper, M., Chance, M.R., Karn, J., 2013. Phosphorylation of CDK9 at Ser175 enhances HIV transcription and is a marker of activated P-TEFb in CD4(+) T lymphocytes. PLoS Pathog. 9, e1003338. https://doi.org/10.1371/journal.ppat.1003338.

McNamara, R.P., McCann, J.L., Gudipaty, S.A., D'Orso, I., 2013. Transcription factors mediate the enzymatic disassembly of promoter-bound 7SK snRNP to locally recruit P-TEFb for transcription elongation. Cell Rep. 5, 1256–1268. https://doi.org/10.1016/j.celrep.2013.11.003.

Mi, S., Lee, X., Li, X., Veldman, G.M., Finnerty, H., Racie, L., LaVallie, E., Tang, X.Y., Edouard, P., Howes, S., Keith, J.C.J., McCoy, J.M., 2000. Syncytin is a captive retroviral envelope protein involved in human placental morphogenesis. Nature 403, 785–789. https://doi.org/10.1038/35001608.

Mitchell, R.S., Beitzel, B.F., Schroder, A.R.W., Shinn, P., Chen, H.M., Berry, C.C., Ecker, J.R., Bushman, F.D., 2004. Retroviral DNA integration: ASLV, HIV, and MLV show distinct target site preferences. PLoS Biol. 2, 1127–1137.

Nabel, G., Baltimore, D., 1987. An inducible transcription factor activates expression of human immunodeficiency virus in T cells. Nature 326, 711–713. https://doi.org/10.1038/326711a0.

Naville, M., Volff, J.-N., 2016. Endogenous retroviruses in fish genomes: from relics of past infections to evolutionary innovations? Front. Microbiol. 7, 1197. https://doi.org/10.3389/fmicb.2016.01197.

Nguyen, K., Das, B., Dobrowolski, C., Karn, J., 2017. Multiple histone lysine methyltransferases are required for the establishment and maintenance of HIV-1 latency. mBio 8. https://doi.org/10.1128/mBio.00133-17.

Nguyen, V.T., Kiss, T., Michels, A.A., Bensaude, O., 2001. 7SK small nuclear RNA binds to and inhibits the activity of CDK9/cyclin T complexes. Nature 414, 322–325.

Nikolai, B.C., Feng, Q., 2017. HIV latency gets a new histone mark. Cell Host Microbe 21, 549–550. https://doi.org/10.1016/j.chom.2017.04.012.

Nikolai, B.C., Lanz, R.B., York, B., Dasgupta, S., Mitsiades, N., Creighton, C.J., Tsimelzon, A., Hilsenbeck, S.G., Lonard, D.M., Smith, C.L., O'Malley, B.W., 2016. HER2 signaling drives DNA anabolism and proliferation through SRC-3 phosphorylation and E2F1-regulated genes. Cancer Res. https://doi.org/10.1158/0008-5472.CAN-15-2383.

O'Dea, E., Hoffmann, A., 2009. NF-kappaB signaling. Wiley Interdiscip. Rev. Syst. Biol. Med. 1, 107–115. https://doi.org/10.1002/wsbm.30.

Oeckinghaus, A., Hayden, M.S., Ghosh, S., 2011. Crosstalk in NF-kappaB signaling pathways. Nat. Immunol. 12, 695–708. https://doi.org/10.1038/ni.2065.

Okano, M., Xie, S., Li, E., 1998. Dnmt2 is not required for de novo and maintenance methylation of viral DNA in embryonic stem cells. Nucleic Acids Res. 26, 2536–2540.

Pak, V., Eifler, T.T., Jager, S., Krogan, N.J., Fujinaga, K., Peterlin, B.M., 2015. CDK11 in TREX/THOC regulates HIV mRNA 3′ end processing. Cell Host Microbe 18, 560–570. https://doi.org/10.1016/j.chom.2015.10.012.

Pandelo Jose, D., Bartholomeeusen, K., da Cunha, R.D., Abreu, C.M., Glinski, J., da Costa, T.B.F., Bacchi Rabay, A.F.M., Pianowski Filho, L.F., Dudycz, L.W., Ranga, U., Peterlin, B.M., Pianowski, L.F., Tanuri, A., Aguiar, R.S., 2014. Reactivation of latent HIV-1 by new semi-synthetic ingenol esters. Virology 462–463, 328–339. https://doi.org/10.1016/j.virol.2014.05.033.

Peng, J., Marshall, N.F., Price, D.H., 1998a. Identification of a cyclin subunit required for the function of Drosophila P-TEFb. J. Biol. Chem. 273, 13855–13860.

Peng, J., Zhu, Y., Milton, J.T., Price, D.H., 1998b. Identification of multiple cyclin subunits of human P-TEFb. Genes Dev. 12, 755–762.

Perkins, N.D., Felzien, L.K., Betts, J.C., Leung, K., Beach, D.H., Nabel, G.J., 1997. Regulation of NF-kappaB by cyclin-dependent kinases associated with the p300 coactivator. Science 275, 523–527.

Pi, W., Yang, Z., Wang, J., Ruan, L., Yu, X., Ling, J., Krantz, S., Isales, C., Conway, S.J., Lin, S., Tuan, D., 2004. The LTR enhancer of ERV-9 human endogenous retrovirus is active in oocytes and progenitor cells in transgenic zebrafish and humans. Proc. Natl. Acad. Sci. U.S.A. 101, 805–810.

Prud'homme, B., Gompel, N., Carroll, S.B., 2007. Emerging principles of regulatory evolution. Proc. Natl. Acad. Sci. U.S.A. 104 (Suppl. 1), 8605–8612. https://doi.org/10.1073/pnas.0700488104.

Rafati, H., Parra, M., Hakre, S., Moshkin, Y., Verdin, E., Mahmoudi, T., 2011. Repressive LTR nucleosome positioning by the BAF complex is required for HIV latency. PLoS Biol. 9, e1001206.

Ramakrishnan, R., Dow, E.C., Rice, A.P., 2009. Characterization of Cdk9 T-loop phosphorylation in resting and activated CD4(+) T lymphocytes. J. Leukoc. Biol. 86, 1345–1350.

Rice, A.P., Kimata, J.T., 2015. Subversion of cell cycle regulatory mechanisms by HIV. Cell Host Microbe 17, 736–740. https://doi.org/10.1016/j.chom.2015.05.010.

Rimsky, L., Hauber, J., Dukovich, M., Malim, M.H., Langlois, A., Cullen, B.R., Greene, W.C., 1988. Functional replacement of the HIV-1 rev protein by the HTLV-1 rex protein. Nature 335, 738–740. https://doi.org/10.1038/335738a0.

Rosen, C.A., Sodroski, J.G., Haseltine, W.A., 1985. The location of cis-acting regulatory sequences in the human T cell lymphotropic virus type III (HTLV-III/LAV) long terminal repeat. Cell 41, 813–823.

Roulois, D., Loo Yau, H., Singhania, R., Wang, Y., Danesh, A., Shen, S.Y., Han, H., Liang, G., Jones, P.A., Pugh, T.J., O'Brien, C., De Carvalho, D.D., 2015. DNA-demethylating agents target colorectal cancer cells by inducing viral mimicry by endogenous transcripts. Cell 162, 961–973. https://doi.org/10.1016/j.cell.2015.07.056.

Santoni, F.A., Guerra, J., Luban, J., 2012. HERV-H RNA is abundant in human embryonic stem cells and a precise marker for pluripotency. Retrovirology 9, 111. https://doi.org/10.1186/1742-4690-9-111.

Scheidereit, C., Geisse, S., Westphal, H.M., Beato, M., 1983. The glucocorticoid receptor binds to defined nucleotide sequences near the promoter of mouse mammary tumour virus. Nature 304, 749–752.

Schlesinger, S., Goff, S.P., 2012. Silencing of proviruses in embryonic cells: efficiency, stability and chromatin modifications. EMBO Rep. 14, 73–79. https://doi.org/10.1038/embor.2012.182.

Schroder, A.R.W., Shinn, P., Chen, H., Berry, C., Ecker, J.R., Bushman, F., 2002. HIV-1 integration in the human genome favors active genes and local hotspots. Cell 110, 521–529. https://doi.org/10.1016/S0092-8674(02)00864-4.

Schulte, A., Czudnochowski, N., Barboric, M., Schonichen, A., Blazek, D., Peterlin, B.M., Geyer, M., 2005. Identification of a cyclin T-binding domain in Hexim1 and biochemical analysis of its binding competition with HIV-1 Tat. J. Biol. Chem. 280, 24968–24977.

Schulze-Gahmen, U., Lu, H., Zhou, Q., Alber, T., 2014. AFF4 binding to Tat-P-TEFb indirectly stimulates TAR recognition of super elongation complexes at the HIV promoter. Elife 3, e02375.

Sedore, S.C., Byers, S.A., Biglione, S., Price, J.P., Maury, W.J., Price, D.H., 2007. Manipulation of P-TEFb control machinery by HIV: recruitment of P-TEFb from the large form by Tat and binding of HEXIM1 to TAR. Nucleic Acids Res. 35, 4347–4358.

Sobhian, B., Laguette, N., Yatim, A., Nakamura, M., Levy, Y., Kiernan, R., Benkirane, M., 2010. HIV-1 Tat assembles a multifunctional transcription elongation complex and stably associates with the 7SK snRNP. Mol. Cell 38, 439–451. https://doi.org/10.1016/j.molcel.2010.04.012.

Sodroski, J., Rosen, C., Wong-Staal, F., Salahuddin, S.Z., Popovic, M., Arya, S., Gallo, R.C., Haseltine, W.A., 1985a. Trans-acting transcriptional regulation of human T-cell leukemia virus type III long terminal repeat. Science 227, 171–173.

Sodroski, J.G., Patarca, R., Rosen, C.A., Haseltine, W.A., 1985b. Location of the trans-activating region on the genome of human T-cell lymphotropic virus III. Science 229, 74–77.

Spivak, A.M., Bosque, A., Balch, A.H., Smyth, D., Martins, L., Planelles, V., 2015. Ex vivo bioactivity and HIV-1 latency reversal by ingenol dibenzoate and panobinostat in resting CD4(+) T cells from aviremic patients. Antimicrob. Agents Chemother. 59, 5984–5991.

Stewart, T.A., Pattengale, P.K., Leder, P., 1984. Spontaneous mammary adenocarcinomas in transgenic mice that carry and express MTV/myc fusion genes. Cell 38, 627–637.

Strasser, K., Masuda, S., Mason, P., Pfannstiel, J., Oppizzi, M., Rodriguez-Navarro, S., Rondon, A.G., Aguilera, A., Struhl, K., Reed, R., Hurt, E., 2002. TREX is a conserved complex coupling transcription with messenger RNA export. Nature 417, 304–308.

Sung, T.-L., Rice, A.P., 2006. Effects of prostratin on Cyclin T1/P-TEFb function and the gene expression profile in primary resting CD4+ T cells. Retrovirology 3, 66. https://doi.org/10.1186/1742-4690-3-66.

Tahirov, T.H., Babayeva, N.D., Varzavand, K., Cooper, J.J., Sedore, S.C., Price, D.H., 2010. Crystal structure of HIV-1 Tat complexed with human P-TEFb. Nature 465, 747–751.

van Nuland, R., van Schaik, F.M., Simonis, M., van Heesch, S., Cuppen, E., Boelens, R., Timmers, H.M., van Ingen, H., 2013. Nucleosomal DNA binding drives the recognition of H3K36-methylated nucleosomes by the PSIP1-PWWP domain. Epigenet. Chromatin 6, 12. https://doi.org/10.1186/1756-8935-6-12.

van Opijnen, T., Jeeninga, R.E., Boerlijst, M.C., Pollakis, G.P., Zetterberg, V., Salminen, M., Berkhout, B., 2004. Human immunodeficiency virus type 1 subtypes have a distinct long terminal repeat that determines the replication rate in a host-cell-specific manner. J. Virol. 78, 3675–3683.

Varmus, H.E., 1990. Nobel lecture. Retroviruses and oncogenes. I. Biosci. Rep. 10, 413–430.

Verdin, E., Paras, P.J., Van Lint, C., 1993. Chromatin disruption in the promoter of human immunodeficiency virus type 1 during transcriptional activation. EMBO J. 12, 3249–3259.

Wada, T., Takagi, T., Yamaguchi, Y., Ferdous, A., Imai, T., Hirose, S., Sugimoto, S., Yano, K., Hartzog, G.A., Winston, F., Buratowski, S., Handa, H., 1998. DSIF, a novel transcription elongation factor that regulates RNA polymerase II processivity, is composed of human Spt4 and Spt5 homologs. Genes Dev. 12, 343–356.

Wei, P., Garber, M.E., Fang, S.M., Fischer, W.H., Jones, K.A., 1998. A novel CDK9-associated C-type cyclin interacts directly with HIV-1 Tat and mediates its high-affinity, loop-specific binding to TAR RNA. Cell 92, 451–462.

Weinberger, C., Hollenberg, S.M., Ong, E.S., Harmon, J.M., Brower, S.T., Cidlowski, J., Thompson, E.B., Rosenfeld, M.G., Evans, R.M., 1985. Identification of human glucocorticoid receptor complementary DNA clones by epitope selection. Science 228, 740–742.

Wolf, D., Goff, S.P., 2009. Embryonic stem cells use ZFP809 to silence retroviral DNAs. Nature 458, 1201–1204. https://doi.org/10.1038/nature07844.

Wolf, D., Goff, S.P., 2007. TRIM28 mediates primer binding site-targeted silencing of murine leukemia virus in embryonic cells. Cell 131, 46–57. https://doi.org/10.1016/j.cell.2007.07.026.

Wu, X.L., Li, Y., Crise, B., Burgess, S.M., 2003. Transcription start regions in the human genome are favored targets for MLV integration. Science 300, 1749–1751.

Yamaguchi, Y., Takagi, T., Wada, T., Yano, K., Furuya, A., Sugimoto, S., Hasegawa, J., Handa, H., 1999. NELF, a multisubunit complex containing RD, cooperates with DSIF to repress RNA polymerase II elongation. Cell 97, 41–51.

Yang, X., Gold, M.O., Tang, D.N., Lewis, D.E., Aguilar-Cordova, E., Rice, A.P., Herrmann, C.H., 1997. TAK, an HIV Tat-associated kinase, is a member of the cyclin-dependent family of protein kinases and is induced by activation of peripheral blood lymphocytes and differentiation of promonocytic cell lines. Proc. Natl. Acad. Sci. U.S.A. 94, 12331–12336.

Yang, Z., Yik, J.H.N., Chen, R., He, N., Jang, M.K., Ozato, K., Zhou, Q., 2005. Recruitment of P-TEFb for stimulation of transcriptional elongation by the bromodomain protein Brd4. Mol. Cell 19, 535–545. https://doi.org/10.1016/j.molcel.2005.06.029.

Yang, Z., Zhu, Q., Luo, K., Zhou, Q., 2001. The 7SK small nuclear RNA inhibits the CDK9/cyclin T1 kinase to control transcription. Nature 414, 317–322 JID: 0410462.

Zhang, S., Cui, X., Li, J., Liang, Z., Qiao, W., Tan, J., 2016. Lysine residues K66, K109, and K110 in the bovine foamy virus transactivator protein are required for transactivation and viral replication. Virol. Sin. 31, 142–149. https://doi.org/10.1007/s12250-015-3652-x.

Zhao, L.J., Giam, C.Z., 1992. Human T-cell lymphotropic virus type I (HTLV-I) transcriptional activator, Tax, enhances CREB binding to HTLV-I 21-base-pair repeats by protein-protein interaction. Proc. Natl. Acad. Sci. U.S.A. 89, 7070–7074.

Zhou, Q., Li, T., Price, D.H., 2012. RNA polymerase II elongation control. Annu. Rev. Biochem. 81, 119–143. https://doi.org/10.1146/annurev-biochem-052610-095910.

Zhu, J., Gaiha, G.D., John, S.P., Pertel, T., Chin, C.R., Gao, G., Qu, H., Walker, B.D., Elledge, S.J., Brass, A.L., 2012. Reactivation of latent HIV-1 by inhibition of BRD4. Cell Rep. 2, 807–816.

Zhu, Y., Pe'ery, T., Peng, J., Ramanathan, Y., Marshall, N.F., Marshall, T.A.B., Amendt, B., Mathews, M.B., Price, D.H., 1997. Transcriptional elongation factor p-TEFb is required for HIV-1 Tat transactivation in vitro. Genes Dev. 11, 2622–2632.

Zou, J.X., Luciw, P.A., 1996. The transcriptional transactivator of simian foamy virus 1 binds to a DNA target element in the viral internal promoter. Proc. Natl. Acad. Sci. U.S.A. 93, 326–330.

Chapter 6

Teetering on the Edge: The Critical Role of RNA Processing Control During HIV-1 Replication

Ahalya Balachandran, Liang Ming, Alan Cochrane
University of Toronto, Toronto, ON, Canada

Chapter Outline

Introduction	229
Role of hnRNPs in the Regulation of HIV-1 RNA Processing	231
Role of SR Proteins in the Regulation of HIV-1 RNA Processing	236
Manipulation of HIV-1 RNA Processing With Small Molecules	240
Acknowledgments	245
References	245

INTRODUCTION

As an obligate parasite, HIV-1 is dependent on the cell for multiple metabolic processes needed for the production and assembly of progeny virions. This point is particularly evident in the regulation of HIV-1 RNA processing. While coding for 15 proteins, the integrated HIV-1 provirus generates only a single 9 kb transcript. To achieve production of all encoded proteins, this transcript undergoes alternative splicing to generate the mRNAs required to synthesize all viral proteins. Although initial studies indicated that 47 different mRNAs were produced from the primary transcript (Purcell and Martin, 1993), recent next-generation sequencing has identified 109 mRNAs, of which 69 are conserved among all HIV-1 strains analyzed to date (Emery et al., 2017; Ocwieja et al., 2012). These transcripts fall into four classes based on size of the mRNA produced: 9 kb unspliced (US) RNAs used to produce Gag and Gagpol; 4 kb singly spliced (SS) RNAs encoding Vif, Vpr, Vpu, and Env; 1.8 kb multiply spliced (MS) RNAs that express Tat, Rev, and Nef; and a 1 kb MS RNA used to produce a Rev–Nef fusion protein (Emery et al., 2017; Ocwieja et al., 2012). Northern blots indicate

Retrovirus-Cell Interactions. https://doi.org/10.1016/B978-0-12-811185-7.00006-6

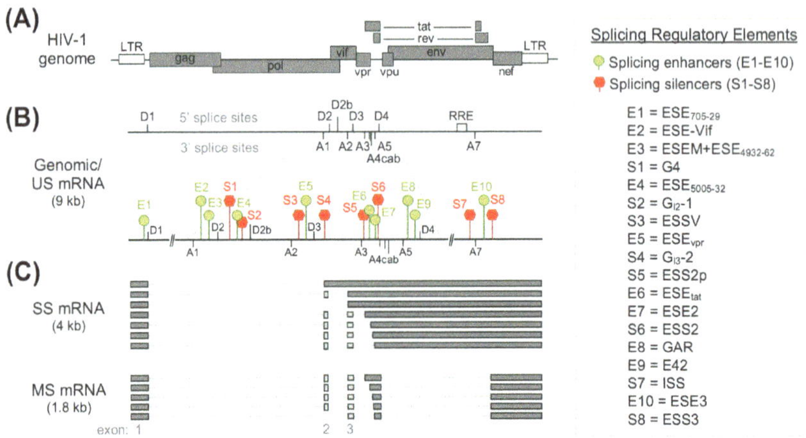

FIGURE 6.1 Summary of HIV-1 RNA processing and cis-elements that regulate splice site use. (A) Schematic of the HIV-1 provirus indicating the location of protein reading frames. (B) Position of HIV-1 splice donor (D) and splice acceptor (A) sites within the primary 9kb transcript as well as the location of splicing enhancer (E) and silencer (S) sequences (Widera et al., 2013, 2014; Zahler et al., 2004; Kammler et al., 2001; Caputi et al., 2004; Stoltzfus and Madsen, 2006; Erkelenz et al., 2014; Domsic et al., 2003; Bilodeau et al., 2001; Si et al., 1997; Amendt et al., 1994; Caputi et al., 1999; Asang et al., 2008; Tange et al., 2001; Amendt et al., 1995; Si et al., 1998). (C) Schematic of spliced HIV-1 RNAs within the SS (4kb) and MS (1.8kb) RNA class. *Open boxes* indicate alternative exons found in some of the viral RNAs. *MS*, multiply spliced; *SS*, singly spliced; *US*, unspliced. *(Adapted from Erkelenz, S., Hillebrand, F., Widera, M., Theiss, S., Fayyaz, A., Degrandi, D., Pfeffer, K., Schaal, H., 2015. Balanced splicing at the Tat-specific HIV-1 3′ss A3 is critical for HIV-1 replication. Retrovirology 12, 29 and Stoltzfus, C.M., Madsen, J.M., 2006. Role of viral splicing elements and cellular RNA binding proteins in regulation of HIV-1 alternative RNA splicing. Curr. HIV Res. 4, 43–55.)*

that the first three classes are the most abundant generated and their composition is outlined in Fig. 6.1. Unlike most host RNA transcripts that remove all introns prior to transport into the cytoplasm and translation, the requirement for HIV-1 to generate key proteins from incompletely processed RNAs has necessitated the evolution of various strategies to regulate the quantity of each RNA produced and facilitate their export to the cytoplasm. The investigation of how HIV-1 RNA splicing is controlled has not only provided important insights into mechanisms involved in the regulation of RNA processing but the recent identification of small molecules that can modulate these events also offers the possibility of novel strategies for the control of this infection.

To regulate the quantity of each mRNA produced, HIV-1 has evolved a complex interplay between *cis*-acting sequences and *trans*-acting host factors to determine the frequency of individual splice site use (Stoltzfus, 2009). In general, inefficient splicing of HIV-1 RNA is attributed to the sequestration of splice sites in RNA secondary structure (Mueller et al., 2014, 2015a,b, 2016; Saliou et al., 2009) or the presence of suboptimal sequences (either disrupted polypyrimidine tracts, nonconsensus branchpoint sequences, or both) within the splice acceptors, A, (Stoltzfus, 2009; Kammler et al., 2006; O'Reilly et al., 1995; Damier et al.,

1997; Dyhr-Mikkelsen and Kjems, 1995). Use of specific splice sites is further regulated by the presence of sequences that either enhance (splicing enhancers, E) or suppress (splicing silencers, S) splice site recognition, in general through the recruitment of serine–arginine (SR) proteins and heterogenous nuclear ribonucleoproteins (hnRNPs), respectively. Fig. 6.1B outlines the position of the splicing silencer and enhancer sequences identified to date. Control of HIV-1 RNA splicing is key to virus replication as evidenced by the finding that inactivation of the exon splicing silencer ESSV or the exon splicing enhancer ESE_{tat} results in marked alterations in viral RNA accumulation and a sharp reduction in virus fitness (Erkelenz et al., 2015; Madsen and Stoltzfus, 2005; Widera et al., 2014). Consequently, identifying the host proteins that regulate HIV-1 RNA processing and understanding their mechanism of action provide important insights for the development of novel approaches to control HIV-1 replication.

The regulation of HIV-1 RNA processing is not limited to splicing as the virus requires a strategy to bypass the host cell's restriction to the export of incompletely spliced RNAs to the cytoplasm. The solution for HIV-1 is Rev. Synthesized from viral MS RNA, Rev shuttles between the nucleus and cytoplasm, mediating the export of HIV-1 US and SS RNAs to the cytoplasm through its interaction with the Rev response element (RRE) within both classes of RNA (Hope, 1999; Pollard and Malim, 1998; McLaren et al., 2008) (see Fig. 6.1B). Consequently, HIV-1 protein expression is biphasic, with expression of MS RNA encoded proteins (Tat, Rev, and Nef) occurring first as these RNAs are exported to the cytoplasm by NXF1, the same export pathway used by host mRNAs (Cullen, 2003). Once sufficient Rev is produced, it mediates the export of viral US and SS RNAs through the CRM1 pathway to allow Gag, Gagpol, Vif, Vpr, Vpu, and Env synthesis. Subsequent research has determined that use of the CRM1 export pathway does affect events within the cytoplasm (Marsh et al., 2008; Swanson et al., 2004, 2010). Such effects could be attributable to differences in the protein composition of the viral US and SS RNPs that allow these RNAs to bypass the modulation of the host protein translation machinery by HIV-1 (Sharma et al., 2012). The dependence on Rev for production of the viral structural proteins and the distinct nature of the viral US and SS RNPs provide another target for potential therapeutic intervention.

ROLE OF hnRNPs IN THE REGULATION OF HIV-1 RNA PROCESSING

Studies investigating the control of HIV-1 RNA splicing have highlighted the importance of hnRNPs (see Table 6.1) with particular emphasis on the role of hnRNP A1 and H in mediating the activity of several of the splicing regulatory elements mapped to date (Stoltzfus, 2009).

In the case of hnRNP A1, studies have demonstrated interaction with ESSV (S3), ESS2 (S6), ISS (S7), and ESS3 (S8) to reduce assembly of spliceosome complexes on the upstream splice acceptor (A2, A3, A7, respectively). In the context of splice acceptor A7, inhibition of splice site recognition by hnRNP A1

TABLE 6.1 Role of hnRNP Proteins in HIV-1 RNA Processing

hnRNP	Binding Sites	Overexpression Effects	Depletion Effects
hnRNP A1	ESSV (S3) ESS2 (S6) ISS (S7) ESS3 (S8)	Increased US RNA level Decreased MS levels	Increased Gag expression Increased use of A2 and A3
hnRNP A2	A2RE-1 (Gag) A2RE-2 (vpr)	Decreased MS RNA levels Decreased Gag synthesis	Redistribution of US RNA to MTOC Increased Gag and Env levels Rev-independent cytoplasmic US RNA accumulation
hnRNP D/ AUF1	ESSV (S3)	p37/p40-reduced US RNA levels, Gag protein p42/p45-increased US RNA levels, Gag protein Increased A1, A2 use	Reduced Gag and Env expression Reduced cytoplasmic levels of US and SS RNAs
hnRNP E1/E2	ND	E1-reduced Gag, Env expression No change in viral RNA levels	E1- or E2-increased Gag and Env levels No change in viral RNA levels
hnRNP H	G_{12}-1 (S2) ESS2p (S5) G_{13}-2 (S4)	Slight increase in US RNA Decreased virion infectivity	Modest change in Gag expression, limited effects on HIV-1 RNA levels
hnRNP I, K	ND	ND	No significant effect in HeLa cells Depletion of K reduced Gag expression in astrocytes
hnRNP R, Q	NH_2 terminus of Rev	Increase expression from Rev-dependent reporter	Reduced Gag expression in astrocytes

HnRNP, heterogeneous nuclear ribonucleoprotein; *MS,* multiply spliced; *ND,* not determined; *US,* unspliced.

binding has been attributed to either looping out A7 due to interaction of hnRNP A1 bound to ISS (S7) and ESS3 (S8) or through oligomerization of hnRNP A1 along the RNA to occlude recognition of the branchpoint and polypyrimidine tracts within A7 (Blanchette and Chabot, 1999; Asai et al., 2003; Zhu et al., 2001; Damgaard et al., 2002; Marchand et al., 2002; Okunola and Krainer,

2009; Nasim et al., 2002). For many of the silencers bound by hnRNP A1, interaction results in the occlusion of overlapping splicing enhancer sequences ESEvpr (E5), ESE2 (E7), or ESE3 (E10), thus preventing their ability to promote use of the adjacent splice site (Marchand et al., 2002; Staffa and Cochrane, 1995; Erkelenz et al., 2013; Zahler et al., 2004). Many of these mechanistic studies were performed in the context of cell extracts or transfection assays using subgenomic HIV-1 constructs. Studies using complete HIV-1 provirus have determined that modulation of hnRNP A1 expression can significantly affect viral RNA accumulation and processing. Overexpression of hnRNP A1 increased US RNA abundance while decreasing MS RNA accumulation, putatively through enhancing the activity of multiple splicing silencers throughout the viral genome (Jablonski and Caputi, 2009). Similar studies investigating the effect of hnRNP A1 depletion (Jablonski and Caputi, 2009; Lund et al., 2012) showed an increase in Gag expression due to higher levels of HIV-1 US RNA accumulation and alterations in splice site usage consistent with reduced activity of splicing silencers ESSV (S3) and ESS2 (S6). Whereas hnRNP A1 depletion shifted the patterns of HIV-1 RNA processing, it did not result in radical changes in viral RNA processing, and virus production and infectivity were increased (Jablonski and Caputi, 2009), suggesting there might be redundancy in the viral regulatory system or the host factors interacting with the splicing control elements. Of note, work by Jean-Philippe et al. (2014) determined that mutants of hnRNP A1 lacking the C-terminal RGG domain have comparable activity to full length protein in modulating splice site A3 usage using model constructs, reducing viral RNA accumulation and infectivity of progeny virus upon cotransfection with a complete HIV-1 provirus. These observations suggest that only binding of hnRNP A1 to ESS2 is required to suppress recognition of the A3 splice site. Given that usage of A3 is required for Tat production, the loss of Tat resulting from hnRNP A1 overexpression results in the failure to increase the elongation efficiency of RNA polymerase II initiating at the viral promoter, leading to the reduced viral RNA accumulation observed in these experiments (Karn and Stoltzfus, 2012).

In contrast to the major role of hnRNP A1 in regulating use of multiple HIV-1 splice sites, hnRNP A2 has been implicated in other aspects of HIV-1 RNA metabolism. Initial studies by Beriault et al. (2004) identified two conserved sequences within the Gag (A2RE-1) and Vpr (A2RE-2) coding regions of HIV-1 that interact with hnRNP A2, and mutation of either decreased association of hnRNP A2 with viral US RNA whereas only mutation of A2RE-2 dramatically reduced virus replication. Although mutation of the A2RE-2 site did not affect Gag synthesis, it altered US RNA subcellular localization (becoming mostly nuclear), consistent with the reduced level of US RNA encapsidation observed. Subsequent studies by the same group (Levesque et al., 2006) revealed that hnRNP A2 depletion altered cytoplasmic distribution of HIV-1 US RNA, from being located throughout the cytoplasm in the presence of hnRNP A2 to accumulating at microtubule-organizing centers in its absence, consistent with a role of this factor in RNA trafficking. Similar studies by other groups (Jablonski and Caputi, 2009; Lund et al., 2012; Ohta et al., 2015) observed increased Gag/Env expression and US RNA abundance upon hnRNP A2 depletion with limited changes in splice site usage. Subsequent work by Gordon et al. (2013) determined that depletion of hnRNP A2 was able to rescue cytoplasmic accumulation of HIV-1 US RNA in the absence of Rev, implicating this factor in the nuclear sequestration of incompletely spliced viral RNAs.

Although known for its involvement in regulating mRNA stability (Barreau et al., 2005; Brewer, 1991), studies investigating the effect of hnRNP D/AUF1 depletion on HIV-1 gene expression indicated a role in viral RNA transport to the cytoplasm. Using HeLa cell lines stably transduced with HIV-1 provirus, Lund et al. (2012) determined that hnRNP D depletion reduced Gag and Env synthesis without changes in the corresponding mRNA abundance or alterations in splice site use. Rather, loss of hnRNP D was associated with a decrease in cytoplasmic accumulation of US and SS viral RNAs. Investigation of the four different isoforms of hnRNP D (p45, p42, p40, and p37) revealed that they had opposing effects on HIV-1 gene expression; p45 or 42 increased Gag protein production whereas overexpression of p40 or p37 reduced Gag synthesis. Consistent with this observation, selective depletion of p45/p42 reduced Gag and Env synthesis. Recent work by Hillebrand et al. (2017) has also implicated hnRNP D in the modulation of A1/A2 splice site usage through interaction with overlapping sequences in ESSV (S3). In the context of a truncated HIV-1 reporter vector, overexpression of the different hnRNP D isoforms increased usage of A1 and A2, with p40 yielding the greatest effect, possibly by displacing hnRNP A1 from ESSV.

Studies investigating the function of ESS3 (S8) identified an interaction with hnRNP E1/E2 (Woolaway et al., 2007). Although subsequent experiments did not confirm a role for either factor in ESS3 function, hnRNP E1 but not E2 overexpression markedly reduced HIV-1 Gag and Env expression from several Rev-dependent, subgenomic reporter constructs, as well as from the HIV-1 provirus (Woolaway et al., 2007). Complementing these observations, depletion of either hnRNP E1 or E2 enhanced Gag and Env synthesis from an HIV-1 proviral vector. In both instances, overexpression and depletion, changes in viral protein expression were not accompanied by any significant alteration in HIV-1 RNA accumulation or a shift in splice site usage. The inhibitory activity of hnRNP E1 was dependent on the KH2 and KH3 domains of the protein and resulted in the accumulation of the protein in the nucleus (Woolaway et al., 2007).

Characterization of sequences regulating HIV-1 splice site usage identified multiple G-rich sequences (S1, S2, S4, S5) that are putative-binding sites for hnRNP H (Widera et al., 2013, 2014; Schaub et al., 2007). Using in vitro splicing assays, Schaub et al., 2007 examined the role of multiple G-rich sequences adjacent to the HIV-1 D1 splice site. Of the seven sequences examined, mutation of only one was shown to negatively affect D1 utilization and its ability to promote splicing was dependent on hnRNP H interaction. Together, these findings suggest a positive role for hnRNP H in D1 recognition. However, subsequent tests in the context of the complete HIV-1 provirus failed to detect a similar requirement for this sequence (Asang et al., 2012). Similar in vitro experiments indicated that ESS2p (S5) suppression of A3 use involved interaction with hnRNP H, indicating that this factor was acting as a suppressor (Jacquenet et al., 2001). A G-rich sequence (G_{I2}-1, S2) proximal to D2 inhibited function of the adjacent splicing enhancer (Exline et al., 2008). hnRNP H binds the sequence (Widera et al., 2013), and introduction of a mutation disrupting hnRNP H interaction within the context of the HIV-1 provirus results in a marked increase in Vif mRNA and protein levels without a significant change in the relative abundance of US, SS, and MS viral RNAs (Widera et al., 2013). Analysis of other G-rich sequences (Widera et al., 2014) downstream of D3 revealed that mutants of G_{I3}-2 (S4) that decreased hnRNP H binding also markedly reduced HIV-1 replication. This effect was associated with decreased US RNA abundance and increased usage of A2, resulting in decreased A1 use and Vif expression. Overexpression and depletion studies investigating the role of hnRNP H in modulating HIV-1 RNA processing have resulted in mixed results. Experiments reported by Jablonski and Caputi (2009) using transfection of HIV-1 proviral DNA into HEK 293 cells determined that hnRNP H overexpression induced a modest increase in Gag release but reduced virion infectivity. The effect is associated with slight shifts in viral RNA accumulation consisting of increased US and SS RNAs with decreased levels of MS RNAs coding for Tat, Rev, and Nef. In contrast, depletion of hnRNP H had more modest effects. Reduced hnRNP H levels resulted in a slight enhancement of Gag release into the media and modest increase in virion infectivity linked with increases in the level of all viral RNAs with the exception of Tat mRNA, which decreased slightly. Of note was the observation that modulation of hnRNP H levels also affected viral RNA subcellular distribution. Increased hnRNP H expression decreased the nuclear/cytoplasmic ratio of US RNA while increasing the ratio for SS RNA. Depletion of hnRNP H had the opposite effect, increasing the nuclear/cytoplasmic ratio for US RNA while decreasing the ratio for SS RNA. Subsequent studies by Lund et al. (2012) using HeLa cells stably transduced with an HIV-1 provirus determined that hnRNP H depletion had little effect on HIV-1 RNA accumulation, splice site usage, or Gag/Env expression.

In contrast to the extensive work outlined above, data on the role of other hnRNPs in regulating HIV-1 gene expression are more limited. In the context of the complete HIV-1 provirus in HeLa cells, depletion of hnRNP I or K has no effect on HIV-1 Gag/Env synthesis, viral RNA accumulation, or splice site usage (Lund et al., 2012). Additional studies have detected an association between the amino terminus of Rev and multiple hnRNP proteins (hnRNP A1, Q, K, R, and U) (Hadian et al., 2009). Depletion of hnRNP Q, R, and K within chronically infected astrocytes reduced Gag expression, with loss of hnRNP Q and R having the greatest effect. Subsequent work (Vincendeau et al., 2013) determined that hnRNP Q overexpression could enhance expression of a Rev-dependent reporter by ~twofold without affecting the abundance or cellular distribution of the target RNA, suggesting that it might be affecting events at the level of translation.

ROLE OF SR PROTEINS IN THE REGULATION OF HIV-1 RNA PROCESSING

Given the suboptimal nature of the splice acceptors and, in most instances, the presence of adjacent splicing silencers, promotion of splice site usage has necessitated the presence of splicing enhancers. The enhancers, in general, operate via binding members of the SR protein family, each member consisting of one or two RRM domains to mediate contact with the RNA and an arginine/serine-rich (RS) C-terminal domain involved in RNA and protein–protein interaction (Long and Caceres, 2009; Sanford et al., 2005). Although several SR proteins have been implicated in RNA export and translation of select mRNAs (Huang et al., 2003; Sanford et al., 2004) and the modulation of HIV-1 RNA expression in the context of mouse cell lines (Swanson et al., 2010), for the purposes of this review, focus will be on their role in the control of HIV-1 RNA splicing (see Table 6.2).

The first report of an SR protein regulating HIV-1 RNA processing and gene expression was by Powell et al. (1997) demonstrating that overexpression of SRSF1 (SF2/ASF) reduced HIV-1 gene expression by decreasing the accumulation of US RNA. Although the initial report (Powell et al., 1997) suggested an interaction with the RRE, subsequent studies have implicated SRSF1 in promoting usage of splice sites A1 (ESEM, E3), A2, A5 (GAR, E8), and A7 (ESE3, E10) (Kammler et al., 2001, 2006; Caputi et al., 2004). Using in vitro splicing assays, several groups exploring the regulation of A7 use have shown that SRSF1 stimulates splicing through interaction with ESE3 (E10) and enhancement of U2AF65 binding to the adjacent polypyrimidine tract (Tange and Kjems, 2001; Zhu and Krainer, 2000). Surprising, stimulation by SRSF1 does not require the RS domain, but the RS domain is required to block binding of hnRNP A1 to the overlapping ESS3 (Zhu et al., 2001; Tange and Kjems, 2001; Zhu and Krainer, 2000). Similar in vitro assays using subgenomic substrates determined that SRSF1 could enhance use of A1 and A2 (Ropers et al., 2004) as well the activity of the GAR ESE (E8) in a heterologous context (Caputi et al., 2004). Function of the GAR ESE is not only required for promoting A4–A5 splice site use but also for recognition of D4 as inactivation of SRSF1-binding sites within GAR resulted in loss of Env expression and strongly promoted the production of nef1 mRNA generated by the joining of D1 to A7 (Caputi et al., 2004). Complementing the in vitro assays, multiple groups have examined the effect of SRSF1 overexpression on HIV-1 protein expression and RNA accumulation (Jablonski and Caputi, 2009; Ropers et al., 2004; Jacquenet et al., 2005). Using transient transfection into HeLa or HEK 293 cells with HIV-1 proviral constructs, SRSF1 overexpression significantly enhanced A2 usage, increasing

TABLE 6.2 Role of Serine–Arginine (SR) Proteins in HIV-1 RNA Processing

SR Protein	Binding Site	Overexpression Effects	Depletion Effects
SRSF1	ESEM (E3) GAR (E8) ESE3 (E10)	Enhanced Vpr levels Reduced Tat, Gag, and Env levels Inhibition of Tat transactivation	Increased levels of all viral RNAs, Gag expression
SRSF2	ESE_{tat} (E6) ESE2 (E7)	Enhances A3 usage	Decreased A3 usage Increased Gag expression
SRSF3	ND	Decreased US/SS RNA levels Decreased Gag/Env expression	ND
SRSF4	ESE_{vif} (E2) GAR (E8)	Increased Gag expression Increased Vpr RNA levels	Loss of vpr RNA
SRSF5	GAR (E8)	Reduced A4–A5 usage, increased A3 usage Reduced Gag levels	Reduced Gag levels Reduced virion infectivity
SRSF6	ESE_{tat} (E6) GAR (E8)	Enhanced A2, A3 usage, reduced A4–5. Promotes D1–A7 products Reduced levels of all HIV-1 RNAs	Reduced A2 usage Decreased Vpr expression
SRSF7	ND	Enhanced use of A3 Decreased Env protein/RNA levels	ND
Tra2α/β	ESE_{vpr} (E5)	Reduced Gag/Env expression Inhibition of US RNA transport Increased Vpr RNA levels	Slight increase in MS RNA levels Twofold reduction in Env expression

MS, multiply spliced; *ND*, not determined; *SS*, singly spliced; *US*, unspliced.

vpr mRNA and protein levels as well as exon 3 inclusion in viral MS RNAs. Consistent with the shift in splice usage and the consequent reduction of Tat expression (due to reduced A3 use), SRSF1 overexpression induces an overall loss in accumulation of all HIV-1 RNAs due to the loss of transcriptional activation by Tat and reduction in Gag and Env protein production (Jablonski and Caputi, 2009; Ropers et al., 2004; Jacquenet et al., 2005). Similar to its effect on A7, the activation of A2 usage by SRSF1 does not require the RS domain

(Ropers et al., 2004). SRSF1 has also been implicated in ESEM (E3) function based on RNA affinity assays, but mutants of ESEM reducing A1 usage had limited effect on virus replication (Kammler et al., 2006). Complementing the overexpression studies, siRNA depletion of SRSF1 in HEK 293 cells was found to have very modest effects, increasing levels of all HIV-1 RNAs by less than twofold, shifting splice usage to some degree, and increasing Gag protein production (Jablonski and Caputi, 2009). More recent studies have also highlighted a more direct role for SRSF1 in regulating HIV-1 transcription, with overexpression of SRSF1 inhibiting Tat transactivation through competition with Tat for binding to TAR (Paz et al., 2014, 2015). In addition to the full-length protein, overexpression of the SRSF1 RRM2 domain alone or a variant lacking the RS domain suppressed viral gene expression even when Tat was provided *in trans* (Paz et al., 2015).

Parallel studies have also highlighted an important role for SRSF2 (SC35) in regulating HIV-1 RNA processing and expression. Initial work using in vitro splicing extracts determined that SRSF2 could modulate A7 usage but, in contrast to SRSF1, its activity within this assay required depletion of hnRNP A1 from the extracts. This finding suggests that the interaction of hnRNP A1 with ESS3 (S8) blocks SRSF2 association with the RNA (Zhu et al., 2001). Similar in vitro assays also indicated that SRSF2 could promote use of A1 and A3 (Ropers et al., 2004). Subsequent work identified SRSF2 as interacting with ESE2 (E7), an association that was blocked by hnRNP A1 binding to the overlapping ESS2 (S6) (Zahler et al., 2004; Hallay et al., 2006). Studies also identified a second binding site for SRSF2 adjacent to A3 designated ESE_{tat} (E6) (Erkelenz et al., 2015). In transient transfection assays with HIV-1 proviral DNA, SRSF2 overexpression strongly enhanced A3 usage resulting in a marked reduction of RNAs using A4–A5 and a dramatic decrease in virus production/infectivity (Erkelenz et al., 2015; Jablonski and Caputi, 2009; Ropers et al., 2004; Jacquenet et al., 2005). Whereas mutations within ESE2 that disrupt SRSF2 interaction have a limited effect on HIV-1 RNA processing and protein expression, mutation of ESE_{tat} significantly reduced usage of A3 and decreased Tat levels as well as accumulation of all viral RNAs and proteins, an effect that could be rescued by expression of Tat *in trans* (Erkelenz et al., 2015). In contrast to the overexpression results, SRSF2 depletion has a more limited effect on HIV-1 RNA accumulation but does cause a marked decrease in A3 usage and an increase in Gag production (correlated with increased US RNA accumulation) and infectivity of the virions produced (Jablonski and Caputi, 2009). Additional SRSF2 binding sites have been identified that overlap with D1. However, whereas a mutation that disrupts SRSF2 binding affects D1 usage in the context of reporter vectors, the same mutations have a limited effect in the context of the complete HIV-1 provirus (Asang et al., 2012). Of note, HIV-1 infection of either macrophages or T-cell lines does result in a transient ~twofold increase in SRSF2 levels, suggesting that the virus may modulate expression/activity of specific SR proteins (Dowling et al., 2008; Maldarelli et al., 1998).

Of the remaining SR proteins, the next most intensively studied in the context of HIV-1 are SRSF5 (SRp40), SRSF6 (SRp55), and SRSF7 (9G8). Experiments using splicing extracts and model RNA substrates indicated weak activation of A1, A2, and A3 by SRSF5, but this factor did promote GAR ESE function (Caputi et al., 2004; Ropers et al., 2004). However, mutations within GAR that reduced SRSF5 binding had limited effect in other assays (Caputi et al., 2004). Experiments in HeLa or HEK 293 cells using HIV-1 proviral DNA determined that SFSR5 overexpression significantly reduced A4–A5 usage, resulting in loss of Nef synthesis, while increasing production of viral RNAs using A3, a response that ultimately reduced Gag protein production and infectivity of the progeny viruses (Jablonski and Caputi, 2009; Ropers et al., 2004; Tranell et al., 2010). Depletion of SRSF5 within the same assays had a limited effect on viral RNA accumulation and splicing but reduced Gag expression by ~two-fold and impaired infectivity of progeny viruses (Jablonski and Caputi, 2009). Studies examining SRSF6 have reported interaction with both ESE_{tat} and GAR ESE (Erkelenz et al., 2015; Tranell et al., 2010) as well as differences in the effect of its overexpression on viral RNA processing. In work by Erkelenz et al. (2015), overexpression of SRSF6 enhanced A2 and A3 usage, reduced A4–5 use, and promoted D1–A7 splicing. SFSF6 also induced a significant reduction in accumulation of all viral RNAs, an effect that is independent of ESE_{tat} and not affected by deletion of the RS domain of the protein. In contrast, only RRM1 and the adjacent C-terminal linker region of SRSF6 was required for its enhancement of A3 usage. Studies by Tranell et al. (2010) using a modified HIV-1 proviral DNA template observed that SRSF6 overexpression inhibited D3 and A3–A5 usage, promoting A2 inclusion and increasing Vpr expression. In contrast to Erkelenz et al. (2015), increased accumulation of vpr mRNA by SRSF6 required the RS domain (Tranell et al., 2010). A report by Tranell et al. indicates that SRSF6 may act, in part, by inhibiting GAR ESE, an activity consistent with the observed loss of A4–A5 usage and stimulation of D1–A7 joining. Complementing the overexpression assays, depletion of SRSF6 reduced A2 usage by reducing vpr mRNA accumulation. Similar studies examining the role of SRSF7 using splicing extracts or transient transfection showed limited effect on HIV-1 RNA processing, while enhancing the use of A3 and decreasing env mRNA/proteins levels with limited effect on Gag expression (Ropers et al., 2004; Jacquenet et al., 2005). As with several other SR proteins, effects of SRSF7 overexpression on HIV-1 RNA processing are not dependent on the RS domain (Ropers et al., 2004).

More limited data are available on the role of other SR and SR-related proteins in modulating HIV-1 RNA processing. Initial experiments observed that SRSF4 (SRp75) overexpression promoted HIV-1 Gag production (Fukuhara et al., 2006) and subsequent assays have implicated a role in the function of ESE_{vif} (E2) (Exline et al., 2008). SRSF4 overexpression had many effects similar to those induced by SRSF6, promoting the accumulation of vpr mRNA putatively by reducing usage of A3–A5 and blocking D3 use in part through inhibition of GAR ESE function (Tranell et al., 2010). Similar to SRSF6, SRSF4 depletion resulted in loss of vpr mRNA (Tranell et al., 2010). In the context of HeLa cell lines containing integrated HIV-1 proviral DNA, overexpression of either SRSF3 or Tra2α/β induced a marked reduction in HIV-1 Gag expression associated with reduced accumulation of HIV-1 US and SS RNAs, as well as increased usage of A3 (and reduced A4cab) in the case of SRSF3 (Wong et al., 2013b). Effects of Tra2α/β overexpression on viral protein synthesis can be attributed, in part, to the inhibition of HIV-1 US export to the cytoplasm, an effect that can be relieved by expression of Rev *in trans* (Platt et al., 2015). In contrast to the marked alterations in HIV-1 gene expression on Tra2α/β overexpression, depletion of Tra2β had more limited effects, with a modest (~1.5 fold) increase in viral MS accumulation and a twofold reduction in Env synthesis (Platt et al., 2015). Other work demonstrated that Tra2α/β can modulate Vpr expression through interaction with ESE_{vpr}, overexpression of Tra2α/β resulting in increased exon 3 inclusion within the SS group of viral RNAs (Erkelenz et al., 2013).

In addition to the role of host proteins in regulating HIV-1 RNA processing, studies have revealed that HIV-1 Tat, while modulating transcription from the viral LTR, also affects splicing of viral RNA (Jablonski et al., 2010). Tat overexpression significantly reduced US RNA accumulation and altered splice site usage (reducing A3 and increasing A4–A5 use), an effect that is independent of its modulation of RNA polymerase II activity and requires GAR ESE function as well as interaction with Tat-SF1 and CA150 (Jablonski et al., 2010). The capacity of Tat to both promote viral RNA production and alter RNA processing to suppress synthesis of its mRNA indicates evolution of a strategy to carefully regulate the abundance of this critical viral factor.

MANIPULATION OF HIV-1 RNA PROCESSING WITH SMALL MOLECULES

The complexity of HIV-1 RNA processing and its regulation suggests that small perturbations in the activity or abundance of key regulatory factors could significantly impair virus replication. Most experiments have used overexpression or sh/siRNA depletion to investigate host factor function but therapeutic strategies to alter HIV-1 RNA processing will require other means of modulating their activity. Research over the last decade has identified a number of small molecules that alter function of key splicing factors through direct effects on their activity or their regulatory kinases. Alternatively, other compounds perturb HIV-1 Rev function. Here, we provide an overview of the small molecules that have been investigated for their effects on HIV-1 RNA processing (summarized in Table 6.3).

To identify small molecules that function by directly effecting SR protein function, in vitro assays were used to assess compound effects on exon splicing enhancer function. Soret et al. (2005) identified multiple indole derivatives

TABLE 6.3 Structure and Function of Small Molecule Modulators of HIV-1 RNA Processing

Compound Name	Compound Structure	Proposed Mechanism	Activity Against HIV	References
IDC16	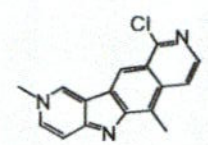	Topo 1, SRSF1	IC_{50} = 0.1 μM (PBMCs)	Bakkour et al. (2007)
ABX464	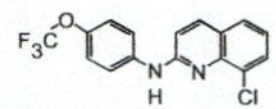	HIV-1 Rev[a]	IC_{50} = 0.1–1 μM (PBMCs)	Campos et al. (2015)
1C8	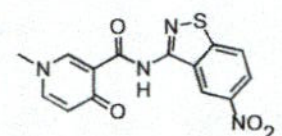	SRSF10, Tra2β	EC_{50} = 0.6–1.5 μM (PBMCs)	Shkreta et al. (2016)
Digoxin	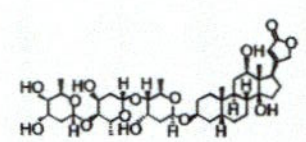	SRSF3	IC_{50} = 1 nM (PBMCs), 40 nM (HeLa)	Wong et al. (2013b)
SRPIN340	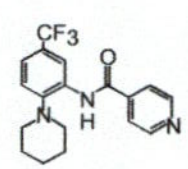	SRPK1, SRPK2	Variable effects	Fukuhara et al. (2006)
TG003	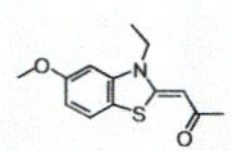	CLK1	Not active at 10 μM	Wong et al. (2011) and Muraki et al (2004)
Chlorhexidine	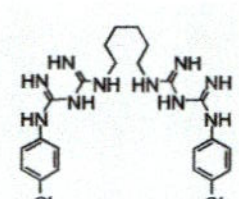	CLK2, CLK3, CLK4	IC_{50} = 1–2.5 μM (PBMCs), 2.5 μM (HeLa)	Wong et al. (2011)
8-azaguanine	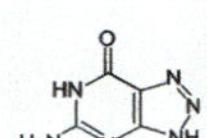	HIV-1 Rev function	IC_{50} = 0.5–1 μM (PBMCs), 10 μM (T-cell line), 8 μM (HeLa)	Wong et al. (2013a)

Continued

TABLE 6.3 Structure and Function of Small Molecule Modulators of HIV-1 RNA Processing—cont'd

Compound Name	Compound Structure	Proposed Mechanism	Activity Against HIV	References
5350150	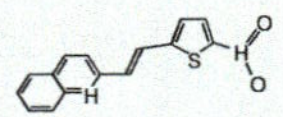	HIV-1 Rev function	IC_{50} = 0.1 μM (PBMCs), 2 μM (T-cell line), 0.5 μM (HeLa)	Wong et al. (2013a)
Clomiphene	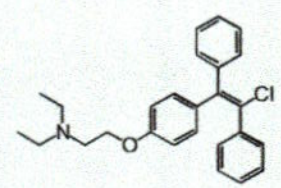	HIV-1 Rev-RRE, LTR-dependent transcription	EC_{50} = 4.3 μM (293T cells)	Prado et al. (2016)
9147791	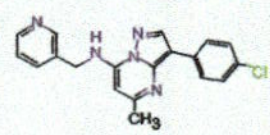	HIV-1 Rev/Tat abundance[a]	IC_{50} = 1–5 μM (PBMCs), 26 μM (T-cell line), 8.2 μM (HeLa)	Balachandran et al. (2017)
5227833	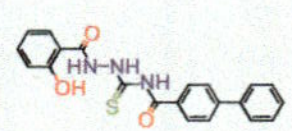	HIV-1 Rev/Tat abundance[a]	IC_{50} = 1–6 μM (PBMCs), 3.5 μM (T-cell line), 0.6 μM (HeLa)	Balachandran et al. (2017)
5193892	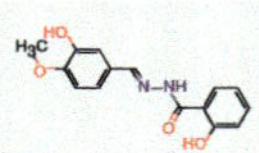	HIV-1 Rev/Tat abundance[a]	IC_{50} = 2 μM (PBMCs), 1.8 μM (HeLa)	Balachandran et al. (2017)

LTR, long terminal repeat; *RRE*, Rev response element.
[a]*Mechanism of action unknown.*

(pyridocarbazoles, benzopyridoindoles, or pyrido-pyrrolo-isoquinolines) that inhibited splicing by blocking assembly of early spliceosome complexes on the RNA substrate, putatively by interacting with SRSF1 and interfering with its function. Compounds within this family were found to have selectivity for specific SR proteins (Soret et al., 2005). Subsequent testing of indole derivatives (i.e., IDC16) identified several that altered HIV-1 RNA processing and suppressed virus replication (Soret et al., 2005; Bakkour et al., 2007). Experiments using IDC16 suggested that it suppressed HIV-1 replication through inhibition of MS RNA production, with limited effect on US RNA accumulation, which was correlated with an ability to reduce SRSF1 function (Bakkour et al., 2007). The capacity of indole derivatives to intercalate into DNA prompted the examination of structural variants that lacked this activity. Two such compounds, ABX464 and 1C8, were recently described (Campos et al., 2015; Shkreta et al., 2016) as potent inhibitors of HIV-1 replication. However, in contrast to IDC16, addition of ABX464 mildly enhanced (~1.5 fold) viral MS RNA accumulation. ABX464 impaired Rev function as evidenced by the retention of viral RNA at the site of transcription (Campos et al., 2015). One striking feature of ABX464 treatment was the failure to observe HIV-1 rebound within humanized mice upon its withdrawal, unlike the response seen with drugs used in existing HAART treatment (Campos et al., 2015). 1C8 was also shown to be a potent inhibitor of HIV-1 replication, but in this instance the compound induced a marked reduction in accumulation of all viral RNAs, an effect that is linked with a selective reduction in SRSF10 function (but not SRSF1 or SRSF9) due in part to SRSF10 dephosphorylation (Shkreta et al., 2016; Cheung et al., 2016).

Another strategy to modulate SR protein function is through targeting the kinases responsible for the phosphorylation of the RS domain. Multiple kinases (SRPK1 and 2, CLK1-4, DYRK1a, and topoisomerase 1) phosphorylate the RS domain of SR proteins, a modification that significantly impacts the capacity of these factors to promote RNA splicing (Zhou and Fu, 2013; Stamm, 2008). Studies to explore the role of SR protein kinases in the regulation of HIV-1 RNA processing and virus replication have shown mixed results. While overexpression of SRPK2 significantly enhanced HIV-1 production in transient transfection assays, the SRPK inhibitor SRPIN340 did not inhibit virus replication (Fukuhara et al., 2006). Studies examining the effect of the CLK group of kinases in the context of HIV-1 have revealed divergent effects; overexpression of CLK1 enhanced HIV-1 Gag expression, CLK2 inhibited Gag synthesis, and CLK3/4 had limited effects (Wong et al., 2011). Subsequent testing of the CLK1 and 4 inhibitor TG003 (Muraki et al., 2004) and CLK2, 3, and 4 inhibitor chlorhexidine (Younis et al., 2010) revealed that only chlorhexidine suppressed HIV-1 gene expression and replication by altering viral RNA accumulation (Wong et al., 2011). The role of CLKs in regulating HIV-1 RNA was further confirmed upon examination of the mechanism underlying the repression of HIV-1 replication by digoxin (Wong et al., 2013b). Digoxin treatment significantly reduced expression of Gag, Env, and Rev, a response that correlated with a marked reduction in US and SS RNA accumulation and a significant increase in MS RNA levels, as well as nuclear sequestration of US RNA (Wong et al., 2013b). At concentrations sufficient to suppress HIV-1 gene expression, digoxin-inhibited CLK1-4 function and altered phosphorylation of select SR proteins, promoting modification of SRSF3 and Tra2β while having no detectable effect on SRSF1 or SRSF7 (Wong et al., 2013b).

A third approach to modulating HIV-1 RNA processing is through altering the accumulation and/or activity of the essential HIV-1 regulatory protein Rev (Karn and Stoltzfus, 2012). Prado et al. (2016) screened a library of 1120 small molecule drugs and identified several bioavailable molecules that interfere with the binding of Rev to the RRE. One of the hits, clomiphene, not only inhibited Rev–RRE complex formation in vitro and blocked viral transcription in cellular assays but also altered the splicing pattern of HIV-1 transcripts in a manner consistent with inhibition of Rev function (Prado et al., 2016). Other screens for compounds impacting HIV-1 RNA processing have identified several structurally unrelated compounds that function by changing Rev function or its accumulation (Balachandran et al., 2017; Wong et al., 2013a). Both 8-azaguanine and 5350150 (2-(2-(5-nitro-2-thienyl)vinyl)quinolone) induced a decrease in US viral RNA accumulation as well as loss of Gag and Env synthesis without altering Rev expression (Wong et al., 2013a). Both compounds induced nuclear retention of HIV-1 US RNA consistent with a loss of Rev function, attributable to the cytoplasmic accumulation of Rev observed upon addition of either compound. Another group of compounds (designated 9147791, 5227833 and 5183892) was recently described that also altered HIV-1 RNA accumulation (reducing US and SS RNA accumulation with limited effect on MS RNA abundance) but, in this instance, not only was Gag and Env synthesis reduced but there was also a marked decrease in expression of Tat and Rev (Balachandran et al., 2017). The ability of the proteasome inhibitor MG132 to restore Tat expression in the presence of the compounds suggests that the effect seen is due to increased turnover of Tat and Rev. Additional tests (Balachandran et al., 2017) suggest that 9147791 functions by reducing levels of Nap1, a Tat/Rev chaperone previously shown to prevent Rev aggregation and modulate Tat function (Cochrane et al., 2009; Vardabasso et al., 2008). Hence, by reducing Nap1 levels, 9147791 may increase the availability of Rev or Tat for degradation by the proteasome.

The structural diversity of the small molecules identified to date that alter HIV-1 RNA processing and inhibit virus replication, coupled with their diverse mechanisms of action, offer the promise that effective therapies targeting this stage of the virus life cycle can be developed. Several of the compounds described are already in human use (chlorhexidine, digoxin, and clomiphene) and have antiviral activity at concentrations at or below those currently used. Of particular note, many of the compounds have been shown to induce very limited alterations in host RNA processing (Campos et al., 2015; Shkreta et al., 2016;

Balachandran et al., 2017; Wong et al., 2013a; Anderson et al., 2012) indicating that HIV-1 is highly sensitive to small perturbations in the activity/abundance of host RNA processing factors. In light of the complexity and the need to carefully balance the extent of HIV-1 RNA processing, such sensitivity is not unexpected. More detailed understanding of how these compounds elicit their effect on HIV-1 RNA processing will likely provide further insight into the host factors involved and help to refine the strategy to enhance antiviral activity while minimizing side effects for the host. The difference in mechanism of action from existing therapeutics to control HIV-1 infection indicates that these compounds could be used to enhance existing treatments or serve as salvage therapy if/when resistance does emerge. Of note, additional studies (Grosso et al., 2017, Cochrane, unpublished observations) have determined that replication of other, unrelated viruses is strongly inhibited by several of the compounds that alter HIV-1 RNA processing, suggesting that the same or similar strategy could be used to control other viral pathogens, offering the potential for generation of a broad spectrum or new class of antiviral therapeutics.

ACKNOWLEDGMENTS

Preparation of this manuscript was supported by grants to A.C. from the Canadian Institutes of Health Research (CIHR); HOP-134065 and CanCURE.

REFERENCES

Amendt, B.A., Hesslein, D., Chang, L.-J., Stoltzfus, C.M., 1994. Presence of negative and positive cis-acting RNA splicing elements within and flanking the first tat coding exon of human immunodeficiency virus type 1. Mol. Cell. Biol. 14, 3960–3970.

Amendt, B., Si, Z., Stoltzfus, C.M., 1995. Presence of exon splicing silencers within human immunodeficiency virus type 1 tat exon 2 and tat-rev exon 3: evidence for inhibition mediated by cellular factors. Mol. Cell. Biol. 15, 4606–4615.

Anderson, E., Lin, C.H., Xiao, X., Stoilov, P., Burge, C.B., Black, D.L., 2012. The cardiotonic steroid digitoxin regulates alternative splicing through depletion of the splicing factors SRSF3 and TRA2B. RNA. 18, 1041–1049.

Asai, K., Platt, C., Cochrane, A., 2003. Control of HIV-1 env RNA splicing and transport: investigating the role of hnRNP A1 in exon splicing silencer (ESS3a) function. Virology 314, 229–242.

Asang, C., Hauber, I., Schaal, H., 2008. Insights into the selective activation of alternatively used splice acceptors by the human immunodeficiency virus type-1 bidirectional splicing enhancer. Nucleic Acids Res. 36, 1450–1463.

Asang, C., Erkelenz, S., Schaal, H., 2012. The HIV-1 major splice donor D1 is activated by splicing enhancer elements within the leader region and the p17-inhibitory sequence. Virology 432, 133–145.

Bakkour, N., Lin, Y.L., Maire, S., Ayadi, L., Mahuteau-Betzer, F., Nguyen, C.H., Mettling, C., Portales, P., Grierson, D., Chabot, B., Jeanteur, P., Branlant, C., Corbeau, P., Tazi, J., 2007. Small-molecule inhibition of HIV pre-mRNA splicing as a novel antiretroviral therapy to overcome drug resistance. PLoS Pathog. 3, 1530–1539.

Balachandran, A., Wong, R., Stoilov, P., Pan, S., Blencowe, B., Cheung, P., Harrigan, P.R., Cochrane, A., 2017. Identification of small molecule modulators of HIV-1 Tat and Rev protein accumulation. Retrovirology 14, 7.

Barreau, C., Paillard, L., Osborne, H.B., 2005. AU-rich elements and associated factors: are there unifying principles? Nucleic Acids Res. 33, 7138–7150.

Beriault, V., Clement, J.F., Levesque, K., Lebel, C., Yong, X., Chabot, B., Cohen, E.A., Cochrane, A.W., Rigby, W.F., Mouland, A.J., 2004. A late role for the association of hnRNP A2 with the HIV-1 hnRNP A2 response elements in genomic RNA, Gag, and Vpr localization. J. Biol. Chem. 279, 44141–44153.

Bilodeau, P.S., Domsic, J.K., Mayeda, A., Krainer, A.R., Stoltzfus, C.M., 2001. RNA splicing at human immunodeficiency virus type 1 3′ splice site A2 is regulated by binding of hnRNP A/B proteins to an exonic splicing silencer element. J. Virol. 75, 8487–8497.

Blanchette, M., Chabot, B., 1999. Modulation of exon skipping by high affinity hnRNP A1-binding sites and by intron elements that repress splice site utilization. EMBO J. 18, 1939–1952.

Brewer, G., 1991. An A + U-rich element RNA-binding factor regulates c-myc mRNA stability in vitro. Mol. Cell. Biol. 11, 2460–2466.

Campos, N., Myburgh, R., Garcel, A., Vautrin, A., Lapasset, L., Nadal, E.S., Mahuteau-Betzer, F., Najman, R., Fornarelli, P., Tantale, K., Basyuk, E., Seveno, M., Venables, J.P., Pau, B., Bertrand, E., Wainberg, M.A., Speck, R.F., Scherrer, D., Tazi, J., 2015. Long lasting control of viral rebound with a new drug ABX464 targeting Rev – mediated viral RNA biogenesis. Retrovirology 12, 30.

Caputi, M., Mayeda, A., Krainer, A.R., Zahler, A.M., 1999. hnRNP A/B proteins are required for inhibition of HIV-1 pre-mRNA splicing. EMBO J. 18, 4060–4067.

Caputi, M., Freund, M., Kammler, S., Asang, C., Schaal, H., 2004. A bidirectional SF2/ASF- and SRp40-dependent splicing enhancer regulates human immunodeficiency virus type 1 rev, env, vpu, and nef gene expression. J. Virol. 78, 6517–6526.

Cheung, P.K., Horhant, D., Bandy, L.E., Zamiri, M., Rabea, S.M., Karagiosov, S.K., Matloobi, M., McArthur, S., Harrigan, P.R., Chabot, B., Grierson, D.S., 2016. A parallel synthesis approach to the identification of novel diheteroarylamide-based compounds blocking HIV replication: potential inhibitors of HIV-1 pre-mRNA alternative splicing. J. Med. Chem. 59, 1869–1879.

Cochrane, A., Murley, L.L., Gao, M., Wong, R., Clayton, K., Brufatto, N., Canadien, V., Mamelak, D., Chen, T., Richards, D., Zeghouf, M., Greenblatt, J., Burks, C., Frappier, L., 2009. Stable complex formation between HIV Rev and the nucleosome assembly protein, NAP1, affects Rev function. Virology 388, 103–111.

Cullen, B.R., 2003. Nuclear mRNA export: insights from virology. Trends Biochem. Sci. 28, 419–424.

Damgaard, C.K., Tange, T.O., Kjems, J., 2002. hnRNP A1 controls HIV-1 mRNA splicing through cooperative binding to intron and exon splicing silencers in the context of a conserved secondary structure. RNA 8, 1401–1415.

Damier, L., Domenjoud, L., Branlant, C., 1997. The D1-A2 and D2-A2 pairs of splice sites from human immunodeficiency virus type 1 are highly efficient in vitro, in spite of an unusual branch site. Biochem. Biophys. Res. Commun. 237, 182–187.

Domsic, J.K., Wang, Y., Mayeda, A., Krainer, A.R., Stoltzfus, C.M., 2003. Human immunodeficiency virus type 1 hnRNP A/B-dependent exonic splicing silencer ESSV antagonizes binding of U2AF65 to viral polypyrimidine tracts. Mol. Cell. Biol. 23, 8762–8772.

Dowling, D., Nasr-Esfahani, S., Tan, C.H., O'Brien, K., Howard, J.L., Jans, D.A., Purcell, D.F., Stoltzfus, C.M., Sonza, S., 2008. HIV-1 infection induces changes in expression of cellular splicing factors that regulate alternative viral splicing and virus production in macrophages. Retrovirology 5, 18.

Dyhr-Mikkelsen, H., Kjems, J., 1995. Inefficient spliceosome assembly and abnormal branch site selection in splicing of an HIV-1 transcript in vitro. J. Biol. Chem. 270, 24060–24066.

Emery, A., Zhou, S., Pollom, E., Swanstrom, R., 2017. Characterizing HIV-1 splicing by using next-generation sequencing. J. Virol. 91, e02515–02516.

Erkelenz, S., Poschmann, G., Theiss, S., Stefanski, A., Hillebrand, F., Otte, M., Stuhler, K., Schaal, H., 2013. Tra2-mediated recognition of HIV-1 5′ splice site D3 as a key factor in the processing of vpr mRNA. J. Virol. 87, 2721–2734.

Erkelenz, S., Theiss, S., Otte, M., Widera, M., Peter, J.O., Schaal, H., 2014. Genomic HEXploring allows landscaping of novel potential splicing regulatory elements. Nucleic Acids Res. 42, 10681–10697.

Erkelenz, S., Hillebrand, F., Widera, M., Theiss, S., Fayyaz, A., Degrandi, D., Pfeffer, K., Schaal, H., 2015. Balanced splicing at the Tat-specific HIV-1 3′ss A3 is critical for HIV-1 replication. Retrovirology 12, 29.

Exline, C.M., Feng, Z., Stoltzfus, C.M., 2008. Negative and positive mRNA splicing elements act competitively to regulate human immunodeficiency virus type 1 vif gene expression. J. Virol. 82, 3921–3931.

Fukuhara, T., Hosoya, T., Shimizu, S., Sumi, K., Oshiro, T., Yoshinaka, Y., Suzuki, M., Yamamoto, N., Herzenberg, L.A., Herzenberg, L.A., Hagiwara, M., 2006. Utilization of host SR protein kinases and RNA-splicing machinery during viral replication. Proc. Natl. Acad. Sci. U.S.A. 103, 11329–11333.

Gordon, H., Ajamian, L., Valiente-Echeverria, F., Levesque, K., Rigby, W.F., Mouland, A.J., 2013. Depletion of hnRNP A2/B1 overrides the nuclear retention of the HIV-1 genomic RNA. RNA Biol. 10, 1714–1725.

Grosso, F., Stoilov, P., Lingwood, C., Brown, M., Cochrane, A., 2017. Suppression of adenovirus replication by cardiotonic steroids. J. Virol. 91.

Hadian, K., Vincendeau, M., Mausbacher, N., Nagel, D., Hauck, S.M., Ueffing, M., Loyter, A., Werner, T., Wolff, H., Brack-Werner, R., 2009. Identification of a heterogeneous nuclear ribonucleoprotein-recognition region in the HIV Rev protein. J. Biol. Chem. 284, 33384–33391.

Hallay, H., Locker, N., Ayadi, L., Ropers, D., Guittet, E., Branlant, C., 2006. Biochemical and NMR study on the competition between proteins SC35, SRp40, and heterogeneous nuclear ribonucleoprotein A1 at the HIV-1 Tat exon 2 splicing site. J. Biol. Chem. 281, 37159–37174.

Hillebrand, F., Peter, J., Brillen, A., Otte, M., Schaal, H., Erkelenz, S., 2017. Differential hnRNP D isoform incorporation may confer plasticity to the ESSV-mediated repressive state across HIV-1 exon 3. Biochim. Biophys. Acta 1860, 205–217.

Hope, T.J., 1999. The ins and outs of HIV Rev. Arch. Biochem. Biophy. 365, 186–191.

Huang, Y., Gattoni, R., Stevenin, J., Steitz, J.A., 2003. SR splicing factors serve as adapter proteins for TAP-dependent mRNA export. Mol. Cell 11, 837–843.

Jablonski, J.A., Caputi, M., 2009. Role of cellular RNA processing factors in human immunodeficiency virus type 1 mRNA metabolism, replication, and infectivity. J. Virol. 83, 981–992.

Jablonski, J.A., Amelio, A.L., Giacca, M., Caputi, M., 2010. The transcriptional transactivator Tat selectively regulates viral splicing. Nucleic Acids Res. 38, 1249–1260.

Jacquenet, S., Mereau, A., Bilodeau, P.S., Damier, L., Stoltzfus, C., Branlant, C., 2001. A second exon splicing silencer within the human immunodeficiency virus type 1 tat exon 2 represses splicing of tat mRNA and binds protein hnRNP H. J. Biol. Chem. 276, 40464–40475.

Jacquenet, S., Decimo, D., Muriaux, D., Darlix, J.L., 2005. Dual effect of the SR proteins ASF/SF2, SC35 and 9G8 on HIV-1 RNA splicing and virion production. Retrovirology 2, 33.

Jean-Philippe, J., Paz, S., Lu, M.L., Caputi, M., 2014. A truncated hnRNP A1 isoform, lacking the RGG-box RNA binding domain, can efficiently regulate HIV-1 splicing and replication. Biochim. Biophys. Acta 1839, 251–258.

Kammler, S., Leurs, C., Freund, M., Krummheuer, J., Seidel, K., Tange, T.O., Lund, M.K., Kjems, J., Scheid, A., Schaal, H., 2001. The sequence complementarity between HIV-1 5′ splice site SD4 and U1 snRNA determines the steady-state level of an unstable env pre-mRNA. RNA 7, 421–434.

Kammler, S., Otte, M., Hauber, I., Kjems, J., Hauber, J., Schaal, H., 2006. The strength of the HIV-1 3′ splice sites affects Rev function. Retrovirology 3, 89.

Karn, J., Stoltzfus, C.M., 2012. Transcriptional and posttranscriptional regulation of HIV-1 gene expression. Cold Spring Harb. Perspect. Med. 2, a006916.

Levesque, K., Halvorsen, M., Abrahamyan, L., Chatel-Chaix, L., Poupon, V., Gordon, H., DesGroseillers, L., Gatignol, A., Mouland, A.J., 2006. Trafficking of HIV-1 RNA is mediated by heterogeneous nuclear ribonucleoprotein A2 expression and impacts on viral assembly. Traffic 7, 1177–1193.

Long, J.C., Caceres, J.F., 2009. The SR protein family of splicing factors: master regulators of gene expression. Biochem. J. 417, 15–27.

Lund, N., Milev, M.P., Wong, R., Sanmuganantham, T., Woolaway, K., Chabot, B., Abou Elela, S., Mouland, A.J., Cochrane, A., 2012. Differential effects of hnRNP D/AUF1 isoforms on HIV-1 gene expression. Nucleic Acids Res. 40, 3663–3675.

Madsen, J.M., Stoltzfus, C.M., 2005. An exonic splicing silencer downstream of the 3′ splice site A2 is required for efficient human immunodeficiency virus type 1 replication. J. Virol. 79, 10478–10486.

Maldarelli, F., Xiang, C., Chamoun, G., Zeichner, S.L., 1998. The expression of the essential nuclear splicing factor SC35 is altered by human immunodeficiency virus infection. Virus Res. 53, 39–51.

Marchand, V., Mereau, A., Jacquenet, S., Thomas, D., Mougin, A., Gattoni, R., Stevenin, J., Branlant, C., 2002. A Janus splicing regulatory element modulates HIV-1 tat and rev mRNA production by coordination of hnRNP A1 cooperative binding. J. Mol. Biol. 323, 629–652.

Marsh, K., Soros, V., Cochrane, A., 2008. Selective translational repression of HIV-1 RNA by Sam68DeltaC occurs by altering PABP1 binding to unspliced viral RNA. Retrovirology 5, 97.

McLaren, M., Marsh, K., Cochrane, A., 2008. Modulating HIV-1 RNA processing and utilization. Front. Biosci. 13, 5693–5707.

Mueller, N., van Bel, N., Berkhout, B., Das, A.T., 2014. HIV-1 splicing at the major splice donor site is restricted by RNA structure. Virology 468–470, 609–620.

Mueller, N., Berkhout, B., Das, A.T., 2015a. HIV-1 splicing is controlled by local RNA structure and binding of splicing regulatory proteins at the major 5′ splice site. J. Gen. Virol. 96, 1906–1917.

Mueller, N., Klaver, B., Berkhout, B., Das, A.T., 2015b. Human immunodeficiency virus type 1 splicing at the major splice donor site is controlled by highly conserved RNA sequence and structural elements. J. Gen. Virol. 96, 3389–3395.

Mueller, N., Das, A.T., Berkhout, B., 2016. A phylogenetic survey on the structure of the HIV-1 leader RNA domain that encodes the splice donor signal. Viruses 8.

Muraki, M., Ohkawara, B., Hosoya, T., Onogi, H., Koizumi, J., Koizumi, T., Sumi, K., Yomoda, J., Murray, M.V., Kimura, H., Furuichi, K., Shibuya, H., Krainer, A.R., Suzuki, M., Hagiwara, M., 2004. Manipulation of alternative splicing by a newly developed inhibitor of Clks. J. Biol. Chem. 279, 24246–24254.

Nasim, F.U., Hutchison, S., Cordeau, M., Chabot, B., 2002. High-affinity hnRNP A1 binding sites and duplex-forming inverted repeats have similar effects on 5′ splice site selection in support of a common looping out and repression mechanism. RNA 8, 1078–1089.

Ocwieja, K.E., Sherrill-Mix, S., Mukherjee, R., Custers-Allen, R., David, P., Brown, M., Wang, S., Link, D.R., Olson, J., Travers, K., Schadt, E., Bushman, F.D., 2012. Dynamic regulation of HIV-1 mRNA populations analyzed by single-molecule enrichment and long-read sequencing. Nucleic Acids Res. 40, 10345–10355.

Ohta, H., Takamune, N., Kishimoto, N., Shoji, S., Misumi, S., 2015. *N*-myristoyltransferase 1 enhances human immunodeficiency virus replication through regulation of viral RNA expression level. Biochem. Biophys. Res. Commun. 463, 988–993.

Okunola, H.L., Krainer, A.R., 2009. Cooperative-binding and splicing-repressive properties of hnRNP A1. Mol. Cell. Biol. 29, 5620–5631.

O'Reilly, M., McNally, M., Beemon, K., 1995. Two strong 5′ splice sites and competing 3′ splice sites involved in alternative splicing of human immunodeficiency virus type 1 RNA. Virology 213, 373–385.

Paz, S., Krainer, A.R., Caputi, M., 2014. HIV-1 transcription is regulated by splicing factor SRSF1. Nucleic Acids Res. 42, 13812–13823.

Paz, S., Lu, M.L., Takata, H., Trautmann, L., Caputi, M., 2015. SRSF1 RNA recognition motifs are strong inhibitors of HIV-1 replication. J. Virol. 89, 6275–6286.

Platt, C., Calimano, M., Nemet, J., Bubenik, J., Cochrane, A., 2015. Differential effects of Tra2ss isoforms on HIV-1 RNA processing and expression. PLoS One 10, e0125315.

Pollard, V., Malim, M., 1998. The HIV-1 Rev protein. Annu. Rev. Microbiol. 52, 491–532.

Powell, D., Amaral, M., Wu, J., Maniatis, T., Greene, W., 1997. HIV Rev-dependent binding of SF2/ASF to the Rev response element: possible role in Rev-mediated inhibition of HIV RNA splicing. Proc. Natl. Acad. Sci. U.S.A. 94, 973–978.

Prado, S., Beltran, M., Coiras, M., Bedoya, L.M., Alcami, J., Gallego, J., 2016. Bioavailable inhibitors of HIV-1 RNA biogenesis identified through a Rev-based screen. Biochem. Pharmacol. 107, 14–28.

Purcell, D., Martin, M.A., 1993. Alternative splicing of human immunodeficiency virus type 1 mRNA modulates viral protein expression, replication, and infectivity. J. Virol. 67, 6365–6378.

Ropers, D., Ayadi, L., Gattoni, R., Jacquenet, S., Damier, L., Branlant, C., Stevenin, J., 2004. Differential effects of the SR proteins 9G8, SC35, ASF/SF2 and SRp40 on the utilization of the A1 to A5 splicing sites of HIV-1 RNA. J. Biol. Chem. 279, 29963–29973.

Saliou, J.M., Bourgeois, C.F., Ayadi-Ben Mena, L., Jacquenet, S., Marchand, V., Stevenin, J., Branlant, C., 2009. Role of RNA structure and protein factors in the control of HIV-1 splicing. Front. Biosci. 14, 2714–2729.

Sanford, J.R., Gray, N.K., Beckmann, K., Caceres, J.F., 2004. A novel role for shuttling SR proteins in mRNA translation. Genes Dev. 18, 755–768.

Sanford, J.R., Ellis, J., Caceres, J.F., 2005. Multiple roles of arginine/serine-rich splicing factors in RNA processing. Biochem. Soc. Trans. 33, 443–446.

Schaub, M.C., Lopez, S.R., Caputi, M., 2007. Members of the heterogeneous nuclear ribonucleoprotein H family activate splicing of an HIV-1 splicing substrate by promoting formation of ATP-dependent spliceosomal complexes. J. Biol. Chem. 282, 13617–13626.

Sharma, A., Yilmaz, A., Marsh, K., Cochrane, A., Boris-Lawrie, K., 2012. Thriving under stress: selective translation of HIV-1 structural protein mRNA during vpr-mediated impairment of eIF4E translation activity. PLoS Pathog. 8, e1002612.

Shkreta, L., Blanchette, M., Toutant, J., Wilhelm, E., Bell, B., Story, B.A., Balachandran, A., Cochrane, A., Cheung, P.K., Harrigan, P.R., Grierson, D.S., Chabot, B., 2016. Modulation of the splicing regulatory function of SRSF10 by a novel compound that impairs HIV-1 replication. Nucleic Acids Res. https://doi.org/10.1093/nar/gkw1223.

Si, Z., Amendt, B.A., Stoltzfus, C.M., 1997. Splicing efficiency of human immunodeficiency virus type 1 tat RNA is determined by both a suboptimal 3′ splice site and a 10 nucleotide exon splicing silencer element located within tat exon 2. Nucleic Acids Res. 25, 861–867.

Si, Z.-H., Rauch, D., Stoltzfus, M., 1998. The exon splicing silencer in the human immunodeficiency virus type 1 tat exon 3 is bipartite and acts early in spliceosome assembly. Mol. Cell. Biol. 18, 5404–5413.

Soret, J., Bakkour, N., Maire, S., Durand, S., Zekri, L., Gabut, M., Fic, W., Divita, G., Rivalle, C., Dauzonne, D., Nguyen, C.H., Jeanteur, P., Tazi, J., 2005. Selective modification of alternative splicing by indole derivatives that target serine-arginine-rich protein splicing factors. Proc. Natl. Acad. Sci. U.S.A. 102, 8764–8769.

Staffa, A., Cochrane, A., 1995. Identification of positive and negative splicing regulatory elements within the terminal tat-rev exon of human immunodeficiency virus type 1. Mol. Cell. Biol. 15, 4597–4605.

Stamm, S., 2008. Regulation of alternative splicing by reversible protein phosphorylation. J. Biol. Chem. 283, 1223–1227.

Stoltzfus, C.M., Madsen, J.M., 2006. Role of viral splicing elements and cellular RNA binding proteins in regulation of HIV-1 alternative RNA splicing. Curr. HIV Res. 4, 43–55.

Stoltzfus, C.M., 2009. Chapter 1. Regulation of HIV-1 alternative RNA splicing and its role in virus replication. Adv. Virus Res. 74, 1–40.

Swanson, C.M., Puffer, B.A., Ahmad, K.M., Doms, R.W., Malim, M.H., 2004. Retroviral mRNA nuclear export elements regulate protein function and virion assembly. EMBO J. 23, 2632–2640.

Swanson, C.M., Sherer, N.M., Malim, M.H., 2010. SRp40 and SRp55 promote the translation of unspliced human immunodeficiency virus type 1 RNA. J. Virol. 84, 6748–6759.

Tange, T.O., Kjems, J., 2001. SF2/ASF binds to a splicing enhancer in the third HIV-1 tat exon and stimulates U2AF binding independently of the RS domain. J. Mol. Biol. 312, 649–662.

Tange, T.O., Damgaard, C.K., Guth, S., Valcarcel, J., Kjems, J., 2001. The hnRNP A1 protein regulates HIV-1 tat splicing via a novel intron silencer element. EMBO J. 20, 5748–5758.

Tranell, A., Fenyo, E.M., Schwartz, S., 2010. Serine- and arginine-rich proteins 55 and 75 (SRp55 and SRp75) induce production of HIV-1 vpr mRNA by inhibiting the 5′-splice site of exon 3. J. Biol. Chem. 285, 31537–31547.

Vardabasso, C., Manganaro, L., Lusic, M., Marcello, A., Giacca, M., 2008. The histone chaperone protein Nucleosome Assembly Protein-1 (hNAP-1) binds HIV-1 Tat and promotes viral transcription. Retrovirology 5, 8.

Vincendeau, M., Nagel, D., Brenke, J.K., Brack-Werner, R., Hadian, K., 2013. Heterogenous nuclear ribonucleoprotein Q increases protein expression from HIV-1 Rev-dependent transcripts. Virol. J. 10, 151.

Widera, M., Erkelenz, S., Hillebrand, F., Krikoni, A., Widera, D., Kaisers, W., Deenen, R., Gombert, M., Dellen, R., Pfeiffer, T., Kaltschmidt, B., Munk, C., Bosch, V., Kohrer, K., Schaal, H., 2013. An intronic G run within HIV-1 intron 2 is critical for splicing regulation of vif mRNA. J. Virol. 87, 2707–2720.

Widera, M., Hillebrand, F., Erkelenz, S., Vasudevan, A.A., Munk, C., Schaal, H., 2014. A functional conserved intronic G run in HIV-1 intron 3 is critical to counteract APOBEC3G-mediated host restriction. Retrovirology 11, 72.

Wong, R., Balachandran, A., Mao, A.Y., Dobson, W., Gray-Owen, S., Cochrane, A., 2011. Differential effect of CLK SR Kinases on HIV-1 gene expression: potential novel targets for therapy. Retrovirology 8, 47.

Wong, R.W., Balachandran, A., Haaland, M., Stoilov, P., Cochrane, A., 2013a. Characterization of novel inhibitors of HIV-1 replication that function via alteration of viral RNA processing and rev function. Nucleic Acids Res. 41, 9471–9483.

Wong, R.W., Balachandran, A., Ostrowski, M.A., Cochrane, A., 2013b. Digoxin suppresses HIV-1 replication by altering viral RNA processing. PLoS Pathog. 9, e1003241.

Woolaway, K., Asai, K., Emili, A., Cochrane, A., 2007. hnRNP E1 and E2 have distinct roles in modulating HIV-1 gene expression. Retrovirology 4, 28.

Younis, I., Berg, M., Kaida, D., Dittmar, K., Wang, C., Dreyfuss, G., 2010. Rapid-response splicing reporter screens identify differential regulators of constitutive and alternative splicing. Mol. Cell. Biol. 30, 1718–1728.

Zahler, A.M., Damgaard, C.K., Kjems, J., Caputi, M., 2004. SC35 and heterogeneous nuclear ribonucleoprotein A/B proteins bind to a juxtaposed exonic splicing enhancer/exonic splicing silencer element to regulate HIV-1 tat exon 2 splicing. J. Biol. Chem. 279, 10077–10084.

Zhou, Z., Fu, X.D., 2013. Regulation of splicing by SR proteins and SR protein-specific kinases. Chromosoma 122, 191–207.

Zhu, J., Krainer, A.R., 2000. Pre-mRNA splicing in the absence of an SR protein RS domain. Genes Dev. 14, 3166–3178.

Zhu, J., Mayeda, A., Krainer, A., 2001. Exon identity established through differential antagonism between exonic splicing silencer-bound hnRNP A1 and enhancer-bound SR proteins. Mol. Cell 8, 1351–1361.

Chapter 7

Cellular RNA Helicases Support Early and Late Events in Retroviral Replication

Gatikrushna Singh[1], Xiao Heng[2], Kathleen Boris-Lawrie[1]
[1]*University of Minnesota, Saint Paul, MN, United States;* [2]*University of Missouri, Columbia, MO, United States*

Chapter Outline

RNA Helicase and Retroviruses in the Advent of "Omics Technology" 258
Early Events: Reverse Transcription and Integration 259
DHX9/RNA Helicase A Activity in the Genomic RNP 259
MOV10 Activity in Virions Remains Undefined 262
Late Events: Provirus Transcription, Primary RNA Processing, Export, Translation, Formation of Genomic RNP 262
  DHX9, Bridging RNA Polymerase to Transcription Coactivators 262
  Shuttling RNA Helicase in Nucleocytoplasmic Transport of Retroviral RNA 264
Nuclear Cap–Binding Proteins and RNA Helicase: Translation Evading Nonsense RNA–Mediated Decay 265
Steady-State Translation: Switching 5′Cap–Binding Proteins to Gain eIF4E 266
Therapeutic Targeting at the Interface of RNA Helicase and Cognate Retroviral RNA 267
References 268

RNA helicases play necessary roles in the growth of eukaryotic cells and the innate responsiveness of multicellular organisms. They support all the steps in gene expression and sense infectious agents by virtue of their RNA binding domains and ATP-dependent catalytic activity. HIV-1 and other retroviruses command the supportive activity of cell-endogenous RHs, whereas influenza, poliovirus, and other RNA viruses invoke the antiviral activity of these enzymes. Instead of registering as an invading pathogen, the retrovirus is construed an endogenous gene whose expression requires the supportive activities of RNA helicases. By contrast, RNA

Retrovirus-Cell Interactions. https://doi.org/10.1016/B978-0-12-811185-7.00007-8

TABLE 7.1 Cytoplasmic RNA Viruses Encode Helicase Domains

Open Reading Frame	Virus Family/ Example Virus	Activity in Virus Replication
NS3	*Flaviviridae* Dengue virus, West nile virus, Yellow fever virus, Japanese encephalitis virus	Unwinding of duplex RNA structures and required for an efficient viral genomic RNA synthesis (Kwong et al., 2005; Ranji and Boris-Lawrie, 2010)
NS3	*Flaviviridae* Hepatitis C virus Hepatitis G virus	Essential for viral RNA replication, electrostatic binding to RNA and assist in packaging of the viral RNA to virions (Kadare and Haenni, 1997; Kwong et al., 2005; Ranji and Boris-Lawrie, 2010). Essential for cleavage of host and viral proteins (Kwong et al., 2005).
2C	*Picornaviridae* Hepatitis A virus	Essential for viral replication with membrane and RNA-binding properties (Kwong et al., 2005).
2C	*Picornaviridae* Poliovirus	Essential for ATPase activity and interact with other elements of the viral replication apparatus (Ranji and Boris-Lawrie, 2010).
ORF-1	*Hepeviridae* Hepatitis E virus	ORF1 encodes several nonstructural proteins required for HEV replication and protein processing (Kadare and Haenni, 1997).
Nsp2	*Togaviridae (Alphavirus)* Semliki forest virus	Essential for inhibition of cellular transcription by inducing rapid degradation of RPB1, a catalytic subunit of the RNAPII complex (Kadare and Haenni, 1997; Kwong et al., 2005).
Nsp13	*Nidovirales* SARS-coronavirus, Human coronavirus 229E	Pathogenicity determinant; essential for nascent RNA hydrolysis to yield 5′-diphosphate RNA (Kwong et al., 2005).
E1	*Papillomaviridae* Human papilloma virus	Interact and recruit host cell replication proteins to viral origin, including DNA polymerase alpha and replication protein A for viral replication (Kwong et al., 2005).
NPH-I NPH-II	*Poxviridae* Vaccinia virus	Phosphohydrolase and helicase activities for virus replication (Kwong et al., 2005; Ranji and Boris-Lawrie, 2010).
Nsp10	*Arteriviridae* Equine arteritis virus	Important for early step of mRNA transcription by stabilizing subgenomic RNA synthesis (Tijms et al., 2007).

viruses replicating in cytoplasmic factories maintain virus-encoded RNA helicases (Table 7.1), which may antagonize the antiviral activity of cellular RNA helicases (Kadare and Haenni, 1997; Kwong et al., 2005; Ranji and Boris-Lawrie, 2010).

Since the advent of "omics technology," the roles of RNA helicases in cell biology have been avidly characterized (Table 7.2). In the context of retroviruses, particularly HIV-1, candidate RNA helicase cofactors were identified by tandem-affinity chromatography-coupled proteomics using tagged retroviral proteins and genome-wide screens using RNA helicase-directed siRNA (Brass et al., 2008; Konig et al., 2008; Jäger et al., 2012). Overlap in the screens and validation of select RNA helicase activities has produced an initial comprehensive picture of the retrovirus–RNA helicase interface (Fig. 7.1).

Whereas RNA helicases support a diversity of activities in cells and the antiviral response to pathogens, a consistent theme is binding and rearranging of nucleotide–nucleotide pairings. At least 85 human proteins encode the core helicase domain, which is designated by highly conserved amino acid motifs, e.g., DEAD, DEIH, or DExH (x = any). Despite the extreme conservation of the core helicase domain, RNA helicases operate in all compartments of eukaryotic cells and exhibit diverse activities that are attributable to ancillary domains. RNA helicases shuttle between the nucleus and cytoplasm and participate in mRNA biogenesis, nucleus–cytoplasm transport, pioneer-round translation, cotranslational decay, and steady-state translation by contributing to the formation of dynamic ribonucleoprotein complexes (RNPs) (Fig. 7.1 and Table 7.2). RIGi and related RNA helicases demonstrate affinity for replication intermediates of RNA viruses that reproduce in the cytoplasm and activate innate response.

The RNA helicase superfamily (SF) is categorized into three subfamilies (SF 1, 2, and 3) according to combination of invariant amino acids in the core helicase motif: Asp-Glu (DE) with two other conserved amino acids; Asp-Glu-Ala-Asp, abbreviated as DEAD, DEAH, or DExH (x = any) residues in the core motif. Besides the namesake domain (DEAD box, DDX or DHX proteins), ancillary domains in RNA helicase SF members vary widely and engender interaction with specific substrate RNAs and protein cofactors (Fig. 7.2).

Functionally conserved across the plant and animal kingdoms, the helicase core promotes native RNA folding by rearranging of duplex nucleotide substrates. By an allosteric mechanism, nucleotide triphosphate (NTP) binding at the helicase core induces conformational change, followed by NTP hydrolysis (Fairman-Williams et al., 2010). Versatile as an RH is, rearranging activity is common without NTP-dependent helicase activity. Changing components of RNA–protein complexes, designated RNPase activity, is associated with RNA binding, rather than NTP hydrolysis, and is categorized a chaperone activity.

Structurally, the catalytic helicase core within each SF is almost identical, consisting of two similar recombination protein RecA domains with highly conserved residues that coordinate binding and hydrolysis of the NTP. The helicase domains share nine sequence motifs: Q is solely present in DEAD box helicase; grouping I, also called Walker A: grouping Ia, Ib; II, also called Walker B;

TABLE 7.2 Cellular RNA Helicase Activity in Gene Expression

RNA Helicase	Molecular Activity
DHX9/DNA Helicase II/ RNA Helicase A (RHA)	Tethers RNA polymerase II to chromatin-bound DNA-binding proteins (e.g., BRCA1, P300/CBP beta, p16ink) (Nakajima et al., 1997). Component of 7SK RNP that regulates availability of PCAF and P-TEF (Van Herreweghe et al., 2007). Binds and is necessary for translation of select RNAs: 5′UTR of junD proto-oncogene, (Hartman et al., 2006) type I collagen mRNAs (Manojlovic and Stefanovic, 2012). Tethers Lin28 to polysomes (Jin et al., 2011).
DDX1	Rearrangement of RNA secondary structure attributed to ATP hydrolysis. SPRY domain facilitates association with cellular RNPs for translational initiation, nuclear export, ribosome, and spliceosome assembly (Fang et al., 2005, 2004).
DDX3	Essential for CRM1-dependent nuclear export of viral RNA, remodel of exporting mRNPs (Yedavalli et al., 2004). Interacts with NF-kB to suppress p65-mediated transcription (Xiang et al., 2016). Associated with eIF4E-dependent mRNA translation (Soto-Rifo et al., 2013).
DDX5 and DDX17	Bind GC-rich sequences, modulate alternative splicing during cell differentiation (Dardenne et al., 2014). Coregulators of estrogen and androgen signaling pathway (Samaan et al., 2014).
UPF1/RENT1	Essential for 5′ to 3′ RNA unwinding activity and factor for nonsense-mediated mRNA decay (NMD) (Gregersen et al., 2014).
DHX30	Nucleocytoplasmic shuttling RH that is enriched in mitochondria; associates with mitoribosome (Lessel et al., 2018). Cofactor of ZAP in NMD (Ye et al., 2010).
DDX19A/DBP5	Associates with cytoplasmic face of nuclear pore complex, remodel of exporting mRNPs (Diot et al., 2016).
UAP56/DDX39B	Contributes ATPase and helicase activity essential for spliceosome assembly (Wisskirchen et al., 2011). Modulates the coupling of mRNA splicing with nucleocytoplasmic transport of mature mRNAs (Delaleau and Borden, 2015).

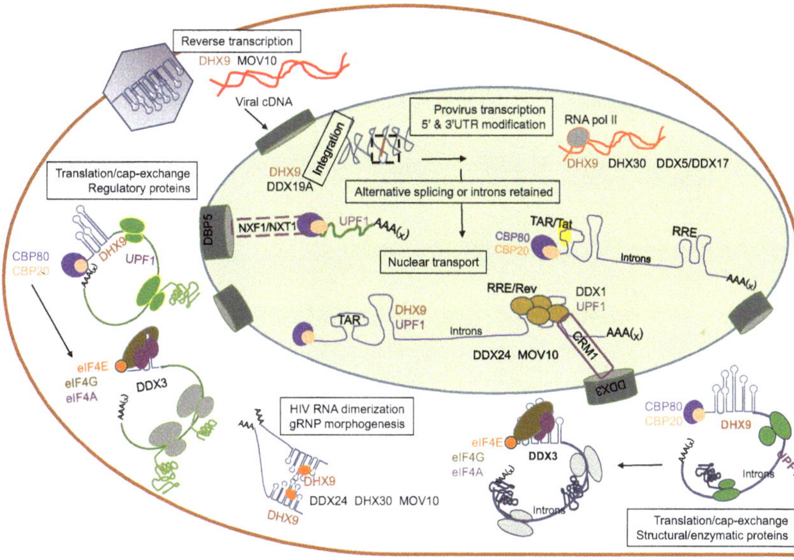

FIGURE 7.1 **Critical events in HIV-1 replication are supported by RNA helicases** Early steps in HIV-1 replication involve reverse transcription in the cytosol of capsid-associated viral RNA (n=2) to double-stranded cDNA that transits the nuclear pore and integrates into the host chromosome (*blue lines*) to form a provirus (*red line* surrounded by *black dotted line*) with the involvement of at least two RHs (DHX9, DDX19A). Late events involve provirus transcription by RNA polymerase II and associated RHs (DHX9, DHX30, DDX5, DDX17); RNA processing by ribonucleoprotein components (i.e., UPF1, UAP56); Rev oligomerization at the Rev responsive element (RRE) of intron-containing HIV RNA fostered by DDX1 and UPF1. Rev-dependent RRE-containing transcripts undergo CRM1-dependent nuclear export assisted by DDX3 on cytoplasmic face of the nuclear pore. Completely spliced viral transcripts are transported by NXF1/NXT1 nuclear export receptor fostered by DBP5 positioned on the inner side of the nuclear membrane. Polysomal CBP80-bound HIV RNA is associated with DHX9; polysomal eIF4E-bound HIV RNA associated with DDX3, eIF4G, eIF4A. UPF1 RNA surveillance cofactors, DDX17 and DHX30, influence HIV RNA dimerization; dimer RNA packaging with viral structural/enzymatic proteins is modulated by MOV10, DDX24, and DHX30 in context of stoichiometric amounts of DHX9 (n=2).

groupings III and IV, V, and VI. SF 1 and 2 members contain conserved Walker A and B motifs in the active site providing NTPase binding and hydrolysis activity, respectively (Del Campo et al., 2009).

Whereas nonspecific RNA binding is attributable to core helicase domain, specific recognition of cognate RNA is achievable by double-stranded RNA-binding domains (dsRBDs), described below. Retroviruses, by mimicking 5′UTR features in cognate cell-derived mRNA targets, commandeer RNA helicases to promote native folding of cis-acting RNA elements. Retrovirus RNA becoming cloaked in cellular RNPs is advantageous to evade surveillance and recognition of this type of virus infection. Retroviral antagonism of innate response remains a controversial area intertwined with biology of restriction factors.

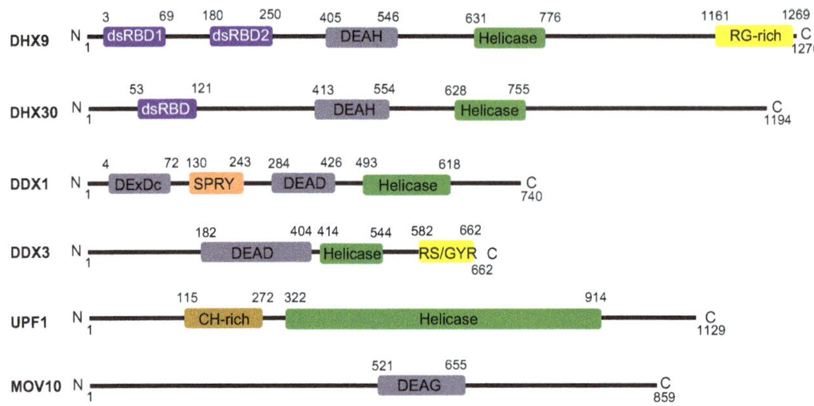

FIGURE 7.2 **Domain structure of selected RNA helicases modulating critical events in retroviral replication.** Variable ancillary domains flanking the helicase core and amino acid coordinates are shown. Canonical double-stranded RNA-binding domains differing in spacing between the motifs are designated as dsRBDI or dsRBDII. Amino acid motifs are: DEAD, DEIH, SPRY, RG (arginine-glycine), and RSGY and cysteine-histidine (CH)-rich region.

RNA HELICASE AND RETROVIRUSES IN THE ADVENT OF "OMICS TECHNOLOGY"

Current understanding of RNA helicase activities in retroviruses has been propelled by results of genome-wide screens. Tandem-affinity purification by Liang and colleagues used epitope-tagged viral proteins and mass spectrophotometry in the first proteomic screen for HIV-associated candidate RNA helicases. The tandem-affinity purification of epitope-tagged HIV Gag in cells isolated Gag binding partners that included DHX9, RNA helicase Gu(a)/DDX36, DDX18, DDX24 (Roy et al., 2006). Affinity purification of HIV proteins and mass spectrometry interaction statistics identified DDX6, DDX20, and DDX49 (Jäger et al., 2012).

Genome-wide short interfering RNA (siRNA) screens to identify host proteins involved in HIV replication identified eight helicases including DHX9, DHX30, DDX3, and DDX1 (Brass et al., 2008; Konig et al., 2008). Williams et al. performed a siRNA library screening targeting 59 cellular helicases (Williams et al., 2015). HeLa cells were cotransfected by specific siRNA to target the helicase and to generate vesicular stomatitis virus G glycoprotein (VSV-G) pseudotyped HIV-1. The culture supernatants were analyzed for virion production by Gag p24 ELISA, and the infectivity of progeny virions was determined by luciferase reporter assay. HIV infectivity was significantly reduced by downregulation of DDX3, DDX5, DDX10, DDX11, DDX17, DDX18, DDX21, DDX24, DDX28, and DDX52 and DHX9; the results indicating direct or indirect activity of the helicase in facilitating virion infectivity (Lorgeoux et al., 2012; Williams et al., 2015).

Ensuing investigations validated selected RNA helicases that promote human retrovirus replication (HIV-1, human T-cell lymphotropic virus type 1) by ectopic overexpression and downregulation by RNA interference and CRISPR technology in cell-based assays, biophysical approaches, and traditional biochemical approaches. An ultimate goal for virologists and structural biologists alike is antiviral targeting of pertinent RNA helicases. A daunting task because retroviral RNA biology is intertwined with nuclear RNA helicases at every step, genetic, biophysical, and structural studies have defined RNA helicases important to early and late steps in HIV replication. The deepening knowledge of RNA helicase ancillary domains and cognate viral nucleotide–nucleotide pairings subject to helicase unwinding is critical information to eventually segregate retroviral RNP targets from cellular RNPs necessary to cell viability. The following sections trace the steps in retroviral replication in context of cell biology and prospective antiviral activity of select RNA helicases (Table 7.3).

EARLY EVENTS: REVERSE TRANSCRIPTION AND INTEGRATION

Cellular RNA helicases are sitting at the center of current research to elucidate the composition and function of retroviral RNPs in retrovirus biology. A number of studies have demonstrated that exogenous overexpression of RNA helicases and/or downregulation by RNA interference diminishes the abundance and infectivity of progeny HIV-1, often in cell type–specific manner (e.g., MOV10, DHX9, DHX30). These observations have been attributed to aberrant reverse transcription of the incoming genomic RNA in an unseemly relationship to the posttranscriptional life of primary retroviral transcripts. A key to interpreting RNA helicase activity in reverse transcription of retroviral RNA has been the determination of stoichiometry in virions and cognate viral RNA structure, as determined for DHX9 (Sharma et al., 2012; Boeras et al., 2016).

DHX9/RNA HELICASE A ACTIVITY IN THE GENOMIC RNP

In 2006, Roy et al. identified HIV-1 virions deficient in DHX9 were poorly infectious and poorly executed reverse transcription. Downregulation of DHX9 by siRNA in producer cells did not affect viral genomic RNA dimerization or packaging, but those virions were poorly infectious, as determined by activation of a galactosidase indicator (MAGI) in Hela-T4 cells (Roy et al., 2006). Similar virion infectivity loss was also observed by infection of primary cells (Bolinger et al., 2010). The $tRNA^{Lys3}$–viral RNA complexes extracted from DHX9-deficient virions were less efficient in base extension in a primer extension assay (Roy et al., 2006; Xing et al., 2011). In addition to reverse transcription initiation, the strand transfer step was also impaired in the absence of DHX9 (Roy et al., 2006). Importantly, a gain-of-function experiment demonstrated that exogenously expressed DHX9 packaged into HIV-1 virions rescues

TABLE 7.3 RNA Helicases Modulate Critical Events in Retroviral Replication

RNA Helicase/ Alternative Names	Modulation of HIV Activity
Reverse Transcription and Integration to Form Provirus	
DHX9 RNA Helicase A Nuclear DNA Helicase II	Component of virions, stoichiometry in proportion to genomic RNA (n=2) (Sharma and Boris-Lawrie, 2012). Promotes initiation phase of HIV-1 reverse transcription (Roy et al., 2006). Binds to topoisomerase II alpha complex that remodels chromatin structure for HIV-1 integration (Raghavendra et al., 2010). Modulates HIV-1 preintegration complex activity through host factors LEDGF/p75 and Integrase Interactor 1 (Lorgeoux et al., 2012).
Provirus Transcription Initiation and Tat/TAR Transactivation of Transcription Elongation	
DHX9	Mediates the association of CREB-binding protein (CBP) with RNA Pol II to promote HIV-1 transcription (Aratani et al., 2001; Nakajima et al., 1997). Binds to HIV-1 TAR RNA through double-standard RNA-binding domains (dsRBDs) (Fujii et al., 2001). Promotes recruitment of PCAF and P-TEFb to TAR/Tat-bound transcription complex (Fujii et al., 2001).
Nucleocytoplasmic RNA Trafficking	
DBP5/DDX19A DDX1 DDX3 DDX5 and DDX17	Promotes nuclear export of the unspliced RNA of avian sarcoma virus through two direct repeats in 3′UTR (LeBlanc et al., 2007). Facilitates Rev oligomerization at Rev responsive element (RRE), enhancing CRM1-dependent nuclear export of HIV-1 transcripts (Fang et al., 2005). Facilitates remodel and release of Rev/RRE-dependent HIV RNPs from nuclear pore (Yedavalli et al., 2004). Interact with HIV-1 Rev protein; may modulate Rev subcellular localization between nucleolus, nucleoplasm, and cytoplasm (Williams et al., 2015).
Translation-Dependent mRNA Surveillance	
UPF1/RENT1	Stabilizes viral unspliced RNA in nuclear and cytoplasmic RNPs; necessary to Gag protein production (Ajamian et al., 2008).
MOV10/UPF-1 like	Facilitates HIV-1 RNA stability; binds to nucleocapsid (NC) region of Gag and packaged into viral particle (Abudu et al., 2012).

TABLE 7.3 RNA Helicases Modulate Critical Events in Retroviral Replication—cont'd

RNA Helicase/ Alternative Names	Modulation of HIV Activity
Translation of Viral mRNA	
eIF4A/DDX2	Component of eIF4F translation initiation complex; ATP-dependent unwinding of 5′UTR structure (Sonenberg, 1993).
DDX3	Recruited to eIF4E-bound HIV RNP, associated with eIF4F translation activity (Soto-Rifo et al., 2013).
DHX9	Facilitates viral protein synthesis of HIV-1 and HTLV-1 by increasing the polysome loading (Bolinger et al., 2010; Hartman et al., 2006). Binds the 5′UTR of HIV-1, HTLV-1, spleen necrosis virus, bovine leukemia virus, Mason Pfizer monkey virus, feline leukemia virus and promote Rev/RRE-independent HIV Gag production (Bolinger et al., 2007).
Dimerization, Packaging of HIV Genomic RNA	
DHX9	Promotes dimerization in-solution; recruited to virion by binding the contiguous 5′UTR in sequences between the primer activation sequence and primer-binding site (Boeras et al., 2016).
DHX30	Regulates production of Gag and inhibits genomic RNA packaging and infectivity (Zhou et al., 2008). Promotes dimerization of Rous sarcoma virus 5′UTR in-solution (Stake et al., 2015).

reverse transcription initiation when the endogenous DHX9 has been silenced (Xing et al., 2012).

DHX9 associates with HIV-1 Gag and is recruited into virus particles in an RNA-dependent manner (Roy et al., 2006). In 2012, the observed stoichiometry of DHX9 in virions was determined proportional to the genomic RNA, $n=2$ (Sharma and Boris-Lawrie, 2012). Recently, isothermal titration calorimetry confirmed that the dimeric HIV-1 5′UTR conformation binds one DHX9 molecule per RNA strand, in a manner independent of binding to the nucleocapsid (NC) domain of Gag. Nuclear magnetic resonance spectra using a deuterium-labeling approach resolved the structure of the dimeric 5′UTR in complex with the N-terminal dsRBDs of DHX9. The structure of the large molecular mass complex was found to be dependent on DHX9 binding to a double-stranded region of the primer-binding site (PBS) segment of the 5′UTR. These results are consistent with measurement of DHX9-depleted virus particles of reduced level of $tRNA^{Lys3}$ that annealed onto the PBS (Roy et al., 2006). In the viral RNA, a single A-to-C substitution was sufficient to disrupt the conformation of the PBS segment and attenuate virion infectivity in cells. Therefore, the contribution of DHX9 to virion infectivity is attributable to its structure-dependent binding at the PBS segment of the HIV 5′UTR during virus assembly (Boeras et al., 2016), consistent with prior data positing DHX9 fosters conformations favoring the action of viral reverse transcriptase (Xing et al., 2011).

Validated cellular cofactors in reverse transcription are few: tRNA synthetase and cognate tRNA and DHX9 that together engender PBS conformation for tRNA annealing and extension by reverse transcriptase. Proteomics results indicate DHX9 is also a component of HIV preintegration complex, conceivably activating chromatin to initiate the integration process (Raghavendra et al., 2010). Once integrated, nuclear RH associates with provirus, supporting transcription and viral RNA processing, export, translation, dimerization, and genomic RNP assembly. The intimate association between the provirus and cellular RNPs mimics that of a cellular gene, and conceivably those nuclear marks shield viral genomes in newly infected cells from recognition by DHX58/RIGi and other innate sensors that detect cytoplasmic viral replication intermediates and trigger interferon response.

MOV10 ACTIVITY IN VIRIONS REMAINS UNDEFINED

The Up-frameshift protein 1 (UPF1)–like SF helicase member MOV10 has a DEAG helicase motif and ATP-dependent 5′ to 3′ directional RNA helicase activity (Gregersen et al., 2014). MOV10 is associated with Gag nucleocapsid RNA–binding domain and is detectable in HIV virions (Abudu et al., 2012), but its activity is controversial. Biochemically, MOV10 interacts with Rev in an RNA-independent manner, and Rev/Rev response element–dependent nuclear export was reduced by dominant-negative helicase mutant (Fig. 7.1). Exogenous expression of MOV10 diminishes production of virions and those virions exhibit deficient HIV reverse transcription and poor infectivity, positing MOV10 as an antiviral factor (Goodier et al., 2012). In unresolved controversy, independent research identified siRNA downregulation of endogenous MOV10 did not change virion production and considered endogenous MOV10 acts as cofactor (Arjan-Odedra et al., 2012). In aggregate, modulation of MOV10 incurs an imbalance in HIV RNP activity, consistent with indirect and possibly beneficial effect on biogenesis of infectious HIV (Huang et al., 2015).

LATE EVENTS: PROVIRUS TRANSCRIPTION, PRIMARY RNA PROCESSING, EXPORT, TRANSLATION, FORMATION OF GENOMIC RNP

DHX9, Bridging RNA Polymerase to Transcription Coactivators

DHX9 is an essential gene product and mouse knockout is lethal to embryos. The dhx9 gene lays in the 1q25 tumor susceptibility locus associated with neoplasms of prostate, lung, and breast. Roles for DHX9 in tumor biology involve dysregulation of DNA damage response related to gene expression and DNA genotoxic stress. A nucleocytoplasmic shuttling DEIH helicase, DHX9 bridges chromatin-bound transcription coactivator proteins (e.g., EP300/CBP) to RNA polymerase II (Table 7.2). The tethering requires DHX9's N-terminus and flanking residues of a minimal transactivation domain (Aratani et al., 2001; Nakajima et al., 1997). By connecting coactivators with RNA polymerase II through a DHX9 bridge, this RNA helicase activates transcription of specific genes. Susceptible coactivators, in addition to CBP, include NFkB p65, BRCA1 breast cancer–specific tumor suppressor BRCA1, the Zic2 protein, EWS-FLI1 oncoprotein, p16INK4a tumor suppressor and multidrug resistance 1 protein (MDR1).

Moreover, DHX9 is component of a 7SK storage RNP that sequesters the positive transcription elongation factor b (P-TEFb), a rate-limiting cofactor of TFIIH that regulates efficiency transcription elongation of many genes, including HIV (Van Herreweghe et al., 2007). Because of limited P-TEFb, RNA polymerase II stalling of HIV transcription complex occurs within the R region of the HIV long terminal repeat yielding truncated transcripts of ~100 nucleotides that are identified as Tat transactivation responsive element, TAR (Table 7.2). Tat/TAR interaction facilitates 5′end capping of nascent RNA necessary to recruitment of nuclear cap–binding proteins CBP80/CBP20 (Zhou et al., 2003). In the absence of Tat, DHX9 exhibits biochemical affinity for TAR, an observation that may explain DHX9 overexpression activates transcription of HIV TAR-luc reporter RNA (Fujii et al., 2001). Overexpression of DHX9 is posited to tip 7SK-P-TEFb equilibrium toward dissociation of P-TEFb, ultimately facilitating Tat/TAR transactivation of HIV. Released from 7SK RNP, P-TEFb stimulates HIV Tat/TAR-dependent transcriptional elongation through a mechanism that involves both phosphorylation of the C-terminal domain of RNA polymerase II and by interfering with the activity of negative elongation factors (Table 7.3). Speculation is DHX9 loading to TAR during basal transcription modulates 5′UTR RNA conformation and the retention of CBP80 in later stages of HIV expression (see below).

DHX9 overexpression increased Gag protein and virion production in HEK293 cells (Fujii et al., 2001) and altered balanced splicing of primary HIV RNA (Tang et al., 1999). Moving beyond exogenous expression assays that disrupt the balance between RNP components, RNA interference proved invaluable to modulate an essential gene product like DHX9. Validation studies in COS cells demonstrated siRNA downregulation of this RH reduced gag, vif, rev, and nef RNA expression (Bolinger et al., 2010; Roy et al., 2006). COS, a simian fibroblast line, did not recapitulate reduced HIV steady-state HIV RNA and instead demonstrated modulation of gag RNA accumulation in polysomes that is proportional to reduction in viral protein production (Hartman et al., 2006). In contrast to results in HEK293 cells, DHX9 downregulation in COS cells did not reduce either HIV RNA abundance or nucleocytoplasmic transport, positing perturbation in RNP components between the transformed cells (Hartman et al., 2006).

Paralogs, DHX30 and DHX9, have nearly identical amino acid conservation but accumulate in different subcellular compartments (Fig. 7.2). DHX9 accumulates in nucleus and DHX30 accumulates in mitochondrion. Attributable to truncation of one dsRBD at N-terminus and ~100 arginine-rich amino acids at the C-terminus, DHX30 functions in the assembly of the large mitochondrial ribosomal mutation of DHX30 reduces mitochondrial localization (Lessel et al., 2018). Overexpression of DHX30 indicates a supportive role in HIV replication: DHX30 overexpression increases production of viral Gag proteins and yield of virus particles by 2- to 3-fold and severely inhibits the packaging of HIV-1 RNA into virus particles, upregulating defective interfering particles by 10-fold (Zhou et al., 2008), consistent with change in RNP dynamics.

Zhou et al. (2008) determined DHX30 interacts with the HIV-1 5′UTR, potentially catalyzing native refolding important for genomic RNA dimerization similar to DHX9. In-solution assays with recombinant proteins demonstrated DHX9 and DHX30 promoted dimerization of the HIV 5′UTR. DHX30, but not DHX9, fostered dimerization of the Rous sarcoma virus (RSV) 5′UTR (Stake et al., 2015). This trend follows the observation that the *Gallus gallus* genome encodes DHX30, but not DHX9, and suggests that DHX30 supports RNP dynamics attributable to higher order conformation of the avian retrovirus 5′UTR and possibly distinct chaperone activity of RSV Gag nucleocapsid (Stake et al., 2015).

Shuttling RNA Helicase in Nucleocytoplasmic Transport of Retroviral RNA

Two decades ago, study of genetically simple retrovirus Mason–Pfizer monkey virus (MPMV) uncovered existence of the cis-acting constitutive transport element (CTE) and its necessity for cytoplasmic expression of simian retrovirus unspliced RNA (SRV1). Synthetic CTE transcripts and cell RNA–binding proteins were incubated with Hela cell lysates and in stringent conditions coisolating RNA-binding proteins were identified, e.g., DHX9 and DHX9-cofactor HAP95; ALY/REF and RH UAP56/DDX39B cofactors that tether the NXF1–NXT1 complex (also known as TAP nuclear export receptor). The study of CTE/TAP activity significantly expanded knowledge of nucleocytoplasmic transport of cell mRNAs and regulation by many virus RNAs. Ten years ago and in context RSV, a genetically simple avian retrovirus, DDX19B/yeast DBP5 was determined to facilitate nuclear export of genome-length unspliced RNA (LeBlanc et al., 2007). Drawing on extensive knowledge of the nuclear pore complex in yeast, DBP5 is positioned on the cytoplasmic face of the nuclear pore complex and catalyzes disassembly of the export RNP (Fig. 7.1).

Whereas SRV1/MPMV CTE connects TAP nuclear export receptor, genetically more complex retroviruses encode a cis-acting viral CTE-like element (RRE) and viral transacting protein (Rev) to activate nuclear export of unspliced viral RNA. Yeast two-hybrid screening identified Rev binding RH, DDX1, and

DDX3 (Fang et al., 2004). Reporter assays and RNA export measurements validated these host cofactors facilitate Rev/RRE-dependent gene expression. DDX1 is a DEAD box family member containing SPRY domain and interacting with poly(A) RNA. Both yeast and mammalian two-hybrid assays demonstrated strong interaction of DDX1 helicase with HIV-1 Rev protein. DDX1 (amino acids 189–333) directly interacts with the nuclear inhibitory signal of Rev at amino acids 10–24. Downregulation of DDX1 reduces the Rev/RRE-dependent luciferase reporter activity and also impairs the nuclear export of unspliced mRNA (Edgcomb et al., 2012; Fang et al., 2005).

DDX3 is a nucleocytoplasmic shuttling protein that localizes to the cytoplasmic side of the nuclear pore (Fig. 7.1). Yedavalli et al. demonstrated siRNA-mediated downregulation of DDX3 reduces Rev/RRE/CRM1-mediated HIV-1 RNA export. Based on the transdominant activity of DDX3 ATPase mutant, DDX3 was postulated to facilitate the final release from the nuclear pore of the CRM1-bound Rev/RRE RNP (Yedavalli et al., 2004). Mechanistically, DDX3 is posited in binding CRM1 molecules bound to the exposed nuclear export signal (NES) of RRE-bound Rev oligomers expelling CRM1/Rev/RRE-containing HIV-1 RNA cargo into the cytoplasm (Lorgeoux et al., 2012). In this remodel of the HIV RNP, DDX3 activity is similar in helicase-driven release by DBP5/NFX1/NXF1 nuclear export receptor activity of mRNA cargo (Fig. 7.1).

DDX1 and DDX3 coordinate Rev/RRE-dependent gene expression. DDX1 binding to RRE modulates RRE conformation, promotes binding of monomeric Rev, resulting in Rev oligomerization (Lamichhane et al., 2017). Rev independently interacts with nuclear export receptor CRM1 through the leucine-rich NES (Cullen, 2003), tethering the RRE/Rev-CRM1 RNP with affinity for DDX3 at the nuclear pore (Lorgeoux et al., 2012) (Fig. 7.1). Roles of other candidate RH are poorly understood, specifically DDX5, DDX17, DDX21, DHX36, and DHX9, in RRE/Rev-mediated nuclear export (Zhou et al., 2013).

NUCLEAR CAP-BINDING PROTEINS AND RNA HELICASE: TRANSLATION EVADING NONSENSE RNA-MEDIATED DECAY

In retrospect, an obvious yet unanticipated precedent came to light through characterizing RNPs deposited at exon–exon junctions; nuclear history of metazoan mRNA facilitates its translational activity in the cytoplasm (Le Hir et al., 2001). These lessons involve RNA helicase exchanges that retroviruses sometimes recapitulate from activity of cellular transcripts, i.e., UPF1/Staufen (Dugre-Brisson et al., 2005), DHX9/junD (Hartman et al., 2006), DHX36 (RHAU)/guanine-quadruplex-rich RNA (Chalupníková et al., 2008) (Table 7.2). Recruited to nascent mRNAs in the nucleus, these RHs remain associated in cytoplasmic RNPs that become polysomes or RNA granules that are translationally-incompetent, dramatically distinct fates attributable to distinct RNP compositions.

UPF1, also known as DDX39B, is an RNA-binding ATPase that contains Zn-coordinating finger motifs, acidic motifs, and basic amino acid clusters, and it is a well-established cofactor in the cotranslational process of nonsense-mediated decay, or NMD (Nicholson et al., 2010). A component of the exon junction complex that binds nascent RNA, UPF1 shuttles between the nucleus and cytoplasm using the CRM1 pathway, and activates RNA surveillance at exon–exon junctions through a series of phosphorylation events during pioneer-round translation. In association with nuclear cap-binding proteins CBP80/CBP20, UPF1 is a sentinel in the process of NMD. NMD parses nonsense-free mRNAs from nonsense-containing transcripts that have potential to translate polypeptide deleterious to the cell. UPF1, UPF2, and UPF3 activate cotranslational RNA decay when scanning ribosome encounter a stop codon that is aberrantly positioned within an open reading frame. However, UPF1 bound to retrovirus unspliced RNAs is responsible for alternative activity: subverting NMD at the premature stop codon laying within the polymerase open reading frame of unspliced transcripts (Fig. 7.1).

Despite its canonical role in activating NMD, UPF1/DHX39A stabilizes HIV unspliced RNA that harbors a premature stop codon and lacks exon junction complex. UPF1 stabilizes primary retroviral RNA in both nucleoplasm and cytoplasm. UPF1 downregulation by siRNA decreases HIV RNA and thereby diminishes the synthesis of virion structural proteins. While residual levels of virions are observed, they are poorly infectious, positing beneficial role of UPF1 in balancing HIV RNP compositions (Ajamian et al., 2008).

STEADY-STATE TRANSLATION: SWITCHING 5′CAP–BINDING PROTEINS TO GAIN eIF4E

The molecular switch triggering cap exchange is poorly understood. Canonically, nuclear cap-binding proteins CBP80/CBP20 are replaced by the cytoplasmic cap-binding protein, eIF4E. eIF4E is considered the rate-limiting factor for steady-state translation and engages eIF4G and the helicase eIF4A to form the eIF4F mRNP. Historically, translation control has been viewed through the prism of eIF4E and the requisite RNA helicase activity of eIF4A. Discovered decades ago, eIF4A provides critical unwinding activity to 5′UTR secondary structures, recruits the 40S ribosomal subunit, and sends the activated ribosome scanning the 5′UTR (Sonenberg, 1993).

Recent studies used antibodies to the nuclear cap-binding protein CBP80 and the cytoplasmic counterpart eIF4E to isolate HIV mRNPs from the cytoplasm (Sharma et al., 2012). HIV-specific RT-qPCR assessing relative abundance of HIV transcripts confirmed the completely spliced nef transcript was in the eIF4E RNP, at levels similar to the cellular RNA control. By comparison, unspliced HIV RNA was poorly enriched in the eIF4 RNP and was instead retained in the CBP80-RNP. Despite the retention of nuclear cap–binding proteins, viral protein synthesis remained robust. Polysome analysis verified Gag was translated from the CBP80-RNP, a noncanonical RNP associated with pioneer-round translation.

RNA granules are attributed to downregulation of eIF4E activity during oxidative stress, heat shock, amino acid deprivation, or other physiological stress. Mouland et al., (2000) characterized downregulation of eIF4E activity and HIV RNA accumulation in RNA granules. In 2013, Soto-Rifo et al., identified DDX3-bound HIV RNA accumulating with eIF4G and PABPC1 in cytoplasmic granules comprised of translation-incompetent mRNPs (Soto-Rifo et al., 2013). Physiological downregulation of eIF4E did not shut down HIV protein synthesis and HIV RNA remained on polysomes (Sharma et al., 2012) and associated with DHX9 (Fritz, Singh, Boris-Lawrie, in preparation). Ongoing studies evaluate the possibility that CBP80-DHX9 RNPs contribute an alternative strategy for cap-dependent translation.

THERAPEUTIC TARGETING AT THE INTERFACE OF RNA HELICASE AND COGNATE RETROVIRAL RNA

RNA helicases serve as drug targets in a variety of settings. In peripheral blood mononuclear cells (PBMCs), a small molecule inhibitor of DDX3 efficiently reduced HIV replication (Radi et al., 2012). Given the importance of several cellular RNA helicases in retrovirus replication, targeted disruption of RNA helicases may attenuate latent provirus reactivation. The prospects in this setting are small molecules targeting cognate RNAs, particularly viral RNA elements. A prospect buoyed by growth in knowledge of structural biology, critical information is complete understanding of three-dimensional features of RNA helicases and cognate RNA. A stealth approach, similar in principle to highly active antiretroviral therapy, is targeting two or more RNA helicase–viral RNA interfaces necessary to distinct events in the viral replication cycle. The potential increases to curb emergence of drug-resistant viruses by leveraging essential RNA helicase activities in cellular cofactor RNAs.

In closing, retroviruses use cellular gene expression machinery, maintaining the interaction with host RNA helicases at every step of viral replication: reverse transcription and integration to form the provirus; transcription of the provirus and posttranscriptional regulation; and dimerization and packaging of the genomic RNP (Fig. 7.1). Reconciling the beneficial activity of shuttling RNA helicases to retroviruses with the antiviral activity of cytosolic helicases that oppose other RNA viruses remains to be fully understood. From the vantage point of a retrovirus, RNA helicases are cofactors required to transcribe provirus DNA, stimulate transcriptional elongation and RNA processing, and facilitate the nucleocytoplasmic transport of retroviral RNPs to become translated. From the vantage point of a cell, RNA helicases are antiviral sensors, detecting non–self-RNA in the cytoplasm and triggering an antiviral state. However, retroviruses circumvent these cellular RNA sensors, instead co-opting nuclear RNA helicases and using them in a manner analogous to cellular transcripts, dissipating the restrictive activity of RNA helicases that actively thwarts cytoplasmic replication of many other RNA viruses.

REFERENCES

Abudu, A., Wang, X., Dang, Y., Zhou, T., Xiang, S.H., Zheng, Y.H., 2012. Identification of molecular determinants from Moloney leukemia virus 10 homolog (MOV10) protein for virion packaging and anti-HIV-1 activity. J. Biol. Chem. 287, 1220–1228. https://doi.org/10.1074/jbc.M111.309831.

Ajamian, L., Abrahamyan, L., Milev, M., Ivanov, P.V., Kulozik, A.E., Gehring, N.H., Mouland, A.J., 2008. Unexpected roles for UPF1 in HIV-1 RNA metabolism and translation. RNA 14, 914–927. https://doi.org/10.1261/rna.829208.

Aratani, S., Fujii, R., Oishi, T., Fujita, H., Amano, T., Ohshima, T., Hagiwara, M., Fukamizu, A., Nakajima, T., 2001. Dual roles of RNA helicase A in CREB-dependent transcription. Mol. Cell. Biol. 21, 4460–4469. https://doi.org/10.1128/MCB.21.14.4460-4469.2001.

Arjan-Odedra, S., Swanson, C.M., Sherer, N.M., Wolinsky, S.M., Malim, M.H., 2012. Endogenous MOV10 inhibits the retrotransposition of endogenous retroelements but not the replication of exogenous retroviruses. Retrovirology 9, 53. https://doi.org/10.1186/1742-4690-9-53.

Boeras, I., Song, Z., Moran, A., Franklin, J., Brown, W.C., Johnson, M., Boris-Lawrie, K., Heng, X., 2016. DHX9/RHA binding to the PBS-segment of the genomic RNA during HIV-1 assembly bolsters virion infectivity. J. Mol. Biol. 428, 2418–2429. https://doi.org/10.1016/j.jmb.2016.04.011.

Bolinger, C., Sharma, A., Singh, D., Yu, L., Boris-Lawrie, K., 2010. RNA helicase A modulates translation of HIV-1 and infectivity of progeny virions. Nucleic Acids Res. 38, 1686–1696. https://doi.org/10.1093/nar/gkp1075.

Bolinger, C., Yilmaz, A., Hartman, T.R., Kovacic, M.B., Fernandez, S., Ye, J., Forget, M., Green, P.L., Boris-Lawrie, K., 2007. RNA helicase A interacts with divergent lymphotropic retroviruses and promotes translation of human T-cell leukemia virus type 1. Nucleic Acids Res. https://doi.org/10.1093/nar/gkm124.

Brass, A.L., Dykxhoorn, D.M., Benita, Y., Yan, N., Engelman, A., Xavier, R.J., Lieberman, J., Elledge, S.J., 2008. Identification of host proteins required for HIV infection through a functional genomic screen. Science (80-.) 319, 921–926. https://doi.org/10.1126/science.1152725.

Chalupníková, K., Lattmann, S., Selak, N., Iwamoto, F., Fujiki, Y., Nagamine, Y., 2008. Recruitment of the RNA helicase RHAU to stress granules via a unique RNA-binding domain. J. Biol. Chem. 283, 35186–35198. https://doi.org/10.1074/jbc.M804857200.

Cullen, B.R., 2003. Nuclear mRNA export: insights from virology. Trends Biochem. Sci. https://doi.org/10.1016/S0968-0004(03)00142-7.

Dardenne, E., PolayEspinoza, M., Fattet, L., Germann, S., Lambert, M.P., Neil, H., Zonta, E., Mortada, H., Gratadou, L., Deygas, M., Chakrama, F.Z., Samaan, S., Desmet, F.O., Tranchevent, L.C., Dutertre, M., Rimokh, R., Bourgeois, C.F., Auboeuf, D., 2014. RNA helicases DDX5 and DDX17 dynamically orchestrate transcription, miRNA, and splicing programs in cell differentiation. Cell Rep. https://doi.org/10.1016/j.celrep.2014.05.010.

Del Campo, M., Mohr, S., Jiang, Y., Jia, H., Jankowsky, E., Lambowitz, A.M., 2009. Unwinding by local strand separation is critical for the function of DEAD-box proteins as RNA chaperones. J. Mol. Biol. 389, 674–693. https://doi.org/10.1016/j.jmb.2009.04.043.

Delaleau, M., Borden, K.L.B., 2015. Multiple export mechanisms for mRNAs. Cells. https://doi.org/10.3390/cells4030452.

Diot, C., Fournier, G., Dos Santos, M., Magnus, J., Komarova, A., Van Der Werf, S., Munier, S., Naffakh, N., 2016. Influenza A virus polymerase recruits the RNA helicase DDX19 to promote the nuclear export of viral mRNAs. Sci. Rep. https://doi.org/10.1038/srep33763.

Dugre-Brisson, S., Elvira, G., Boulay, K., Chatel-Chaix, L., Mouland, A.J., DesGroseillers, L., 2005. Interaction of Staufen1 with the 5′ end of mRNA facilitates translation of these RNAs. Nucleic Acids Res. 33, 4797–4812 33/15/4797.

Edgcomb, S.P., Carmel, A.B., Naji, S., Ambrus-Aikelin, G., Reyes, J.R., Saphire, A.C., Gerace, L., Williamson, J.R., 2012. DDX1 is an RNA-dependent ATPase involved in HIV-1 Rev function and virus replication. J. Mol. Biol. 415, 61–74. https://doi.org/10.1016/j.jmb.2011.10.032.

Fairman-Williams, M.E., Guenther, U.P., Jankowsky, E., 2010. SF1 and SF2 helicases: family matters. Curr. Opin. Struct. Biol. 20, 313–324. https://doi.org/10.1016/j.sbi.2010.03.011.

Fang, J., Acheampong, E., Dave, R., Wang, F., Mukhtar, M., Pomerantz, R.J., 2005. The RNA helicase DDX1 is involved in restricted HIV-1 Rev function in human astrocytes. Virology 336, 299–307 S0042-6822(05)00166-2 [pii].

Fang, J., Kubota, S., Yang, B., Zhou, N., Zhang, H., Godbout, R., Pomerantz, R.J., 2004. A DEAD box protein facilitates HIV-1 replication as a cellular co-factor of Rev. Virology 330, 471–480 S0042-6822(04)00645-2 [pii].

Fujii, R., Okamoto, M., Aratani, S., Oishi, T., Ohshima, T., Taira, K., Baba, M., Fukamizu, A., Nakajima, T., 2001. A role of RNA helicase a in cis-acting transactivation response element-mediated transcriptional regulation of human immunodeficiency virus type 1. J. Biol. Chem. 276, 5445–5451. https://doi.org/10.1074/jbc.M006892200.

Goodier, J.L., Cheung, L.E., Kazazian Jr., H.H., 2012. MOV10 RNA helicase is a potent inhibitor of retrotransposition in cells. PLoS Genet. 8, e1002941. https://doi.org/10.1371/journal.pgen.1002941.

Gregersen, L.H., Schueler, M., Munschauer, M., Mastrobuoni, G., Chen, W., Kempa, S., Dieterich, C., Landthaler, M., 2014. MOV10 is a 5′ to 3′ RNA helicase contributing to UPF1 mRNA target degradation by translocation along 3′ UTRs. Mol. Cell 54, 573–585. https://doi.org/10.1016/j.molcel.2014.03.017.

Hartman, T.R., Qian, S., Bolinger, C., Fernandez, S., Schoenberg, D.R., Boris-Lawrie, K., 2006. RNA helicase A is necessary for translation of selected messenger RNAs. Nat. Struct. Mol. Biol. 13, 509–516.

Huang, F., Zhang, J., Zhang, Y., Geng, G., Liang, J., Li, Y., Chen, J., Liu, C., Zhang, H., 2015. RNA helicase MOV10 functions as a co-factor of HIV-1 Rev to facilitate Rev/RRE-dependent nuclear export of viral mRNAs. Virology 486, 15–26. https://doi.org/10.1016/j.virol.2015.08.026.

Jäger, S., Cimermancic, P., Gulbahce, N., Johnson, J.R., McGovern, K.E., Clarke, S.C., Shales, M., Mercenne, G., Pache, L., Li, K., Hernandez, H., Jang, G.M., Roth, S.L., Akiva, E., Marlett, J., Stephens, M., D'Orso, I., Fernandes, J., Fahey, M., Mahon, C., O'Donoghue, A.J., Todorovic, A., Morris, J.H., Maltby, D.A., Alber, T., Cagney, G., Bushman, F.D., Young, J.A., Chanda, S.K., Sundquist, W.I., Kortemme, T., Hernandez, R.D., Craik, C.S., Burlingame, A., Sali, A., Frankel, A.D., Krogan, N.J., 2012. Global landscape of HIV-human protein complexes. Nature 481, 365–370. https://doi.org/10.1038/nature10719.

Jin, J., Jing, W., Lei, X.X., Feng, C., Peng, S., Boris-Lawrie, K., Huang, Y., 2011. Evidence that Lin28 stimulates translation by recruiting RNA helicase A to polysomes. Nucleic Acids Res. https://doi.org/10.1093/nar/gkq1350.

Kadare, G., Haenni, A., 1997. Virus-encoded RNA helicases. J. Virol.

Konig, R., Zhou, Y., Elleder, D., Diamond, T.L., Bonamy, G.M.C., Irelan, J.T., Chiang, C.Y., Tu, B.P., De Jesus, P.D., Lilley, C.E., Seidel, S., Opaluch, A.M., Caldwell, J.S., Weitzman, M.D., Kuhen, K.L., Bandyopadhyay, S., Ideker, T., Orth, A.P., Miraglia, L.J., Bushman, F.D., Young, J.A., Chanda, S.K., 2008. Global analysis of host-pathogen interactions that regulate early-stage HIV-1 replication. Cell 135, 49–60. https://doi.org/10.1016/j.cell.2008.07.032.

Kwong, A.D., Rao, B.G., Jeang, K.T., 2005. Viral and cellular RNA helicases as antiviral targets. Nat. Rev. Drug Discov. https://doi.org/10.1038/nrd1853.

Lamichhane, R., Hammond, J.A., Pauszek, R.F., Anderson, R.M., Pedron, I., Van Der Schans, E., Williamson, J.R., Millar, D.P., 2017. A DEAD-box protein acts through RNA to promote HIV-1 Rev-RRE assembly. Nucleic Acids Res. 45, 4632–4641. https://doi.org/10.1093/nar/gkx206.

Le Hir, H., Gatfield, D., Izaurralde, E., Moore, M.J., 2001. The exon-exon junction complex provides a binding platform for factors involved in mRNA export and nonsense-mediated mRNA decay. EMBO J. 20, 4987–4997. https://doi.org/10.1093/emboj/20.17.4987.

LeBlanc, J.J., Uddowla, S., Abraham, B., Clatterbuck, S., Beemon, K.L., 2007. Tap and Dbp5, but not Gag, are involved in DR-mediated nuclear export of unspliced Rous sarcoma virus RNA. Virology 363, 376–386. https://doi.org/10.1016/j.virol.2007.01.026.

Lessel, D., Schob, C., Küry, S., Reijnders, M.R.F., Harel, T., Eldomery, M.K., Coban-Akdemir, Z., Denecke, J., Edvardson, S., Colin, E., Stegmann, A.P.A., Gerkes, E.H., Tessarech, M., Bonneau, D., Barth, M., Besnard, T., Cogné, B., Revah-Politi, A., Strom, T.M., Rosenfeld, J.A., Yang, Y., Posey, J.E., Immken, L.D., Oundjian, N., Helbig, K.L., Meeks, N., Zegar, K., Morton, J., the DDD study, Schieving, J.H., Claasen, A., Huentelman, M., Narayanan, V., Ramsey, K., Brunner, H.G., Elpeleg, O., Mercier, S., Bézieau, S., Kubisch, C., Kleefstra, T., Kindler, S., Lupski, J.R., Kreienkamp, H.J., 2018. Erratum: De novo missense mutations in DHX30 impair global translation and cause a neurodevelopmental disorder. Am. J. Hum. Genet. 101 (5), 716–724. https://doi.org/10.1016/j.ajhg.2017.09.014. 2017, pii: S000292971730383X, https://doi.org/j.ajhg.2017.12.016.

Lorgeoux, R.P., Guo, F., Liang, C., 2012. From promoting to inhibiting: diverse roles of helicases in HIV-1 replication. Retrovirology 9, 79. https://doi.org/10.1186/1742-4690-9-79.

Manojlovic, Z., Stefanovic, B., 2012. A novel role of RNA helicase A in regulation of translation of type I collagen mRNAs. RNA. https://doi.org/10.1261/rna.030288.111.

Mouland, A.J., Mercier, J., Luo, M., Bernier, L., DesGroseillers, L., Cohen, E.A., 2000. The double-stranded RNA-binding protein Staufen is incorporated in human immunodeficiency virus type 1: evidence for a role in genomic RNA encapsidation. J. Virol. 74, 5441–5451.

Nakajima, T., Uchida, C., Anderson, S.F., Lee, C.G., Hurwitz, J., Parvin, J.D., Montminy, M., 1997. RNA helicase A mediates association of CBP with RNA polymerase II. Cell 90, 1107–1112, pii: S0092-8674(00)80376-1.

Nicholson, P., Yepiskoposyan, H., Metze, S., Zamudio Orozco, R., Kleinschmidt, N., Muhlemann, O., 2010. Nonsense-mediated mRNA decay in human cells: mechanistic insights, functions beyond quality control and the double-life of NMD factors. Cell. Mol. Life Sci. 67, 677–700. https://doi.org/10.1007/s00018-009-0177-1.

Radi, M., Falchi, F., Garbelli, A., Samuele, A., Bernardo, V., Paolucci, S., Baldanti, F., Schenone, S., Manetti, F., Maga, G., Botta, M., 2012. Discovery of the first small molecule inhibitor of human DDX3 specifically designed to target the RNA binding site: towards the next generation HIV-1 inhibitors. Bioorg. Med. Chem. Lett. 22, 2094–2098. https://doi.org/10.1016/j.bmcl.2011.12.135.

Raghavendra, N.K., Shkriabai, N., Graham, R.L.J., Hess, S., Kvaratskhelia, M., Wu, L., 2010. Identification of host proteins associated with HIV-1 preintegration complexes isolated from infected CD4+ cells. Retrovirology 7, 66. https://doi.org/10.1186/1742-4690-7-66.

Ranji, A., Boris-Lawrie, K., 2010. RNA helicases: Emerging roles in viral replication and the host innate response. RNA Biol. https://doi.org/10.4161/rna.7.6.14249.

Roy, B.B., Hu, J., Guo, X., Russell, R.S., Guo, F., Kleiman, L., Liang, C., 2006. Association of RNA helicase a with human immunodeficiency virus type 1 particles. J. Biol. Chem. 281, 12625–12635. https://doi.org/10.1074/jbc.M510596200.

Samaan, S., Tranchevent, L.C., Dardenne, E., Polay Espinoza, M., Zonta, E., Germann, S., Gratadou, L., Dutertre, M., Auboeuf, D., 2014. The Ddx5 and Ddx17 RNA helicases are cornerstones in the complex regulatory array of steroid hormone-signaling pathways. Nucleic Acids Res. https://doi.org/10.1093/nar/gkt1216.

Sharma, A., Boris-Lawrie, K., 2012. Determination of host RNA helicases activity in viral replication. Methods Enzymol. 511, 405–435. https://doi.org/10.1016/B978-0-12-396546-2.00019-X.

Sharma, A., Yilmaz, A., Marsh, K., Cochrane, A., Boris-Lawrie, K., 2012. Thriving under stress: selective translation of HIV-1 structural protein mRNA during Vpr-mediated impairment of eIF4E translation activity. PLoS Pathog. 8, e1002612. https://doi.org/10.1371/journal.ppat.1002612.

Sonenberg, N., 1993. Remarks on the mechanism of ribosome binding to eukaryotic mRNAs. Gene Expr. 3, 317–323.

Soto-Rifo, R., Rubilar, P.S., Ohlmann, T., 2013. The DEAD-box helicase DDX3 substitutes for the cap-binding protein eIF4E to promote compartmentalized translation initiation of the HIV-1 genomic RNA. Nucleic Acids Res. 41, 6286–6299. https://doi.org/10.1093/nar/gkt306.

Stake, M., Singh, D., Singh, G., Marcela, H.J., Kaddis, M.R., Parent, L.J., Boris-Lawrie, K., 2015. HIV-1 and two avian retroviral 5′untranslated regions bind orthologous human and chicken RNA binding proteins. Virology 486, 307–320.

Tang, H., McDonald, D., Middlesworth, T., Hope, T.J., Wong-Staal, F., 1999. The carboxyl terminus of RNA helicase A contains a bidirectional nuclear transport domain. Mol. Cell. Biol. 19, 3540–3550.

Tijms, M.A., Nedialkova, D.D., Zevenhoven-Dobbe, J.C., Gorbalenya, A.E., Snijder, E.J., 2007. Arterivirus subgenomic mRNA synthesis and virion biogenesis depend on the multifunctional nsp1 autoprotease. J. Virol. https://doi.org/10.1128/JVI.00683-07.

Van Herreweghe, E., Egloff, S., Goiffon, I., Jády, B.E., Froment, C., Monsarrat, B., Kiss, T., 2007. Dynamic remodelling of human 7SK snRNP controls the nuclear level of active P-TEFb. EMBO J. https://doi.org/10.1038/sj.emboj.7601783.

Williams, C.A., Abbink, T.E., Jeang, K.T., Lever, A.M., 2015. Identification of RNA helicases in human immunodeficiency virus 1 (HIV-1) replication – a targeted small interfering RNA library screen using pseudotyped and WT HIV-1. J. Gen. Virol. 96, 1484–1489. https://doi.org/10.1099/vir.0.000092.

Wisskirchen, C., Ludersdorfer, T.H., Muller, D.A., Moritz, E., Pavlovic, J., 2011. The cellular RNA helicase UAP56 is required for prevention of double-stranded RNA formation during influenza A virus infection. J. Virol. https://doi.org/10.1128/JVI.02559-10.

Xing, L., Liang, C., Kleiman, L., 2011. Coordinate roles of Gag and RNA helicase A in promoting the annealing of formula to HIV-1 RNA. J. Virol. 85, 1847–1860. https://doi.org/10.1128/JVI.02010-10.

Xing, L., Niu, M., Kleiman, L., 2012. In vitro and in vivo analysis of the interaction between RNA helicase A and HIV-1 RNA. https://doi.org/10.1128/JVI.01993-12.

Ye, P., Liu, S., Zhu, Y., Chen, G., Gao, G., 2010. DEXH-box protein DHX30 is required for optimal function of the zinc-finger antiviral protein. Protein Cell. https://doi.org/10.1007/s13238-010-0117-8.

Yedavalli, V.S., Neuveut, C., Chi, Y.H., Kleiman, L., Jeang, K.T., 2004. Requirement of DDX3 DEAD box RNA helicase for HIV-1 Rev-RRE export function. Cell 119, 381–392, pii: S0092867404008360.

Zhou, M., Deng, L., Kashanchi, F., Brady, J.N., Shatkin, A.J., Kumar, A., 2003. The Tat/TAR-dependent phosphorylation of RNA polymerase II C-terminal domain stimulates cotranscriptional capping of HIV-1 mRNA. Proc. Natl. Acad. Sci. U.S.A. 100, 12666–12671. https://doi.org/10.1073/pnas.1835726100.

Zhou, X., Luo, J., Mills, L., Wu, S., Pan, T., Geng, G., Zhang, J., Luo, H., Liu, C., Zhang, H., 2013. DDX5 facilitates HIV-1 replication as a cellular co-factor of Rev. PLoS One 8. https://doi.org/10.1371/journal.pone.0065040.

Zhou, Y., Ma, J., Bushan Roy, B., Wu, J.Y., Pan, Q., Rong, L., Liang, C., 2008. The packaging of human immunodeficiency virus type 1 RNA is restricted by overexpression of an RNA helicase DHX30. Virology 372, 97–106, pii: S0042-6822(07)00714-3.

Chapter 8

Role of Host Factors in the Subcellular Trafficking of Gag Proteins and Genomic RNA Leading to Virion Assembly

Eunice C. Chen, Leslie J. Parent
Penn State College of Medicine, Hershey, PA, United States

Chapter Outline

Introduction 274
Nuclear Export of Viral RNAs 275
Unspliced Viral RNA Export in Simple Retroviruses 277
The Constitutive Transport Element of MMTV 277
Export of Unspliced RNA in Rous Sarcoma Virus 278
Spumaretrovirus RNA Nuclear Export 278
Retroviral Genome Trafficking in the Cytoplasm 279
Relationship Between Viral RNA Export and Virus Assembly 279
Transport of Unspliced Viral RNA Through the Cytoplasm to the Assembly Site 279
Dimerization of Retroviral Genomic RNA 281
Retroviral Gag Protein Trafficking 281
Nuclear Trafficking of Retroviral Gag Proteins 284
Foamy Virus Gag Association With Chromatin 285
Transport of Gag Proteins in the Cytoplasm 286
Role of Cytoskeletal Proteins and Microtubule-Organizing Center in Gag Transport 288
Multivesicular Bodies and Endosomal Proteins in Gag Trafficking 289
Role of Myristoylation and Lipid Binding in Gag Membrane Targeting 291
Multivesicular Body and Endosomal Sorting Complexes Required for Transport Pathway Interactions With Gag Late Domains 291
Nedd4 and E3 Ubiquitin Ligases 292
Tsg101 as a Central Player in Wide Range of Gag Late Domain Interactions 292

Retrovirus-Cell Interactions. https://doi.org/10.1016/B978-0-12-811185-7.00008-X

Alix/AIP-1 Interactions With HIV Gag Proteins 294
Late Domains in EIAV, MLV, and Foamy Virus Gag Proteins 294
Endophilin 2 in Murine Leukemia Virus Gag Trafficking 295
Association of Gag With tRNA Synthetases 296
Factors That Interact With Gag or Viral RNA to Restrict Virus Replication 296
Host Restriction Factors That Interfere With Gag Functions Early or Late in Replication 296
Concluding Remarks 298
Acknowledgments 299
References 299

INTRODUCTION

The retroviral genome consists of a single-positively stranded RNA that forms a noncovalent dimer when packaged into newly formed virions. All retroviral genomes contain three major genes: *gag*, which encodes the structural protein, Gag; *pol*, containing protease, reverse transcriptase, RNAseH, and integrase activities; and *env*, which produces Env, the envelope glycoprotein (Vogt, 1997). These proteins are required for infectivity and completion of the replication cycle. Following binding and entry of a retrovirus, the genome undergoes reverse transcription and the double-stranded DNA intermediate is integrated into the cellular DNA. This proviral DNA is transcribed by host RNA polymerase II, and the resulting retroviral RNA has three potential fates: (1) it can be spliced to form subgenomic mRNAs that are transported into the cytoplasm where they encode viral proteins, such as Env; (2) it can remain unspliced, and after transport into the cytoplasm, serve as the template for Gag and Gag-Pol protein expression; or (3) it can be maintained as full-length, unspliced viral RNA genomes for packaging into nascent virus particles (Rabson and Graves, 1997). Each of these options requires the viral RNA to be exported from the nucleus into the cytoplasm to facilitate efficient expression of viral proteins or assembly of new virus particles. Complex retroviruses such as human immunodeficiency virus (HIV), human T-cell lymphotropic virus type I (HTLV-I), and mouse mammary tumor virus (MMTV) generate additional spliced subgenomic mRNA species that encode accessory proteins to achieve greater control over the replication cycle. However, simple retroviruses do not encode accessory proteins and therefore must rely more heavily on cellular factors to complete their replication cycle (Coffin et al., 1997).

Once the unspliced viral mRNA arrives in the cytoplasm, it is translated on free ribosomes to produce the Gag protein. The Gag-Pol fusion protein is synthesized from the same mRNA template as a result of either ribosomal frameshifting or termination suppression, depending on the retrovirus. After translation, the Gag protein must identify and bind the viral RNA genome for encapsidation, necessitating that Gag and the viral RNA traffic to the same location within the same cellular compartment. The subcellular site of this initial Gag–genomic RNA interaction is not well understood and remains a fundamental question in the field. Moreover, it is not clear how Gag locates the unspliced viral RNA genome, which constitutes only about 1% of the cellular RNA pool and appears identical to cellular mRNAs, as both are modified with a 5′ trimethyl cap and a 3′ polyadenylated tail (Furuichi et al., 1975; Keith and Fraenkel-Conrat, 1975; Lai and Duesberg, 1972; Gillespie et al., 1972). The selectivity and specificity of genomic RNA packaging requires the presence of a structured, high affinity–binding site for Gag within the 5′ untranslated region (UTR) of the genomic RNA, known as "psi" for packaging signal (Linial and Miller, 1990; Rein, 1994). Gag and the genomic RNA form a ribonucleoprotein complex (RNP) that is transported from the site of initial contact to the assembly site in the cytoplasm or at the inner leaflet of the plasma membrane depending on the virus (Swanstrom and Wills, 1997). Failure of either Gag or the viral RNA genome to traffic to the correct subcellular site disrupts the production of infectious virus particles.

In this chapter, we will discuss the host factors that facilitate trafficking of the viral RNA genome from the site of transcription into the virion; we will also discuss the cellular proteins, RNAs, and lipids that interact with Gag during its journey to the plasma membrane, where virus particles are released by budding. Gag and genomic RNA co-opt a variety of cellular factors in the nucleus, nuclear pore complex (NPC), microtubule-organizing center (MTOC), cytoskeleton, cytoplasmic RNA granules, multivesicular bodies, and phospholipids. Importantly, the cell also mounts resistance to virus infection, and those factors that restrict virus assembly and release are important to understand as the battle between the invader and the host defense system takes place.

NUCLEAR EXPORT OF VIRAL RNAs

Expression of the integrated provirus is mediated by the cellular enzyme RNA polymerase II and numerous transcription factors influence the efficiency of transcription, and in the case of lentiviruses, silencing of transcription leading to latency (see Chapter 6). Posttranscription processing of the viral RNA occurs in parallel to host RNAs, with the association of capping and polyadenylation machinery (Schrom et al., 2013). Splicing factors bind to 5′ splice donor and 3′ splice acceptor sites to mediate removal of introns, yet a portion of the viral RNA does not undergo splicing and instead remains full length to be used as mRNA or genomic RNA. Inhibition of splicing occurs through multiple mechanisms, including *cis*-acting negative regulators of splicing that serve as decoys for splicing factors and suboptimal splice sites in the viral RNA (Giles and Beemon, 2005; Hibbert et al., 1999; McNally et al., 1991; Ogert et al., 1996; Karn and Stoltzfus, 2012; Stoltzfus, 2009; Stoltzfus and Madsen, 2006).

Export of the unspliced viral RNA from the nucleus poses a unique challenge for retroviruses to overcome because incompletely processed cellular mRNAs are normally retained or degraded in the nucleus (Aguilera, 2005; Bjork and Wieslander, 2017; Brodsky and Silver, 2000; Cullen, 2002; Erkmann and Kutay, 2004; Ghosh and Jacobson, 2010; Jackson, 2004; Neumann et al., 2003;

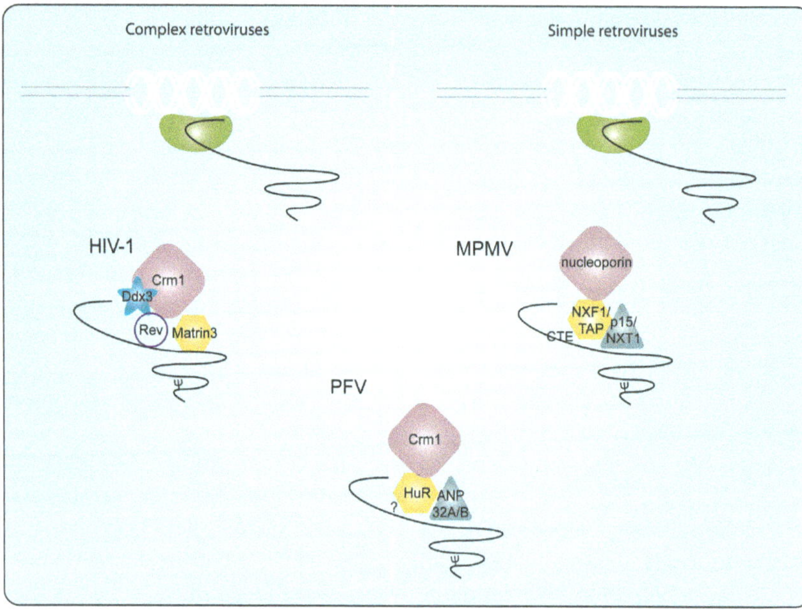

FIGURE 8.1 RNA trafficking in simple versus complex retroviruses. Human immunodeficiency virus type 1 (HIV-1) is an example of a complex retrovirus, which utilizes the viral accessory protein Rev and the host protein Ddx3 to bind the nuclear export receptor Crm1 and facilitate nuclear export of the unspliced viral genome. The simple retrovirus Mason–Pfizer monkey virus (MPMV) relies solely on host proteins NFX1/TAP and p15/NXT1 for nuclear export of its unspliced viral genome. In contrast, prototype foamy virus (PFV) uses a mechanism that combines aspects of simple and complex retroviruses.

Rondon and Proudfoot, 2008; Saguez et al., 2005; Sommer and Nehrbass, 2005; Stutz and Izaurralde, 2003; Swanson et al., 2004; Zenklusen and Stutz, 2001). In complex retroviruses, virally encoded accessory proteins mediate nuclear export of unspliced or incompletely spliced viral RNAs (Cullen, 2002; Harris and Hope, 2000). However, simple retroviruses co-opt host nuclear export proteins to facilitate this step (Fig. 8.1). The complex retroviruses HIV-1, HTLV, and MMTV encode *trans*-acting viral proteins Rev (Rosen et al., 1988; Malim et al., 1989; Felber et al., 1989), Rex (Hidaka et al., 1988), and Rem (Mertz et al., 2005), respectively, to bridge viral interactions with the nuclear export receptor Crm1 (also known as exportin1/XPO1) to facilitate nuclear egress of their viral unspliced RNAs (Fornerod et al., 1997a; Nakielny and Dreyfuss, 1999; Neville et al., 1997). Crm1 associates with nucleoporins to bring RNA complexes to the NPC for nuclear export (Fornerod et al., 1997a,b). Additional host proteins have been identified to aid these viral-encoded proteins. For example in HIV-1, Matrin 3 (MATR3), a highly conserved component of the nuclear matrix that has been identified as an interacting factor of the NPC (Belgrader et al., 1991; Hibino et al., 2006; Nakayasu and Berezney, 1991; Cronshaw et al., 2002), and DDX3, a DEAD box RNA helicase and ATPase, aids the nuclear export of unspliced and partially spliced viral RNA transcripts via interactions with Rev (Kula et al., 2011, 2013; Yedavalli and Jeang, 2011; Yedavalli et al., 2004). DDX3 binds Crm1 and Rev and travels to the outer surface of the nuclear pore and specifically facilitates Rev/RRE-mediated viral genome export (Yedavalli et al., 2004). MATR3 associates with and stabilizes RRE-containing spliced and unspliced viral transcripts in a Rev-dependent manner (Yedavalli and Jeang, 2011).

Unspliced Viral RNA Export in Simple Retroviruses

Simple retroviruses such as Rous sarcoma virus (RSV), Mason–Pfizer monkey virus (MPMV), and MLV do not encode Rev-like proteins and instead co-opt the function of the host protein NFX1/TAP and its cofactor p15/NXT1 to facilitate nuclear export of spliced host mRNAs (Santos-Rosa et al., 1998; Herold et al., 2001; Jin et al., 2003; Pessel-Vivares et al., 2015). In MLV, the genomic RNA contains a *cis*-acting cytoplasmic accumulation element (CAE) located at the end of *pol* (Sakuma et al., 2014; Pessel-Vivares et al., 2014). As shown in Fig. 8.1, this sequence interacts with NXF1/TAP to enhance cytoplasmic accumulation of both spliced and unspliced viral RNA transcripts and increase viral protein expression, without altering splicing or stability of the viral RNA transcripts (Pessel-Vivares et al., 2014; Sakuma et al., 2014). The efficacy of NFX1/TAP-mediated export is regulated by p15/NXT1: expression of p15/NXT1 increases the affinity of NFX1/TAP for nucleoporin proteins in in vitro experiments and increases NFX1/TAP-dependent RNA export in vivo (Levesque et al., 2001; Guzik et al., 2001). There effects are sequence specific, and when the CAE is introduced into a Rev-dependent HIV-1 Gag construct, an increase in Gag expression is observed in a Rev-independent manner (Sakuma et al., 2014). Additionally, the cellular RNA export factor UAP56 plays a role in export of MLV spliced and unspliced RNAs, and THOC5 and THOC7, members of the TREX export complex, also contribute to efficiency of export of MLV unspliced transcripts, with the final common pathway for both species of RNAs being NXF1 (Sakuma et al., 2014).

The Constitutive Transport Element of MMTV

MPMV utilizes a similar pathway for nuclear export of its full-length viral RNA. The viral genome contains a constitutive transport element (CTE) within the 3′UTR, which directly interacts with the amino terminal of NXF1/TAP to facilitate nuclear export of viral RNA (Bray et al., 1994; Ernst et al., 1997; Braun et al., 1999). Microinjection or transfection of NFX1/TAP results in nuclear export of CTE-containing RNA in oocytes and quail cells (Gruter et al., 1998; Kang and Cullen, 1999; Guzik et al., 2001). Similarly, the presence of one or more copies of the CTE sequence allows HIV-1 RNA to be exported in a Rev-independent fashion (Bray et al., 1994; Wodrich et al., 2000; Swanson et al., 2004).

Export of Unspliced RNA in Rous Sarcoma Virus

In RSV, the mechanism underlying nuclear export of unspliced viral transcripts is a bit more complicated. The RSV genome contains two direct repeats (DRs) flanking the *src* gene that have been proposed, but not widely accepted, to be involved in the nuclear export of viral full-length unspliced RNAs (Ogert et al., 1996; Simpson et al., 1997, 1998; Ogert and Beemon, 1998; Yang and Cullen, 1999; Paca et al., 2000; LeBlanc et al., 2007). LeBlanc et al. (2007) utilized dominant-negative constructs of NXF1/Tap and DEAD box helicase (Dbp5), host factors involved in cellular mRNA export, to demonstrate their role in export of DR-containing RSV unspliced viral RNA. Furthermore, the authors showed that insertion of DR1 into a heterologous RNA leads to the translation of a reporter while the presence of Ψ in the heterologous RNA did not result in translation. However, the findings did not demonstrate a direct interaction between the host factors NXF1/Tap or Dbp5 and the DR-repeats, suggesting a possible adapter protein could be important for export.

Subsequently, it was demonstrated that the RSV Gag protein undergoes nucleocytoplasmic trafficking, and reduction of Gag nuclear localization leads to a significant reduction in packaging of the viral RNA (see section on Retroviral Gag Protein Trafficking) (Garbitt-Hirst et al., 2009). Later studies suggested that the nuclear binding of RSV Gag to viral RNA via ψ could trigger exposure of a nuclear export signal (NES) in the Gag protein, facilitating the nuclear export of the viral RNP in a Crm1-dependent mechanism (Gudleski et al., 2010). Biochemical and imaging studies suggest that once RSV Gag binds to the unspliced viral RNA in the nucleus, the viral RNP is exported into the cytoplasm via Crm1 (Gudleski et al., 2010; Scheifele et al., 2002). Taken together, these data suggest that the unspliced RSV RNA is likely exported from the nucleus by two distinct pathways, using NXF1/Tap and Dbp5 to sort the unspliced viral mRNA for translation and Gag to export the unspliced genomic viral RNA for packaging.

Spumaretrovirus RNA Nuclear Export

Nuclear export of the viral genome in spumaretroviruses, such as prototype foamy virus (PFV), utilizes a mechanism that is a mix of that of simple and complex retroviruses (Fig. 8.1). Nuclear export in PFV is Crm1-dependent, as evidenced by decreased Gag expression in infectious provirus–transfected 293T cells following leptomycin B (LMB) treatment (Bodem et al., 2011). However, PFV does not encode a viral protein that interacts with Crm1. Instead, PFV relies on a cellular RNA–binding protein HuR and adapter proteins ANP32A/B to bridge the interaction between the viral genome and Crm1. This interaction is believed to be a direct interaction, as it was identified by RNA-protein cross-linking and immunoprecipitation techniques between the viral genome and tagged proteins. However, the mechanisms of this interaction are not fully clarified, as the secondary RNA structure commonly required for HuR–RNA interactions has not been identified in the PFV RNA (Bodem et al., 2011; Prechtel et al., 2006).

RETROVIRAL GENOME TRAFFICKING IN THE CYTOPLASM

Once the retrovirus full-length genome has entered the cytoplasm, it is used as either a template for translation of Gag and Gag-Pol or it travels through the cytoplasm with Gag to the plasma membrane for budding. There is some debate whether the same pool of unspliced viral RNA can be used for translation and packaging. Some studies suggest that the same population of RNA can be used for either purpose or if they exist as distinct pools with specific fates (LeBlanc and Beemon, 2004; Levin et al., 1974; Levin and Rosenak, 1976; Bender et al., 1987; Swanson and Malim, 2006; Butsch and Boris-Lawrie, 2002).

Relationship Between Viral RNA Export and Virus Assembly

It was recently proposed that the pathway by which the full-length HIV-1 genome is exported from the nucleus targets it toward the site of virus assembly. In this study, researchers used imaging techniques to compare Rev/RRE-dependent export of HIV-1 viral RNA or the NXF1(TAP)-NXT1/CTE-dependent nuclear export of the MPMV viral RNA. The HIV-1 RRE-containing RNA showed "*en masse*" nonlocalized nuclear export into the cytoplasm, whereas the MMTV CTE-tagged RNA accumulated specifically near the MTOC just outside the nuclear membrane (Pocock et al., 2016). When the RRE sequence of HIV-1 was replaced with the CTE sequence of MPMV, nuclear export of HIV-1 viral RNA became Rev-independent and was "reprogrammed" to accumulate near the MTOC and nuclear membrane, as observed for MPMV CTE-RNA (Bray et al., 1994; Wodrich et al., 2000; Swanson et al., 2004; Pocock et al., 2016). The presence of additional CTEs further enhanced targeting of the HIV-1 RNA to the MTOC. Similar patterns in RNA trafficking were observed when CTE and RRE sequences were placed into heterologous, nonviral RNAs (Pocock et al., 2016). The researchers concluded that the MTOC and cytoplasmic RNA trafficking patterns observed were intrinsic to the nuclear export pathway utilized, possibly due to an interaction between the CTE sequence and NXF1/TAP-NXT1 molecules. A distinction between nuclear export pathways would allow each retroviral RNA genome to be efficiently targeted to the site of virus assembly and could potentially facilitate downstream genomic RNA packaging steps.

Transport of Unspliced Viral RNA Through the Cytoplasm to the Assembly Site

A few host factors have also been identified that chaperone the viral genome throughout the cytoplasm to facilitate trafficking to sites of viral assembly and budding, including many that interact with the host cell's cytoskeletal network. Members of the heterogeneous nuclear ribonucleoprotein (hnRNP) family have been identified to play a role in the trafficking of HIV-1 genomic RNA. hnRNP A and B proteins are involved in multiple aspects of posttranscriptional gene regulation in the host cell, including splicing, metabolism, transport, and translation (Shyu and Wilkinson, 2000). The function of A2RE sequences and hnRNP A2 in targeted trafficking of RNA from the nucleus into the cytoplasm occurs in a microtubule-dependent fashion and has been characterized in a variety of organisms (Carson and Barbarese, 2005; Kosturko et al., 2005). In HIV-1, hnRNP A1 has been proposed to facilitate expression of Gag from the full-length viral RNA, although the data are controversial (Monette et al., 2009). The hnRNP A2 protein interacts with two hnRNP A2 response element (A2RE)–like sequences located in the HIV-1 genome, one within the major homology region coding region of the *gag* gene (A2RE-1) and the other between the *vpr* and *tat* genes (A2RE-2) (Mouland et al., 2001). By microinjecting fluorescently labeled viral RNAs containing wild-type and deleted A2RE sequences into oligodendrocytes and following their movement, researchers demonstrated that RNAs containing the *gag* and A2RE-1 sequence or *vpr* and A2RE-2 sequence were transported in a majority of injected cells, whereas deletion of either A2RE-like sequence reduced RNA transport significantly. Additionally, researchers demonstrated the function of A2RE-1 and -2 in HIV-1 RNA transport is sequence-dependent, the inclusion of either sequence into a GFP mRNA allowed for mRNA transport at levels equivalent to that of HIV-1 *gag* or *vpr* RNA.

Consistent with the function of hnRNP A2 in host cells, hnRNP A2-mediated HIV-1 RNA transport appears to be dependent on microtubules. Treatment with nocodazole, a disrupter of microtubules, results in decreased trafficking and reduced viral infectivity (Mouland et al., 2001). These A2RE-like sequences are highly conserved in the genome and contain fewer sequence polymorphisms as compared to surrounding genome regions in HIV-1 and other lentiviruses. Of interest, the position of these sequences relative to splicing sites means they are differentially expressed on unspliced full-length genomic RNAs compared with spliced subgenomic mRNAs, strengthening the argument that they play an important role in the trafficking of genomic viral RNA for assembly and packaging.

Additionally, the human Staufen (hStau) protein, a double-stranded RNA-binding protein that traffics along the microtubule pathway (Ferrandon et al., 1994), has been implicated in packaging of HIV-1. Staufen is packaged into HIV-1, HIV-2, and MLV virions, although the packaging of Staufen is specific to the viral RNA as it was not observed in HIV-1 VLPs. Researchers demonstrated the significance of Staufen's role in viral genome trafficking by showing that overexpression of hStau enhanced the number of copies of genomic viral RNA present in virions (Mouland et al., 2000).

An important contribution to RNA trafficking is the structural state of the unspliced viral RNA during its passage from the nucleus through the cytoplasm to sites of assembly and budding. In addition to the evidence that different unspliced viral RNA species are bound by distinct RNA-binding proteins to form RNP complexes destined for either translation or packaging, it is likely that the unspliced viral RNA adopts a different structure depending on whether it is being used for translation of viral proteins or packaged as genome (Stake et al., 2015; Keane et al., 2015, 2016; Keane and Summers, 2016; Boeras et al., 2017).

Dimerization of Retroviral Genomic RNA

To be infectious, the viral genome must form a noncovalent dimer with a second copy of the genomic RNA (Greatorex, 2004; Rein, 1994; Paillart et al., 1996). It is not yet clear whether genomic RNA dimerization occurs within the cell or after the RNA is encapsidated into a virus particle, and perhaps the site of dimer formation varies for different retroviruses (Jouvenet et al., 2011). In MLV, it has been suggested that genome dimerization occurs in the nucleus and in a cotranscriptional manner (Maurel and Mougel, 2010). This finding is consistent with the observation that MLV has a low recombination rate presumably because two identical unspliced RNAs are copackaged (Kharytonchyk et al., 2005). For HIV, which has a higher recombination rate (Hu et al., 2003; Jetzt et al., 2000; Onafuwa et al., 2003), it has been suggested that genome dimerization occurs in the cytoplasm and at the plasma membrane in the presence of the HIV-1 Gag protein (Chen et al., 2016). However, the mechanism and subcellular location of dimer formation are not known for other retroviruses. For most retroviruses, it remains to be determined whether the genome travels from the nucleus, through the cytoplasm, and to the plasma membrane as a genome monomer or dimer, and whether distinct RNA-binding proteins facilitate the transport in each case.

RETROVIRAL GAG PROTEIN TRAFFICKING

The correct localization of Gag proteins is critical for the replication of retroviruses, as Gag directs the assembly of new virions. Studies have demonstrated that the expression of Gag alone, without other viral proteins, results in the formation of spherical virus-like particles (VLPs) released from cells (Dickson et al., 1984; Delchambre et al., 1989; Gheysen et al., 1989; Smith et al., 1990; Wills et al., 1989; Shioda and Shibuta, 1990; Madisen et al., 1987; Haffar et al., 1990; Luo et al., 1990; Freed, 1998). The location of assembly and budding varies depending on the type of retrovirus. Type B and D retroviruses (for example, MMTV and MPMV, respectively) assemble immature capsids in the perinuclear space and subsequently transport them to the plasma membrane for budding. Conversely type C retroviruses (MLV, HIV-1, and RSV) assemble their capsids at or near the plasma membrane in a mechanism that may occur simultaneously with budding (Swanstrom and Wills, 1997). Mutations in the MA domain that prevent plasma membrane targeting/binding of Gag result in reduced production of virions and decreased infectivity (REF). Interestingly, complementation of MA mutants with wild-type Gag is possible in only a subset of retroviruses (RSV and HIV-1), but not in MLV or SNV, suggesting that Gag–Gag interactions prior to localization at the plasma membrane may be important for assembly and budding in certain retroviruses (Wills et al., 1991; Morikawa et al., 1996; Yuan et al., 1993; Parent et al., 1996; Schultz and Rein, 1989; Weaver and Panganiban, 1990; Bennett and Wills, 1999). The host factors involved in Gag trafficking described in this chapter are summarized in Table 8.1.

TABLE 8.1 Host Factors That Influence Retroviral Gag Trafficking in the Host Cell

		Nucleus/Nucleolus Proteins	MTOC/ Cytoskeleton Proteins	Cytoplasmic Proteins	Plasma Membrane Proteins
Alpharetrovirus	RSV	Crm1 (Scheifele et al., 2002) Importin-α, importin-β, importin-11, TNPO3 (Butterfield-Gerson et al., 2006; Gudleski et al., 2010; Scheifele et al., 2002) SC35, SF2 (Rice et al., 2015)	KIF-4 (Tang et al., 1999)	Nedd4 (Kikonyogo et al., 2001)	
Betaretrovirus	MMTV	RPL9 (Beyer et al., 2013)			
	MPMV		KIF-4 (Tang et al., 1999) Ubc9 (Weldon et al., 2003)	Dynein/dynactin (Petit et al., 2003; Saib et al., 1997) Nedd4, Tsg 101 (Gottwein et al., 2003; Yasuda and Hunter, 1998) *Rab11* (Sfakianos and Hunter, 2003)	
Gammaretrovirus	MLV	C-terminus of nucleolin (Bacharach et al., 2000)	IQGAP1 (Leung et al., 2006) KIF-4 (Kim et al., 1998)	Endophilin 2 (Wang et al., 2003) Nedd4, Tsg101, Alix/AIP-1 (Segura-Morales et al., 2005)	

Lentivirus	HIV-1	nucleolin (Gao et al., 2014)	KIF-4 (Tang et al., 1999; Martinez et al., 2008) Ubc9 (Mo and Moschos, 2005)	Alix/AIP-1 (Garrus et al., 2001; Strack et al., 2003) Annexin[b] (Ma et al., 2004; Ryzhova et al., 2006) AP-3 (Dong et al., 2005) hPOSH (Alroy et al., 2005) Rab9, TIP47, p40, PIKfye, Rab11A (Murray et al., 2005) SOCS1 (Ryo et al., 1999, 2008) Tsg101 (VerPlank et al., 2001)	ABCE1 (Zimmerman et al., 2002; Dooher et al., 2007) Annexin[b] (Ma et al., 2004; Ryzhova et al., 2006) $PI(4,5)P_2$ (Tang et al., 2004; Saad et al., 2006)
	EIAV			AP-50/AP-2 (Puffer et al., 1998) Alix/AIP-1 (Strack et al., 2003)	
	SIV		KIF-4 (Tang et al., 1999)		
	FIV	Crm1[c] (Kemler et al., 2012)		Nedd4, Tsg101 (Luttge et al., 2008; Calistri et al., 2009)	
Spumavirus	PFV	Crm1β (Renault et al., 2011) H2A-H2B/nucleosomes[a] (Saib et al., 1997) PLKs (Zurnic et al., 2016)	Light chain 8/dynein[a] (Petit et al., 2003; Saib et al., 1997)	Tsg101 (Stange et al., 2005)	

Retroviral Gag proteins utilize host factors to efficiently traffic through the cell during infection. A list of factors and their subcellular location are shown in this table. *EIAV*, equine infectious anemia virus; *FIV*, feline immunodeficiency virus; *HIV-1*, human immunodeficiency virus type 1; *MLV*, murine leukemia virus; *MPMV*, Mason–Pfizer monkey virus; *PFV*, prototype foamy virus; *RSV*, Rous sarcoma virus; *SIV*, simian immunodeficiency virus.

[a] *Trafficking of preintegration complex.*

[b] *Conflicting data.*

[c] *LMB-sensitive, not necessarily a direct interaction with Crm1.*

Nuclear Trafficking of Retroviral Gag Proteins

In the traditional view of C-type retrovirus assembly, the Gag protein is translated in the cytoplasm where it binds the viral RNA genome and then travels directly to the plasma membrane. However, the observations from numerous research groups that multiple retroviral Gag proteins have some degree of nuclear localization have challenged that dogma. For example, the RSV Gag protein transiently traffics through the nucleus, and viral mutants that are reduced in nuclear localization packaged less genomic RNA; restoration of nuclear packaging rescues genomic RNA levels in particles. Bidirectional nuclear trafficking is mediated by host importins and exportins, soluble nuclear transport receptors belonging to the importinβ/karyopherinβ family, and components of the NPC (Terry et al., 2007) RSV Gag enters the nucleus via two nuclear localization signals (NLS), a canonical NLS located in nucleocapsid protein (NC), which interacts directly with importin-α and recruits importin-β, and a noncanonical NLS located in MA, which interacts with importin-11 and transportin-SR (transportin-3, TNPO3) (Gudleski et al., 2010; Scheifele et al., 2002; Butterfield-Gerson et al., 2006). Studies have demonstrated that overexpression of importin-11 or importin-beta alone is each sufficient to drive NLS-mediated RSV Gag nuclear import (Gudleski et al., 2010). While in the nucleus, RSV Gag localizes near sites of active transcription, as demonstrated by imaging studies demonstrating colocalization of Gag and phosphorylated splicing speckle components SC35 and SF2, with the hypothesis that Gag binds newly transcribed unspliced viral RNA for packaging as genomes (Rice et al., 2015). RSV Gag exit from the nucleus is mediated by an NES present in the p10 domain, which functions in a host export factor Crm1-dependent mechanism. Treatment with LMB, a drug that covalently binds Crm1 to prevent its interaction with a cargo NES or mutation of any of the hydrophobic residues in the p10 NES sequence traps RSV Gag in the nucleus (Scheifele et al., 2002).

Localization of Gag to the nucleus has been observed in other retroviruses, including HIV, feline immunodeficiency virus (FIV), MTMV, and MPMV (Fig. 8.2) (Kemler et al., 2010, 2012; Nash et al., 1993; Dupont et al., 1999), although the biological relevance of Gag nuclear trafficking has not been fully characterized. HIV-1 Gag has been observed in the nucleus and nucleolus (Bukrinsky et al., 1993; Haffar et al., 2000; Dupont et al., 1999; Kemler et al., 2012; Baluyot et al., 2012; Grewe et al., 2012), and an NLS in NC has been identified (Yu et al., 2016; Lochmann et al., 2013). A mechanism for nuclear export of HIV-1 has not been convincingly identified, and there appears to be little effect of Crm1 inhibition on HIV-1 subcellular localization (Lochmann et al., 2013; Baluyot et al., 2012). By contrast, FIV Gag shuttles through the nucleus in a Crm1-dependent fashion. Expression of an FIV Gag-CFP molecule results in a diffuse cytoplasmic pattern; however, treatment with LMB results in a partial accumulation of FIV Gag-CFP in the nucleus and nucleoli. Motifs suggestive of NLSs and NESs were identified but they did not prove to be functional, instead

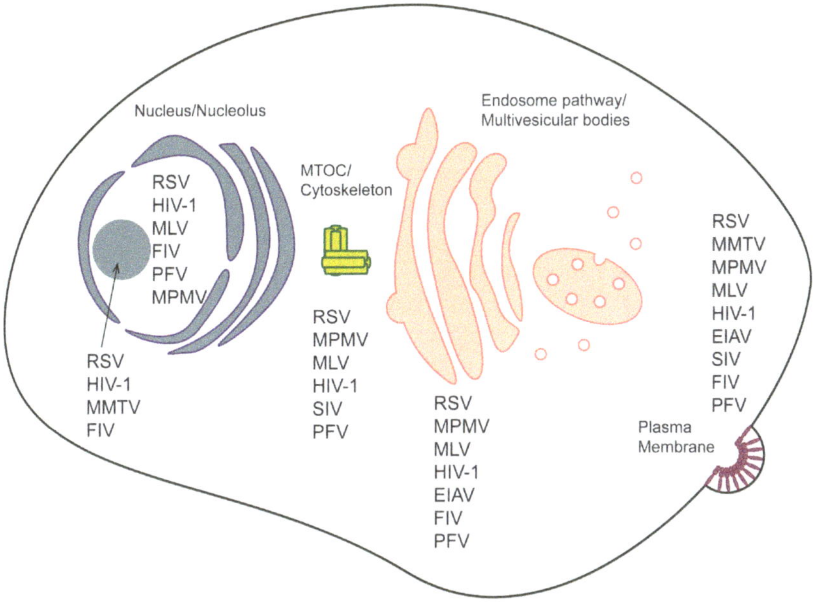

FIGURE 8.2 Subcellular location of retroviral Gag proteins. Retroviral Gag proteins traffic through numerous subcellular compartments during the replication cycle: the nucleus and nucleolus, the microtubule-organizing center (MTOC) and components of the cell cytoskeleton, the endosomal pathway and multivesicular body/endosomal sorting complexes required for transport components, and the plasma membrane. *EIAV*, equine infectious anemia virus; *FIV*, feline immunodeficiency virus; *HIV-1*, human immunodeficiency virus type 1; *MLV*, murine leukemia virus; *MPMV*, Mason–Pfizer monkey virus; *PFV*, prototype foamy virus; *RSV*, Rous sarcoma virus; *SIV*, simian immunodeficiency virus.

the nuclear shuttling properties were attributed to the CA-NCp2 portion of Gag (Kemler et al., 2012). MMTV Gag was found to be nucleolar-localized in association with ribosomal protein L9 (RPL9), and knockdown of RPL9 reduces MMTV particle production (Beyer et al., 2013).

Foamy Virus Gag Association With Chromatin

Nuclear localization of PFV Gag was characterized with an immunofluorescence assay in foamy virus-infected cells using sera from infected hosts (Schliephake and Rethwilm, 1994). PFV Gag initially enters the nucleus as part of the preintegration complex (PIC); Gag directly interacts with light chain 8 of the microtubule-dependent dynein–dynactin motor to transport the PIC to the MTOC (Petit et al., 2003; Saib et al., 1997). PFV Gag contains an NLS in the glycine–arginine-rich motifs (GR boxes) in the NC domain, mutation of the second motif results in loss of nuclear translocation of PFV Gag although it does not affect MTOC targeting (Schliephake and Rethwilm, 1994; Petit et al., 2003;

Yu et al., 1996). During mitosis and nuclear envelope breakdown, PFV Gag in the PIC associates with host core histones H2A–H2B via a chromatin-binding site (CBS) present in the C-terminus of Gag to facilitate integration of the retrovirus genome (Petit et al., 2003; Saib et al., 1997; Tobaly-Tapiero et al., 2008). A subsequent study demonstrated that the CBS is essential for PFV nuclear targeting, and therefore, the NLS may not contribute or may not be functional (Mullers et al., 2011). The crystal structure of PFV Gag directly interacting with a mononucleosome via the CBS was recently solved and reveals the importance of an arginine anchor motif in the CBS for chromatin binding (Lesbats et al., 2017). A similar chromatin-tethering sequence is present in the C-terminus of MLV p12 and is important for nuclear retention of the MLV PIC in mitotic chromosomes (Schneider et al., 2013). In later stages of the replication cycle, PFV Gag utilizes an LMB-sensitive NES located within the N-terminus of the protein to shuttle out of the nucleus for virus assembly (Schliephake and Rethwilm, 1994; Renault et al., 2011). Renault et al. (2011) demonstrated that mutation of the NES (GagG110V) resulted in decreased capsid assembly, without any effect on Gag expression or processing.

While in the nucleus PFV Gag also interacts with polo-like kinases (PLKs), a family of host proteins involved in cell cycle regulation and entry into mitosis (Zitouni et al., 2014). Using full-length PFV Gag as bait, Zurnic et al. (2016) performed yeast two-hybrid assays to demonstrate PFV Gag interacted with full-length human PLK2 and PLK3. Follow-up confocal microscopy studies demonstrated mCherry-tagged PFV Gag recruited eGFP-tagged human PLK1, PLK2, and to a lesser degree PLK3 to mitotic chromosomes in 293T cells. This recruitment was dependent on the presence of a consensus S-(T/S)-P-binding motif on PFV Gag (Elia et al., 2003). Mutation of the STP motif or abrogation of the PFV Gag–PLK interaction resulted in reduced efficiency of provirus integration, the number of integrated genomes was reduced to 45% as compared with wild type, without affecting other steps of retrovirus replication (Zurnic et al., 2016).

Transport of Gag Proteins in the Cytoplasm

Efficient and accurate trafficking of Gag within the cytoplasm to the plasma membrane is critical for assembly of viral particles, the presence of Gag is sufficient to package genomic viral RNA into virions (Shields et al., 1978; Sakalian et al., 1994). It is not yet determined how Gag is targeted to the proper subcellular location for assembly and budding, although researchers have proposed three possible mechanisms: by free diffusion of the Gag protein (Tang et al., 1999); hitching a ride on transport vesicles traveling between the Golgi apparatus and plasma membrane (Krausslich and Welker, 1996; Hansen et al., 1990); or co-opting the host cell's cytoskeleton proteins (Tang et al., 1999; Edbauer and Naso, 1983; Kim et al., 1998). Experimental evidence does not support free diffusion as a mechanism for Gag trafficking. A single tryptophan to arginine mutation in the MA domain of MPMV Gag can functionally switch the virus from type D to type C assembly, and Gag targeting is redirected from the cytoplasm to plasma membrane with no negative effect on virus assembly (Rhee and Hunter, 1990). Additionally, MPMV mutant viruses encoding a Gag that cannot be myristylated to accumulate A-type capsids within the cytoplasm that are unable to bud (Rhee and Hunter, 1987). These findings suggest that capsid assembly in MPMV requires Gag to accumulate above a concentration threshold at the perinuclear region allowing for self-assembly and then trafficking of the virus to the plasma membrane for budding, which depends on targeting of Gag to a specific cellular compartment. Similarly, Gag trafficking via transport vesicles is not supported by the experimental data. First, this model could not explain trafficking in type B and D retroviruses, which assemble at the perinuclear space and then travel to the plasma membrane (Hansen et al., 1990). Second, budding studies in HIV-1 and RSV failed to demonstrate sensitivity to brefeldin A treatment, a drug that interferes with transport by the secretory pathway between the ER and the Golgi apparatus (Pal et al., 1991; Krishna et al., 1996).

Gag relies on additional host factors to facilitate the proper assembly of virions. Previous studies demonstrated that retroviruses assembly occurred in an ATP-dependent mechanism following translation of the Gag polyprotein (Lingappa et al., 1997). HIV-1 Gag is not known to interact with ATP, which prompted Zimmerman et al. to search for a host protein to serve as an adapter. ABCE1, also known as HP68 and RNaseL inhibitor, was identified by performing a coimmunoprecipitation screen using antibodies targeting host ATP-binding molecular chaperones and Gag (Zimmerman et al., 2002). The screen was performed in a cell-free system using wheat germ extract as a source of cellular factors. The interaction between Gag and ABCE1 was confirmed by co-IP, imaging studies, and functional studies using a dominant-negative clone. Subsequent studies demonstrated that ABCE1 initially associates with p55 Gag in cytoplasm following translation and remains associated throughout the assembly process, as demonstrated by pulse-chase experiments and the radiolabeling of increasingly higher density subcellular fractions that contain Gag (Zimmerman et al., 2002; Dooher et al., 2007). Experiments using immunogold labeling and cryoEM show an interaction between HIV-1 Gag and ABCE1 at sites of budding at the plasma membrane and within newly formed virions (Dooher et al., 2007).

The nucleolar protein nucleolin has also been suggested to play a role in retrovirus assembly. Using a yeast two-hybrid screen, Bacharach et al. identified a C-terminal fragment of nucleolin as an interacting host factor for MLV Gag and weakly for MPMV and HIV-1 (Bacharach et al., 2000). Overexpression of this fragment resulted in a reduction in MLV virion release, in a dose-dependent manner, and infecitivity. This function was specific to the C-terminus fragment, as full-length nucleolin did not have a similar effect. Gao et al. also identified nucleolin as an interacting host factor for HIV-1 Gag using tandem-immunoprecipitation and mass spectrometry (Gao et al., 2014). These authors demonstrated that expression of HIV-1 Gag increased the nucleolin levels present in the cytoplasm and the presence of Gag:nucleolin complexes. Overexpression of nucleolin enhanced release of Gag in the supernatant, suggesting nucleolin facilitates HIV-1 assembly.

Role of Cytoskeletal Proteins and Microtubule-Organizing Center in Gag Transport

There is strong evidence that host cytoskeletal proteins and cellular transport proteins may facilitate retrovirus Gag trafficking (Fig. 8.2). MLV Gag associates with a host cytoskeleton protein IQGAP1, an actin and microtubule-binding factor, and utilizes this interaction to traffic through the cytoplasm. The interaction was initially discovered through a yeast two-hybrid screen in a mouse macrophage cell line using the MLV MA domain as bait (Leung et al., 2006). IQGAP1 has been demonstrated to be an important regulator of the cytoskeleton and controls directional migration of host proteins, thus it is proposed that the IQGAP1/Gag interaction facilitates trafficking of Gag during assembly. As expected, subsequent studies using alanine mutagenesis and siRNA knockdown demonstrated the importance of IQGAP1 to the replication and infectivity of MLV (Leung et al., 2006).

A host cytoskeletal protein identified to interact with multiple retroviruses is KIF-4, a plus-end microtubule motor protein belonging to the kinesin superfamily initially characterized by Sekine et al. while studying organelle transport along the length of axons. Immunohistochemistry studies demonstrated the expression of KIF-4 localizes in the cytoplasm and with membranous organelles in axons (Sekine et al., 1994). Yeast two-hybrid studies demonstrated interactions between KIF-4 and the Gag proteins of RSV, HIV-1, MuMLV, MPMV, and SIV. The interaction has been mapped to occur between the MA domain of Gag and the C-terminus of KIF-4 (Tang et al., 1999; Kim et al., 1998). KIF-4 has been demonstrated to play a role early in infection, by stabilizing microtubules to allow for trafficking of viral particles into the nucleus (Sabo et al., 2013), and late in infection during assembly. Martinez et al. demonstrated that disruption of KIF-4 expression, through the use of a dominant-negative construct or by siRNA-mediated knockdown, reduced the production of HIV-1 VLPs and increased Gag degradation by facilitating the turnover of intracellular Gag levels (Martinez et al., 2008). Live-cell imaging of cells coexpressing RFP-tagged dominant-negative KIF-4 with HIV-1 Gag-GFP demonstrated that KIF4 facilitated HIV-1 Gag trafficking, and the presence of a dominant-negative mutant reduced trafficking to the plasma membrane as compared with coexpression with wild-type KIF-4 (Martinez et al., 2008).

When KIF-4 expression is disrupted, HIV-1 Gag-GFP molecules are visualized to accumulate in the perinuclear region near MTOCs where it accumulates with Ubc9, an E2 SUMO-conjugating enzyme that is associated with the nucleus and nuclear pore (Mo and Moschos, 2005). Ubc9 and SUMO conjugase activity has been demonstrated to facilitate trafficking to the nucleus and mitotic spindle (Bhaskar et al., 2000; Rangasamy et al., 2000; Duprez et al., 1999; Joseph et al., 2002; Saitoh et al., 1997, 1998), mediate protein–protein interactions (Subramanian et al., 2003), and assist in the function of other proteins (Desterro et al., 1998; Giorgino et al., 2000; Rodriguez et al., 1999; Tashiro et al., 1997). Interestingly, Ubc9 has been demonstrated to interact with MPMV Gag in the perinuclear region. Weldon et al. initially identified hUbc9 during a yeast two-hybrid screen using MPMV full-length Gag (Pr78) as bait (Weldon et al., 2003). The interaction was confirmed with in vitro pulldowns using lysates from transfected cells and Gag proteins purified from virions. Imaging studies demonstrated colocalization between Gag and GST-tagged Ubc9 in the nucleus and in the cytoplasm near the nuclear membrane, where MPMV virion assembly occurs. As Ubc9 is known to facilitate nuclear trafficking of other proteins, the authors posit that Ubc9 targets MPMV Gag into the nucleus to facilitate Gag–viral RNA interactions (similar to RSV Gag, see below) for assembly. The authors admit that trafficking of MPMV Gag into the nucleus may not be exclusive to the action of Ubc9, as depletion of intracellular ATP levels with sodium azide and 2-deoxy-d-glucose similarly results in an accumulation of nuclear MPMV Gag (Weldon et al., 1998). Taken together, these findings led researchers to hypothesize that the perinuclear compartment played a role as an intermediate location in the intracellular trafficking of HIV-1 Gag, in a mechanism similar to assembly of MPMV (Martinez et al., 2008). Subsequent imaging studies in support of this idea demonstrated colocalization of HIV-1 Gag with the viral genomic RNA in perinuclear regions and with centrosomes (Poole et al., 2005).

Multivesicular Bodies and Endosomal Proteins in Gag Trafficking

Other host factors identified to be involved in trafficking of HIV-1 Gag to sites of assembly include the δ subunit of the AP-3 complex, which interacts directly with the HIV-1 Gag MA domain to target Gag to the late endosome/multivesicular bodies (MVBs) to facilitate HIV-1 virion assembly and budding (Dong et al., 2005); Plenty of SH3s (hPOSH), an ubiquitin ligase associated with the trans-Golgi network, facilitates targeting of HIV-1 Gag to the plasma membrane (Alroy et al., 2005); and suppressor of cytokine signaling 1 (SOCS1), a protein upregulated in response to HIV-1 infection, stabilizes Gag expression intracellularly and in the supernatant and targets Gag to the plasma membrane (Ryo et al., 1999, 2008). For these proteins, knockdown of the host factor results in mistargeting of HIV-1 Gag to another cellular compartment, resulting in decreased virion formation and infectivity, further demonstrating that the correct expression and targeting of HIV-1 Gag to the plasma membrane is critical for maintaining infectivity (Dong et al., 2005; Alroy et al., 2005; Ryo et al., 1999, 2008). Experiments using primary cell lines derived from patients with AP-3 deficiency demonstrate reduced HIV-1 assembly and virion release that is rescued by exogenous expression of the β3A subunit of AP-3 (Liu et al., 2012). Using microscopy, Liu L. et al. showed that deficiency of AP-3 decreased localization of HIV-1 Gag-CFP to the plasma membrane and treatment with endocytosis inhibitors demonstrated that HIV-1 Gag was not trapped in the clathrin-mediated endocytic pathway (Liu et al., 2012).

These findings suggest that Gag may take advantage of vesicular trafficking factors to facilitate the assembly and budding of virions (Fig. 8.2), a hypothesis supported by the accumulation of HIV-1 Gag in intracellular vesicles in macrophage cells (Raposo et al., 2002). Annexin 2 (Anx2), a protein involved in endosomal transport and actin (Gerke and Moss, 2002; Mollenhauer, 1997) was investigated for its potential role in HIV-1 replication. Ma et al. (2004) initially identified Anx2 (referred to as annexin II) as a host cofactor involved in HIV-1 entry in macrophages. Subsequently, Anx2 was detected by a mass spectrometry proteomics study as a host factor that interacts with HIV-1 $p55^{Gag}$ in activated infected monocyte-derived macrophages (MDMs), and immunofluorescence studies showed colocalization of Anx2 and HIV-1 Gag at late endosomal membranes (CD63-positive) in infected MDMs (Ryzhova et al., 2006). Depletion of Anx2 expression, either siRNA-mediated or by antibody treatment, resulted in decreased virus particle production (Ma et al., 2004; Ryzhova et al., 2006; Woodham et al., 2016). Ryzhoa et al. suggest the budding defect was due to incomplete Gag processing and virus assembly in the absence of Anx2. Consistent with this conclusion, cotransfection of Anx2 and HIV-1_{YU2} proviral construct in 293T cells resulted in increased Gag processing and virus particle production (Harrist et al., 2009). Membrane flotation and bimolecular fluorescence complementation imaging studies showed Anx2–Gag interactions at plasma membranes, specifically in regions enriched in the phospholipid $PI(4,5)P_2$ (see Chapter 11) (Harrist et al., 2009). Further studies found conflicting results, and it was proposed that the role of Anx2 in HIV-1 replication could be cell-type specific (Rai et al., 2010) or might involve an interaction with HIV-1 Env rather than Gag (Woodham et al., 2016).

Similarly, factors involved in vesicle transport between late endosomes, the *trans*-Golgi and the plasma membrane have been implicated in HIV-1 replication (Murray et al., 2005). Using siRNAs, Murray et al. (2005) reduced expression of Rab9, TIP47, p40, PIKfyve, and Rab11A resulting in significant inhibition of HIV-1 replication. Murray et al. attribute the block in virus replication to abnormal intracellular accumulation of HIV-1 Gag. When a Rab9 siRNA stable cell line was transfected with Gag-EGFP the scientists observed higher intracellular Gag expression, defined by the presence of more than 10 Gag-EGFP puncta, in Rab9-siRNA cells as compared with wild-type cells (75% vs. 25%, respectively). However, a subsequent study proposes that Gag seen in endosomes may result from endocytosis of assembled virions from the plasma membrane (Neil et al., 2006). Colocalization has also been demonstrated between Rab11 and MPMV Gag; although the secretory pathway was shown to not be involved, Rab11 mutants interfered with MPMV Gag trafficking (Sfakianos and Hunter, 2003).

Role of Myristoylation and Lipid Binding in Gag Membrane Targeting

The MA domain of HIV-1 Gag also contains a myristylation site, which has been characterized to be important but not sufficient for targeting of Gag to the plasma membrane (Freed and Martin, 1995), and a bipartite signal consisting of myristate plus a cluster of basic amino acids was found to constitute the HIV-1 Gag membrane–binding signal (Zhou et al., 1994). HIV-1 Gag membrane–binding is regulated by a mechanism known as the "myristyl switch," in which the myristic acid moiety is postulated to buried in a hydrophobic pocket until its exposure is triggered when Gag approaches the plasma membrane (Spearman et al., 1997; Zhou and Resh, 1996; Paillart and Gottlinger, 1999; Ono et al., 2000; Resh, 2004; Bryant and Ratner, 1990; Zhou et al., 1994; Veronese et al., 1988). Myristate exposure occurs either due to Gag multimerization or binding of Gag to PI(4,5)P_2, which is preferentially localized to the inner leaflet in the plasma membrane (Tang et al., 2004; Saad et al., 2006). However, kinetic simulations suggest myristate exposure and insertion into the plasma membrane can occur independently of these two events (Charlier et al., 2014). HIV-1 Gag mutants that are not myristylated are unable to assemble into virions and fail to bud from the host cell (Bryant and Ratner, 1990). However, myristoylation is not required for all retroviruses because RSV and EIAV are not N-terminally myristylated. Once at the plasma membrane, Gag proteins have been shown to interact with lipid rafts (see Chapter 11 for a more extensive discussion of this topic).

Multivesicular Body and Endosomal Sorting Complexes Required for Transport Pathway Interactions With Gag Late Domains

At the plasma membrane, Gag relies on host proteins to facilitate release of newly formed virions from infected cells. To accomplish this retroviruses hijack the budding processes used in the MVBs pathway, in which ubiquitinated proteins are delivered to lysosomes for degradation (Fig. 8.2). The pathway utilizes endosomal sorting complexes required for transport (ESCRT) machinery, the components of which were initially characterized in yeast and are highly conserved in vertebrates (Babst et al., 2002a,b; Bache et al., 2003; Katzmann et al., 2001; Martin-Serrano et al., 2003b; Raiborg et al., 2003; Schmidt and Teis, 2012). In the MVB pathway, more than 15 proteins work together in three main complexes: ESCRT-I and -II are involved in identification and sorting of cargo and recruitment of ESCRT-III, which is the core complex involved in budding and fission in the release of vesicles. There are also additional host proteins that interact with the complexes and assist in the pathway (Babst et al., 2002a,b; Katzmann et al., 2001). In the case of retroviruses, the Gag polyprotein directly interacts with the ESCRT machinery via a series of short peptide motifs identified as "late" (L) domains (Freed, 2002; Parent et al., 1995; Wills et al., 1994). Three classes of motifs have been identified: Pro-Thr/Ser-Ala-Pro [P(T/SA)AP], Tyr-Pro-Asp-Leu (YPDL/LYPx_nLxxL), and Pro-Pro-Pro-Tyr (PPPY). Each of these motifs is responsible for the interaction of Gag with a specific ESCRT or ESCRT-related protein. The function of L domains is independent of the position and interchangeable between viruses, shifting the location or swapping the sequence has no effect on virion budding, suggesting the motif mediates direct protein binding (Xiang et al., 1996; Parent et al., 1995; Wills et al., 1994; Yuan et al., 2000; Li et al., 2002; Accola et al., 2000; Stange et al., 2005). Mutation of the PTAP domain in HIV-1 renders the virus noninfectious, this defect can be rescued by *trans* expression of HIV-1 Gag containing a PTAP or YPDL L domain (Martin-Serrano et al., 2001).

Nedd4 and E3 Ubiquitin Ligases

Nedd4 is a family of E3 ubiquitin–protein ligases that are involved in sorting of targets for internalization or endocytosis (Marmor and Yarden, 2004; Staub et al., 2000). In retroviruses, Nedd4 directly binds the PPPY motif on the Gag protein of RSV, MLV, HTLV, and MPMV (Martin-Serrano et al., 2003b; Strack et al., 2000; Blot et al., 2004; Bouamr et al., 2003; Freed, 2002; Garnier et al., 1996; Gottwein et al., 2003; Heidecker et al., 2004; Kikonyogo et al., 2001; Segura-Morales et al., 2005; Sakurai et al., 2004; Yasuda et al., 2002; Yasuda and Hunter, 1998; Yuan et al., 1999; Stange et al., 2005). The PPPY domain is located between the MA and CA domains in the Gag polyprotein of MLV, RSV, and MPMV (Xiang et al., 1996; Yasuda and Hunter, 1998; Yuan et al., 1999). Expression of the WW domains, the protein-interaction domain of Nedd4 that specifically interacts with the PPPY motif, result in inhibition of budding in RSV, HTLV, and MPMV in a dominant-negative manner (Kikonyogo et al., 2001; Patnaik and Wills, 2002; Sakurai et al., 2004; Yasuda et al., 2002). Overexpression of Nedd4 in 293T cells expressing HTLV results in increased budding with no effect on HTLV Gag synthesis or stability (Sakurai et al., 2004). Similarly, deletion of the PPPY motif in MLV abolished p27 in supernatant, as detected by pulse-chase, in comparison to wild type. Reintroduction of the PPPY motif rescued expression of p27 in the supernatant, even without the pp16 protein (Yasuda and Hunter, 1998). As expected, the Gag proteins are ubiquitinated during budding and can be found in virions as mono- or diubiquitinated proteins and ubiquitination plays a critical role in retrovirus budding (Strack et al., 2000; Ott et al., 1998, 2000; Patnaik et al., 2000; Blot et al., 2004; Schubert et al., 2000).

Tsg101 as a Central Player in Wide Range of Gag Late Domain Interactions

Tsg101 was initially identified in a genome-wide RNAi screen as a tumor susceptibility gene and later characterized to be a core component of the ESCRT-I complex, critical for recruitment of ESCRT components, and function in the vacuolar protein sorting (Vps) pathway involved in trafficking targets for lysosomal degradation (Raiborg et al., 2003; Li and Cohen, 1996; Babst et al., 2000; Slagsvold et al., 2006; Lemmon and Traub, 2000). In retrovirus budding, Tsg101 binds the PSAP motif in FIV, MPMV, PFV, and MLV and the PTAP motif in HIV-1, HIV-2, and HTLV (Bouamr et al., 2003; Gottwein et al., 2003; Heidecker et al., 2004; Segura-Morales et al., 2005; Garrus et al., 2001; Myers and Allen, 2002; VerPlank et al., 2001; Gottlinger et al., 1991; Huang et al., 1995; Luttge et al., 2008; Stange et al., 2005). The PTAP motif–Tsg101 interaction in retroviruses occurs through the N-terminal ubiquitin E2 variant domain (UEV) of Tsg101, which cannot mediate ubiquitin transfer but can bind ubiquitin and function as a cofactor (Garrus et al., 2001; VerPlank et al., 2001; Pornillos et al., 2002, 2003; Martin-Serrano et al., 2001; Koonin and Abagyan, 1997; Hofmann and Pickart, 1999). Overexpression of the 5′ end of Tsg101 inhibits HIV-1 virion production. This effect is specific to the specific to the PTAP motif of HIV-1, as a similar result is not seen in MLV or MPMV, which contains PPPY-type motifs (Demirov et al., 2002). Tsg101 was identified as a direct binding partner of HIV-1 p6 by yeast two-hybrid studies (VerPlank et al., 2001). Knockdown of Tsg101 in vivo drastically reduced HIV-1 virus production and infectivity; defects were rescued by expression of a siRNA-resistant Tsg101 (Garrus et al., 2001; VerPlank et al., 2001; Martin-Serrano et al., 2001). Overexpression of the Vps pathway, Tsg101 functions with Vps4, an ATPase that hydrolyzes ATP to disassemble endosome complexes to facilitate endosome recycling. Expression of an ATPase-defective Vps4 results in the formation of defective, enlarged endosomes and mutation of Vps4 disrupts normal Tsg101 trafficking (Bishop and Woodman, 2000, 2001; Martin-Serrano et al., 2003a). Coexpression of ATPase-defective Vps4 with a proviral HIV-1 plasmid reduced HIV-1 particle production in a dose-dependent fashion, suggesting a link between retrovirus budding and the Vps pathway (Garrus et al., 2001).

In HTLV, the PPPY and PTAP motifs are located adjacent to each other in the C terminus of MA (PPPYEPTAP) and successful virion budding requires the complementary actions of Nedd4 and Tsg101 (Blot et al., 2004; Bouamr et al., 2003). Blot et al. (2004) demonstrated that the PPPY/Nedd4 and PTAP/Tsg101 interactions occur in distinct intracellular compartments in a specific order; knockout of either cellular host factor resulted in arrest of Gag trafficking and virion budding. Similarly, in MPMV, the PPPY and PSAP are located adjacent to each other and function in a complementary manner. Gottwein et al. (2003) used EM to visualize the role of each motif in MPMV: mutation of the PPPY motif resulted in accumulation of early stage immature procapsids near the plasma membrane while mutation of the PSAP motif resulted in accumulation of late stage immature procapsids linked together at the plasmid membrane and outside of the host cell. Similarly, FIV contains a PSAP motif near the C terminus of Gag which interacts with Tsg101. Knockdown of Tsg101 by siRNA decreases release of FIV virions in feline cells and overexpression of the 5′ UEV domain has a dominant-negative effect (Luttge et al., 2008). Although FIV Gag does not contain a PPPY domain, Calistri et al. (2009) demonstrated that ubiquitination of FIV Gag by Nedd4-like proteins will rescue a PSAP-mutant virus. A similar experiment in HIV-1 had the same result (Chung et al., 2008). These experiments further corroborate the importance of ubiquitination in retrovirus budding. Ubiquitination of HIV-1 p6 increases the affinity of p6-Tsg101 binding by 10-fold (Garrus et al., 2001).

Alix/AIP-1 Interactions With HIV Gag Proteins

Alix/AIP-1 interacts with the motif in LYPx_nLxxL HIV-1 and HIV-2 and the YPDL motif in EIAV and MLV (Strack et al., 2003; Segura-Morales et al., 2005; Puffer et al., 1997; Martin-Serrano et al., 2003a). Alix/AIP-1 contains a Bro1 domain, which can recruit and bind to ESCRT-III proteins and also bind membranes to stabilize curvature (Babst et al., 2002a; Bissig et al., 2013; Kim et al., 2005). Alix/AIP-1 was proposed to be involved in HIV-1 budding because it was packaged into HIV-1 VLPs in a p6-dependent manner (Strack et al., 2003). The LYPLxxLF motif was subsequently identified in HIV-1 and HIV-2 by looking for sequence homology in the respective p6 domains. Alix/AIP-1-LYPLxxLF binding was confirmed by GST-pulldown assays, the introduction of point mutations abolished binding and reduced VLP production (Strack et al., 2003). As previously mentioned, Tsg101 is critical for HIV-1 budding leading Strack et al. to suggest that Alix/AIP-1 contributes to budding by functioning in a complex with Tsg101 and HIV-1. A screen to identify the human protein interaction network involved in formation of MVBs demonstrated that Alix/AIP-1 and Tsg101 directly interacted, linking ESCRT-1 and ESCRT-III proteins (von Schwedler et al., 2003).

Late Domains in EIAV, MLV, and Foamy Virus Gag Proteins

EIAV p9 was originally identified to contain an L domain in experiments that demonstrated EIAV p9 could replace the L domain in RSV and HIV (Parent et al., 1995). A proline-rich peptide was not readily identifiable in EIAV p9, but progressive C-terminal truncations and alanine-scanning experiments identified the YPDL motif (Puffer et al., 1997). The EIAV YPDL motif was initially identified to interact with the medium chain (AP-50 subunit) of the AP-2 clathrin-associated adaptor protein complex. Coimmunoprecipitation experiments demonstrated the AP-50 interaction was specific to the EIAV YPDL motif, and not the HIV-1 PTAPP or PPPPY RSV motifs. The introduction of individual alanine mutations reduced, but did not completely abrogate, AP-50 binding (Puffer et al., 1998). YPXL/I was later identified as a peptide motif recognized by PalA, a component of the *Aspergillus nidulans* ambient pH signal transduction pathway, and its human homolog Alix/AIP-1(Vincent et al., 2003). Coimmunoprecipitation, alanine mutation screens, and GST-pulldown assays confirmed a direct interaction between EIAV p9 YPDL and Alix/AIP-1 (von Schwedler et al., 2003; Strack et al., 2003; Puffer et al., 1997, 1998). Strack et al. (2003) demonstrated that Alix/AIP-1 bound EIAV p9 more tightly than HIV-1 p6 and mutations in the YPDL motif completely reduced Alix/AIP-1 binding, which led them to conclude that Alix/AIP-1 was the primary binding partner of the YPDL motif. Additionally, mutation of the YPDL motif in EIAV and shRNA knockdown of Alix in HeLa cells reduced particle release 12-fold and 10-fold, respectively (Fisher et al., 2007).

MLV was the first retrovirus Gag protein identified that contains three functional L domain motifs; yeast two-hybrid and coimmunoprecipitation assays to confirm interactions among all three proteins and demonstrated that all three motifs were functional (Segura-Morales et al., 2005). Disruption of the PPPY motif had the greatest impact on budding, which could not be rescued by the interactions at the other two motifs, suggesting the PPPY–Nedd4 interaction was critical in MLV budding. However, all three motifs were required for optimal budding. Interestingly, overexpression of Alix increases MLV budding, suggesting additional host-binding partners of Alix may also participate to facilitate particle release (Segura-Morales et al., 2005).

PFV Gag similarly contains three L domain motifs: a PSAP motif, a PPPI motif, and a YEIL motif. The introduction of mutations to these motifs suggests that the PSAP motif is the primary motif for particle release: the introduction of alanines reduces extracellular infectivity over 1000-fold, as compared with wild type, while mutation of the PPPI and YEIL motifs only reduced infectivity by 10-fold and 4-fold, respectively (Stange et al., 2005). To confirm the role of PSAP–Tsg101 interactions, the overexpression of the Tsg101 5′ region reduces PFV infectivity and virion release, similar to the effect seen in HIV-1, as mentioned earlier.

Endophilin 2 in Murine Leukemia Virus Gag Trafficking

The mouse homolog of human endophilin 2 was also identified as a direct binding partner of MLV Gag, by using the Gag of an MLV isolate as bait in a yeast two-hybrid assay in a mouse T-lymphoma library (Wang et al., 2003). Endophilin 2 is a protein involved in clathrin-mediated endocytosis and both exogenously and endogenously expressed endophilin 2 are found in MLV virions (Huttner and Schmidt, 2000; Wang et al., 2003). However, the role of endophilin 2 in MLV budding is unclear, as overexpression of endophilin 2 functions in a dominant-negative manner to inhibit MLV virion production but siRNA-mediated knockdown of endophilin 2 has no apparent effect on virion production (Wang et al., 2003). The researchers repeated the experiments with HIV-1 Gag and found no effect of endophilin 2 overexpression on formation of HIV-1 VLPs. The authors propose that endophilin 2 may function as part of a larger complex in vesicular trafficking, as AIP/Alix-1, which interacts with HIV-1 Gag as mentioned earlier, has been shown to interact with endophilin (Wang et al., 2003).

Association of Gag With tRNA Synthetases

In some cases, host factors are trafficked by Gag to be packaged into newly formed virions. For example, lysyl-tRNA synthetase (LysRS) is packaged into HIV-1 virions due to association with HIV-1 Gag (Halwani et al., 2004; Cen et al., 2001). The Gag–LysRS interaction has been mapped to occur between the C-terminal 54 amino acids of HIV-1 CA and the region between amino acids 207 and 259 in LysRS (Javanbakht et al., 2003). It is hypothesized that LysRS serves as a signal to facilitate packaging of $tRNA^{Lys}$ into HIV-1 virions (Halwani et al., 2004; Cen et al., 2001, 2004; Javanbakht et al., 2003; Gabor et al., 2002).

FACTORS THAT INTERACT WITH GAG OR VIRAL RNA TO RESTRICT VIRUS REPLICATION

The host cell factors described earlier in this chapter are co-opted by retroviruses to facilitate virus replication. However, there are also restriction factors, proteins that negatively regulate virus trafficking and assembly to limit virus replication in host cells. These factors can interact with Gag to interfere with its normal function, interact with the viral RNA to facilitate RNA genome degradation, or prevent the final steps in virus release.

Host Restriction Factors That Interfere With Gag Functions Early or Late in Replication

Fv1 and TRIM5α function during early stages of virus replication by interacting with CA, although a direct interaction between Fv1 and CA is presumed but not yet demonstrated (Boone et al., 1983, 1988; Kozak and Chakraborti, 1996). Fv1 is a mouse protein initially identified to restrict replication of Friend leukemia virus and later MLV (Lilly, 1967; Pincus et al., 1971a,b; Bishop et al., 2001). Fv1 functions by inhibiting virus replication at a step between reverse transcription and integration (Jolicoeur and Baltimore, 1976; Yang et al., 1980). The block is hypothesized to occur during the nuclear translocation of the PIC, as viable PICs can be isolated from infected cells (Pryciak and Varmus, 1992). Similarly, TRIM5α interacts with CA to block replication after entry and before reverse transcription (Himathongkham and Luciw, 1996; Shibata et al., 1995). TRIM5α was identified in a cDNA screen in primary rhesus monkey lung fibroblast cells to determine the cause of HIV-1 resistance in Old World monkeys (Stremlau et al., 2004). The B30.2 domain in the C-terminus of TRIM5α mediates the interaction with CA, and the sequence of this domain dictates restriction of a specific retrovirus (Hatziioannou et al., 2004; Perez-Caballero et al., 2005; Nakayama et al., 2005; Sebastian and Luban, 2005; Stremlau et al., 2005, 2006; Yap et al., 2005). Fv1 and TRIM5α are not involved in assembly of the virus, so no further details in regard to these proteins will be discussed in this chapter.

The zinc finger antiviral protein (ZAP) was identified as a restriction factor for MLV during a cDNA screen in Rat2 fibroblast cells, which lack detectable thymidine kinase (Topp, 1981). MLV vectors were modified to express thymidine kinase and transduced in Rat2 cells containing a cDNA library (Gao et al., 2002). The cells were treated with trifluorothymidine to select for cells that were nonpermissive to infection and did not express thymidine kinase and the selected cells were assayed to determine which cDNA clone conferred resistance to infection. ZAP contains four CCCH-type zinc fingers, which were previously identified in a family of RNA-binding proteins that negatively regulate mRNAs by binding the 3′UTR of the target mRNA via ARE motifs and recruiting exosomes to degrade the mRNA (Lai et al., 1999, 2000; Carballo et al., 2000; Chen et al., 2001). The second and fourth CCCH zinc fingers allow ZAP to directly bind the MLV 3′UTR; introduction of mutations to disrupt the zinc fingers (C88R and H191R, respectively) eliminate ZAP's ability to inhibit virus replication (Guo et al., 2004). ZAP interferes with the MLV replication cycle by blocking accumulation of viral RNAs in the cytoplasm, possibly by recruiting the exosome to degrade the viral RNA (Gao et al., 2002; Guo et al., 2004). Gao et al. (2002) demonstrated a 30-fold reduction in the amount of viral RNA in the cytoplasm of cells overexpressing ZAP as compared with control cells, with no change in the mRNA levels of housekeeping genes.

Restriction factors can also function by preventing the release of newly packaged virions from the host cell by tethering the virions to surface of the host cell. Tetherin (BST-2/HM1.24/CD317) was identified by scientists seeking to understand the function of the HIV-1 accessory protein Vpu (Neil et al., 2008). Initial experiments with HIV-1 strains lacking Vpu showed accumulation of virions within intracellular vacuoles and along the surface of the host cell (Klimkait et al., 1990; Gottlinger et al., 1993). The requirement of Vpu to facilitate release of virions was shown to be cell type–dependent (Geraghty et al., 1994; Gottlinger et al., 1993; Sakai et al., 1993) and induced by interferon-α (IFN-α) (Neil et al., 2007). Using this information, microarray analysis was performed to determine mRNA expression changes in HIV-1 infected cells with and without Vpu and IFN- α, and tetherin was identified as a host factor that antagonizes virus release (Neil et al., 2008). In single-cycle replication studies, stable expression of tetherin reduced Vpu-negative HIV-1 virion production in two different cell lines (more than 100-fold reduction in 293T cells and approximately 80-fold reduction in HT1080 cells) with minimal effect on wild-type HIV-1 virion production (approximately twofold reduction) (Neil et al., 2008). Similarly, a negative effect on virion production was observed when BST-2 was coexpressed with Vpu-negative HIV-1, and the results of the experiment demonstrated that the ability of Vpu to counteract the effects of BST-2 was saturable (Van Damme et al., 2008). Venkatesh and Bieniasz (2013) demonstrated that tetherin traps virions by forming homodimers aligned along the long axis with one pair of membrane domains in the viral envelope and the other pair in the membrane of the host cell. The negative effects of tetherin on virus budding were also seen in MLV, PFV, and EIAV although these viruses are not known to encode a Vpu-like protein (Yin et al., 2014; Xu et al., 2011; Neil et al., 2008).

A second host factor, UBP (Vpu-binding protein) was found to interfere with HIV-1 particle release and was initially identified in a yeast two-hybrid assay using HIV-1 Vpu as bait in a CD4-negative B lymphocyte cDNA library (Callahan et al., 1998). Overexpression of UBP reduced HIV-1 virion release in a Vpu-dependent manner. GST-pulldown studies showed a direct interaction between UBP and HIV-1 Gag; however, coexpression of HIV-1 Gag, Vpu, and UBP resulted in destabilization of the Gag–UBP interaction in favor of Vpu-UBP (Callahan et al., 1998). These findings suggest an important role of UBP in HIV-1 virion release and suggest that the interaction is through Vpu rather than Gag.

CONCLUDING REMARKS

This chapter summarizes the numerous host factors involved in retroviral genomic RNA and Gag intracellular transport on the journey to the plasma membrane where viruses are released. Although retroviruses encode a varying number of viral proteins to assist in the replication cycle, they have relatively limited genomes and therefore, they are dependent on the host cell to complete their replication cycle. There are three instances of RNA and protein trafficking in retroviruses described: (1) the nuclear export of viral RNA, (2) trafficking of the viral genome through the cytoplasm to the sites of assembly and budding, and (3) trafficking of Gag to sites of assembly and budding. Although the exact host protein involved in trafficking in each of these instances varies, there are certain similarities. For example, simple retroviruses utilize a host protein that bridges viral RNA to the Crm1 nuclear export factor to facilitate export of unspliced viral genomes. In contrast, complex retroviruses rely on a combination of viral accessory proteins and host proteins. It is astonishing that analogous proteins perform the same function considering the variety of host species that retroviruses infect.

There are also specific host protein interactions that are distinctive to a specific class of retrovirus. For example, type B and D retroviruses assemble their virions in the cytoplasm, whereas type C retroviruses assemble their virions at the plasma membrane. Accordingly, differences exist in the trafficking patterns of viral genomes and Gag proteins to facilitate these variations in assembly. These differences in trafficking appear to be equally directed by the host protein being co-opted and by the viral RNA and protein being trafficked.

The host proteins described in this chapter are hijacked by retroviruses for purposes other than their intended cellular role. Retroviruses have adapted over time to co-opt these factors to benefit virus replication. It should be noted that the host factors described in this chapter include currently known proteins, lipids, and RNAs involved in Gag and genomic viral RNA transport and virus particle assembly; many more host interacting factors have to yet to be identified and characterized. The continued study of host factors hijacked by retroviruses provides opportunities to further understand the intricate mechanisms governing the retroviral replication cycle and may provide novel targets for therapeutic intervention in the future.

ACKNOWLEDGMENTS

We are grateful to have received support from the National Institutes of Health awards R01CA76534 (LJP) and F30 CA214010 (ECC) and the Penn State College of Medicine Medical Scientist Training Program and the Department of Medicine.

REFERENCES

Accola, M.A., Strack, B., Gottlinger, H.G., 2000. Efficient particle production by minimal Gag constructs which retain the carboxy-terminal domain of human immunodeficiency virus type 1 capsid-p2 and a late assembly domain. J. Virol. 74, 5395–5402.

Aguilera, A., 2005. Cotranscriptional mRNP assembly: from the DNA to the nuclear pore. Curr. Opin. Cell Biol. 17, 242–250.

Alroy, I., Tuvia, S., Greener, T., Gordon, D., Barr, H.M., Taglicht, D., Mandil-Levin, R., Ben-Avraham, D., Konforty, D., Nir, A., Levius, O., Bicoviski, V., Dori, M., Cohen, S., Yaar, L., Erez, O., Propheta-Meiran, O., Koskas, M., Caspi-Bachar, E., Alchanati, I., Sela-Brown, A., Moskowitz, H., Tessmer, U., Schubert, U., Reiss, Y., 2005. The trans-Golgi network-associated human ubiquitin-protein ligase POSH is essential for HIV type 1 production. Proc. Natl. Acad. Sci. U.S.A. 102, 1478–1483.

Babst, M., Katzmann, D.J., Estepa-Sabal, E.J., Meerloo, T., Emr, S.D., 2002a. Escrt-III: an endosome-associated heterooligomeric protein complex required for mvb sorting. Dev. Cell 3, 271–282.

Babst, M., Katzmann, D.J., Snyder, W.B., Wendland, B., Emr, S.D., 2002b. Endosome-associated complex, ESCRT-II, recruits transport machinery for protein sorting at the multivesicular body. Dev. Cell 3, 283–289.

Babst, M., Odorizzi, G., Estepa, E.J., Emr, S.D., 2000. Mammalian tumor susceptibility gene 101 (TSG101) and the yeast homologue, Vps23p, both function in late endosomal trafficking. Traffic 1, 248–258.

Bacharach, E., Gonsky, J., Alin, K., Orlova, M., Goff, S.P., 2000. The carboxy-terminal fragment of nucleolin interacts with the nucleocapsid domain of retroviral gag proteins and inhibits virion assembly. J. Virol. 74, 11027–11039.

Bache, K.G., Brech, A., Mehlum, A., Stenmark, H., 2003. Hrs regulates multivesicular body formation via ESCRT recruitment to endosomes. J. Cell Biol. 162, 435–442.

Baluyot, M.F., Grosse, S.A., Lyddon, T.D., Janaka, S.K., Johnson, M.C., 2012. CRM1-dependent trafficking of retroviral Gag proteins revisited. J. Virol. 86, 4696–4700.

Belgrader, P., Dey, R., Berezney, R., 1991. Molecular cloning of matrin 3. A 125-kilodalton protein of the nuclear matrix contains an extensive acidic domain. J. Biol. Chem. 266, 9893–9899.

Bender, M.A., Palmer, T.D., Gelinas, R.E., Miller, A.D., 1987. Evidence that the packaging signal of Moloney murine leukemia virus extends into the gag region. J. Virol. 61, 1639–1646.

Bennett, R.P., Wills, J.W., 1999. Conditions for copackaging Rous sarcoma virus and murine leukemia virus Gag proteins during retroviral budding. J. Virol. 73, 2045–2051.

Beyer, A.R., Bann, D.V., Rice, B., Pultz, I.S., Kane, M., Goff, S.P., Golovkina, T.V., Parent, L.J., 2013. Nucleolar trafficking of the mouse mammary tumor virus gag protein induced by interaction with ribosomal protein L9. J. Virol. 87, 1069–1082.

Bhaskar, V., Valentine, S.A., Courey, A.J., 2000. A functional interaction between dorsal and components of the Smt3 conjugation machinery. J. Biol. Chem. 275, 4033–4040.

Bishop, K.N., Bock, M., Towers, G., Stoye, J.P., 2001. Identification of the regions of Fv1 necessary for murine leukemia virus restriction. J. Virol. 75, 5182–5188.

Bishop, N., Woodman, P., 2000. ATPase-defective mammalian VPS4 localizes to aberrant endosomes and impairs cholesterol trafficking. Mol. Biol. Cell 11, 227–239.

Bishop, N., Woodman, P., 2001. TSG101/mammalian VPS23 and mammalian VPS28 interact directly and are recruited to VPS4-induced endosomes. J. Biol. Chem. 276, 11735–11742.

Bissig, C., Lenoir, M., Velluz, M.C., Kufareva, I., Abagyan, R., Overduin, M., Gruenberg, J., 2013. Viral infection controlled by a calcium-dependent lipid-binding module in ALIX. Dev. Cell 25, 364–373.

Bjork, P., Wieslander, L., 2017. Integration of mRNP formation and export. Cell. Mol. Life Sci. 74, 2875–2897.

Blot, V., Perugi, F., Gay, B., Prevost, M.C., Briant, L., Tangy, F., Abriel, H., Staub, O., Dokhelar, M.C., Pique, C., 2004. Nedd4.1-mediated ubiquitination and subsequent recruitment of Tsg101 ensure HTLV-1 Gag trafficking towards the multivesicular body pathway prior to virus budding. J. Cell Sci. 117, 2357–2367.

Bodem, J., Schied, T., Gabriel, R., Rammling, M., Rethwilm, A., 2011. Foamy virus nuclear RNA export is distinct from that of other retroviruses. J. Virol. 85, 2333–2341.

Boeras, I., Seufzer, B., Brady, S., Rendahl, A., Heng, X., Boris-Lawrie, K., 2017. The basal translation rate of authentic HIV-1 RNA is regulated by 5′UTR nt-pairings at junction of R and U5. Sci. Rep. 7, 6902.

Boone, L.R., Glover, P.L., Innes, C.L., Niver, L.A., Bondurant, M.C., Yang, W.K., 1988. Fv-1 N- and B-tropism-specific sequences in murine leukemia virus and related endogenous proviral genomes. J. Virol. 62, 2644–2650.

Boone, L.R., Myer, F.E., Yang, D.M., Ou, C.Y., Koh, C.K., Roberson, L.E., Tennant, R.W., Yang, W.K., 1983. Reversal of Fv-1 host range by in vitro restriction endonuclease fragment exchange between molecular clones of N-tropic and B-tropic murine leukemia virus genomes. J. Virol. 48, 110–119.

Bouamr, F., Melillo, J.A., Wang, M.Q., Nagashima, K., de Los Santos, M., Rein, A., Goff, S.P., 2003. PPPYVEPTAP motif is the late domain of human T-cell leukemia virus type 1 Gag and mediates its functional interaction with cellular proteins Nedd4 and Tsg101 [corrected]. J. Virol. 77, 11882–11895.

Braun, I.C., Rohrbach, E., Schmitt, C., Izaurralde, E., 1999. TAP binds to the constitutive transport element (CTE) through a novel RNA-binding motif that is sufficient to promote CTE-dependent RNA export from the nucleus. EMBO J. 18, 1953–1965.

Bray, M., Prasad, S., Dubay, J.W., Hunter, E., Jeang, K.T., Rekosh, D., Hammarskjold, M.L., 1994. A small element from the Mason-Pfizer monkey virus genome makes human immunodeficiency virus type 1 expression and replication Rev-independent. Proc. Natl. Acad. Sci. U.S.A. 91, 1256–1260.

Brodsky, A.S., Silver, P.A., 2000. Pre-mRNA processing factors are required for nuclear export. RNA 6, 1737–1749.

Bryant, M., Ratner, L., 1990. Myristoylation-dependent replication and assembly of human immunodeficiency virus 1. Proc. Natl. Acad. Sci. U.S.A. 87, 523–527.

Bukrinsky, M.I., Haggerty, S., Dempsey, M.P., Sharova, N., Adzhubel, A., Spitz, L., Lewis, P., Goldfarb, D., Emerman, M., Stevenson, M., 1993. A nuclear localization signal within HIV-1 matrix protein that governs infection of non-dividing cells. Nature 365, 666–669.

Butsch, M., Boris-Lawrie, K., 2002. Destiny of unspliced retroviral RNA: ribosome and/or virion? J. Virol. 76, 3089–3094.

Butterfield-Gerson, K.L., Scheifele, L.Z., Ryan, E.P., Hopper, A.K., Parent, L.J., 2006. Importin-beta family members mediate alpharetrovirus gag nuclear entry via interactions with matrix and nucleocapsid. J. Virol. 80, 1798–1806.

Calistri, A., Del Vecchio, C., Salata, C., Celestino, M., Celegato, M., Gottlinger, H., Palu, G., Parolin, C., 2009. Role of the feline immunodeficiency virus L-domain in the presence or absence of Gag processing: involvement of ubiquitin and Nedd4-2s ligase in viral egress. J. Cell. Physiol. 218, 175–182.

Callahan, M.A., Handley, M.A., Lee, Y.H., Talbot, K.J., Harper, J.W., Panganiban, A.T., 1998. Functional interaction of human immunodeficiency virus type 1 Vpu and Gag with a novel member of the tetratricopeptide repeat protein family. J. Virol. 72, 5189–5197.

Carballo, E., Lai, W.S., Blackshear, P.J., 2000. Evidence that tristetraprolin is a physiological regulator of granulocyte-macrophage colony-stimulating factor messenger RNA deadenylation and stability. Blood 95, 1891–1899.

Carson, J.H., Barbarese, E., 2005. Systems analysis of RNA trafficking in neural cells. Biol. Cell 97, 51–62.

Cen, S., Javanbakht, H., Niu, M., Kleiman, L., 2004. Ability of wild-type and mutant lysyl-tRNA synthetase to facilitate tRNA(Lys) incorporation into human immunodeficiency virus type 1. J. Virol. 78, 1595–1601.

Cen, S., Khorchid, A., Javanbakht, H., Gabor, J., Stello, T., Shiba, K., Musier-Forsyth, K., Kleiman, L., 2001. Incorporation of lysyl-tRNA synthetase into human immunodeficiency virus type 1. J. Virol. 75, 5043–5048.

Charlier, L., Louet, M., Chaloin, L., Fuchs, P., Martinez, J., Muriaux, D., Favard, C., Floquet, N., 2014. Coarse-grained simulations of the HIV-1 matrix protein anchoring: revisiting its assembly on membrane domains. Biophys. J. 106, 577–585.

Chen, C.Y., Gherzi, R., Ong, S.E., Chan, E.L., Raijmakers, R., Pruijn, G.J., Stoecklin, G., Moroni, C., Mann, M., Karin, M., 2001. AU binding proteins recruit the exosome to degrade ARE-containing mRNAs. Cell 107, 451–464.

Chen, J., Rahman, S.A., Nikolaitchik, O.A., Grunwald, D., Sardo, L., Burdick, R.C., Plisov, S., Liang, E., Tai, S., Pathak, V.K., Hu, W.S., 2016. HIV-1 RNA genome dimerizes on the plasma membrane in the presence of Gag protein. Proc. Natl. Acad. Sci. U.S.A. 113, E201–E208.

Chung, H.Y., Morita, E., von Schwedler, U., Muller, B., Krausslich, H.G., Sundquist, W.I., 2008. NEDD4L overexpression rescues the release and infectivity of human immunodeficiency virus type 1 constructs lacking PTAP and YPXL late domains. J. Virol. 82, 4884–4897.

Coffin, J.M., Hughes, S.H., Varmus, H.E., 1997a. The interactions of retroviruses and their hosts. In: Coffin, J.M., Hughes, S.H., Varmus, H.E. (Eds.), Retroviruses Cold Spring Harbor, NY.

Cronshaw, J.M., Krutchinsky, A.N., Zhang, W., Chait, B.T., Matunis, M.J., 2002. Proteomic analysis of the mammalian nuclear pore complex. J. Cell Biol. 158, 915–927.

Cullen, B.R., 2002. Using retroviruses to study the nuclear export of mRNA. Results Probl. Cell Differ. 35, 151–168.

Delchambre, M., Gheysen, D., Thines, D., Thiriart, C., Jacobs, E., Verdin, E., Horth, M., Burny, A., Bex, F., 1989. The GAG precursor of simian immunodeficiency virus assembles into virus-like particles. EMBO J. 8, 2653–2660.

Demirov, D.G., Ono, A., Orenstein, J.M., Freed, E.O., 2002. Overexpression of the N-terminal domain of TSG101 inhibits HIV-1 budding by blocking late domain function. Proc. Natl. Acad. Sci. U.S.A. 99, 955–960.

Desterro, J.M., Rodriguez, M.S., Hay, R.T., 1998. SUMO-1 modification of IkappaBalpha inhibits NF-kappaB activation. Mol. Cell 2, 233–239.

Dickson, C., Eisenman, R., Fan, H., Hunter, E., Teich, N., 1984. Protein biosynthesis and assembly. RNA Tumor Viruses 1, 513–648.

Dong, X., Li, H., Derdowski, A., Ding, L., Burnett, A., Chen, X., Peters, T.R., Dermody, T.S., Woodruff, E., Wang, J.J., Spearman, P., 2005. AP-3 directs the intracellular trafficking of HIV-1 Gag and plays a key role in particle assembly. Cell 120, 663–674.

Dooher, J.E., Schneider, B.L., Reed, J.C., Lingappa, J.R., 2007. Host ABCE1 is at plasma membrane HIV assembly sites and its dissociation from Gag is linked to subsequent events of virus production. Traffic 8, 195–211.

Dupont, S., Sharova, N., Dehoratius, C., Virbasius, C.M., Zhu, X., Bukrinskaya, A.G., Stevenson, M., Green, M.R., 1999. A novel nuclear export activity in HIV-1 matrix protein required for viral replication. Nature 402, 681–685.

Duprez, E., Saurin, A.J., Desterro, J.M., Lallemand-Breitenbach, V., Howe, K., Boddy, M.N., Solomon, E., de The, H., Hay, R.T., Freemont, P.S., 1999. SUMO-1 modification of the acute promyelocytic leukaemia protein PML: implications for nuclear localisation. J. Cell Sci. 112 (Pt 3), 381–393.

Edbauer, C.A., Naso, R.B., 1983. Cytoskeleton-associated Pr65gag and retrovirus assembly. Virology 130, 415–426.

Elia, A.E., Cantley, L.C., Yaffe, M.B., 2003. Proteomic screen finds pSer/pThr-binding domain localizing Plk1 to mitotic substrates. Science 299, 1228–1231.

Erkmann, J.A., Kutay, U., 2004. Nuclear export of mRNA: from the site of transcription to the cytoplasm. Exp. Cell Res. 296, 12–20.

Ernst, R.K., Bray, M., Rekosh, D., Hammarskjold, M.L., 1997. Secondary structure and mutational analysis of the Mason-Pfizer monkey virus RNA constitutive transport element. RNA 3, 210–222.

Felber, B.K., Hadzopoulou-Cladaras, M., Cladaras, C., Copeland, T., Pavlakis, G.N., 1989. rev protein of human immunodeficiency virus type 1 affects the stability and transport of the viral mRNA. Proc. Natl. Acad. Sci. U.S.A. 86, 1495–1499.

Ferrandon, D., Elphick, L., Nusslein-Volhard, C., St Johnston, D., 1994. Staufen protein associates with the 3′UTR of bicoid mRNA to form particles that move in a microtubule-dependent manner. Cell 79, 1221–1232.

Fisher, R.D., Chung, H.Y., Zhai, Q., Robinson, H., Sundquist, W.I., Hill, C.P., 2007. Structural and biochemical studies of ALIX/AIP1 and its role in retrovirus budding. Cell 128, 841–852.

Fornerod, M., Ohno, M., Yoshida, M., Mattaj, I.W., 1997a. CRM1 is an export receptor for leucine-rich nuclear export signals. Cell 90, 1051–1060.

Fornerod, M., van Deursen, J., van Baal, S., Reynolds, A., Davis, D., Murti, K.G., Fransen, J., Grosveld, G., 1997b. The human homologue of yeast CRM1 is in a dynamic subcomplex with CAN/Nup214 and a novel nuclear pore component Nup88. EMBO J. 16, 807–816.

Freed, E.O., 1998. HIV-1 gag proteins: diverse functions in the virus life cycle. Virology 251, 1–15.

Freed, E.O., 2002. Viral late domains. J. Virol. 76, 4679–4687.

Freed, E.O., Martin, M.A., 1995. Virion incorporation of envelope glycoproteins with long but not short cytoplasmic tails is blocked by specific, single amino acid substitutions in the human immunodeficiency virus type 1 matrix. J. Virol. 69, 1984–1989.

Furuichi, Y., Shatkin, A.J., Stavnezer, E., Bishop, J.M., 1975. Blocked, methylated 5′-terminal sequence in avian sarcoma virus RNA. Nature 257, 618–620.

Gabor, J., Cen, S., Javanbakht, H., Niu, M., Kleiman, L., 2002. Effect of altering the tRNA(Lys)(3) concentration in human immunodeficiency virus type 1 upon its annealing to viral RNA, GagPol incorporation, and viral infectivity. J. Virol. 76, 9096–9102.

Gao, G., Guo, X., Goff, S.P., 2002. Inhibition of retroviral RNA production by ZAP, a CCCH-type zinc finger protein. Science 297, 1703–1706.

Gao, G., Li, M., Zhang, J., 2014. Tandem immunoprecipitation approach to identify HIV-1 Gag associated host factors. J. Virol. Methods 203, 116–119.

Garbitt-Hirst, R., Kenney, S.P., Parent, L.J., 2009. Genetic evidence for a connection between Rous sarcoma virus gag nuclear trafficking and genomic RNA packaging. J. Virol. 83, 6790–6797.

Garnier, L., Wills, J.W., Verderame, M.F., Sudol, M., 1996. WW domains and retrovirus budding. Nature 381, 744–745.

Garrus, J.E., von Schwedler, U.K., Pornillos, O.W., Morham, S.G., Zavitz, K.H., Wang, H.E., Wettstein, D.A., Stray, K.M., Cote, M., Rich, R.L., Myszka, D.G., Sundquist, W.I., 2001. Tsg101 and the vacuolar protein sorting pathway are essential for HIV-1 budding. Cell 107, 55–65.

Geraghty, R.J., Talbot, K.J., Callahan, M., Harper, W., Panganiban, A.T., 1994. Cell type-dependence for Vpu function. J. Med. Primatol. 23, 146–150.

Gerke, V., Moss, S.E., 2002. Annexins: from structure to function. Physiol. Rev. 82, 331–371.

Gheysen, D., Jacobs, E., de Foresta, F., Thiriart, C., Francotte, M., Thines, D., De Wilde, M., 1989. Assembly and release of HIV-1 precursor Pr55gag virus-like particles from recombinant baculovirus-infected insect cells. Cell 59, 103–112.

Ghosh, S., Jacobson, A., 2010. RNA decay modulates gene expression and controls its fidelity. Wiley Interdiscip. Rev. RNA 1, 351–361.

Giles, K.E., Beemon, K.L., 2005. Retroviral splicing suppressor sequesters a 3′ splice site in a 50S aberrant splicing complex. Mol. Cell. Biol. 25, 4397–4405.

Gillespie, D., Gallo, R.C., Marshall, S., 1972. RNA of RNA tumour viruses contains poly A. Nat. N. Biol. 236, 227–231.

Giorgino, F., de Robertis, O., Laviola, L., Montrone, C., Perrini, S., Mccowen, K.C., Smith, R.J., 2000. The sentrin-conjugating enzyme mUbc9 interacts with GLUT4 and GLUT1 glucose transporters and regulates transporter levels in skeletal muscle cells. Proc. Natl. Acad. Sci. U.S.A. 97, 1125–1130.

Gottlinger, H.G., Dorfman, T., Cohen, E.A., Haseltine, W.A., 1993. Vpu protein of human immunodeficiency virus type 1 enhances the release of capsids produced by gag gene constructs of widely divergent retroviruses. Proc. Natl. Acad. Sci. U.S.A. 90, 7381–7385.

Gottlinger, H.G., Dorfman, T., Sodroski, J.G., Haseltine, W.A., 1991. Effect of mutations affecting the p6 gag protein on human immunodeficiency virus particle release. Proc. Natl. Acad. Sci. U.S.A. 88, 3195–3199.

Gottwein, E., Bodem, J., Muller, B., Schmechel, A., Zentgraf, H., Krausslich, H.G., 2003. The Mason-Pfizer monkey virus PPPY and PSAP motifs both contribute to virus release. J. Virol. 77, 9474–9485.

Greatorex, J., 2004. The retroviral RNA dimer linkage: different structures may reflect different roles. Retrovirology 1, 22.

Grewe, B., Hoffmann, B., Ohs, I., Blissenbach, M., Brandt, S., Tippler, B., Grunwald, T., Uberla, K., 2012. Cytoplasmic utilization of human immunodeficiency virus type 1 genomic RNA is not dependent on a nuclear interaction with gag. J. Virol. 86, 2990–3002.

Gruter, P., Tabernero, C., von Kobbe, C., Schmitt, C., Saavedra, C., Bachi, A., Wilm, M., Felber, B.K., Izaurralde, E., 1998. TAP, the human homolog of Mex67p, mediates CTE-dependent RNA export from the nucleus. Mol. Cell 1, 649–659.

Gudleski, N., Flanagan, J.M., Ryan, E.P., Bewley, M.C., Parent, L.J., 2010. Directionality of nucleocytoplasmic transport of the retroviral gag protein depends on sequential binding of karyopherins and viral RNA. Proc. Natl. Acad. Sci. U.S.A. 107, 9358–9363.

Guo, X., Carroll, J.W., Macdonald, M.R., Goff, S.P., Gao, G., 2004. The zinc finger antiviral protein directly binds to specific viral mRNAs through the CCCH zinc finger motifs. J. Virol. 78, 12781–12787.

Guzik, B.W., Levesque, L., Prasad, S., Bor, Y.-C., Black, B.E., Paschal, B.M., Rekosh, D., Hammarskjöld, M.-L., 2001. NXT1 (p15) is a crucial cellular cofactor in TAP-dependent export of intron-containing RNA in mammalian cells. Mol. Cell. Biol. 21, 2545–2554.

Haffar, O., Garrigues, J., Travis, B., Moran, P., Zarling, J., Hu, S.L., 1990. Human immunodeficiency virus-like, nonreplicating, gag-env particles assemble in a recombinant vaccinia virus expression system. J. Virol. 64, 2653–2659.

Haffar, O.K., Popov, S., Dubrovsky, L., Agostini, I., Tang, H., Pushkarsky, T., Nadler, S.G., Bukrinsky, M., 2000. Two nuclear localization signals in the HIV-1 matrix protein regulate nuclear import of the HIV-1 pre-integration complex. J. Mol. Biol. 299, 359–368.

Halwani, R., Cen, S., Javanbakht, H., Saadatmand, J., Kim, S., Shiba, K., Kleiman, L., 2004. Cellular distribution of Lysyl-tRNA synthetase and its interaction with Gag during human immunodeficiency virus type 1 assembly. J. Virol. 78, 7553–7564.

Hansen, M., Jelinek, L., Whiting, S., Barklis, E., 1990. Transport and assembly of gag proteins into Moloney murine leukemia virus. J. Virol. 64, 5306–5316.

Harris, M.E., Hope, T.J., 2000. RNA export: insights from viral models. Essays Biochem. 36, 115–127.

Harrist, A.V., Ryzhova, E.V., Harvey, T., Gonzalez-Scarano, F., 2009. Anx2 interacts with HIV-1 Gag at phosphatidylinositol (4,5) bisphosphate-containing lipid rafts and increases viral production in 293T cells. PLoS One 4, e5020.

Hatziioannou, T., Perez-Caballero, D., Yang, A., Cowan, S., Bieniasz, P.D., 2004. Retrovirus resistance factors Ref1 and Lv1 are species-specific variants of TRIM5alpha. Proc. Natl. Acad. Sci. U.S.A. 101, 10774–10779.

Heidecker, G., Lloyd, P.A., Fox, K., Nagashima, K., Derse, D., 2004. Late assembly motifs of human T-cell leukemia virus type 1 and their relative roles in particle release. J. Virol. 78, 6636–6648.

Herold, A., Klymenko, T., Izaurralde, E., 2001. NXF1/p15 heterodimers are essential for mRNA nuclear export in Drosophila. RNA 7, 1768–1780.

Hibbert, C.S., Gontarek, R.R., Beemon, K.L., 1999. The role of overlapping U1 and U11 5′ splice site sequences in a negative regulator of splicing. RNA 5, 333–343.

Hibino, Y., Usui, T., Morita, Y., Hirose, N., Okazaki, M., Sugano, N., Hiraga, K., 2006. Molecular properties and intracellular localization of rat liver nuclear scaffold protein P130. Biochim. Biophys. Acta 1759, 195–207.

Hidaka, M., Inoue, J., Yoshida, M., Seiki, M., 1988. Post-transcriptional regulator (rex) of HTLV-1 initiates expression of viral structural proteins but suppresses expression of regulatory proteins. EMBO J. 7, 519–523.

Himathongkham, S., Luciw, P.A., 1996. Restriction of HIV-1 (subtype B) replication at the entry step in rhesus macaque cells. Virology 219, 485–488.

Hofmann, R.M., Pickart, C.M., 1999. Noncanonical MMS2-encoded ubiquitin-conjugating enzyme functions in assembly of novel polyubiquitin chains for DNA repair. Cell 96, 645–653.

Hu, W.S., Rhodes, T., Dang, Q., Pathak, V., 2003. Retroviral recombination: review of genetic analyses. Front. Biosci. 8, d143–d155.

Huang, M., Orenstein, J.M., Martin, M.A., Freed, E.O., 1995. p6Gag is required for particle production from full-length human immunodeficiency virus type 1 molecular clones expressing protease. J. Virol. 69, 6810–6818.

Huttner, W.B., Schmidt, A., 2000. Lipids, lipid modification and lipid–protein interaction in membrane budding and fission—insights from the roles of endophilin A1 and synaptophysin in synaptic vesicle endocytosis. Curr. Opin. Neurobiol. 10, 543–551.

Jackson, D.A., 2004. Regulating gene expression in mammalian cells: how nuclear architecture influences mRNA synthesis and export to the cytoplasm. Symp. Soc. Exp. Biol. 135–155.

Javanbakht, H., Halwani, R., Cen, S., Saadatmand, J., Musier-Forsyth, K., Gottlinger, H., Kleiman, L., 2003. The interaction between HIV-1 Gag and human lysyl-tRNA synthetase during viral assembly. J. Biol. Chem. 278, 27644–27651.

Jetzt, A.E., Yu, H., Klarmann, G.J., Ron, Y., Preston, B.D., Dougherty, J.P., 2000. High rate of recombination throughout the human immunodeficiency virus type 1 genome. J. Virol. 74, 1234–1240.

Jin, L., Guzik, B.W., Bor, Y.C., Rekosh, D., Hammarskjold, M.L., 2003. Tap and NXT promote translation of unspliced mRNA. Genes Dev. 17, 3075–3086.

Jolicoeur, P., Baltimore, D., 1976. Effect of Fv-1 gene product on proviral DNA formation and integration in cells infected with murine leukemia viruses. Proc. Natl. Acad. Sci. U.S.A. 73, 2236–2240.

Joseph, J., Tan, S.H., Karpova, T.S., Mcnally, J.G., Dasso, M., 2002. SUMO-1 targets RanGAP1 to kinetochores and mitotic spindles. J. Cell Biol. 156, 595–602.

Jouvenet, N., Laine, S., Pessel-Vivares, L., Mougel, M., 2011. Cell biology of retroviral RNA packaging. RNA Biol. 8, 572–580.

Kang, Y., Cullen, B.R., 1999. The human Tap protein is a nuclear mRNA export factor that contains novel RNA-binding and nucleocytoplasmic transport sequences. Genes Dev. 13, 1126–1139.

Karn, J., Stoltzfus, C.M., 2012. Transcriptional and posttranscriptional regulation of HIV-1 gene expression. Cold Spring Harb. Perspect. Med. 2, a006916.

Katzmann, D.J., Babst, M., Emr, S.D., 2001. Ubiquitin-dependent sorting into the multivesicular body pathway requires the function of a conserved endosomal protein sorting complex, ESCRT-I. Cell 106, 145–155.

Keane, S.C., Heng, X., Lu, K., Kharytonchyk, S., Ramakrishnan, V., Carter, G., Barton, S., Hosic, A., Florwick, A., Santos, J., Bolden, N.C., Mccowin, S., Case, D.A., Johnson, B.A., Salemi, M., Telesnitsky, A., Summers, M.F., 2015. RNA structure. Structure of the HIV-1 RNA packaging signal. Science 348, 917–921.

Keane, S.C., Summers, M.F., 2016. NMR studies of the structure and function of the HIV-1 5′-leader. Viruses 8.

Keane, S.C., Van, V., Frank, H.M., Sciandra, C.A., Mccowin, S., Santos, J., Heng, X., Summers, M.F., 2016. NMR detection of intermolecular interaction sites in the dimeric 5'-leader of the HIV-1 genome. Proc. Natl. Acad. Sci. U.S.A. 113, 13033–13038.

Keith, J., Fraenkel-Conrat, H., 1975. Identification of the 5′ end of Rous sarcoma virus RNA. Proc. Natl. Acad. Sci. U.S.A. 72, 3347–3350.

Kemler, I., Meehan, A., Poeschla, E.M., 2010. Live-cell coimaging of the genomic RNAs and Gag proteins of two lentiviruses. J. Virol. 84, 6352–6366.

Kemler, I., Saenz, D., Poeschla, E., 2012. Feline immunodeficiency virus Gag is a nuclear shuttling protein. J. Virol. 86, 8402–8411.

Kharytonchyk, S.A., Kireyeva, A.I., Osipovich, A.B., Fomin, I.K., 2005. Evidence for preferential copackaging of Moloney murine leukemia virus genomic RNAs transcribed in the same chromosomal site. Retrovirology 2, 3.

Kikonyogo, A., Bouamr, F., Vana, M.L., Xiang, Y., Aiyar, A., Carter, C., Leis, J., 2001. Proteins related to the Nedd4 family of ubiquitin protein ligases interact with the L domain of Rous sarcoma virus and are required for gag budding from cells. Proc. Natl. Acad. Sci. U.S.A. 98, 11199–11204.

Kim, J., Sitaraman, S., Hierro, A., Beach, B.M., Odorizzi, G., Hurley, J.H., 2005. Structural basis for endosomal targeting by the Bro1 domain. Dev. Cell 8, 937–947.

Kim, W., Tang, Y., Okada, Y., Torrey, T.A., Chattopadhyay, S.K., Pfleiderer, M., Falkner, F.G., Dorner, F., Choi, W., Hirokawa, N., Morse 3rd, H.C., 1998. Binding of murine leukemia virus Gag polyproteins to KIF4, a microtubule-based motor protein. J. Virol. 72, 6898–6901.

Klimkait, T., Strebel, K., Hoggan, M.D., Martin, M.A., Orenstein, J.M., 1990. The human immunodeficiency virus type 1-specific protein vpu is required for efficient virus maturation and release. J. Virol. 64, 621–629.

Koonin, E.V., Abagyan, R.A., 1997. TSG101 may be the prototype of a class of dominant negative ubiquitin regulators. Nat. Genet. 16, 330–331.

Kosturko, L.D., Maggipinto, M.J., D'sa, C., Carson, J.H., Barbarese, E., 2005. The microtubule-associated protein tumor overexpressed gene binds to the RNA trafficking protein heterogeneous nuclear ribonucleoprotein A2. Mol. Biol. Cell 16, 1938–1947.

Kozak, C.A., Chakraborti, A., 1996. Single amino acid changes in the murine leukemia virus capsid protein gene define the target of Fv1 resistance. Virology 225, 300–305.

Krausslich, H.G., Welker, R., 1996. Intracellular transport of retroviral capsid components. Curr. Top. Microbiol. Immunol. 214, 25–63.

Krishna, N.K., Weldon, R., Wills, J.W., 1996. Transport and processing of the Rous sarcoma virus Gag protein in the endoplasmic reticulum. J. Virol. 70, 1570–1579.

Kula, A., Gharu, L., Marcello, A., 2013. HIV-1 pre-mRNA commitment to Rev mediated export through PSF and Matrin 3. Virology 435, 329–340.

Kula, A., Guerra, J., Knezevich, A., Kleva, D., Myers, M.P., Marcello, A., 2011. Characterization of the HIV-1 RNA associated proteome identifies Matrin 3 as a nuclear cofactor of Rev function. Retrovirology 8, 60.

Lai, M.M., Duesberg, P.H., 1972. Adenylic acid-rich sequence in RNAs of Rous sarcoma virus and Rauscher mouse leukaemia virus. Nature 235, 383–386.

Lai, W.S., Carballo, E., Strum, J.R., Kennington, E.A., Phillips, R.S., Blackshear, P.J., 1999. Evidence that tristetraprolin binds to AU-rich elements and promotes the deadenylation and destabilization of tumor necrosis factor alpha mRNA. Mol. Cell. Biol. 19, 4311–4323.

Lai, W.S., Carballo, E., Thorn, J.M., Kennington, E.A., Blackshear, P.J., 2000. Interactions of CCCH zinc finger proteins with mRNA. Binding of tristetraprolin-related zinc finger proteins to Au-rich elements and destabilization of mRNA. J. Biol. Chem. 275, 17827–17837.

LeBlanc, J.J., Beemon, K.L., 2004. Unspliced Rous sarcoma virus genomic RNAs are translated and subjected to nonsense-mediated mRNA decay before packaging. J. Virol. 78, 5139–5146.

LeBlanc, J.J., Uddowla, S., Abraham, B., Clatterbuck, S., Beemon, K.L., 2007. Tap and Dbp5, but not Gag, are involved in DR-mediated nuclear export of unspliced Rous sarcoma virus RNA. Virology 363, 376–386.

Lemmon, S.K., Traub, L.M., 2000. Sorting in the endosomal system in yeast and animal cells. Curr. Opin. Cell Biol. 12, 457–466.

Lesbats, P., Serrao, E., Maskell, D.P., Pye, V.E., O'reilly, N., Lindemann, D., Engelman, A.N., Cherepanov, P., 2017. Structural basis for spumavirus GAG tethering to chromatin. Proc. Natl. Acad. Sci. U.S.A. 114, 5509–5514.

Leung, J., Yueh, A., Appah Jr., F.S., Yuan, B., De Los Santos, K., Goff, S.P., 2006. Interaction of Moloney murine leukemia virus matrix protein with IQGAP. EMBO J. 25, 2155–2166.

Levesque, L., Guzik, B., Guan, T., Coyle, J., Black, B.E., Rekosh, D., Hammarskjold, M.L., Paschal, B.M., 2001. RNA export mediated by tap involves NXT1-dependent interactions with the nuclear pore complex. J. Biol. Chem. 276, 44953–44962.

Levin, J.G., Grimley, P.M., Ramseur, J.M., Berezesky, I.K., 1974. Deficiency of 60 to 70S RNA in murine leukemia virus particles assembled in cells treated with actinomycin D. J. Virol. 14, 152–161.

Levin, J.G., Rosenak, M.J., 1976. Synthesis of murine leukemia virus proteins associated with virions assembled in actinomycin D-treated cells: evidence for persistence of viral messenger RNA. Proc. Natl. Acad. Sci. U.S.A. 73, 1154–1158.

Li, F., Chen, C., Puffer, B.A., Montelaro, R.C., 2002. Functional replacement and positional dependence of homologous and heterologous L domains in equine infectious anemia virus replication. J. Virol. 76, 1569–1577.

Li, L., Cohen, S.N., 1996. Tsg101: a novel tumor susceptibility gene isolated by controlled homozygous functional knockout of allelic loci in mammalian cells. Cell 85, 319–329.

Lilly, F., 1967. Susceptibility to two strains of Friend leukemia virus in mice. Science 155, 461–462.

Lingappa, J.R., Hill, R.L., Wong, M.L., Hegde, R.S., 1997. A multistep, ATP-dependent pathway for assembly of human immunodeficiency virus capsids in a cell-free system. J. Cell Biol. 136, 567–581.

Linial, M.L., Miller, A.D., 1990. Retroviral RNA packaging: sequence requirements and implications. Curr. Top. Microbiol. Immunol. 157, 125–152.

Liu, L., Sutton, J., Woodruff, E., Villalta, F., Spearman, P., Dong, X., 2012. Defective HIV-1 particle assembly in AP-3-deficient cells derived from patients with Hermansky-Pudlak syndrome type 2. J. Virol. 86, 11242–11253.

Lochmann, T.L., Bann, D.V., Ryan, E.P., Beyer, A.R., Mao, A., Cochrane, A., Parent, L.J., 2013. NC-mediated nucleolar localization of retroviral gag proteins. Virus Res. 171, 304–318.

Luo, L., Li, Y., Kang, C.Y., 1990. Expression of gag precursor protein and secretion of virus-like gag particles of HIV-2 from recombinant baculovirus-infected insect cells. Virology 179, 874–880.

Luttge, B.G., Shehu-Xhilaga, M., Demirov, D.G., Adamson, C.S., Soheilian, F., Nagashima, K., Stephen, A.G., Fisher, R.J., Freed, E.O., 2008. Molecular characterization of feline immunodeficiency virus budding. J. Virol. 82, 2106–2119.

Ma, G., Greenwell-Wild, T., Lei, K., Jin, W., Swisher, J., Hardegen, N., Wild, C.T., Wahl, S.M., 2004. Secretory leukocyte protease inhibitor binds to annexin II, a cofactor for macrophage HIV-1 infection. J. Exp. Med. 200, 1337–1346.

Madisen, L., Travis, B., Hu, S.L., Purchio, A.F., 1987. Expression of the human immunodeficiency virus gag gene in insect cells. Virology 158, 248–250.

Malim, M.H., Hauber, J., Le, S.Y., Maizel, J.V., Cullen, B.R., 1989. The HIV-1 rev trans-activator acts through a structured target sequence to activate nuclear export of unspliced viral mRNA. Nature 338, 254–257.

Marmor, M.D., Yarden, Y., 2004. Role of protein ubiquitylation in regulating endocytosis of receptor tyrosine kinases. Oncogene 23, 2057–2070.

Martin-Serrano, J., Zang, T., Bieniasz, P.D., 2001. HIV-1 and Ebola virus encode small peptide motifs that recruit Tsg101 to sites of particle assembly to facilitate egress. Nat. Med. 7, 1313–1319.

Martin-Serrano, J., Yarovoy, A., Perez-Caballero, D., Bieniasz, P.D., 2003a. Divergent retroviral late-budding domains recruit vacuolar protein sorting factors by using alternative adaptor proteins. Proc. Natl. Acad. Sci. U.S.A. 100, 12414–12419.

Martin-Serrano, J., Zang, T., Bieniasz, P.D., 2003b. Role of ESCRT-I in retroviral budding. J. Virol. 77, 4794–4804.

Martinez, N.W., Xue, X., Berro, R.G., Kreitzer, G., Resh, M.D., 2008. Kinesin KIF4 regulates intracellular trafficking and stability of the human immunodeficiency virus type 1 Gag polyprotein. J. Virol. 82, 9937–9950.

Maurel, S., Mougel, M., 2010. Murine leukemia virus RNA dimerization is coupled to transcription and splicing processes. Retrovirology 7, 64.

McNally, M.T., Gontarek, R.R., Beemon, K., 1991. Characterization of Rous sarcoma virus intronic sequences that negatively regulate splicing. Virology 185, 99–108.

Mertz, J.A., Simper, M.S., Lozano, M.M., Payne, S.M., Dudley, J.P., 2005. Mouse mammary tumor virus encodes a self-regulatory RNA export protein and is a complex retrovirus. J. Virol. 79, 14737–14747.

Mo, Y.Y., Moschos, S.J., 2005. Targeting Ubc9 for cancer therapy. Expert Opin. Ther. Targets 9, 1203–1216.

Mollenhauer, J., 1997. Annexins: what are they good for? Cell. Mol. Life Sci. 53, 506–507.

Monette, A., Ajamian, L., Lopez-Lastra, M., Mouland, A.J., 2009. Human immunodeficiency virus type 1 (HIV-1) induces the cytoplasmic retention of heterogeneous nuclear ribonucleoprotein A1 by disrupting nuclear import: implications for HIV-1 gene expression. J. Biol. Chem. 284, 31350–31362.

Morikawa, Y., Hinata, S., Tomoda, H., Goto, T., Nakai, M., Aizawa, C., Tanaka, H., Omura, S., 1996. Complete inhibition of human immunodeficiency virus Gag myristoylation is necessary for inhibition of particle budding. J. Biol. Chem. 271, 2868–2873.

Mouland, A.J., Mercier, J., Luo, M., Bernier, L., Desgroseillers, L., Cohen, E.A., 2000. The double-stranded RNA-binding protein Staufen is incorporated in human immunodeficiency virus type 1: evidence for a role in genomic RNA encapsidation. J. Virol. 74, 5441–5451.

Mouland, A.J., Xu, H., Cui, H., Krueger, W., Munro, T.P., Prasol, M., Mercier, J., Rekosh, D., Smith, R., Barbarese, E., Cohen, E.A., Carson, J.H., 2001. RNA trafficking signals in human immunodeficiency virus type 1. Mol. Cell. Biol. 21, 2133–2143.

Mullers, E., Stirnnagel, K., Kaulfuss, S., Lindemann, D., 2011. Prototype foamy virus gag nuclear localization: a novel pathway among retroviruses. J. Virol. 85, 9276–9285.

Murray, J.L., Mavrakis, M., Mcdonald, N.J., Yilla, M., Sheng, J., Bellini, W.J., Zhao, L., Le Doux, J.M., Shaw, M.W., Luo, C.C., Lippincott-Schwartz, J., Sanchez, A., Rubin, D.H., Hodge, T.W., 2005. Rab9 GTPase is required for replication of human immunodeficiency virus type 1, filoviruses, and measles virus. J. Virol. 79, 11742–11751.

Myers, E.L., Allen, J.F., 2002. Tsg101, an inactive homologue of ubiquitin ligase e2, interacts specifically with human immunodeficiency virus type 2 gag polyprotein and results in increased levels of ubiquitinated gag. J. Virol. 76, 11226–11235.

Nakayama, E.E., Miyoshi, H., Nagai, Y., Shioda, T., 2005. A specific region of 37 amino acid residues in the SPRY (B30.2) domain of African green monkey TRIM5alpha determines species-specific restriction of simian immunodeficiency virus SIVmac infection. J. Virol. 79, 8870–8877.

Nakayasu, H., Berezney, R., 1991. Nuclear matrins: identification of the major nuclear matrix proteins. Proc. Natl. Acad. Sci. U.S.A. 88, 10312–10316.

Nakielny, S., Dreyfuss, G., 1999. Transport of proteins and RNAs in and out of the nucleus. Cell 99, 677–690.

Nash, M.A., Meyer, M.K., Decker, G.L., Arlinghaus, R.B., 1993. A subset of Pr65gag is nucleus associated in murine leukemia virus-infected cells. J. Virol. 67, 1350–1356.

Neil, S.J., Eastman, S.W., Jouvenet, N., Bieniasz, P.D., 2006. HIV-1 Vpu promotes release and prevents endocytosis of nascent retrovirus particles from the plasma membrane. PLoS Pathog. 2, e39.

Neil, S.J., Sandrin, V., Sundquist, W.I., Bieniasz, P.D., 2007. An interferon-alpha-induced tethering mechanism inhibits HIV-1 and Ebola virus particle release but is counteracted by the HIV-1 Vpu protein. Cell Host Microbe 2, 193–203.

Neil, S.J., Zang, T., Bieniasz, P.D., 2008. Tetherin inhibits retrovirus release and is antagonized by HIV-1 Vpu. Nature 451, 425–430.

Neumann, S., Petfalski, E., Brugger, B., Grosshans, H., Wieland, F., Tollervey, D., Hurt, E., 2003. Formation and nuclear export of tRNA, rRNA and mRNA is regulated by the ubiquitin ligase Rsp5p. EMBO Rep. 4, 1156–1162.

Neville, M., Stutz, F., Lee, L., Davis, L.I., Rosbash, M., 1997. The importin-beta family member Crm1p bridges the interaction between Rev and the nuclear pore complex during nuclear export. Curr. Biol. 7, 767–775.

Ogert, R.A., Beemon, K.L., 1998. Mutational analysis of the Rous sarcoma virus DR posttranscriptional control element. J. Virol. 72, 3407–3411.

Ogert, R.A., Lee, L.H., Beemon, K.L., 1996. Avian retroviral RNA element promotes unspliced RNA accumulation in the cytoplasm. J. Virol. 70, 3834–3843.

Onafuwa, A., An, W., Robson, N.D., Telesnitsky, A., 2003. Human immunodeficiency virus type 1 genetic recombination is more frequent than that of Moloney murine leukemia virus despite similar template switching rates. J. Virol. 77, 4577–4587.

Ono, A., Demirov, D., Freed, E.O., 2000. Relationship between human immunodeficiency virus type 1 Gag multimerization and membrane binding. J. Virol. 74, 5142–5150.

Ott, D.E., Coren, L.V., Chertova, E.N., Gagliardi, T.D., Schubert, U., 2000. Ubiquitination of HIV-1 and MuLV Gag. Virology 278, 111–121.

Ott, D.E., Coren, L.V., Copeland, T.D., Kane, B.P., Johnson, D.G., Sowder 2nd, R.C., Yoshinaka, Y., Oroszlan, S., Arthur, L.O., Henderson, L.E., 1998. Ubiquitin is covalently attached to the p6Gag proteins of human immunodeficiency virus type 1 and simian immunodeficiency virus and to the p12Gag protein of Moloney murine leukemia virus. J. Virol. 72, 2962–2968.

Paca, R.E., Ogert, R.A., Hibbert, C.S., Izaurralde, E., Beemon, K.L., 2000. Rous sarcoma virus DR posttranscriptional elements use a novel RNA export pathway. J. Virol. 74, 9507–9514.

Paillart, J.C., Gottlinger, H.G., 1999. Opposing effects of human immunodeficiency virus type 1 matrix mutations support a myristyl switch model of gag membrane targeting. J. Virol. 73, 2604–2612.

Paillart, J.C., Marquet, R., Skripkin, E., Ehresmann, C., Ehresmann, B., 1996. Dimerization of retroviral genomic RNAs: structural and functional implications. Biochimie 78, 639–653.

Pal, R., Mumbauer, S., Hoke, G.M., Takatsuki, A., Sarngadharan, M.G., 1991. Brefeldin A inhibits the processing and secretion of envelope glycoproteins of human immunodeficiency virus type 1. AIDS Res. Hum. Retrovir. 7, 707–712.

Parent, L.J., Bennett, R.P., Craven, R.C., Nelle, T.D., Krishna, N.K., Bowzard, J.B., Wilson, C.B., Puffer, B.A., Montelaro, R.C., Wills, J.W., 1995. Positionally independent and exchangeable late budding functions of the Rous sarcoma virus and human immunodeficiency virus Gag proteins. J. Virol. 69, 5455–5460.

Parent, L.J., Wilson, C.B., Resh, M.D., Wills, J.W., 1996. Evidence for a second function of the MA sequence in the Rous sarcoma virus Gag protein. J. Virol. 70, 1016–1026.

Patnaik, A., Chau, V., Wills, J.W., 2000. Ubiquitin is part of the retrovirus budding machinery. Proc. Natl. Acad. Sci. U.S.A. 97, 13069–13074.

Patnaik, A., Wills, J.W., 2002. In vivo interference of Rous sarcoma virus budding by cis expression of a WW domain. J. Virol. 76, 2789–2795.

Perez-Caballero, D., Hatziioannou, T., Yang, A., Cowan, S., Bieniasz, P.D., 2005. Human tripartite motif 5alpha domains responsible for retrovirus restriction activity and specificity. J. Virol. 79, 8969–8978.

Pessel-Vivares, L., Ferrer, M., Laine, S., Mougel, M., 2014. MLV requires Tap/NXF1-dependent pathway to export its unspliced RNA to the cytoplasm and to express both spliced and unspliced RNAs. Retrovirology 11, 21.

Pessel-Vivares, L., Houzet, L., Laine, S., Mougel, M., 2015. Insights into the nuclear export of murine leukemia virus intron-containing RNA. RNA Biol. 12, 942–949.

Petit, C., Giron, M.L., Tobaly-Tapiero, J., Bittoun, P., Real, E., Jacob, Y., Tordo, N., De The, H., Saib, A., 2003. Targeting of incoming retroviral Gag to the centrosome involves a direct interaction with the dynein light chain 8. J. Cell Sci. 116, 3433–3442.

Pincus, T., Hartley, J.W., Rowe, W.P., 1971a. A major genetic locus affecting resistance to infection with murine leukemia viruses. I. Tissue culture studies of naturally occurring viruses. J. Exp. Med. 133, 1219–1233.

Pincus, T., Rowe, W.P., Lilly, F., 1971b. A major genetic locus affecting resistance to infection with murine leukemia viruses. II. Apparent identity to a major locus described for resistance to friend murine leukemia virus. J. Exp. Med. 133, 1234–1241.

Pocock, G.M., Becker, J.T., Swanson, C.M., Ahlquist, P., Sherer, N.M., 2016. HIV-1 and M-PMV RNA nuclear export elements program viral genomes for distinct cytoplasmic trafficking behaviors. PLoS Pathog. 12, e1005565.

Poole, E., Strappe, P., Mok, H.P., Hicks, R., Lever, A.M., 2005. HIV-1 Gag-RNA interaction occurs at a perinuclear/centrosomal site; analysis by confocal microscopy and FRET. Traffic 6, 741–755.

Pornillos, O., Alam, S.L., Rich, R.L., Myszka, D.G., Davis, D.R., Sundquist, W.I., 2002. Structure and functional interactions of the Tsg101 UEV domain. EMBO J. 21, 2397–2406.

Pornillos, O., Higginson, D.S., Stray, K.M., Fisher, R.D., Garrus, J.E., Payne, M., He, G.P., Wang, H.E., Morham, S.G., Sundquist, W.I., 2003. HIV Gag mimics the Tsg101-recruiting activity of the human Hrs protein. J. Cell Biol. 162, 425–434.

Prechtel, A.T., Chemnitz, J., Schirmer, S., Ehlers, C., Langbein-Detsch, I., Stulke, J., Dabauvalle, M.C., Kehlenbach, R.H., Hauber, J., 2006. Expression of CD83 is regulated by HuR via a novel cis-active coding region RNA element. J. Biol. Chem. 281, 10912–10925.

Pryciak, P.M., Varmus, H.E., 1992. Fv-1 restriction and its effects on murine leukemia virus integration in vivo and in vitro. J. Virol. 66, 5959–5966.

Puffer, B.A., Parent, L.J., Wills, J.W., Montelaro, R.C., 1997. Equine infectious anemia virus utilizes a YXXL motif within the late assembly domain of the Gag p9 protein. J. Virol. 71, 6541–6546.

Puffer, B.A., Watkins, S.C., Montelaro, R.C., 1998. Equine infectious anemia virus Gag polyprotein late domain specifically recruits cellular AP-2 adapter protein complexes during virion assembly. J. Virol. 72, 10218–10221.

Rabson, A.B., Graves, B.J., 1997. Synthesis and processing of viral RNA. In: Coffin, J.M., Hughes, S.H., Varmus, H.E. (Eds.), Retroviruses Cold Spring Harbor, NY.

Rai, T., Mosoian, A., Resh, M.D., 2010. Annexin 2 is not required for human immunodeficiency virus type 1 particle production but plays a cell type-dependent role in regulating infectivity. J. Virol. 84, 9783–9792.

Raiborg, C., Rusten, T.E., Stenmark, H., 2003. Protein sorting into multivesicular endosomes. Curr. Opin. Cell Biol. 15, 446–455.

Rangasamy, D., Woytek, K., Khan, S.A., Wilson, V.G., 2000. SUMO-1 modification of bovine papillomavirus E1 protein is required for intranuclear accumulation. J. Biol. Chem. 275, 37999–38004.

Raposo, G., Moore, M., Innes, D., Leijendekker, R., Leigh-Brown, A., Benaroch, P., Geuze, H., 2002. Human macrophages accumulate HIV-1 particles in MHC II compartments. Traffic 3, 718–729.

Rein, A., 1994. Retroviral RNA packaging: a review. Arch. Virol. Suppl. 9, 513–522.

Renault, N., Tobaly-Tapiero, J., Paris, J., Giron, M.L., Coiffic, A., Roingeard, P., Saib, A., 2011. A nuclear export signal within the structural Gag protein is required for prototype foamy virus replication. Retrovirology 8, 6.

Resh, M.D., 2004. A myristoyl switch regulates membrane binding of HIV-1 Gag. Proc. Natl. Acad. Sci. U.S.A. 101, 417–418.

Rhee, S.S., Hunter, E., 1987. Myristylation is required for intracellular transport but not for assembly of D-type retrovirus capsids. J. Virol. 61, 1045–1053.

Rhee, S.S., Hunter, E., 1990. A single amino acid substitution within the matrix protein of a type D retrovirus converts its morphogenesis to that of a type C retrovirus. Cell 63, 77–86.

Rice, B.L., Kaddis, R.J., Stake, M.S., Lochmann, T.L., Parent, L.J., 2015. Interplay between the alpharetroviral Gag protein and SR proteins SF2 and SC35 in the nucleus. Front. Microbiol. 6, 925.

Rodriguez, M.S., Desterro, J.M., Lain, S., Midgley, C.A., Lane, D.P., Hay, R.T., 1999. SUMO-1 modification activates the transcriptional response of p53. EMBO J. 18, 6455–6461.

Rondon, A.G., Proudfoot, N.J., 2008. Nuclear roadblocks for mRNA export. Cell 135, 207–208.

Rosen, C.A., Terwilliger, E., Dayton, A., Sodroski, J.G., Haseltine, W.A., 1988. Intragenic cis-acting art gene-responsive sequences of the human immunodeficiency virus. Proc. Natl. Acad. Sci. U.S.A. 85, 2071–2075.

Ryo, A., Suzuki, Y., Ichiyama, K., Wakatsuki, T., Kondoh, N., Hada, A., Yamamoto, M., Yamamoto, N., 1999. Serial analysis of gene expression in HIV-1-infected T cell lines. FEBS Lett. 462, 182–186.

Ryo, A., Tsurutani, N., Ohba, K., Kimura, R., Komano, J., Nishi, M., Soeda, H., Hattori, S., Perrem, K., Yamamoto, M., Chiba, J., Mimaya, J., Yoshimura, K., Matsushita, S., Honda, M., Yoshimura, A., Sawasaki, T., Aoki, I., Morikawa, Y., Yamamoto, N., 2008. SOCS1 is an inducible host factor during HIV-1 infection and regulates the intracellular trafficking and stability of HIV-1 Gag. Proc. Natl. Acad. Sci. U.S.A. 105, 294–299.

Ryzhova, E.V., Vos, R.M., Albright, A.V., Harrist, A.V., Harvey, T., Gonzalez-Scarano, F., 2006. Annexin 2: a novel human immunodeficiency virus type 1 Gag binding protein involved in replication in monocyte-derived macrophages. J. Virol. 80, 2694–2704.

Saad, J.S., Miller, J., Tai, J., Kim, A., Ghanam, R.H., Summers, M.F., 2006. Structural basis for targeting HIV-1 Gag proteins to the plasma membrane for virus assembly. Proc. Natl. Acad. Sci. U.S.A. 103, 11364–11369.

Sabo, Y., Walsh, D., Barry, D.S., Tinaztepe, S., De Los Santos, K., Goff, S.P., Gundersen, G.G., Naghavi, M.H., 2013. HIV-1 induces the formation of stable microtubules to enhance early infection. Cell Host Microbe 14, 535–546.

Saguez, C., Olesen, J.R., Jensen, T.H., 2005. Formation of export-competent mRNP: escaping nuclear destruction. Curr. Opin. Cell Biol. 17, 287–293.

Saib, A., Puvion-Dutilleul, F., Schmid, M., Peries, J., De The, H., 1997. Nuclear targeting of incoming human foamy virus Gag proteins involves a centriolar step. J. Virol. 71, 1155–1161.

Saitoh, H., Pu, R., Cavenagh, M., Dasso, M., 1997. RanBP2 associates with Ubc9p and a modified form of RanGAP1. Proc. Natl. Acad. Sci. U.S.A. 94, 3736–3741.

Saitoh, H., Sparrow, D.B., Shiomi, T., Pu, R.T., Nishimoto, T., Mohun, T.J., Dasso, M., 1998. Ubc9p and the conjugation of SUMO-1 to RanGAP1 and RanBP2. Curr. Biol. 8, 121–124.

Sakai, H., Shibata, R., Sakuragi, J., Sakuragi, S., Kawamura, M., Adachi, A., 1993. Cell-dependent requirement of human immunodeficiency virus type 1 Vif protein for maturation of virus particles. J. Virol. 67, 1663–1666.

Sakalian, M., Wills, J.W., Vogt, V.M., 1994. Efficiency and selectivity of RNA packaging by Rous sarcoma virus Gag deletion mutants. J. Virol. 68, 5969–5981.

Sakuma, T., Davila, J.I., Malcolm, J.A., Kocher, J.P., Tonne, J.M., Ikeda, Y., 2014. Murine leukemia virus uses NXF1 for nuclear export of spliced and unspliced viral transcripts. J. Virol. 88, 4069–4082.

Sakurai, A., Yasuda, J., Takano, H., Tanaka, Y., Hatakeyama, M., Shida, H., 2004. Regulation of human T-cell leukemia virus type 1 (HTLV-1) budding by ubiquitin ligase Nedd4. Microbes Infect. 6, 150–156.

Santos-Rosa, H., Moreno, H., Simos, G., Segref, A., Fahrenkrog, B., Pante, N., Hurt, E., 1998. Nuclear mRNA export requires complex formation between Mex67p and Mtr2p at the nuclear pores. Mol. Cell. Biol. 18, 6826–6838.

Scheifele, L.Z., Garbitt, R.A., Rhoads, J.D., Parent, L.J., 2002. Nuclear entry and CRM1-dependent nuclear export of the Rous sarcoma virus Gag polyprotein. Proc. Natl. Acad. Sci. U.S.A. 99, 3944–3949.

Schliephake, A.W., Rethwilm, A., 1994. Nuclear localization of foamy virus Gag precursor protein. J. Virol. 68, 4946–4954.

Schmidt, O., Teis, D., 2012. The ESCRT machinery. Curr. Biol. 22, R116–R120.

Schneider, W.M., Brzezinski, J.D., Aiyer, S., Malani, N., Gyuricza, M., Bushman, F.D., Roth, M.J., 2013. Viral DNA tethering domains complement replication-defective mutations in the p12 protein of MuLV Gag. Proc. Natl. Acad. Sci. U.S.A. 110, 9487–9492.

Schrom, E.M., Moschall, R., Schuch, A., Bodem, J., 2013. Regulation of retroviral polyadenylation. Adv. Virus Res. 85, 1–24.

Schubert, U., Ott, D.E., Chertova, E.N., Welker, R., Tessmer, U., Princiotta, M.F., Bennink, J.R., Krausslich, H.G., Yewdell, J.W., 2000. Proteasome inhibition interferes with gag polyprotein processing, release, and maturation of HIV-1 and HIV-2. Proc. Natl. Acad. Sci. U.S.A. 97, 13057–13062.

Schultz, A.M., Rein, A., 1989. Unmyristylated Moloney murine leukemia virus Pr65gag is excluded from virus assembly and maturation events. J. Virol. 63, 2370–2373.

Sebastian, S., Luban, J., 2005. TRIM5alpha selectively binds a restriction-sensitive retroviral capsid. Retrovirology 2, 40.

Segura-Morales, C., Pescia, C., Chatellard-Causse, C., Sadoul, R., Bertrand, E., Basyuk, E., 2005. Tsg101 and Alix interact with murine leukemia virus Gag and cooperate with Nedd4 ubiquitin ligases during budding. J. Biol. Chem. 280, 27004–27012.

Sekine, Y., Okada, Y., Noda, Y., Kondo, S., Aizawa, H., Takemura, R., Hirokawa, N., 1994. A novel microtubule-based motor protein (KIF4) for organelle transports, whose expression is regulated developmentally. J. Cell Biol. 127, 187–201.

Sfakianos, J.N., Hunter, E., 2003. M-PMV capsid transport is mediated by Env/Gag interactions at the pericentriolar recycling endosome. Traffic. 4, 671–680.

Shibata, R., Sakai, H., Kawamura, M., Tokunaga, K., Adachi, A., 1995. Early replication block of human immunodeficiency virus type 1 in monkey cells. J. Gen. Virol. 76 (Pt 11), 2723–2730.

Shields, A., Witte, W.N., Rothenberg, E., Baltimore, D., 1978. High frequency of aberrant expression of Moloney murine leukemia virus in clonal infections. Cell 14, 601–609.

Shioda, T., Shibuta, H., 1990. Production of human immunodeficiency virus (HIV)-like particles from cells infected with recombinant vaccinia viruses carrying the gag gene of HIV. Virology 175, 139–148.

Shyu, A.-B., Wilkinson, M.F., 2000. The double lives of shuttling mRNA binding proteins. Cell 102, 135–138.

Simpson, S.B., Guo, W., Winistorfer, S.C., Craven, R.C., Stoltzfus, C.M., 1998. The upstream, direct repeat sequence of Prague A Rous sarcoma virus is deficient in mediating efficient Gag assembly and particle release. Virology 247, 86–96.

Simpson, S.B., Zhang, L., Craven, R.C., Stoltzfus, C.M., 1997. Rous sarcoma virus direct repeat cis elements exert effects at several points in the virus life cycle. J. Virol. 71, 9150–9156.

Slagsvold, T., Pattni, K., Malerod, L., Stenmark, H., 2006. Endosomal and non-endosomal functions of ESCRT proteins. Trends Cell Biol. 16, 317–326.

Smith, A.J., Cho, M.I., Hammarskjold, M.L., Rekosh, D., 1990. Human immunodeficiency virus type 1 Pr55gag and Pr160gag-pol expressed from a simian virus 40 late replacement vector are efficiently processed and assembled into viruslike particles. J. Virol. 64, 2743–2750.

Sommer, P., Nehrbass, U., 2005. Quality control of messenger ribonucleoprotein particles in the nucleus and at the pore. Curr. Opin. Cell Biol. 17, 294–301.

Spearman, P., Horton, R., Ratner, L., Kuli-Zade, I., 1997. Membrane binding of human immunodeficiency virus type 1 matrix protein in vivo supports a conformational myristyl switch mechanism. J. Virol. 71, 6582–6592.

Stake, M., Singh, D., Singh, G., Marcela Hernandez, J., Kaddis Maldonado, R., Parent, L.J., Boris-Lawrie, K., 2015. HIV-1 and two avian retroviral 5′ untranslated regions bind orthologous human and chicken RNA binding proteins. Virology 486, 307–320.

Stange, A., Mannigel, I., Peters, K., Heinkelein, M., Stanke, N., Cartellieri, M., Gottlinger, H., Rethwilm, A., Zentgraf, H., Lindemann, D., 2005. Characterization of prototype foamy virus gag late assembly domain motifs and their role in particle egress and infectivity. J. Virol. 79, 5466–5476.

Staub, O., Abriel, H., Plant, P., Ishikawa, T., Kanelis, V., Saleki, R., Horisberger, J.D., Schild, L., Rotin, D., 2000. Regulation of the epithelial Na^+ channel by Nedd4 and ubiquitination. Kidney Int. 57, 809–815.

Stoltzfus, C.M., 2009. Chapter 1. Regulation of HIV-1 alternative RNA splicing and its role in virus replication. Adv. Virus Res. 74, 1–40.

Stoltzfus, C.M., Madsen, J.M., 2006. Role of viral splicing elements and cellular RNA binding proteins in regulation of HIV-1 alternative RNA splicing. Curr. HIV Res. 4, 43–55.

Strack, B., Calistri, A., Accola, M.A., Palu, G., Gottlinger, H.G., 2000. A role for ubiquitin ligase recruitment in retrovirus release. Proc. Natl. Acad. Sci. U.S.A. 97, 13063–13068.

Strack, B., Calistri, A., Craig, S., Popova, E., Gottlinger, H.G., 2003. AIP1/ALIX is a binding partner for HIV-1 p6 and EIAV p9 functioning in virus budding. Cell 114, 689–699.

Stremlau, M., Owens, C.M., Perron, M.J., Kiessling, M., Autissier, P., Sodroski, J., 2004. The cytoplasmic body component TRIM5alpha restricts HIV-1 infection in Old World monkeys. Nature 427, 848–853.

Stremlau, M., Perron, M., Lee, M., Li, Y., Song, B., Javanbakht, H., Diaz-Griffero, F., Anderson, D.J., Sundquist, W.I., Sodroski, J., 2006. Specific recognition and accelerated uncoating of retroviral capsids by the TRIM5alpha restriction factor. Proc. Natl. Acad. Sci. U.S.A. 103, 5514–5519.

Stremlau, M., Perron, M., Welikala, S., Sodroski, J., 2005. Species-specific variation in the B30.2(SPRY) domain of TRIM5alpha determines the potency of human immunodeficiency virus restriction. J. Virol. 79, 3139–3145.

Stutz, F., Izaurralde, E., 2003. The interplay of nuclear mRNP assembly, mRNA surveillance and export. Trends Cell Biol. 13, 319–327.

Subramanian, L., Benson, M.D., Iniguez-Lluhi, J.A., 2003. A synergy control motif within the attenuator domain of CCAAT/enhancer-binding protein alpha inhibits transcriptional synergy through its PIASy-enhanced modification by SUMO-1 or SUMO-3. J. Biol. Chem. 278, 9134–9141.

Swanson, C.M., Malim, M.H., 2006. Retrovirus RNA trafficking: from chromatin to invasive genomes. Traffic 7, 1440–1450.

Swanson, C.M., Puffer, B.A., Ahmad, K.M., Doms, R.W., Malim, M.H., 2004. Retroviral mRNA nuclear export elements regulate protein function and virion assembly. EMBO J. 23, 2632–2640.

Swanstrom, R., Wills, J.W., 1997. Synthesis, assembly, and processing of viral proteins. In: Coffin, J.M., Hughes, S.H., Varmus, H.E. (Eds.), Retroviruses Cold Spring Harbor, NY.

Tang, C., Loeliger, E., Luncsford, P., Kinde, I., Beckett, D., Summers, M.F., 2004. Entropic switch regulates myristate exposure in the HIV-1 matrix protein. Proc. Natl. Acad. Sci. U.S.A. 101, 517–522.

Tang, Y., Winkler, U., Freed, E.O., Torrey, T.A., Kim, W., Li, H., Goff, S.P., Morse 3rd, H.C., 1999. Cellular motor protein KIF-4 associates with retroviral Gag. J. Virol. 73, 10508–10513.

Tashiro, K., Pando, M.P., Kanegae, Y., Wamsley, P.M., Inoue, S., Verma, I.M., 1997. Direct involvement of the ubiquitin-conjugating enzyme Ubc9/Hus5 in the degradation of IkappaBalpha. Proc. Natl. Acad. Sci. U.S.A. 94, 7862–7867.

Terry, L.J., Shows, E.B., Wente, S.R., 2007. Crossing the nuclear envelope: hierarchical regulation of nucleocytoplasmic transport. Science 318, 1412–1416.

Tobaly-Tapiero, J., Bittoun, P., Lehmann-Che, J., Delelis, O., Giron, M.L., De The, H., Saib, A., 2008. Chromatin tethering of incoming foamy virus by the structural Gag protein. Traffic 9, 1717–1727.

Topp, W.C., 1981. Normal rat cell lines deficient in nuclear thymidine kinase. Virology 113, 408–411.

Van Damme, N., Goff, D., Katsura, C., Jorgenson, R.L., Mitchell, R., Johnson, M.C., Stephens, E.B., Guatelli, J., 2008. The interferon-induced protein BST-2 restricts HIV-1 release and is downregulated from the cell surface by the viral Vpu protein. Cell Host Microbe 3, 245–252.

Venkatesh, S., Bieniasz, P.D., 2013. Mechanism of HIV-1 virion entrapment by tetherin. PLoS Pathog. 9, e1003483.

Veronese, F.D., Copeland, T.D., Oroszlan, S., Gallo, R.C., Sarngadharan, M.G., 1988. Biochemical and immunological analysis of human immunodeficiency virus gag gene products p17 and p24. J. Virol. 62, 795–801.

VerPlank, L., Bouamr, F., Lagrassa, T.J., Agresta, B., Kikonyogo, A., Leis, J., Carter, C.A., 2001. Tsg101, a homologue of ubiquitin-conjugating (E2) enzymes, binds the L domain in HIV type 1 Pr55(Gag). Proc. Natl. Acad. Sci. U.S.A. 98, 7724–7729.

Vincent, O., Rainbow, L., Tilburn, J., Arst Jr., H.N., Penalva, M.A., 2003. YPXL/I is a protein interaction motif recognized by aspergillus PalA and its human homologue, AIP1/Alix. Mol. Cell. Biol. 23, 1647–1655.

Vogt, V.M., 1997. Retroviral virions and genomes. In: Coffin, J.M., Hughes, S.H., Varmus, H.E. (Eds.), Retroviruses Cold Spring Harbor, NY.

von Schwedler, U.K., Stuchell, M., Muller, B., Ward, D.M., Chung, H.Y., Morita, E., Wang, H.E., Davis, T., He, G.P., Cimbora, D.M., Scott, A., Krausslich, H.G., Kaplan, J., Morham, S.G., Sundquist, W.I., 2003. The protein network of HIV budding. Cell 114, 701–713.

Wang, M.Q., Kim, W., Gao, G., Torrey, T.A., Morse 3rd, H.C., De Camilli, P., Goff, S.P., 2003. Endophilins interact with Moloney murine leukemia virus Gag and modulate virion production. J. Biol. 3, 4.

Weaver, T.A., Panganiban, A.T., 1990. N myristoylation of the spleen necrosis virus matrix protein is required for correct association of the Gag polyprotein with intracellular membranes and for particle formation. J. Virol. 64, 3995–4001.

Weldon Jr., R.A., Parker, W.B., Sakalian, M., Hunter, E., 1998. Type D retrovirus capsid assembly and release are active events requiring ATP. J. Virol. 72, 3098–3106.

Weldon Jr., R.A., Sarkar, P., Brown, S.M., Weldon, S.K., 2003. Mason-Pfizer monkey virus Gag proteins interact with the human sumo conjugating enzyme, hUbc9. Virology 314, 62–73.

Wills, J.W., Cameron, C.E., Wilson, C.B., Xiang, Y., Bennett, R.P., Leis, J., 1994. An assembly domain of the Rous sarcoma virus Gag protein required late in budding. J. Virol. 68, 6605–6618.

Wills, J.W., Craven, R.C., Achacoso, J.A., 1989. Creation and expression of myristylated forms of Rous sarcoma virus gag protein in mammalian cells. J. Virol. 63, 4331–4343.

Wills, J.W., Craven, R.C., Weldon Jr., R.A., Nelle, T.D., Erdie, C.R., 1991. Suppression of retroviral MA deletions by the amino-terminal membrane-binding domain of p60src. J. Virol. 65, 3804–3812.

Wodrich, H., Schambach, A., Krausslich, H.G., 2000. Multiple copies of the Mason-Pfizer monkey virus constitutive RNA transport element lead to enhanced HIV-1 Gag expression in a context-dependent manner. Nucleic Acids Res. 28, 901–910.

Woodham, A.W., Sanna, A.M., Taylor, J.R., Skeate, J.G., Da Silva, D.M., Dekker, L.V., Kast, W.M., 2016. Annexin A2 antibodies but not inhibitors of the annexin A2 heterotetramer impair productive HIV-1 infection of macrophages in vitro. Virol. J. 13, 187.

Xiang, Y., Cameron, C.E., Wills, J.W., Leis, J., 1996. Fine mapping and characterization of the Rous sarcoma virus Pr76gag late assembly domain. J. Virol. 70, 5695–5700.

Xu, F., Tan, J., Liu, R., Xu, D., Li, Y., Geng, Y., Liang, C., Qiao, W., 2011. Tetherin inhibits prototypic foamy virus release. Virol. J. 8, 198.

Yang, J., Cullen, B.R., 1999. Structural and functional analysis of the avian leukemia virus constitutive transport element. RNA 5, 1645–1655.

Yang, W.K., Kiggans, J.O., Yang, D.M., Ou, C.Y., Tennant, R.W., Brown, A., Bassin, R.H., 1980. Synthesis and circularization of N- and B-tropic retroviral DNA Fv-1 permissive and restrictive mouse cells. Proc. Natl. Acad. Sci. U.S.A. 77, 2994–2998.

Yap, M.W., Nisole, S., Stoye, J.P., 2005. A single amino acid change in the SPRY domain of human Trim5alpha leads to HIV-1 restriction. Curr. Biol. 15, 73–78.

Yasuda, J., Hunter, E., 1998. A proline-rich motif (PPPY) in the Gag polyprotein of Mason-Pfizer monkey virus plays a maturation-independent role in virion release. J. Virol. 72, 4095–4103.

Yasuda, J., Hunter, E., Nakao, M., Shida, H., 2002. Functional involvement of a novel Nedd4-like ubiquitin ligase on retrovirus budding. EMBO Rep. 3, 636–640.

Yedavalli, V.S., Jeang, K.T., 2011. Matrin 3 is a co-factor for HIV-1 Rev in regulating post-transcriptional viral gene expression. Retrovirology 8, 61.

Yedavalli, V.S., Neuveut, C., Chi, Y.H., Kleiman, L., Jeang, K.T., 2004. Requirement of DDX3 DEAD box RNA helicase for HIV-1 Rev-RRE export function. Cell 119, 381–392.

Yin, X., Hu, Z., Gu, Q., Wu, X., Zheng, Y.H., Wei, P., Wang, X., 2014. Equine tetherin blocks retrovirus release and its activity is antagonized by equine infectious anemia virus envelope protein. J. Virol. 88, 1259–1270.

Yu, K.L., Lee, S.H., Lee, E.S., You, J.C., 2016. HIV-1 nucleocapsid protein localizes efficiently to the nucleus and nucleolus. Virology 492, 204–212.

Yu, S.F., Edelmann, K., Strong, R.K., Moebes, A., Rethwilm, A., Linial, M.L., 1996. The carboxyl terminus of the human foamy virus Gag protein contains separable nucleic acid binding and nuclear transport domains. J. Virol. 70, 8255–8262.

Yuan, B., Campbell, S., Bacharach, E., Rein, A., Goff, S.P., 2000. Infectivity of Moloney murine leukemia virus defective in late assembly events is restored by late assembly domains of other retroviruses. J. Virol. 74, 7250–7260.

Yuan, B., Li, X., Goff, S.P., 1999. Mutations altering the moloney murine leukemia virus p12 Gag protein affect virion production and early events of the virus life cycle. EMBO J. 18, 4700–4710.

Yuan, X., Yu, X., Lee, T.H., Essex, M., 1993. Mutations in the N-terminal region of human immunodeficiency virus type 1 matrix protein block intracellular transport of the Gag precursor. J. Virol. 67, 6387–6394.

Zenklusen, D., Stutz, F., 2001. Nuclear export of mRNA. FEBS Lett. 498, 150–156.

Zhou, W., Parent, L.J., Wills, J.W., Resh, M.D., 1994. Identification of a membrane-binding domain within the amino-terminal region of human immunodeficiency virus type 1 Gag protein which interacts with acidic phospholipids. J. Virol. 68, 2556–2569.

Zhou, W., Resh, M.D., 1996. Differential membrane binding of the human immunodeficiency virus type 1 matrix protein. J. Virol. 70, 8540–8548.

Zimmerman, C., Klein, K.C., Kiser, P.K., Singh, A.R., Firestein, B.L., Riba, S.C., Lingappa, J.R., 2002. Identification of a host protein essential for assembly of immature HIV-1 capsids. Nature 415, 88–92.

Zitouni, S., Nabais, C., Jana, S.C., Guerrero, A., Bettencourt-Dias, M., 2014. Polo-like kinases: structural variations lead to multiple functions. Nat. Rev. Mol. Cell Biol. 15, 433–452.

Zurnic, I., Hutter, S., Rzeha, U., Helbig, R., Stanke, N., Reh, J., Mullers, E., Hamann, M.V., Kern, T., Gerresheim, G.K., Lindel, F., Serrao, E., Lesbats, P., Engelman, A.N., Cherepanov, P., Lindemann, D., 2016. Correction: interactions of prototype foamy virus capsids with host cell polo-like kinases are important for efficient viral DNA integration. PLoS Pathog. 12, e1005956.

Chapter 9

Tumor Suppressor Gene 101: A Virus' Multifunctional Conduit to the ESCRT Trafficking Machinery

Carol A. Carter
Stony Brook University, Stony Brook, NY, United States

Chapter Outline

Introduction 318
The Endosomal Sorting Complex Required for Transport Machinery 318
ESCRT-0 318
ESCRT-I 320
ESCRT-II 320
ESCRT-III 320
Endosomal Sorting Complexes Required for Transport Participation in Cellular Processes 322
Endosomal Sorting Complexes Required for Transport Participation in Virus Budding 322
Tumor Suppressor Gene 101 Structure 323
Tsg101-UEV Domain 325
Tsg101-UEV PTAP-Binding Pocket 325
Tsg101-UEV Ub-Binding Pocket 326
Ub- and PTAP-Binding Pockets: Is There Cross Talk? 327
Tumor Suppressor Gene 101 Participation 328
Tumor Suppressor Gene 101 Participation in the Cellular Endocytic Pathway 328
Tumor Suppressor Gene 101 Participation in Virus Budding 329
Cellular Factors That Work With Tumor Suppressor Gene 101 330
Convergence of Tsg101-Ub-Nedd4-Alix Participation 333
Tumor Suppressor Gene 101 Role in Budding of Other Viruses 335
Questions Remaining 339
Acknowledgments 340
References 340

Retrovirus-Cell Interactions. https://doi.org/10.1016/B978-0-12-811185-7.00009-1

INTRODUCTION

The Endosomal Sorting Complex Required for Transport Machinery

The *e*ndosomal *s*orting *c*omplex *r*equired for *t*ransport (ESCRT) system, discovered relatively recently, is evolutionarily conserved from yeast to man (Field et al., 2011; Wideman et al., 2014). Using yeast as their model system, the Emr, Jenness and Stevens groups identified genes required for sorting of cargo into a *m*ulti*v*esicular *b*ody (MVB) compartment and designated these as genes encoding the class E *v*acuolar *p*rotein *s*orting (Vps) proteins (Babst et al., 1998, 2000; Katzmann et al., 2001, 2003; Li et al., 1999; Rothman et al., 1989). Mammalian cells encode orthologs and, in some case, isoforms of these yeast gene products (*reviewed in* Coonrod and Stevens, 2010), the latter serving to broaden the functional range of the ESCRT complexes that comprise the system.

The machinery includes more than two dozen of unique proteins and protein isoforms organized into four complexes (ESCRT-0, -I, -II, and -III; Fig. 9.1A) that function collectively to transform a single, continuous membrane bilayer into two distinct bilayers, while simultaneously segregating associated cargo away from the cytoplasmic face of the bilayer. To date, the machinery has been established to participate in the biogenesis of the MVB in endolysosomal sorting, the budding of HIV-1 and other enveloped viruses from the plasma membrane of infected cells, and in the membrane abscission step in cytokinesis during mitosis. It is suspected to contribute to microvesicle and exosome biogenesis, repair of plasma membrane defects, extraction of defective nuclear pore complexes, nuclear envelope reformation during mitosis, micro- and macroautophagy, neuronal pruning, and plus-stranded RNA virus replication compartment formation (*reviewed in* Hurley, 2015; Schuh and Audhya, 2014).

ESCRT-0

ESCRT-0 consists of two components, Hrs and signal-transducing adaptor molecule (STAM), which both contain motifs that bind ubiquitin (Ub) with relatively high affinity. These motifs are presumed to foster cargo clustering by permitting cargoes modified with multiple Ub molecules to bind more avidly to ESCRT-0 through multiple simultaneous interactions. When coexpressed recombinantly, the proteins form an elongated 1:1 heterodimeric complex and its multiple Ub-binding domains lie in close proximity, fostering cargo capture. Hrs possesses an FYVE domain that, in the context of the protein, exhibits nanomolar affinity for the phospholipid, phosphatidylinositol 3-phosphate (PI_3P), which is enriched on early endosomes. Interestingly, on membranes, the heterodimer undergoes oligomerization; the resulting complex brings into proximity multiple low-affinity Ub-binding motifs, contributing to the aforementioned avidity. The carboxyl termini of the ESCRT-0 subunits promote interactions with other proteins: STAM possesses SH3 domains that permit

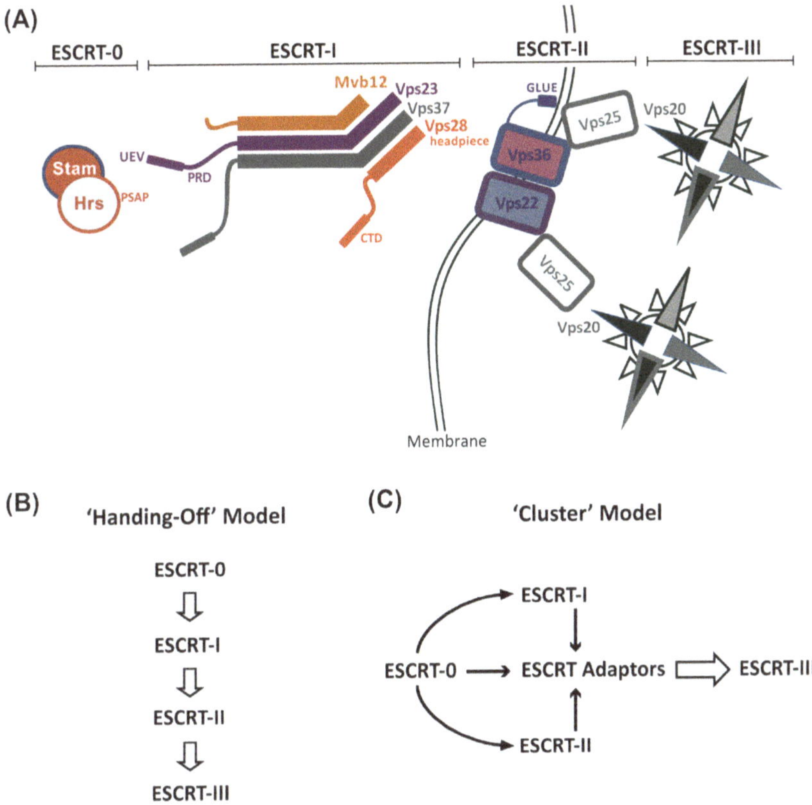

FIGURE 9.1 Panel A, key components of the endosomal sorting complexes required for transport (ESCRT) machinery. For simplicity, in ESCRT-III, only Vps20 (Chmp6), the major factor interacting with ESCRT-II, is labeled. Panel B, A rendition of the "*Handing-Off*" model. Panel C, A rendition of the "Cluster" model.

binding to the deubiquitinating enzymes, AMSH and UBPY. Hrs contains a clathrin-interaction motif that enables the ESCRT-0 complex to associate with flat clathrin lattices assembled on early endosomes. These lattices are proposed to facilitate clustering of ESCRT-0 into microdomains, which ultimately become sites of intraluminal vesicle formation. Deletion of the clathrin-binding box in Hrs results in a more uniform distribution on the endosome surface. Hrs also associates with endocytic adaptors, including the AP-2 complex, which targets ESCRT-0 to clathrin-coated pits at the plasma membrane. The Hrs C-terminus has a PSAP motif that permits it to recruit ESCRT-I by binding to the PT/SAP-binding pocket in the "ubiquitin-E2-variant" (UEV) domain in tumor suppressor gene 101 (Tsg101) (*discussed further below*). The affinity of the interaction is weak and therefore ESCRT-I is not constitutively localized with ESCRT-0 at endosomal subdomains.

ESCRT-I

ESCRT-I is a heterotetramer consisting of Tsg101 (yeast homologue Vps23), Vps28, Vps37 and Mvb12 (isoform name UBAP1) with the components in 1:1:1:1 subunit stoichiometry. Refinement of crystal structures of the yeast core ESCRT-I complex, the UEV domain of Vps23 and the Vps28 C-terminal region using small-angle X-ray scattering, electron paramagnetic resonance, and fluorescence resonance energy transfer spectroscopy reveal the yeast ESCRT-I complex as highly asymmetric, consisting of a headpiece formed by the Vps28 N-terminal region attached to stalk nearly 13 nm long comprises a helical bundle donated by the other three components (Boura et al., 2011; Kostelansky et al., 2006, 2007). Two of the ESCRT-I components, Tsg101 and Mvb12 (UBAP1) can bind Ub (Agromayor et al., 2012; Gill et al., 2007; Pornillos et al., 2002a,b; Shields et al., 2009; Stefani et al., 2011; Teo et al., 2004). The binding affinity for Ub of UBAP1 (~140 μM) is significantly greater than that of Tsg101 (~500 μM). Structural studies place the Ub-binding carboxyl-terminus of Mvb12 (UBAP1) near the UEV domain of Tsg101 suggesting, as with ESCRT-0, that both could contribute to stronger binding to di-Ub-modified cargo (Boura et al., 2011; Stefani et al., 2011). The presence of UBAP1, in combination with Vps37A but not Vps37C, Vps28, and Tsg101, defines an endosome-specific ESCRT-I complex essential for MVB sorting (Stefani et al., 2011). UBAP1 is not required for virus budding or cytokinesis.

ESCRT-II

ESCRT-II forms a stable heterotetramer composed of three unique subunits: Vps22 (mammalian counterpart Eap30), Vps36 (Eap45), and two copies of Vps25 (Eap20). Like ESCRT-I, ESCRT-II has only a few Ub-binding motifs and exhibits weak affinity for Ub and membranes. Both Vps22 and Vps36 bind to membranes: Vps36 via its "*G*ram-*l*ike *Ub*-binding in *E*ap45" or GLUE domain, which exhibits a slight preference for PI_3P yet actually contributes little to ESCRT-II localization (Slagsvold et al., 2005; Teo et al., 2004); and Vps22 through a conserved, basic helical domain that associates with acidic phospholipids nonspecifically but nevertheless is required for ESCRT-II endosomal targeting (Im and Hurley, 2008).

ESCRT-III

The membrane scission activity of the machinery is embodied in the ESCRT-III complex in a subset of four core components, Vps 2, 20, 24, and 32 (Snf7) (mammalian counterparts CHMP space 2, 6, 3, and 4, *respectively*) and several additional factors. In contrast to ESCRT complexes I and II, ESCRT-III assembly occurs only on lipid bilayers and is initiated on the membrane through interaction with ESCRT-II Vps25 (Im et al., 2009). A type I AAA ATPase, Vps4 (SKD1 in mammalian cells) and its cofactor Vta1 (LIP5 in mammalian cells) is responsible for disassembly of the machinery for reutilization. An amino-terminal *m*icrotubule *i*nteracting *and t*ransport (MIT) domain enables the ATPase to bind to nearly all of the ESCRT-III subunits. The features of ESCRT-III and the recycling action of Vps4 and its regulator Vta1 have been described extensively and the reader is referred to several outstanding reviews that focus on these aspects of ESCRT machinery function (McCullough et al., 2015; Christ et al., 2017; Lippincott-Schwartz et al., 2017; Schoneberg et al., 2017).

The four complexes were proposed to function sequentially in a "handing-off" model (Fig. 9.1B), but this was found to occur only in some cases. A major function of ESCRT-0, -I, and -II is to concentrate/sequester Ub-modified cargo through various Ub-binding motifs of ESCRT members on the limiting membrane of the endosome that eventually invaginates (Shields et al., 2009). There are currently no experimental findings that define how this might be accomplished. The "handing-off" model is not supported by structural considerations, as the binding sites in ESCRT-I for the ubiquitinated cargo received from ESCRT-0 and passed on to ESCRT-II are at opposite ends of a modeled 25 nm long ESCRT-I complex (*c.f.,* Fig. 9.1A; *reviewed in* Hurley, 2008). In retroviral assembly, the late domains of Gag are believed to indirectly recruit ESCRT-III. Some models proposed that ESCRT-III factors then polymerize into ring or spiral structures at the bud neck to drive bending and ultimate bud scission (Lippincott-Schwartz et al., 2017; Cashikar et al., 2014; Fabrikant et al., 2009; Hanson et al., 2008). Johnson et al. (2018) investigated when membrane bending occurs relative to the assembly of Gag and the recruitment of ESCRT-III factors. It was concluded that membrane bending occurs throughout assembly of virus-like particles whether or not Gag was full-length or missing its carboxyl terminal p6 domain, which functions to recruit early acting proteins such as ESCRT-I/TSG101 or Alix. This idea is inconsistent with notions that recruitment of ESCRT-III factors subsequent to early ESCRT recruitment facilitates the transition from a flat lattice to a spherical particle. Moreover, if membrane bending occurs prior to the arrival of ESCRT-III, then ESCRT-III factors are also not required to initiate curvature. Johnson et al. suggest the process is promoted by Gag multimerization. Strickland et al. (2017) suggest the involvement of ubiquitination based on the apparent arrest of membrane bending following small molecule disruption of the Ub-binding pockets in Tsg101. The suggestions are not mutually exclusive.

An alternative model based on ESCRT complex proximity in membrane-associated clusters has been proposed (Shields et al., 2009; Fig. 9.1C). Subsequent action of the ESCRT-III complex moves the cargo away from the cytoplasmic side of the endosome/MVB membrane by pinching off the neck of invaginated membrane sections associated with the cargo, forming cargo-containing intraluminal vesicles within the MVB. The AAA ATPase Vps4, also highly conserved, functions to maintain the availability of the ESCRT complex components, essentially promoting the disassembly of the membrane-bound complex and the return of the factors into the intracellular pool. It is now recognized that the ESCRT pathway is involved in membrane fission events at various cellular locations: at the plasma membrane; at the midbody following cell division; at the nuclear membrane; and at the endoplasmic reticulum membrane (Hurley et al., 2010; Olmos et al., 2015; Tabata et al., 2016; Vietri et al., 2015).

Endosomal Sorting Complexes Required for Transport Participation in Cellular Processes

The role for which the ESCRT machinery is mostly known is delivery of cell surface receptors cargo to MVBs for sorting and eventual degradation in the lysosome or the proteasome (Eden et al., 2009; Woodman, 2009). As noted above, ESCRTs participate in a broad spectrum of cellular activities (*reviewed broadly in* Campsteijn et al., 2016; Christ et al., 2017; Hurley, 2015; Olmos and Carlton, 2016). Somewhat more focused reviews are indicated for each activity:

- Cytokinetic abscission during mitotic exit (Kitagawa and Lee, 2015; Nakayama, 2016);
- Reformation of the nuclear envelope (Olmos et al., 2015; Vietri et al., 2015);
- Repair of damaged regions of the plasma membrane where ESCRTs are recruited in a Ca^{2+}-dependent manner by the cellular factor ALG2 (Jimenez et al., 2014; Scheffer et al., 2014);
- Autophagy (Murrow et al., 2015);
- Neuronal pruning (Loncle et al., 2015);
- Release of exosomes (Baietti et al., 2012);
- Polarized release of T cell receptor–enriched microvesicles at the T-cell immunological synapse (Choudhuri et al., 2014). Interestingly, it is speculated that this native pathway in T cells, triggered by antigen recognition to form an immunological synapse, may be co-opted by HIV to mediate polarized retroviral transmission, becoming a virological synapse (Vardhana et al., 2010).

Endosomal Sorting Complexes Required for Transport Participation in Virus Budding

The discovery that ESCRTs participate in virus budding arose from findings that proline-rich motifs in retroviral Gags dictated budding efficiency (*reviewed in* Carter, 2002; Demirov and Freed, 2004). These motifs, designated as late (L) domains in the context of viral-encoded proteins, take three general forms, Pro-Tyr-X-Pro-Pro (the so called *PY motif*), Pro-Thr/Ser-Ala/X-Pro (PTAP), and Tyr-Pro/Xn-Leu (*YXXL motif*). They are (*somewhat*) positionally independent and functionally interchangeable: these observations provided an early clue that proteins bearing these motifs were interacting with *trans*-acting adaptors engaged in the same or parallel-acting trafficking pathways. The first to be identified was Tsg101, as the binding partner for the PT/SAP motif in HIV-1 structural precursor polyprotein, Gag (Garrus et al., 2001; Martin-Serrano et al., 2001; VerPlank et al., 2001). Early on, two major groups of Gag partners were described: proteins involved in sorting of cargo into the MVB and proteins involved in clathrin adaptor complexes (Batonick et al., 2005; Camus et al., 2007; Dong et al., 2005). Interestingly, some of the cellular factors used for budding by the *e*quine *i*nfectious *a*nemia *v*irus (EIAV) and by HIV-1 (*both members of the lentivirus subgroup of Retroviridae*) such as the adaptor protein complexes, AP-1, -2, and -3, participate in endo- and exocytic machines (Chen et al., 2005). AP-1 and AP-3 could function on the same pathway that leads to Gag release via ESCRTs as it has been reported that AP-1 interacts with Tsg101 and Nedd4.1, an E3 Ub ligase (Camus et al., 2007). Both proteins are involved in budding of several retroviruses (Bouamr et al., 2003; Kikonyogo et al., 2001; Mercenne et al., 2015; Segura-Morales et al., 2005). Although Gag contains all of the determinants necessary for assembly and budding of (immature) virus particle from the plasma membrane (Freed, 2015), interaction with cellular cofactors involved in endosomal sorting and trafficking appears to be crucial. Precisely how the viruses might recruit them and avoid nonproductive endocytosis is not known. Depletion studies suggest that ESCRT-0 is required for HIV-1 Vpu-mediated BST-2/tetherin restriction of released particles (Janvier et al., 2011). ESCRT-I and -III have been shown to function in HIV-1 Gag trafficking; the involvement of ESCRT-II is controversial having been found to be both dispensable (Langelier et al., 2006; Pincetic et al., 2008) and indispensable (Ghoujal et al., 2012; Meng et al., 2015). Interestingly, we have found a cellular protein, Sprouty 2 (Spry2), whose depletion diminished HIV-1 release efficiency and which is capable of disrupting ESCRT-I and ESCRT-II association by sequestering the ESCRT-II component Eap20 (Medina et al., 2011). Spry 2 had been shown earlier to interfere with Hrs–Tsg101 interaction (Kim et al., 2007), indicating the protein's potential to disengage ESCRT-0 and ESCRT-I association as well. What may look like conflicting findings may be a reflection of the more complex nature of the participation of the ESCRT machinery in virus budding. At least generally, ESCRT machinery participation in virus budding can be differentiated from its behavior in other functions by the lack of sequential action of the ESCRT complexes.

At this point, it is clear that ESCRT budding functions are manifestations of topologically equivalent events occurring at other cellular locations: Thus, MVB sorting and retroviral egress both involve budding of vesicles and enveloped assemblages *away from* cytoplasm that is taking place in the cell interior and at the plasma membrane, *respectively*. They are the topological opposite of endocytosis, which involves the budding of vesicles *into* cytoplasm that takes place at the cell periphery and which uses different cellular machinery.

TUMOR SUPPRESSOR GENE 101 STRUCTURE

The highly conserved Tsg101 protein consists of four recognizable domains, designated as *U*b-*E*2-*v*ariant (UEV), pro-rich, coiled-coil, and steadiness-box domains (Li and Cohen, 1996). Fig. 9.2 shows the location of the domains

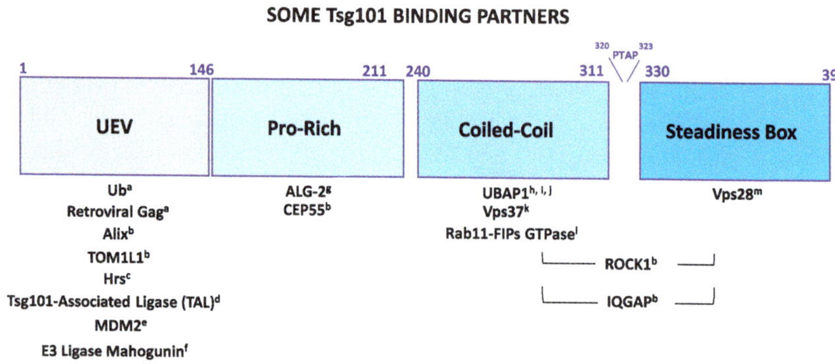

FIGURE 9.2 Schematic showing tumor suppressor gene 101 (Tsg101) domains and some of the binding partners mapped to these regions. References are [a]Pornillos, O., Sundquist, W.I., 2002a,b. EMBO; [b]Morita, E., Sandrin, V., Chung, H.Y., Morham, S.G., Gygi, S.P., Rodesch, C.K., Sundquist, W.I., 2007. Human ESCRT and ALIX proteins interact with proteins of the midbody and function in cytokinesis. EMBO J. 26, 4215–4227; [c]Lu, Q., Hope, L.W., Brasch, M., Reinhard, C., Cohen, S.N., 2003. Tsg101 interaction with Hrs mediates endosomal trafficking and receptor downregulation. Proc. Natl. Acad. Sci. U.S.A 100, 7626–7631; [d]McDonald, B., Martin-Serrano, J., 2008. Regulation of Tsg101 expression by the steadiness box: a role of Tsg101-associated ligase. Mol. Biol. Cell. 19, 754–763; [e]Li, L., Liao, J., Ruland, J., Mak, T.W., Cohen, S.N., 2001. A Tsg101/MDM2 regulatory loop modulates MDM2 degradation and MDM2/p53 feedback control. Proc. Natl. Acad. Sci. U.S.A. 98, 1619–1624; [f]Kim, B.Y., Olzmann, J.A., Barsh, G.S., Chin, L.S., Li, L., 2007. Spongiform neurodegeneration-associated E3 ligase Mahogunin ubiquitylates Tsg101 and regulates endosomal trafficking. Mol. Biol. Cell. 18, 1129–1142; [g]Katoh, K., Suzuki, H., Terasawa, Y., Mizuno, T., Yasuda, J., Shibata, H., Maki, M., 2005. The penta-EF-hand protein ALG-2 interacts directly with the ESCRT-I component Tsg101, and Ca^{2+}-dependently co-localizes to aberrant endosomes with dominant-negative AAA ATPase SKD1/Vps4B. Biochem. J. 391, 677–685; [h]Stefani, F., Zhang, L., Taylor, S., Donovan, J., Rollinson, S., Doyotte, A., Brownhill, K., Bennion, J., Pickering-Brown S, Woodman, P., 2011. UBAP1 is a component of an endosome-specific ESCRT-I complex that is essential for MVB sorting. Curr. Biol. 21, 1245–1250; [i]Agromayor, M., Soler, N., Caballe, A., Kueck, T., Freund, S.M., Allen, M.D., Bycroft, M., Perisic, O., Ye, Y., McDonald, B., Scheel, H., Hofmann, K., Neil, S.J., Martin-Serrano, J., Williams, R.L., 2012. The UBAP1 subunit of ESCRT-I interacts with ubiquitin via a SOUBA domain. Structure 20, 414–428; [j]Morita, E., Sandrin, V., Alam, S.L., Eckert, D.M., Gygi, S.P., Sundquist, W.I., 2007. Identification of human MVB12 proteins as ESCRT-I subunits that function in HIV budding. Cell Host Microbe 2, 41–53; [k]Eastman, S.W., Martin-Serrano, J., Chung, W., Zang, T., Bieniasz, P.D., 2005. Identification of human Vps37C, a component of endosomal sorting complex required for transport-I important for viral budding. J. Biol. Chem. 280, 628–636; [l]Horgan, C.P., Hanscom, S.R., Kelly, E.E., McCaffrey, M.W. Tumor susceptibility gene 101 (Tsg101) is a novel binding-partner for the class II Rab11-FIPs. PLoS One. 7, e32030; [m]Bishop, N., Woodman, P., 2001. Tsg101/mammalian Vps23 and mammalian Vps28 interact directly and are recruited to Vps4-induced endosomes. J. Biol. Chem. 276, 11,735–11,742.

within the protein and the interaction region of some of the known binding partners. Only the UEV domain has been structurally characterized in depth (Im et al., 2010; Pornillos et al., 2002a,b; Sundquist et al., 2004; Teo et al., 2004). The discussion below focuses on this region.

Tsg101-UEV Domain

The N-terminal ~150 amino acids in the Tsg101 protein fold into the structure described above as the UEV domain. The core of the structure (amino acids ~53 to 138) resembles that found in Ub E2 conjugases and in Mms2, the only other structurally characterized UEV protein (Eddins et al., 2006; Lewis et al., 2006). In conjunction with E3 Ub ligases, E2 Ub conjugases catalyze the linkage of "acceptor" and "donor" Ub molecules in the active site region where a Cys residue forms a transient thioester bond with the donor Ub. In Tsg101, this region is defunct: the Cys is replaced by a Tyr residue and the protein is unable to form the transient thioester bond with the C-terminus of Ub (Koonin and Abagyan, 1997; Ponting et al., 1997). The region is designated as the "vestigial active–site loop" because it is analogous to the region containing the active-site cysteine in authentic E2 conjugases. Outside of this region, there are additional important differences between the Tsg101-UEV domain, UEV proteins and enzymatically active E2 conjugases that have functional consequences for Gag. In Tsg101, residues from five different secondary structural elements come together to create a PTAP-binding "groove" (Teo et al., 2004). The analogous sites in E2 and UEV proteins are buried. In both the Tsg101-UEV and the yeast Vps23 homologue, but not E2, two beta strands, S1 and S2, extend to form a so-called "β-hairpin tongue." While this tongue is the most prominent feature of the Ub-binding site in Tsg101, as described below, it is distinct from the Ub-binding surface in enzymatically active E2 enzymes (Sundquist et al., 2004). In the latter case, the C-terminal tail of Ub binds near the active-site Cys and these residues form the major contacts with the E2 enzymes (Hamilton et al., 2001; Miura et al., 1999). This is not the case for the Tsg101 UEV, nevertheless, residues in the vestigial active–site region affect Gag binding (VerPlank et al., 2001).

Tsg101-UEV PTAP-Binding Pocket

The Tsg101-UEV domain residues exhibiting the greatest chemical shift changes on PTAP binding surround the rim of the aforementioned groove. Ala substitutions in three of these residues (Tyr_{63}, Met_{95}, and Val_{141}) all reduced p6 binding significantly (Pornillos et al., 2002a,b). In vitro binding of a nine-residue peptide that includes the core PTAP motif (PE**PTAP**PEE) or binding of the full-length Gag p6 sequence (~55 amino acids) that includes the PTAP motif produced very similar chemical shift changes and bound the Tsg101-UEV with similar affinity (K_d ~3 to 30 μM under nuclear magnetic resonance solution conditions). This permits the conclusion that the PTAP peptide contains all of the most important residues for Tsg101 binding. It should be noted that, while the UEV domains of Vps23 and Tsg101 both bind to the C terminus of Vps27 and Hrs, *respectively*, the interactions are not identical. Vps27 possesses two PSDP motifs that directly interact with Vps23 (Bilodeau et al., 2003); Tsg101 has a similar interaction with the PTAP motif in the HIV-1 Gag protein and with the PSAP motif in Hrs (Pornillos et al., 2002a,b). The structural basis of the Tsg101–PTAP interaction is known, as described above. However, it is not known for the Vps23–UEV-[PSDP-X_n-PSDP]–Vps27 interaction. Thus, although the Vps23 UEV accommodates a PSDP motif in a manner analogous to the Tsg101/P(S/T)AP interaction, they recognize related, but distinct PXXP motifs. Thus, the PTAP-binding pocket of Tsg101 is not conserved in Vps23.

Tsg101-UEV Ub-Binding Pocket

Chemical shift mapping was used to define the region in the UEV domain with Ub-binding activity (Pornillos et al., 2002b). Ub is a 76 amino acid protein that can be covalently attached to proteins through the formation of an isopeptide bond between the C-terminal Gly residue of Ub and the ε-amino group of Lys residues on target proteins (*reviewed in* MacGurn et al., 2012). The attachment of Ub to substrate proteins is carried out through an ATP-dependent mechanism that requires the sequential activity of three enzymes: E1 (Ub-activating enzyme), E2 (Ub-conjugating enzyme), and E3 (Ub ligase). The E3 ligase recognizes the substrate protein and provides specificity to the reaction. Ub contains seven Lys residues that can serve as acceptor sites for additional Ub molecules. The N-terminus of Ub can also join to the C-terminus. The fate of Ub-modified proteins is determined in part by the length of the Ub chain and on the configuration of the Ub–Ub linkages. Chains of four or more Ub moieties that are linked through Ub Lys48 are efficiently targeted to the 26S proteasome for degradation. By contrast, chains that are Ub Lys63-linked have different consequences for the target protein, e.g., endocytic sorting, perhaps even to the lysosome, but not proteasomal degradation.

Structural analysis shows two patches on the Ub molecule, a hydrophobic Leu_8-Ile_{44}-Val_{70} region that makes contact with the β-hairpin tongue and a hydrophilic Asn_{60}-Gln_{62}-Glu_{64} that contacts the vestigial active–site loop in the UEV. Mutations in Leu_8-Ile_{44}-Val_{70} in the hydrophobic patch interfere with endocytosis of monoubiquitinated proteins, proteasomal degradation of polyubiquitinated proteins, and interactions with Ub-binding modules in proteins (NZF, CUE, UBA, and Ub-interacting motif [UIM] modules). In contrast, the hydrophilic patch was suggested to be specific for ESCRT-I functions because mutations in Gln_{62}–Glu_{64} did not affect Hrs/ESCRT-0-mediated functions (Bilodeau et al., 2003). Based on the structure of the Ub–Tsg101-UEV complex (Pornillos et al., 2002b), the hydrophilic patch bears the Lys63 residue, which is expected to be surface exposed when Tsg101 binds Ub (Fig. 9.3). On the Tsg101-UEV side, the residues undergoing the greatest chemical shift changes on Ub binding are located throughout the β-tongue projection and in the so-called "lip" within the loop of the vestigial active–site region (Tsg101-UEV residues ~ Lys90-Tyr113). These two features flank a large hydrophobic patch on the UEV surface. Alanine substitutions at positions Val43, Asn45, and Asp46 in the β-tongue and at residue Phe88 in a hydrophobic patch residue significantly reduced binding of Tsg101-UEV to both p6–Ub and Ub alone

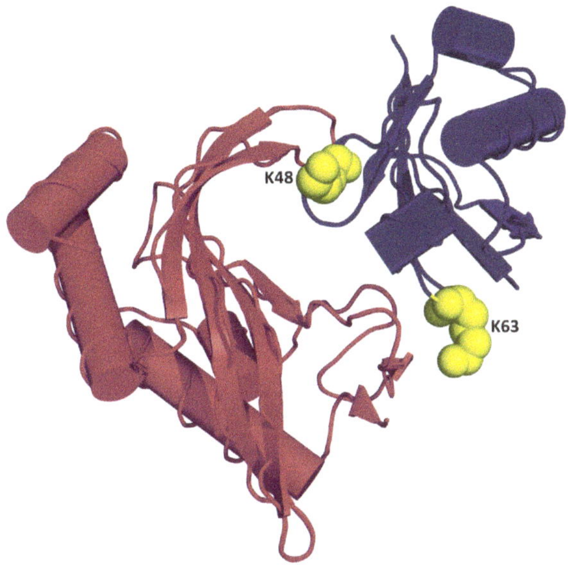

FIGURE 9.3 Solution structure of the ubiquitin (Ub)–Tsg101–UEV complex. The orientation of Ub (blue) in the Tsg101 UEV (purple) Ub-binding pocket is such that Lys63 is exposed. Lys48 and Lys63 are represented as *yellow spheres*. The molecular rendering of Protein Data Bank file 1S1Q was made using PyMOL.

(>threefold; Pornillos et al., 2002b). The orientation of Ub has been suggested to preclude polyubiquitylation via Lys48 linkages (Pornillos et al., 2002b). As part of the characterization of how small molecule inhibitors target the UEV domain (Strickland et al., 2017), we confirmed the solution structure of the Tsg101-UEV in complex with a mono-Ub moiety. Our analysis of chemical shift perturbations, docking predictions, and solvent saturation transfer to proteins (SSTP, Mishra et al., 2017) indicate that, beyond mono-Ub, the UEV can accommodate at least one additional Ub moiety in di-Ub linkage (Strickland, Fushman, Carter, Tjandra, *unpublished*).

Ub- and PTAP-Binding Pockets: Is There Cross Talk?

Titration with Ub produces chemical shift changes in a subset of the Tsg101-UEV amides that are distinct from those perturbed by titration with p6, consistent with the idea that PTAP and Ub bind to completely different sites (Pornillos et al., 2002a,b; Teo et al., 2004). Several observations suggest that the two ligands bind cooperatively. Direct fusion of Ub to p6 was found to enhance the binding affinity over that of either ligand alone (Pornillos et al., 2002b). As the PTAP peptide binds, the binding groove of Tsg101-UEV closes significantly, enveloping the core residues of the motif (Sundquist et al., 2004; Teo et al., 2004). Binding induces the flexible C-terminal UEV residues to become

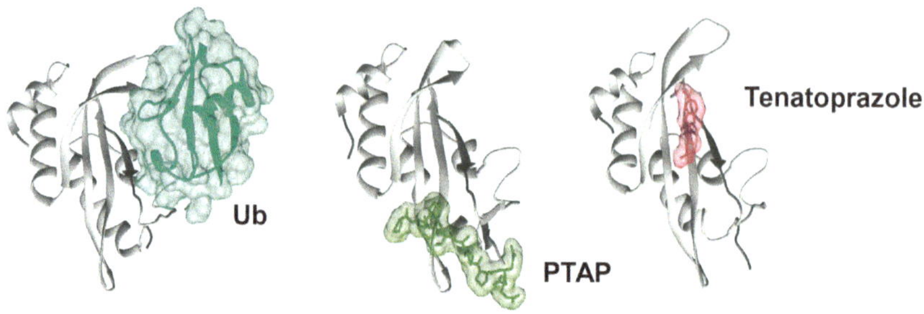

FIGURE 9.4 Binding locations of ubiquitin (Ub) (left, shown in same orientation as in Fig. 9.3), a PTAP peptide (center) and a small molecule inhibitor that targets the Ub-binding region (right).

ordered, wrapping around one side of the PTAP motif. The N-terminal third of the vestigial active–site loop changes its orientation, leading to an increase in order (based on a 30% increase in intramolecular NOEs in this region; Teo et al., 2004). Furthermore, we previously demonstrated that mutations in residues unique to the Tsg101 vestigial active–site region (Lys108, Tyr110, Tyr113, Trp117, Lys118) interfered with Tsg101 interaction with the SP2-p6 region of HIV-1 Gag in the yeast two-hybrid assay (VerPlank et al., 2001). This finding is consistent with observations that (1) PTAP binding perturbs residues in the N-terminal third of this loop (Pornillos et al., 2002a); Tsg101 residues 89–92; 94–96 undergo changes 1–1.5 standard deviations from zero (Strickland, Carter, Tjandra, *unpublished observation*) and (2) Ub binding perturbs residues in the C-terminal third of this loop (amino acids 106, 107, and 113; Pornillos et al., 2002a; Strickland, Carter, Tjandra, *unpublished*). This apparent structural cross talk may underlie the influence that Ub binding exerts on Gag occupancy of the PTAP-binding pocket. In Strickland et al. (2017), we demonstrate that adduct formation with Cys73 by the sulfenamide derivative of tenatoprazole, a small molecule in the prazole family, disrupts UEV Ub binding and arrests budding at an early stage of membrane evagination. The PTAP-binding pocket was not perturbed. Therefore, the two binding pockets indeed operate independently (i.e., Ub-binding pocket occupancy does not influence PTAP pocket occupancy) albeit cooperatively. Fig. 9.4 compares the binding locations of Ub (*left*, shown in same orientation as in Fig. 9.3), a PTAP peptide (*center*) and a small molecule inhibitor that targets the Ub-binding region (*right*).

TUMOR SUPPRESSOR GENE 101 PARTICIPATION

Tumor Suppressor Gene 101 Participation in the Cellular Endocytic Pathway

Studies, initially in the yeast system (Babst et al., 2000; Li et al., 1999), revealed that Vps23/Tsg101 is a key player in the biogenesis of MVBs and performs an important role in selecting which proteins enter the MVB lumen and which remain on the limiting membrane. Indeed, Vps23 was recognized to serve this function years before the discovery that it functioned in conjunction with Vps27, the orthologue of Hrs in mammalian cells. Vps23 was found early on to bind Ub (Katzmann et al., 2001) and it was proposed that its role was to recognize ubiquitinated cargo. As noted above, the Ub-binding pocket in Tsg101 binds Ub with very weak affinity (~500 μM; Pornillos et al., 2002b) and is not required for its function in ligand-induced downregulation of internalized growth factor receptors that are tagged by Ub and destined for degradation. Rather, other components of ESCRT-I and factors in ESCRT-0 and -II possess Ub-binding domains that bind Ub with greater affinity and likely play a more significant role in this downregulation process (Bilodeau et al., 2003; Ren and Hurley, 2010; Shields et al., 2009; Shih et al., 2002). It is interesting that Tsg101 participates in both ligand-induced and constitutive epidermal growth factor receptor (EGFR) recycling (Rush and Ceresa, 2013): the aforementioned small molecule was found to interfere with constitutive but not ligand-induced EGFR trafficking (Strickland et al., 2017), implicating the UEV Ub–binding capacity in one but not the other function. As noted above, ESCRT-I can interact directly with the ESCRT-0 complex (via a Tsg101–PSAP–Hrs interaction, Lu et al., 2003) and with the ESCRT-II complex (via Vps23 C-terminal domain binding to Vps28 which, in turn, binds to a composite surface that includes regions of Vps36 Gill et al., 2007). By serving to bridge ESCRT-0 and ESCRT-II, the Tsg101 molecule is the integral component that could ensure the sequential action of ESCRT-0, -I, -II, and -III complexes when required in this context.

It is interesting that although HIV-1 budding and EGFR downregulation both require Tsg101 (Babst et al., 2000; Garrus et al., 2001; Lu et al., 2003; Martin-Serrano et al., 2001), budding is much more dependent on the Gag-PT/SAP–Tsg101 interaction than is downregulation on the Hrs–PSAP–Tsg101 binding. Indeed, loss of the PSDP motifs in the yeast Vps27-ESCRT-0 that are required to bind to yeast Vps23-ESCRT-I reduces, but does not completely eliminate, colocalization of ESCRT-0 and -I (Wollert and Hurley, 2010). Some reports suggest that other parts of the ESCRT-0 and -I complexes contact with one another (Bouamr et al., 2007; Eastman et al., 2005; Stuchell et al., 2004).

Tumor Suppressor Gene 101 Participation in Virus Budding

The budding requirement for Tsg101 interaction is cell type–dependent: In the case of HIV-1, quantifiably different impacts were observed, e.g., budding from Jurkat cells was reported to be on the order of 1 magnitude less PTAP-dependent than release from HeLa cells (Demirov et al., 2002). In contrast, in the case of the vesicular stomatitis virus (VSV), mutations in the PSAP motif or small interfering RNA-mediated depletion of the endogenous Tsg101 had no detectable effect in BHK-21 cells (Irie et al., 2004). However, in chicken embryo–related Alger clone cells, PSAP mutation impaired VSV budding (Obiang et al., 2012). Also, recombinant viruses with mutations in and around

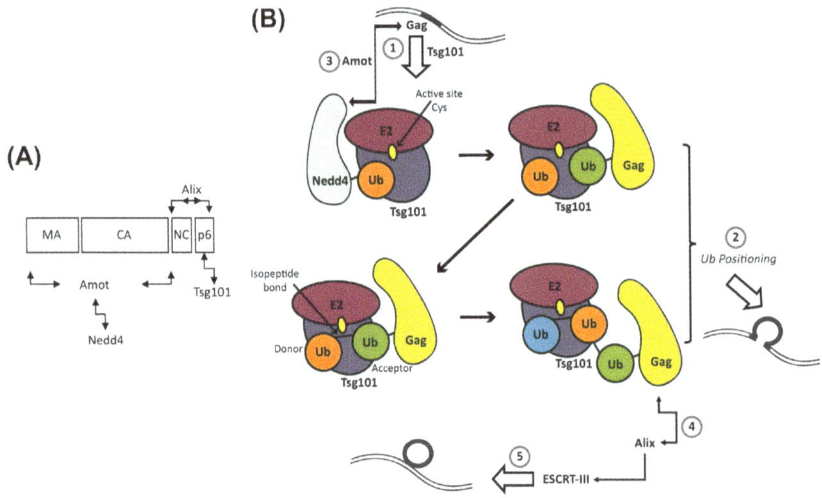

FIGURE 9.5 A model for Tsg101-ubiquitin (Ub)-Nedd4-Alix cooperation in HIV-1 budding. Panel A, schematic of HIV-1 Gag showing location of Tsg101, Nedd4 adaptor Amot, and Alix binding regions. Panel B, cartoon representation of the Tsg101 (blue), Ub (orange), E2 (purple)/ Nedd4 (light blue), Amot, and Alix proteins working together to position Ub to promote formation of Lys63-specific linkages. (1) Gag binds Tsg101 through the PT/SAP motif in the Gag protein. (2) The Ub-binding capability of Tsg101 combined with its similarity to authentic E2 Ub conjugases permit it to influence the interaction of (unknown) E2 enzyme(s) with the E3 Ub ligase, Nedd4. This influence positions the Lys63 residue in an acceptor Ub for proximity to the C-terminal Gly residue of a donor Ub. (3) HIV-1 Gag is linked to Nedd4 through the adaptor molecule, Amot; the active E2/E3 enzymes mediate formation of the Lys63 di-Ub linkage. (4) The Ub functions in Alix activation, which is required for (5) ESCRT-III recruitment, bud-host membrane scission and ultimate separation of the viral particle. *(The concept for this model is adapted from Eddins, M.J., Carlile, C.M., Gomez, K.M., Pickart, C.M., Wolberger, C., 2006. Mms2-Ubc13 covalently bound to ubiquitin reveals the structural basis of linkage-specific polyubiquitin chain formation. Nat. Struct. Mol. Biol. 13, 915–920.)*

the PSAP motif were attenuated in both cell culture and mice (Irie et al., 2007, 2012; Irie and Harty, 2005). Both HIV-1 and VSV, indeed most viruses, have conserved more than one type of late domain, permitting plasticity in budding pathway utilization. Doubtlessly, such plasticity reflects the abundance, stability, and/or compartmentalization of cell dependency factors. In the case of HIV-1, some of the cellular factors that potentiate Tsg101 utilization have been identified.

CELLULAR FACTORS THAT WORK WITH TUMOR SUPPRESSOR GENE 101

Tsg101 and Alix. The ability of Tsg101 to interact with the ESCRT adaptor Alix permits the protein to function independently of ESCRT-0 and -II and nevertheless link its partners to the ESCRT-III membrane scission machinery.

The HIV-1 Gag protein bears two binding sites for Alix: one in the NC domain and the other in the p6 domain (Fig. 9.5A). These cooperate in promoting the virus release function but are functionally redundant to the Tsg101 role. Interestingly, the PT/SAP motif has been intact in all variants isolated to date from infected individuals and archived in the Los Alamos National Laboratory database of HIV-1, -2, and recombinant viral sequences. However, variants containing mutations in the LYPXnL motif recognized by Alix were detected with high frequency in an investigation of a small cohort (Watanabe et al., 2013). The Alix protein can bind the ESCRT-III factor CHMP4B (Katoh et al., 2004), as well as Gag and Tsg101 (Martin-Serrano et al., 2003; Strack et al., 2003; von Schwedler et al., 2003), and could thereby provide a link to critical membrane-scission machinery in the absence of ESCRT-II involvement. However, disruption of the Tsg101-binding site in Alix, mutation of Alix-binding sites in HIV-1 Gag, or depletion of the endogenous pool of Alix have minimal impact on virus production when the Tsg101-binding motif in Gag is intact. Whether Alix is transported to the plasma membrane with HIV-1 Gag (Jouvenet et al., 2011), or later (Ku et al., 2014), is also controversial. However, there is general agreement that its function is performed late in the assembly process (Bleck et al., 2014; Jouvenet et al., 2011; Ku et al., 2014). How is this temporal control achieved? The native form of ALIX has a closed conformation due to intramolecular interactions between its N-terminal (Bro1) domain and its C-terminal (proline-rich) domain that precludes interaction with CHMP4B unless the protein is activated to the open conformation. Differences correlated to the respective affinities of Alix binding to EIAV Gag versus HIV-1-Gag dictate the ease of ALIX activation and its dependency on Alix multimerization (Zhou et al., 2008). In the absence of Tsg101 binding, budding can be rescued by Alix overexpression (Fisher et al., 2007; Fujii et al., 2009; Usami et al., 2007) indicating that Alix can link Gag to an alternative budding pathway. This alternative pathway requires the enzymatic activity of E3 Ub ligase Nedd4-1 (Sette et al., 2010).

Tsg101 and Ub. An estimated 2%–5% of Gag proteins in the retrovirion are monoubiquitylated (Ott et al., 1998, 2000; Putterman et al., 1990). Treatment of HIV- or VSV-infected cells with proteasome inhibitors, which decreases the intracellular concentration of free Ub, inhibits virus release (Harty et al., 2001; Schubert et al., 2000), suggesting a requirement for a ubiquitination event in the budding process. Consistent with this notion, fusion of the herpes simplex virus–encoded UL36 deubiquitination enzyme to Gag severely inhibited budding (Sette et al., 2013). The inhibitory effect of UL36 fusion was abolished when its catalytic activity was ablated. The target of the required ubiquitination event and when it is required remains unclear. Preventing Gag modification by mutation of acceptor Lys residues was not deleterious unless the mutations were cumulative (Gottwein et al., 2006; Gottwein and Krausslich, 2005; Ott et al., 2000). Studies demonstrate that partial alleviation of budding defects can be achieved by targeting ligases or their *h*omologous to the

*E*6-AP *c*arboxyl *t*erminus (HECT) domains to Gag, by Ub overexpression, or by appending Ub to Gag (Chung et al., 2008; Harty et al., 2000; Jadwin et al., 2010; Joshi et al., 2008; Kikonyogo et al., 2001; Martin-Serrano et al., 2004, 2005; Patnaik et al., 2000, 2002; Schubert et al., 2000; Strack et al., 2000; Vogt, 2000; Weiss et al., 2010; Zhadina and Bieniasz, 2010). The targeting of Gag itself for Ub modification as a requirement for budding has been suggested for *a*vian *s*arcoma *v*irus (ASV) budding (Vana et al., 2004), EIAV and HIV-1 budding, (Sette et al., 2013) and Nedd4-stimulated HIV-1 budding (Weiss et al., 2010). Our laboratory observed that deletion of a cluster of residues (aa 41–43) in the Tsg101-UEV domain that would perturb the β-hairpin tongue reduced budding efficiency (Goff et al., 2003). Mutation of single residues (e.g., aa 42 or 110) located in the β-hairpin tongue or the vestigial active site, *respectively*, also prevented Gag–Tsg101 interaction and accumulation on the plasma membrane (Watanabe and Carter, *unpublished observations*). Also, as noted above, small molecule interference with Ub binding to Tsg101 inhibited virus production and arrested budding at an early stage of membrane evagination (Strickland et al., 2017). Collectively, these observations support the notion that the Ub-binding pocket in Tsg101 contributes to the overall facilitating role of Ub in budding.

Tsg101 and Nedd4. The Nedd4 family of proteins typically possess an N-terminal C2 domain, three or four central WW domains that recognize pro-rich motifs and a C-terminal HECT domain that houses the catalytic Cys residue responsible for enzymatic activity (*reviewed in* Boase and Kumar, 2015). Several retroviruses encode PY motifs in the Gag protein that interact with the WW domains and thereby recruit the enzyme; mutational disruption of these sequences inhibits budding. A notable exception is HIV-1, which lacks a recognizable PY motif in Gag and whose release is not inhibited by depletion of any one of the several Nedd4 family members (Alroy et al., 2005). Nevertheless, HIV-1 egress can be rescued in the context of disrupted Tsg101 and Alix binding, by the Nedd4 family member Nedd4-2s (Chung et al., 2008). Our laboratory reported detection of Nedd4 in immune precipitates containing Tsg101 and HIV-1 Gag (Medina et al., 2005). The association was blocked by expression of a Nedd4-related fragment containing the C2–WW domains, suggesting that Nedd4 was interacting specifically. Nedd4 interaction with Gag can be bridged through the adaptor protein angiomotin p130 (Mercenne et al., 2015). Amot overexpression stimulated HIV-1 virus production in a manner that is dependent on Amot's ability to bind Nedd4L; Amot depletion inhibited virus production. Interestingly, the manifestation of early budding arrest in Amot-depleted cells (Mercenne et al., 2015) bears striking similarity to the early budding arrest following treatment with inhibitors that disrupt Tsg101 Ub–binding activity (Strickland et al., 2017). However, Martin-Serrano et al. (2004) demonstrated that for HIV-1, a variety of PPXY motifs were unable to replace the L-domain function that is normally provided by the PTAP motif even though they were all able to stimulate Gag ubiquitination. These findings support the notion that for budding Nedd4 functions in conjunction with Tsg101.

Convergence of Tsg101-Ub-Nedd4-Alix Participation

It has been suggested that Nedd4-mediated ubiquitination might occur upstream of bud neck scission and that three early-acting ESCRT-associated factors recruited by HIV-1 Gag, i.e., Nedd4, Tsg101 itself, and Alix, could work together in a stepwise fashion to facilitate virus budding (*reviewed in* Votteler and Sundquist, 2013). It is, therefore, an intriguing possibility that Tsg101-Ub-Nedd4-Alix participation is all about obtaining a Lys63-Ub-specific modification of Gag in a spatiotemporal manner (Fig. 9.4B). The key events in this model for budding would be (1) that ubiquitination of Gag is Lys63-specific and (2) that Tsg101 Ub binding plays a critical role in making sure that the ubiquitination event is Lys63-specific. The following observations make this speculation particularly compelling: (1) Synthesis of Lys63-linked Ub chains was observed to be critical for Ub ligase-mediated HIV-1 release (Weiss et al., 2010). This suggests that Ub contributes in a highly specific manner. (2) The HECT E3 ligase family that is implicated in HIV-1 budding is Lys63-specific Ub ligases (Kim and Huibregtse, 2009; Maspero et al., 2011, 2013). (3) The formation of Lys63 linkages employs E2 enzymes bound to UEVs (Eddins et al., 2006; Hofmann and Pickart, 1999; Maspero et al., 2013). (4) It is the binding of Mms2, a UEV protein, such as Tsg101, to the active E2 ligase Ubc13 that positions the E2 for Lys63-specific catalysis (Eddins et al., 2006). (5) Similarly, Tsg101 UEV-binding positions Ub Lys63 to be the accessible E2 ligase target (Pornillos et al., 2002b; Fig. 9.3). (6) Considering that Alix is now known to prefer Lys63 Ub modification and that this affinity is important for its role in budding (Dowlatshahi et al., 2012; Keren-Kaplan et al., 2013), it seems especially pertinent that Nedd4L was found to build Lys63-linked Ub chains on Gag.

Lys63 Ub modification is implicated in a variety of cellular functions (*reviewed in* Erpapazoglou et al., 2014). Lys63-linked di-Ub chains are known to play a role in endocytosis as part of a process that involves a budding event from the plasma membrane (*reviewed in* Hicke, 2001). Such modification on Gag might (1) be recognized as a sorting signal by the Ub-binding domain of Tsg101; (2) function in control of actin polymerization necessary for budding; or (3) function in Alix activation for ESCRT-III recruitment. The scheme in Fig. 9.5B suggests Alix activation (stage 4); this would not preclude Alix-mediated control of actin.

Tsg101 and PI(4,5)P$_2$. PI(4,5)P$_2$ is an anionic lipid that is found primarily in the cytoplasmic leaflet of the PM. For HIV-1 and murine leukemia virus Gag, PI(4,5)P$_2$ facilitates proper membrane targeting and myristate exposure (Chukkapalli and Ono, 2011; Fledderman et al., 2010; Hamard-Peron et al., 2010; Ono et al., 2004; Saad et al., 2006). The contribution of PI(4,5)P$_2$ to membrane targeting is less clear in the case of other retroviruses, such as EIAV and ASV. The Gag protein of EIAV appears to be less dependent on this particular lipid for targeting or stable membrane association (Fernandes et al., 2011). ASV Gag protein appears to be similar to HIV-1 in its $PI(4,5)P_2$ dependency (Nadaraia-Hoke et al., 2013; Vlach, Watanabe, Carter, and Saad, *unpublished observations*), however, this is controversial (Chan et al., 2011). In the case of HIV-1, $PI(4,5)P_2$ is involved in several Gag assembly events, including plasma membrane targeting and calcium signaling necessary for Gag accumulation on the plasma membrane (*reviewed in* Chukkapalli and Ono, 2011; Ehrlich and Carter, 2012; Mariani et al., 2014; Mucksch et al., 2017; Rocha-Perugini et al., 2014). When $PI(4,5)P_2$ was depleted from cells using siRNA directed against a kinase critical for $PI(4,5)P_2$ biogenesis, or through use of a lipid phosphatase, Gag accumulated in the cell interior, with resulting reductions in virus release (Ono et al., 2004). We found that budding also requires the hydrolysis of $PI(4,5)P_2$ (Ehrlich et al., 2010, 2011a,b). Thus, HIV-1 budding requires both $PI(4,5)P_2$ formation and turnover.

Because $PI(4,5)P_2$ is mostly at the plasma membrane (Watt et al., 2002), targeting Gag to this lipid (Ono et al., 2004) can be viewed as providing a basis for the plasma membrane as the site of HIV-1 assembly (reviewed in Ono, 2010). Our laboratory observed that the membrane envelope of VLPs assembled from a chimeric ASV GagΔp2b–Tsg101 protein shared properties with the anionic lipid–enriched plasma membrane microdomain that generated the HIV-1 Gag VLPs rather than the neutral lipid–enriched membranes from which the parental ASV VLPs were released (Medina et al., 2008). This finding suggests that "sensing" of the assembly site on the plasma membrane was dictated by the fused Tsg101. Tsg101 binding implies recruitment of other members of the ESCRT-I complex. ESCRT-I member Mvb12A and B subunits contain a so-called "β-prism" domain that binds anionic lipid bilayers (Boura and Hurley, 2012). The Mvb12-directed lipid binding exhibits no specificity for $PI(4,5)P_2$. However, possibly, Tsg101-mediated recruitment of Mvb12 might have been responsible for the apparent delivery of the assemblage to a membrane domain more reflective of the ESCRT-I propensities than that of the native ASV membrane–targeting determinants. Collectively, the notion that Tsg101/ESCRT-I influences the virus membrane localization properties supports a current hypothesis that Gag actively directs formation of its assembly platform rather than budding from a preexisting membrane domain (Hogue et al., 2012; Kerviel et al., 2013).

Tsg101 and Calcium. The presence of an intact PTAP motif in Gag was directly correlated to a 1.5- to 2-fold increase in the cytosolic Ca^{2+} concentration, inferring that Tsg101 binding directly or indirectly influenced Ca^{2+} influx and/or intracellular Ca^{2+} store release (Ehrlich et al., 2011a, 2014). Consistent with this finding, we observed that possession of an intact PTAP motif conferred resistance to 2-aminoethoxydiphenyl-borate (2-APB)–mediated inhibition of budding (Ehrlich et al., 2011a). 2-APB is a modulator of IP3R- and TRP-gated calcium influx channels. It is known that elevation of the intracellular Ca^{2+} level is required for multiple membrane fusion events, including fusion of membrane vesicles with the plasma membrane. Whether Gag assembles in intracellular vesicles or is subject to endocytosis from the cell periphery, Ca^{2+} mobilization sustained through Tsg101/ESCRT-I-influence on phospholipase C (PLC)–mediated $PI(4,5)P_2$ hydrolysis would, in effect, either rescue virus by promoting retention of Gag assemblages at the plasma membrane (i.e., maintaining $PI(4,5)P_2$ intact) or induce Ca^{2+} fluxes that facilitate membrane fusion or protein activation events (i.e., promoting PLC-mediated $PI(4,5)P_2$ hydrolysis to IP3, signaling release of IP3R-gated Ca^{2+} stores). Reciprocal co-immunoprecipitation experiments suggested that the proximity of Gag to IP3R, a resident protein on the intracellular Ca^{2+} ER store, was favored when the PTAP motif in Gag was intact (Ehrlich et al., 2014). The cellular factor ALG-2 is implicated in repair of damaged regions of the plasma membrane (Jimenez et al., 2014; Scheffer et al., 2014) to which it is recruited in a Ca^{2+}-dependent manner. Possibly, if budding indeed occurs near plasma membrane-ER junctions (Ehrlich et al., 2014), ALG-2 which binds Tsg101 and Alix in a Ca^{2+}-dependent manner (Maki et al., 2016) and limits their availability for budding (Maki et al., 2016; Ehrlich and Carter, *unpublished observations*) might serve to impose a necessary "brake" on the rate of virus production to permit intermittent repairs and thus minimize induction of nonproductive cellular responses signaled by membrane disruption.

TUMOR SUPPRESSOR GENE 101 ROLE IN BUDDING OF OTHER VIRUSES

Until relatively recently, involvement of Tsg101 and the ESCRT machinery in budding of a particular virus group was based on identification of a PT/SAP motif in a viral-encoded protein and/or a demonstration that viral release was inhibited by inactivation of the enzymatic activity of the AAA ATPase Vps4. These expectations were based on inviolate observations made with retroviruses, filoviruses (Ebola virus, Marburg virus), rhabdoviruses (VSV), reoviruses (bluetongue virus), and arena viruses (lymphochoriomenigitis virus, LCM, and Lassa fever virus). Nevertheless, it is now clear that these criteria are not universal. The Influenza A virus was the first well-established virus group to challenge this thinking (Chen et al., 2008; Rossman et al., 2010; Yondola and Carter, 2011). Collectively, the several observations made to date with viruses outside the Retroviridae reveal that vesiculation into the ER lumen have different requirements from vesiculation into the MVB lumen or into the extracellular space. Although the details of the virus, ESCRT, and adaptor interplay are not yet clear in most cases, the differences apparent in diverse contexts will be broadly informative, e.g., elucidation of the ESCRT role in *E*pstein–*B*arr *V*irus envelopment and release will most likely also provide novel insights relevant to how ESCRT machinery can function at the nuclear envelope. A few additional

TABLE 9.1 **Viruses that Co-opt Tsg101 during virus replication**

Virus	Tsg101: Viral Protein Link	Function
RNA Viruses		
Picornaviruses		
Hepatitis A virus	$YPXXLX_nYPXXL$: VP2	Budding (Feng et al., 2013)
Bunyaviruses		
Crimean–Congo hemorrhagic fever virus	Tsg101: ?[a]	Entry (Shtanko et al., 2014)
Filoviruses		
Ebola virus	PTAPPEY: VP40	Budding (Licata et al., 2003; Timmins et al., 2003; Dolnik et al., 2010)
Marburg virus	PSAP: NP	Budding (Licata et al., 2003; Timmins et al., 2003; Dolnik et al., 2010)
Hepatitis Viruses		
Hepatitis C virus	Tsg101:?	Budding (Ariumi et al., 2011)
Hepatitis E virus	PSAP: ORF3	Egress (Kenneya et al., 2010)
Flaviviruses		
Dengue	Tsg101: NS2A, NS3[a]	Budding from internal membrane (Tabata et al., 2016)
Japanese Encephalitis virus	Tsg101: NS2A, NS3[a]	Budding from internal membrane (Tabata et al., 2016)
Paramyxoviruses		
Nipah virus	Tsg101: C protein	Budding (Park et al., 2016)
Reoviruses		
Bluetongue virus	PPRYAPSAP: NS3	Budding from MVB? (Wirblich et al., 2006)
Retroviruses		
Avian sarcoma leukosis virus	$ASAPPPPYX_nYPXL$: Gag	Budding (Kikonyogo et al., 2001)
Equine infectious anemia virus	YXXL: Gag	Budding (Martin-Serrano et al., 2003; Strack et al., 2003)

TABLE 9.1 Viruses that Co-opt Tsg101 during virus replication—cont'd

Virus	Tsg101: Viral Protein Link	Function
Human immunodeficiency virus-1/2	PT/SAPX$_n$YPXL: Gag	Budding (Garrus et al., 2001; Martin-Serrano et al., 2001; VerPlank et al., 2001)
Human T-Cell Leukemia virus	PPPYX$_n$PTAP: Gag	Budding (Blot et al., 2004; Bouamr et al., 2003; Heidecker et al., 2004; Wang et al., 2004)
Mason–Pfizer Monkey virus	PPPYX$_n$PSAP: Gag	Budding (Gottwein et al., 2003)
Murine leukemia virus	PSAPX$_n$YPXLX$_n$PPPY: Gag	Budding (Segura-Morales et al., 2005)
Rhabdoviruses		
Vesicular stomatitis virus	PPPYX$_n$PSAP:M	Budding (Obiang et al., 2012)
DNA Viruses		
Hepatitis Viruses		
Hepatitis B virus	ESCRT-II/III: ?[a]	RNA transport; Budding from intracellular membranes (Stieler and Prange, 2014)
Herpes Viruses		
Epstein–Barr virus	Tsg101: Rta N-terminal region	Rta transactivation (Chua et al., 2007)
Herpes simplex virus-1	PT/SAP:?; Tsg101 UL36 (VP1/2)	Budding (Calistri et al., 2015; Pawliczek and Crump, 2009)
Kaposi sarcoma–associated herpes virus	Tsg101:?	Nuclear viral genome delivery; Cytoplasmic trafficking (Kumar and Chandran, 2016)
Orthopoxviruses		
Vaccinia virus	Tsg101/Alix: F13L	Egress (Honeychurch et al., 2007)
Papova Viruses		
Papilloma virus	Tsg101: L2	Not yet determined (Broniarczyk et al., 2014)

Tsg101, tumor suppressor gene 101.
[a]*Depletion of Tsg101 is inhibitory, however, no Tsg101 binding motif identified; :?, denotes no binding partner identified to date.*

illustrative systems are summarized below. Table 9.1 lists animal viruses for which direct protein–protein interaction with Tsg101 or evidence of dysfunction following Tsg101 depletion has been reported. Included is the unique case wherein the naked picornavirus, hepatitis A virus, exploits an adaptor to the host ESCRT system to hijack host cell membranes and noncytolytically exit infected cells, thus mimicking true enveloped viruses.

Orthomyxoviruses: Influenza A virus. Influenza A viruses are believed to employ an ESCRT-independent mode of budding, but the exact mechanism by which Influenza A virus exits infected cells is not clear. On one hand, the surface glycoproteins HA and NA associate with the viral matrix protein M1 at lipid microdomains then the M2 protein, an ion channel capable of altering membrane curvature, is recruited (Chen et al., 2008; Rossman et al., 2010). Mutations introduced into the amphipathic helix of M2 abolish membrane scission and virus release (Rossman et al., 2010; Rossman and Lamb, 2011). Yet, although there is no apparent PTAP motif encoded in any of the viral proteins and budding does neither require ESCRT-I component Vps28 nor the AAA ATPase Vps4, M1 binds to the ESCRT-I complex (Bruce et al., 2009; Watanabe and Lamb, 2010). Tsg101-deficient cells are defective in the surface display of HA and for infectious virion release (Sanyal et al., 2013). Interestingly, HA traffic from Golgi to plasma membrane but not its transport from ER to Golgi was affected. Precisely how Tsg101 contributes is not clear. Possibly, proteins destined for the extracellular environment from the secretory pathway are delivered by vesicular transport from the ER to the Golgi and then, via transport vesicles generated in a Tsg101-dependent fashion, from the Golgi network to the cell surface. Rab11, a protein enriched in the TGN and in recycling endosomes/MVBs, is required for Influenza virus budding (Bruce et al., 2010; Cox et al., 2000; Eisfeld et al., 2011) and also is capable of interacting with Tsg101 through the coiled-coil region in both proteins (Horgan et al., 2012).

Flaviviruses: Japanese Encephalitis Virus (JEV) and Dengue Virus. Several ESCRT factors are recruited to the site of replication of these viruses on the ER. Of these, only a unique subset that includes Tsg101 are necessary for the ultimate membrane deformation required for formation and release of the viral particle (Tabata et al., 2016). Particles that bud into the ER are secreted into the extracellular environment through the classical secretory pathway or through vesicles that fuse with the Golgi complex and exit through the secretory pathway from that intersection. Interestingly, like Influenza virus but unlike HIV-1, release was unaffected by depletion of Vps4. Thus, budding associated with membranes distinct from the plasma membrane appears to be less dependent on reutilization requiring Vps4-dependent disengagement of the ESCRT factors from the membrane. Also interesting, although depletion of Tsg101 significantly reduces the budding of JEV, dengue and HIV, only the latter requires the Ub-binding function of Tsg101 for budding (Tabata et al., 2016). Supporting this, dengue virus replication was tenatoprazole-resistant (Stauft, Wimmer and

Carter, *unpublished observations*). Again, this difference in ESCRT requirements would appear to suggest the existence of plasma membrane versus ER-specific needs for ESCRT-mediated membrane remodeling and implies that Ub provides a key feature when budding is from the plasma membrane. Identification of such features, if they exist, could provide invaluable insights pertinent to drug design.

Hepatitis C Virus (HCV): Like the Influenza virus A and the Flaviviruses described above, HCV lacks a defined PT/SAP motif and buds into or must transport structural proteins out of intracellular compartments (Barouch-Bentov et al., 2016). However, HCV appears to be unique even when compared with these viruses because it employs K63-linked polyubiquitinated Lys residues in its NS2 protein to bind to the UIM in the Hrs component of the ESCRT-0 complex. Through this means, it gains access to ESCRT-III machinery. As we suggested above, the Ub-binding pocket in Tsg101 could provide HIV-1 with access to K63-linked Ub for the purpose of communication with Alix and thereby ESCRT-III. If so, the "gate-keeping" role of K63-linked Ub might explain the broad conservation of Ub and E3 ligases in virus budding generally.

QUESTIONS REMAINING

Although much of our understanding was brought to light more than a decade ago, precisely how Tsg101 functions in budding remains a question. The precise roles of Tsg101 itself and its apparent partners, Nedd4, Ub, and enzymes that remove Ub are unclear. Whether or not routing through the cell interior in endosomes is required for virus budding from the plasma membrane is still controversial. The N-terminal UEV domain of Tsg101 contains a Ub-binding pocket and a PT/SAP-binding pocket. How do viral proteins responsible for Tsg101 interaction recruit the Tsg101 protein from its site of cellular accumulation to their assembly platforms? Indeed, is the Tsg101 accumulation site a domain on endosomes, as suggested by Welsch et al. (2006) or cytosolic, as initially believed? Do these two cellular compartments present Tsg101 in different forms (i.e., uncomplexed vs. complexed)? Tsg101-binding partners remain to be identified and the roles in budding of the known partners need clarification. For example, the same C-terminal region of Tsg101 binds ROCK, IQGAP, and Vps28 (Morita et al., 2007). How are their respective interactions controlled? Currently, some viruses, e.g., the togavirus Semliki Forest virus and the respiratory syncytial virus (a paramyxovirus) are thought to exit cells through a totally ESCRT-independent mechanism (Taylor et al., 2007; Utley et al., 2008). The budding of Nipah virus, a member of the Henipavirus genus within the Paramyxoviridae, was previously presumed to be independent of the ESCRT pathway because the viral matrix protein could induce budding when expressed alone in cells, but subsequently it was found that the virus C protein, an antagonist of innate immunity, recruits the ESCRT pathway, including

Tsg101, to promote efficient virus release (Park et al., 2016). Enveloped hepatitis virus and some other enveloped viruses utilize some ESCRT factors, but not Tsg101 (Feng et al., 2013). Will this status change as additional cell types are investigated? Mammalian cells contain soluble factors that selectively bind to K63-linked Ub chains and inhibit or prevent their association with proteasomes (Nathan et al., 2013). Rather, they are delivered instead to the endosomal-lysosomal pathway. Are there "still-to-be-discovered" cellular factors that determine whether a virus is delivered to the plasma membrane or the lysosome? Identification of these would provide novel antiviral targets. Are the diverse virus budding processes to which Tsg101 can contribute and at disparate subcellular locations an impediment or an advantage to efforts to target the protein for antiviral drug development?

Currently, there are no pharmacological inhibitors available that target Tsg101 or the virus budding event (reviewed in Jiang et al., 2011; Waheed and Freed, 2008). Interestingly, cyclic peptides (Tavassoli et al., 2008) were found to differentially inhibit HIV-1 budding while leaving EGFR downregulation unimpaired. As the former requires ESCRT-I and -III functions while the latter requires the participation of all four ESCRT complexes, the observation suggests that the peptides blocked some event that is unique to the budding pathway. This supposition, in turn, infers the existence of distinct surfaces that can be targeted for specific disruption of ESCRT specific–dependent steps in ubiquitinated cargo sorting. Compounds capable of similar differential inhibitory activity were recently identified through random screening of a library of small molecules (Strickland et al., 2017) suggesting that antiviral therapy based on the targeting of Tsg101 is feasible.

ACKNOWLEDGMENTS

We thank Drs. L. Ehrlich, M. Strickland, N. Tjandra, and S. Watanabe for their much appreciated comments and assistance. We thank Dr. Robert C. Rizzo and Mr. Yuchen Zhou for kindly providing the rendering in Fig. 9.4. We sincerely apologize to all those authors whose important work could not be cited in this overview due to space restrictions.

REFERENCES

Agromayor, M., Soler, N., Caballe, A., Kueck, T., Freund, S.M., Allen, M.D., Bycroft, M., Perisic, O., Ye, Y., McDonald, B., Scheel, H., Hofmann, K., Neil, S.J., Martin-Serrano, J., Williams, R.L., 2012. The UBAP1 subunit of ESCRT-I interacts with ubiquitin via a SOUBA domain. Structure 20, 414–428.

Alroy, I., Tuvia, S., Greener, T., Gordon, D., Barr, H.M., Taglicht, D., Mandil-Levin, R., Ben-Avraham, D., Konforty, D., Nir, A., Levius, O., Bicoviski, V., Dori, M., Cohen, S., Yaar, L., Erez, O., Propheta-Meiran, O., Koskas, M., Caspi-Bachar, E., Alchanati, I., Sela-Brown, A., Moskowitz, H., Tessmer, U., Schubert, U., Reiss, Y., 2005. The trans-Golgi network-associated human ubiquitin-protein ligase POSH is essential for HIV type 1 production. Proc. Natl. Acad. Sci. U.S.A. 102, 1478–1483.

Ariumi, Y., Kuroki, M., Maki, M., Ikeda, M., Dansako, H., Wakita, T., Kato, N., 2011. The ESCRT system is required for hepatitis C virus production. PLoS One 6, e14517.

Babst, M., Wendland, B., Estepa, E.J., Emr, S.D., 1998. The Vps4p AAA ATPase regulates membrane association of a Vps protein complex required for normal endosome function. EMBO J. 17, 2982–2993.

Babst, M., Odorizzi, G., Estepa, E.J., Emr, S.D., 2000. Mammalian tumor susceptibility gene 101 (Tsg101) and the yeast homologue, Vps23p, both function in late endosomal trafficking. Traffic 1, 248–258.

Baietti, M.F., Zhang, Z., Mortier, E., Melchior, A., Degeest, G., Geeraerts, A., Ivarsson, Y., Depoortere, F., Coomans, C., Vermeiren, E., Zimmermann, P., David, G., 2012. Syndecan-syntenin-ALIX regulates the biogenesis of exosomes. Nat. Cell Biol. 14, 677–685.

Barouch-Bentov, R., Neveu, G., Xiao, F., Beer, M., Bekerman, E., Schor, S., Campbell, J., Boonyaratanakornkit, J., Lindenbach, B., Lu, A., Jacob, Y., Einav, S., 2016. Hepatitis C virus proteins interact with the endosomal sorting complex required for transport (ESCRT) machinery via ubiquitination to facilitate viral envelopment. mBio 7.

Batonick, M., Favre, M., Boge, M., Spearman, P., Honing, S., Thali, M., 2005. Interaction of HIV-1 Gag with the clathrin-associated adaptor AP-2. Virology 342, 190–200.

Bilodeau, P.S., Winistorfer, S.C., Kearney, W.R., Robertson, A.D., Piper, R.C., 2003. Vps27-Hse1 and ESCRT-I complexes cooperate to increase efficiency of sorting ubiquitinated proteins at the endosome. J. Cell Biol. 163, 237–243.

Bleck, M., Itano, M.S., Johnson, D.S., Thomas, V.K., North, A.J., Bieniasz, P.D., Simon, S.M., 2014. Temporal and spatial organization of ESCRT protein recruitment during HIV-1 budding. Proc. Natl. Acad. Sci. U.S.A. 111, 12211–12216.

Blot, V., Perugi, F., Gay, B., Prévost, M.-C., Briant, L., Tangy, F., Abriel, H., Staub, O., Dokhélar, M.-C., Pique, C., 2004. Nedd4.1-mediated ubiquitination and subsequent recruitment of Tsg101 ensure HTLV-1 Gag trafficking towards the multivesicular body pathway prior to virus budding. J. Cell Sci. 117, 2357–2367.

Boase, N.A., Kumar, S., 2015. NEDD4: the founding member of a family of ubiquitin-protein ligases. Gene 557, 113–122.

Bouamr, F., Melillo, J.A., Wang, M.Q., Nagashima, K., de Los Santos, M., Rein, A., Goff, S.P., 2003. PPPYVEPTAP motif is the late domain of human T-cell leukemia virus type 1 Gag and mediates its functional interaction with cellular proteins Nedd4 and Tsg101 [corrected]. J. Virol. 77, 11882–11895.

Bouamr, F., Houck-Loomis, B.R., De Los Santos, M., Casaday, R.J., Johnson, M.C., Goff, S.P., 2007. The C-terminal portion of the Hrs protein interacts with Tsg101 and interferes with human immunodeficiency virus type 1 Gag particle production. J. Virol. 81, 2909–2922.

Boura, E., Hurley, J.H., 2012. Structural basis for membrane targeting by the MVB12-associated beta-prism domain of the human ESCRT-I MVB12 subunit. Proc. Natl. Acad. Sci. U.S.A. 109, 1901–1906.

Boura, E., Rozycki, B., Herrick, D.Z., Chung, H.S., Vecer, J., Eaton, W.A., Cafiso, D.S., Hummer, G., Hurley, J.H., 2011. Solution structure of the ESCRT-I complex by small-angle X-ray scattering, EPR, and FRET spectroscopy. Proc. Natl. Acad. Sci. U.S.A. 108, 9437–9442.

Broniarczyk, J., Bergant, M., Goździcka-Józefiak, A., Banks, L., 2014. Human papillomavirus infection requires the Tsg101 component of the ESCRT machinery. Virology 460–461, 83–90.

Bruce, E.A., Medcalf, L., Crump, C.M., Noton, S.L., Stuart, A.D., Wise, H.M., Elton, D., Bowers, K., Digard, P., 2009. Budding of filamentous and non-filamentous influenza A virus occurs via a Vps4 and VPS28-independent pathway. Virology 390, 268–278.

Bruce, E.A., Digard, P., Stuart, A.D., 2010. The Rab11 pathway is required for influenza A virus budding and filament formation. J. Virol. 84, 5848–5859.

Caillat, C., Macheboeuf, P., Wu, Y., McCarthy, A.A., Boeri-Erba, E., Effantin, G., Gottlinger, H.G., Weissenhorn, W., Renesto, P., 2015. Asymmetric ring structure of Vps4 required for ESCRT-III disassembly. Nat. Commun. 6, 8781.

Calistri, A., Munegato, D., Toffoletto, M., Celestino, M., Franchin, E., Comin, A., Sartori, E., Salata, C., Parolin, C., Palù, G., 2015. Functional interaction between the ESCRT-I component Tsg101 and the HSV-1 tegument ubiquitin specific protease. J. Cell. Physiol. 230, 1794–1806.

Campsteijn, C., Vietri, M., Stenmark, H., 2016. Novel ESCRT functions in cell biology: spiraling out of control? Curr. Opin. Cell Biol. 41, 1–8.

Camus, G., Segura-Morales, C., Molle, D., Lopez-Verges, S., Begon-Pescia, C., Cazevieille, C., Schu, P., Bertrand, E., Berlioz-Torrent, C., Basyuk, E., 2007. The clathrin adaptor complex AP-1 binds HIV-1 and MLV Gag and facilitates their budding. Mol. Biol. Cell 18, 3193–3203.

Carter, C.A., 2002. Tsg101: HIV-1's ticket to ride. Trends Microbiol. 10, 203–205.

Cashikar, A.G., Shim, S., Roth, R., Maldazys, M.R., Heuser, J.E., Hanson, P.I., May 30, 2014. Structure of cellular ESCRT-III spirals and their relationship to HIV budding. Elife 3. https://doi.org/10.7554/eLife.02184. PubMed PMID: 24878737; PubMed Central PMCID: PMC4073282.

Chan, J., Dick, R.A., Vogt, V.M., 2011. Rous sarcoma virus gag has no specific requirement for phosphatidylinositol-(4,5)-bisphosphate for plasma membrane association in vivo or for liposome interaction in vitro. J. Virol. 85, 10851–10860.

Chen, C., Vincent, O., Jin, J., Weisz, O.A., Montelaro, R.C., 2005. Functions of early (AP-2) and late (AIP1/ALIX) endocytic proteins in equine infectious anemia virus budding. J. Biol. Chem. 280, 40474–40480.

Chen, B.J., Leser, G.P., Jackson, D., Lamb, R.A., 2008. The influenza virus M2 protein cytoplasmic tail interacts with the M1 protein and influences virus assembly at the site of virus budding. J. Virol. 82, 10059–10070.

Choudhuri, K., Llodra, J., Roth, E.W., Tsai, J., Gordo, S., Wucherpfennig, K.W., Kam, L.C., Stokes, D.L., Dustin, M.L., 2014. Polarized release of T-cell-receptor-enriched microvesicles at the immunological synapse. Nature 507, 118–123.

Christ, L., Raiborg, C., Wenzel, E.M., Campsteijn, C., Stenmark, H., 2017. Cellular functions and molecular mechanisms of the ESCRT membrane-scission machinery. Trends Biochem. Sci. 42, 42–56.

Chua, H.H., Lee, H.H., Chang, S.S., Lu, C.C., Yeh, T.H., Hsu, T.Y., Cheng, T.H., Cheng, J.T., Chen, M.R., Tsai, C.H., 2007. Role of the Tsg101 gene in Epstein-Barr virus late gene transcription. J. Virol. 2007 (81), 2459–2471.

Chukkapalli, V., Ono, A., 2011. Molecular determinants that regulate plasma membrane association of HIV-1 Gag. J. Mol. Biol. 410, 512–524.

Chung, H.Y., Morita, E., von Schwedler, U., Muller, B., Krausslich, H.G., Sundquist, W.I., 2008. NEDD4L overexpression rescues the release and infectivity of human immunodeficiency virus type 1 constructs lacking PTAP and YPXL late domains. J. Virol. 82, 4884–4897.

Coonrod, E.M., Stevens, T.H., 2010. The yeast vps class E mutants: the beginning of the molecular genetic analysis of multivesicular body biogenesis. Mol. Biol. Cell 21, 4057–4060.

Cox, D., Lee, D.J., Dale, B.M., Calafat, J., Greenberg, S., 2000. A Rab11-containing rapidly recycling compartment in macrophages that promotes phagocytosis. Proc. Natl. Acad. Sci. U.S.A. 97, 680–685.

Demirov, D.G., Freed, E.O., 2004. Retrovirus budding. Virus Res. 106, 87–102.

Demirov, D.G., Orenstein, J.M., Freed, E.O., 2002. The late domain of human immunodeficiency virus type 1 p6 promotes virus release in a cell type-dependent manner. J. Virol. 76, 105–117.

Dolnik, O., Kolesnikova, L., Stevermann, L., Becker, S., 2010. Tsg101 is recruited by a Late domain of the nucleocapsid protein to support budding of Marburg virus-like particles. J. Virol. 84, 7847–7856.

Dong, X., Li, H., Derdowski, A., Ding, L., Burnett, A., Chen, X., Peters, T.R., Dermody, T.S., Woodruff, E., Wang, J.J., Spearman, P., 2005. AP-3 directs the intracellular trafficking of HIV-1 Gag and plays a key role in particle assembly. Cell 120, 663–674.

Dowlatshahi, D.P., Sandrin, V., Vivona, S., Shaler, T.A., Kaiser, S.E., Melandri, F., Sundquist, W.I., Kopito, R.R., 2012. ALIX is a Lys63-specific polyubiquitin binding protein that functions in retrovirus budding. Dev. Cell 23, 1247–1254.

Eastman, S.W., Martin-Serrano, J., Chung, W., Zang, T., Bieniasz, P.D., 2005. Identification of human Vps37C, a component of endosomal sorting complex required for transport-I important for viral budding. J. Biol. Chem. 280, 628–636.

Eddins, M.J., Carlile, C.M., Gomez, K.M., Pickart, C.M., Wolberger, C., 2006. Mms2-Ubc13 covalently bound to ubiquitin reveals the structural basis of linkage-specific polyubiquitin chain formation. Nat. Struct. Mol. Biol. 13, 915–920.

Eden, E.R., White, I.J., Futter, C.E., 2009. Down-regulation of epidermal growth factor receptor signalling within multivesicular bodies. Biochem. Soc. Trans. 37, 173–177.

Ehrlich, L.S., Carter, C.A., 2012. HIV assembly and budding: Ca(2+) signaling and non-ESCRT proteins set the stage. Mol. Biol. Int. 2012, 851670.

Ehrlich, L.S., Medina, G.N., Khan, M.B., Powell, M.D., Mikoshiba, K., Carter, C.A., 2010. Activation of the inositol (1,4,5)-triphosphate calcium gate receptor is required for HIV-1 Gag release. J. Virol. 84, 6438–6451.

Ehrlich, L.S., Medina, G.N., Carter, C.A., 2011a. ESCRT machinery potentiates HIV-1 utilization of the PI(4,5)P(2)-PLC-IP3R-Ca(2+) signaling cascade. J. Mol. Biol. 413, 347–358.

Ehrlich, L.S., Medina, G.N., Carter, C.A., 2011b. Sprouty2 regulates PI(4,5)P2/Ca2+ signaling and HIV-1 Gag release. J. Mol. Biol. 410, 716–725.

Ehrlich, L.S., Medina, G.N., Photiadis, S., Whittredge, P.B., Watanabe, S., Taraska, J.W., Carter, C.A., 2014. Tsg101 regulates PI(4,5)P2/Ca(2+) signaling for HIV-1 Gag assembly. Front. Microbiol. 5, 234.

Eisfeld, A.J., Kawakami, E., Watanabe, T., Neumann, G., Kawaoka, Y., 2011. RAB11A is essential for transport of the influenza virus genome to the plasma membrane. J. Virol. 85, 6117–6126.

Erpapazoglou, Z., Walker, O., Haguenauer-Tsapis, R., 2014. Versatile roles of k63-linked ubiquitin chains in trafficking. Cells 3, 1027–1088.

Fabrikant, G., Lata, S., Riches, J.D., Briggs, J.A., Weissenhorn, W., Kozlov, M.M., Nov, 2009. Computational model of membrane fission catalyzed by ESCRT-III. PLoS Comput. Biol. 5 (11), e1000575. https://doi.org/10.1371/journal.pcbi.1000575. Epub Nov 20, 2009. PubMed PMID: 19936052; PubMed Central PMCID: PMC2773929.

Feng, Z., Hensley, L., McKnight, K.L., Hu, F., Madden, V., Ping, L., Jeong, S.H., Walker, C., Lanford, R.E., Lemon, S.M., 2013. A pathogenic picornavirus acquires an envelope by hijacking cellular membranes. Nature 496, 367–371.

Fernandes, F., Chen, K., Ehrlich, L.S., Jin, J., Chen, M.H., Medina, G.N., Symons, M., Montelaro, R., Donaldson, J., Tjandra, N., Carter, C.A., 2011. Phosphoinositides direct equine infectious anemia virus gag trafficking and release. Traffic 12, 438–451.

Field, M.C., Sali, A., Rout, M.P., 2011. Evolution: on a bender–BARs, ESCRTs, COPs, and finally getting your coat. J. Cell Biol. 193, 963–972.

Fisher, R.D., Chung, H.Y., Zhai, Q., Robinson, H., Sundquist, W.I., Hill, C.P., 2007. Structural and biochemical studies of ALIX/AIP1 and its role in retrovirus budding. Cell 128, 841–852.

Fledderman, E.L., Fujii, K., Ghanam, R.H., Waki, K., Prevelige, P.E., Freed, E.O., Saad, J.S., 2010. Myristate exposure in the human immunodeficiency virus type 1 matrix protein is modulated by pH. Biochemistry 49, 9551–9562.

Freed, E.O., 2015. HIV-1 assembly, release and maturation. Nat. Rev. Microbiol. 13, 484–496.

Fujii, K., Munshi, U.M., Ablan, S.D., Demirov, D.G., Soheilian, F., Nagashima, K., Stephen, A.G., Fisher, R.J., Freed, E.O., 2009. Functional role of Alix in HIV-1 replication. Virology 391, 284–292.

Garrus, J.E., von Schwedler, U.K., Pornillos, O.W., Morham, S.G., Zavitz, K.H., Wang, H.E., Wettstein, D.A., Stray, K.M., Cote, M., Rich, R.L., Myszka, D.G., Sundquist, W.I., 2001. Tsg101 and the vacuolar protein sorting pathway are essential for HIV-1 budding. Cell 107, 55–65.

Ghoujal, B., Milev, M.P., Ajamian, L., Abel, K., Mouland, A.J., 2012. ESCRT-II's involvement in HIV-1 genomic RNA trafficking and assembly. Biol. Cell 104, 706–721.

Gill, D.J., Teo, H., Sun, J., Perisic, O., Veprintsev, D.B., Emr, S.D., Williams, R.L., 2007. Structural insight into the ESCRT-I/-II link and its role in MVB trafficking. EMBO J. 26, 600–612.

Goff, A., Ehrlich, L.S., Cohen, S.N., Carter, C.A., 2003. Tsg101 control of human immunodeficiency virus type 1 Gag trafficking and release. J. Virol. 77, 9173–9182.

Gottwein, E., Krausslich, H.G., 2005. Analysis of human immunodeficiency virus type 1 Gag ubiquitination. J. Virol. 79, 9134–9144.

Gottwein, E., Bodem, J., Müller, B., Schmechel, A., Zentgraf, H., Kräusslich, H.-G., 2003. The Mason-Pfizer Monkey Virus PPPY and PSAP motifs both contribute to virus release. J. Virol. 77, 9474–9485.

Gottwein, E., Jager, S., Habermann, A., Krausslich, H.G., 2006. Cumulative mutations of ubiquitin acceptor sites in human immunodeficiency virus type 1 gag cause a late budding defect. J. Virol. 80, 6267–6275.

Hamard-Peron, E., Juillard, F., Saad, J.S., Roy, C., Roingeard, P., Summers, M.F., Darlix, J.L., Picart, C., Muriaux, D., 2010. Targeting of murine leukemia virus gag to the plasma membrane is mediated by PI(4,5)P2/PS and a polybasic region in the matrix. J. Virol. 84, 503–515.

Hamilton, K.S., Ellison, M.J., Barber, K.R., Williams, R.S., Huzil, J.T., McKenna, S., Ptak, C., Glover, M., Shaw, G.S., 2001. Structure of a conjugating enzyme-ubiquitin thiolester intermediate reveals a novel role for the ubiquitin tail. Structure 9, 897–904.

Hanson, P.I., Roth, R., Lin, Y., Heuser, J.E., Jan 28, 2008. Plasma membrane deformation by circular arrays of ESCRT-III protein filaments. J. Cell. Biol. 180 (2), 389–402. https://doi.org/10.1083/jcb.200707031. Epub Jan 21, 2008. PubMed PMID: 18209100; PubMed Central PMCID: PMC2213594.

Harty, R.N., Brown, M.E., Wang, G., Huibregtse, J., Hayes, F.P., 2000. A PPxY motif within the VP40 protein of Ebola virus interacts physically and functionally with a ubiquitin ligase: implications for filovirus budding. Proc. Natl. Acad. Sci. U.S.A. 97, 13871–13876.

Harty, R.N., Brown, M.E., McGettigan, J.P., Wang, G., Jayakar, H.R., Huibregtse, J.M., Whitt, M.A., Schnell, M.J., 2001. Rhabdoviruses and the cellular ubiquitin-proteasome system: a budding interaction. J. Virol. 75, 10623–10629.

Heidecker, G., Lloyd, P.A., Fox, K., Nagashima, K., Derse, D., 2004. Late assembly motifs of Human T-Cell Leukemia Virus Type 1 and their relative roles in particle release. J. Virol. 78, 6636–6648.

Hicke, L., 2001. A new ticket for entry into budding vesicles-ubiquitin. Cell 106, 527–530.

Hofmann, R.M., Pickart, C.M., 1999. Noncanonical MMS2-encoded ubiquitin-conjugating enzyme functions in assembly of novel polyubiquitin chains for DNA repair. Cell 96, 645–653.

Hogue, I.B., Llewellyn, G.N., Ono, A., 2012. Dynamic association between HIV-1 gag and membrane domains. Mol. Biol. Int. 2012, 979765.

Honeychurch, K.M., Yang, G., Jordan, R., Hruby, D.E., 2007. The Vaccinia Virus F13L YPPL motif is required for efficient release of extracellular enveloped virus. J. Virol. 81, 7310–7315.

Horgan, C.P., Hanscom, S.R., Kelly, E.E., McCaffrey, M.W., 2012. Tumor susceptibility gene 101 (Tsg101) is a novel binding-partner for the class II Rab11-FIPs. PLoS One 7, e32030.

Hurley, J.H., 2008. ESCRT complexes and the biogenesis of multivesicular bodies. Curr. Opin. Cell Biol. 20, 4–11.

Hurley, J.H., 2015. ESCRTs are everywhere. EMBO J. 34, 2398–2407.

Hurley, J.H., Boura, E., Carlson, L.A., Rozycki, B., 2010. Membrane budding. Cell 143, 875–887.

Im, Y.J., Hurley, J.H., 2008. Integrated structural model and membrane targeting mechanism of the human ESCRT-II complex. Dev. Cell 14, 902–913.

Im, Y.J., Wollert, T., Boura, E., Hurley, J.H., 2009. Structure and function of the ESCRT-II-III interface in multivesicular body biogenesis. Dev. Cell 17, 234–243.

Im, Y.J., Kuo, L., Ren, X., Burgos, P.V., Zhao, X.Z., Liu, F., Burke Jr., T.R., Bonifacino, J.S., Freed, E.O., Hurley, J.H., 2010. Crystallographic and functional analysis of the ESCRT-I/HIV-1 Gag PTAP interaction. Structure 18, 1536–1547.

Irie, T., Harty, R.N., 2005. L-domain flanking sequences are important for host interactions and efficient budding of vesicular stomatitis virus recombinants. J. Virol. 79, 12617–12622.

Irie, T., Licata, J.M., McGettigan, J.P., Schnell, M.J., Harty, R.N., 2004. Budding of PPxY-containing rhabdoviruses is not dependent on host proteins TGS101 and Vps4A. J. Virol. 78, 2657–2665.

Irie, T., Carnero, E., Okumura, A., Garcia-Sastre, A., Harty, R.N., 2007. Modifications of the PSAP region of the matrix protein lead to attenuation of vesicular stomatitis virus in vitro and in vivo. J. Gen. Virol. 88, 2559–2567.

Irie, T., Liu, Y., Drolet, B.S., Carnero, E., Garcia-Sastre, A., Harty, R.N., 2012. Cytopathogenesis of vesicular stomatitis virus is regulated by the PSAP motif of M protein in a species-dependent manner. Viruses 4, 1605–1618.

Jadwin, J.A., Rudd, V., Sette, P., Challa, S., Bouamr, F., 2010. Late domain-independent rescue of a release-deficient Moloney murine leukemia virus by the ubiquitin ligase itch. J. Virol. 84, 704–715.

Janvier, K., Pelchen-Matthews, A., Renaud, J.B., Caillet, M., Marsh, M., Berlioz-Torrent, C., 2011. The ESCRT-0 component Hrs is required for HIV-1 Vpu-mediated BST-2/tetherin down-regulation. PLoS Pathog. 7, e1001265.

Jiang, Y., Liu, X., De Clercq, E., 2011. New therapeutic approaches targeted at the late stages of the HIV-1 replication cycle. Curr. Med. Chem. 18, 16–28.

Jimenez, A.J., Maiuri, P., Lafaurie-Janvore, J., Divoux, S., Piel, M., Perez, F., 2014. ESCRT machinery is required for plasma membrane repair. Science 343, 1247136.

Johnson, D.S., Bleck, M., Simon, S.M., Jul 4, 2018. Timing of ESCRT-III protein recruitment and membrane scission during HIV-1 assembly. Elife 7. https://doi.org/10.7554/eLife.36221. pii: e36221. PubMed PMID: 29972351.

Joshi, A., Munshi, U., Ablan, S.D., Nagashima, K., Freed, E.O., 2008. Functional replacement of a retroviral late domain by ubiquitin fusion. Traffic 9, 1972–1983.

Jouvenet, N., Zhadina, M., Bieniasz, P.D., Simon, S.M., 2011. Dynamics of ESCRT protein recruitment during retroviral assembly. Nat. Cell Biol. 13, 394–401.

Katoh, K., Shibata, H., Hatta, K., Maki, M., 2004. CHMP4b is a major binding partner of the ALG-2-interacting protein Alix among the three CHMP4 isoforms. Arch. Biochem. Biophys. 421, 159–165.

Katzmann, D.J., Babst, M., Emr, S.D., 2001. Ubiquitin-dependent sorting into the multivesicular body pathway requires the function of a conserved endosomal protein sorting complex, ESCRT-I. Cell 106, 145–155.

Katzmann, D.J., Stefan, C.J., Babst, M., Emr, S.D., 2003. Vps27 recruits ESCRT machinery to endosomes during MVB sorting. J. Cell Biol. 162, 413–423.

Kenneya, S.P., Wentworth, J., Heffrona, C.L., Meng, X.-J., 2010. Replacement of the Hepatitis E Virus ORF3 protein PxxP motif with heterologous Late domain motifs affects virus release via interaction with Tsg101. Virology 486, 198–208.

Keren-Kaplan, T., Attali, I., Estrin, M., Kuo, L.S., Farkash, E., Jerabek-Willemsen, M., Blutraich, N., Artzi, S., Peri, A., Freed, E.O., Wolfson, H.J., Prag, G., 2013. Structure-based in silico identification of ubiquitin-binding domains provides insights into the ALIX-V:ubiquitin complex and retrovirus budding. EMBO J. 32, 538–551.

Kerviel, A., Thomas, A., Chaloin, L., Favard, C., Muriaux, D., 2013. Virus assembly and plasma membrane domains: which came first? Virus Res. 171, 332–340.

Kikonyogo, A., Bouamr, F., Vana, M.L., Xiang, Y., Aiyar, A., Carter, C., Leis, J., 2001. Proteins related to the Nedd4 family of ubiquitin protein ligases interact with the L domain of Rous sarcoma virus and are required for gag budding from cells. Proc. Natl. Acad. Sci. U.S.A. 98, 11199–11204.

Kim, H.C., Huibregtse, J.M., 2009. Polyubiquitination by HECT E3s and the determinants of chain type specificity. Mol. Cell. Biol. 29, 3307–3318.

Kim, H.J., Taylor, L.J., Bar-Sagi, D., 2007. Spatial regulation of EGFR signaling by Sprouty2. Curr. Biol. 17, 455–461.

Kitagawa, M., Lee, S.H., 2015. The chromosomal passenger complex (CPC) as a key orchestrator of orderly mitotic exit and cytokinesis. Front. Cell Dev. Biol. 3, 14.

Koonin, E.V., Abagyan, R.A., 1997. Tsg101 may be the prototype of a class of dominant negative ubiquitin regulators. Nat. Genet. 16, 330–331.

Kostelansky, M.S., Sun, J., Lee, S., Kim, J., Ghirlando, R., Hierro, A., Emr, S.D., Hurley, J.H., 2006. Structural and functional organization of the ESCRT-I trafficking complex. Cell 125, 113–126.

Kostelansky, M.S., Schluter, C., Tam, Y.Y., Lee, S., Ghirlando, R., Beach, B., Conibear, E., Hurley, J.H., 2007. Molecular architecture and functional model of the complete yeast ESCRT-I heterotetramer. Cell 129, 485–498.

Ku, P.I., Bendjennat, M., Ballew, J., Landesman, M.B., Saffarian, S., 2014. ALIX is recruited temporarily into HIV-1 budding sites at the end of gag assembly. PLoS One 9, e96950.

Kumar, B., Chandran, B., 2016. KSHV entry and trafficking in target cells—hijacking of cell signal pathways, actin and membrane dynamics. Viruses 8, 305–334.

Langelier, C., von Schwedler, U.K., Fisher, R.D., De Domenico, I., White, P.L., Hill, C.P., Kaplan, J., Ward, D., Sundquist, W.I., 2006. Human ESCRT-II complex and its role in human immunodeficiency virus type 1 release. J. Virol. 80, 9465–9480.

Lewis, M.J., Saltibus, L.F., Hau, D.D., Xiao, W., Spyracopoulos, L., 2006. Structural basis for non-covalent interaction between ubiquitin and the ubiquitin conjugating enzyme variant human MMS2. J. Biomol. NMR 34, 89–100.

Li, L., Cohen, S.N., 1996. Tsg101: a novel tumor susceptibility gene isolated by controlled homozygous functional knockout of allelic loci in mammalian cells. Cell 85, 319–329.

Li, Y., Kane, T., Tipper, C., Spatrick, P., Jenness, D.D., 1999. Yeast mutants affecting possible quality control of plasma membrane proteins. Mol. Cell. Biol. 19, 3588–3599.

Licata, J.M., Simpson-Holley, M., Wright, N.T., Han, Z., Paragas, J., Harty, R.N., 2003. Overlapping motifs (PTAP and PPEY) within the Ebola virus VP40 protein function independently as late budding domains: involvement of host proteins Tsg101 and Vps4. J. Virol. 77, 1812–1819.

Lippincott-Schwartz, J., Freed, E.O., van Engelenburg, S.B., 2017. A Consensus View of ESCRT-Mediated Human Immunodeficiency Virus Type 1 Abscission. Ann. Rev. Virol. 4, 309–325.

Loncle, N., Agromayor, M., Martin-Serrano, J., Williams, D.W., 2015. An ESCRT module is required for neuron pruning. Sci. Rep. 5, 8461.

Lu, Q., Hope, L.W., Brasch, M., Reinhard, C., Cohen, S.N., 2003. Tsg101 interaction with Hrs mediates endosomal trafficking and receptor down-regulation. Proc. Natl. Acad. Sci. U.S.A. 100, 7626–7631.

MacGurn, J.A., Hsu, P.C., Emr, S.D., 2012. Ubiquitin and membrane protein turnover: from cradle to grave. Annu. Rev. Biochem. 81, 231–259.

Maki, M., Takahara, T., Shibata, H., 2016. Multifaceted roles of ALG-2 in Ca(2+)-regulated membrane trafficking. Int. J. Mol. Sci. 17.

Mariani, C., Desdouits, M., Favard, C., Benaroch, P., Muriaux, D.M., 2014. Role of Gag and lipids during HIV-1 assembly in CD4(+) T cells and macrophages. Front. Microbiol. 5, 312.

Martin-Serrano, J., Zang, T., Bieniasz, P.D., 2001. HIV-1 and Ebola virus encode small peptide motifs that recruit Tsg101 to sites of particle assembly to facilitate egress. Nat. Med. 7, 1313–1319.

Martin-Serrano, J., Yarovoy, A., Perez-Caballero, D., Bieniasz, P.D., 2003. Divergent retroviral late-budding domains recruit vacuolar protein sorting factors by using alternative adaptor proteins. Proc. Natl. Acad. Sci. U.S.A. 100, 12414–12419.

Martin-Serrano, J., Perez-Caballero, D., Bieniasz, P.D., 2004. Context-dependent effects of L domains and ubiquitination on viral budding. J. Virol. 78, 5554–5563.

Martin-Serrano, J., Eastman, S.W., Chung, W., Bieniasz, P.D., 2005. HECT ubiquitin ligases link viral and cellular PPXY motifs to the vacuolar protein-sorting pathway. J. Cell Biol. 168, 89–101.

Maspero, E., Mari, S., Valentini, E., Musacchio, A., Fish, A., Pasqualato, S., Polo, S., 2011. Structure of the HECT:ubiquitin complex and its role in ubiquitin chain elongation. EMBO Rep. 12, 342–349.

Maspero, E., Valentini, E., Mari, S., Cecatiello, V., Soffientini, P., Pasqualato, S., Polo, S., 2013. Structure of a ubiquitin-loaded HECT ligase reveals the molecular basis for catalytic priming. Nat. Struct. Mol. Biol. 20, 696–701.

McCullough, J., Clippinger, A.K., Talledge, N., Skowyra, M.L., Saunders, M.G., Naismith, T.V., Colf, L.A., Afonine, P., Arthur, C., Sundquist, W.I., Hanson, P.I., Frost, A., 2015. Structure and membrane remodeling activity of ESCRT-III helical polymers. Science 350, 1548–1551.

Medina, G., Zhang, Y., Tang, Y., Gottwein, E., Vana, M.L., Bouamr, F., Leis, J., Carter, C.A., 2005. The functionally exchangeable L domains in RSV and HIV-1 Gag direct particle release through pathways linked by Tsg101. Traffic 6, 880–894.

Medina, G., Pincetic, A., Ehrlich, L.S., Zhang, Y., Tang, Y., Leis, J., Carter, C.A., 2008. Tsg101 can replace Nedd4 function in ASV Gag release but not membrane targeting. Virology 377, 30–38.

Medina, G.N., Ehrlich, L.S., Chen, M.H., Khan, M.B., Powell, M.D., Carter, C.A., 2011. Sprouty 2 binds ESCRT-II factor Eap20 and facilitates HIV-1 gag release. J. Virol. 85, 7353–7362.

Meng, B., Ip, N.C., Prestwood, L.J., Abbink, T.E., Lever, A.M., 2015. Evidence that the endosomal sorting complex required for transport-II (ESCRT-II) is required for efficient human immunodeficiency virus-1 (HIV-1) production. Retrovirology 12, 72.

Mercenne, G., Alam, S.L., Arii, J., Lalonde, M.S., Sundquist, W.I., 2015. Angiomotin functions in HIV-1 assembly and budding. Elife 4.

Mishra, P., Barnes, C.A., Strickland, M., Tjandra, N., 2017. Solvent saturation transfer to proteins (SSTP) for structural and functional characterization of proteins. J. Biomol. NMR. 70, 11–20.

Miura, T., Klaus, W., Gsell, B., Miyamoto, C., Senn, H., 1999. Characterization of the binding interface between ubiquitin and class I human ubiquitin-conjugating enzyme 2b by multidimensional heteronuclear NMR spectroscopy in solution. J. Mol. Biol. 290, 213–228.

Morita, E., Sandrin, V., Chung, H.Y., Morham, S.G., Gygi, S.P., Rodesch, C.K., Sundquist, W.I., 2007. Human ESCRT and ALIX proteins interact with proteins of the midbody and function in cytokinesis. EMBO J. 26, 4215–4227.

Mucksch, F., Laketa, V., Muller, B., Schultz, C., Krausslich, H.G., 2017. Synchronized HIV assembly by tunable PIP2 changes reveals PIP2 requirement for stable Gag anchoring. Elife 6.

Murrow, L., Malhotra, R., Debnath, J., 2015. ATG12-ATG3 interacts with Alix to promote basal autophagic flux and late endosome function. Nat. Cell Biol. 17, 300–310.

Nadaraia-Hoke, S., Bann, D.V., Lochmann, T.L., Gudleski-O'Regan, N., Parent, L.J., 2013. Alterations in the MA and NC domains modulate phosphoinositide-dependent plasma membrane localization of the Rous sarcoma virus Gag protein. J. Virol. 87, 3609–3615.

Nakayama, K., 2016. Regulation of cytokinesis by membrane trafficking involving small GTPases and the ESCRT machinery. Crit. Rev. Biochem. Mol. Biol. 51, 1–6.

Nathan, J.A., Kim, H.T., Ting, L., Gygi, S.P., Goldberg, A.L., 2013. Why do cellular proteins linked to K63-polyubiquitin chains not associate with proteasomes? EMBO J. 32, 552–565.

Obiang, L., Raux, H., Ouldali, M., Blondel, D., Gaudin, Y., 2012. Phenotypes of vesicular stomatitis virus mutants with mutations in the PSAP motif of the matrix protein. J. Gen. Virol. 93, 857–865.

Olmos, Y., Carlton, J.G., 2016. The ESCRT machinery: new roles at new holes. Curr. Opin. Cell Biol. 38, 1–11.

Olmos, Y., Hodgson, L., Mantell, J., Verkade, P., Carlton, J.G., 2015. ESCRT-III controls nuclear envelope reformation. Nature 522, 236–239.

Ono, A., 2010. Relationships between plasma membrane microdomains and HIV-1 assembly. Biol. Cell 102, 335–350.

Ono, A., Ablan, S.D., Lockett, S.J., Nagashima, K., Freed, E.O., 2004. Phosphatidylinositol (4,5) bisphosphate regulates HIV-1 Gag targeting to the plasma membrane. Proc. Natl. Acad. Sci. U.S.A. 101, 14889–14894.

Ott, D.E., Coren, L.V., Copeland, T.D., Kane, B.P., Johnson, D.G., Sowder, R.C., Yoshinaka, Y., Oroszlan, S., Arthur, L.O., Henderson, L.E., 1998. Ubiquitin is covalently attached to the p6Gag proteins of human immunodeficiency virus type 1 and simian immunodeficiency virus and to the p12Gag protein of Moloney murine leukemia virus. J. Virol. 72, 2962–2968.

Ott, D.E., Coren, L.V., Chertova, E.N., Gagliardi, T.D., Schubert, U., 2000. Ubiquitination of HIV-1 and MuLV Gag. Virology 278, 111–121.

Park, A., Yun, T., Vigant, F., Pernet, O., Won, S.T., Dawes, B.E., Bartkowski, W., Freiberg, A.N., Lee, B., 2016. Nipah Virus C protein recruits Tsg101 to promote the efficient release of virus in an ESCRT-dependent pathway. PLoS Pathog. 12, e1005659. https://doi.org/10.1371/journal.ppat.1005659.

Patnaik, A., Chau, V., Wills, J.W., 2000. Ubiquitin is part of the retrovirus budding machinery. Proc. Natl. Acad. Sci. U.S.A. 97, 13069–13074.

Patnaik, A., Chau, V., Li, F., Montelaro, R.C., Wills, J.W., 2002. Budding of equine infectious anemia virus is insensitive to proteasome inhibitors. J. Virol. 76, 2641–2647.

Pawliczek, T., Crump, C.M., 2009. Herpes Simplex Virus type 1 production requires a functional ESCRT-III complex but is independent of Tsg101 and ALIX expression. J. Virol. 83, 11254–11264.

Pincetic, A., Medina, G., Carter, C., Leis, J., 2008. Avian sarcoma virus and human immunodeficiency virus, type 1 use different subsets of ESCRT proteins to facilitate the budding process. J. Biol. Chem. 283, 29822–29830.

Ponting, C.P., Cai, Y.D., Bork, P., 1997. The breast cancer gene product Tsg101: a regulator of ubiquitination? J. Mol. Med. (Berl.) 75, 467–469.

Pornillos, O., Alam, S.L., Davis, D.R., Sundquist, W.I., 2002a. Structure of the Tsg101 UEV domain in complex with the PTAP motif of the HIV-1 p6 protein. Nat. Struct. Biol. 9, 812–817.

Pornillos, O., Alam, S.L., Rich, R.L., Myszka, D.G., Davis, D.R., Sundquist, W.I., 2002b. Structure and functional interactions of the Tsg101 UEV domain. EMBO J. 21, 2397–2406.

Putterman, D., Pepinsky, R.B., Vogt, V.M., 1990. Ubiquitin in avian leukosis virus particles. Virology 176, 633–637.

Ren, X., Hurley, J.H., 2010. VHS domains of ESCRT-0 cooperate in high-avidity binding to polyubiquitinated cargo. EMBO J. 29, 1045–1054.

Rocha-Perugini, V., Gordon-Alonso, M., Sanchez-Madrid, F., 2014. PIP2: choreographer of actin-adaptor proteins in the HIV-1 dance. Trends Microbiol. 22, 379–388.

Rossman, J.S., Lamb, R.A., 2011. Influenza virus assembly and budding. Virology 411, 229–236.

Rossman, J.S., Jing, X., Leser, G.P., Balannik, V., Pinto, L.H., Lamb, R.A., 2010. Influenza virus m2 ion channel protein is necessary for filamentous virion formation. J. Virol. 84, 5078–5088.

Rothman, J.H., Yamashiro, C.T., Kane, P.M., Stevens, T.H., 1989. Protein targeting to the yeast vacuole. Trends Biochem. Sci. 14, 347–350.

Rush, J.S., Ceresa, B.P., 2013. RAB7 and Tsg101 are required for the constitutive recycling of unliganded EGFRs via distinct mechanisms. Mol. Cell. Endocrinol. 381, 188–197.

Saad, J.S., Miller, J., Tai, J., Kim, A., Ghanam, R.H., Summers, M.F., 2006. Structural basis for targeting HIV-1 Gag proteins to the plasma membrane for virus assembly. Proc. Natl. Acad. Sci. U.S.A. 103, 11364–11369.

Sanyal, S., Ashour, J., Maruyama, T., Altenburg, A.F., Cragnolini, J.J., Bilate, A., Avalos, A.M., Kundrat, L., Garcia-Sastre, A., Ploegh, H.L., 2013. Type I interferon imposes a Tsg101/ISG15 checkpoint at the Golgi for glycoprotein trafficking during influenza virus infection. Cell Host Microbe 14, 510–521.

Scheffer, L.L., Sreetama, S.C., Sharma, N., Medikayala, S., Brown, K.J., Defour, A., Jaiswal, J.K., 2014. Mechanism of Ca(2)(+)-triggered ESCRT assembly and regulation of cell membrane repair. Nat. Commun. 5, 5646.

Schoneberg, J., Lee, I.H., Iwasa, J.H., Hurley, J.H., 2017. Reverse-topology membrane scission by the ESCRT proteins. Nat. Rev. Mol. Cell Biol. 18, 5–17.

Schubert, U., Ott, D.E., Chertova, E.N., Welker, R., Tessmer, U., Princiotta, M.F., Bennink, J.R., Krausslich, H.G., Yewdell, J.W., 2000. Proteasome inhibition interferes with gag polyprotein processing, release, and maturation of HIV-1 and HIV-2. Proc. Natl. Acad. Sci. U.S.A. 97, 13057–13062.

Schuh, A.L., Audhya, A., 2014. The ESCRT machinery: from the plasma membrane to endosomes and back again. Crit. Rev. Biochem. Mol. Biol. 49, 242–261.

Segura-Morales, C., Pescia, C., Chatellard-Causse, C., Sadoul, R., Bertrand, E., Basyuk, E., 2005. Tsg101 and Alix interact with murine leukemia virus Gag and cooperate with Nedd4 ubiquitin ligases during budding. J. Biol. Chem. 280, 27004–27012.

Sette, P., Jadwin, J.A., Dussupt, V., Bello, N.F., Bouamr, F., 2010. The ESCRT-associated protein Alix recruits the ubiquitin ligase Nedd4-1 to facilitate HIV-1 release through the LYPXnL L domain motif. J. Virol. 84, 8181–8192.

Sette, P., Nagashima, K., Piper, R.C., Bouamr, F., 2013. Ubiquitin conjugation to Gag is essential for ESCRT-mediated HIV-1 budding. Retrovirology 10, 79.

Shields, S.B., Oestreich, A.J., Winistorfer, S., Nguyen, D., Payne, J.A., Katzmann, D.J., Piper, R., 2009. ESCRT ubiquitin-binding domains function cooperatively during MVB cargo sorting. J. Cell Biol. 185, 213–224.

Shih, S.C., Katzmann, D.J., Schnell, J.D., Sutanto, M., Emr, S.D., Hicke, L., 2002. Epsins and Vps27p/Hrs contain ubiquitin-binding domains that function in receptor endocytosis. Nat. Cell Biol. 4, 389–393.

Shtanko, O., Nikitina, R.A., Altuntas, C.Z., Chepurnov, A.A., Davey, R.A., 2014. Crimean-Congo hemorrhagic fever virus entry into host cells occurs through the multivesicular body and requires ESCRT regulators. PLoS Pathog. 10, e1004390.

Slagsvold, T., Aasland, R., Hirano, S., Bache, K.G., Raiborg, C., Trambaiolo, D., Wakatsuki, S., Stenmark, H., 2005. Eap45 in mammalian ESCRT-II binds ubiquitin via a phosphoinositide-interacting GLUE domain. J. Biol. Chem. 280, 19600–19606.

Stefani, F., Zhang, L., Taylor, S., Donovan, J., Rollinson, S., Doyotte, A., Brownhill, K., Bennion, J., Pickering-Brown, S., Woodman, P., 2011. UBAP1 is a component of an endosome-specific ESCRT-I complex that is essential for MVB sorting. Curr. Biol. 21, 1245–1250.

Stieler, J.T., Prange, R., 2014. Involvement of ESCRT-II in hepatitis B virus morphogenesis. PLoS One 9, e91279.

Strack, B., Calistri, A., Accola, M.A., Palu, G., Gottlinger, H.G., 2000. A role for ubiquitin ligase recruitment in retrovirus release. Proc. Natl. Acad. Sci. U.S.A. 97, 13063–13068.

Strack, B., Calistri, A., Craig, S., Popova, E., Göttlinger, H.G., 2003. AIP1/ALIX is a binding partner for HIV-1 p6 and EIAV p9 functioning in virus budding. Cell 114, 689–699.

Strickland, M., Ehrlich, L.S., Watanabe, S., Khan, M., Strub, M.P., Luan, C.H., Powell, M.D., Leis, J., Tjandra, N., Carter, C.A., 2017. Tsg101 chaperone function revealed by HIV-1 assembly inhibitors. Nat. Commun. 8, 1391.

Stuchell, M.D., Garrus, J.E., Muller, B., Stray, K.M., Ghaffarian, S., McKinnon, R., Krausslich, H.G., Morham, S.G., Sundquist, W.I., 2004. The human endosomal sorting complex required for transport (ESCRT-I) and its role in HIV-1 budding. J. Biol. Chem. 279, 36059–36071.

Sundquist, W.I., Schubert, H.L., Kelly, B.N., Hill, G.C., Holton, J.M., Hill, C.P., 2004. Ubiquitin recognition by the human Tsg101 protein. Mol. Cell 13, 783–789.

Tabata, K., Arimoto, M., Arakawa, M., Nara, A., Saito, K., Omori, H., Arai, A., Ishikawa, T., Konishi, E., Suzuki, R., Matsuura, Y., Morita, E., 2016. Unique requirement for ESCRT factors in Flavivirus particle formation on the endoplasmic reticulum. Cell Rep. 16, 2339–2347.

Tavassoli, A., Lu, Q., Gam, J., Pan, H., Benkovic, S.J., Cohen, S.N., 2008. Inhibition of HIV budding by a genetically selected cyclic peptide targeting the Gag-Tsg101 interaction. ACS Chem. Biol. 3, 757–764.

Taylor, G.M., Hanson, P.I., Kielian, M., 2007. Ubiquitin depletion and dominant-negative Vps4 inhibit rhabdovirus budding without affecting alphavirus budding. J. Virol. 81, 13631–13639.

Teo, H., Veprintsev, D.B., Williams, R.L., 2004. Structural insights into endosomal sorting complex required for transport (ESCRT-I) recognition of ubiquitinated proteins. J. Biol. Chem. 279, 28689–28696.

Timmins, J., Schoehn, G., Ricard-Blum, S., Scianimanico, S., Vernet, T., Ruigrok, R.W., Weissenhorn, W., 2003. Ebola virus matrix protein VP40 interaction with human cellular factors Tsg101 and Nedd4. J. Mol. Biol. 326, 493–502.

Usami, Y., Popov, S., Gottlinger, H.G., 2007. Potent rescue of human immunodeficiency virus type 1 late domain mutants by ALIX/AIP1 depends on its CHMP4 binding site. J. Virol. 81, 6614–6622.

Utley, T.J., Ducharme, N.A., Varthakavi, V., Shepherd, B.E., Santangelo, P.J., Lindquist, M.E., Goldenring, J.R., Crowe Jr., J.E., 2008. Respiratory syncytial virus uses a Vps4-independent budding mechanism controlled by Rab11-FIP2. Proc. Natl. Acad. Sci. U.S.A. 105, 10209–10214.

Vana, M.L., Tang, Y., Chen, A., Medina, G., Carter, C., Leis, J., 2004. Role of Nedd4 and ubiquitination of Rous sarcoma virus Gag in budding of virus-like particles from cells. J. Virol. 78, 13943–13953.

Vardhana, S., Choudhuri, K., Varma, R., Dustin, M.L., 2010. Essential role of ubiquitin and Tsg101 protein in formation and function of the central supramolecular activation cluster. Immunity 32, 531–540.

VerPlank, L., Bouamr, F., LaGrassa, T.J., Agresta, B., Kikonyogo, A., Leis, J., Carter, C.A., 2001. Tsg101, a homologue of ubiquitin-conjugating (E2) enzymes, binds the L domain in HIV type 1 Pr55(Gag). Proc. Natl. Acad. Sci. U.S.A. 98, 7724–7729.

Vietri, M., Schink, K.O., Campsteijn, C., Wegner, C.S., Schultz, S.W., Christ, L., Thoresen, S.B., Brech, A., Raiborg, C., Stenmark, H., 2015. Spastin and ESCRT-III coordinate mitotic spindle disassembly and nuclear envelope sealing. Nature 522, 231–235.

Vogt, V.M., 2000. Ubiquitin in retrovirus assembly: actor or bystander? Proc. Natl. Acad. Sci. U.S.A. 97, 12945–12947.

von Schwedler, U.K., Stuchell, M., Muller, B., Ward, D.M., Chung, H.Y., Morita, E., Wang, H.E., Davis, T., He, G.P., Cimbora, D.M., Scott, A., Krausslich, H.G., Kaplan, J., Morham, S.G., Sundquist, W.I., 2003. The protein network of HIV budding. Cell 114, 701–713.

Votteler, J., Sundquist, W.I., 2013. Virus budding and the ESCRT pathway. Cell Host Microbe 14, 232–241.

Waheed, A.A., Freed, E.O., 2008. Peptide inhibitors of HIV-1 egress. ACS Chem. Biol. 3, 745–747.

Wang, H., Machesky, N.J., Mansky, L.M., 2004. Both the PPPY and PTAP motifs are involved in human T cell leukemia virus type 1 particle release. J. Virol. 78, 1503–1512.

Watanabe, R., Lamb, R.A., 2010. Influenza virus budding does not require a functional AAA+ ATPase, Vps4. Virus Res. 153, 58–63.

Watanabe, S.M., Chen, M.H., Khan, M., Ehrlich, L., Kemal, K.S., Weiser, B., Shi, B., Chen, C., Powell, M., Anastos, K., Burger, H., Carter, C.A., 2013. The S40 residue in HIV-1 Gag p6 impacts local and distal budding determinants, revealing additional late domain activities. Retrovirology 10, 143.

Watt, S.A., Kular, G., Fleming, I.N., Downes, C.P., Lucocq, J.M., 2002. Subcellular localization of phosphatidylinositol 4,5-bisphosphate using the pleckstrin homology domain of phospholipase C delta1. Biochem. J. 363, 657–666.

Weiss, E.R., Popova, E., Yamanaka, H., Kim, H.C., Huibregtse, J.M., Gottlinger, H., 2010. Rescue of HIV-1 release by targeting widely divergent NEDD4-type ubiquitin ligases and isolated catalytic HECT domains to Gag. PLoS Pathog. 6, e1001107.

Welsch, S., Habermann, A., Jager, S., Muller, B., Krijnse-Locker, J., Krausslich, H.G., 2006. Ultrastructural analysis of ESCRT proteins suggests a role for endosome-associated tubular-vesicular membranes in ESCRT function. Traffic 7, 1551–1566.

Wideman, J.G., Leung, K.F., Field, M.C., Dacks, J.B., 2014. The cell biology of the endocytic system from an evolutionary perspective. Cold Spring Harb. Perspect. Biol. 6, a016998.

Wirblich, C., Bhattacharya, B., Roy, P., 2006. Nonstructural protein 3 of bluetongue virus assists virus release by recruiting ESCRT-I protein Tsg101. J. Virol. 80, 460–473.

Wollert, T., Hurley, J.H., 2010. Molecular mechanism of multivesicular body biogenesis by ESCRT complexes. Nature 464, 864–869.

Woodman, P., 2009. ESCRT proteins, endosome organization and mitogenic receptor down-regulation. Biochem. Soc. Trans. 37, 146–150.

Yondola, M., Carter, C., 2011. Un-"ESCRT"-ed budding. Viruses 3, 26–31.

Zhadina, M., Bieniasz, P.D., 2010. Functional interchangeability of late domains, late domain cofactors and ubiquitin in viral budding. PLoS Pathog. 6, e1001153.

Zhou, X., Pan, S., Sun, L., Corvera, J., Lin, S.H., Kuang, J., 2008. The HIV-1 p6/EIAV p9 docking site in Alix is autoinhibited as revealed by a conformation-sensitive anti-Alix monoclonal antibody. Biochem. J. 414, 215–220.

Chapter 10

The Role of Lipids in Retroviral Replication

Abdul A. Waheed, Eric O. Freed
National Cancer Institute, Frederick, MD, United States

Chapter Outline

Introduction 353
Composition of Cellular Membranes 354
Composition of Viral Membranes 357
Membrane Microdomains 358
Lipids in Retroviral Entry 360
Role of Cholesterol in Retroviral Entry 360
Role of Sphingolipids in Retroviral Entry 362
Role of Phospholipids in Retroviral Entry 363
Retroviral Assembly 363
Trafficking of Gag and Sites of Retroviral Assembly 365
Gag–Membrane Interactions 366
Lipids in Retroviral Assembly 368
Role of Cholesterol in Retroviral Assembly 368
Role of $PI(4,5)P_2$ in Retroviral Assembly 370
Gag Binding to Giant Unilamellar Vesicles 374
Lipids in Retroviral Cell–Cell Transfer 375
Retroviral Accessory Proteins and Lipid Microdomains 377
Inhibition of Retroviral Replication With Lipid-Modifying Agents 379
Cholesterol Modifying Agents 379
Inhibitors of Sphingolipid Synthesis 380
Concluding Remarks 381
References 382

INTRODUCTION

Retroviruses are enveloped, single-stranded RNA viruses belonging to the family *Retroviridae*. The major human retroviruses are human immunodeficiency virus (HIV), the causative agent of AIDS, and human T-cell leukemia virus (HTLV), the causative agent of adult T-cell leukemia/lymphoma, and HTLV-associated myelopathy/tropical spastic paraparesis. Retroviruses use the enzyme reverse transcriptase (RT) to convert their RNA genomes into linear, double-stranded DNA. The viral enzyme integrase (IN) then catalyzes the integration of the newly synthesized viral DNA into the host cell genome. The integrated

Retrovirus-Cell Interactions. https://doi.org/10.1016/B978-0-12-811185-7.00010-8

proviral DNA is transcribed into full-length (unspliced) genomic RNA and spliced mRNAs, which are transported from the nucleus to the cytoplasm where they are translated into viral proteins, and, in the case of full-length, unspliced viral RNA, packaged into the next generation of virus particles.

Replication-competent retroviruses contain three genes: *gag*, *pol*, and *env*. The *gag* gene encodes the Gag precursor protein, which contains matrix (MA), capsid (CA), and nucleocapsid (NC) domains. A small amount of a GagPol precursor is expressed during Gag translation; GagPol contains, in addition to major Gag domains, the protease (PR), RT, and IN domains. The *env* open reading frame encodes the envelope (Env) glycoproteins, which are embedded in the viral membrane during assembly. Retroviral Env glycoprotein complexes comprise a heterotrimer of surface (SU) and transmembrane (TM) subunits. In addition to *gag, pol*, and *env*, complex retroviruses, such as HIV, encode a number of additional regulatory and accessory genes.

Retroviral particles measure ~80–120nm in diameter. Based on their morphology and genome organization, retroviruses are classified into seven genera: alpharetroviruses, betaretroviruses, gammaretroviruses, deltaretroviruses, epsilonretroviruses, lentiviruses, and spumaviruses. In many respects, the spumaviruses differ significantly from the other genera.

As is the case for other enveloped viruses, retroviruses must cross the host cell membrane both early (viral entry) and late (assembly and release) in their replication cycle. The plasma membrane (PM) is a central element in the interaction of virus with the host. The entry process begins when the SU subunit of the Env glycoprotein binds to receptor(s) on the cell surface followed by the fusion of viral and cell membranes, triggered by the TM subunit (Freed and Martin, 2013). The generation of new viral particles takes place when a subset of viral proteins such as Gag, GagPol, and Env traffic to the PM of the host cell where they assemble and bud to form new immature viral particles (reviewed in Balasubramaniam and Freed, 2011; Freed, 2015). Thus, retroviruses acquire their lipid Env by budding through the PM of an infected cell. Nonetheless, the lipid composition of the viral membrane is often considerably different from that of the host cell PM from which it is derived. Much of this chapter will focus on HIV-1 and other retroviruses that assemble at, and bud from, the PM. It should be noted, however, that some retroviruses (e.g., the betaretrovirus Mason–Pfizer monkey virus [M-PMV] and mouse mammary tumor virus [MMTV]) assemble in the cytosol before trafficking to the PM, whereas others (e.g., the spumaretroviruses) assemble in the cytosol and bud at the endoplasmic reticulum (ER) and PM in an Env-dependent fashion.

Composition of Cellular Membranes

The lipid composition of the PM is very different from that of other membranes in the cell, and changes in lipid composition have been implicated in a number of diseases (Holthuis and Menon, 2014; van Meer, 2005; van Meer et al.,

2008). The predominant classes of lipids in mammalian cell membranes are the glycerophospholipids (GPLs), sphingolipids, and sterols (reviewed in Munro, 2003; Simons and Vaz, 2004; van Meer et al., 2008; Waheed and Freed, 2009, 2010). The GPLs that contribute to forming lipid bilayers comprise a head group attached to two acyl chains via glycerol. Two acyl chains together with glycerol, known as diacylglycerol, constitute the hydrophobic portion of GPLs. Variations in the polar head group lead to different GPLs such as phosphatidylcholine (PC), phosphatidylethanolamine (PE), phosphatidylserine (PS), phosphatidylinositol (PI), phosphatidic acid (PA), phosphatidylglycerol, and cardiolipin (Fig. 10.1) (Li et al., 2015; Waheed and Freed, 2010). The length of the apolar moiety also differs between GPLs, e.g., dimyristoyl, dipalmitoyl, distearoyl, and dioleoyl. The acyl chains can be saturated (e.g., dipalmitoyl or distearoyl) or unsaturated (e.g., dioleoyl) (Li et al., 2015). Some GPLs (e.g., the lysophospholipids) have only a single acyl chain. In some GPLs, the fatty acyl chains are linked to glycerol with an ether linkage at the sn-1 position and an ester linkage at the sn-2 position as observed in plasmalogens (Li et al., 2015). PC is the most abundant phospholipid in mammalian membranes and contributes ~50% of total cellular phospholipid content. PE, the lipid that imposes membrane curvature and plays a critical role in budding, fission and fusion of cell membranes, is the second most abundant GPL (Marrink et al., 2009; Marsh, 2007; van Meer et al., 2008). Sphingolipids, which are predominantly enriched in the PM, consist of two parts: a hydrophobic ceramide (Cer) backbone and a polar head group. The head group of sphingolipids is either PC (sphingomyelin, SM) or a carbohydrate moiety (glycosphingolipids, GSLs). GSLs are highly diverse, and in mammals hundreds of GSLs exist with differing carbohydrate modifications, such as glucose, galactose, mannose, fucose, xylose, sialic acid, *N*-acetylglucosamine, *N*-acetylgalactosamine, and *N*-acetylmannosamine. The simplest GSLs are glucose-containing glucosylceramide, galactose-containing galactosylceramide (GalCer), and sialic acid-containing gangliosides (GM). The acyl chains of sphingolipids are usually long (~24 carbons) and saturated, in some cases transunsaturated; these lipids are thus more closely packed than GPLs, resulting in a solid gel phase. The third major class of lipids, the sterols, is amphipathic. Cholesterol is the principle sterol in many biological membranes. Cholesterol has a tetracyclic structure with a polar hydroxyl group in the equatorial position on its first A ring, and an eight-member branched acyl chain on the pentameric ring D (Fig. 10.1). Because of its tetracyclic structure, cholesterol increases the rigidity of membranes in which it is embedded. Sphingolipids and sterols are enriched in the PM and PM-derived endosomal membranes compared with intracellular membranes (Hao et al., 2002; van Meer, 1998; van Meer et al., 2008).

The key organelle in the cell for lipid biosynthesis is the ER, which is the site of synthesis for GPLs, sterols, and sphingolipid precursors such as Cer and GalCer (Bell et al., 1981; Sprong et al., 1998). Although the ER is the main site of sterol biosynthesis, sterols are rapidly transported to other cellular organelles. Sphingolipids are initially synthesized in the ER and later modified in the Golgi

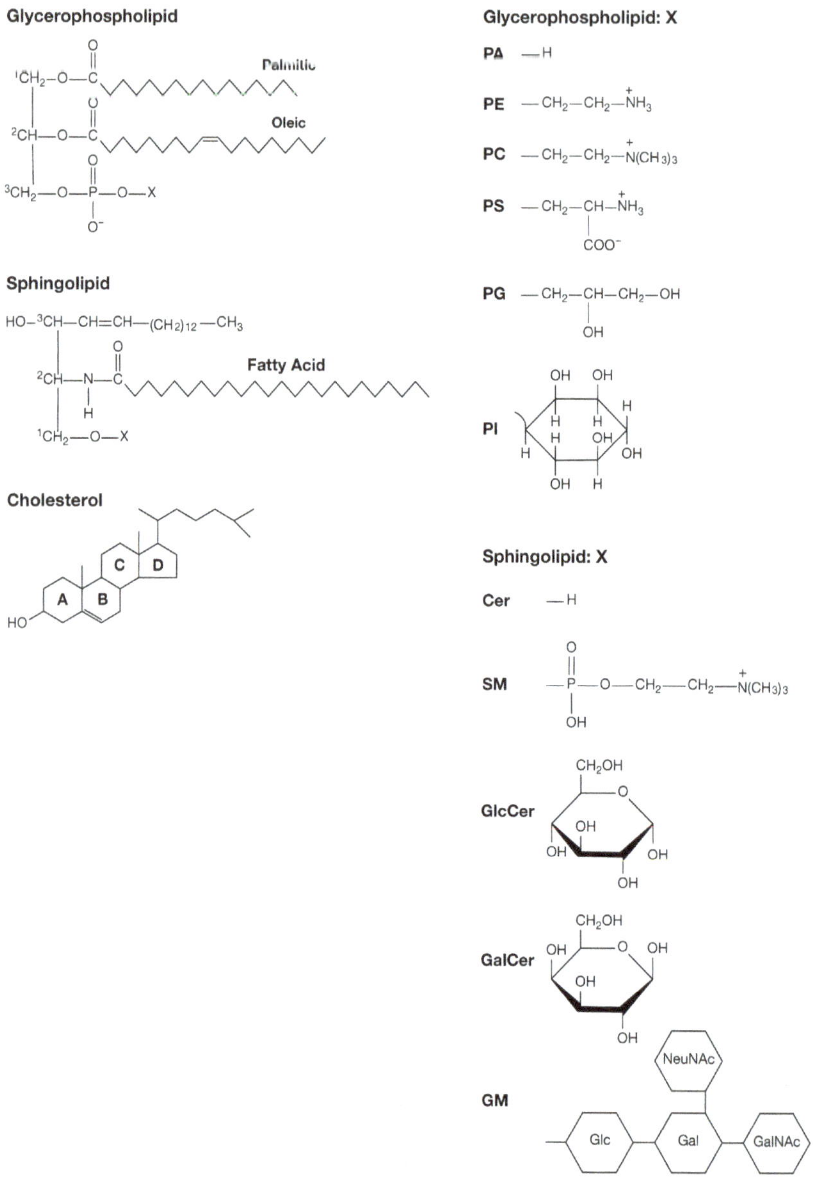

FIGURE 10.1 Structures of principle classes of lipids in eukaryotic cell membranes. Glycerophospholipids contain diacylglycerol in which two fatty acids are ester-linked to glycerol at C-1 and C-2, and a polar head group is attached to C-3. The acyl chains are commonly 16–18 carbons long and contain a saturated fatty acid at C-1 and an unsaturated fatty acid at C-2. The head groups vary among the phospholipids; e.g., ethanolamine in phosphatidylethanolamine (PE), choline in phosphatidylcholine (PC), serine in phosphatidylserine (PS), glycerol in phosphatidylglycerol (PG), and inositol in phosphatidylinositol (PI). Sphingolipids are derivatives of sphingosine. Ceramide, the parent compound for sphingolipids, bears a long, saturated acyl chain of 16–26 carbons attached to the C-2 amino group by an amide linkage. The polar head group attached to C-1 differs, with a phosphocholine in sphingomyelin (SM) and a sugar in glycosphingolipids. Depending on the type of sugar, glycosphingolipids are further subclassified into several types as listed. A third class of membrane lipids is sterols, with cholesterol being a major sterol in mammalian cells. The cholesterol core consists of four fused rings, with a polar hydroxyl group on ring A and a nonpolar eight-carbon hydrocarbon side chain on pentameric ring D.

apparatus as enzymes involved in sphingolipid synthesis are localized in the ER and the Golgi (Futerman and Riezman, 2005).

More than 1000 different lipids make up the cellular lipidome (van Meer, 2005; Wenk, 2010). Lipids typically constitute ~50% of the mass of cellular membranes, although this percentage is influenced by a variety of factors, including species of origin, cell type, and stage of the cell cycle. Lipids in the PM are distinct from those in intracellular membranes, and even among intracellular membranes, the composition differs depending on the organelle. The approximate abundance of common lipids in cellular membranes is PC (45%–55%), PE (15%–25%), PI (10%–15%), PS (5%–10%), PA (1%–2%), SM (5%–10%), and cholesterol (10%–20%) (Balasubramanian and Schroit, 2003; Feigenson, 2007; van Meer et al., 2008). The composition of the inner and outer leaflet of membrane bilayers also differs; in the PM of mammalian cells, PC, SM, and GM are located predominately in the outer leaflet, whereas PE, PS, PI, and phosphoinositides are primarily in the inner leaflet. In certain pathological conditions, such as apoptosis and viral infections, PS is relocalized to the outer leaflet due to the action of the flippase enzyme, scramblase (Marino and Kroemer, 2013; Suzuki et al., 2013). Although cholesterol displays preferential interactions with sphingolipids relative to phospholipids, it is present in both inner and outer leaflets and may flip-flop between them (Hamilton, 2003). The relative content of cholesterol in mammalian cell membranes is as follows: PM > trans-Golgi network > cis-Golgi > ER (Watson, 2015). Due to this gradient in cholesterol concentration, the trans-Golgi and PM bilayers are slightly thicker than the ER membrane, allowing them to accommodate longer hydrophobic transmembrane domains (Watson, 2015).

Composition of Viral Membranes

Enveloped viruses typically acquire their lipid Envs from the host cell membranes in which particle assembly and budding take place. The composition of retroviral lipid Envs can thus provide valuable information regarding the site of assembly in an infected cell. In most cases, retroviruses assemble at the PM; however, the lipid composition of retroviral particles differs significantly from that of cellular PMs, with some lipids reduced and others enriched in the viral Env (Aloia et al., 1993; Brugger et al., 2006; Chan et al., 2008; Lorizate et al., 2013). The lipids that are enriched over twofold in purified HIV-1 and MLV virions compared with producer cell PM include cholesterol, SM, PS, plasmenylethanolamine (pl-PE), GM1, and PI 4,5-bisphosphate [$PI(4,5)P_2$] (Aloia et al., 1988, 1993; Brugger et al., 2006; Chan et al., 2008; Kawabata et al., 2009; Lorizate et al., 2009, 2013; Nguyen and Hildreth, 2000). In contrast, lipids such as PC, Cer, PE, and PI are reduced in virions relative to host cell PM (Brugger et al., 2006; Chan et al., 2008; Lorizate et al., 2013). These studies support the hypothesis that enveloped viruses assemble in specific PM microdomains.

The lipid composition of the HIV-1 and MLV membranes is very similar to that of lipid raft membranes, also known as detergent-resistant membranes (DRMs), suggesting that the HIV-1 and MLV membranes are derived from lipid rafts (see below) (Brugger et al., 2006; Chan et al., 2008).

Membrane Microdomains

Membrane microdomains, often referred to as lipid rafts, are specific subdomains in cellular membranes that are enriched in cholesterol and sphingolipids containing saturated acyl chains (Brown and London, 2000; Simons and Toomre, 2000). These microdomains arise due to interactions of cholesterol with saturated acyl chains that result in tight packing of these lipids in comparison to phospholipids, which are rich in kinked, unsaturated acyl chains (Brown, 2006; Brown and London, 2000). This differential lipid packing induces the formation of a liquid-ordered (lo) phase and a more loosely packed liquid-disordered (ld) phase (Brown, 2006; Brown and London, 2000). Cholesterol plays a critical role in regulating phase behavior; above a threshold concentration, the lo phase is favored (Brown and London, 2000; Simons and Toomre, 2000; Simons and Vaz, 2004). The formation of the lo phase is also enhanced by intermolecular hydrogen bonds that form between sphingolipids (Li et al., 2001; Ramstedt and Slotte, 2002). The enrichment of sphingolipids in the outer leaflet of the PM results in lo domain that "floats" in the surrounding ld domain, hence the term "lipid rafts" (Fig. 10.2). These lipid microdomains, due to their insolubility in nonionic detergent (typically Triton X-100) at low temperatures, can be biochemically isolated as DRM. Because of their high lipid content, DRM complexes float to low density in sucrose density gradients (Brown and London, 2000; Simons and Toomre, 2000).

DRM isolation has been widely used to identify proteins associated with lipid rafts. The proteins that tend to reside in these microdomains include singly or multiply acylated proteins such as G-proteins and certain tyrosine kinases that generally reside in the inner leaflet and glycosylphosphatidylinositol (GPI)-anchored proteins, which are attached to the outer leaflet of membrane microdomains (Brown and London, 2000; Simons and Toomre, 2000). A major caveat to the DRM isolation approach is that cold detergent treatment can cause aggregation of membrane domains. Hence, the isolated lipid rafts can represent artifacts of detergent treatment rather than membrane microdomains that are present in native, living cell membranes (Mayor and Maxfield, 1995; Munro, 2003). To overcome this technical limitation, methods that visualize copatching of a protein of interest with well-known raft markers (e.g., GM1) by immunofluorescence microscopy are often coupled with DRM analysis to identify the association of proteins with lipid rafts (Harder et al., 1998). Because native lipid raft microdomains are defined as being 10–200 nm in diameter (Pike, 2006), advanced microscopy approaches such as fluorescence resonance energy transfer (FRET), single-particle tracking, fluorescence recovery after photobleaching

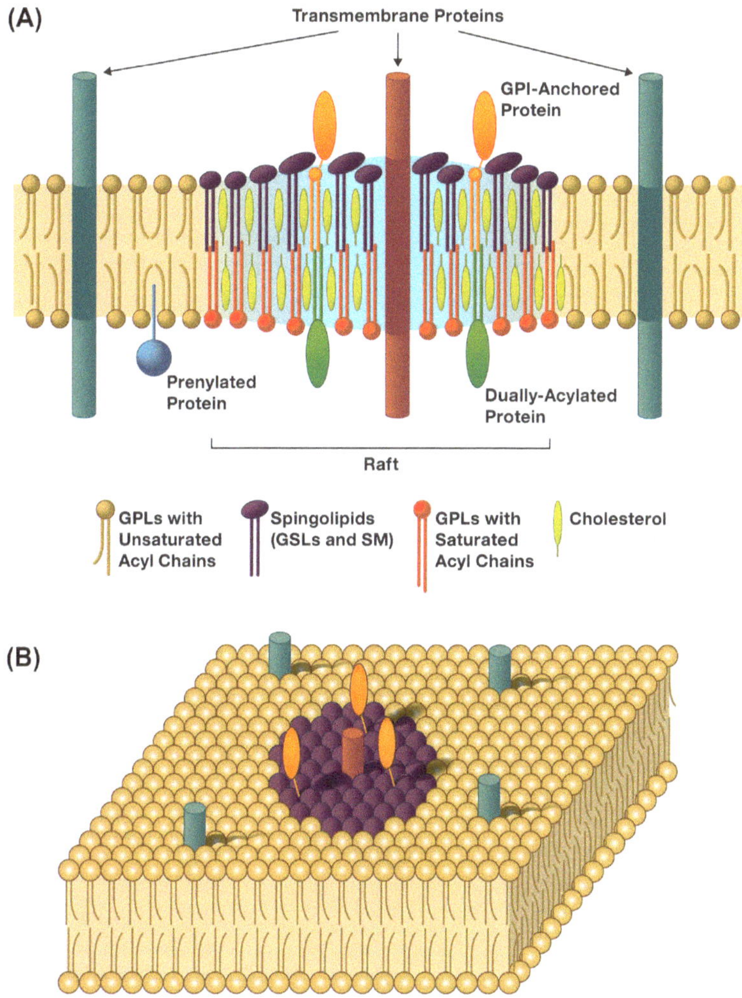

FIGURE 10.2 **A simplified model of lipid raft structure in cell membranes.** The phospholipids (brown and red) and cholesterol (yellow) are distributed in both the leaflets, whereas sphingolipids (violet) are enriched in the outer leaflet of the bilayer. The acyl chains of raft lipids are generally long and saturated (violet and red), whereas those in nonraft domains are shorter and contain one or two unsaturated acyl chains (brown). Raft domains contain concentrations of dually acylated (green) and glycosylphosphatidylinositol (GPI)–anchored (orange) proteins, whereas transmembrane (blue) and prenylated (blue) proteins are usually nonraft associated. Panel A depicts a lipid raft in cross section; panel B shows a 3D view, looking at 45 degrees from the top of the membrane. *GPL*, glycerophospholipid; *GSL*, glycosphingolipid; *SM*, sphingomyelin.

(FRAP), immunoelectron microscopy, and two-photon microscopy have been applied to characterize lipid rafts in living cells (Gaus et al., 2003; Kenworthy, 2007; Kenworthy and Edidin, 1998; Kim et al., 2008; Loura et al., 2009; Loura and Prieto, 2011; Prior et al., 2003; Ritchie and Kusumi, 2003; Sharma et al., 2004; Varma and Mayor, 1998; Wilson et al., 2001; Wu et al., 2016).

LIPIDS IN RETROVIRAL ENTRY

The entry of retroviral particles is initiated by interactions between the viral Env glycoproteins on the surface of the virion and specific receptors (and, in some cases, coreceptors) on the target cell surface. The cellular receptors are generally either glycoproteins or glycolipids, and the specific interactions between the Env glycoproteins and cellular receptors determine the host range, or tropism, of the virus. After binding to cell surface receptor(s), structural rearrangements take place in both the SU and TM subunits of the Env glycoprotein complex that ultimately result in exposure of a hydrophobic fusion peptide in TM. Insertion of the fusion peptide into the target cell membrane results in lipid mixing and fusion of viral and target cell membranes. Depending on the retrovirus, and potentially other factors (perhaps including target cell type, receptor/coreceptor density, etc.), fusion can take place directly at the PM or in low-pH endosomes following endocytic uptake of the virus particle. The fusion of viral and cell membranes leads to penetration of the viral core into the host cell cytoplasm where postentry events can proceed. It is well established that lipids and lipid microdomains on the PM of the target cell and on the viral membrane play important roles in retroviral entry (Lorizate and Krausslich, 2011; Waheed and Freed, 2010).

Role of Cholesterol in Retroviral Entry

Several studies indicate that cholesterol is enriched in the lipid Env of many retroviruses (Aloia et al., 1988, 1993; Brugger et al., 2006; Chan et al., 2008; Lorizate et al., 2009). The enrichment of cholesterol is important for viral entry, as treating HIV-1 and simian immunodeficiency virus (SIV) virions with cholesterol-extracting agents such as methyl-β-cyclodextrin (MβCD) or 2-hydroxypropyl-β-cyclodextrin (2OHpβCD) significantly impairs particle infectivity (Campbell et al., 2002; Graham et al., 2003; Guyader et al., 2002; Liao et al., 2003). 2OHpβCD is considered to be a milder cholesterol-sequestering agent than MβCD as, unlike MβCD, 2OHpβCD does not induce the formation of holes in the viral membrane and loss of viral proteins; nonetheless, infectivity is markedly reduced by 2OHpβCD treatment (Graham et al., 2003; Liao et al., 2003). Removal of virion cholesterol also increases the buoyant density of viral particles (Campbell et al., 2002). This loss of infectivity can be rescued by replenishment of cholesterol, but not its structural analog cholestenone, although cholestenone could restore the buoyant density of cholesterol-depleted virions (Campbell et al., 2002).

Cholesterol in both the virus-producing cell and the target cell also contributes to efficient viral entry. Several studies demonstrated that depleting cholesterol from virus-producing cells resulted in HIV-1 particles with low cholesterol content that were poorly infectious relative to those obtained from normal control cells (Liao et al., 2001; Ono and Freed, 2001; Pickl et al., 2001).

Depletion of cholesterol from the target cells also has a significant impact on HIV-1 infectivity (Graham et al., 2003; Liao et al., 2001; Manes et al., 2000; Nguyen and Taub, 2002; Popik et al., 2002). It has been hypothesized that this inhibition is due to perturbation of lipid raft association of CD4 and CCR5 (Kozak et al., 2002; Popik et al., 2002), although it is worth noting that studies from other groups indicate that raft localization of CD4 and CCR5 is dispensable for HIV-1 entry (Percherancier et al., 2003; Popik and Alce, 2004). Nonetheless, Zhukovsky et al. (2013) identified putative cholesterol-binding sites in CXCR4 and CCR5 that are responsible for the raft localization of these coreceptors. Several studies have proposed that a highly conserved motif (LWYIK), located near the membrane-spanning domain of gp41, binds to cholesterol and is essential for HIV-1 infectivity (Chen et al., 2009; Epand, 2006; Salzwedel et al., 1999; Schwarzer et al., 2014; Vincent et al., 2002). Besides cyclodextrins, other cholesterol-sequestering agents such as filipin and nystatin that modify cholesterol organization without affecting its overall concentration in treated membranes inhibited the entry of HIV-1 into macrophages (Carter et al., 2009). Inhibition of HIV-1 and SIV entry was observed with cholesterol-binding compounds such as amphotericin B methyl ester (AME), a water-soluble derivative of amphotericin B (Pleskoff et al., 1995; Waheed et al., 2006, 2007, 2010). Furthermore, 3-hydroxy-3-methylglutaryl coenzyme (HMG-CoA) reductase inhibitors (statins) that inhibit cholesterol biosynthesis impair HIV-1 entry in macrophages (Carter et al., 2009). In this study, the inhibition of HIV-1 infectivity is correlated with a reduction in cell-surface expression of CD4 and CCR5 (Carter et al., 2009).

HTLV-1 entry is also dependent on cholesterol, as its depletion disrupts binding of HTLV-1 virions to T cells (Hague et al., 2003) and inhibits syncytium formation (Niyogi and Hildreth, 2001). However, another study suggested that this inhibition of HTLV-1 entry is independent of virion binding to its receptors (Wielgosz et al., 2005). The HTLV-1 receptors GLUT-1 and neuropilin-1 have been shown to cluster and colocalize with raft-associated markers (Barnes et al., 2004; Wielgosz et al., 2005) and monoclonal antibodies that bind to raft-associated proteins inhibit HTLV-1 entry and syncytium formation (Niyogi and Hildreth, 2001; Wielgosz et al., 2005). These results suggest that lipid rafts play a significant role in HTLV-1 entry. MLVs are thought to enter cells predominantly via direct fusion at the PM, though some reports have suggested that ecotropic MLV (eco-MLV) may enter via a low-pH endocytic route (Beer et al., 2005; Katen et al., 2001; Lee et al., 1999; McClure et al., 1990). Nevertheless, both the eco-MLV and amphotropic MLV receptors, cationic amino acid transporter (CAT1) and Pit2, respectively, directly associate with caveolin-1, an essential component of caveolae, and are detected in DRM fractions (Beer et al., 2005; Lu and Silver, 2000). Silver et al. showed that cholesterol depletion from target cells, but not from the viral membrane, inhibited eco-MLV entry, suggesting that cholesterol plays an important role in caveolae-mediated endocytosis of MLV (Lu et al., 2002). Equine infectious anemia virus

(EIAV), which reportedly uses a low-pH endosomal entry pathway, requires virion cholesterol, but target cell cholesterol appears to be dispensable for particle infectivity (Brindley and Maury, 2008).

Role of Sphingolipids in Retroviral Entry

A growing literature suggests that GSLs play a role in HIV-1 entry, and several GSLs on the target cell membrane have been proposed as alternate or accessory receptors. These include GalCer, galactosylceramide-3-sulphate (SGalCer), globotriaosylceramide (Gb3), and monosialoganglioside (GM3). It has been proposed that the V3 loop of gp120 interacts with GalCer and promotes infection in cells that lack CD4 (Bhat et al., 1991, 1993; Fantini et al., 1997; Hammache et al., 1998). A role for GalCer in HIV-1 entry is supported by the observations that antibodies against GalCer, and soluble analogues of GalCer, inhibit HIV-1 entry and fusion (Fantini et al., 1993, 1997; Harouse et al., 1991). In addition to gp120, gp41 residues 650–685 also bind to GalCer (Alfsen and Bomsel, 2002). GM3 and Gb3 have been shown to interact with the Env glycoproteins of HIV-1 and HIV-2 and enhance viral fusion and entry into CD4+ T lymphocytes (Hammache et al., 1998, 1999; Hug et al., 2000). Among these two GSLs, Gb3 showed preferential interaction with CXCR4-tropic gp120, whereas GM3 interacted exclusively with CCR5-tropic gp120 (Hammache et al., 1999; Puri et al., 1998). These observations are supported by the finding that inhibition of GSL biosynthesis by 1-phenyl-2-hexadecanoylamino-3-morpholino-1-propanol (PPMP) blocks infection of primary T cells (Hug et al., 2000; Puri et al., 2004). In contrast, Rawat et al. (2004) showed that high-level expression of GM3 rendered a mouse melanoma cell line (B16) resistant to HIV-1 Env-mediated membrane fusion. This block could be relieved by treating the cells with the GSL inhibitor PPMP. High expression of GM3 could disrupt fusion by acting as a barrier to the lateral mobility of CD4 in B16 cells (Rawat et al., 2008). Similarly, a synthetic analogue of Gb3 was also shown to inhibit HIV-1 infection (Harrison et al., 2010; Ramkumar et al., 2009). SM was shown to play a role in HIV-1 infection; cotreating cells with sphingomyelinase (SMase) and cholesterol oxidase inhibited HIV-1 infection by restricting lateral diffusion of CD4 (Finnegan et al., 2007; Nguyen and Taub, 2003). Ceramide, the hydrophobic backbone of sphingolipids, impairs HIV-1 infection, as enhancing ceramide levels using pharmacological agents such as *N*-(4-hydroxyphenyl)-retinamide or treating cells with SMase or exogenous addition of long-chain ceramide inhibited viral infection in CD4+ lymphocytes and monocyte-derived macrophages (MDMs) (Finnegan et al., 2004). In addition, several studies also showed the role of GSLs, in particular GM3, in regulating the signaling and activation of CD4+ T cells (Nagafuku et al., 2012; Zhu et al., 2011). These studies suggest that GSLs at normal levels can enhance HIV-1 fusion by stabilizing the association of gp120 with CD4 and coreceptors, but at higher concentrations could interfere with HIV-1 fusion by impairing the lateral mobility of receptors or restricting fusion pore formation.

Role of Phospholipids in Retroviral Entry

Increasing evidence suggests that some phospholipids, in particular the negatively charged PS, enhance the infection of a number of enveloped viruses. PS, generally localized in the cytoplasmic leaflet of the PM, flips to the outer leaflet on the onset of apoptosis (Marino and Kroemer, 2013; Verhoven et al., 1995). Virus infection can also cause a redistribution of PS to the outer leaflet of the PM (Soares et al., 2008). Furthermore, mature macrophages, but not monocytes, express PS in the outer leaflet of their PM and thus incorporate PS into the outer leaflet of HIV-1 virions (Callahan et al., 2000, 2003). As mentioned in the Introduction (Composition of Viral Membranes section), PS is one of the phospholipids that is enriched in HIV-1 and MLV virions by more than twofold relative to the producer cell PM (Aloia et al., 1993; Brugger et al., 2006; Chan et al., 2008). Callahan et al. (2003) reported that PS promotes macrophage infection by stabilizing virus–cell interactions and enhancing virus fusion. Treating target cells that contain functional receptors with PS-containing liposomes markedly enhances infection by multiple enveloped viruses (Coil and Miller, 2005a), and PS treatment overcomes the block to retroviral infection of cells expressing nonfunctional receptors (Coil and Miller, 2005b). This enhanced infection is not due to altered virus receptors or Env binding but rather to enhancement of viral fusion by PS-containing liposomes (Coil and Miller, 2005a,b). These studies suggest that PS plays a role in virus fusion, although the detailed mechanism remains to be elucidated.

RETROVIRAL ASSEMBLY

The assembly of retroviral particles is a multistep process driven by the Gag precursor protein (Fig. 10.3) (reviewed in Bieniasz, 2009; Freed, 2015; Sundquist and Krausslich, 2012). The retroviral Gag precursor polyprotein is necessary and sufficient to mediate all essential steps in particle assembly, as demonstrated by the early observation that its expression alone in recombinant baculovirus–infected insect cells led to the production of virus-like particles (Gheysen et al., 1989). In addition to the Gag precursor, retroviruses (with the exception of the spumaviruses) also express a GagPol polyprotein, which includes domains for the viral enzymes PR, RT, and IN. Concomitant with particle release, PR cleaves Gag and GagPol to produce mature Gag proteins MA, CA, NC, spacer peptides (SP1 and SP2), and p6, and mature enzymes PR, RT, and IN. In addition to Gag and GagPol, Env glycoproteins (SU and TM) on the surface of the virions are necessary for the production of infectious virus particles. Most retroviruses assemble at the PM of infected cells, though some retroviruses (e.g., the betaretroviruses M-PMV and MMTV) assemble in the cytosol before preformed CAs are transported to the PM. Many, but not all, retroviral Gag proteins are modified at their N-termini by a covalently attached myristic acid. The myristate moiety promotes membrane binding by inserting into the

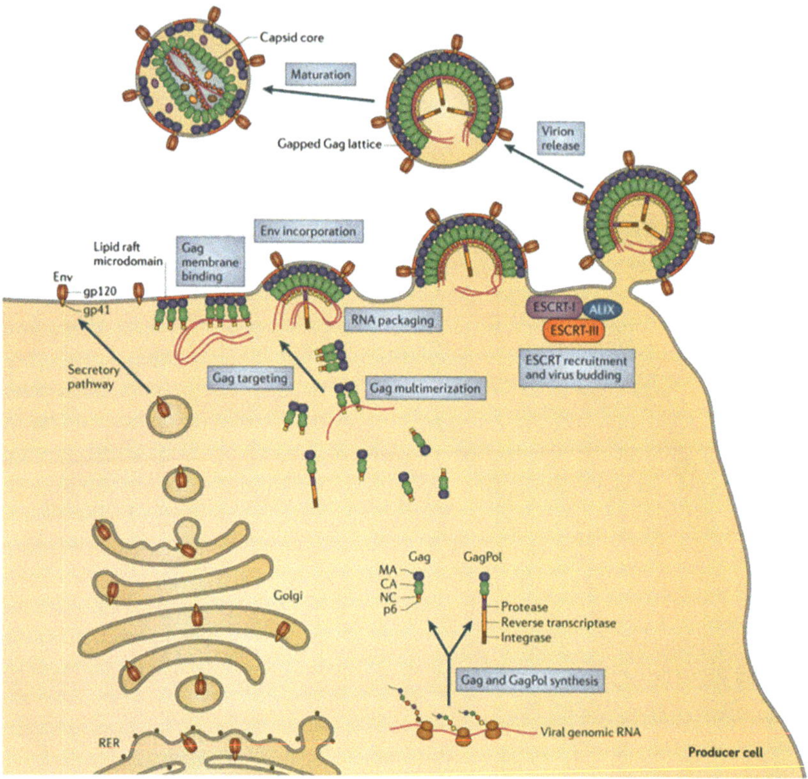

FIGURE 10.3 **The assembly and release of HIV-1 virions.** HIV-1 Env glycoproteins are synthesized as a polyprotein precursor, gp160, which is cleaved by cellular furin to gp120 and gp41, and transported to the plasma membrane through the secretory pathway. The Gag (containing matrix (MA), capsid (CA), nucleocapsid (NC), and p6 domains) and GagPol (containing MA, CA, NC, PR, RT, and IN domains) precursor polyproteins are synthesized in the cytosol from the full-length viral RNA, the latter as the result of a programmed frameshift event during the translation of Gag-encoding viral RNA. The amino-terminal myristate moiety and direct interaction of MA with PI(4,5)P_2 promote Gag binding to lipid raft microdomains in the plasma membrane. Gag recruits the viral genomic RNA, which initiates the multimerization of Gag. The viral Env glycoproteins on the plasma membrane are incorporated into assembling virions. The assembling particle recruits endosomal sorting complex required for transport (ESCRT-I) and ESCRT-associated factor ALG2-interacting protein X (ALIX) via a direct interaction with the specific motifs in the p6 domain of Gag resulting in virion release. The release of virions takes place by a membrane scission reaction driven by ESCRT-III and vacuolar protein sorting 4 (VPS4) complexes. The viral PR cleaves Gag and GagPol polyprotein precursors leading to maturation of virions that result in the formation of the conical capsid core. *(Reprinted with permission from Freed, E.O., 2015. HIV-1 assembly, release and maturation. Nat. Rev. Microbiol. 13, 484–496. Copyright 2015.)*

lipid bilayer. Gag–Gag interactions that drive particle assembly are promoted by CA, SP1, and NC domains. NC binds to viral genomic RNA and recruits it into the assembling particle. Short peptide motifs in Gag, known as "late domains," recruit the cellular endosomal sorting complex required for transport (ESCRT) machinery that is responsible for the membrane scission reaction that allows the

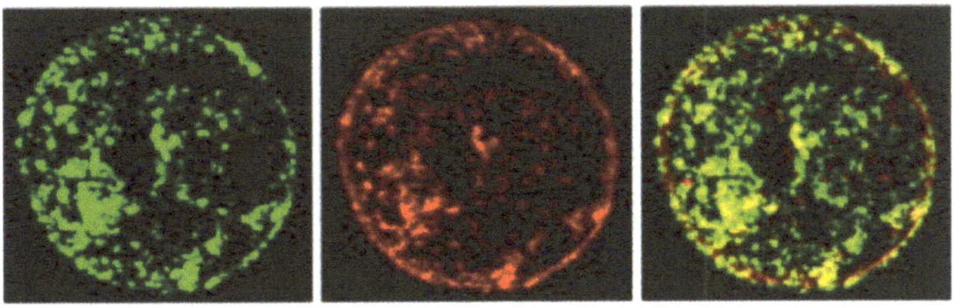

FIGURE 10.4 **Copatching of HIV-1 Gag proteins with the raft marker GM1.** Jurkat cells expressing HIV-1 Gag were labeled with cholera toxin B subunit to stain GM1 (red) before fixation. Cells were then fixed, permeabilized, and Gag proteins were detected with an anti-p17 MA antibody (green). Merged images of Gag and GM1 show colocalization, as shown in yellow. *(Reprinted from Waheed, A.A., Freed, E.O., 2010. The role of lipids in retrovirus replication. Viruses 2, 1146–1180. Copyright 2010, MDPI open access.)*

virion to bud off from the PM (Freed, 2015; Lippincott-Schwartz et al., 2017; Schoneberg et al., 2017). Biochemical assays showing that HIV-1 Gag proteins associate with DRMs, colocalization studies in which Gag copatches with raft markers such as the glycosphingolipid GM1 and GPI-anchored proteins, and the raftlike composition of HIV-1 envelopes together demonstrate that assembly takes place in lipid raft microdomains (Fig. 10.4) (Harrist et al., 2009; Hogue et al., 2011; Holm et al., 2003; Krementsov et al., 2010; Nguyen and Hildreth, 2000; Ono and Freed, 2001, 2005).

Trafficking of Gag and Sites of Retroviral Assembly

The retroviral Gag and GagPol polyproteins are synthesized in the cytosol and then traffic to the site of assembly. Gag proteins have been reported to interact with a number of host cell factors, and in some cases disruption of these interactions causes defects in Gag trafficking to the membrane. However, the mechanism and itinerary of Gag trafficking are incompletely defined (Balasubramaniam and Freed, 2011).

While most retroviruses assemble at the PM, some genera of retroviruses assemble in either the cytosol (e.g., the betaretroviruses) or the ER (e.g., the spumaviruses). Assembly at the PM facilitates rapid and efficient release of cell-free virions to the extracellular space or to adjacent cells through virological synapses. Previously, endosomes were thought to be a native assembly sites based on the observation that HIV-1 Gag was often observed in intracellular endosomal compartments. However, in most cases, endosome-associated particles are likely to have been internalized from the PM (Chu et al., 2009; Finzi et al., 2013; Jouvenet et al., 2006; Welsch et al., 2007). Gag mutants that are directed to endosomal compartments produce extracellular particles much less efficiently than WT when expressed in laboratory cell lines such as HeLa, 293T, and COS (Freed et al., 1994; Joshi et al., 2009; Ono and Freed, 2004; Ono et al., 2000; Yuan et al., 1993). However, in some T-cell lines and in primary T cells, mutants targeted to late endosome are released efficiently, suggesting an alternative pathway to particle release (Joshi et al., 2009).

In primary MDMs, HIV-1 assembly takes place in an intracellular compartment that contains markers for late endosomes or multivesicular bodies (MVBs) such as CD63, CD81, LAMP1, and major histocompatibility class II (MHC II) (Nguyen et al., 2003; Ono and Freed, 2004; Pelchen-Matthews et al., 2003; Raposo et al., 2002). This observation led to the hypothesis that MVBs are the site of HIV-1 assembly in macrophages. However, a number of studies using a variety of techniques demonstrated that these virus-containing, MVB-like compartments are actually connected to the extracellular space through thin, extended channels and thus represented PM invaginations rather than endosomes (Bennett et al., 2009; Deneka et al., 2007; Gousset et al., 2008; Jouve et al., 2007; Mlcochova et al., 2013; Nkwe et al., 2016; Welsch et al., 2011). Virus assembling in these apparently internal, virus-containing compartments move rapidly to the site of cell–cell contact when T cells are added to the cultures of infected macrophages (Gousset et al., 2008). In polarized T cells, Gag accumulates at the lagging end, known as the uropod, and the virological synapse in these cells is enriched in uropod markers. NC-mediated Gag multimerization and the highly basic region (HBR) of MA contribute to uropod localization (Llewellyn et al., 2010, 2013). Analysis of Gag colocalization with uropod markers suggests that assembling Gag reorganizes uropod microdomains in a fashion that may promote formation of the virological synapse and cell–cell transfer. This accumulation of Gag and assembling virions is likely to facilitate cell–cell transfer of virus through virological synapses (Llewellyn et al., 2010, 2013).

Gag–Membrane Interactions

Several factors have been identified that govern the binding of Gag to the membrane at the site of assembly. The covalent attachment of myristic acid, a 14-carbon fatty acid, to the N-terminal Gly of Gag, is an important factor for Gag binding to membrane. The MA domain of many, but not all, retroviral Gag proteins is myristylated. Examples of myristylated Gag proteins include the lentiviruses HIV-1, SIV, and feline immunodeficiency virus (FIV); the betaretrovirus M-PMV; the deltaretrovirus HTLV; and the gammaretrovirus MLV. Nonmyristylated Gag proteins include those of the lentiviruses EIAV and visna virus; the alpharetrovirus avian leukosis virus; and the spumavirus human foamy virus. The N-terminal myristate promotes the interaction of Gag with membranes by direct insertion into the lipid bilayer (Saad et al., 2006; Tang et al., 2004; Vlach and Saad, 2013). Mutating the N-terminal Gly of myristylated Gag proteins completely abrogates the binding of Gag to membrane and blocks virus assembly (Bryant and Ratner, 1990; Gottlinger et al., 1989; Schultz and Rein, 1989). It has been proposed that the membrane binding of HIV-1 Gag is regulated by a myristyl switch mechanism whereby the myrsitate can adopt either a sequestered or an exposed conformation (Hermida-Matsumoto and Resh, 1999; Resh, 2004). When sequestered, the myristate is unable to bind membrane; when exposed, it can insert into the lipid bilayer (Freed, 2006). The myristyl switch can be triggered by Gag multimerization (Tang et al., 2004). Mutations near the N-terminus of HIV-1 MA just downstream from the residues required for myristylation block the myristyl switch and disrupt virus assembly (Ono and Freed, 1999; Saad et al., 2007).

The second factor that plays a critical role in Gag binding to membrane is electrostatic interactions of MA with negatively charged lipids. MA comprises five major helices and a number of basic residues on one surface form a basic patch (the HBR) that interacts with negatively charged lipids in the inner leaflet (Hill et al., 1996; Murray et al., 2005; Zhou et al., 1994). While most retroviral MA proteins have a basic patch, the net charge differs among retroviruses (Murray et al., 2005). Because RSV Gag is not myristylated, it provides a useful tool to study the role of basic residues in membrane binding. Mutating one or two basic residues in RSV MA resulted in loss of Gag binding to membrane and virus production, and additional mutations elsewhere in MA that resulted in +3 net surface charge restored virus release (Callahan and Wills, 2000). Converting two acidic residues to basic amino acids in RSV Gag–enhanced virus release, suggesting that the placement of basic residues is less important than the net charge for membrane binding (Callahan and Wills, 2000). Similarly, for HIV-1, basic residue mutations such as K29T/K31T resulted in a threefold reduction in virus release, whereas mutations that increased the overall positive charge (e.g., I18K/L20K) increased membrane binding relative to WT (Ono et al., 2000). The role of specific lipids, such as $PI(4,5)P_2$, in Gag–PM binding is discussed in detail in Role of $PI(4,5)P_2$ in Retroviral Assembly section. Using a tethered lipid bilayer membrane system, HIV-1 Gag was shown to be compact, with both MA and NC domains interacting with the lipid bilayer (Datta et al., 2011). In the presence of short oligonucleotides, NC prefers to bind to the nucleic acids and MA binds to the lipid bilayer, resulting in extension of the Gag protein (Datta et al., 2011). Similarly, RSV Gag adopts a more compact structure on a supported bilayer compared with its extended and flexible form in solution (Dick et al., 2015). Although RSV Gag binds to liposomes containing acidic lipids, and binding is enhanced by physiological levels of $PI(4,5)P_2$ and cholesterol, it is insensitive to acyl chain saturation, in contrast to HIV-1 Gag (Dick et al., 2015). Under some conditions, Gag multimerization contributes to membrane targeting. At low concentrations, HIV-1 Gag is localized to the cytoplasm, and when its concentration increases the amount of Gag on the PM increases, due to an increase in lower-order multimerization (Fogarty et al., 2014). Conversely, Gag interaction with the PM promotes Gag multimerization. Furthermore, the PM-localized Gag undergoes higher-order multimerization to facilitate virus assembly (Kutluay and Bieniasz, 2010). Whereas dimerization of HIV-1 Gag is sufficient for membrane binding, RSV Gag dimerization is not sufficient but rather requires hexamerization for PM association (Dick et al., 2013).

Lipids in Retroviral Assembly

As discussed above, the Envs of many retroviruses are enriched in specific lipids such as cholesterol, SM, PS, pl-PE, GM1, and PI(4,5)P_2 compared with the host cell PM, and the lipid composition of some retroviral membranes is very similar to that of lipid rafts (Aloia et al., 1993; Brugger et al., 2006; Chan et al., 2008; Lorizate et al., 2013). These observations are consistent with the hypothesis that retroviral budding takes place in lipid raft microdomains. This hypothesis is supported by the observation that the Gag proteins of several retroviruses, including those of HIV-1 (Ding et al., 2003; Halwani et al., 2003; Holm et al., 2003; Lindwasser and Resh, 2001; Nguyen and Hildreth, 2000; Ono and Freed, 2001; Ono et al., 2007; Pickl et al., 2001), MLV (Beer and Pedersen, 2006), and HTLV-1 (Feng et al., 2003) associate with DRMs.

Retroviral Gag interactions with specific lipids have been extensively studied in vitro by using liposome-binding assays. Liposomes, spherical vesicles made up of lipids, can be formed as unilamellar or multilamellar vesicles with specific compositions. Liposome diameters vary as follows: small unilamellar vesicles, ~25 nm; large unilamellar vesicles (LUVs), ~100 nm; giant unilamellar vesicles (GUVs), >1 μm; large multilamellar vesicles (200 nm–3 μm); and multivesicular vesicles, 200 nm–3 μm.

Role of Cholesterol in Retroviral Assembly

As mentioned above, the observation that the Envs of several retroviruses contain higher amounts of cholesterol than the host cell PM suggests that retroviral budding takes place in cholesterol-enriched membrane microdomains (Aloia et al., 1992, 1993; Brugger et al., 2006; Chan et al., 2008; Lorizate et al., 2013; Quigley et al., 1972; Slosberg and Montelaro, 1982). Cholesterol is known to regulate protein and lipid packing by reducing the intermolecular electrostatic repulsive forces among phosphoinositides and the cholesterol hydroxyl group participates in hydrogen-bonding networks that stabilize the membrane (Jiang et al., 2014). The importance of cholesterol and lipid rafts in retroviral assembly was demonstrated by using cholesterol-depleting or sequestering agents. Cholesterol-depleting agents used to disrupt lipid rafts include MβCD, a cyclic sugar molecule that extracts cholesterol from the PM into its hydrophobic cavity, and the statins, which inhibit the activity of HMG-CoA reductase, a key enzyme in cholesterol biosynthesis (Ono and Freed, 2001; Pickl et al., 2001). Agents that sequester cholesterol without depleting it from the cell membrane (e.g., filipin and nystatin), alter the composition and microorganization of the PM, and disrupt the functions of membrane-associated proteins (Bolard, 1986; Peyron et al., 2000; Smart and Anderson, 2002; Subtil et al., 1999). The third class of compounds that alter cholesterol function in membranes is cholesterol-binding compounds such as the fungal antibiotic AME (Waheed et al., 2006, 2008). Several studies have shown that treating virus-producing cells with the above-mentioned cholesterol-disrupting agents significantly reduces virus production

(Finzi et al., 2007; Ono and Freed, 2001; Pickl et al., 2001; Waheed et al., 2006, 2008). In cells, cholesterol depletion was shown to decrease the mobility of Gag in the membrane (Gomez and Hope, 2006). Ono et al. (2007) showed that the inhibition of HIV-1 production from cholesterol-depleted cells is due to reduced Gag–membrane binding and impaired higher-order Gag multimerization. Treating virus-producing cells with a nonraft targeting, unsaturated myristic acid analog impaired HIV-1 particle production by reducing Gag association with DRMs (Lindwasser and Resh, 2002). Assembly of HTLV-1 also reportedly depends on lipid rafts; treating virus-producing cells with α-interferon inhibited its assembly and release by preventing Gag association with lipid rafts (Feng et al., 2003), though the basis for this observation is not known.

Env glycoprotein incorporation into virions is essential for virus particle infectivity (Checkley et al., 2011). If cholesterol-enriched membrane microdomains are sites for virus assembly, then viral Env glycoproteins should also be preferentially targeted to these domains. The Env glycoproteins of HIV-1 (Nguyen and Hildreth, 2000; Pickl et al., 2001; Rousso et al., 2000) and MLV (Beer et al., 2005; Li et al., 2002; Pickl et al., 2001) are reported to be associated with lipid rafts. Treating cells with MβCD blocked HIV-1 Env association with DRM (Bhattacharya et al., 2006). Truncations and mutations in the cytoplasmic tail (CT) of HIV-1 gp41 can disrupt Env cellular localization and incorporation into virions (Bhattacharya et al., 2004; Byland et al., 2007; Freed and Martin, 1996; Lopez-Verges et al., 2006; Murakami and Freed, 2000a,b; Rousso et al., 2000; Wyss et al., 2001; Yang et al., 2010). Truncating the gp41 CT inhibits Env incorporation in the majority of T-cell lines and in primary cells but has little effect when virions are produced in HeLa or 293T cells (Murakami and Freed, 2000b). Distinct motifs in the gp41 CT, such as the membrane-interacting lentivirus lytic peptides (LLP1, LLP2, and LLP3) or two gp41 CT cysteines that are sites of palmitoylation, have been reported to target HIV-1 Env to lipid rafts (Bhattacharya et al., 2004; Rousso et al., 2000; Yang et al., 2010). Although Rousso et al. (2000) reported that palmitoylation of the CT Cys residues is critical for Env incorporation into virions, Bhattacharya et al. (2004) suggested that palmitoylation is dispensable for Env incorporation. In contrast, Chan et al. (2005) reported that Cys mutants replicate with kinetics similar to those of WT, and Cys palmitoylation is not critical for either DRM association or Env incorporation. Subsequent studies suggested that Gag–Env interactions are required for Env association with lipid rafts, as mutations in MA or the gp41 CT that block the putative Gag–gp41 interaction prevent Env association with DRM (Bhattacharya et al., 2006). Palmitoylation of Env glycoprotein was reported for other viruses such as SIV (Yang et al., 1995), MLV (Li et al., 2002; Olsen and Andersen, 1999; Yang and Compans, 1996), RSV (Ochsenbauer-Jambor et al., 2001), and FIV (Gonzalez et al., 2012). Mutating the palmitoylation site in MLV Env abrogated DRM association (Li et al., 2002) but had no effect on syncytium formation, Env incorporation, or virus replication (Yang and Compans, 1996). In contrast, mutating the palmitoylation sites in the FIV TM protein abrogated both Env fusion activity and Env incorporation into virions (Gonzalez et al., 2012). Using scanning electron microscopy, Jorgenson et al. (2009) showed that the glycoproteins of HIV-1, VSV, and MLV are actively recruited to budding sites formed by either HIV-1 or RSV Gag, and among these Envs, RSV Env is more selective and exhibited recruitment only to RSV budding sites. Another motif in gp41 that is critical for raft association is the cholesterol-binding amino acid sequence LWYIK (Chen et al., 2009; Epand et al., 2006; Salzwedel et al., 1999; Schwarzer et al., 2014; Vincent et al., 2002). This cholesterol-binding motif is embedded in the nonpolar portion of the membrane-proximal external region of gp41. Mutating this cholesterol-binding motif significantly reduced Env-mediated fusion and virus infectivity (Salzwedel et al., 1999).

Role of $PI(4,5)P_2$ in Retroviral Assembly

$PI(4,5)P_2$ belongs to the family of acidic lipids known as phosphoinositides. Although all phosphoinositides have an inositol ring in their head group, the number and positions of phosphates on the inositol ring differ among its members (McLaughlin et al., 2002). Each phosphoinositide has a preferred subcellular localization. For example, in most cell types, $PI(4,5)P_2$ is concentrated primarily in the inner leaflet of the PM, $PI(3,5)P_2$ in late endosomes, and PI(3) P in early endosomes (De Matteis and Godi, 2004; Hurley and Meyer, 2001). A number of cellular proteins interact directly and electrostatically with phosphoinositides. Many of these proteins have basic domains such as pleckstrin homology (PH) and epsin N-terminal homology domains (Lemmon, 2003; McLaughlin et al., 2002). These basic domains bind preferentially to one phosphoinositide; as a result, these domains confer specific subcellular targeting. For example, the PH domain of phospholipase C delta 1 (PLCδ 1) binds specifically to $PI(4,5)P_2$; PLCδ 1 is thus targeted to the PM where $PI(4,5)P_2$ is enriched (Garcia et al., 1995). $PI(4,5)P_2$ is primarily synthesized by the phosphorylation of PI 4-phosphate by type I PI 4-phosphate 5-kinase, and its activity is regulated by a number of factors including the small G protein ADP-ribosylation factor 6 (Arf6) (reviewed in Sasaki et al., 2009). Overexpression of a constitutively active Arf6 mutant (Arf6/Q67L) translocates $PI(4,5)P_2$ from the PM to $PI(4,5)P_2$-enriched endosomal structures (Aikawa and Martin, 2003; Brown et al., 2001). $PI(4,5)P_2$ levels can also be modulated by phosphatases such as PI 5-phosphatase IV (5ptaseIV) that hydrolyzes the phosphate at position 5 thereby reducing $PI(4,5)P_2$ levels at the PM (Kisseleva et al., 2000).

The role of $PI(4,5)P_2$ in HIV-1 assembly was initially investigated by disrupting $PI(4,5)P_2$ levels by two different approaches: overexpressing 5ptaseIV or expressing the Arf6/Q67L mutant. When 5ptaseIV was overexpressed in virus-producing HeLa cells, Gag was mislocalized to MVB compartments and the total amount of Gag associated with membrane was reduced (Chukkapalli et al., 2008; Ono et al., 2004). In contrast, expressing the Arf6/Q67L mutant retargeted Gag to $PI(4,5)P_2$-enriched endosomal compartments (Ono et al., 2004). In both cases, due to altered Gag localization, HIV-1 particle production was severely reduced (Chukkapalli et al., 2008; Ono et al., 2004). The phenotype of Gag-MVB localization resulting from overexpression of 5ptaseIV (Ono et al., 2004) was similar to that observed for MA HBR mutants (Freed et al., 1994; Ono and Freed, 2004); these observations led to the suggestion that basic residues of MA might interact directly with PI(4,5)P_2 (Ono et al., 2004).

The direct interaction of MA and with PI(4,5)P_2 was demonstrated by several groups using different approaches. Summers et al. used a nuclear magnetic resonance (NMR)–based approach to show direct binding of soluble, short-acyl-chain forms of PI(4,5)P_2 to HIV-1 MA. This interaction was found to induce conformational changes within MA that promote myristate exposure. The solution structure of the complex between MA and truncated, soluble derivatives of PI(4,5)P_2 also suggested that the 2′-unsaturated acyl chain of PI(4,5)P_2 binds to a hydrophobic cleft within the globular core of MA, an interaction that would require extrusion of the 2′-acyl chain from the lipid bilayer (Saad et al., 2006). However, in a recent study, Mercredi et al. (2016) showed that PI(4,5)P_2 molecules with truncated acyl chains also bind to nonmembrane proteins such as HIV-1 CA, and, using native PI(4,5)P_2, they showed that the MA domain of Gag directly binds to PI(4,5)P_2 head groups with both PI(4,5)P_2 acyl chains anchored in the membrane (Fig. 10.5). Thus, it appears that the initial observations of acyl chain contacts within MA were an artifact of using PI(4,5)P_2 derivatives that were not membrane-associated. In an independent study, Shkriabai et al. (2006) used a mass spectrometric protein footprinting approach to identify MA residues 29 and 31 as being important for the MA–PI(4,5)P_2 interaction.

Mutating the basic residues 29 and 31 mistargets Gag to MVBs (Gousset et al., 2008; Joshi et al., 2009; Ono and Freed, 2004; Ono et al., 2000), supporting a role for these residues in PI(4,5)P_2 interactions. The incorporation and enrichment of PI(4,5)P_2 in HIV-1 virions require the MA domain, as mutants of MA that lack the highly basic globular head incorporate significantly less PI(4,5)P_2 than the WT (Chan et al., 2008). These observations suggest that not only in vitro but also in cells, HIV-1 MA interacts with PI(4,5)P_2.

The MA domains of retroviruses other than HIV-1 also contain basic residues implicated in PI(4,5)P_2 binding. The structure of HIV-2 MA is very similar to that of HIV-1 MA. HIV-1 and HIV-2 MA bind to PI(4,5)P_2 in a similar manner, except that HIV-2 MA is less sensitive to factors that modulate myristate exposure. As with HIV-1 (Ono et al., 2004), PI(4,5)P_2 disruption retargets HIV-2 Gag assembly (Saad et al., 2008). The role of PI(4,5)P_2 in MLV Gag assembly has been somewhat controversial; Muriaux et al. reported that MLV Gag targeting to the PM is mediated by PI(4,5)P_2 and PS (Hamard-Peron et al., 2010), whereas Inlora et al. (2014), using Gag-leucine zipper constructs (GagLZ), showed that PI(4,5)P_2 is not required for MLV Gag binding to the PM. Whether RSV Gag binds to PI(4,5)P_2 is also not fully resolved. As discussed earlier, RSV Gag lacks a myristyl group; its interaction with the PM is therefore primarily driven by electrostatic interactions with negatively charged lipids (Dalton et al., 2005). Studies from the Vogt laboratory showed that the negatively charged

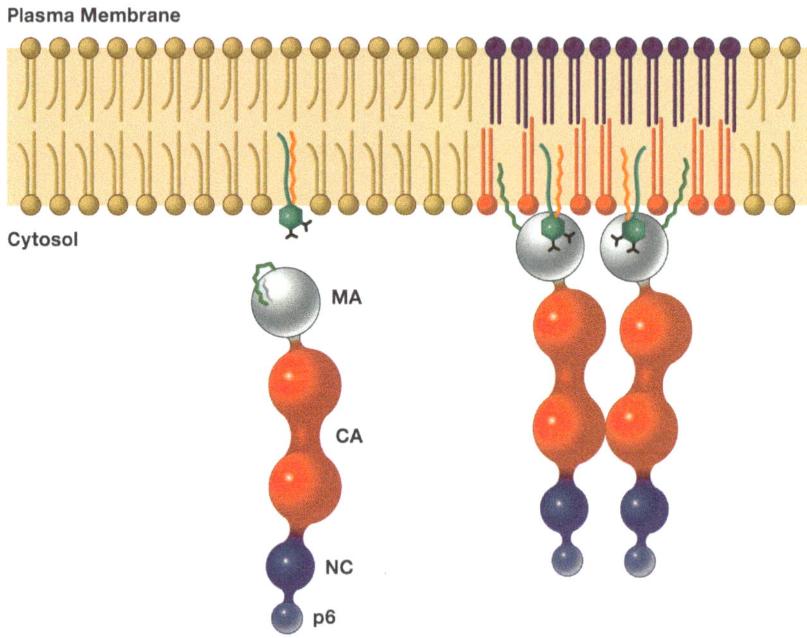

FIGURE 10.5 **HIV-1 Gag association with $PI(4,5)P_2$ at the plasma membrane.** $PI(4,5)P_2$ is embedded in the inner leaflet of the plasma membrane with both of its acyl chains (green and orange) in the lipid bilayer and the inositol ring (green) on the inner leaflet of the bilayer with two phosphate groups (black) at positions 4 and 5. The amino-terminal myristic acid moiety (light green) of matrix (MA) is sequestered in its unbound form (left). On binding of MA with $PI(4,5)P_2$ the myristate moiety flips out and is exposed, leading to its insertion into the lipid bilayer. Gag binding to lipid raft domains of the membrane promotes the multimerization of Gag molecules. *CA*, capsid; *NC*, nucleocapsid.

lipid PS, but not $PI(4,5)P_2$, is required for RSV Gag binding to the PM in cells or in vitro liposome binding assays (Chan et al., 2011). However, Parent and colleagues demonstrated that depletion of $PI(4,5)P_2$ and phosphatidylinositol-(3,4,5)-triphosphate [$PI(3,4,5)P_2$] interferes with localization of RSV Gag to the PM (Nadaraia-Hoke et al., 2013). A subsequent study from Ono's laboratory using GagLZ constructs supported the latter result, showing that, like HIV-1 MA, RSV-MA binding to the PM and to liposomes is dependent on PI(4,5) P_2 and is subject to RNA-mediated inhibition (Inlora et al., 2014). EIAV Gag, which is also nonmyristylated, was reported to interact with a soluble analogue of $PI(4,5)P_2$ (Chen et al., 2008); however, $PI(4,5)P_2$ perturbation by 5ptas-eIV overexpression was found to have no significant effect on EIAV release (Fernandes et al., 2011). EIAV Gag localizes not only to the PM but also to intracellular compartments enriched in phosphatidylinositol-(3,5)-bisphosphate [$PI(3,5)P_2$], and NMR studies indicate that EIAV MA binds to PI(3)P more tightly than to $PI(4,5)P_2$. EIAV Gag has also been reported to assemble in the ER and Golgi network (Zhang et al., 2015). These observations suggest that

EIAV Gag membrane targeting occurs via a mechanism that is different from that of HIV-1 (Fernandes et al., 2011; Zhang et al., 2015). HTLV-1 Gag localizes to the PM and to intracellular compartments, and its subcellular localization and virus release are not sensitive to perturbation of PI(4,5)P_2. These results suggest that membrane binding of HTLV-1 Gag is independent of PI(4,5)P_2 (Inlora et al., 2011, 2014). M-PMV, a betaretrovirus, is dependent on PI(4,5)P_2 for Gag binding to the PM, based on the observation that depletion of PI(4,5)P_2 significantly reduces particle release (Stansell et al., 2007). Prchal et al. (2012) reported that M-PMV MA directly interacts with soluble PI(4,5)P_2. FIV Gag assembly is also misdirected by PI(4,5)P_2 disruption. These studies suggest that many retroviruses, in addition to HIV-1, exploit MA-PI(4,5)P_2 interactions for virus assembly and release.

The Ono, Barklis, and Vogt laboratories showed that PI(4,5)P_2 significantly enhances WT HIV-1 Gag binding to liposomes, and the HBR of HIV-1 MA, together with the N-terminal myristate, is essential for PI(4,5)P_2-dependent liposome binding (Alfadhli et al., 2009a, 2011; Chukkapalli et al., 2008, 2010; Dick et al., 2012). Although PI(3,5)P_2, a phosphoinositide with the same charge as PI(4,5)P_2, also enhances Gag binding to liposomes, this enhancement is consistently lower compared to that conferred by PI(4,5)P_2 (Chukkapalli et al., 2008). Barklis et al. showed that HIV-1 MA binds preferentially to PI(4,5)P_2-containing liposomes relative to those containing PC, PS, or PI, and PI(4,5)P_2 successfully outcompetes nucleic acid for binding to MA (Alfadhli et al., 2009b, 2011). This observation was complemented by studies from Ono et al. in which RNAase treatment of in vitro translated Gag reduced the selectivity of Gag binding to PI(4,5)P_2-containing liposomes (Chukkapalli et al., 2010; Chukkapalli and Ono, 2011). These studies suggest that MA–RNA binding increases the specificity of HIV-1 Gag for PI(4,5)P_2-containing membranes (i.e., the PM) by preventing nonspecific binding of MA to inappropriate membranes (Alfadhli et al., 2009b, 2011; Chukkapalli et al., 2010, 2013). Interestingly, mutants in which lysine and arginines were exchanged (denoted "RK switch") are defective for binding to PI(4,5)P_2-containing liposomes, suggesting that it is not simply the overall charge but rather the specific amino acid sequence in the HBR that is necessary for the Gag–PI(4,5)P_2 interaction (Llewellyn et al., 2013; Olety and Ono, 2014). Mutation of the basic amino acids of MLV MA reduced Gag binding to liposomes containing PI(4,5)P_2 or PS (Hamard-Peron et al., 2010). As mentioned above, RSV Gag binding to liposomes in vitro was driven by electrostatic interactions with PS and displayed no specific association with PI(4,5)P_2 in one study (Chan et al., 2011), although in a later study, RSV MA binding to liposomes was dependent on PI(4,5)P_2 (Inlora et al., 2014).

Liposomes have also been used to evaluate the contribution of cholesterol and lipid acyl chains to Gag–membrane binding. Barklis et al. reported that saturation of PI(4,5)P_2 acyl chains has no significant effect on the interaction between MA and PI(4,5)P_2-containing liposomes (Alfadhli et al., 2009b). In contrast, the Ono lab reported that PI(4,5)P_2 with two saturated acyl chains [dipalmitoyl PI(4,5)P_2] is defective in Gag binding compared with PI(4,5)P_2 containing at least one unsaturated acyl chain (Olety et al., 2015). Vogt et al. reported that HIV-1 Gag senses the hydrophobic component of the bilayer, as Gag binds to PS- and PC-containing liposomes more strongly when both acyl chains are unsaturated than when one acyl chain is unsaturated (Dick et al., 2012). Cholesterol in liposomes also influences Gag binding, as demonstrated in studies showing that HIV-1 Gag binding to liposomes increases with increasing cholesterol concentration (Alfadhli et al., 2009b; Dick et al., 2012). In a later study, Barros et al. (2016) observed that above 30 mol% cholesterol, membrane binding of myristylated HIV-1 MA increased, although binding of nonmyristylated HIV-1 MA was insensitive to changes in cholesterol concentration. They also observed that myristylated MA binds more strongly to liposomes containing PS than those containing PI, although both lipids have the same charge (Barros et al., 2016). Yandrapalli et al. (2016) reported that HIV-1 Gag self-assembly on LUVs or supported lipid bilayers generates the formation of PI(4,5)P_2 lipid nanoclusters that are enriched in cholesterol but not SM. This clustering of cholesterol and PI(4,5)P_2 is independent of surrounding lipids and is driven by full-length HIV-1 Gag but not by the MA domain alone. These studies indicate that not only the polar head group but also the nature of the PI(4,5) P_2 acyl chains, and the cholesterol content in the membrane, play a critical role in HIV-1 Gag binding to membrane.

Gag Binding to Giant Unilamellar Vesicles

One of the limitations of liposomes is that they are very small and therefore have highly curved lipid bilayers relative to the cell PM, which is nearly planar. To overcome this potential caveat, GUVs have been used to understand the mechanism and kinetics of HIV-1 budding. GUVs are very large vesicles with single-lipid bilayers, with diameters of up to a few microns. Due to their large size, the binding of fluorescently tagged or labeled Gag proteins to these vesicles can be visualized by optical microscopy. Schwille et al. showed that HIV-1 Gag binding to GUVs is dependent on myristylation, Gag multimerization, and the presence of PI(4,5)P_2 in the membrane (Keller et al., 2013). They also observed that multimerized MA preferentially resides in Ld domains that resemble the nonraft fraction of the PM. In another study, Mohideen and colleagues showed that interaction of HIV-1 Gag with lipids and RNA in GUVs, as well as budding of minivesicles on the inner surface of the GUVs, occurs even in the absence of cellular proteins using GUVs composed of 1-palmitoyl-2-oleoyl-sn-glycero-3-phosphocholine + 1-palmitoyl-2-oleoylphosphatidylserine (Gui et al., 2015).

Recapitulating the physiological membrane concentration of cholesterol and PS by combining POPS + cholesterol + brain PI(4,5)P_2 at a molar ratio 4:2:3:1, Olety et al. (2016) showed robust and efficient binding of HIV-1 Gag-YFP to GUVs in real time. Carlson and Hurley (2012) used a GUV system and purified ESCRT components to understand the sequential order of recruitment of ESCRT proteins in HIV-1 Gag assembly, and the minimal set of components needed to complete ESCRT assembly at the site of viral budding. Recently, studies were undertaken to show that other retroviral Gag proteins also bind to GUVs containing specific lipids. Wen et al. (2016) showed that RSV Gag binding to GUVs depends on membrane charge provided by PS but has no preference for membrane order. These studies demonstrate the utility of GUVs as a model system for the study of interactions between retroviral Gag and lipid bilayers.

LIPIDS IN RETROVIRAL CELL–CELL TRANSFER

Retroviruses can spread to uninfected cells by two distinct modes, either as cell-free virions or by cell–cell transfer (Igakura et al., 2003; McDonald et al., 2003; Mothes et al., 2010; Phillips, 1994; Sato et al., 1992; Sattentau, 2011; Zhong et al., 2013). Whereas cell-free virus infection can take place at a distance through diffusion of virus particle within or between hosts, cell–cell transfer is more efficient and can overcome multiple donor and target cell barriers such as antibody-mediated neutralization, particle degradation, and innate cellular defense (Sattentau, 2008). Many studies have demonstrated that cell–cell transmission takes place at intercellular adhesive structures known as the virological synapse, which, in many respects, resembles the immunological synapse involved in antigen presentation (Arrighi et al., 2004; Fackler et al., 2007; Gousset et al., 2008; Hubner et al., 2009; Igakura et al., 2003; Jolly et al., 2004; Jolly and Sattentau, 2004; McDonald et al., 2003; Piguet and Sattentau, 2004). The virological synapse is formed by interactions between the viral Env and adhesion molecules in the infected cell and the receptor (CD4) and coreceptor (CXCR4 and CCR5) on target cells (Fig. 10.6) (Jolly et al., 2004; Jolly and Sattentau, 2005). Although the formation of the HIV-1 virological synapse between T cells is mediated by Env and receptor interactions, the recruitment of Gag to macrophage–macrophage or macrophage–T-cell synapses does not require Env expression (Gousset et al., 2008). HTLV-1 cell-free particles are poorly infectious, so HTLV-1 spreads between cells almost exclusively by cell–cell contact (Bangham, 2003; Igakura et al., 2003). MLV spreads efficiently in tissue culture by polarizing virus assembly to cell–cell contact regions, and the basic residues within MA are critical for targeting of Gag to the virological synapse (Jin et al., 2009; Li et al., 2013; Sherer et al., 2007). Live-cell imaging has demonstrated that viral transfer can occur through filopodia and nanotubes (Lehmann et al., 2005; Sherer et al., 2007; Sowinski et al., 2008). These nanotubes physically connect T cells over a long distance, and transfer of viral particles is mediated by actin filaments (Sowinski et al., 2008).

The colocalization of Gag and Env with raft markers GM1 and CD59 at the T-cell synapse in the infected effector cell, and Env and GM1 in the target cell, suggests that lipid raft–rich microdomains are involved in the transfer of virus particles across the virological synapse (Jolly and Sattentau, 2005). Following cholesterol depletion in effector cells, the polarization of HIV-1

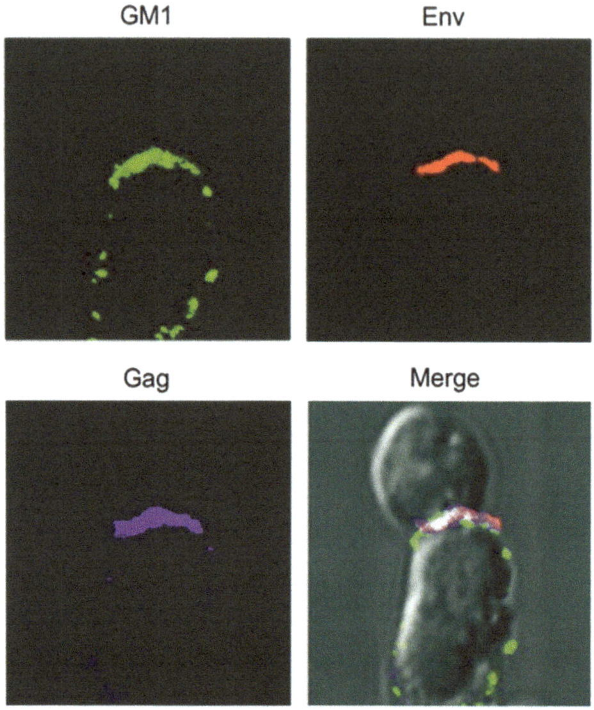

FIGURE 10.6 **HIV-1 Gag and envelope (Env) concentrate in a raft-rich domain at the virological synapse.** HIV-1-infected Jurkat effector cells (lower cell) were incubated with target primary CD4+ T cells (upper cell) for 1 h at 37°C. During the incubation, cells were stained for Env (red), fixed, and stained with cholera toxin B subunit (green), specific for the raft marker GM1. Cells were then permeabilized and Gag proteins were detected with anti-p17 antibody (blue). Images were superimposed; white indicates the colocalization of three colors on the corresponding Nomarski image. *(Reprinted with permission from Jolly, C., Sattentau, Q.J., 2005. Human immunodeficiency virus type 1 virological synapse formation in T cells requires lipid raft integrity. J. Virol. 79, 12088–12094. Copyright 2005, The American Society of Microbiology.)*

Gag and Env, in addition to their colocalization with raft markers, is disrupted, and CD4 polarization in targets cells is prevented, further supporting the role of lipid rafts in cell–cell transfer (Jolly and Sattentau, 2005). Studies from the Derse laboratory showed that HTLV-1 Gag colocalizes with CD82 and other tetraspanins that segregate into patches upon cross-linking with CD82 antibody in T cells (Mazurov et al., 2006). Copatching and segregation were prevented by cholesterol depletion (Mazurov et al., 2006). These studies suggest that cholesterol plays a critical role in cell-to-cell transfer at the virological synapse. As discussed earlier, tetherin inhibits the release of nascent cell-free HIV-1 virions in the absence of Vpu, but some evidence suggests that Vpu-defective HIV-1 spreads more efficiently through cell-to-cell contact between T cells that express low levels of tetherin (Jolly et al., 2010). Similarly, Heidecker et al. showed that whereas release of cell-free HTLV-1 was significantly reduced by tetherin expression, cell-to-cell transmission was less affected (Ilinskaya et al., 2013). Nonetheless, in several studies, it was shown that tetherin can restrict not only cell-free virus but also cell–cell transmission of HIV-1 (Casartelli et al., 2010; Giese and Marsh, 2014; Kuhl et al., 2010).

Retroviral Accessory Proteins and Lipid Microdomains

Nef, a 27-kDa multifunctional accessory protein encoded by HIV-1, HIV-2, and SIV, is responsible for inactivation of several cellular proteins involved in the antiviral immune response and plays an important role in the pathogenesis of these primate lentiviruses. Nef downmodulates CD4, MHC I, CD28, and several other molecules from the cell surface (reviewed in Landi et al., 2011). Nef has also been proposed to be involved in modulating signal transduction pathways required for T-cell activation and regulation of cholesterol trafficking in infected cells (reviewed in Foster and Garcia, 2008). Although Nef is not essential for virus replication in vitro, it enhances the infectivity of cell-free HIV-1 particles (Chowers et al., 1994; Miller et al., 1994). A proposed mechanism by which Nef regulates viral infectivity is through increasing cholesterol content of lipid rafts in virus-producing cells (Zheng et al., 2003). Nef undergoes N-terminal myristylation, a modification that, together with a patch of basic residues, targets Nef to DRM (Giese et al., 2006; Wang et al., 2000; Zheng et al., 2001). Nef expression has been reported to induce a number of genes involved in cholesterol synthesis (van ‘t Wout et al., 2005; Zheng et al., 2003) and facilitate cholesterol delivery to lipid rafts (Zheng et al., 2001, 2003). Because lipid rafts are favored sites for HIV-1 assembly, the cholesterol content in lipid rafts of virus-producing cells could influence the cholesterol content of virions and particle infectivity (Zheng et al., 2001, 2003). It has been reported that HIV-1 Nef impairs cholesterol efflux from infected macrophages by inhibiting the activity of ATP-binding cassette transporter A1 (ABCA1), an integral membrane lipid transporter, thereby increasing the cholesterol content in host cell lipid rafts and virions (Mujawar et al., 2006). This inhibition of ABCA1 activity is due to Nef blocking ABCA1 exit from the ER and subsequent stimulation of its degradation (Brichacek et al., 2014; Cui et al., 2012). According to this model, Nef not only in infected cells but also secreted from infected cells can suppress cholesterol efflux in uninfected macrophages and hepatocytes (Asztalos et al., 2010). In contrast to these studies, Brugger et al. (2007) reported that in virions derived from MT4 cells, Nef enhanced virion-associated SM and decreased PC without affecting virion cholesterol content. Nef associates with microdomains in which Lck, Fyn, and LAT kinases are enriched and activates T cells by interacting with and stimulating p21-activated kinase (PAK) in lipid rafts (Krautkramer et al., 2004; Wang et al., 2000). Although Nef membrane binding is generally required for all Nef activities, its raft localization is required for PAK binding and infectivity enhancement but is apparently dispensable for CD4 downregulation (Giese et al., 2006; Krautkramer et al., 2004). At odds with the hypothesis that Nef association with lipid rafts is important for its functions, Sol-Foulon et al. (2004) reported that CD4 downregulation and virus infectivity are both independent of Nef localization in lipid rafts. Clearly, further studies are required to elucidate the relevance of lipid raft association of Nef for its function. In later studies, it was shown that Nef counteracts members of the SERINC family of host cell restriction factors (primarily SERINC5) in virus-producer cells by reducing SERINC cell-surface expression and incorporation into virions (Rosa et al., 2015; Usami et al., 2015). It now appears likely that the primary mechanism by which Nef enhances cell-free HIV-1 virion infectivity in vitro is through SERINC5 counteraction. Precisely how SERINC5 expression in the virus-producing cell inhibits particle infectivity remains an area of active investigation.

Vpu, a 16-kDa, membrane-associated accessory protein that is not incorporated into virions significantly enhances HIV-1 release in certain cell types. In these cells, in the absence of Vpu, virions are tethered to the PM after completion of the virus budding process (Malim and Emerman, 2008). Two independent studies initially reported that BST2 (CD317/HM1.24/tetherin) restricts the release of budded virions by tethering them to the cell surface. Vpu enhances virus release by counteracting this restriction factor (Neil et al., 2008; Van Damme et al., 2008). Tetherin has an unusual membrane topology, with an N-terminal transmembrane domain and a C-terminal GPI-anchor (Kupzig et al., 2003). Stephens et al. demonstrated that Vpu is associated with lipid rafts, and that Vpu mutants (W22A and IVV19-21AAA) that are defective for raft association have impaired antitetherin activity (Ruiz et al., 2010). However, in an independent study, Fackler et al. showed that another Vpu mutant (V25G/Y29G) that is incapable of raft association could still downmodulate cell-surface tetherin and promote HIV-1 release. This latter result suggested that lipid raft association of Vpu is dispensable for its function in tetherin downmodulation (Fritz et al., 2012). Nonetheless, Banting et al. reported that Vpu expression leads to retargeting of lipid raft–associated endogenous tetherin to nonraft fractions followed by tetherin internalization and lysosomal degradation (Rollason et al., 2013). These studies further support the concept that HIV-1 assembly takes place in cholesterol-enriched membrane microdomains.

Some strains of MLV encode a unique form of glycosylated Gag polyprotein containing an additional 88 N-terminal amino acids. This protein is referred to as $gPr80^{Gag}$ or glyco-Gag. Nitta et al. (2010) reported that glyco-Gag facilitates MLV production by promoting Gag association with lipid rafts, and MLV particles expressed in the absence of glyco-Gag have a lower cholesterol content and are less sensitive to MβCD than particles produced in the presence of glyco-Gag. Recent reports demonstrated that glyco-Gag counteracts the host-restriction factor SERINC5 by promoting its intracellular localization via a clathrin-dependent mechanism, thereby preventing its incorporation into virions (Rosa et al., 2015; Usami et al., 2015). In this respect, glyco-Gag is functionally analogous to primate lentiviral Nef. Depending on the viral glycoprotein that they contain, glyco-Gag enhances, has no effect on, or reduces, the infectivity of MLV particles (Ahi et al., 2016). Again, as mentioned above, it is not yet clear how SERINC expression modulates particle infectivity.

INHIBITION OF RETROVIRAL REPLICATION WITH LIPID-MODIFYING AGENTS

As discussed above, lipids such as cholesterol, sphingolipids, and PS play a critical role in multiple steps of retroviral replication. Modulating the levels of these lipids by inhibiting their biosynthesis, or treating with lipid-binding compounds, could potentially serve as a therapeutic strategy to inhibit retroviral replication.

Cholesterol Modifying Agents

The importance of cholesterol in both early and late phases of the HIV-1 replication suggests that cholesterol could be targeted for antiretroviral therapy. As discussed earlier, treating virus-producing cells, virions, or target cells with cholesterol-depleting agents, such as MβCD, inhibits both HIV-1 particle production and infectivity of the released virions (Campbell et al., 2002, 2004; Graham et al., 2003; Guyader et al., 2002; Liao et al., 2001, 2003; Manes et al., 2000; Nguyen and Taub, 2002; Ono and Freed, 2001; Pickl et al., 2001; Popik et al., 2002). Statins, which inhibit cholesterol biosynthesis by blocking the activity of the enzyme HMG-CoA reductase, also reduce the production of viral particles and the infectivity of released virions (Ono and Freed, 2001; Pickl et al., 2001). Markham et al. used severe combined immunodeficiency (SCID) mice carrying human peripheral blood lymphocytes (Hu-PBL-SCID) in which cell-associated HIV-1, but not cell-free virus, was efficiently transmitted by vaginal inoculation of HIV-1-infected cells (Khanna et al., 2002). They observed that application of 2OHpβCD to the vaginal mucosa prior to inoculation with HIV-1-infected cells reduced HIV-1 transmission without damaging the vaginal epithelial cell membrane (Khanna et al., 2002). In another approach, statins were examined for antiviral activity in vivo by several groups (del Real et al., 2004; Drechsler et al., 2017; Ganesan et al., 2011; Moncunill et al., 2005; Montoya et al., 2009; Negredo et al., 2006; Probasco et al., 2008; Sklar et al., 2005). Studies from the Manes and Rugeles labs showed that in chronically HIV-1-infected patients, and in SCID-Hu-PBMC mice, lovastatin reduced HIV-1 viral loads and increased CD4+ cell counts (del Real et al., 2004; Montoya et al., 2009). The authors concluded that inhibition was due to downregulation of Rho GTPase activity by preventing the prenylation of small G proteins (del Real et al., 2004). In contrast to these initial findings, several other groups failed to confirm any consistent effect of statin treatment on viral loads in HIV-1-infected individuals (Ganesan et al., 2011; Moncunill et al., 2005; Negredo et al., 2006; Probasco et al., 2008; Sklar et al., 2005).

The lack of reproducible antiviral effect of statins in patients led to the evaluation of cholesterol-binding compounds for antiviral activity. Several studies showed that the cholesterol-binding compound AME has a profound effect on the replication of HIV-1 and SIV by inhibiting both viral entry and particle production (Pleskoff et al., 1995; Waheed et al., 2006, 2007, 2008, 2010). Truncating the long CT of either HIV-1 or SIV Env, or pseudotyping HIV-1 particles with Env glycoproteins (for example, those of MLV or VSV) bearing short CTs, conferred resistance to AME (Waheed et al., 2006, 2007). This observation suggests that the binding of AME to cholesterol in the viral lipid bilayer inhibits viral infectivity in a manner that is specific to virions bearing the full-length HIV-1 or SIV Env. Interestingly, single amino acid substitutions in the gp41 CT confer AME resistance (Waheed et al., 2006, 2007, 2010). These mutations create a cleavage site in the gp41 CT for the viral PR, leading to truncation of gp41 during particle maturation after Env is incorporated into virions (Waheed et al., 2007, 2010). The inhibition of virus release by AME is due to impairment of Vpu-mediated enhancement of HIV-1 release (Waheed et al., 2008).

Inhibitors of Sphingolipid Synthesis

Studies from several independent groups have suggested that sphingolipids play a critical role in HIV-1 infection, although the exact functions of this lipid in HIV-1 entry remain elusive (reviewed in Rawat et al., 2005). The levels of sphingolipids or sphingolipid-related compounds can be manipulated by several strategies; for example, inhibiting sphingolipid biosynthesis using L-cycloserine or fumonisin B1, or upregulating sphingolipid metabolites such as ceramide. L-cycloserine inhibits the activity of serine palmitoyltransferase, the first enzyme in the sphingolipid pathway, and treating cells with this compound inhibits HIV-1 replication by reducing virion infectivity (Mizrachi et al., 1996). Another inhibitor of the sphingolipid biosynthesis pathway, fumonisin B1, which blocks the activity of dihydroceramide synthase, was reported to reduce the infectivity of virions produced from MT-4 cells treated with this compound (Brugger et al., 2006). Furthermore, Gummuluru et al. reported that virions produced from fumonisin B1–treated cells were defective in capture by dendritic cells, suggesting that sphingolipids are not only critical for HIV-1 virion infectivity but also play a role in the transfer of HIV-1 by dendritic cells (Hatch et al., 2009). In another approach targeting sphingolipids, the upregulation of the sphingolipid metabolite ceramide reduced HIV-1 infectivity without significant toxicity (Finnegan et al., 2004). In this study, Blumenthal et al. modulated ceramide levels by treating target cells with either *N*-(4-hydroxyphenyl) retinamide or SMase or exogenous addition of long-chain ceramide. The GSL biosynthesis inhibitor PPMP blocks HIV-1 infection in various target cells that expresses CD4, CXCR4, or CCR5 by inhibiting membrane fusion and entry (Hug et al., 2000; Puri et al., 2004). Another glucosyl transferase inhibitor, *N*-butyldeoxynojirimycin, inhibits HIV entry by reducing gp120 levels in virions and preventing conformational changes in Env induced by CD4 binding (Fischer et al., 1995, 1996). However, this compound was reported to have no effect on viral loads in HIV-1-infected patients (Fischl et al., 1994). Because various GSLs are involved in gp120 binding to target cells, these analogues could compete with gp120–GSL interaction and might potentially serve as HIV-1 entry inhibitors (Fantini, 2007). Indeed, several studies showed that GalCer, SGalCer, and Gb3 analogues inhibited HIV-1 replication by blocking the virus–cell fusion of various HIV strains (Augustin et al., 2006; Daoudi et al., 2004; Garg et al., 2008; Kensinger et al., 2004; Lund et al., 2006). These studies indicate that GSL inhibitors possess anti-HIV-1 activity, and if toxicity is minimized these compounds could potentially serve as antiviral drugs.

CONCLUDING REMARKS

Recent technical advances in the identification and quantification of lipids in virions, in concert with the ability to manipulate factors involved in lipid metabolism in cells, have provided important tools for understanding the involvement of lipids in retroviral replication. It is now well established that many enveloped viruses, including HIV-1, have evolved to use lipids such as cholesterol, sphingolipids, and PS in both early and late steps of the replication cycle. The membrane microdomains that are enriched in these lipids have been shown to be sites of residence not only for viral structural and accessory proteins (e.g., Gag, Env, Nef, Vpu) but also for their host cellular partners (e.g., CD4, CXCR4, CCR5, and tetherin). Lipid rafts serve as platforms for virus assembly and for cell–cell transmission. The mechanism by which these microdomains promote the various steps involved in viral replication will be the subject of further investigation. Although the role of lipid microdomains in retroviral replication is well established, currently there are no FDA-approved drugs that target lipids in these microdomains for the treatment of diseases caused by retroviruses. Clearly, toxic side effects of lipid-modifying agents are a formidable barrier to drug development. Topical, rather than systemic, approaches would thus seem to hold the most promise. Studies on the role of lipids in retroviral replication could lead to novel strategies for the treatment of other enveloped viral pathogens, particularly those that lead to acute, rather than chronic, infections in vivo. Further understanding of membrane microdomains that are critical for diverse cellular processes, and their association with viral replication, will be helpful in targeting these domains for antiviral therapies in the future.

REFERENCES

Ahi, Y.S., Zhang, S., Thappeta, Y., Denman, A., Feizpour, A., Gummuluru, S., Reinhard, B., Muriaux, D., Fivash, M.J., Rein, A., 2016. Functional interplay between murine leukemia virus glycogag, serinc5, and surface glycoprotein governs virus entry, with opposite effects on gammaretroviral and ebolavirus glycoproteins. mBio 7.

Aikawa, Y., Martin, T.F., 2003. ARF6 regulates a plasma membrane pool of phosphatidylinositol(4,5)bisphosphate required for regulated exocytosis. J. Cell Biol. 162, 647–659.

Alfadhli, A., Barklis, R.L., Barklis, E., 2009a. HIV-1 matrix organizes as a hexamer of trimers on membranes containing phosphatidylinositol-(4,5)-bisphosphate. Virology 387, 466–472.

Alfadhli, A., McNett, H., Tsagli, S., Bachinger, H.P., Peyton, D.H., Barklis, E., 2011. HIV-1 matrix protein binding to RNA. J. Mol. Biol. 410, 653–666.

Alfadhli, A., Still, A., Barklis, E., 2009b. Analysis of human immunodeficiency virus type 1 matrix binding to membranes and nucleic acids. J. Virol. 83, 12196–12203.

Alfsen, A., Bomsel, M., 2002. HIV-1 gp41 envelope residues 650-685 exposed on native virus act as a lectin to bind epithelial cell galactosyl ceramide. J. Biol. Chem. 277, 25649–25659.

Aloia, R.C., Curtain, C.C., Jensen, F.C., 1992. Membrane cholesterol and human immunodeficiency virus infectivity. In: Aloia, R.C., Curtain, C.C. (Eds.), Advances in Membrane Fluidity. Wiley-Liss, Inc., New York, pp. 283–304.

Aloia, R.C., Jensen, F.C., Curtain, C.C., Mobley, P.W., Gordon, L.M., 1988. Lipid composition and fluidity of the human immunodeficiency virus. Proc. Natl. Acad. Sci. U.S.A. 85, 900–904.

Aloia, R.C., Tian, H., Jensen, F.C., 1993. Lipid composition and fluidity of the human immunodeficiency virus envelope and host cell plasma membranes. Proc. Natl. Acad. Sci. U.S.A. 90, 5181–5185.

Arrighi, J.F., Pion, M., Garcia, E., Escola, J.M., van Kooyk, Y., Geijtenbeek, T.B., Piguet, V., 2004. DC-SIGN-mediated infectious synapse formation enhances X4 HIV-1 transmission from dendritic cells to T cells. J. Exp. Med. 200, 1279–1288.

Asztalos, B.F., Mujawar, Z., Morrow, M.P., Grant, A., Pushkarsky, T., Wanke, C., Shannon, R., Geyer, M., Kirchhoff, F., Sviridov, D., Fitzgerald, M.L., Bukrinsky, M., Mansfield, K.G., 2010. Circulating Nef induces dyslipidemia in simian immunodeficiency virus-infected macaques by suppressing cholesterol efflux. J. Infect. Dis. 202, 614–623.

Augustin, L.A., Fantini, J., Mootoo, D.R., 2006. C-glycoside analogues of beta-galactosylceramide with a simple ceramide substitute: synthesis and binding to HIV-1 gp120. Bioorg. Med. Chem. 14, 1182–1188.

Balasubramaniam, M., Freed, E.O., 2011. New insights into HIV assembly and trafficking. Physiology (Bethesda) 26, 236–251.

Balasubramanian, K., Schroit, A.J., 2003. Aminophospholipid asymmetry: a matter of life and death. Annu. Rev. Physiol. 65, 701–734.

Bangham, C.R., 2003. The immune control and cell-to-cell spread of human T-lymphotropic virus type 1. J. Gen. Virol. 84, 3177–3189.

Barnes, K., Ingram, J.C., Bennett, M.D., Stewart, G.W., Baldwin, S.A., 2004. Methyl-beta-cyclodextrin stimulates glucose uptake in Clone 9 cells: a possible role for lipid rafts. Biochem. J. 378, 343–351.

Barros, M., Heinrich, F., Datta, S.A., Rein, A., Karageorgos, I., Nanda, H., Losche, M., 2016. Membrane binding of HIV-1 matrix protein: dependence on bilayer composition and protein lipidation. J. Virol. 90, 4544–4555.

Beer, C., Pedersen, L., 2006. Amphotropic murine leukemia virus is preferentially attached to cholesterol-rich microdomains after binding to mouse fibroblasts. Virol. J. 3, 21.

Beer, C., Pedersen, L., Wirth, M., 2005. Amphotropic murine leukaemia virus envelope protein is associated with cholesterol-rich microdomains. Virol. J. 2, 36.

Bell, R.M., Ballas, L.M., Coleman, R.A., 1981. Lipid topogenesis. J. Lipid Res. 22, 391–403.

Bennett, A.E., Narayan, K., Shi, D., Hartnell, L.M., Gousset, K., He, H., Lowekamp, B.C., Yoo, T.S., Bliss, D., Freed, E.O., Subramaniam, S., 2009. Ion-abrasion scanning electron microscopy reveals surface-connected tubular conduits in HIV-infected macrophages. PLoS Pathog. 5, e1000591.

Bhat, S., Mettus, R.V., Reddy, E.P., Ugen, K.E., Srikanthan, V., Williams, W.V., Weiner, D.B., 1993. The galactosyl ceramide/sulfatide receptor binding region of HIV-1 gp120 maps to amino acids 206-275. AIDS Res. Hum. Retrovir. 9, 175–181.

Bhat, S., Spitalnik, S.L., Gonzalez-Scarano, F., Silberberg, D.H., 1991. Galactosyl ceramide or a derivative is an essential component of the neural receptor for human immunodeficiency virus type 1 envelope glycoprotein gp120. Proc. Natl. Acad. Sci. U.S.A. 88, 7131–7134.

Bhattacharya, J., Peters, P.J., Clapham, P.R., 2004. Human immunodeficiency virus type 1 envelope glycoproteins that lack cytoplasmic domain cysteines: impact on association with membrane lipid rafts and incorporation onto budding virus particles. J. Virol. 78, 5500–5506.

Bhattacharya, J., Repik, A., Clapham, P.R., 2006. Gag regulates association of human immunodeficiency virus type 1 envelope with detergent-resistant membranes. J. Virol. 80, 5292–5300.

Bieniasz, P.D., 2009. The cell biology of HIV-1 virion genesis. Cell Host Microbe 5, 550–558.

Bolard, J., 1986. How do the polyene macrolide antibiotics affect the cellular membrane properties? Biochim. Biophys. Acta 864, 257–304.

Brichacek, B., Darwish, C., Popratiloff, A., Dubrovsky, L., Bukrinsky, M., 2014. HIV-1 infection of macrophages induces retention of cholesterol transporter ABCA1 in the endoplasmic reticulum. AIDS Res. Hum. Retrovir. 30, 947–948.

Brindley, M.A., Maury, W., 2008. Equine infectious anemia virus entry occurs through clathrin-mediated endocytosis. J. Virol. 82, 1628–1637.

Brown, D.A., 2006. Lipid rafts, detergent-resistant membranes, and raft targeting signals. Physiology (Bethesda) 21, 430–439.

Brown, D.A., London, E., 2000. Structure and function of sphingolipid- and cholesterol-rich membrane rafts. J. Biol. Chem. 275, 17221–17224.

Brown, F.D., Rozelle, A.L., Yin, H.L., Balla, T., Donaldson, J.G., 2001. Phosphatidylinositol 4,5-bisphosphate and Arf6-regulated membrane traffic. J. Cell Biol. 154, 1007–1017.

Brugger, B., Glass, B., Haberkant, P., Leibrecht, I., Wieland, F.T., Krausslich, H.G., 2006. The HIV lipidome: a raft with an unusual composition. Proc. Natl. Acad. Sci. U.S.A. 103, 2641–2646.

Brugger, B., Krautkramer, E., Tibroni, N., Munte, C.E., Rauch, S., Leibrecht, I., Glass, B., Breuer, S., Geyer, M., Krausslich, H.G., Kalbitzer, H.R., Wieland, F.T., Fackler, O.T., 2007. Human immunodeficiency virus type 1 Nef protein modulates the lipid composition of virions and host cell membrane microdomains. Retrovirology 4, 70.

Bryant, M., Ratner, L., 1990. Myristoylation-dependent replication and assembly of human immunodeficiency virus 1. Proc. Natl. Acad. Sci. U.S.A. 87, 523–527.

Byland, R., Vance, P.J., Hoxie, J.A., Marsh, M., 2007. A conserved dileucine motif mediates clathrin and AP-2-dependent endocytosis of the HIV-1 envelope protein. Mol. Biol. Cell 18, 414–425.

Callahan, E.M., Wills, J.W., 2000. Repositioning basic residues in the M domain of the Rous sarcoma virus gag protein. J. Virol. 74, 11222–11229.

Callahan, M.K., Popernack, P.M., Tsutsui, S., Truong, L., Schlegel, R.A., Henderson, A.J., 2003. Phosphatidylserine on HIV envelope is a cofactor for infection of monocytic cells. J. Immunol. 170, 4840–4845.

Callahan, M.K., Williamson, P., Schlegel, R.A., 2000. Surface expression of phosphatidylserine on macrophages is required for phagocytosis of apoptotic thymocytes. Cell Death Differ. 7, 645–653.

Campbell, S., Gaus, K., Bittman, R., Jessup, W., Crowe, S., Mak, J., 2004. The raft-promoting property of virion-associated cholesterol, but not the presence of virion-associated Brij 98 rafts, is a determinant of human immunodeficiency virus type 1 infectivity. J. Virol. 78, 10556–10565.

Campbell, S.M., Crowe, S.M., Mak, J., 2002. Virion-associated cholesterol is critical for the maintenance of HIV-1 structure and infectivity. AIDS 16, 2253–2261.

Carlson, L.A., Hurley, J.H., 2012. In vitro reconstitution of the ordered assembly of the endosomal sorting complex required for transport at membrane-bound HIV-1 Gag clusters. Proc. Natl. Acad. Sci. U.S.A. 109, 16928–16933.

Carter, G.C., Bernstone, L., Sangani, D., Bee, J.W., Harder, T., James, W., 2009. HIV entry in macrophages is dependent on intact lipid rafts. Virology 386, 192–202.

Casartelli, N., Sourisseau, M., Feldmann, J., Guivel-Benhassine, F., Mallet, A., Marcelin, A.G., Guatelli, J., Schwartz, O., 2010. Tetherin restricts productive HIV-1 cell-to-cell transmission. PLoS Pathog. 6, e1000955.

Chan, J., Dick, R.A., Vogt, V.M., 2011. Rous sarcoma virus gag has no specific requirement for phosphatidylinositol-(4,5)-bisphosphate for plasma membrane association in vivo or for liposome interaction in vitro. J. Virol. 85, 10851–10860.

Chan, R., Uchil, P.D., Jin, J., Shui, G., Ott, D.E., Mothes, W., Wenk, M.R., 2008. Retroviruses human immunodeficiency virus and murine leukemia virus are enriched in phosphoinositides. J. Virol. 82, 11228–11238.

Chan, W.E., Lin, H.H., Chen, S.S., 2005. Wild-type-like viral replication potential of human immunodeficiency virus type 1 envelope mutants lacking palmitoylation signals. J. Virol. 79, 8374–8387.

Checkley, M.A., Luttge, B.G., Freed, E.O., 2011. HIV-1 envelope glycoprotein biosynthesis, trafficking, and incorporation. J. Mol. Biol. 410, 582–608.

Chen, K., Bachtiar, I., Piszczek, G., Bouamr, F., Carter, C., Tjandra, N., 2008. Solution NMR characterizations of oligomerization and dynamics of equine infectious anemia virus matrix protein and its interaction with PIP2. Biochemistry 47, 1928–1937.

Chen, S.S., Yang, P., Ke, P.Y., Li, H.F., Chan, W.E., Chang, D.K., Chuang, C.K., Tsai, Y., Huang, S.C., 2009. Identification of the LWYIK motif located in the human immunodeficiency virus type 1 transmembrane gp41 protein as a distinct determinant for viral infection. J. Virol. 83, 870–883.

Chowers, M.Y., Spina, C.A., Kwoh, T.J., Fitch, N.J., Richman, D.D., Guatelli, J.C., 1994. Optimal infectivity in vitro of human immunodeficiency virus type 1 requires an intact nef gene. J. Virol. 68, 2906–2914.

Chu, H., Wang, J.J., Spearman, P., 2009. Human immunodeficiency virus type-1 gag and host vesicular trafficking pathways. Curr. Top. Microbiol. Immunol. 339, 67–84.

Chukkapalli, V., Hogue, I.B., Boyko, V., Hu, W.S., Ono, A., 2008. Interaction between the human immunodeficiency virus type 1 Gag matrix domain and phosphatidylinositol-(4,5)-bisphosphate is essential for efficient gag membrane binding. J. Virol. 82, 2405–2417.

Chukkapalli, V., Inlora, J., Todd, G.C., Ono, A., 2013. Evidence in support of RNA-mediated inhibition of phosphatidylserine-dependent HIV-1 Gag membrane binding in cells. J. Virol. 87, 7155–7159.

Chukkapalli, V., Oh, S.J., Ono, A., 2010. Opposing mechanisms involving RNA and lipids regulate HIV-1 Gag membrane binding through the highly basic region of the matrix domain. Proc. Natl. Acad. Sci. U.S.A. 107, 1600–1605.

Chukkapalli, V., Ono, A., 2011. Molecular determinants that regulate plasma membrane association of HIV-1 Gag. J. Mol. Biol. 410, 512–524.
Coil, D.A., Miller, A.D., 2005a. Enhancement of enveloped virus entry by phosphatidylserine. J. Virol. 79, 11496–11500.
Coil, D.A., Miller, A.D., 2005b. Phosphatidylserine treatment relieves the block to retrovirus infection of cells expressing glycosylated virus receptors. Retrovirology 2, 49.
Cui, H.L., Grant, A., Mukhamedova, N., Pushkarsky, T., Jennelle, L., Dubrovsky, L., Gaus, K., Fitzgerald, M.L., Sviridov, D., Bukrinsky, M., 2012. HIV-1 Nef mobilizes lipid rafts in macrophages through a pathway that competes with ABCA1-dependent cholesterol efflux. J. Lipid Res. 53, 696–708.
Dalton, A.K., Murray, P.S., Murray, D., Vogt, V.M., 2005. Biochemical characterization of rous sarcoma virus MA protein interaction with membranes. J. Virol. 79, 6227–6238.
Daoudi, J.M., Greiner, J., Aubertin, A.M., Vierling, P., 2004. New bicyclam-GalCer analogue conjugates: synthesis and in vitro anti-HIV activity. Bioorg. Med. Chem. Lett. 14, 495–498.
Datta, S.A., Heinrich, F., Raghunandan, S., Krueger, S., Curtis, J.E., Rein, A., Nanda, H., 2011. HIV-1 Gag extension: conformational changes require simultaneous interaction with membrane and nucleic acid. J. Mol. Biol. 406, 205–214.
De Matteis, M.A., Godi, A., 2004. PI-loting membrane traffic. Nat. Cell Biol. 6, 487–492.
del Real, G., Jimenez-Baranda, S., Mira, E., Lacalle, R.A., Lucas, P., Gomez-Mouton, C., Alegret, M., Pena, J.M., Rodriguez-Zapata, M., Alvarez-Mon, M., Martinez, A.C., Manes, S., 2004. Statins inhibit HIV-1 infection by down-regulating Rho activity. J. Exp. Med. 200, 541–547.
Deneka, M., Pelchen-Matthews, A., Byland, R., Ruiz-Mateos, E., Marsh, M., 2007. In macrophages, HIV-1 assembles into an intracellular plasma membrane domain containing the tetraspanins CD81, CD9, and CD53. J. Cell Biol. 177, 329–341.
Dick, R.A., Datta, S.A., Nanda, H., Fang, X., Wen, Y., Barros, M., Wang, Y.X., Rein, A., Vogt, V.M., 2015. Hydrodynamic and membrane binding properties of purified rous sarcoma virus gag protein. J. Virol. 89, 10371–10382.
Dick, R.A., Goh, S.L., Feigenson, G.W., Vogt, V.M., 2012. HIV-1 Gag protein can sense the cholesterol and acyl chain environment in model membranes. Proc. Natl. Acad. Sci. U.S.A. 109, 18761–18766.
Dick, R.A., Kamynina, E., Vogt, V.M., 2013. Effect of multimerization on membrane association of Rous sarcoma virus and HIV-1 matrix domain proteins. J. Virol. 87, 13598–13608.
Ding, L., Derdowski, A., Wang, J.J., Spearman, P., 2003. Independent segregation of human immunodeficiency virus type 1 Gag protein complexes and lipid rafts. J. Virol. 77, 1916–1926.
Drechsler, H., Ayers, C., Cutrell, J., Maalouf, N., Tebas, P., Bedimo, R., 2017. Current use of statins reduces risk of HIV rebound on suppressive HAART. PLoS One 12, e0172175.
Epand, R.F., Thomas, A., Brasseur, R., Vishwanathan, S.A., Hunter, E., Epand, R.M., 2006. Juxtamembrane protein segments that contribute to recruitment of cholesterol into domains. Biochemistry 45, 6105–6114.
Epand, R.M., 2006. Cholesterol and the interaction of proteins with membrane domains. Prog. Lipid Res. 45, 279–294.
Fackler, O.T., Alcover, A., Schwartz, O., 2007. Modulation of the immunological synapse: a key to HIV-1 pathogenesis? Nat. Rev. Immunol. 7, 310–317.
Fantini, J., 2007. Interaction of proteins with lipid rafts through glycolipid-binding domains: biochemical background and potential therapeutic applications. Curr. Med. Chem. 14, 2911–2917.
Fantini, J., Cook, D.G., Nathanson, N., Spitalnik, S.L., Gonzalez-Scarano, F., 1993. Infection of colonic epithelial cell lines by type 1 human immunodeficiency virus is associated with cell surface expression of galactosylceramide, a potential alternative gp120 receptor. Proc. Natl. Acad. Sci. U.S.A. 90, 2700–2704.

Fantini, J., Hammache, D., Delezay, O., Yahi, N., Andre-Barres, C., Rico-Lattes, I., Lattes, A., 1997. Synthetic soluble analogs of galactosylceramide (GalCer) bind to the V3 domain of HIV-1 gp120 and inhibit HIV-1-induced fusion and entry. J. Biol. Chem. 272, 7245–7252.
Feigenson, G.W., 2007. Phase boundaries and biological membranes. Annu. Rev. Biophys. Biomol. Struct. 36, 63–77.
Feng, X., Heyden, N.V., Ratner, L., 2003. Alpha interferon inhibits human T-cell leukemia virus type 1 assembly by preventing Gag interaction with rafts. J. Virol. 77, 13389–13395.
Fernandes, F., Chen, K., Ehrlich, L.S., Jin, J., Chen, M.H., Medina, G.N., Symons, M., Montelaro, R., Donaldson, J., Tjandra, N., Carter, C.A., 2011. Phosphoinositides direct equine infectious anemia virus gag trafficking and release. Traffic 12, 438–451.
Finnegan, C.M., Rawat, S.S., Cho, E.H., Guiffre, D.L., Lockett, S., Merrill Jr., A.H., Blumenthal, R., 2007. Sphingomyelinase restricts the lateral diffusion of CD4 and inhibits human immunodeficiency virus fusion. J. Virol. 81, 5294–5304.
Finnegan, C.M., Rawat, S.S., Puri, A., Wang, J.M., Ruscetti, F.W., Blumenthal, R., 2004. Ceramide, a target for antiretroviral therapy. Proc. Natl. Acad. Sci. U.S.A. 101, 15452–15457.
Finzi, A., Orthwein, A., Mercier, J., Cohen, E.A., 2007. Productive human immunodeficiency virus type 1 assembly takes place at the plasma membrane. J. Virol. 81, 7476–7490.
Finzi, A., Perlman, M., Bourgeois-Daigneault, M.C., Thibodeau, J., Cohen, E.A., 2013. Major histocompatibility complex class-II molecules promote targeting of human immunodeficiency virus type 1 virions in late endosomes by enhancing internalization of nascent particles from the plasma membrane. Cell. Microbiol. 15, 809–822.
Fischer, P.B., Collin, M., Karlsson, G.B., James, W., Butters, T.D., Davis, S.J., Gordon, S., Dwek, R.A., Platt, F.M., 1995. The alpha-glucosidase inhibitor *N*-butyldeoxynojirimycin inhibits human immunodeficiency virus entry at the level of post-CD4 binding. J. Virol. 69, 5791–5797.
Fischer, P.B., Karlsson, G.B., Dwek, R.A., Platt, F.M., 1996. *N*-butyldeoxynojirimycin-mediated inhibition of human immunodeficiency virus entry correlates with impaired gp120 shedding and gp41 exposure. J. Virol. 70, 7153–7160.
Fischl, M.A., Resnick, L., Coombs, R., Kremer, A.B., Pottage Jr., J.C., Fass, R.J., Fife, K.H., Powderly, W.G., Collier, A.C., Aspinall, R.L., et al., 1994. The safety and efficacy of combination *N*-butyl-deoxynojirimycin (SC-48334) and zidovudine in patients with HIV-1 infection and 200-500 CD4 cells/mm^3. J. Acquir. Immune Defic. Syndr. 7, 139–147.
Fogarty, K.H., Berk, S., Grigsby, I.F., Chen, Y., Mansky, L.M., Mueller, J.D., 2014. Interrelationship between cytoplasmic retroviral Gag concentration and Gag-membrane association. J. Mol. Biol. 426, 1611–1624.
Foster, J.L., Garcia, J.V., 2008. HIV-1 Nef: at the crossroads. Retrovirology 5, 84.
Freed, E.O., 2006. HIV-1 Gag: flipped out for PI(4,5)P(2). Proc. Natl. Acad. Sci. U.S.A. 103, 11101–11102.
Freed, E.O., 2015. HIV-1 assembly, release and maturation. Nat. Rev. Microbiol. 13, 484–496.
Freed, E.O., Martin, M.A., 1996. Domains of the human immunodeficiency virus type 1 matrix and gp41 cytoplasmic tail required for envelope incorporation into virions. J. Virol. 70, 341–351.
Freed, E.O., Martin, M.A., 2013. Human Immunodeficiency Virus: Replication. Lippincott Williams & Wilkins.
Freed, E.O., Orenstein, J.M., Buckler-White, A.J., Martin, M.A., 1994. Single amino acid changes in the human immunodeficiency virus type 1 matrix protein block virus particle production. J. Virol. 68, 5311–5320.
Fritz, J.V., Tibroni, N., Keppler, O.T., Fackler, O.T., 2012. HIV-1 Vpu's lipid raft association is dispensable for counteraction of the particle release restriction imposed by CD317/Tetherin. Virology 424, 33–44.

Futerman, A.H., Riezman, H., 2005. The ins and outs of sphingolipid synthesis. Trends Cell Biol. 15, 312–318.

Ganesan, A., Crum-Cianflone, N., Higgins, J., Qin, J., Rehm, C., Metcalf, J., Brandt, C., Vita, J., Decker, C.F., Sklar, P., Bavaro, M., Tasker, S., Follmann, D., Maldarelli, F., 2011. High dose atorvastatin decreases cellular markers of immune activation without affecting HIV-1 RNA levels: results of a double-blind randomized placebo controlled clinical trial. J. Infect. Dis. 203, 756–764.

Garcia, P., Gupta, R., Shah, S., Morris, A.J., Rudge, S.A., Scarlata, S., Petrova, V., McLaughlin, S., Rebecchi, M.J., 1995. The pleckstrin homology domain of phospholipase C-delta 1 binds with high affinity to phosphatidylinositol 4,5-bisphosphate in bilayer membranes. Biochemistry 34, 16228–16234.

Garg, H., Francella, N., Tony, K.A., Augustine, L.A., Barchi Jr., J.J., Fantini, J., Puri, A., Mootoo, D.R., Blumenthal, R., 2008. Glycoside analogs of beta-galactosylceramide, a novel class of small molecule antiviral agents that inhibit HIV-1 entry. Antivir. Res. 80, 54–61.

Gaus, K., Gratton, E., Kable, E.P., Jones, A.S., Gelissen, I., Kritharides, L., Jessup, W., 2003. Visualizing lipid structure and raft domains in living cells with two-photon microscopy. Proc. Natl. Acad. Sci. U.S.A. 100, 15554–15559.

Gheysen, D., Jacobs, E., de Foresta, F., Thiriart, C., Francotte, M., Thines, D., De Wilde, M., 1989. Assembly and release of HIV-1 precursor Pr55gag virus-like particles from recombinant baculovirus-infected insect cells. Cell 59, 103–112.

Giese, S., Marsh, M., 2014. Tetherin can restrict cell-free and cell-cell transmission of HIV from primary macrophages to T cells. PLoS Pathog. 10, e1004189.

Giese, S.I., Woerz, I., Homann, S., Tibroni, N., Geyer, M., Fackler, O.T., 2006. Specific and distinct determinants mediate membrane binding and lipid raft incorporation of HIV-1(SF2) Nef. Virology 355, 175–191.

Gomez, C.Y., Hope, T.J., 2006. Mobility of human immunodeficiency virus type 1 Pr55Gag in living cells. J. Virol. 80, 8796–8806.

Gonzalez, S.A., Paladino, M.G., Affranchino, J.L., 2012. Palmitoylation of the feline immunodeficiency virus envelope glycoprotein and its effect on fusion activity and envelope incorporation into virions. Virology 428, 1–10.

Gottlinger, H.G., Sodroski, J.G., Haseltine, W.A., 1989. Role of capsid precursor processing and myristoylation in morphogenesis and infectivity of human immunodeficiency virus type 1. Proc. Natl. Acad. Sci. U.S.A. 86, 5781–5785.

Gousset, K., Ablan, S.D., Coren, L.V., Ono, A., Soheilian, F., Nagashima, K., Ott, D.E., Freed, E.O., 2008. Real-time visualization of HIV-1 GAG trafficking in infected macrophages. PLoS Pathog. 4, e1000015.

Graham, D.R., Chertova, E., Hilburn, J.M., Arthur, L.O., Hildreth, J.E., 2003. Cholesterol depletion of human immunodeficiency virus type 1 and simian immunodeficiency virus with beta-cyclodextrin inactivates and permeabilizes the virions: evidence for virion-associated lipid rafts. J. Virol. 77, 8237–8248.

Gui, D., Gupta, S., Xu, J., Zandi, R., Gill, S., Huang, I.C., Rao, A.L., Mohideen, U., 2015. A novel minimal in vitro system for analyzing HIV-1 Gag-mediated budding. J. Biol. Phys. 41, 135–149.

Guyader, M., Kiyokawa, E., Abrami, L., Turelli, P., Trono, D., 2002. Role for human immunodeficiency virus type 1 membrane cholesterol in viral internalization. J. Virol. 76, 10356–10364.

Hague, B.F., Zhao, T.M., Kindt, T.J., 2003. Binding of HTLV-1 virions to T cells occurs by a temperature and calcium-dependent process and is blocked by certain type 2 adenosine receptor antagonists. Virus Res. 93, 31–39.

Halwani, R., Khorchid, A., Cen, S., Kleiman, L., 2003. Rapid localization of Gag/GagPol complexes to detergent-resistant membrane during the assembly of human immunodeficiency virus type 1. J. Virol. 77, 3973–3984.

Hamard-Peron, E., Juillard, F., Saad, J.S., Roy, C., Roingeard, P., Summers, M.F., Darlix, J.L., Picart, C., Muriaux, D., 2010. Targeting of murine leukemia virus gag to the plasma membrane is mediated by PI(4,5)P2/PS and a polybasic region in the matrix. J. Virol. 84, 503–515.

Hamilton, J.A., 2003. Fast flip-flop of cholesterol and fatty acids in membranes: implications for membrane transport proteins. Curr. Opin. Lipidol. 14, 263–271.

Hammache, D., Pieroni, G., Yahi, N., Delezay, O., Koch, N., Lafont, H., Tamalet, C., Fantini, J., 1998. Specific interaction of HIV-1 and HIV-2 surface envelope glycoproteins with monolayers of galactosylceramide and ganglioside GM3. J. Biol. Chem. 273, 7967–7971.

Hammache, D., Yahi, N., Maresca, M., Pieroni, G., Fantini, J., 1999. Human erythrocyte glycosphingolipids as alternative cofactors for human immunodeficiency virus type 1 (HIV-1) entry: evidence for CD4-induced interactions between HIV-1 gp120 and reconstituted membrane microdomains of glycosphingolipids (Gb3 and GM3). J. Virol. 73, 5244–5248.

Hao, M., Lin, S.X., Karylowski, O.J., Wustner, D., McGraw, T.E., Maxfield, F.R., 2002. Vesicular and non-vesicular sterol transport in living cells. The endocytic recycling compartment is a major sterol storage organelle. J. Biol. Chem. 277, 609–617.

Harder, T., Scheiffele, P., Verkade, P., Simons, K., 1998. Lipid domain structure of the plasma membrane revealed by patching of membrane components. J. Cell Biol. 141, 929–942.

Harouse, J.M., Bhat, S., Spitalnik, S.L., Laughlin, M., Stefano, K., Silberberg, D.H., Gonzalez-Scarano, F., 1991. Inhibition of entry of HIV-1 in neural cell lines by antibodies against galactosyl ceramide. Science 253, 320–323.

Harrison, A.L., Olsson, M.L., Jones, R.B., Ramkumar, S., Sakac, D., Binnington, B., Henry, S., Lingwood, C.A., Branch, D.R., 2010. A synthetic globotriaosylceramide analogue inhibits HIV-1 infection in vitro by two mechanisms. Glycoconj. J. 27, 515–524.

Harrist, A.V., Ryzhova, E.V., Harvey, T., Gonzalez-Scarano, F., 2009. Anx2 interacts with HIV-1 Gag at phosphatidylinositol (4,5) bisphosphate-containing lipid rafts and increases viral production in 293T cells. PLoS One 4, e5020.

Hatch, S.C., Archer, J., Gummuluru, S., 2009. Glycosphingolipid composition of human immunodeficiency virus type 1 (HIV-1) particles is a crucial determinant for dendritic cell-mediated HIV-1 trans-infection. J. Virol. 83, 3496–3506.

Hermida-Matsumoto, L., Resh, M.D., 1999. Human immunodeficiency virus type 1 protease triggers a myristoyl switch that modulates membrane binding of Pr55(gag) and p17MA. J. Virol. 73, 1902–1908.

Hill, C.P., Worthylake, D., Bancroft, D.P., Christensen, A.M., Sundquist, W.I., 1996. Crystal structures of the trimeric human immunodeficiency virus type 1 matrix protein: implications for membrane association and assembly. Proc. Natl. Acad. Sci. U.S.A. 93, 3099–3104.

Hogue, I.B., Grover, J.R., Soheilian, F., Nagashima, K., Ono, A., 2011. Gag induces the coalescence of clustered lipid rafts and tetraspanin-enriched microdomains at HIV-1 assembly sites on the plasma membrane. J. Virol. 85, 9749–9766.

Holm, K., Weclewicz, K., Hewson, R., Suomalainen, M., 2003. Human immunodeficiency virus type 1 assembly and lipid rafts: Pr55(gag) associates with membrane domains that are largely resistant to Brij98 but sensitive to Triton X-100. J. Virol. 77, 4805–4817.

Holthuis, J.C., Menon, A.K., 2014. Lipid landscapes and pipelines in membrane homeostasis. Nature 510, 48–57.

Hubner, W., McNerney, G.P., Chen, P., Dale, B.M., Gordon, R.E., Chuang, F.Y., Li, X.D., Asmuth, D.M., Huser, T., Chen, B.K., 2009. Quantitative 3D video microscopy of HIV transfer across T cell virological synapses. Science 323, 1743–1747.

Hug, P., Lin, H.M., Korte, T., Xiao, X., Dimitrov, D.S., Wang, J.M., Puri, A., Blumenthal, R., 2000. Glycosphingolipids promote entry of a broad range of human immunodeficiency virus type 1 isolates into cell lines expressing CD4, CXCR4, and/or CCR5. J. Virol. 74, 6377–6385.

Hurley, J.H., Meyer, T., 2001. Subcellular targeting by membrane lipids. Curr. Opin. Cell Biol. 13, 146–152.

Igakura, T., Stinchcombe, J.C., Goon, P.K., Taylor, G.P., Weber, J.N., Griffiths, G.M., Tanaka, Y., Osame, M., Bangham, C.R., 2003. Spread of HTLV-I between lymphocytes by virus-induced polarization of the cytoskeleton. Science 299, 1713–1716.

Ilinskaya, A., Derse, D., Hill, S., Princler, G., Heidecker, G., 2013. Cell-cell transmission allows human T-lymphotropic virus 1 to circumvent tetherin restriction. Virology 436, 201–209.

Inlora, J., Chukkapalli, V., Derse, D., Ono, A., 2011. Gag localization and virus-like particle release mediated by the matrix domain of human T-lymphotropic virus type 1 Gag are less dependent on phosphatidylinositol-(4,5)-bisphosphate than those mediated by the matrix domain of HIV-1 Gag. J. Virol. 85, 3802–3810.

Inlora, J., Collins, D.R., Trubin, M.E., Chung, J.Y., Ono, A., 2014. Membrane binding and subcellular localization of retroviral Gag proteins are differentially regulated by MA interactions with phosphatidylinositol-(4,5)-bisphosphate and RNA. mBio 5, e02202.

Jiang, Z., Redfern, R.E., Isler, Y., Ross, A.H., Gericke, A., 2014. Cholesterol stabilizes fluid phosphoinositide domains. Chem. Phys. Lipids 182, 52–61.

Jin, J., Sherer, N.M., Heidecker, G., Derse, D., Mothes, W., 2009. Assembly of the murine leukemia virus is directed towards sites of cell-cell contact. PLoS Biol. 7, e1000163.

Jolly, C., Booth, N.J., Neil, S.J., 2010. Cell-cell spread of human immunodeficiency virus type 1 overcomes tetherin/BST-2-mediated restriction in T cells. J. Virol. 84, 12185–12199.

Jolly, C., Kashefi, K., Hollinshead, M., Sattentau, Q.J., 2004. HIV-1 cell to cell transfer across an Env-induced, actin-dependent synapse. J. Exp. Med. 199, 283–293.

Jolly, C., Sattentau, Q.J., 2004. Retroviral spread by induction of virological synapses. Traffic 5, 643–650.

Jolly, C., Sattentau, Q.J., 2005. Human immunodeficiency virus type 1 virological synapse formation in T cells requires lipid raft integrity. J. Virol. 79, 12088–12094.

Jorgenson, R.L., Vogt, V.M., Johnson, M.C., 2009. Foreign glycoproteins can be actively recruited to virus assembly sites during pseudotyping. J. Virol. 83, 4060–4067.

Joshi, A., Ablan, S.D., Soheilian, F., Nagashima, K., Freed, E.O., 2009. Evidence that productive human immunodeficiency virus type 1 assembly can occur in an intracellular compartment. J. Virol. 83, 5375–5387.

Jouve, M., Sol-Foulon, N., Watson, S., Schwartz, O., Benaroch, P., 2007. HIV-1 buds and accumulates in "nonacidic" endosomes of macrophages. Cell Host Microbe 2, 85–95.

Jouvenet, N., Neil, S.J., Bess, C., Johnson, M.C., Virgen, C.A., Simon, S.M., Bieniasz, P.D., 2006. Plasma membrane is the site of productive HIV-1 particle assembly. PLoS Biol. 4, e435.

Katen, L.J., Januszeski, M.M., Anderson, W.F., Hasenkrug, K.J., Evans, L.H., 2001. Infectious entry by amphotropic as well as ecotropic murine leukemia viruses occurs through an endocytic pathway. J. Virol. 75, 5018–5026.

Kawabata, A., Tang, H., Huang, H., Yamanishi, K., Mori, Y., 2009. y Human herpesvirus 6 envelope components enriched in lipid rafts: evidence for virion-associated lipid rafts. Virol. J. 6, 127.

Keller, H., Krausslich, H.G., Schwille, P., 2013. Multimerizable HIV Gag derivative binds to the liquid-disordered phase in model membranes. Cell. Microbiol. 15, 237–247.

Kensinger, R.D., Catalone, B.J., Krebs, F.C., Wigdahl, B., Schengrund, C.L., 2004. Novel polysulfated galactose-derivatized dendrimers as binding antagonists of human immunodeficiency virus type 1 infection. Antimicrob. Agents Chemother. 48, 1614–1623.

Kenworthy, A.K., 2007. Fluorescence recovery after photobleaching studies of lipid rafts. Methods Mol. Biol. 398, 179–192.

Kenworthy, A.K., Edidin, M., 1998. Distribution of a glycosylphosphatidylinositol-anchored protein at the apical surface of MDCK cells examined at a resolution of <100 A using imaging fluorescence resonance energy transfer. J. Cell Biol. 142, 69–84.

Khanna, K.V., Whaley, K.J., Zeitlin, L., Moench, T.R., Mehrazar, K., Cone, R.A., Liao, Z., Hildreth, J.E., Hoen, T.E., Shultz, L., Markham, R.B., 2002. Vaginal transmission of cell-associated HIV-1 in the mouse is blocked by a topical, membrane-modifying agent. J. Clin. Investig. 109, 205–211.

Kim, H.M., Jeong, B.H., Hyon, J.Y., An, M.J., Seo, M.S., Hong, J.H., Lee, K.J., Kim, C.H., Joo, T., Hong, S.C., Cho, B.R., 2008. Two-photon fluorescent turn-on probe for lipid rafts in live cell and tissue. J. Am. Chem. Soc. 130, 4246–4247.

Kisseleva, M.V., Wilson, M.P., Majerus, P.W., 2000. The isolation and characterization of a cDNA encoding phospholipid-specific inositol polyphosphate 5-phosphatase. J. Biol. Chem. 275, 20110–20116.

Kozak, S.L., Heard, J.M., Kabat, D., 2002. Segregation of CD4 and CXCR4 into distinct lipid microdomains in T lymphocytes suggests a mechanism for membrane destabilization by human immunodeficiency virus. J. Virol. 76, 1802–1815.

Krautkramer, E., Giese, S.I., Gasteier, J.E., Muranyi, W., Fackler, O.T., 2004. Human immunodeficiency virus type 1 Nef activates p21-activated kinase via recruitment into lipid rafts. J. Virol. 78, 4085–4097.

Krementsov, D.N., Rassam, P., Margeat, E., Roy, N.H., Schneider-Schaulies, J., Milhiet, P.E., Thali, M., 2010. HIV-1 assembly differentially alters dynamics and partitioning of tetraspanins and raft components. Traffic 11, 1401–1414.

Kuhl, B.D., Sloan, R.D., Donahue, D.A., Bar-Magen, T., Liang, C., Wainberg, M.A., 2010. Tetherin restricts direct cell-to-cell infection of HIV-1. Retrovirology 7, 115.

Kupzig, S., Korolchuk, V., Rollason, R., Sugden, A., Wilde, A., Banting, G., 2003. Bst-2/HM1.24 is a raft-associated apical membrane protein with an unusual topology. Traffic 4, 694–709.

Kutluay, S.B., Bieniasz, P.D., 2010. Analysis of the initiating events in HIV-1 particle assembly and genome packaging. PLoS Pathog. 6, e1001200.

Landi, A., Iannucci, V., Nuffel, A.V., Meuwissen, P., Verhasselt, B., 2011. One protein to rule them all: modulation of cell surface receptors and molecules by HIV Nef. Curr. HIV Res. 9, 496–504.

Lee, S., Zhao, Y., Anderson, W.F., 1999. Receptor-mediated Moloney murine leukemia virus entry can occur independently of the clathrin-coated-pit-mediated endocytic pathway. J. Virol. 73, 5994–6005.

Lehmann, M.J., Sherer, N.M., Marks, C.B., Pypaert, M., Mothes, W., 2005. Actin- and myosin-driven movement of viruses along filopodia precedes their entry into cells. J. Cell Biol. 170, 317–325.

Lemmon, M.A., 2003. Phosphoinositide recognition domains. Traffic 4, 201–213.

Li, F., Jin, J., Herrmann, C., Mothes, W., 2013. Basic residues in the matrix domain and multimerization target murine leukemia virus Gag to the virological synapse. J. Virol. 87, 7113–7126.

Li, J., Wang, X., Zhang, T., Wang, C., Huang, Z., Luo, X., Deng, Y., 2015. A review on phospholipids and their main applications in drug delivery systems. Asian J. Pharm. Sci. 10, 81–98.

Li, M., Yang, C., Tong, S., Weidmann, A., Compans, R.W., 2002. Palmitoylation of the murine leukemia virus envelope protein is critical for lipid raft association and surface expression. J. Virol. 76, 11845–11852.

Li, X.M., Momsen, M.M., Smaby, J.M., Brockman, H.L., Brown, R.E., 2001. Cholesterol decreases the interfacial elasticity and detergent solubility of sphingomyelins. Biochemistry 40, 5954–5963.

Liao, Z., Cimakasky, L.M., Hampton, R., Nguyen, D.H., Hildreth, J.E., 2001. Lipid rafts and HIV pathogenesis: host membrane cholesterol is required for infection by HIV type 1. AIDS Res. Hum. Retrovir. 17, 1009–1019.

Liao, Z., Graham, D.R., Hildreth, J.E., 2003. Lipid rafts and HIV pathogenesis: virion-associated cholesterol is required for fusion and infection of susceptible cells. AIDS Res. Hum. Retrovir. 19, 675–687.

Lindwasser, O.W., Resh, M.D., 2001. Multimerization of human immunodeficiency virus type 1 Gag promotes its localization to barges, raft-like membrane microdomains. J. Virol. 75, 7913–7924.

Lindwasser, O.W., Resh, M.D., 2002. Myristoylation as a target for inhibiting HIV assembly: unsaturated fatty acids block viral budding. Proc. Natl. Acad. Sci. U.S.A. 99, 13037–13042.

Lippincott-Schwartz, J.L., Freed, E.O., van Engelenburg, S.B., 2017. A consensus view of ESCRT-mediated HIV-1 abscission. Ann. Rev. Virol. 4, 309–325.

Llewellyn, G.N., Grover, J.R., Olety, B., Ono, A., 2013. HIV-1 Gag associates with specific uropod-directed microdomains in a manner dependent on its MA highly basic region. J. Virol. 87, 6441–6454.

Llewellyn, G.N., Hogue, I.B., Grover, J.R., Ono, A., 2010. Nucleocapsid promotes localization of HIV-1 gag to uropods that participate in virological synapses between T cells. PLoS Pathog. 6, e1001167.

Lopez-Verges, S., Camus, G., Blot, G., Beauvoir, R., Benarous, R., Berlioz-Torrent, C., 2006. Tail-interacting protein TIP47 is a connector between Gag and Env and is required for Env incorporation into HIV-1 virions. Proc. Natl. Acad. Sci. U.S.A. 103, 14947–14952.

Lorizate, M., Brugger, B., Akiyama, H., Glass, B., Muller, B., Anderluh, G., Wieland, F.T., Krausslich, H.G., 2009. Probing HIV-1 membrane liquid order by Laurdan staining reveals producer cell-dependent differences. J. Biol. Chem. 284, 22238–22247.

Lorizate, M., Krausslich, H.G., 2011. Role of lipids in virus replication. Cold Spring Harb. Perspect. Biol. 3, a004820.

Lorizate, M., Sachsenheimer, T., Glass, B., Habermann, A., Gerl, M.J., Krausslich, H.G., Brugger, B., 2013. Comparative lipidomics analysis of HIV-1 particles and their producer cell membrane in different cell lines. Cell. Microbiol. 15, 292–304.

Loura, L.M., de Almeida, R.F., Silva, L.C., Prieto, M., 2009. FRET analysis of domain formation and properties in complex membrane systems. Biochim. Biophys. Acta 1788, 209–224.

Loura, L.M., Prieto, M., 2011. FRET in membrane biophysics: an overview. Front. Physiol. 2, 82.

Lu, X., Silver, J., 2000. Ecotropic murine leukemia virus receptor is physically associated with caveolin and membrane rafts. Virology 276, 251–258.

Lu, X., Xiong, Y., Silver, J., 2002. Asymmetric requirement for cholesterol in receptor-bearing but not envelope-bearing membranes for fusion mediated by ecotropic murine leukemia virus. J. Virol. 76, 6701–6709.

Lund, N., Branch, D.R., Mylvaganam, M., Chark, D., Ma, X.Z., Sakac, D., Binnington, B., Fantini, J., Puri, A., Blumenthal, R., Lingwood, C.A., 2006. A novel soluble mimic of the glycolipid, globotriaosyl ceramide inhibits HIV infection. AIDS 20, 333–343.

Malim, M.H., Emerman, M., 2008. HIV-1 accessory proteins–ensuring viral survival in a hostile environment. Cell Host Microbe 3, 388–398.

Manes, S., del Real, G., Lacalle, R.A., Lucas, P., Gomez-Mouton, C., Sanchez-Palomino, S., Delgado, R., Alcami, J., Mira, E., Martinez, A.C., 2000. Membrane raft microdomains mediate lateral assemblies required for HIV-1 infection. EMBO Rep. 1, 190–196.

Marino, G., Kroemer, G., 2013. Mechanisms of apoptotic phosphatidylserine exposure. Cell Res. 23, 1247–1248.

Marrink, S.J., de Vries, A.H., Tieleman, D.P., 2009. Lipids on the move: simulations of membrane pores, domains, stalks and curves. Biochim. Biophys. Acta 1788, 149–168.

Marsh, D., 2007. Lateral pressure profile, spontaneous curvature frustration, and the incorporation and conformation of proteins in membranes. Biophys. J. 93, 3884–3899.

Mayor, S., Maxfield, F.R., 1995. Insolubility and redistribution of GPI-anchored proteins at the cell surface after detergent treatment. Mol. Biol. Cell 6, 929–944.

Mazurov, D., Heidecker, G., Derse, D., 2006. HTLV-1 Gag protein associates with CD82 tetraspanin microdomains at the plasma membrane. Virology 346, 194–204.

McClure, M.O., Sommerfelt, M.A., Marsh, M., Weiss, R.A., 1990. The pH independence of mammalian retrovirus infection. J. Gen. Virol. 71 (Pt 4), 767–773.

McDonald, D., Wu, L., Bohks, S.M., KewalRamani, V.N., Unutmaz, D., Hope, T.J., 2003. Recruitment of HIV and its receptors to dendritic cell-T cell junctions. Science 300, 1295–1297.

McLaughlin, J.N., Thulin, C.D., Bray, S.M., Martin, M.M., Elton, T.S., Willardson, B.M., 2002. Regulation of angiotensin II-induced G protein signaling by phosducin-like protein. J. Biol. Chem. 277, 34885–34895.

Mercredi, P.Y., Bucca, N., Loeliger, B., Gaines, C.R., Mehta, M., Bhargava, P., Tedbury, P.R., Charlier, L., Floquet, N., Muriaux, D., Favard, C., Sanders, C.R., Freed, E.O., Marchant, J., Summers, M.F., 2016. Structural and molecular determinants of membrane binding by the HIV-1 matrix protein. J. Mol. Biol. 428, 1637–1655.

Miller, M.D., Warmerdam, M.T., Gaston, I., Greene, W.C., Feinberg, M.B., 1994. The human immunodeficiency virus-1 nef gene product: a positive factor for viral infection and replication in primary lymphocytes and macrophages. J. Exp. Med. 179, 101–113.

Mizrachi, Y., Lev, M., Harish, Z., Sundaram, S.K., Rubinstein, A., 1996. L-cycloserine, an inhibitor of sphingolipid biosynthesis, inhibits HIV-1 cytopathic effects, replication, and infectivity. J. Acquir. Immune Defic. Syndr. Hum. Retrovirol. 11, 137–141.

Mlcochova, P., Pelchen-Matthews, A., Marsh, M., 2013. Organization and regulation of intracellular plasma membrane-connected HIV-1 assembly compartments in macrophages. BMC Biol. 11, 89.

Moncunill, G., Negredo, E., Bosch, L., Vilarrasa, J., Witvrouw, M., Llano, A., Clotet, B., Este, J.A., 2005. Evaluation of the anti-HIV activity of statins. AIDS 19, 1697–1700.

Montoya, C.J., Jaimes, F., Higuita, E.A., Convers-Paez, S., Estrada, S., Gutierrez, F., Amariles, P., Giraldo, N., Penaloza, C., Rugeles, M.T., 2009. Antiretroviral effect of lovastatin on HIV-1-infected individuals without highly active antiretroviral therapy (The LIVE study): a phase-II randomized clinical trial. Trials 10, 41.

Mothes, W., Sherer, N.M., Jin, J., Zhong, P., 2010. Virus cell-to-cell transmission. J. Virol. 84, 8360–8368.

Mujawar, Z., Rose, H., Morrow, M.P., Pushkarsky, T., Dubrovsky, L., Mukhamedova, N., Fu, Y., Dart, A., Orenstein, J.M., Bobryshev, Y.V., Bukrinsky, M., Sviridov, D., 2006. Human immunodeficiency virus impairs reverse cholesterol transport from macrophages. PLoS Biol. 4, e365.

Munro, S., 2003. Lipid rafts: elusive or illusive? Cell 115, 377–388.

Murakami, T., Freed, E.O., 2000a. Genetic evidence for an interaction between human immunodeficiency virus type 1 matrix and alpha-helix 2 of the gp41 cytoplasmic tail. J. Virol. 74, 3548–3554.

Murakami, T., Freed, E.O., 2000b. The long cytoplasmic tail of gp41 is required in a cell type-dependent manner for HIV-1 envelope glycoprotein incorporation into virions. Proc. Natl. Acad. Sci. U.S.A. 97, 343–348.

Murray, P.S., Li, Z., Wang, J., Tang, C.L., Honig, B., Murray, D., 2005. Retroviral matrix domains share electrostatic homology: models for membrane binding function throughout the viral life cycle. Structure 13, 1521–1531.

Nadaraia-Hoke, S., Bann, D.V., Lochmann, T.L., Gudleski-O'Regan, N., Parent, L.J., 2013. Alterations in the MA and NC domains modulate phosphoinositide-dependent plasma membrane localization of the Rous sarcoma virus Gag protein. J. Virol. 87, 3609–3615.

Nagafuku, M., Okuyama, K., Onimaru, Y., Suzuki, A., Odagiri, Y., Yamashita, T., Iwasaki, K., Fujiwara, M., Takayanagi, M., Ohno, I., Inokuchi, J., 2012. CD4 and CD8 T cells require different membrane gangliosides for activation. Proc. Natl. Acad. Sci. U.S.A. 109, E336–E342.

Negredo, E., Clotet, B., Puig, J., Perez-Alvarez, N., Ruiz, L., Romeu, J., Molto, J., Rey-Joly, C., Blanco, J., 2006. The effect of atorvastatin treatment on HIV-1-infected patients interrupting antiretroviral therapy. AIDS 20, 619–621.

Neil, S.J., Zang, T., Bieniasz, P.D., 2008. Tetherin inhibits retrovirus release and is antagonized by HIV-1 Vpu. Nature 451, 425–430.

Nguyen, D.G., Booth, A., Gould, S.J., Hildreth, J.E., 2003. Evidence that HIV budding in primary macrophages occurs through the exosome release pathway. J. Biol. Chem. 278, 52347–52354.

Nguyen, D.H., Hildreth, J.E., 2000. Evidence for budding of human immunodeficiency virus type 1 selectively from glycolipid-enriched membrane lipid rafts. J. Virol. 74, 3264–3272.

Nguyen, D.H., Taub, D., 2002. CXCR4 function requires membrane cholesterol: implications for HIV infection. J. Immunol. 168, 4121–4126.

Nguyen, D.H., Taub, D.D., 2003. Inhibition of chemokine receptor function by membrane cholesterol oxidation. Exp. Cell Res. 291, 36–45.

Nitta, T., Kuznetsov, Y., McPherson, A., Fan, H., 2010. Murine leukemia virus glycosylated Gag (gPr80gag) facilitates interferon-sensitive virus release through lipid rafts. Proc. Natl. Acad. Sci. U.S.A. 107, 1190–1195.

Niyogi, K., Hildreth, J.E., 2001. Characterization of new syncytium-inhibiting monoclonal antibodies implicates lipid rafts in human T-cell leukemia virus type 1 syncytium formation. J. Virol. 75, 7351–7361.

Nkwe, D.O., Pelchen-Matthews, A., Burden, J.J., Collinson, L.M., Marsh, M., 2016. The intracellular plasma membrane-connected compartment in the assembly of HIV-1 in human macrophages. BMC Biol. 14, 50.

Ochsenbauer-Jambor, C., Miller, D.C., Roberts, C.R., Rhee, S.S., Hunter, E., 2001. Palmitoylation of the Rous sarcoma virus transmembrane glycoprotein is required for protein stability and virus infectivity. J. Virol. 75, 11544–11554.

Olety, B., Ono, A., 2014. Roles played by acidic lipids in HIV-1 Gag membrane binding. Virus Res. 193, 108–115.

Olety, B., Veatch, S.L., Ono, A., 2015. Phosphatidylinositol-(4,5)-bisphosphate acyl chains differentiate membrane binding of HIV-1 Gag from that of the phospholipase Cdelta1 pleckstrin homology domain. J. Virol. 89, 7861–7873.

Olety, B., Veatch, S.L., Ono, A., 2016. Visualization of HIV-1 Gag binding to giant unilamellar vesicle (GUV) membranes. J. Vis. Exp. 113, 54293.

Olsen, K.E., Andersen, K.B., 1999. Palmitoylation of the intracytoplasmic R peptide of the transmembrane envelope protein in Moloney murine leukemia virus. J. Virol. 73, 8975–8981.

Ono, A., Ablan, S.D., Lockett, S.J., Nagashima, K., Freed, E.O., 2004. Phosphatidylinositol (4,5) bisphosphate regulates HIV-1 Gag targeting to the plasma membrane. Proc. Natl. Acad. Sci. U.S.A. 101, 14889–14894.

Ono, A., Freed, E.O., 1999. Binding of human immunodeficiency virus type 1 Gag to membrane: role of the matrix amino terminus. J. Virol. 73, 4136–4144.

Ono, A., Freed, E.O., 2001. Plasma membrane rafts play a critical role in HIV-1 assembly and release. Proc. Natl. Acad. Sci. U.S.A. 98, 13925–13930.

Ono, A., Freed, E.O., 2004. Cell-type-dependent targeting of human immunodeficiency virus type 1 assembly to the plasma membrane and the multivesicular body. J. Virol. 78, 1552–1563.

Ono, A., Freed, E.O., 2005. Role of lipid rafts in virus replication. Adv. Virus Res. 64, 311–358.
Ono, A., Orenstein, J.M., Freed, E.O., 2000. Role of the Gag matrix domain in targeting human immunodeficiency virus type 1 assembly. J. Virol. 74, 2855–2866.
Ono, A., Waheed, A.A., Freed, E.O., 2007. Depletion of cellular cholesterol inhibits membrane binding and higher-order multimerization of human immunodeficiency virus type 1 Gag. Virology 360, 27–35.
Pelchen-Matthews, A., Kramer, B., Marsh, M., 2003. Infectious HIV-1 assembles in late endosomes in primary macrophages. J. Cell Biol. 162, 443–455.
Percherancier, Y., Lagane, B., Planchenault, T., Staropoli, I., Altmeyer, R., Virelizier, J.L., Arenzana-Seisdedos, F., Hoessli, D.C., Bachelerie, F., 2003. HIV-1 entry into T-cells is not dependent on CD4 and CCR5 localization to sphingolipid-enriched, detergent-resistant, raft membrane domains. J. Biol. Chem. 278, 3153–3161.
Peyron, P., Bordier, C., N'Diaye, E.N., Maridonneau-Parini, I., 2000. Nonopsonic phagocytosis of *Mycobacterium kansasii* by human neutrophils depends on cholesterol and is mediated by CR3 associated with glycosylphosphatidylinositol-anchored proteins. J. Immunol. 165, 5186–5191.
Phillips, D.M., 1994. The role of cell-to-cell transmission in HIV infection. AIDS 8, 719–731.
Pickl, W.F., Pimentel-Muinos, F.X., Seed, B., 2001. Lipid rafts and pseudotyping. J. Virol. 75, 7175–7183.
Piguet, V., Sattentau, Q., 2004. Dangerous liaisons at the virological synapse. J. Clin. Investig. 114, 605–610.
Pike, L.J., 2006. Rafts defined: a report on the Keystone Symposium on lipid rafts and cell function. J. Lipid Res. 47, 1597–1598.
Pleskoff, O., Seman, M., Alizon, M., 1995. Amphotericin B derivative blocks human immunodeficiency virus type 1 entry after CD4 binding: effect on virus-cell fusion but not on cell-cell fusion. J. Virol. 69, 570–574.
Popik, W., Alce, T.M., 2004. CD4 receptor localized to non-raft membrane microdomains supports HIV-1 entry. Identification of a novel raft localization marker in CD4. J. Biol. Chem. 279, 704–712.
Popik, W., Alce, T.M., Au, W.C., 2002. Human immunodeficiency virus type 1 uses lipid raft-colocalized CD4 and chemokine receptors for productive entry into CD4(+) T cells. J. Virol. 76, 4709–4722.
Prchal, J., Srb, P., Hunter, E., Ruml, T., Hrabal, R., 2012. The structure of myristoylated Mason-Pfizer monkey virus matrix protein and the role of phosphatidylinositol-(4,5)-bisphosphate in its membrane binding. J. Mol. Biol. 423, 427–438.
Prior, I.A., Muncke, C., Parton, R.G., Hancock, J.F., 2003. Direct visualization of Ras proteins in spatially distinct cell surface microdomains. J. Cell Biol. 160, 165–170.
Probasco, J.C., Spudich, S.S., Critchfield, J., Lee, E., Lollo, N., Deeks, S.G., Price, R.W., 2008. Failure of atorvastatin to modulate CSF HIV-1 infection: results of a pilot study. Neurology 71, 521–524.
Puri, A., Hug, P., Jernigan, K., Barchi, J., Kim, H.Y., Hamilton, J., Wiels, J., Murray, G.J., Brady, R.O., Blumenthal, R., 1998. The neutral glycosphingolipid globotriaosylceramide promotes fusion mediated by a CD4-dependent CXCR4-utilizing HIV type 1 envelope glycoprotein. Proc. Natl. Acad. Sci. U.S.A. 95, 14435–14440.
Puri, A., Rawat, S.S., Lin, H.M., Finnegan, C.M., Mikovits, J., Ruscetti, F.W., Blumenthal, R., 2004. An inhibitor of glycosphingolipid metabolism blocks HIV-1 infection of primary T-cells. AIDS 18, 849–858.
Quigley, J.P., Rifkin, D.B., Reich, E., 1972. Lipid studies of Rous sarcoma virus and host cell membranes. Virology 50, 550–557.

Ramkumar, S., Sakac, D., Binnington, B., Branch, D.R., Lingwood, C.A., 2009. Induction of HIV-1 resistance: cell susceptibility to infection is an inverse function of globotriaosyl ceramide levels. Glycobiology 19, 76–82.

Ramstedt, B., Slotte, J.P., 2002. Membrane properties of sphingomyelins. FEBS Lett. 531, 33–37.

Raposo, G., Moore, M., Innes, D., Leijendekker, R., Leigh-Brown, A., Benaroch, P., Geuze, H., 2002. Human macrophages accumulate HIV-1 particles in MHC II compartments. Traffic 3, 718–729.

Rawat, S.S., Gallo, S.A., Eaton, J., Martin, T.D., Ablan, S., KewalRamani, V.N., Wang, J.M., Blumenthal, R., Puri, A., 2004. Elevated expression of GM3 in receptor-bearing targets confers resistance to human immunodeficiency virus type 1 fusion. J. Virol. 78, 7360–7368.

Rawat, S.S., Johnson, B.T., Puri, A., 2005. Sphingolipids: modulators of HIV-1 infection and pathogenesis. Biosci. Rep. 25, 329–343.

Rawat, S.S., Zimmerman, C., Johnson, B.T., Cho, E., Lockett, S.J., Blumenthal, R., Puri, A., 2008. Restricted lateral mobility of plasma membrane CD4 impairs HIV-1 envelope glycoprotein mediated fusion. Mol. Membr. Biol. 25, 83–94.

Resh, M.D., 2004. A myristoyl switch regulates membrane binding of HIV-1 Gag. Proc. Natl. Acad. Sci. U.S.A. 101, 417–418.

Ritchie, K., Kusumi, A., 2003. Single-particle tracking image microscopy. Methods Enzymol. 360, 618–634.

Rollason, R., Dunstan, K., Billcliff, P.G., Bishop, P., Gleeson, P., Wise, H., Digard, P., Banting, G., 2013. Expression of HIV-1 Vpu leads to loss of the viral restriction factor CD317/Tetherin from lipid rafts and its enhanced lysosomal degradation. PLoS One 8, e75680.

Rosa, A., Chande, A., Ziglio, S., De Sanctis, V., Bertorelli, R., Goh, S.L., McCauley, S.M., Nowosielska, A., Antonarakis, S.E., Luban, J., Santoni, F.A., Pizzato, M., 2015. HIV-1 Nef promotes infection by excluding SERINC5 from virion incorporation. Nature 526, 212–217.

Rousso, I., Mixon, M.B., Chen, B.K., Kim, P.S., 2000. Palmitoylation of the HIV-1 envelope glycoprotein is critical for viral infectivity. Proc. Natl. Acad. Sci. U.S.A. 97, 13523–13525.

Ruiz, A., Hill, M.S., Schmitt, K., Stephens, E.B., 2010. Membrane raft association of the Vpu protein of human immunodeficiency virus type 1 correlates with enhanced virus release. Virology 408, 89–102.

Saad, J.S., Ablan, S.D., Ghanam, R.H., Kim, A., Andrews, K., Nagashima, K., Soheilian, F., Freed, E.O., Summers, M.F., 2008. Structure of the myristylated human immunodeficiency virus type 2 matrix protein and the role of phosphatidylinositol-(4,5)-bisphosphate in membrane targeting. J. Mol. Biol. 382, 434–447.

Saad, J.S., Loeliger, E., Luncsford, P., Liriano, M., Tai, J., Kim, A., Miller, J., Joshi, A., Freed, E.O., Summers, M.F., 2007. Point mutations in the HIV-1 matrix protein turn off the myristyl switch. J. Mol. Biol. 366, 574–585.

Saad, J.S., Miller, J., Tai, J., Kim, A., Ghanam, R.H., Summers, M.F., 2006. Structural basis for targeting HIV-1 Gag proteins to the plasma membrane for virus assembly. Proc. Natl. Acad. Sci. U.S.A. 103, 11364–11369.

Salzwedel, K., West, J.T., Hunter, E., 1999. A conserved tryptophan-rich motif in the membrane-proximal region of the human immunodeficiency virus type 1 gp41 ectodomain is important for Env-mediated fusion and virus infectivity. J. Virol. 73, 2469–2480.

Sasaki, T., Takasuga, S., Sasaki, J., Kofuji, S., Eguchi, S., Yamazaki, M., Suzuki, A., 2009. Mammalian phosphoinositide kinases and phosphatases. Prog. Lipid Res. 48, 307–343.

Sato, H., Orenstein, J., Dimitrov, D., Martin, M., 1992. Cell-to-cell spread of HIV-1 occurs within minutes and may not involve the participation of virus particles. Virology 186, 712–724.

Sattentau, Q., 2008. Avoiding the void: cell-to-cell spread of human viruses. Nat. Rev. Microbiol. 6, 815–826.

Sattentau, Q.J., 2011. The direct passage of animal viruses between cells. Curr. Opin. Virol. 1, 396–402.

Schoneberg, J., Lee, I.H., Iwasa, J.H., Hurley, J.H., 2017. Reverse-topology membrane scission by the ESCRT proteins. Nat. Rev. Mol. Cell Biol. 18, 5–17.

Schultz, A.M., Rein, A., 1989. Unmyristylated Moloney murine leukemia virus Pr65gag is excluded from virus assembly and maturation events. J. Virol. 63, 2370–2373.

Schwarzer, R., Levental, I., Gramatica, A., Scolari, S., Buschmann, V., Veit, M., Herrmann, A., 2014. The cholesterol-binding motif of the HIV-1 glycoprotein gp41 regulates lateral sorting and oligomerization. Cell. Microbiol. 16, 1565–1581.

Sharma, P., Varma, R., Sarasij, R.C., Ira, Gousset, K., Krishnamoorthy, G., Rao, M., Mayor, S., 2004. Nanoscale organization of multiple GPI-anchored proteins in living cell membranes. Cell 116, 577–589.

Sherer, N.M., Lehmann, M.J., Jimenez-Soto, L.F., Horensavitz, C., Pypaert, M., Mothes, W., 2007. Retroviruses can establish filopodial bridges for efficient cell-to-cell transmission. Nat. Cell Biol. 9, 310–315.

Shkriabai, N., Datta, S.A., Zhao, Z., Hess, S., Rein, A., Kvaratskhelia, M., 2006. Interactions of HIV-1 Gag with assembly cofactors. Biochemistry 45, 4077–4083.

Simons, K., Toomre, D., 2000. Lipid rafts and signal transduction. Nat. Rev. 1, 31–39.

Simons, K., Vaz, W.L., 2004. Model systems, lipid rafts, and cell membranes. Annu. Rev. Biophys. Biomol. Struct. 33, 269–295.

Sklar, P.A., Masur, H., Grubb, J.R., Voell, J., Witek, J., Ono, A., Freed, E.O., Maldarelli, F., 2005. Pravastatin does not have a consistent antiviral effect in chronically HIV-infected individuals on antiretroviral therapy. AIDS 19, 1109–1111.

Slosberg, B.N., Montelaro, R.C., 1982. A comparison of the mobilities and thermal transitions of retrovirus lipid envelopes and host cell plasma membranes by electron spin resonance spectroscopy. Biochim. Biophys. Acta 689, 393–402.

Smart, E.J., Anderson, R.G., 2002. Alterations in membrane cholesterol that affect structure and function of caveolae. Methods Enzymol. 353, 131–139.

Soares, M.M., King, S.W., Thorpe, P.E., 2008. Targeting inside-out phosphatidylserine as a therapeutic strategy for viral diseases. Nat. Med. 14, 1357–1362.

Sol-Foulon, N., Esnault, C., Percherancier, Y., Porrot, F., Metais-Cunha, P., Bachelerie, F., Schwartz, O., 2004. The effects of HIV-1 Nef on CD4 surface expression and viral infectivity in lymphoid cells are independent of rafts. J. Biol. Chem. 279, 31398–31408.

Sowinski, S., Jolly, C., Berninghausen, O., Purbhoo, M.A., Chauveau, A., Kohler, K., Oddos, S., Eissmann, P., Brodsky, F.M., Hopkins, C., Onfelt, B., Sattentau, Q., Davis, D.M., 2008. Membrane nanotubes physically connect T cells over long distances presenting a novel route for HIV-1 transmission. Nat. Cell Biol. 10, 211–219.

Sprong, H., Kruithof, B., Leijendekker, R., Slot, J.W., van Meer, G., van der Sluijs, P., 1998. UDP-galactose:ceramide galactosyltransferase is a class I integral membrane protein of the endoplasmic reticulum. J. Biol. Chem. 273, 25880–25888.

Stansell, E., Apkarian, R., Haubova, S., Diehl, W.E., Tytler, E.M., Hunter, E., 2007. Basic residues in the Mason-Pfizer monkey virus gag matrix domain regulate intracellular trafficking and capsid-membrane interactions. J. Virol. 81, 8977–8988.

Subtil, A., Gaidarov, I., Kobylarz, K., Lampson, M.A., Keen, J.H., McGraw, T.E., 1999. Acute cholesterol depletion inhibits clathrin-coated pit budding. Proc. Natl. Acad. Sci. U.S.A. 96, 6775–6780.

Sundquist, W.I., Krausslich, H.G., 2012. HIV-1 assembly, budding, and maturation. Cold Spring Harb. Perspect. Med. 2, a006924.

Suzuki, J., Denning, D.P., Imanishi, E., Horvitz, H.R., Nagata, S., 2013. Xk-related protein 8 and CED-8 promote phosphatidylserine exposure in apoptotic cells. Science 341, 403–406.

Tang, C., Loeliger, E., Luncsford, P., Kinde, I., Beckett, D., Summers, M.F., 2004. Entropic switch regulates myristate exposure in the HIV-1 matrix protein. Proc. Natl. Acad. Sci. U.S.A. 101, 517–522.

Usami, Y., Wu, Y., Gottlinger, H.G., 2015. SERINC3 and SERINC5 restrict HIV-1 infectivity and are counteracted by Nef. Nature 526, 218–223.

van 't Wout, A.B., Swain, J.V., Schindler, M., Rao, U., Pathmajeyan, M.S., Mullins, J.I., Kirchhoff, F., 2005. Nef induces multiple genes involved in cholesterol synthesis and uptake in human immunodeficiency virus type 1-infected T cells. J. Virol. 79, 10053–10058.

Van Damme, N., Goff, D., Katsura, C., Jorgenson, R.L., Mitchell, R., Johnson, M.C., Stephens, E.B., Guatelli, J., 2008. The interferon-induced protein BST-2 restricts HIV-1 release and is downregulated from the cell surface by the viral Vpu protein. Cell Host Microbe 3, 245–252.

van Meer, G., 1998. Lipids of the Golgi membrane. Trends Cell Biol. 8, 29–33.

van Meer, G., 2005. Cellular lipidomics. EMBO J. 24, 3159–3165.

van Meer, G., Voelker, D.R., Feigenson, G.W., 2008. Membrane lipids: where they are and how they behave. Nat. Rev. Mol. Cell Biol. 9, 112–124.

Varma, R., Mayor, S., 1998. GPI-anchored proteins are organized in submicron domains at the cell surface. Nature 394, 798–801.

Verhoven, B., Schlegel, R.A., Williamson, P., 1995. Mechanisms of phosphatidylserine exposure, a phagocyte recognition signal, on apoptotic T lymphocytes. J. Exp. Med. 182, 1597–1601.

Vincent, N., Genin, C., Malvoisin, E., 2002. Identification of a conserved domain of the HIV-1 transmembrane protein gp41 which interacts with cholesteryl groups. Biochim. Biophys. Acta 1567, 157–164.

Vlach, J., Saad, J.S., 2013. Trio engagement via plasma membrane phospholipids and the myristoyl moiety governs HIV-1 matrix binding to bilayers. Proc. Natl. Acad. Sci. U.S.A. 110, 3525–3530.

Waheed, A.A., Ablan, S.D., Mankowski, M.K., Cummins, J.E., Ptak, R.G., Schaffner, C.P., Freed, E.O., 2006. Inhibition of HIV-1 replication by amphotericin B methyl ester: selection for resistant variants. J. Biol. Chem. 281, 28699–28711.

Waheed, A.A., Ablan, S.D., Roser, J.D., Sowder, R.C., Schaffner, C.P., Chertova, E., Freed, E.O., 2007. HIV-1 escape from the entry-inhibiting effects of a cholesterol-binding compound via cleavage of gp41 by the viral protease. Proc. Natl. Acad. Sci. U.S.A. 104, 8467–8471.

Waheed, A.A., Ablan, S.D., Soheilian, F., Nagashima, K., Ono, A., Schaffner, C.P., Freed, E.O., 2008. Inhibition of human immunodeficiency virus type 1 assembly and release by the cholesterol-binding compound amphotericin B methyl ester: evidence for Vpu dependence. J. Virol. 82, 9776–9781.

Waheed, A.A., Ablan, S.D., Sowder, R.C., Roser, J.D., Schaffner, C.P., Chertova, E., Freed, E.O., 2010. Effect of mutations in the human immunodeficiency virus type 1 protease on cleavage of the gp41 cytoplasmic tail. J. Virol. 84, 3121–3126.

Waheed, A.A., Freed, E.O., 2009. Lipids and membrane microdomains in HIV-1 replication. Virus Res. 143, 162–176.

Waheed, A.A., Freed, E.O., 2010. The role of lipids in retrovirus replication. Viruses 2, 1146–1180.

Wang, J.K., Kiyokawa, E., Verdin, E., Trono, D., 2000. The Nef protein of HIV-1 associates with rafts and primes T cells for activation. Proc. Natl. Acad. Sci. U.S.A. 97, 394–399.

Watson, H., 2015. Biological membranes. Essays Biochem. 59, 43–69.

Welsch, S., Groot, F., Krausslich, H.G., Keppler, O.T., Sattentau, Q.J., 2011. Architecture and regulation of the HIV-1 assembly and holding compartment in macrophages. J. Virol. 85, 7922–7927.

Welsch, S., Keppler, O.T., Habermann, A., Allespach, I., Krijnse-Locker, J., Krausslich, H.G., 2007. HIV-1 buds predominantly at the plasma membrane of primary human macrophages. PLoS Pathog. 3, e36.

Wen, Y., Dick, R.A., Feigenson, G.W., Vogt, V.M., 2016. Effects of membrane charge and order on membrane binding of the retroviral structural protein Gag. J. Virol. 90, 9518–9532.

Wenk, M.R., 2010. Lipidomics: new tools and applications. Cell 143, 888–895.

Wielgosz, M.M., Rauch, D.A., Jones, K.S., Ruscetti, F.W., Ratner, L., 2005. Cholesterol dependence of HTLV-I infection. AIDS Res. Hum. Retrovir. 21, 43–50.

Wilson, B.S., Pfeiffer, J.R., Surviladze, Z., Gaudet, E.A., Oliver, J.M., 2001. High resolution mapping of mast cell membranes reveals primary and secondary domains of Fc(epsilon)RI and LAT. J. Cell Biol. 154, 645–658.

Wu, H.M., Lin, Y.H., Yen, T.C., Hsieh, C.L., 2016. Nanoscopic substructures of raft-mimetic liquid-ordered membrane domains revealed by high-speed single-particle tracking. Sci. Rep. 6, 20542.

Wyss, S., Berlioz-Torrent, C., Boge, M., Blot, G., Honing, S., Benarous, R., Thali, M., 2001. The highly conserved C-terminal dileucine motif in the cytosolic domain of the human immunodeficiency virus type 1 envelope glycoprotein is critical for its association with the AP-1 clathrin adaptor [correction of adapter]. J. Virol. 75, 2982–2992.

Yandrapalli, N., Lubart, Q., Tanwar, H.S., Picart, C., Mak, J., Muriaux, D., Favard, C., 2016. Self assembly of HIV-1 Gag protein on lipid membranes generates PI(4,5)P2/Cholesterol nanoclusters. Sci. Rep. 6, 39332.

Yang, C., Compans, R.W., 1996. Palmitoylation of the murine leukemia virus envelope glycoprotein transmembrane subunits. Virology 221, 87–97.

Yang, C., Spies, C.P., Compans, R.W., 1995. The human and simian immunodeficiency virus envelope glycoprotein transmembrane subunits are palmitoylated. Proc. Natl. Acad. Sci. U.S.A. 92, 9871–9875.

Yang, P., Ai, L.S., Huang, S.C., Li, H.F., Chan, W.E., Chang, C.W., Ko, C.Y., Chen, S.S., 2010. The cytoplasmic domain of human immunodeficiency virus type 1 transmembrane protein gp41 harbors lipid raft association determinants. J. Virol. 84, 59–75.

Yuan, X., Yu, X., Lee, T.H., Essex, M., 1993. Mutations in the N-terminal region of human immunodeficiency virus type 1 matrix protein block intracellular transport of the Gag precursor. J. Virol. 67, 6387–6394.

Zhang, Z., Ma, J., Zhang, X., Su, C., Yao, Q.C., Wang, X., 2015. Equine infectious anemia virus Gag assembly and export are directed by matrix protein through trans-Golgi networks and cellular vesicles. J. Virol. 90, 1824–1838.

Zheng, Y.-H., Plemenitas, A., Linneman, T., Fackler, O.T., Peterlin, B.M., 2001. Nef increases infectivity of HIV via lipid rafts. Curr. Biol. 11, 875–879.

Zheng, Y.H., Plemenitas, A., Fielding, C.J., Peterlin, B.M., 2003. Nef increases the synthesis of and transports cholesterol to lipid rafts and HIV-1 progeny virions. Proc. Natl. Acad. Sci. U.S.A. 100, 8460–8465.

Zhong, P., Agosto, L.M., Munro, J.B., Mothes, W., 2013. Cell-to-cell transmission of viruses. Curr. Opin. Virol. 3, 44–50.

Zhou, W., Parent, L.J., Wills, J.W., Resh, M.D., 1994. Identification of a membrane-binding domain within the amino-terminal region of human immunodeficiency virus type 1 Gag protein which interacts with acidic phospholipids. J. Virol. 68, 2556–2569.

Zhu, Y., Gumlaw, N., Karman, J., Zhao, H., Zhang, J., Jiang, J.L., Maniatis, P., Edling, A., Chuang, W.L., Siegel, C., Shayman, J.A., Kaplan, J., Jiang, C., Cheng, S.H., 2011. Lowering glycosphingolipid levels in CD4+ T cells attenuates T cell receptor signaling, cytokine production, and differentiation to the Th17 lineage. J. Biol. Chem. 286, 14787–14794.

Zhukovsky, M.A., Lee, P.H., Ott, A., Helms, V., 2013. Putative cholesterol-binding sites in human immunodeficiency virus (HIV) coreceptors CXCR4 and CCR5. Proteins 81, 555–567.

Chapter 11

Cellular Immune Responses to Retroviruses

Susan R. Ross
UIC College of Medicine, Chicago, IL, United States

Chapter Outline

Introduction 401
Steps of the Retrovirus Infection Pathway Targeted by the Host Intrinsic/Innate Response 402
Host Pathways Implicated in Cellular Control of Retrovirus Infection 406
Subversion of Cellular Immune Responses by Retroviruses 412
Conclusions 414
References 414

INTRODUCTION

Many retroviruses have been in their mammalian hosts for millions of years, as evidenced by the "fossil record" of endogenous proviruses found in the genomes of a variety of species (Baillie et al., 2004; Mager and Stoye, 2015; Maksakova et al., 2006). As such, the host has evolved many ways to keep infection under control. This defense occurs particularly at the cellular level and at the early stages of infection, because once infection is established, integration of the retroviral DNA into the host chromosome occurs. This creates a pool of latently infected cells, at which point elimination of the virus is difficult. The host restriction factors that control the early stages of infection are generally referred to as "intrinsic" if they function in many cell types and "innate" if their expression is limited to cells of the immune system, which includes sentinel cells such as macrophages and dendritic cells (DCs) and lymphocytes. Viruses that persist as active exogenous viruses in a species do so because they develop the means to counteract host restriction factors, either by limiting accessibility of these factors to the virus or by the expression of viral proteins that specifically counteract their function.

Most exogenous retroviruses are associated with pathogenic diseases, such as cancers and immunodeficiencies. Retroviruses that have persisted in their hosts as exogenous infectious agents in an evolutionary sense are generally not highly pathogenic, e.g., relative to the species' life span, they cause diseases

Retrovirus-Cell Interactions. https://doi.org/10.1016/B978-0-12-811185-7.00011-X

that only minimally affect reproductive capacity. This is due in large part to the ability of host restriction factors to control the early stages of infection, thereby limiting virus load. For example, the murine retrovirus mouse mammary tumor virus (MMTV), which has existed in mice for millions of years, is naturally acquired by newborns from their mothers via milk and typically causes breast cancer in mice over 6 months of age (Baillie et al., 2004; Dudley et al., 2016). This allows multiple litters to be produced prior to the onset of disease and for the virus to be passed on to offspring. Similarly, human T cell leukemia virus I, which is also naturally transmitted through milk, is estimated to have been in human populations for >100,000 years and is associated with leukemias that occur in the fourth or fifth decade of life (Panfil et al., 2016).

In contrast, the introduction of new retrovirus infections from one species to another, termed zoonosis, usually has one of two outcomes. If the host restriction factors act on the virus, it cannot establish infection, thus preventing zoonosis. This has been demonstrated for example, for xenotropic murine leukemia virus (XMLV) and xenotropic-like murine retrovirus infection of human cells as well as for human immunodeficiency virus-1 (HIV-1) infection of nonhuman primates (Groom et al., 2010; Harris et al., 2012; Kirchhoff, 2010). In contrast, successful zoonotic infections arise when host restriction factors cannot act on the virus, as has occurred for HIV-1. HIV-1 is derived from simian immunodeficiency virus (SIV) found in chimpanzees and successfully zoonosed into humans within the past 100 years (Sharp and Hahn, 2011). This may have also occurred with viruses such as koala retrovirus, which resulted from zoonotic infection of koalas by a rodent retrovirus that is estimated to have occurred around 200 years ago (Denner, 2016).

Understanding the initial immune response is thus not only critical to developing means to control viruses that are endemic within a species but also to determine what allows trans-species viral transmission. This chapter will focus on the cellular proteins involved in the intrinsic and innate response to mammalian retroviruses at the initial stages of infection, as well as the means by which these viruses subvert the response and establish successful infection.

STEPS OF THE RETROVIRUS INFECTION PATHWAY TARGETED BY THE HOST INTRINSIC/INNATE RESPONSE

The steps of the retrovirus infection/replication pathway are described in detail in other chapters of this volume. Here I will briefly describe the known steps in the pathway targeted by the cellular intrinsic/innate immune response.

Retroviruses first bind to cellular receptors on the cell surface via the viral envelope (Env) glycoproteins incorporated into their membranes. These include the entry receptors that mediate virus–cell fusion and entry, either from the plasma membrane or from within an endosomal compartment (Fig. 11.1). The presence or absence of entry receptors on cells determines whether they are susceptible to infection. The entry receptors themselves, such as the chemokine

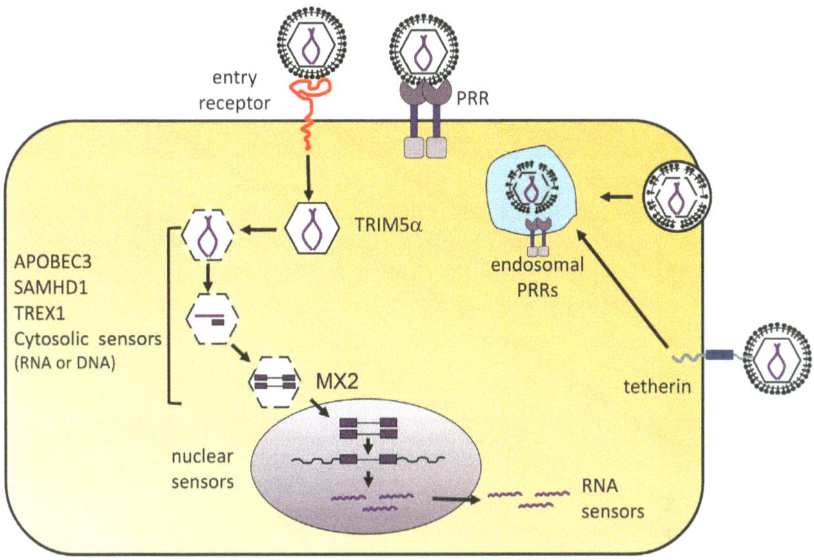

FIGURE 11.1 Steps at which different host genes restrict retrovirus infection. Viruses can enter via the plasma membrane after binding to a cell surface receptor or after internalization into an acidic compartment. Entry is determined by the binding specificity of the virus Env for receptor. Binding of virus to cell surface pattern recognition receptors (PRRs), such as TLR2, TLR4, or DC-SIGN can also trigger activation of antiviral signal transduction pathways. There are also endosomal PRRs activated by viral RNA; this occurs when defective virions are endocytosed, when phagocytic cells ingest retrovirus-infected cells, or when molecules such as tetherin cause the internalization of virions trapped on the cell surface. During productive infection, virions enter the cell and the capsid, which forms a lattice-like structure, can be recognized by host cell factors such as TRIM5α. After or during capsid partial or total dissociation, factors such as virus-packaged or cell-intrinsic APOBEC3 proteins can mutate viral DNA by deaminating cytidine residues or block reverse transcription, whereas cell-intrinsic SAMHD1 can deplete nucleotide pools needed for reverse transcription. Reverse transcription products, including single-stranded viral DNA, DNA/RNA hybrid molecules, and double-stranded viral DNA can be recognized by cytosolic sensors such as cGAS, DDX41, and members of the absent in melanoma 2–like receptor family. Some sensors are also present in the nucleus, although there have not yet been reports that they can recognize retroviral DNA prior to integration. For HIV-1, it has also been reported that early after infection, short polyA- transcripts are generated and recognized by RNA sensors such as DDX3, which in turn signal through the MAVS restriction factor (see Fig. 11.2).

coreceptors used by HIV, or additional virus-binding receptors such as the C-type lectin pattern recognition receptor (PRR) DC-specific intercellular adhesion molecule-3-grabbing nonintegrin (DC-SIGN), also transduce intracellular signals on retrovirus binding and trigger responses that can be pro- or antiviral (Collman et al., 2000; Gringhuis et al., 2009). Some retroviruses interact with additional PRRs on the cell surface, such as toll-like receptor 2 (TLR2) or TLR4, either via direct interaction with viral glycoproteins that present conserved structural features, termed pathogen-associated molecular patterns (PAMPs), or because the virions themselves bind to bacterial ligands that serve as PAMPs (Del Corno et al., 2016; Nazli et al., 2013; Rassa et al., 2002; Wilks et al., 2015) (Fig. 11.1).

After fusion of the virus–cell membrane, the retroviral capsid enters the cytoplasm. The regular lattice-like structure of the capsid protein in retroviral virions can serve as a PAMP. For example, it has been reported that the HIV capsid is recognized by cellular restriction factors such as TRIM5α, leading to the induction of innate immune responses in nonhuman cells; recent studies suggest that TRIM5α may function as a cell-type specific PRR in humans Langerhans DCs (Pertel et al., 2011; Ribeiro et al., 2016) (Figs. 11.1 and 11.2). Another antiviral restriction factor in mice, termed Fv1, also interacts with incoming capsid, but it is not known if this triggers innate immunity (Sanz-Ramos and Stoye, 2013). Interestingly, not all species have functional TRIM5-like proteins.

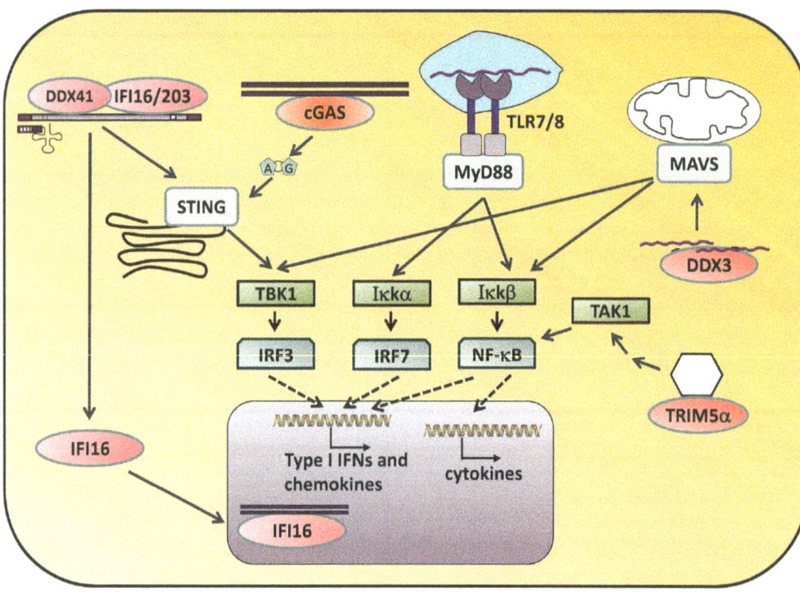

FIGURE 11.2 Pathways activated by viral gene products, resulting in innate/intrinsic immune responses to retroviruses. During infection, viral RNA/DNA hybrids are recognized by DDX41 and the ALRs IFI16 (humans) or IFI203 (mice), which then activate STING), localized in the endosome. STING activation causes phosphorylation ofTBK1 and in turn IRF3, which enters the nucleus and induces transcription of type 1 IFNs and chemokines such as CXCL10. Some ALRs, such as IFI16, have also been shown to sense nuclear DNA, although this has not been shown for retroviruses. It has also been demonstrated that truncated polyA- transcripts generated after integration of HIV provirus into the host chromosome are recognized by DDX3, which in turn activates the MAVS sensing pathway. Virion RNA that ends up in endosomal compartments, as shown in Fig. 11.1, can bind to TLR7 or TLR8 and trigger signaling through the MyD88 adaptor molecule, resulting in IκK phosphorylation and activation of the transcription factors IRF7 and NF-κB, which traffic to the nucleus and activate transcription of type 1 IFNs and cytokines, respectively. It has also been reported that the dsRNA PRR TLR3 may recognize retroviral RNA in endosomes. TRIM5α recognition of HIV-1 capsid has also been reported to activate the NF-κB pathway, via phosphorylation of the cellular kinase TAK1.

For example, the *TRIM5α* gene in cats is truncated, making it unable to interact with feline immunodeficiency (FIV) capsid (Zielonka and Munk, 2011).

During the course of reverse transcription in the cytoplasm, several forms of nucleic acid are generated, all of which may be perceived as "foreign" by the cell. These include RNA/DNA hybrids, single-stranded and double-stranded viral DNA (ssDNA and dsDNA, respectively) (Figs. 11.2 and 11.3). Moreover, all retroviruses have a structurally unusual replication intermediate consisting of a tRNA-bound DNA; the tRNA primer is removed by the RNAase H activity of reverse transcriptase during DNA synthesis (Fig. 11.3). Reverse transcription in the cytoplasm likely initiates within the capsid structure and thus the capsid serves to protect the nucleic acids from recognition by host cytosolic factors that recognize foreign nucleic acid (Lahaye et al., 2013; Stavrou et al., 2013).

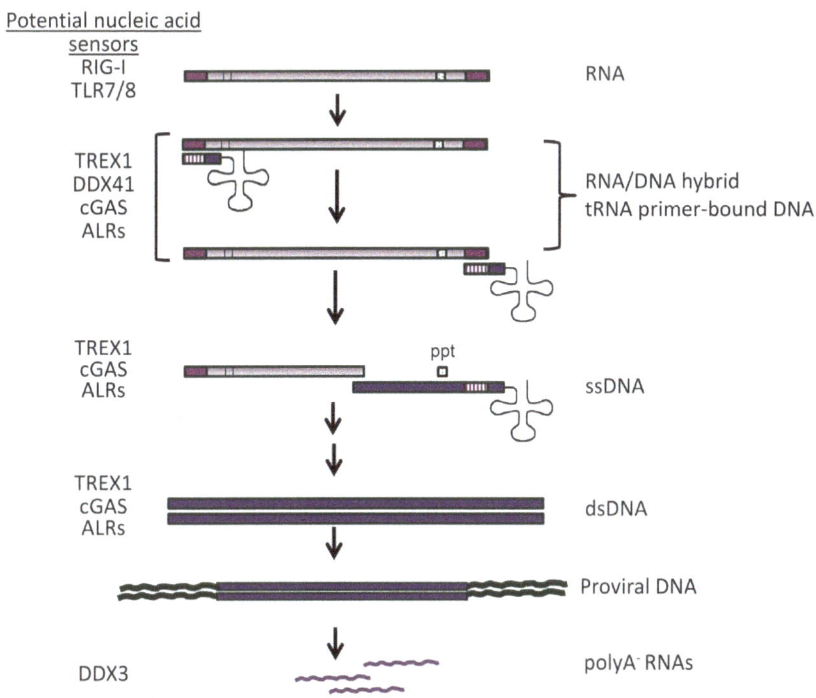

FIGURE 11.3 Nucleic acid structures generated during retrovirus replication that are recognized by host sensors. Light purple structures indicate RNA and dark purple DNA. The specific tRNA packaged in virions serves as the primer for reverse transcription. Different retroviruses use different tRNA as primers. The RNaseH function of reverse transcription degrades viral RNA as the DNA is synthesized, although it leaves behind a small piece of RNA termed the polypyrimidine tract (ppt) to serve as a primer for plus strand DNA synthesis. The RNaseH activity also eventually degrades the tRNA primer. After integration in the genome, viral genomic RNA and mRNAs are likely not recognized by host sensor. However, it has been suggested that HIV produces short polyA- transcripts until activation of the TEFb elongation factor by NF-κB occurs and that these are sensed by the DDX3 RNA helicase (Barboric et al., 2001; Gringhuis et al., 2017; Ribeiro et al., 2016). *dsDNA*, double-stranded viral DNA; *ssDNA*, single-stranded viral DNA; RIG-I, retinoic acid-inducible gene I.

Eventually, the capsid structure dissociates, and the dsDNA enters the nucleus, where it could also be recognized by nuclear DNA sensors, although this has not been reported (Fig. 11.1). Once integrated into the genome, viral RNA is transcribed and resembles host mRNAs. However, it has also been reported that at least for HIV-1, truncated viral transcripts that lack polyA tails are produced from integrated proviruses early in infection and that these are sensed by host RNA sensors (Gringhuis et al., 2017).

Incomplete or defective retroviruses may also enter cells via endocytosis into endosomal compartments, where PRRs that recognize single-stranded RNA (ssRNA) such as TLR7/8 reside (Beignon et al., 2005; Browne, 2011; Heil et al., 2004; Kane et al., 2011). Similarly, host factors such as bone-marrow stromal cell antigen 2/tetherin, which tethers virions to the cell surface, may lead to their endocytosis and subsequent degradation within compartments where these PRRs reside (Fig. 11.1) (Li et al., 2016) Thus, retroviral proteins, nucleic acid replication intermediates, and transcripts all potentially serve as recognition elements for host nucleic acid sensors.

HOST PATHWAYS IMPLICATED IN CELLULAR CONTROL OF RETROVIRUS INFECTION

There are many host PRRs that respond to viral PAMPs. In this section, I will describe the known PRRs that play a role in retrovirus sensing, the PAMPs they recognize, and the signal transduction pathways that they activate. Different cell types are infected by retroviruses during in vivo infection and this may influence the steps of infection that are recognized by these PRRs. Innate/intrinsic immune responses, particularly by sentinel cells such as macrophages and DCs, are now widely recognized as providing the first line of defense against incoming pathogens. Because most retroviruses are acquired via mucosal tissue, resident DCs and macrophages are likely the cell type of first encounter. In addition to providing cell-intrinsic signals that block infection, these cells secrete cytokines, chemokines, and interferons (IFNs) that activate antiviral pathways in bystander cells.

Macrophages and DCs express retrovirus entry receptors, including the HIV receptor CD4 and the chemokine coreceptors as well as mouse cationic amino acid transporter 1 (MCAT1/Slc7a1) and transferrin receptor 1, the MLV and MMTV receptors, respectively (Koppensteiner et al., 2012; Manches et al., 2014; Overbaugh et al., 2001; Ross et al., 2002). What are the initial in vivo target cells for HIV infection remains controversial. While some studies suggest that only T cells become infected at early times after HIV is acquired, macrophages and DCs are productively infected. In addition, conventional DCs bind and uptake HIV-1, depositing it into a nondegradative endocytic compartment (Koppensteiner et al., 2012; Manches et al., 2014). Virus delivery to uninfected T cells is either by productively infected or carrier macrophages, and DCs can occur via release of cell-free virus or by cell–cell transmission through what is called the virological synapse (Chen, 2012; Koppensteiner et al., 2012; Manches et al., 2014). This mode of transmission has also been reported to occur with MLV (Sewald et al., 2015). It may be that some of these endocytosed particles also reach a degradative compartment in macrophages and DCs, thereby activating endosomal PRRs.

Because these cells express high levels of most if not all of the PRRs discussed here, the outcome of DC or macrophage infection is increased secretion of IFNs, cytokines, and chemokines, which play a role in autocrine and paracrine induction of IFN-stimulated genes (ISG) encoding antiviral factors. In addition, the secretion of these factors leads to the influx and maturation of antigen-presenting cells and lymphocytes needed for the adaptive immune response. However, because of this activation, the level of DC and macrophage infection is potentially low. Moreover, retroviruses encode proteins that counteract the antiviral responses in sentinel cells, which should allow them to avoid the immune response and to replicate (see next section) and thus it may be difficult to detect early sentinel infection in in vivo settings. Nonetheless, for the murine retroviruses MLV and MMTV, it is known that productive infection of DCs occurs; these most likely are the first cells to be infected in vivo, followed by infection of lymphocytes (Burzyn et al., 2004; Courreges et al., 2007; Low et al., 2009).

There are a number of cell surface receptors that interact with retroviral glycoproteins and transduce signals that activate antiviral pathways as a result of these interactions. One of the earliest examples of this type of activation is binding of gp55, a truncated form of mouse retrovirus spleen focus-forming virus Env, to the erythropoietin receptor, a member of the cytokine receptor family on hematopoietic stem cells. Binding of gp55 triggers proliferation of these stem cells, eventually leading to oncogenesis (Li et al., 1990). Similar pathways of activation have been described when soluble gp120 Env present on HIV-1 virions binds to the CD4 entry receptor or chemokine coreceptors, initiating a signal transduction cascade leading to the expression of antiviral cytokine and chemokine genes, although it has not been demonstrated in vivo that this plays a role in the innate immune response (Del Corno et al., 2001; Liu et al., 2000; Popik et al., 1998). These receptors are typically G-coupled proteins that signal through src family and other cellular kinases. Other animal retroviruses, such as XMLV, also use G-protein–coupled entry receptors (Battini et al., 1999; Overbaugh et al., 2001). Whether the entry receptors also signal on XMLV binding in cells of the immune system has not been studied.

TLR2 and TLR4, the cell surface receptors for the bacterial PAMPs peptidoglycan and lipopolysaccharide, respectively, have also been implicated in the innate immune responses to retroviruses (Fig. 11.1). Several groups have reported direct binding of retroviral glycoproteins to both PRRs, although it has been suggested that incorporation of bacterial ligands into the viral membrane is responsible for the induction of IFNs and cytokines, at least in the case of the murine retrovirus MMTV (Nazli et al., 2013; Rassa et al., 2002; Wilks et al., 2015).

TLRs signal through cytosolic adaptor molecules such as MyD88, TRAM, and TRIF, ultimately resulting in the phosphorylation and migration into the nucleus of the transcription factors interferon-regulatory factor (IRF) 7 and nuclear factor (NF)-κB (Fitzgerald et al., 2003) (Fig. 11.2). These transcription factors induce genes involved in cell activation, which may be required for efficient sentinel cell infection by retroviruses, as well as for inducing ISGs involved in both innate and adaptive immunity (Burzyn et al., 2004; Courreges et al., 2007; Jude et al., 2003; Wilks et al., 2015).

In addition to the cell surface TLRs, other members of this family that recognize viral nucleic acids are found in endosomal compartments. These include TLR3, TLR7 and TLR8, and TLR9, which recognize dsRNA, ssRNA, and dsDNA, respectively (O'Neill, 2000) (Fig. 11.2). Purified HIV-1 RNA induces a TLR7-mediated immune response, although whether this is biologically important has not been shown (Meier et al., 2007). Moreover, germline knockout of both *Tlr3* and *Tlr7* as well as the downstream adaptor, *Myd88*, results in mice that are more highly infected by MLV and MMTV than are wild-type mice (Browne, 2011; Browne and Littman, 2009; Gibbert et al., 2014; Kane et al., 2011). Because retroviral replication occurs in the cytoplasm, access by these TLRs to either incoming viral RNA genomes or replication intermediates during productive infection is unlikely to occur. Instead, the ssRNA found in capsids or other viral nucleic acid replication intermediates only access this compartment after uptake of defective particles or perhaps by endocytosis via tetherin (Fig. 11.1), or when phagocytic cells such as macrophages engulf already-infected cells, respectively (Beignon et al., 2005; Galao et al., 2012; Gringhuis et al., 2010; Li et al., 2016). Mice with targeted knockout of *Myd88*, *Tlr3*, and *Tlr7* have defects in their ability to develop adaptive immune responses, which contributes to their higher infection levels (Browne, 2011; Browne and Littman, 2009; Gibbert et al., 2014; Kane et al., 2011). Interestingly, knockout of *Myd88* in B cells but not DCs affected infection by friend MLV (Browne, 2011). This supports the idea that particular pathways involved in the innate immune response are cell-type specific. Given that B cells infection by MLV is likely secondary to DC infection in vivo, this also suggests that TLR-mediated IFN/cytokine production is not critical for controlling infection in the initial in vivo targets but instead occurs after virus transmission to the lymphoid compartment. Interestingly, TLR9 has not been implicated in the innate immune response to retroviral infection. This may be because the dsDNA generated during reverse transcription rapidly moves into the nucleus and integrates into the host chromosomes; this would limit the amount of dsDNA that is released into the endosome after, for example, phagocytosis of infected cells.

The glycoproteins of many different enveloped viruses, including HIV-1 gp120, interact with the C-type lectin receptor DC-SIGN, in many cases using it to enhance virus entry (Van Breedam et al., 2014). DC-SIGN has also been implicated in HIV-1's ability to activate DCs. HIV-1 binding to DC-SIGN activates RAF-1 kinase, which triggers a signal transcription cascade that ultimately blocks the ability of the RNA helicase DDX3 to sense short aborted HIV-1 transcripts (see below) (Gringhuis et al., 2009, 2017).

There are also host factors in the cytoplasm that recognize various viral nucleic acids (Fig. 11.1). Notably, apolipoprotein B mRNA editing enzyme, catalytic polypeptide-like 3 (APOBEC3) proteins bind to both retroviral RNA and nucleocapsid protein in virus-producing cells and are thereby packaged into virions (Stavrou and Ross, 2015). APOBEC3 proteins belong to a family of RNA and DNA cytidine deaminases whose expression is induced by IFNs and that show strong evidence of positive selection. For example, humans have 7 *APOBEC3* genes, while mice and rats have a single gene. During reverse transcription, APOBEC3-mediated cytidine deamination causes mutations of reversed-transcribed viral DNA, leading to highly mutated viruses. APOBEC3 proteins also block reverse transcription (Koning et al., 2011; Holmes et al., 2007; MacMillan et al., 2013; Stavrou et al., 2013). APOBEC3 proteins expressed in target cells also have the ability to block reverse transcription by incoming virus (Gillick et al., 2013; Okeoma et al., 2009; Stavrou et al., 2015).

Another more recently discovered IFN-inducible host restriction factor, SAM domain and HD domain-containing protein 1 (SAMHD1), also acts at this step of replication to block incoming virus replication in quiescent myeloid cells and possibly resting T cells as well (Baldauf et al., 2017; Laguette et al., 2011). SAMHD1 is a phosphohydrolase that degrades deoxynucleoside triphosphates (dNTPs) and depletes the dNTP pools available for viral DNA synthesis during reverse transcription (Berger et al., 2011; Lahouassa et al., 2012). Interestingly, while APOBEC3 knockout mice are more susceptible to infection by a number of different murine retroviruses, SAMHD1 mice show little if any difference in susceptibility to MLV (Low et al., 2009; Okeoma et al., 2007; Rehwinkel et al., 2013; Santiago et al., 2008; Takeda et al., 2008).

Retroviral replication, which largely occurs in the cytoplasm, generates a number of nucleic acid intermediates that are recognized by these cellular nucleic acid sensors (Fig. 11.3 and Table 11.1). It is evident that DNA is an important PAMP recognized by host sensors because of the depletion or loss of the cytosolic DNA exonuclease Three prime repair exonuclease 1 (TREX1) leads to increase IFN production after infection with HIV or MLV, particularly in sentinel cells (Stavrou et al., 2013; Yan et al., 2010). Although TREX1 preferentially binds dsDNA, it also degrades the other reverse transcription intermediate (ssDNA, dsDNA, DNA in RNA/DNA hybrids) (Lindahl et al., 1969; Yan et al., 2010; Yuan et al., 2015). Nucleic acid that is not degraded by TREX1 can be recognized by several sensors upstream of the downstream mediator stimulator of interferon genes (STING). STING, when activated, translocates from the endoplasmic reticulum to a perinuclear compartment, where it activates tank-binding kinase 1 (TBK1) and as a result, phosphorylation of the transcription factor IRF3 (Fig. 11.2). IRF3 then enters the nucleus and induces transcription of type 1 IFNs and chemokines such as CXCL10 (Bhat and Fitzgerald, 2014; Ishikawa et al., 2009; Mukai et al., 2016; Saitoh et al., 2009).

TABLE 11.1 Viral Proteins That Subvert the Host Innate Immune Response

Virus	Virus Gene	Host Gene(s)	ISG
MLV	Capsid	APOBEC3	Yes
HIV, SIV	Vif	APOBEC3F, 3G, 3H[a]	Yes
MLV	Capsid	FV-1	Unknown
HIV-1	Capsid	TRIM5α	Yes
HIV-1	Capsid	MX2	Yes
HIV-1, MLV	Capsid	cGAS	Yes
HIV-1, MLV	Capsid	DDX41	No
HIV-1	Capsid	IFI16	Yes
MLV	Capsid	IFI203	Yes
HIV-1	Env	DDX3	Unknown
HIV, SIV	Vpu	Tetherin	Yes
SIV	Vpx	SAMHD1	Yes

ISG, IFN-stimulated genes.
[a]*Only some APOBEC3H haplotypes are counteracted by Vif.*

The most widely studied of the cytosolic sensors is cyclic guanosine monophosphate-adenosine monophosphate (cGAMP) synthase (cGAS), which on binding DNA generates cGAMP, which in turn activates STING (Cai et al., 2014; Gao et al., 2013) (Fig. 11.2). cGAS binds several forms of DNA, although dsDNA is the most effective at inducing this response (Sun et al., 2013). Depletion of cGAS, either by si- or shRNA-mediated knockdown or gene knockout, dramatically reduces the production of cGAMP and IFN production that occurs after HIV infection via the STING pathway (Cai et al., 2014; Luecke and Paludan, 2016). Similarly, depletion of cGAS also dramatically reduces the STING-mediated IFN response to MLV in human and mouse cells (Gao et al., 2013; Stavrou et al., 2013, 2015). Interestingly, the response to MLV occurs within 2 h of infection, coincident with the peak of reverse transcription, whereas the cGAS-dependent response to HIV takes over 20 h; it is not clear if HIV reverse transcription is much slower than MLV's or if HIV but not MLV reverse transcripts persist in the cytoplasm after the dsDNA viral genome has migrated to the nucleus and integrated into the host chromosomes. Recently, another protein polyglutamine binding protein 1 has been shown to bind dsDNA and interact with cGAS and thereby facilitate IFN induction (Yoh et al., 2015).

Several other molecules have been shown to participate in the IFN response to retroviral infection, including members of the IFN-inducible absent in melanoma 2–like receptor (ALR) family. The ALR genes encode proteins with PYRIN-PAAD-DAPIN (PYD) and hematopoietic interferon-inducible nuclear antigens with 200 amino acid repeats (HIN) domains, which bind to the PYDs of other proteins and dsDNA, respectively (Albrecht et al., 2005; Asefa et al., 2004). The *ALR* genes or interferon-inducible (IFI) proteins, are under strong positive selection. For example, humans have 4 genes whereas mice have 12 or more genes, tandemly duplicated at the same genetic locus (Brunette et al., 2012; Cridland et al., 2012; Nakaya et al., 2017). The human IFI16 and mouse IFI203 ALRs are required for the maximal response to retrovirus infection via the STING pathway in human and mouse cells, respectively (Jakobsen et al., 2013; Stavrou et al., 2015) (Fig. 11.2). Recently it has been suggested that IFI16 amplifies cGAS's ability to produce cGAMP, as well as to facilitate the recruitment of TBK1 to STING (Almine et al., 2017; Jonsson et al., 2017). Whether this also occurs with IFI203 in mouse cells has not been demonstrated.

Another sensor that participates in the cytosolic response to both DNA and DNA viruses is the dead-box helicase DDX41, which belongs to a larger family of RNA helicases most commonly associated with RNA binding (Zhang et al., 2011). Interestingly, DDX41 is highly conserved among mammals and is not induced by IFNs. Recently it has been shown that both familial and sporadic mutations in human DDX41 lead to acute myeloblastic leukemia and myelodysplastic syndromes, suggesting that it also functions as a tumor suppressor (Lewinsohn et al., 2015; Polprasert et al., 2015). Nonetheless, DDX41 depletion dampens STING-dependent IFN induction by HIV and MLV in mouse cells (Stavrou et al., 2015) (Figs. 11.2 and 11.3). DDX41 and IFI203 form a complex, and both bind STING but not cGAS (Stavrou et al., 2015). Importantly, both DDX41 and IFI203 appear to act independently of cGAS because addition of cGAMP to cells rescues cGAS but not IFI203 or DDX41 deficiency (Stavrou et al., 2018). Our data also suggest that DDX41 preferentially senses RNA/DNA hybrids in retrovirus-infected mouse cells, whereas cGAS senses dsDNA (Stavrou et al., 2018). Interestingly, we found that targeted loss of DDX41 in DCs but not macrophages rendered mice more susceptible to MLV infection than wild type mice, suggesting that DCs play a more important role in modulating the in vivo innate immune response. Moreover, mice with DC-specific loss of DDX41 were as susceptible to MLV infection as cGAS knockout mice, whereas mice deficient for the downstream effector STING were the most highly infected (Stavrou et al., 2018). Thus, multiple sensors of retroviral replication intermediates are required for maximal induction of the IFN response in vivo.

Myxovirus Resistance 2 (MX2) is another recently discovered IFN-induced host factor, which specifically blocks HIV and not MLV infection at the step of entry of proviral DNA into the nucleus (Goujon et al., 2013; Kane et al., 2013; Liu et al., 2013) (Fig. 11.1). MX2 is related to a well-known ISG, MX1, which is a GTPase of the dynamin superfamily and has potent antiviral activity against a number of viruses (Haller and Kochs, 2011). MX2 interacts with the HIV-1 capsid protein and blocks its interaction with the cellular proteins need for importation of the viral reverse transcription complex (RTC) into the nucleus. MX2's inability to inhibit MLV is likely due to both its inability to interact with the capsid protein and perhaps because there are different host factors required for nuclear import of HIV and MLV viral dsDNA. The MX2 blockade of nuclear entry could enhance the cytosolic sensor/STING-mediated response to HIV infection; by blocking entrance of proviral DNA into the nucleus, it potentially would both increase cGAS/IFI16 ligand levels as well as the time needed for the interaction of sensor and nucleic acid.

The dead-box helicase, DDX3, has also been implicated in the innate immune response to HIV infection. In the absence of the transcription elongation factor pTEF-b, newly integrated HIV-1 proviruses produce short HIV-1 transcripts that lack polyA tails. These transcripts are exported to the cytoplasm, where they encounter DDX3, which in turn binds to the adaptor protein mitochondrial antiviral signaling (MAVS), a well-known player in the response to cytosolic RNA sensors (Gringhuis et al., 2017; Johnson and Gale, 2006) (Figs. 11.2 and 11.3). MAVS-mediated signaling results in increased transcription of both IRF3 and NF-κB–dependent IFNs, chemokines, and cytokines (Fig. 11.2). Other cytosolic sensors of viral RNA, such as the RIG-I-like receptors, are probably not involved in recognition of retroviral RNA (Lee et al., 2013).

SUBVERSION OF CELLULAR IMMUNE RESPONSES BY RETROVIRUSES

As discussed above, viruses that successfully infect hosts must have the means to evade the intrinsic/innate responses. Retroviruses have evolved multiple means to avoid the host response. Table 11.1 lists some of the viral proteins and the host factors that they subvert.

Although there is some disagreement in the field, it is thought that reverse transcription and capsid disassembly are coordinated and that the capsid remains associated with the RTC until it docks at the nuclear pore and the viral dsDNA enters the nucleus (Ambrose and Aiken, 2014). One of the major ways that viruses can avoid detection, particularly by host cytosolic restriction factors, is through limiting accessibility. Viruses achieve this by maintaining their capsid structure. For example, gammaretroviruses make a protein termed glycosylated Gag, which stabilizes the capsid and protects the RTC from both APOBEC3 proteins and cytosolic nucleic acid sensors (Kolokithas et al., 2010; Rosales Gerpe et al., 2015; Stavrou et al., 2013, 2015; Renner et al., 2018). The HIV-1 capsid forms a complex with a cellular proteins, CypA and CPSF6, which mask viral nucleic acid from interaction with cGAS and other cytosolic sensors (Lahaye et al., 2013; Manel et al., 2010). In contrast, infection of sentinel cells by HIV-2 induces an innate immune response because the capsid allows reverse transcripts to be detected by these sensors (Lahaye et al., 2013; Manel et al., 2010). Amino acid changes in the HIV-1 capsid protein also render it resistant to sensing by TRIM5α (Kratovac et al., 2008). It is not known if glycoGag interacts with specific host proteins to block access to sensors.

Capsid stability appears to be the major means by which MLV blocks access of APOBEC3 proteins to the RTC; whether this is also the case for lentiviruses, or other retroviruses restricted by APOBEC3 proteins has not been tested. Lentiviruses encode a protein termed viral infectivity factor (Vif), however, which binds APOBEC3 proteins in virus producer cells and targets these proteins for ubiquitinylation and degradation in the proteosome, thereby preventing packaging into virions (Malim, 2009). Because Vif affects APOBEC3 proteins in virus producer cells, it would promote HIV spread in already-infected individuals rather that acting at the initial steps of infection. Nonetheless, as expression of APOBEC3 proteins is induced by type I IFNs, and these factors are one of the most effective host antiviral factors, increasing virus spread is clearly important to retrovirus persistence (Stavrou et al., 2015). There is also species-specific degradation of human APOBEC3 proteins by different Vifs.

As discussed in the preceding section, short aborted polyA(-) RNAs transcribed from newly integrated HIV-1 proviruses are recognized by DDX3 and activate the MAVS signaling pathway. HIV-1 Env gp120 interaction with DC-SIGN blocks IFN production via this pathway by suppressing MAVS signaling (Gringhuis et al., 2017). Not much is known about the interaction of other retroviruses and cell surface lectin receptors. MLV, as well as HIV, interacts with the I-type lectin CD169 on the surface of DCs and uses this for transporting endocytosed virions to lymphoid organs (Sewald et al., 2015). Whether this provides an intracellular signal or whether other retroviruses also generate abortive transcripts has not been studied.

Lentiviruses also encode additional proteins that counteract the other restriction factors. Many SIVs and HIV-2, but not HIV-1, encode a protein called Vpx, which binds to SAMHD1 and targets it for degradation (Etienne et al., 2013). Vpx is packaged into virions and thus could affect the initial infection of quiescent sentinel cells. In contrast, the lentiviral proteins that counteract tetherin likely act at later stages of infection in already-infected, virus-producing cells. These include the Vpu protein encoded in HIV-1 and some strains of SIV, Nef protein found in other strains of SIV, and the Env proteins found in HIV-2, FIV, and equine infectious anemia virus (Harris et al., 2012; Yin et al., 2014; Zielonka and Munk, 2011). While tetherin has clearly been shown to play a significant role in restricting a variety of retroviruses in vitro, in mouse knockout models it has only been shown to have modest effects on infection by MLV and MMTV (Jones et al., 2012; Liberatore and Bieniasz, 2011). Both MLV and MMTV induce low levels of type I IFNs in vivo, the result of APOBEC3-mediated inhibition of reverse transcription and the inability of host nucleic acid sensors to readily access the RTC. Because tetherin expression is low in the absence of type I IFNs, it likely is not a major factor in controlling infection by these murine viruses in vivo.

CONCLUSIONS

Many host factors contribute to resistance to retrovirus infection, including those that confer intrinsic and innate immunity. The wide variety of proteins and the mechanisms by which they block infection reflect the different cell types that are infected during the course of an in vivo infection as well as the need to have multiple means to protect the organism at multiple steps of the virus infection pathway. In addition to directly blocking the early steps of infection, the signal transduction cascades initiated by host recognition of the different retroviral proteins and nucleic acids results in antiviral autocrine and paracrine signals that prime bystander cells to resist infection, as well as to recruit the cells needed to generate adaptive immunity. Conversely, retroviruses have developed multiple means for overcoming these factors. Determining which host proteins play major roles in the initial antiviral response, as well as how viruses counteract these defenses, is critical to developing means to block the latent infection by retroviruses that leads to pathogenesis.

REFERENCES

Albrecht, M., Choubey, D., Lengauer, T., 2005. The HIN domain of IFI-200 proteins consists of two OB folds. Biochem. Biophys. Res. Commun. 327, 679–687.

Almine, J.F., O'Hare, C.A., Dunphy, G., Haga, I.R., Naik, R.J., Atrih, A., Connolly, D.J., Taylor, J., Kelsall, I.R., Bowie, A.G., Beard, P.M., Unterholzner, L., 2017. IFI16 and cGAS cooperate in the activation of STING during DNA sensing in human keratinocytes. Nat. Commun. 8, 14392.

Ambrose, Z., Aiken, C., 2014. HIV-1 uncoating: connection to nuclear entry and regulation by host proteins. Virology 454–455, 371–379.

Asefa, B., Klarmann, K.D., Copeland, N.G., Gilbert, D.J., Jenkins, N.A., Keller, J.R., 2004. The interferon-inducible p200 family of proteins: a perspective on their roles in cell cycle regulation and differentiation. Blood Cells Mol. Dis. 32, 155–167.

Baillie, G.J., van de Lagemaat, L.N., Baust, C., Mager, D.L., 2004. Multiple groups of endogenous betaretroviruses in mice, rats and other mammals. J. Virol. 78, 5784–5798.

Baldauf, H.M., Stegmann, L., Schwarz, S.M., Ambiel, I., Trotard, M., Martin, M., Burggraf, M., Lenzi, G.M., Lejk, H., Pan, X., Fregoso, O.I., Lim, E.S., Abraham, L., Nguyen, L.A., Rutsch, F., Konig, R., Kim, B., Emerman, M., Fackler, O.T., Keppler, O.T., 2017. Vpx overcomes a SAMHD1-independent block to HIV reverse transcription that is specific to resting CD4 T cells. Proc. Natl. Acad. Sci. U.S.A. 114, 2729–2734.

Barboric, M., Nissen, R.M., Kanazawa, S., Jabrane-Ferrat, N., Peterlin, B.M., 2001. NF-kappaB binds P-TEFb to stimulate transcriptional elongation by RNA polymerase II. Mol. Cell 8, 327–337.

Battini, J.L., Rasko, J.E., Miller, A.D., 1999. A human cell-surface receptor for xenotropic and polytropic murine leukemia viruses: possible role in G protein-coupled signal transduction. Proc. Natl. Acad. Sci. U.S.A. 96, 1385–1390.

Beignon, A.S., McKenna, K., Skoberne, M., Manches, O., DaSilva, I., Kavanagh, D.G., Larsson, M., Gorelick, R.J., Lifson, J.D., Bhardwaj, N., 2005. Endocytosis of HIV-1 activates plasmacytoid dendritic cells via Toll-like receptor-viral RNA interactions. J. Clin. Investig. 115, 3265–3275.

Berger, A., Sommer, A.F., Zwarg, J., Hamdorf, M., Welzel, K., Esly, N., Panitz, S., Reuter, A., Ramos, I., Jatiani, A., Mulder, L.C., Fernandez-Sesma, A., Rutsch, F., Simon, V., Konig, R., Flory, E., 2011. SAMHD1-deficient CD14+ cells from individuals with Aicardi-Goutieres syndrome are highly susceptible to HIV-1 infection. PLoS Pathog. 7, e1002425.

Bhat, N., Fitzgerald, K.A., 2014. Recognition of cytosolic DNA by cGAS and other STING-dependent sensors. Eur. J. Immunol. 44, 634–640.

Browne, E.P., 2011. Toll-like receptor 7 controls the anti-retroviral germinal center response. PLoS Pathog. 7, e1002293.

Browne, E.P., Littman, D.R., 2009. Myd88 is required for an antibody response to retroviral infection. PLoS Pathog. 5, e1000298.

Brunette, R.L., Young, J.M., Whitley, D.G., Brodsky, I.E., Malik, H.S., Stetson, D.B., 2012. Extensive evolutionary and functional diversity among mammalian AIM2-like receptors. J. Exp. Med. 209, 1969–1983.

Burzyn, D., Rassa, J.C., Kim, D., Nepomnaschy, I., Ross, S.R., Piazzon, I., 2004. Toll-like receptor 4-dependent activation of dendritic cells by a retrovirus. J. Virol. 78, 576–584.

Cai, X., Chiu, Y.H., Chen, Z.J., 2014. The cGAS-cGAMP-STING pathway of cytosolic DNA sensing and signaling. Mol. Cell 54, 289–296.

Chen, B.K., 2012. T cell virological synapses and HIV-1 pathogenesis. Immunol. Res. 54, 133–139.

Collman, R.G., Yi, Y., Liu, Q.H., Freedman, B.D., 2000. Chemokine signaling and HIV-1 fusion mediated by macrophage CXCR4: implications for target cell tropism. J. Leukoc. Biol. 68, 318–323.

Courreges, M.C., Burzyn, D., Nepomnaschy, I., Piazzon, I., Ross, S.R., 2007. Critical role of dendritic cells in mouse mammary tumor virus *in vivo* infection. J. Virol. 81, 3769–3777.

Cridland, J.A., Curley, E.Z., Wykes, M.N., Schroder, K., Sweet, M.J., Roberts, T.L., Ragan, M.A., Kassahn, K.S., Stacey, K.J., 2012. The mammalian PYHIN gene family: phylogeny, evolution and expression. BMC Evol. Biol. 12, 140.

Del Corno, M., Liu, Q.H., Schols, D., de Clercq, E., Gessani, S., Freedman, B.D., Collman, R.G., 2001. HIV-1 gp120 and chemokine activation of Pyk2 and mitogen-activated protein kinases in primary macrophages mediated by calcium-dependent, pertussis toxin-insensitive chemokine receptor signaling. Blood 98, 2909–2916.

Del Corno, M., Cappon, A., Donninelli, G., Varano, B., Marra, F., Gessani, S., 2016. HIV-1 gp120 signaling through TLR4 modulates innate immune activation in human macrophages and the biology of hepatic stellate cells. J. Leukoc. Biol. 100, 599–606.

Denner, J., 2016. Transspecies transmission of gammaretroviruses and the origin of the Gibbon Ape Leukaemia Virus (GaLV) and the Koala Retrovirus (KoRV). Viruses 8.

Dudley, J.P., Golovkina, T.V., Ross, S.R., 2016. Lessons learned from mouse mammary tumor virus in animal models. ILAR J. 57, 12–23.

Etienne, L., Hahn, B.H., Sharp, P.M., Matsen, F.A., Emerman, M., 2013. Gene loss and adaptation to hominids underlie the ancient origin of HIV-1. Cell Host Microbe 14, 85–92.

Fitzgerald, K.A., Rowe, D.C., Barnes, B.J., Caffrey, D.R., Visintin, A., Latz, E., Monks, B., Pitha, P.M., Golenbock, D.T., 2003. LPS-TLR4 signaling to IRF-3/7 and NF-kappaB involves the toll adapters TRAM and TRIF. J. Exp. Med. 198, 1043–1055.

Galao, R.P., Le Tortorec, A., Pickering, S., Kueck, T., Neil, S.J., 2012. Innate sensing of HIV-1 assembly by Tetherin induces NFkappaB-dependent proinflammatory responses. Cell Host Microbe 12, 633–644.

Gao, D., Wu, J., Wu, Y.T., Du, F., Aroh, C., Yan, N., Sun, L., Chen, Z.J., 2013. Cyclic GMP-AMP synthase is an innate immune sensor of HIV and other retroviruses. Science 341, 903–906.

Rosales Gerpe, M.C., Renner, T.M., Belanger, K., Lam, C., Aydin, H., Langlois, M.A., 2015. N-linked glycosylation protects gammaretroviruses against deamination by APOBEC3 proteins. J. Virol. 89, 2342–2357.

Gibbert, K., Francois, S., Sigmund, A.M., Harper, M.S., Barrett, B.S., Kirchning, C.J., Lu, M., Santiago, M.L., Dittmer, U., 2014. Friend retrovirus drives cytotoxic effectors through toll-like receptor 3. Retrovirology 11, 126.

Gillick, K., Pollpeter, D., Phalora, P., Kim, E.Y., Wolinsky, S.M., Malim, M.H., 2013. Suppression of HIV-1 infection by APOBEC3 proteins in primary human CD4(+) T cells is associated with inhibition of processive reverse transcription as well as excessive cytidine deamination. J. Virol. 87, 1508–1517.

Goujon, C., Moncorge, O., Bauby, H., Doyle, T., Ward, C.C., Schaller, T., Hue, S., Barclay, W.S., Schulz, R., Malim, M.H., 2013. Human MX2 is an interferon-induced post-entry inhibitor of HIV-1 infection. Nature 502, 559–562.

Gringhuis, S.I., den Dunnen, J., Litjens, M., van der Vlist, M., Geijtenbeek, T.B., 2009. Carbohydrate-specific signaling through the DC-SIGN signalosome tailors immunity to *Mycobacterium tuberculosis*, HIV-1 and *Helicobacter pylori*. Nat. Immunol. 10, 1081–1088.

Gringhuis, S.I., van der Vlist, M., van den Berg, L.M., den Dunnen, J., Litjens, M., Geijtenbeek, T.B., 2010. HIV-1 exploits innate signaling by TLR8 and DC-SIGN for productive infection of dendritic cells. Nat. Immunol. 11, 419–426.

Gringhuis, S.I., Hertoghs, N., Kaptein, T.M., Zijlstra-Willems, E.M., Sarrami-Fooroshani, R., Sprokholt, J.K., van Teijlingen, N.H., Kootstra, N.A., Booiman, T., van Dort, K.A., Ribeiro, C.M., Drewniak, A., Geijtenbeek, T.B., 2017. HIV-1 blocks the signaling adaptor MAVS to evade antiviral host defense after sensing of abortive HIV-1 RNA by the host helicase DDX3. Nat. Immunol. 18, 225–235.

Groom, H.C., Yap, M.W., Galao, R.P., Neil, S.J., Bishop, K.N., 2010. Susceptibility of xenotropic murine leukemia virus-related virus (XMRV) to retroviral restriction factors. Proc. Natl. Acad. Sci. U.S.A. 107, 5166–5171.

Haller, O., Kochs, G., 2011. Human MxA protein: an interferon-induced dynamin-like GTPase with broad antiviral activity. J. Interferon Cytokine Res. 31, 79–87.

Harris, R.S., Hultquist, J.F., Evans, D.T., 2012. The restriction factors of human immunodeficiency virus. J. Biol. Chem. 287, 40875–40883.

Heil, F., Hemmi, H., Hochrein, H., Ampenberger, F., Kirschning, C., Akira, S., Lipford, G., Wagner, H., Bauer, S., 2004. Species-specific recognition of single-stranded RNA via toll-like receptor 7 and 8. Science (New York, N.Y.) 303, 1526–1529.

Holmes, R.K., Koning, F.A., Bishop, K.N., Malim, M.H., 2007. APOBEC3F can inhibit the accumulation of HIV-1 reverse transcription products in the absence of hypermutation: comparisons with APOBEC3G. J. Biol. Chem. 282, 2587–2595.

Ishikawa, H., Ma, Z., Barber, G.N., 2009. STING regulates intracellular DNA-mediated, type I interferon-dependent innate immunity. Nature 461, 788–792.

Jakobsen, M.R., Bak, R.O., Andersen, A., Berg, R.K., Jensen, S.B., Tengchuan, J., Laustsen, A., Hansen, K., Ostergaard, L., Fitzgerald, K.A., Xiao, T.S., Mikkelsen, J.G., Mogensen, T.H., Paludan, S.R., 2013. IFI16 senses DNA forms of the lentiviral replication cycle and controls HIV-1 replication. Proc. Natl. Acad. Sci. U.S.A. 110, E4571–E4580.

Johnson, C.L., Gale Jr., M., 2006. CARD games between virus and host get a new player. Trends Immunol. 27, 1–4.

Jones, P.H., Mehta, H.V., Maric, M., Roller, R.J., Okeoma, C.M., 2012. Bone marrow stromal cell antigen 2 (BST-2) restricts mouse mammary tumor virus (MMTV) replication in vivo. Retrovirology 9, 10.

Jonsson, K.L., Laustsen, A., Krapp, C., Skipper, K.A., Thavachelvam, K., Hotter, D., Egedal, J.H., Kjolby, M., Mohammadi, P., Prabakaran, T., Sorensen, L.K., Sun, C., Jensen, S.B., Holm, C.K., Lebbink, R.J., Johannsen, M., Nyegaard, M., Mikkelsen, J.G., Kirchhoff, F., Paludan, S.R., Jakobsen, M.R., 2017. IFI16 is required for DNA sensing in human macrophages by promoting production and function of cGAMP. Nat. Commun. 8, 14391.

Jude, B.A., Pobezinskaya, Y., Bishop, J., Parke, S., Medzhitov, R.M., Chervonsky, A.V., Golovkina, T.V., 2003. Subversion of the innate immune system by a retrovirus. Nat. Immunol. 4, 573–578.

Kane, M., Case, L.K., Wang, C., Yurkovetskiy, L., Dikiy, S., Golovkina, T.V., 2011. Innate immune sensing of retroviral infection via Toll-like receptor 7 occurs upon viral entry. Immunity 35, 135–145.

Kane, M., Yadav, S.S., Bitzegeio, J., Kutluay, S.B., Zang, T., Wilson, S.J., Schoggins, J.W., Rice, C.M., Yamashita, M., Hatziioannou, T., Bieniasz, P.D., 2013. MX2 is an interferon-induced inhibitor of HIV-1 infection. Nature 502, 563–566.

Kirchhoff, F., 2010. Immune evasion and counteraction of restriction factors by HIV-1 and other primate lentiviruses. Cell Host Microbe 8, 55–67.

Kolokithas, A., Rosenke, K., Malik, F., Hendrick, D., Swanson, L., Santiago, M.L., Portis, J.L., Hasenkrug, K.J., Evans, L.H., 2010. The glycosylated Gag protein of a murine leukemia virus inhibits the antiretroviral function of APOBEC3. J. Virol. 84, 10933–10936.

Koning, F.A., Goujon, C., Bauby, H., Malim, M.H., 2011. Target cell-mediated editing of HIV-1 cDNA by APOBEC3 proteins in human macrophages. J. Virol. 85, 13448–13452.

Koppensteiner, H., Brack-Werner, R., Schindler, M., 2012. Macrophages and their relevance in human immunodeficiency virus type I infection. Retrovirology 9, 82.

Kratovac, Z., Virgen, C.A., Bibollet-Ruche, F., Hahn, B.H., Bieniasz, P.D., Hatziioannou, T., 2008. Primate lentivirus capsid sensitivity to TRIM5 proteins. J. Virol. 82, 6772–6777.

Laguette, N., Sobhian, B., Casartelli, N., Ringeard, M., Chable-Bessia, C., Segeral, E., Yatim, A., Emiliani, S., Schwartz, O., Benkirane, M., 2011. SAMHD1 is the dendritic- and myeloid-cell-specific HIV-1 restriction factor counteracted by Vpx. Nature 474, 654–657.

Lahaye, X., Satoh, T., Gentili, M., Cerboni, S., Conrad, C., Hurbain, I., El Marjou, A., Lacabaratz, C., Lelievre, J.D., Manel, N., 2013. The capsids of HIV-1 and HIV-2 determine immune detection of the viral cDNA by the innate sensor cGAS in dendritic cells. Immunity 39, 1132–1142.

Lahouassa, H., Daddacha, W., Hofmann, H., Ayinde, D., Logue, E.C., Dragin, L., Bloch, N., Maudet, C., Bertrand, M., Gramberg, T., Pancino, G., Priet, S., Canard, B., Laguette, N., Benkirane, M., Transy, C., Landau, N.R., Kim, B., Margottin-Goguet, F., 2012. SAMHD1 restricts the replication of human immunodeficiency virus type 1 by depleting the intracellular pool of deoxynucleoside triphosphates. Nat. Immunol. 13, 223–228.

Lee, H., Komano, J., Saitoh, Y., Yamaoka, S., Kozaki, T., Misawa, T., Takahama, M., Satoh, T., Takeuchi, O., Yamamoto, N., Matsuura, Y., Saitoh, T., Akira, S., 2013. Zinc-finger antiviral protein mediates retinoic acid inducible gene I-like receptor-independent antiviral response to murine leukemia virus. Proc. Natl. Acad. Sci. U.S.A. 110, 12379–12384.

Lewinsohn, M., Brown, A.L., Weinel, L.M., Phung, C., Rafidi, G., Lee, M.K., Schreiber, A.W., Feng, J., Babic, M., Chong, C.E., Lee, Y., Yong, A., Suthers, G.K., Poplawski, N., Altree, M., Phillips, K., Jaensch, L., Fine, M., D'Andrea, R.J., Lewis, I.D., Medeiros, B.C., Pollyea, D.A., King, M.C., Walsh, T., Keel, S., Shimamura, A., Godley, L.A., Hahn, C.N., Churpek, J.N., Scott, H.S., 2015. Novel germline DDX41 mutations define families with a lower age of MDS/AML onset, and lymphoid malignancies. Blood 127, 1017–1023.

Li, J.P., D'Andrea, A., Lodish, H., Baltimore, D., 1990. Activation of cell growth by binding of Friend spleen focus-forming virus gp55 glycoprotein to the erythropoietin receptor. Nature 343, 762–764.

Li, S.X., Barrett, B.S., Guo, K., Santiago, M.L., 2016. Tetherin/BST-2: restriction factor or immunomodulator? Curr. HIV Res. 14, 235–246.

Liberatore, R.A., Bieniasz, P.D., 2011. Tetherin is a key effector of the antiretroviral activity of type I interferon in vitro and in vivo. Proc. Natl. Acad. Sci. U.S.A. 108, 18097–18101.

Lindahl, T., Gally, J.A., Edelman, G.M., 1969. Properties of deoxyribonuclease 3 from mammalian tissues. J. Biol. Chem. 244, 5014–5019.

Liu, Q.H., Williams, D.A., McManus, C., Baribaud, F., Doms, R.W., Schols, D., De Clercq, E., Kotlikoff, M.I., Collman, R.G., Freedman, B.D., 2000. HIV-1 gp120 and chemokines activate ion channels in primary macrophages through CCR5 and CXCR4 stimulation. Proc. Natl. Acad. Sci. U.S.A. 97, 4832–4837.

Liu, Z., Pan, Q., Ding, S., Qian, J., Xu, F., Zhou, J., Cen, S., Guo, F., Liang, C., 2013. The interferon-inducible MxB protein inhibits HIV-1 infection. Cell Host Microbe 14, 398–410.

Low, A., Okeoma, C.M., Lovsin, N., de las Heras, M., Taylor, T.H., Peterlin, B.M., Ross, S.R., Fan, H., 2009. Enhanced replication and pathogenesis of Moloney murine leukemia virus in mice defective in the murine APOBEC3 gene. Virology 385, 455–463.

Luecke, S., Paludan, S.R., 2016. Molecular requirements for sensing of intracellular microbial nucleic acids by the innate immune system. Cytokine 98, 4–14.

MacMillan, A.L., Kohli, R.M., Ross, S.R., 2013. APOBEC3 inhibition of mouse mammary tumor virus infection: the role of cytidine deamination versus inhibition of reverse transcription. J. Virol. 87, 4808–4817.

Mager, D.L., Stoye, J.P., 2015. Mammalian endogenous retroviruses. Microbiol. Spectr. 3 MDNA3-0009-2014.

Maksakova, I.A., Romanish, M.T., Gagnier, L., Dunn, C.A., van de Lagemaat, L.N., Mager, D.L., 2006. Retroviral elements and their hosts: insertional mutagenesis in the mouse germ line. PLoS Genet. 2, e2.

Malim, M.H., 2009. APOBEC proteins and intrinsic resistance to HIV-1 infection. Philos. Trans. R. Soc. Lond. B Biol. Sci. 364, 675–687.

Manches, O., Frleta, D., Bhardwaj, N., 2014. Dendritic cells in progression and pathology of HIV infection. Trends Immunol. 35, 114–122.

Manel, N., Hogstad, B., Wang, Y., Levy, D.E., Unutmaz, D., Littman, D.R., 2010. A cryptic sensor for HIV-1 activates antiviral innate immunity in dendritic cells. Nature 467, 214–217.

Meier, A., Alter, G., Frahm, N., Sidhu, H., Li, B., Bagchi, A., Teigen, N., Streeck, H., Stellbrink, H.J., Hellman, J., van Lunzen, J., Altfeld, M., 2007. MyD88-dependent immune activation mediated by human immunodeficiency virus type 1-encoded Toll-like receptor ligands. J. Virol. 81, 8180–8191.

Mukai, K., Konno, H., Akiba, T., Uemura, T., Waguri, S., Kobayashi, T., Barber, G.N., Arai, H., Taguchi, T., 2016. Activation of STING requires palmitoylation at the Golgi. Nat. Commun. 7, 11932.

Nakaya, Y., Lilue, J., Stavrou, S., Moran, E., Ross, S.R., 2017. AIM2-like receptors positively and negatively regulate the interferon response induced by cytosolic DNA. MBio. 8, 4. https://doi.org/10.1128/mBio.00944-17.

Nazli, A., Kafka, J.K., Ferreira, V.H., Anipindi, V., Mueller, K., Osborne, B.J., Dizzell, S., Chauvin, S., Mian, M.F., Ouellet, M., Tremblay, M.J., Mossman, K.L., Ashkar, A.A., Kovacs, C., Bowdish, D.M., Snider, D.P., Kaul, R., Kaushic, C., 2013. HIV-1 gp120 induces TLR2- and TLR4-mediated innate immune activation in human female genital epithelium. J. Immunol. 191, 4246–4258.

O'Neill, L.A.J., 2000. The interleukin-1 receptor/toll-like receptor superfamily: signal transduction during inflammation and host defense. Sci. STKE 1–11.

Okeoma, C.M., Lovsin, N., Peterlin, B.M., Ross, S.R., 2007. APOBEC3 inhibits mouse mammary tumour virus replication in vivo. Nature 445, 927–930.

Okeoma, C.M., Low, A., Bailis, W., Fan, H.Y., Peterlin, B.M., Ross, S.R., 2009. Induction of APOBEC3 in vivo causes increased restriction of retrovirus infection. J. Virol. 83, 3486–3495.

Overbaugh, J., Miller, A.D., Eiden, M.V., 2001. Receptors and entry cofactors for retroviruses include single and multiple transmembrane-spanning proteins as well as newly described glycophosphatidylinositol-anchored and secreted proteins. Microbiol. Mol. Biol. Rev. 65, 371–389.

Panfil, A.R., Martinez, M.P., Ratner, L., Green, P.L., 2016. Human T-cell leukemia virus-associated malignancy. Curr. Opin. Virol. 20, 40–46.

Pertel, T., Hausmann, S., Morger, D., Zuger, S., Guerra, J., Lascano, J., Reinhard, C., Santoni, F.A., Uchil, P.D., Chatel, L., Bisiaux, A., Albert, M.L., Strambio-De-Castillia, C., Mothes, W., Pizzato, M., Grutter, M.G., Luban, J., 2011. TRIM5 is an innate immune sensor for the retrovirus capsid lattice. Nature 472, 361–365.

Polprasert, C., Schulze, I., Sekeres, M.A., Makishima, H., Przychodzen, B., Hosono, N., Singh, J., Padgett, R.A., Gu, X., Phillips, J.G., Clemente, M., Parker, Y., Lindner, D., Dienes, B., Jankowsky, E., Saunthararajah, Y., Du, Y., Oakley, K., Nguyen, N., Mukherjee, S., Pabst, C., Godley, L.A., Churpek, J.E., Pollyea, D.A., Krug, U., Berdel, W.E., Klein, H.U., Dugas, M., Shiraishi, Y., Chiba, K., Tanaka, H., Miyano, S., Yoshida, K., Ogawa, S., Muller-Tidow, C., Maciejewski, J.P., 2015. Inherited and somatic defects in DDX41 in myeloid neoplasms. Canc. Cell 27, 658–670.

Popik, W., Hesselgesser, J.E., Pitha, P.M., 1998. Binding of human immunodeficiency virus type 1 to CD4 and CXCR4 receptors differentially regulates expression of inflammatory genes and activates the MEK/ERK signaling pathway. J. Virol. 72, 6406–6413.

Rassa, J.C., Meyers, J.L., Zhang, Y., Kudaravalli, R., Ross, S.R., 2002. Murine retroviruses activate B cells via interaction with Toll-like receptor 4. Proc. Natl. Acad. Sci. U.S.A. 99, 2281–2286.

Renner, T.M., Bélanger, K., Lam, C., Gerpe, M.C.R., McBane, J.E., Langlois, M.A., 2018. Full-length glycosylated Gag of murine leukemia virus can associate with the viral envelope as a type I integral membrane protein. J. Virol. 92, e01530–17.

Rehwinkel, J., Maelfait, J., Bridgeman, A., Rigby, R., Hayward, B., Liberatore, R.A., Bieniasz, P.D., Towers, G.J., Moita, L.F., Crow, Y.J., Bonthron, D.T., Reis e Sousa, C., 2013. SAMHD1-dependent retroviral control and escape in mice. EMBO J. 32, 2454–2462.

Ribeiro, C.M., Sarrami-Forooshani, R., Setiawan, L.C., Zijlstra-Willems, E.M., van Hamme, J.L., Tigchelaar, W., van der Wel, N.N., Kootstra, N.A., Gringhuis, S.I., Geijtenbeek, T.B., 2016. Receptor usage dictates HIV-1 restriction by human TRIM5alpha in dendritic cell subsets. Nature 540, 448–452.

Ross, S.R., Schofield, J.J., Farr, C.J., Bucan, M., 2002. Mouse transferrin receptor 1 is the cell entry receptor for mouse mammary tumor virus. Proc. Natl. Acad. Sci. U.S.A. 99, 12386–12390.

Saitoh, T., Fujita, N., Hayashi, T., Takahara, K., Satoh, T., Lee, H., Matsunaga, K., Kageyama, S., Omori, H., Noda, T., Yamamoto, N., Kawai, T., Ishii, K., Takeuchi, O., Yoshimori, T., Akira, S., 2009. Atg9a controls dsDNA-driven dynamic translocation of STING and the innate immune response. Proc. Natl. Acad. Sci. U.S.A. 106, 20842–20846.

Santiago, M.L., Montano, M., Benitez, R., Messer, R.J., Yonemoto, W., Chesebro, B., Hasenkrug, K.J., Greene, W.C., 2008. Apobec3 encodes Rfv3, a gene influencing neutralizing antibody control of retrovirus infection. Science 321, 1343–1346.

Sanz-Ramos, M., Stoye, J.P., 2013. Capsid-binding retrovirus restriction factors: discovery, restriction specificity and implications for the development of novel therapeutics. J. Gen. Virol. 94, 2587–2598.

Sewald, X., Ladinsky, M.S., Uchil, P.D., Beloor, J., Pi, R., Herrmann, C., Motamedi, N., Murooka, T.T., Brehm, M.A., Greiner, D.L., Shultz, L.D., Mempel, T.R., Bjorkman, P.J., Kumar, P., Mothes, W., 2015. Retroviruses use CD169-mediated trans-infection of permissive lymphocytes to establish infection. Science 350, 563–567.

Sharp, P.M., Hahn, B.H., 2011. Origins of HIV and the AIDS pandemic. Cold Spring Harb. Perspect. Med. 1, a006841.

Stavrou, S., Ross, S.R., 2015. APOBEC3 proteins in viral immunity. J. Immunol. 195, 4565–4570.

Stavrou, S., Nitta, T., Kotla, S., Ha, D., Nagashima, K., Rein, A.R., Fan, H., Ross, S.R., 2013. Murine leukemia virus glycosylated Gag blocks apolipoprotein B editing complex 3 and cytosolic sensor access to the reverse transcription complex. Proc. Natl. Acad. Sci. U.S.A. 110, 9078–9083.

Stavrou, S., Blouch, K., Kotla, S., Bass, A., Ross, S.R., 2015. Nucleic acid recognition orchestrates the anti-viral response to retroviruses. Cell Host Microbe 17, 478–488.

Stavrou, S., Aguilera, A., Blouch, K., Ross, S.R., 2018. DDX41 senses retroviral RNA/DNA hybrids generated during reverse transcription and is critical for the control of *in vivo* MLV infection. mBIO. pii: e00923–18. https://doi.org/10.1128/mBio.00923-18.

Sun, L., Wu, J., Du, F., Chen, X., Chen, Z.J., 2013. Cyclic GMP-AMP synthase is a cytosolic DNA sensor that activates the type I interferon pathway. Science 339, 786–791.

Takeda, E., Tsuji-Kawahara, S., Sakamoto, M., Langlois, M.A., Neuberger, M.S., Rada, C., Miyazawa, M., 2008. Mouse APOBEC3 restricts friend leukemia virus infection and pathogenesis in vivo. J. Virol. 82, 10998–11008.

Van Breedam, W., Pohlmann, S., Favoreel, H.W., de Groot, R.J., Nauwynck, H.J., 2014. Bitter-sweet symphony: glycan-lectin interactions in virus biology. FEMS Microbiol. Rev. 38, 598–632.

Wilks, J., Lien, E., Jacobson, A.N., Fischbach, M.A., Qureshi, N., Chervonsky, A.V., Golovkina, T.V., 2015. Mammalian lipopolysaccharide receptors incorporated into the retroviral envelope augment virus transmission. Cell Host Microbe 18, 456–462.

Yan, N., Regalado-Magdos, A.D., Stiggelbout, B., Lee-Kirsch, M.A., Lieberman, J., 2010. The cytosolic exonuclease TREX1 inhibits the innate immune response to human immunodeficiency virus type 1. Nat. Immunol. 11, 1005–1013.

Yin, X., Hu, Z., Gu, Q., Wu, X., Zheng, Y.H., Wei, P., Wang, X., 2014. Equine tetherin blocks retrovirus release and its activity is antagonized by equine infectious anemia virus envelope protein. J. Virol. 88, 1259–1270.

Yoh, S.M., Schneider, M., Seifried, J., Soonthornvacharin, S., Akleh, R.E., Olivieri, K.C., De Jesus, P.D., Ruan, C., de Castro, E., Ruiz, P.A., Germanaud, D., des Portes, V., Garcia-Sastre, A., Konig, R., Chanda, S.K., 2015. PQBP1 is a proximal sensor of the cGAS-dependent innate response to HIV-1. Cell 161, 1293–1305.

Yuan, F., Dutta, T., Wang, L., Song, L., Gu, L., Qian, L., Benitez, A., Ning, S., Malhotra, A., Deutscher, M.P., Zhang, Y., 2015. Human DNA exonuclease TREX1 is also an exoribonuclease that acts on single-stranded RNA. J. Biol. Chem. 290, 13344–13353.

Zhang, Z., Yuan, B., Bao, M., Lu, N., Kim, T., Liu, Y.-J., 2011. The helicase DDX41 senses intracellular DNA mediated by the adaptor STING in dendritic cells. Nat. Immunol. 12, 959–965.

Zielonka, J., Munk, C., 2011. Cellular restriction factors of feline immunodeficiency virus. Viruses 3, 1986–2005.

Chapter 12

Noncoding RNAs in Retrovirus Replication

Alice A. Duchon, Karin Musier-Forsyth
The Ohio State University, Columbus, OH, United States

Chapter Outline

Introduction 421
Host RNAs Packaged Into Virions 423
Transfer RNAs 423
7SL RNA 430
U Small Nuclear RNA 431
Y RNA 432
Vault RNA 432
High-Throughput RNA-Sequencing Studies 433
Long Noncoding RNA and Endogenous Retroviruses 434
NRON: HIV-1 434
NEAT1: HIV-1 435
Large Intergenic RNA–p21: HIV-1 435
Antisense Protein RNA: HIV-1 436
Long Noncoding RNA 00173: HIV-1 436
ASP RNA: HTLV-1 436
Bic: ALV and His-1: MLV 437
Endogenous Retroviruses 437
Functional Transactivating Response Element RNAs 439
RNA Interference and miRNAs 440
RNAi, miRNA, and siRNA Production 441
Cellular MicroRNAs Made in Response to Viral Infections 444
Viral MicroRNAs Encoded Within Genomes 446
Viral Suppressors of RNA Silencing 450
Therapeutics 450
Concluding Remarks 453
Acknowledgments 453
References 453

INTRODUCTION

Prior to the Human Genome Project, it was estimated that human genomes contained roughly 100,000 genes encoding proteins (Taft et al., 2007). In actuality, humans possess only ~20,000 protein-coding genes, which is 10,000 more than the fruit fly, *Drosophila melanogaster*, and 15,000 fewer than rice, negating the hypothesis that the number of genes correlated with complexity of the organism (Mattick, 2001). It is now apparent that only 2% of the human genome codes for proteins, whereas the remaining 98% is composed of endogenous retroviruses (ERVs) and noncoding RNA (ncRNA) (Frias-Lasserre, 2012).

Retrovirus-Cell Interactions. https://doi.org/10.1016/B978-0-12-811185-7.00012-1

Although a limited percentage of the genome produces proteins, approximately 90% is transcribed, indicating important roles for ncRNA (Veneziano et al., 2016; Birney et al., 2007; Elgar and Vavouri, 2008). The term "ncRNA" applies to any RNA that is not translated into protein (Mattick and Makunin, 2006). Generally, ncRNAs are divided into groups based on their size and activity, although other classification systems are sometimes used (Mattick and Makunin, 2006; Geisler and Coller, 2013; St Laurent et al., 2015). Ribosomal RNA (rRNA) and transfer RNA (tRNA) are the most well-characterized species of ncRNA, which are both integral components of mRNA translation (Mattick and Makunin, 2006). The untranslated regions (UTRs) of mRNAs can also be classified as ncRNA, and play critical roles in the regulation of mRNA translation, as well as viral life cycles (Mattick and Makunin, 2006). Small ncRNAs (generally <200 nt) include small nuclear RNA (snRNA), microRNA (miRNA), tRNA-derived fragments (tRFs), piwi-interacting RNA (piRNA), repeat-associated small interfering RNA, and Y RNAs (Veneziano et al., 2016; Mattick and Makunin, 2006; Berkhout and Jeang, 2007). Long ncRNAs (lncRNAs) are longer than 200 nts, more loosely classified, and less well understood (Geisler and Coller, 2013; Mercer et al., 2009; Rinn and Chang, 2012). For more in-depth analysis of small ncRNA and lncRNA nomenclature and functions see reviews by Mattick and Makunin (2006) and St Laurent et al. (2015).

Because viral genomes are compact, the evolutionary conservation of proteins and ncRNA underscores their importance in the viral life cycle (Tycowski et al., 2015). The first viral ncRNAs were identified in 1966 and were isolated from the adenovirus (Reich et al., 1966). These RNAs were found to counteract protein kinase R (PKR) through competitive binding with viral dsRNA transcripts, blocking the host's ability to sense infection (de Haro et al., 1996). ncRNAs have been found to have functions at nearly every step of viral replication cycles (Cullen, 2009). For example, many viruses use internal ribosomal entry sites (IRESs) to preferentially translate viral mRNA over host mRNA (Pelletier and Sonenberg, 1988; Jang et al., 1988). Some retroviruses and all coronaviruses exploit RNA structures to cause ribosomal frameshifting resulting in the Gag-Pol and 1a/1b polyproteins, respectively (Brierley and Dos Ramos, 2006). Specific RNA structures within viral RNA can also enhance transcription and exportation of viral RNA from the nucleus (Fisher et al., 1986; Kao et al., 1987; Feng and Holland, 1988; Malim et al., 1989). RNA interference (RNAi), thought to be an early version of the innate immune response to viral infection, can still be observed in plants and invertebrates today, as discussed in more detail later in this chapter (Cullen, 2009; Voinnet, 2001). Many viruses produce miRNAs to counteract host responses and defenses (Tycowski et al., 2015), and retroviruses and long terminal repeat (LTR) retrotransposons use host tRNAs to prime reverse transcription (Waters and Mullin, 1977; Mak and Kleiman, 1997). This chapter will focus on ncRNAs of both host and viral origin and their roles in retroviral replication cycles.

HOST RNAs PACKAGED INTO VIRIONS

It is estimated that RNA constitutes approximately 50% of the mass of retroviral particles; roughly half of retroviral RNA by weight is cellular RNA (Bonar et al., 1967; Berkowitz et al., 1996; Bonar and Beard, 1959; Eckwahl et al., 2016b). The most commonly found RNAs packaged into retrovirus virions, excluding viral RNA and tRNAs, are host mRNA, 7SL RNA, U snRNA, Y RNA, and vault RNA (typically abbreviated vtRNA or vRNA, but here it is represented by vltRNA to avoid confusion with viral tRNAs and viral RNA), a majority of which are ncRNA (Bonar et al., 1967; Telesnitsky and Wolin, 2016; Onafuwa-Nuga et al., 2005, 2006; Eckwahl et al., 2015; Bishop et al., 1970; Giles et al., 2004; Garcia et al., 2009; Ali et al., 2016). Host mRNAs are a minority species of RNA packaged into retrovirus virions (Onafuwa-Nuga et al., 2006; Harris et al., 2003; Muckenfuss et al., 2006). Generally, the amount of packaged mRNA is proportional to the concentration within the host cell in the absence of viral genomic RNA (Kiss, 2004). In addition, mRNAs packaged in place of genomic RNA tend to have longer 3′ UTRs, a common feature of retroviral genomic RNA, suggesting the length of the 3′UTRs may be a determinant for packaging (Kiss, 2004; Matera et al., 2007). Initially, it was hypothesized that packaging of cellular RNAs was either random or a contamination of the purified virions, and that these RNAs did not play a significant role in the retrovirus life cycle (Bonar et al., 1967; Wollmann and Kirsten, 1968; Harel et al., 1965). However, as more roles for ncRNA in the cell began to emerge it became clear that these RNAs may indeed participate in the viral life cycle. Below, the most common RNAs packaged into virions will be described and potential packaging mechanisms will be briefly discussed.

Transfer RNAs

During reverse transcription, retroviruses, retrotransposons, and plant viruses known as pararetroviruses, which replicate through an RNA intermediate and are also reverse transcribed, use host tRNAs as primers for cDNA synthesis (Mak and Kleiman, 1997; Guilley et al., 1983; Medberry et al., 1990; Hay et al., 1991; Marquet et al., 1995). Each virus contains a primer-binding site (PBS) in their 5′ UTR that is complementary to 8–18 nucleotides (nt) of a host tRNA (Mak and Kleiman, 1997; Marquet et al., 1995) (Table 12.1). Generally, the 3′ acceptor stem of the tRNA is unwound and annealed to the PBS (Marquet et al., 1995). However, for some members of the retrotransposon Ty1/copia and Ty5 groups, priming begins at an internal site of the tRNA overlapping the anticodon stem loop (Voytas and Boeke, 1993). Interestingly, T1 retrotransposons found in *D. melanogaster* were directly shown to be primed by a fragment of $tRNA_i^{Met}$ instead of the full-length tRNA that is thought to be the primer in most other cases (Marquet et al., 1995; Voytas and Boeke, 1993; Sandmeyer and Menees, 1996). It has been suggested that a fragment of $tRNA^{Pro}$ (tRF-3019) could also be substituted for the primer tRNA in human T-cell leukemia virus (HTLV) (Ruggero et al., 2014; Seiki et al., 1982). tRF-3019 produced from the

TABLE 12.1 Proposed Host Transfer RNAs (tRNAs) Used as Primers for Retroviral Reverse Transcription

Genera	Virus	Host Species	Primer	References
Alpharetrovirus	ALV/RSV	Avian	$tRNA^{Trp}$	Sawyer and Dahlberg (1973)
Betaretrovirus	MMTV	Murine	$tRNA^{Lys3a}$	Majors and Varmus (1983)
	Mason–Pfizer monkey virus	Macaca	$tRNA^{Lys1,2}$	Leis (1993)
Gammaretrovirus	MLV	Murine	$tRNA^{Pro,\ Glu}$	Taylor (1977) and Colicelli and Goff (1986)
	GALV	Primate	$tRNA^{Pro}$	Delassus et al. (1989)
	FLV	Feline	$tRNA^{Pro}$	Laprevotte et al. (1984)
Deltaretrovirus	BLV	Bovine	$tRNA^{Pro}$	Couez et al. (1984)
	HTLV	*H. sapiens*	$tRNA^{Pro}$	Seiki et al. (1983)
Epsilonretrovirus	Walleye dermal sarcoma	*Sander vitreus*	$tRNA^{His}$	Holzschu et al. (1995)
Lentivirus	HIV-1, HIV-2	*H. sapiens*	$tRNA^{Lys3}$	Wain-Hobson et al. (1985)
	SIV	Primate	$tRNA^{Lys3}$	Berkhout (1997)
	EIAV	Equine	$tRNA^{Lys3}$	Kawakami et al. (1987)
	FIV	Feline	$tRNA^{Lys3}$	Miller et al. (2001)
Spumavirus	SFV	Primate	$tRNA^{Lys1,2}$	Rabin et al. (1976) and Schnitzer (1981)
	PFV	Primate	$tRNA^{Lys1}$	Maurer et al. (1988)
	EFV	Equine	$tRNA^{Lys1}$	Tobaly-Tapiero et al. (2000)
	FFV	Feline	$tRNA^{Lys1}$	Riggs et al. (1969)
	BFV	Bovine	$tRNA^{Lys1}$	Malmquist et al. (1969)

Genera	Virus	Host Species	Primer	References
Endogenous	CHIA34	Chinese Hamster	$tRNA^{Phe}$	Lie et al. (1994)
	PERV	Porcine	$tRNA^{Gly, Pro}$	Akiyoshi et al. (1998) and Bartosch et al. (2002)
	Squirrel Monkey	Saimirinae	$tRNA^{Lys1,2}$	Chiu and Skuntz (1986)
	Jaagsiekte sheep	*Ovis aries*	$tRNA^{Lys1,2}$	Cann (2001)
	HERV	*H. sapiens*	$tRNA^{Lysb}$	Tristem (2000)
	AKV	AKR mice	$tRNA^{Glu}$	Nikbakht et al. (1985)
	KoRV	Phascolarctos	$tRNA^{Pro}$	Hanger et al. (2000)
	MINERVa	Miniopteridae	$tRNA_{Pro}$	Farkasova et al. (2017)
	Rhinolophus ferrumequinum retrovirus (RfRV)	*R. ferrumequinum*	$tRNA^{Pro}$	Cui et al. (2012)

Continued

TABLE 12.1 Proposed Host Transfer RNAs (tRNAs) Used as Primers for Retroviral Reverse Transcription—cont'd

Genera	Virus	Host Species	Primer	References
Retrotransposons	Ty1, Ty2	*Saccharomyces cerevisiae*	$tRNA_i^{Met}$	Friant et al. (1997)
	tom	*Drosophila*	$tRNA^{Ser}$	Stefanov et al. (2012)
	Ty4	*S. cerevisiae*	$tRNA^{Asn}$	Voytas and Boeke (1993)
	Ty5	*S. cerevisiae*	$tRNA_i^{Met}$-fragment	Voytas and Boeke (1993)
	Copia	*Drosophila*	$tRNA_i^{Met}$-fragment	Kikuchi et al. (1986)
	Gypsy	*Drosophila*	$tRNA^{Lys}$	Sandmeyer and Menees (1996) and Springer and Britten (1993)
	mdg1 and 412	*Drosophila*	$tRNA^{Arg}$	Tubio et al. (2004)
	Elements 297 and 17.6	*Drosophila*	$tRNA^{Ser}$	Marquet et al. (1995)
	mdg.3	*Drosophila*	$tRNA^{Leu}$	Marquet et al. (1995)
	Ty3	*S. cerevisiae*	$tRNA_i^{Met}$	Pochart et al. (1993)

ALV, avian leukosis virus; *BFV*, bovine foamy virus; *BLV*, bovine leukemia virus; *EFV*, equine foamy virus; *EIAV*, equine infectious anemia virus; *FFV*, feline foamy virus; *FIV*, feline immunodeficiency virus; *FLV*, feline leukemia virus; *GALV*, gibbon ape leukemia virus; *HERV*, human endogenous retrovirus; *HTLV*, human T-cell leukemia lymphoma virus; *KoRV*, Koala retrovirus; *MLV*, murine leukemia virus; *RSV*, Rous sarcoma virus; *SIV*, simian immunodeficiency virus; *SFV*, simian foamy virus; *PFV*, prototype foamy virus.

[a] *Mammalian $tRNA^{Lys}$ has two isoacceptors, CUU and UUU. The human genome encodes at least 44 putative $tRNA^{Lys}$ species (Lowe and Eddy, 1997-http://gtrnadb.ucsc.edu) that have slight variation outside of the anticodon, i.e., isodecoders. Therefore, $tRNA^{Lys1,2}$ represents $tRNA^{Lys,CUU}$ with 24 different isodecoders and $tRNA^{Lys3}$ represents $tRNA^{Lys,UUU}$ with 20 different isodecoders. Whether all of these are expressed and function as tRNAs in translation or as primers is unknown.*

[b] *tRNA primer varies depending on specific HERV species, however, many use $tRNA^{Lys}$ isoacceptors.*

3′ acceptor stem was found to prime RT in vitro and was detected in viral particles collected from HTLV-infected cells (Ruggero et al., 2014).

Low levels of tRFs are present under homeostatic conditions, however specific cleavage of tRNAs occurs under stress conditions (oxidative stress, heat shock, UV radiation, and viral infection) (Anderson and Ivanov, 2014; Sobala and Hutvagner, 2011). The biological function of at least some tRFs is thought to involve the RNA silencing pathway, and exogenous expression of the 5′ acceptor stem fragments results in a 10%–15% reduction in global translation (Anderson and Ivanov, 2014; Emara et al., 2010). Production of tRFs is also found in human cancers and may play a role in neurodegenerative diseases (Anderson and Ivanov, 2014). Recently, two tRFs were characterized across several mouse stem cell lines (Schorn et al., 2017). tRFs (18-nt) derived from the 3′ terminus of $tRNA^{Lys3}$ were found to interfere with reverse transcription of ERVs, maintaining viral expression but limiting the mobility of the retroelements (Schorn et al., 2017). Additionally, 22-nt tRFs also derived from the 3′ end of $tRNA^{Lys3}$ resulted in posttranscriptional silencing of viral RNA and proteins (Schorn et al., 2017). These tRFs allow for fine tuning of ERV expression, many of which are thought to have gained regulatory roles in the host cells as described below.

tRNA maturation is highly regulated and involves processing 3′ and 5′ leader sequences, excision of introns, if present, and extensive nucleotide modification (Phizicky and Hopper, 2010). The process of tRNA maturation is further complicated by several nuclear export and import steps to proofread some tRNA species (Hopper, 2013). Intriguingly, some pre-tRNAs are packaged into MLV and HIV-1 virions, although the function of these RNAs is not known (Eckwahl et al., 2015). Limiting the expression of a transport protein known to export pre-tRNAs, exportin-5, blocked packaging of pre-tRNAs into MLV and HIV-1 virions (Eckwahl et al., 2015). This finding suggests that pre-tRNAs, and potentially other RNAs packaged into retroviral virions, are recruited shortly after transcription.

The evolutionary selection of one primer tRNA over another is not well understood (Marquet et al., 1995). In some cases, a subpopulation of cellular tRNAs is packaged into virions or virus-like particles (VLPs) at ratios that differ from the host pool, suggesting that these tRNA species are selectively packaged (Waters and Mullin, 1977; Marquet et al., 1995; Keith and Heyman, 1990; Mak et al., 1994). For example, $tRNA^{Trp}$, the primer for avian leucosis virus (ALV), accounts for 32% of the tRNA packaged into virions, in contrast to 1.4% of the tRNA found in cells (Waters and Mullin, 1977). Additionally, human $tRNA^{Lys3}$ serves as the primer in HIV-1, and $tRNA^{Lys}$ isoacceptors represent approximately 50%–60% of the low-molecular weight RNA within the HIV-1 virion, but only 5%–6% within the cell (Mak et al., 1994). A similar trend, albeit less striking, is observed for Moloney murine leukemia virus (MLV), where primer $tRNA^{Pro}$ constitutes 5%–6% of cellular low-molecular weight RNA and 12%–24% of low-molecular weight vRNA (Waters and Mullin, 1977). Retrotransposons also package their primer tRNAs into VLPs; the relative concentration of $tRNA_i^{Met}$ increases 10- to 40-fold in Ty1 VLPs (Voytas and Boeke, 1993; Prats et al., 1988).

The mechanism by which tRNA primer enrichment occurs is not known. However, in some instances the cognate aminoacyl-tRNA synthetase is also packaged (Cen et al., 2001, 2002). In the case of HIV-1, human lysyl-tRNA synthetase (LysRS) is packaged into HIV-1 virions (Cen et al., 2001). Knockdown of LysRS, which interacts with the CA domain of HIV-1 Gag, reduces the levels of $tRNA^{Lys}$ packaged into virions, as well as viral infectivity, suggesting that primer $tRNA^{Lys3}$ packaging is facilitated through LysRS (Cen et al., 2001, 2004; Halwani et al., 2004; Dewan et al., 2012; Guo et al., 2003; Javanbakht et al., 2003). Intriguingly, upstream of the PBS in the NL4-3 (subtype B) HIV-1 gRNA there is a stem loop that mimics both the secondary and tertiary structure of the anticodon stem loop of $tRNA^{Lys3}$ and contains exposed uridine residues critical to LysRS binding (Jones et al., 2013, 2014; Liu et al., 2016). Mutation of the uridine residues in the tRNA-like element (TLE) to adenines reduced LysRS binding, as well as primer placement of $tRNA^{Lys3}$ on the PBS, inhibiting reverse transcription and viral infection (Jones et al., 2013). This tRNA mimicry is also found in the MAL isolate (subtype A-like) of HIV-1, suggesting conservation of TLEs across distinct HIV subtypes (Comandur et al., 2017). Overall, HIV-1 Gag is proposed to bind $tRNA^{Lys}$-bound LysRS and recruit the complex to sites of virion assembly where the TLE effectively competes for LysRS binding, causing the release of the primer and facilitating efficient annealing (Jones et al., 2013; Kleiman et al., 2010).

Mutagenesis studies have been carried out to determine the importance of the specific PBS sequence in retroviral replication (Whitcomb et al., 1995; Das et al., 1995; Lund et al., 1993; Rhim et al., 1991; Wakefield et al., 1996). The PBS of HIV-1 is the best characterized of any retrovirus. Deletions of the entire PBS or nt 1–9 did not produce infectious virus (Rhim et al., 1991). However, deletion of nt 10–18, 7–18, or replacement of nt 7–18 with random nt resulted in delayed production of functionally active virus. Upon sequencing, these virions were found to contain an 18-nt region that was once again fully complementary to $tRNA^{Lys3}$. Nucleotide insertions and deletions were found 3′ to the PBS suggesting that $tRNA^{Lys3}$ did in fact anneal to the PBS resulting in the reversion to the wild-type (WT), $tRNA^{Lys3}$ specific, PBS (Rhim et al., 1991). Multiple studies mutating the PBS region of HIV-1 to become complementary to different tRNAs produced infectious virions utilizing the substituted tRNA (Das et al., 1995; Wakefield et al., 1995). However, every virus reverted back to a PBS specific for $tRNA^{Lys3,}$ indicating a selective advantage in the use of this particular tRNA. Interestingly, an HIV-1 virus with a PBS complementary to $tRNA^{His}$ was maintained by mutating residues within the HIV-1 genome that are thought to further stabilize interactions with the tRNA outside of the 3′ 18-nt, indicating that residues outside of the PBS may play a role in tRNA use (Wakefield et al., 1996; Berkhout, 1997). Furthermore, packaging of $tRNA^{Lys}$

occurs independently of HIV-1 genomic RNA packaging, which could contribute to the selective pressure in maintaining a WT PBS (Huang et al., 1994).

Similarly to HIV-1, mutations to the PBS of ALV allowing for the use of $tRNA^{Pro}$, $tRNA^{Lys}$, $tRNA^{Met}$, $tRNA^{Ile}$, $tRNA^{Phe}$, and $tRNA^{Ser}$ resulted in short-term maintenance of the mutated PBS and delayed infection followed by reversion to the WT PBS-Trp, suggesting that although other tRNAs can be used, there is a distinct infectivity advantage for the WT PBS (Whitcomb et al., 1995). Interestingly, MLV appears to be more permissive to PBS mutations. Mutation of the PBS matching $tRNA^{Pro}$ to sequences accommodating $tRNA^{Glu}$ or $tRNA^{Lys}$ produced infectious single-cycle replication-competent virus, suggesting that the PBS did not require perfect nt base pairing for efficient priming (Lund et al., 1993). Additionally, using a randomized PBS it was found that MLV replicated efficiently with $tRNA^{Arg(CUC)}$, $tRNA^{Phe(GAA)}$, and $tRNA^{Ser(CGA)}$ (Lund et al., 2000); however, it is possible, based on studies with HIV-1 and ALV, that after a longer period, the sequence would revert back to WT, which was not tested.

Another function of host tRNAs in the retroviral lifecycle is Gag membrane targeting. Phosphatidylinositol (4,5) bisphosphate ($PI(4,5)P_2$)is associated with the cytosolic inner leaflet of the plasma membrane (PM) and acts as a targeting mechanism for PM-localized proteins (Veneziano et al., 2016; Birney et al., 2007; van Meer et al., 2008; Ono et al., 2004). In some retroviruses, including HIV-1, the matrix (MA) domain of Gag binds to $PI(4,5)P_2$, targeting Gag to the plasma membrane (Ono et al., 2004; Olety et al., 2015; Inlora et al., 2011, 2014; Chukkapalli et al., 2008). Plasma membrane targeting of HIV-1 Gag is mediated through a bipartite signal consisting of a myristic acid moiety added to the second amino acid of MA, which inserts into the plasma membrane-anchoring Gag (Waheed and Freed, 2009), in addition to a group of positively charged residues within MA (Zhou et al., 1994). HIV-1 MA contains a highly basic patch that sequesters the myristate prior to membrane binding and is also capable of binding RNA (Waheed and Freed, 2009). Interestingly, cellular RNAs block binding to nonspecific liposomes but are not able to compete for binding to specific PI(4,5)P2-containing membranes, thus enhancing HIV-1 Gag selectivity for $PI(4,5)P_2$-rich membranes (Inlora et al., 2014; Chukkapalli et al., 2008, 2010, 2013; Alfadhli et al., 2009). A cross-linking immunoprecipitation sequencing (CLIP-Seq) study revealed that HIV-1 MA binds almost exclusively to a subset of tRNAs in the cytoplasm (Kutluay et al., 2014) and recently, specific species of tRNA were shown to block nonspecific Gag membrane binding (Todd et al., 2017). Surprisingly, removal of the nucleocapsid (NC) domain of HIV-1 Gag abolished selective binding, suggesting NC may also play a role in membrane targeting (Todd et al., 2017).

tRNAs may also contribute to the nuclear localization of the preintegration complex (PIC) (Zaitseva et al., 2006). Following reverse transcription, the newly transcribed proviral DNA must be imported into the nucleus for integration of the viral genome into the cellular chromosome (Hu and Hughes, 2012). The CA protein is thought to assist in translocation of the PIC into the nucleus in nondividing cells (Dismuke and Aiken, 2006; Lee et al., 2010; Yamashita and Emerman, 2004). Intriguingly, when HeLa cell cytosolic fractions containing high amounts of tRNA were added back to purified PICs and nuclei, nuclear accumulation of the PICs was observed (Zaitseva et al., 2006). Once imported into the nucleus, tRNAs and CA must dissociate from the HIV-1 proviral DNA to allow integration into the host genome (Dismuke and Aiken, 2006). Transportin 3 (TNPO3) was found to bind and export tRNAs and CA in a RanGTP-dependent manner, facilitating efficient uncoating of the viral genome after nuclear entry (Zhou et al., 2011). TNPO3 is implicated in the nuclear export of defective tRNAs, including those lacking the 3′ CCA end (Zaitseva et al., 2006). The export of defective tRNAs to the cytoplasm is counterproductive to protein translation, suggesting the defective tRNAs may function as tRFs (Zhou et al., 2011).

7SL RNA

7SL RNA is packaged at high concentrations by all retroviruses that have been examined to date including MLV, HIV-1, avian retroviruses, feline leukemia virus, visna virus, and equine infectious anemia virus (Telesnitsky and Wolin, 2016; Bishop et al., 1970; Brian et al., 1975; Cheevers et al., 1977; Duesberg and Robinson, 1966; Levin et al., 1974; Lin and Thormar, 1971; Peters et al., 1977). Besides genomic RNA and tRNAs, 7SL RNA is the most well-characterized ncRNA packaged into virions. In eukaryotes, 7SL is a 300-nt RNA named according to its sedimentation coefficient and was first identified as a prevalent RNA in RSV particles (Bishop et al., 1970; Ullu and Tschudi, 1984). Within the cell, 7SL RNA is part of the signal recognition particle (SRP) ribonucleoprotein complex (RNP), which functions in regulating the secretion of proteins in mammals (Ullu and Tschudi, 1984; Walter and Blobel, 1982). As protein translation occurs, hydrophobic sequences at the N-terminus of nascent peptides are recognized by the SRP (Cross et al., 2009), which is composed of six proteins (SRP9, 14, 19, 54, 68, and 72) and 7SL RNA. Once bound to membrane or secretory proteins, SRP is responsible for targeting these growing proteins to the endoplasmic reticulum (ER) for further distribution (Walter and Blobel, 1982; Cross et al., 2009; Luirink and Sinning, 2004). In addition to acting as scaffolding for the assembly of the SRP proteins, the RNA has roles in accelerating the binding to the SRP receptor on the ER surface and triggering conformational changes that induce GTP hydrolysis by the receptor (Telesnitsky and Wolin, 2016; Cross et al., 2009).

In the case of retroviruses, 7SL RNA is highly enriched in virions (Bishop et al., 1970; Erikson et al., 1973; Walker et al., 1974; Brameier et al., 2013; Wang et al., 2007; Keene and Telesnitsky, 2012). For example, in HIV-1 virions, there are 14 copies of 7SL packaged in comparison with 2 copies of gRNA per HIV-1 virion (Onafuwa-Nuga et al., 2006). Interestingly, although 7SL RNA is predominantly found within the SRP, major SRP components SRP19 and 54 have not been found in virions, while minimal amounts of SPR14 are encapsidated (Eckwahl et al., 2015; Wang et al., 2008). These studies suggest that viral 7SL is newly synthesized and diverted to assembling virions prior to assembly of the larger RNP (Eckwahl et al., 2015). Although the exact mechanism of packaging remains unclear, 7SL RNA was shown to be encapsidated into minimal VLPs containing a mini-Gag construct, which contains the first 12 amino acids of matrix followed by the C-terminal domain of CA and a leucine zipper in the place of NC (Keene et al., 2010). A different report suggested NC was required for 7SL RNA packaging (Wang et al., 2007; Tian et al., 2007). Many questions remain as to the role of cellular ncRNAs that are packaged into virions. 7SL and other packaged cellular RNA may act as nonspecific nucleating factors for initiation of retroviral Gag multimerization in the cytoplasm (Telesnitsky and Wolin, 2016). In a recent study, MS2-stem loops were used to label and track 7SL RNA during HIV-1 infection (Itano et al., 2018). This study revealed that Gag and 7SL RNA arrived at the PM simultaneously, consistent with a model wherein 7SL RNA and lower-order multimers of Gag interact in the cytoplasm prior to assembly at the PM (Itano et al., 2018). Alternatively, 7SL RNA has been proposed to facilitate apolipoprotein B mRNA-editing catalytic polypeptide–like 3G and 3F (APOBEC3G, -3F) packaging (Wang et al., 2007, 2008). APOBEC family members are cytidine deaminases that modify RNA, DNA and, in the case of APOBEC3 family members, the HIV-1 genome (Jarmuz et al., 2002; Sheehy et al., 2003). In virions lacking Vif, which induces the degradation of APOBEC3G, hypermutations of the viral genome are found in the form of C- to U- changes (Harris et al., 2003; Lecossier et al., 2003). APOBEC3C, 3F, 3A, and 3B also appear to possess broad antiviral activities against viruses including simian immunodeficiency virus (SIV), MLV, hepatitis B virus, and some endogenous retroelements (Turelli et al., 2004; Bogerd et al., 2006; Chen et al., 2006; Esnault et al., 2005; Muckenfuss et al., 2006). APOBEC3G immunoprecipitated from 293 and Jurkat cells copurified with Y RNAs, specifically Y1, and Y3, and 7SL RNA (Wang et al., 2007). Overexpression of 7SL binding partner SRP19 in 293T cells reduced levels of packaged 7SL RNA and APOBEC3G, whereas overexpression of 7SL RNA increased inclusion of both components into HIV-1 virions (Wang et al., 2007). Similar results were found with APBOPEC3F, suggesting other ABOBEC3 family members may also interact with 7SL (Wang et al., 2008). The authors suggested that the potentially conserved interaction between APOBEC3G and 7SL could explain the more general antiviral nature of the protein and provide a reason for the abundant 7SL RNA present in retroviruses (Wang et al., 2007).

U Small Nuclear RNA

U snRNAs are associated with pre-mRNA splicing machinery and are between 60 and 450-nt in length (Kiss, 2004). All known U snRNAs are transcribed by RNA polymerase II with the exception of U6, which is transcribed by RNA polymerase III (Matera et al., 2007). Although U1, U2, U3, U4, U5, and U6 snRNAs have been found in MLV virions, U6 is the only one detected at levels that approach those of 7SL RNA, whereas the others are much lower in abundance (Onafuwa-Nuga et al., 2005). Interestingly, in HIV-1 virions, 3′-extended forms of pre-U1 snRNAs were detected at low levels indicating that these RNAs had not been fully processed and are likely packaged as newly transcribed RNAs (Eckwahl et al., 2015, 2016b). RSV virions were found to contain approximately one U6 snRNA per virion and trace amounts of U1 and U2 (Giles et al., 2004). Interestingly, the authors also found products of reverse transcription that matched U6, U1, and U2 RNA suggesting that RT may use packaged cellular RNA to create recombinant viral genomes (Giles et al., 2004). It is surprising that snRNAs are frequently found in retrovirus virions as they are exclusively found in the nucleus and most retroviruses bud from the cytoplasmic membrane (Giles et al., 2004).

Y RNA

Y RNAs are small ~100-nt RNAs that are present in the cytoplasm of most vertebrates and some lower eukaryotes in complex with the Ro Protein (Hendrick et al., 1981; Smith et al., 1984; O'Brien et al., 1993; Chen and Wolin, 2004). The number of different Y RNAs depends on the species; the human genome encodes four Y RNAs denoted as hY1, hY3, hY4, and hY5 (O'Brien and Wolin, 1994). Upon transcription of Y RNAs by RNA polymerase III, Y RNA is transiently bound by the La protein and ultimately forms a complex with the Ro protein. The Y RNA–Ro complex has an important role in unfolded and mutated RNA degradation (Wolin and Steitz, 1984). Several roles for Y RNA have been proposed: (1) the Y RNA-binding site on Ro overlaps the target RNA-binding site, suggesting that Y RNA regulates RNAs, which are bound by Ro, (2) in *Xenopus* oocytes, Ro has been found to bind mutant 5S rRNA and misfolded U2 snRNA, supporting the role of the Ro complex as a scavenger of unfolded RNA, (3) Y RNA-bound Ro is exclusively localized to the cytoplasm, whereas Y RNA-depleted Ro is localized to the nucleus, suggesting Y RNA also regulates Ro trafficking, (4) Y RNA may regulate the ability of Ro to enhance cell survival following UV radiation. Free Y RNA is found in MLV and HIV-1 virions at lower concentrations than 7SL RNA, bolstering the idea that many cellular RNAs are packaged from a newly synthesized pool prior to interaction with their protein complex partners (Onafuwa-Nuga et al., 2005; Garcia et al., 2009; Chen and Wolin, 2004; O'Brien and Wolin, 1994; Reinisch and Wolin, 2007; Sim et al., 2009; Chen et al., 2003). The role of Y RNAs in viral replication is unknown; however, there is an enrichment of mouse Y1 (mY1) and mY3 RNAs even in the absence of Ro, when cellular Y RNAs are degraded, showing there is a selectivity for the virus to package these RNAs (Garcia et al., 2009).

Vault RNA

Within the cytoplasm of eukaryotes, large RNP particles consisting of three proteins [major vault protein (MVP), poly(ADP-ribosyl) polymerase (VPARP), and telomerase/vault-associated protein (TEP1)] and multiple small ncRNAs are known as vaults. These RNPs are roughly three times larger than ribosomes with a molecular mass of 13 MDa (Kedersha and Rome, 1986; Kickhoefer et al., 2001; Stadler et al., 2009). vltRNAs are 80- to 150-nt long and vary considerably in sequence but are characterized by a GC-rich stem–loop 5′ to a box A promoter sequence (Stadler et al., 2009). The purpose of the vaults is not well understood despite the fact that they are conserved across most eukaryotic species except *Drosophila*, *Arabidopsis*, *Caenorhabditis*, and *Saccharomyces cerevisiae* (Stadler et al., 2009). Human vaults are the best characterized and are thought to primarily function in general intracellular transportation and have been implicated as a contributing factor to drug resistance by exporting chemical compounds (Kickhoefer et al., 2001; Stadler et al., 2009; Steiner et al., 2006). However, other roles have been attributed to vaults and vltRNAs, which are able to fold into structures similar to pri-miRNAs and be processed by Dicer into functional miRNA. Upregulation of vaults correlated with drug resistance to cancer therapeutics in some studies, and vltRNAs are also upregulated during Epstein–Barr virus infection (EBV) (Persson et al., 2009; Kitazono et al., 1999, 2001; Nandy et al., 2009). Of relevance to the retrovirus field is the fact that MVP alone can form vaultlike particles (Stephen et al., 2001) reminiscent of VLPs produced by Gag. Additionally, a study of all cellular RNAs packaged into MLV virions identified the presence of previously undetected vltRNAs (Eckwahl et al., 2015). The proteins associated with vaults have not been found in virions to date.

High-Throughput RNA-Sequencing Studies

With advancements in the ability to perform whole-cell RNA sequencing, more recent studies have been conducted to examine ncRNA interactions with viral proteins within the cell and to identify RNAs packaged into virions (Eckwahl et al., 2016a,b; Onafuwa-Nuga et al., 2005; Kutluay et al., 2014). CLIP-Seq was used to determine which cellular and vRNAs are preferentially bound by different domains of Gag (Kutluay et al., 2014). During the initial stages of HIV-1 viral assembly, the NC domain of Gag bound to the genomic RNA packaging signal, psi, and to the Rev response element, as well as to GU-rich mRNA transcripts. However, at the plasma membrane, NC preferentially bound to A-rich regions of both the viral genome and mRNA (Kutluay et al., 2014). Surprisingly, HIV-1 MA almost exclusively bound host tRNAs in the cytoplasm, as mentioned earlier, and did not interact with the genomic RNA (Kutluay et al., 2014). However, HIV-1 MA isolated from membrane fractions showed significantly less bound tRNA (Kutluay et al., 2014), which is in agreement with reports indicating that host tRNAs regulate HIV-1 MA–Gag plasma membrane targeting, as discussed earlier (Chukkapalli et al., 2010, 2013; Todd et al., 2017; Dick et al., 2013). A high-throughput RNA-sequencing (RNA-seq) method has also been applied to identify RNAs packaged into HIV-1 virions, revealing several

RNAs that had not been detected in previous studies, including many unprocessed RNA species, such as pre-tRNAs, mentioned earlier, and pre-snRNAs (Eckwahl et al., 2016b). Future applications of RNA-seq to other retroviruses may uncover RNAs that are universally packaged into virions and those that are selective to specific viruses.

LONG NONCODING RNA AND ENDOGENOUS RETROVIRUSES

lncRNAs are not abundantly packaged into retroviral virions with the exception of 7SL RNA and some long interspersed nuclear elements (Wapinski and Chang, 2011). Classification of lncRNA is typically more difficult than short ncRNAs, which have well-defined features (St Laurent et al., 2015). They are generally split into six groups: pseudogenes (mutated gene sequence that no longer code functional genes); long intronic ncRNA; divergent transcripts (associated with and enhance promoters); large intergenic RNA (occurring between coding regions); circular RNA; and natural antisense transcripts (Veneziano et al., 2016). Although lncRNAs are abundant within the human genome, only a few are well characterized (Geisler and Coller, 2013). Those with known functions include roles in cell cycle regulation, telomerase regulation, chromatin remodeling, and scaffolding for protein complexes. Many have been associated with diseases, including Alzheimer's disease (Mattick, 2001; Frias-Lasserre, 2012; Geisler and Coller, 2013; Groen and Morris, 2013). Several viruses, including several types of herpesviruses, encode lncRNAs known to interact with and, in some cases, inhibit host proteins, as reviewed in Tycowski et al. (2015). The human, murine, and avian lncRNAs described below adopt functions that affect retroviral life cycles in diverse ways.

NRON: HIV-1

HIV-1 latency and maintenance of virally infected reservoirs has been a central conundrum for the development of effective vaccines and antiviral therapeutic development (Groen and Morris, 2013). Several host and viral proteins are thought to regulate viral latency including nuclear factor kappa B, nuclear factor of activated T cells (NFAT) and HIV-1 Tat (Kao et al., 1987; Nabel and Baltimore, 1987; Kinoshita et al., 1998). Although controversial, host miRNAs are proposed to contribute to viral latency (Kiss, 2004). Using a large-scale siRNA screen, knockdown of the lncRNA called nonprotein-coding RNA repressor of NFAT (NRON) was found to enhance HIV-1 replication in activated primary $CD4^+$ T lymphocytes (Li et al., 2016). NRON is ~2700-nt long and its expression was found to be altered over the course of HIV-1 infection; expression is reduced initially due to the viral accessory protein Nef, whereas enhanced expression is observed in the later stages of infection due to Vpu (Imam et al., 2015). NRON was found to affect HIV-1 replication in an NFAT-dependent manner, wherein reduction of NRON resulted in increased activity of NFAT and enhanced viral transcription (Imam et al., 2015). During HIV-1 replication, the viral protein Tat has an essential role in enhancing efficient transcription of the viral genome (Leblanc et al., 2013). NRON was reported to form a complex with components of the ubiquitin/proteasome, ultimately resulting in the degradation of Tat and reduction of viral transcription in an NFAT-independent mechanism (Li et al., 2016). NRON is present at high concentrations in resting $CD4^+$ T lymphocytes and therefore, may contribute to viral latency via viral transcriptional silencing (Li et al., 2016).

NEAT1: HIV-1

Many different types of nuclear and cytoplasmic bodies have been described in cells (Sleeman and Trinkle-Mulcahy, 2014). Paraspeckles are dynamic nuclear bodies, the components of which include the 3200-nt lncRNA nuclear paraspeckle assembly transcript 1 (NEAT1). Paraspeckles respond to stress conditions by sequestering components of transcription and halting the production of specific genes; they also sequester a subset of mRNAs to be edited including those containing Alu repeats (Barichievy et al., 2015; Chen and Carmichael, 2009). Paraspeckles have also been implicated as an innate immune response to viral infections (Imamura et al., 2014). HIV-1 RNA has been reported to be localized to paraspeckles, potentially for A to I editing carried out by ADAR1 (Doria et al., 2009; Gallo and Locatelli, 2012). In one study, the expression profile of over 83 lncRNAs was analyzed during HIV-1 infection of both Jurkat and MT4 cells and NEAT1 expression was found to be increased by 5- to 10-fold (Zhang et al., 2013). Knockdown of NEAT1 resulted in reduced paraspeckle formation and increased cytoplasmic concentration of Rev-dependent instability element-containing HIV-1 RNA, enhancing viral replication (Zhang et al., 2013).

Large Intergenic RNA–p21: HIV-1

During the integration step of the retroviral life cycle, double-strand breaks must be made in the host genome to insert the proviral DNA (Craigie and Bushman, 2012). Once inserted, host machinery must repair the break in the phosphodiester bonds before damage to the host DNA triggers cell death (Houtgraaf et al., 2006). The proapoptotic protein, p53, is activated when DNA damage occurs, which also activates transcription of lincRNA-p21. The 3100-nt large intergenic RNA (lincRNA)–p21 can induce apoptotic pathways by downregulating the transcription of prosurvival genes (Huarte et al., 2010). Expression of lncRNA-p21 is regulated during HIV-1 infection in macrophages to block detection of DNA breaks, allowing HIV-1 to gain control over the MAP2K/ERK2 pathway, one of the targets of lincRNA-p21 that is needed for cell survival (Barichievy et al., 2015).

Antisense Protein RNA: HIV-1

In 1988, HIV-1 was proposed to contain an antisense protein (ASP) produced from the antisense promoter within the 3′ LTR based on computational analysis of the viral RNA (Miller, 1988). The expression of ASP has been detected in patient sera and induces $CD8^+$ T cells (Bet et al., 2015). Possible roles for the ASP RNA include epigenetic silencing of the HIV-1 integrated DNA by recruiting chromatin remodeling proteins to the viral LTR (Zapata et al., 2017). This lncRNA is proposed to suppress viral transcription, favoring latency, by enhancing DNA methylation. It is also possible, given the degree of similarity between the 3′ and 5′ LTRs, that ASP RNA acts by forming a triple helix with the 5′ LTR of the genome (Zapata et al., 2017). ASP RNA can be detected in chronically infected cell lines and resting $CD4^+$ T cells from patients currently taking antiretroviral therapies, demonstrating ASP RNA is expressed in vivo (Zapata et al., 2017).

Long Noncoding RNA 00173: HIV-1

Analysis of two RNA-seq studies revealed that three lncRNAs were differentially expressed during HIV-1 infection; long noncoding RNA 00173 (LINC00173) was the most evolutionarily conserved prompting further investigation (Postler et al., 2017). Intriguingly, when knocked out, LINC00173 did not impact the retroviral life cycle in Jurkat T cells. However, loss of LINC00173 resulted in higher expression of several cytokines, including interferon-γ, which is important in the adaptive immune response (Postler et al., 2017). Although the authors did not find a direct effect of LINC00173 on HIV-1 infection, they speculate that the increased production of interferon-γ could have implications on an organismal level (Postler et al., 2017).

ASP RNA: HTLV-1

The HTLV-1 accessory protein, HTLV-1 basic leucine zipper factor (HBZ) is an ASP whose gene is located near the 3′ viral LTR (Gaudray et al., 2002). The RNA transcript of this protein has been implicated in cell proliferation leading to HTLV-1-associated diseases and chronic infections (Barbeau and Mesnard, 2011). However, it has also been reported that the mRNA transcript functions by activating antiapoptotic pathways and inducing proliferation by IL-2 independent pathways (Satou et al., 2006a). Interestingly, progression to adult T-cell leukemia (ATL) is correlated with increased viral expression (Zhao and Matsuoka, 2012). However, Tax, another accessory protein associated with cell proliferation, is often mutated and inactive in ATL cases, whereas HBZ is expressed, suggesting a role for HBZ protein in chronic infection (Ma et al., 2016). The HBZ mRNA has been shown to be retained within the nucleus indicating a potential functional role in transcription regulation (Rende et al., 2011). The functional HBZ mRNA forms stem–loop structures that function to transcriptionally activate E2F1, a transcription factor upregulated in many cancers, although the mechanism by which this occurs is not understood (Satou et al., 2006b; Matsuoka and Green, 2009). Expression of miRNAs targeting HBZ RNA in ATL cells decreased proliferation of cells (Satou et al., 2006b). In mice, HBZ mRNA has been found to activate the transcription of antiapoptotic genes including *survivin*, which may enhance the progression to ATL (Mitobe et al., 2015).

Bic: ALV and His-1: MLV

ALV infection of birds results in B-cell lymphomas. The virus often integrates at the *bic* locus and is frequently associated with *c-myc* activation (Clurman and Hayward, 1989; Tam et al., 1997). Interestingly, the expression of the proto-oncogenic *c-myc* alone is not sufficient to cause lymphoma in avian systems, suggesting that other factors are required (Adams and Cory, 1992). It was found that insertions at the *bic* locus activated the transcription of a chimeric bic-virus RNA; however, the RNA lacks an open reading frame suggesting it functions as a lncRNA (Tam et al., 1997). In a follow-up study, the expression of bic and c-myc in chicken embryonic fibroblasts induced greater cell growth than activation of *c-myc* alone (Tam et al., 2002). However, the coexpression was not sufficient to induce full transformation of cultured cells, suggesting that other genetic changes are required (Tam et al., 2002). While the exact mechanism of the bic lncRNA in cell regulation is still unclear, this discovery was the first report that an lncRNA can interact with and enhance c-myc activity (Tam et al., 2002).

In mice, MLV commonly inserts into the *His*-1 locus, which does not produce any known coding protein (Askew et al., 1991). His-1 transcripts do not contain an open reading frame and the 3000-nt transcript contains multiple stop codons suggesting that it functions through its RNA (Askew et al., 1994). In murine cell lines and mouse tissue, His-1 expression was undetected; however, in two transformed cell lines with detectable MLV insertion near the *His-1* site, increased RNA expression was observed (Askew et al., 1994). Similar to *bic* in ALV, insertion of the MLV genome into the *His-1* locus served to activate the proto-oncogene and enhance cell survival (Askew et al., 1994).

Endogenous Retroviruses

It is estimated that ~10% of the human genome comprises ERVs, which are considered a type of retrotransposon (Lower et al., 1996; Kurth and Bannert, 2010; Wildschutte et al., 2016). ERVs can arise from the integration of retroviruses in germ cells; inherited viruses are generally unable to replicate due to an accumulation of mutations, frameshifts, and deletions (Gifford and Tristem, 2003). Human ERVs (HERVs) are the best characterized, but many animal species also contain ERVs in their genomes including pigs, plants, fish, rodents, insects, and cats (Boeke and Stoye, 1997; Herniou et al., 1998; Johnson and Coffin, 1999). ERVs are divided into three classes: class I, which resembles gamma- and epsilonretroviruses; class II, which bears similarities to betaretroviruses; and class III, which resembles spumaretroviruses (Lower et al., 1996; Gifford and Tristem, 2003). Alternative methods of classification use the primer tRNAs to categorize ERVs, see Table 12.1 (Gifford and Tristem, 2003). Initially, it was thought that ERVs were simply remnants of ancient retroviruses; however, there is increasing evidence that some ERVs have become essential regulatory elements and provide genetic diversity (Wang et al., 2014; Lu et al., 2014; Lee et al., 2017). Also, although many ERVs do not cause disease, others have been associated with multiple sclerosis, autoimmune disorders, and cancer (Mameli et al., 2007; Krieg et al., 1992). HERV-K is the most intact ERV, with a long open reading frame and ability to produce vRNA and proteins (Lower et al., 1996). HERV-K is also able to illicit a weak immune response, as antibodies to HERV-K proteins are found at very low levels following pregnancies, as well as in patients with leukemias, and certain types of tumors (Lower et al., 1996; Sauter et al., 1995). The presence of ERVs in pigs, porcine endogenous retrovirus (PERVs), is of special interest as pig tissues can be transplanted into humans. While ERVs are generally innocuous, the consequences of cross-species interactions are unknown (Weiss, 2006). A recent study examining the transplantation of porcine islet cells into humans for the treatment of type 1 diabetes found no indication of PERV infection up to 113 weeks postimplantation, lending credence to the safety of porcine transplants (Morozov et al., 2017). Interestingly, the feline ERV, RD114, was initially detected in a human cell line that had been xenografted into fetal kitten brain tissue, suggesting that xenotropic ERVs are possible (McAllister et al., 1972; Achong et al., 1976; Patience et al., 1997).

Two studies have found that ERVs are important for human stem cell function (Wang et al., 2014; Lu et al., 2014). HERV-H silencing in human embryonic stem cells was found to lead to differentiation to fibroblast-like cells (Lu et al., 2014). Using RNA-seq, the levels of HERV-H in human pluripotent stem cells were shown to be elevated compared to embryoid bodies and fibroblasts (Wang et al., 2014). Both studies revealed that HERV-H sites of integration were in regions encoding lncRNAs, prompting the authors to examine the expression of these lncRNAs and adjacent protein-coding regions (Wang et al., 2014; Lu et al., 2014). Knockdown of HERV-H also resulted in the reduced expression of neighboring lncRNAs and genes (Lu et al., 2014). It was determined that the HERV-H fragment was acting as a *cis*-gene regulator during stem cell development suggesting that other functional lncRNAs could be derived from ERVs (Lu et al., 2014). Other HERVs have been shown to play roles in cell fusion in the placenta, and fetal tolerance is linked to expression of HERV *env* genes (Blond et al., 1999; Mangeney et al., 2007). Varying levels of HERV expression are observed across many different types of cancer, prompting the examination of HERV activity in tumorigenesis (Ruprecht et al., 2008). Interestingly, HERVs can be inserted into the genome in the antisense orientation, overlapping coding genes. This orientation results in the production of short RNAs complementary to the overlapping gene sequences, creating an inhibitory RNA able to regulate gene expression (Gogvadze et al., 2009). HERV-K, which is generally present in introns, has a functional 3′ LTR expression and was initially inserted into Old World monkey genomes (Broecker et al., 2016; Dangel et al., 1994; Mack et al., 2004). *DAP3*, a gene that is located within *HERV-K* in HepG2, 293T, and HeLa cells, is a signaling protein that when overexpressed induces apoptosis (Broecker et al., 2016; Kissil et al., 1995). HERV-K antisense RNA fragments were able to reduce levels of DAP3 and, conversely, knockdown of HERV-K transcripts enhanced DAP3 expression to induce apoptosis in HeLa cells (Broecker et al., 2016). This finding suggests HERV-K expression could potentially lead to a favorable outcome for tumor formation (Broecker et al., 2016). HERV-E was found to be differentially expressed in urotherlial carcinoma cells and coincided with a decrease in cytosolic phospholipase A2, which is often dysregulated in tumor cells (Gosenca et al., 2012). Although HERV-E transcripts are found in several cancers, including breast cancer, they are also present in normal cell lines implying more studies will need to be conducted before a link between cancers and HERV-E expression is solidified (Gosenca et al., 2012; Frank et al., 2008). Therefore, although ERVs generally do not cause diseases when inserted into host genomes, transcriptionally active portions of the genomes could inherit functional roles within the cell.

FUNCTIONAL TRANSACTIVATING RESPONSE ELEMENT RNAs

Located in the 5′UTR region of the HIV-1 vRNA, the transactivating response (TAR) element forms a stem–loop structure that specifically binds the viral protein Tat (Frankel and Young, 1998). During transcription, Tat binds to TAR and phosphorylates the C-terminal domain of RNA polymerase II, enhancing transcription of the retroviral genome (Karn and Stoltzfus, 2012). Outside of its essential role in transcription, TAR RNA has also been suggested to have a role in blocking host cellular RNAi responses to HIV-1 infection (Bennasser et al., 2006). TAR RNA–binding protein (TRBP) is known to interact with Dicer, a component of the RNAi pathway discussed in more detail in Therapeutics section, by trafficking siRNA to the multiprotein RNA-induced silencing complex (RISC) (Haase et al., 2005; Forstemann et al., 2005; Chendrimada et al., 2005). TRBP can be sequestered by TAR RNA, reducing the effectiveness of RNAi within the cell (Bennasser et al., 2006).

TAR RNA has also been implicated in both *cis*- and *trans*-inhibition of translation within host cells (Parkin et al., 1988; Edery et al., 1989; SenGupta et al., 1990). TAR RNA can block the ability of the ribosomal initiation factor eIF-4B to effectively recognize the mRNA cap (Parkin et al., 1988). Additionally, TAR RNA can be recognized by interferon inducible protein kinase (PKR), which typically recognizes dsRNA, and by 2-5A synthetase, which produces

2′,5′-oligoadenylates that activate RNase L, which can degrade single-stranded vRNA and induce autophagy in response to viral infection (SenGupta and Silverman, 1989; Roy et al., 1991). At low concentrations of TAR, PKR is activated and at high concentrations PKR is inhibited (Maitra et al., 1994). Presumably to counteract the ability of PKR and 2-5A synthetase to silence translation, when HIV-1 Tat is expressed, both PKR and 2-5A-synthetase activities are blocked (Maitra et al., 1994; Judware et al., 1993).

Most cellular mRNA is translated by a cap-dependent mechanism wherein assembling ribosomes scan along the 5′ UTRs to find start codons (Jackson et al., 2010; Deforges et al., 2017). However, the HIV-1 5′ UTR is highly structured and contains stable stem loops including TAR that are capable of hindering ribosomal scanning (Yilmaz et al., 2006). Within the 5′ UTR of several retroviruses, including HIV-1, a posttranscriptional control element (PCE) helps recruit host proteins essential for efficient translation (Bolinger et al., 2007, 2010; Hull and Boris-Lawrie, 2002; Russell et al., 2001; Roberts and Boris-Lawrie, 2003). Dhx9/RNA helicase A (RHA) is known to interact with PCEs to destabilize structured elements allowing for cap-dependent translation (Hartman et al., 2006). RHA interacts with TAR RNA, is packaged into HIV-1 virions, and is important for efficient translation of the HIV-1 vRNA and progeny virion infectivity (Bolinger et al., 2010). In addition, some cap-dependent cellular translation is downregulated during HIV-1 infection but vRNA is still translated, the possibility of IRESs within the HIV-1 genome has been explored (Agy et al., 1990; Brasey et al., 2003). Indeed, IRES-like elements within the 5′ UTR and in the *gag* coding frame have been characterized (Deforges et al., 2017; Brasey et al., 2003; Buck et al., 2001).

Surprisingly, the TAR element was recently shown to be one of the regions of vRNA to bind to integrase (IN) in HIV-1 virions (Kessl et al., 2016). Previously, allosteric IN inhibitors (ALLINIs) had been shown to inhibit late stages of HIV-1, disrupting viral maturation and resulting in impaired viral infectivity (Jurado et al., 2013; Fontana et al., 2015; Desimmie et al., 2013), although the mechanism was unknown. Using CLIP-seq, IN was found to bind discreet regions of the viral RNA including TAR and RRE. Disruption of IN–viral RNA interactions by mutagenesis or treatment with ALLINIs resulted in eccentric viral core formation and reducing infectivity (Kessl et al., 2016). The functional significance of this interaction is not completely understood, although it is postulated that the IN–viral RNA interaction may ensure the presence of IN in the CA core and thereby facilitate maturation of the core (Kessl et al., 2016).

RNA INTERFERENCE AND miRNAs

Although plants do not have an immune system in the typical mammalian sense, higher-order plants are able to mount targeted immune responses to viral threats using RNAi (Voinnet, 2001; Plasterk, 2002). RNAi was first described in transgenic petunias, wherein introduced genes were silenced through an unknown pathway (van der Krol et al., 1990; Napoli et al., 1990). Eventually, it was found that short 21–23 nt siRNAs were targeting mRNAs within the cell, causing the degradation of the RNA and suppression of protein expression (van der Krol et al., 1990). Over 90% of viruses that infect higher-order plants are transmitted through RNA at some stage of the viral life cycle (Voinnet, 2001). Exogenous viral dsRNA, showing no homology to the host genome, has been shown to trigger the RNAi pathway to suppress vRNA invaders (Elmayan and Vaucheret, 1996). It was observed that viruses containing similar gene sequences to those of the plant cell could cause virus-induced gene silencing (Ruiz et al., 1998; Ratcliff et al., 1999). Essentially, viral ssRNA can catalyze the synthesis of dsRNA by plant RNA-dependent RNA polymerases (Plasterk, 2002). The typical RNAi pathway can then occur, ultimately creating siRNAs targeted to the viral genome (Voinnet, 2001). Importantly, this pathway can lead to the production of secondary siRNAs, effectively amplifying the "immune response" (Plasterk, 2002). As for retroviruses discussed below, plant viruses adapt to the host RNAi pathway by producing virally encoded suppressors of RNA silencing (Pruss et al., 1997). For example, tomato bushy stunt RNA viruses produce a protein called p19, an inhibitor of Dicer, which blocks the host plant cell's immune response to initial viral infection (Voinnet et al., 1999).

RNAi, miRNA, and siRNA Production

The RNAi pathway was initially thought to be unique to plants, but was later identified in many eukaryotic species (Plasterk, 2002; Almeida et al., 2011). Briefly, RNA Pol II and III recognize specific promoters in DNA to begin transcription of long primary miRNA transcripts (pri-miRNA), usually containing a 33-bp stem loop (Lee et al., 2004; Borchert et al., 2006) (Fig 12.1). Pri-miRNAs produced from RNA Pol II are sensitive to alpha-amanitin, allowing for the identification of the polymerase responsible for transcription (Lee et al., 2004). These transcripts are polyadenylated and capped via a similar mechanism to that used for mRNAs, and like mRNAs, can be edited (Cai et al., 2004; Luciano et al., 2004). The nuclear microprocessor complex composed of Drosha, an RNase III enzyme, and DGCR8, which contains two dsRNA-binding motifs, cleaves the pri-miRNA 11 bp from the base of the stem–loop structure and the product of this step is pre-miRNA (Lee et al., 2003; Han et al., 2004). Although most miRNAs are processed through Drosha and DGCR8, some miRNAs are formed after intron splicing and can bypass this step (Ruby et al., 2007). At this stage, the pre-miRNA is exported to the cytoplasm and further processed by the RISC-loading complex, which consists of the RNase Dicer, TRBP, the protein activator of PKR (PACT) and Ago2, the main component of Argonaute-2 (Gregory et al., 2005; Lee et al., 2006). TRBP and PACT help to stabilize the pre-miRNA–protein complex, whereas Ago2 creates a nick in the 3′ arm of the stem loop (Lee et al., 2006; Diederichs and Haber, 2007). Dicer cleaves the remaining loop resulting in a 21–23 nt RNA with 2-nt overhangs on both the

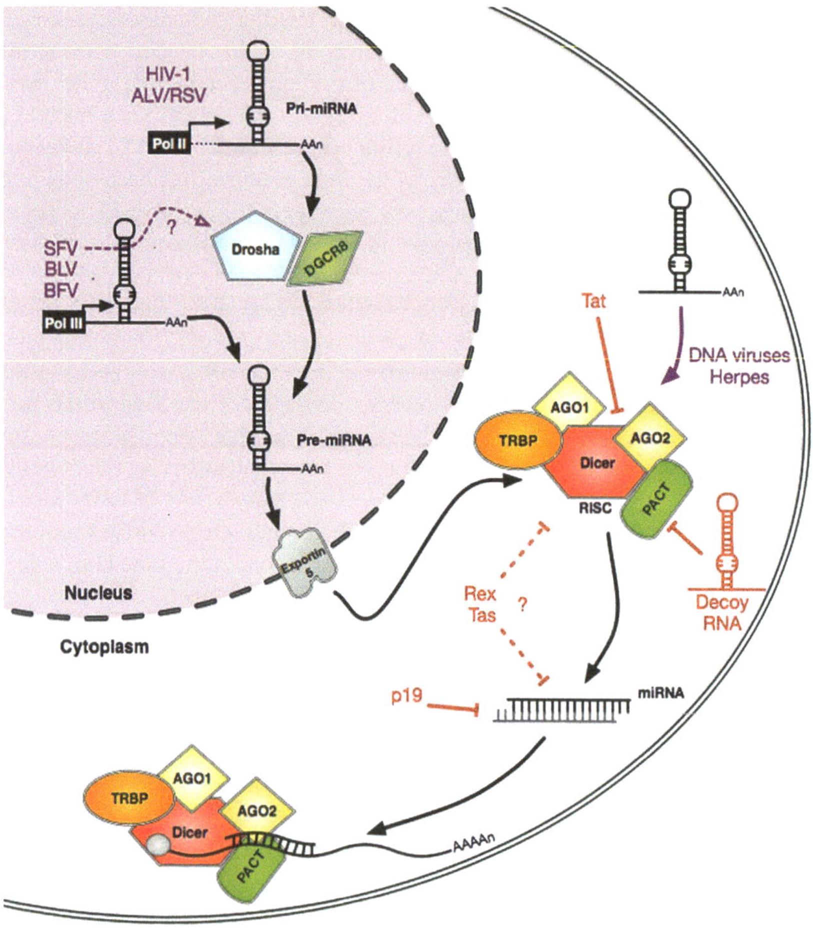

FIGURE 12.1 MicroRNA (miRNA) production pathway and viral interaction with RNA interference (RNAi) machinery. Host pri-miRNAs are transcribed by RNA polymerase II (Pol II) or III (Pol III) and can be processed by Drosha and DGCR8 into pre-miRNAs. Pre-miRNAs are then exported from the nucleus by exportin-5 and cleaved into a 21–22 nt miRNA duplex by Dicer of the RNA-induced silencing complex (RISC) complex. One miRNA strand is loaded onto the Argonaut (AGO)-RISC complex and used to select mRNAs for degradation or translational silencing. This process is depicted by *black arrows*. Viral miRNA entry into the RNAi pathway is shown in *purple text*. Inhibition of the host RNAi pathway by viral proteins or RNA is indicated by *red text* and *arrows* directed towards their proposed inhibitory target. *ALV*, avian leukosis virus; *BFV*, bovine foamy virus; *BLV*, bovine leukemia virus; *RSV*, Rous sarcoma virus; *SFV*, simian foamy virus; *TRBP*, TAR RNA–binding protein. *(Adapted from Winter, J., Jung, S., Keller, S., Gregory, R.I., Diederichs, S.S., 2009. Many roads to maturity: microRNA biogenesis pathways and their regulation. Nat. Cell Biol. 11, 228–234; Trobaugh, D.W., Klimstra, W.B., 2017. MicroRNA regulation of RNA virus replication and pathogenesis. Trends Mol. Med. 23, 80–93; Klase, Z.A., Sampey, G.C., Kashanchi, F., 2013. Retrovirus infected cells contain viral microRNAs. Retrovirology 10, 15; Soifer, H.S., Rossi, J.J., Saetrom, P., 2007. MicroRNAs in disease and potential therapeutic applications. Mol. Ther. 15, 2070–2079.)*

3′ ends (Bernstein et al., 2001). Although both strands could be utilized as the guide RNA for silencing, it is typical that the one with the least stable 5′ end becomes the functional miRNA, whereas the other one is degraded (Schwarz et al., 2003; Khvorova et al., 2003). Generally, miRNAs target the 3′ UTRs of mRNA blocking protein translation (Ouellet and Provost, 2010). A majority of the mRNAs that are completely complementary to the seed sequence, the minimal portion of miRNA required to recognize the target RNA, will be translationally suppressed, whereas targeted mRNAs with partial complementarity to guide RNAs tend to be degraded (Yekta et al., 2004; Mansfield et al., 2004; Lewis et al., 2005). Many of the steps of this process can be bypassed or altered depending on cellular conditions; for a more detailed review see Winter et al. (2009). Processing bodies (P-bodies) within the cytoplasm of cells appear to be concentrated areas of translationally silenced RNAs, mRNA degradation, and nonsense-mediated decay (Liu et al., 2005; Sen and Blau, 2005). Since the first identification of miRNAs, over 5500 human and 130 viral miRNAs have been predicted (Griffiths-Jones et al., 2006, 2008; Londin et al., 2015; Trobaugh and Klimstra, 2017). Expression of miRNAs is tightly regulated within cells and is often tissue specific, leading to the use of miRNAs in diagnostics for various types of cancers, cardiovascular diseases, autoimmune diseases, and neurodegenerative diseases (Almeida et al., 2011).

Effective viral infection requires regulation of various cellular processes including mRNA translation (Barichievy et al., 2015). HIV-1 infection has been found to modulate the expression of over 300 genes including miRNAs (Barichievy et al., 2015; Ouellet and Provost, 2010; van ‘t Wout et al., 2003). The ability of host miRNAs to impact viral replication in mammals has been well characterized. The question remains whether these miRNAs are physiologically capable of directly regulating retroviral replication (Grassmann and Jeang, 2008). Additionally, many different types of viruses have been found to produce miRNAs from their genomes (Tycowski et al., 2015; Ouellet and Provost, 2010; Klase et al., 2013; Umbach and Cullen, 2009). The most widely characterized viral miRNAs are produced from DNA viruses including *Herpesviridae* and mouse cytomegalovirus (Ouellet and Provost, 2010). The discovery of viral miRNAs in RNA viruses, such as EBV, was surprising and remains controversial (Tycowski et al., 2015; Klase et al., 2013; Houzet and Jeang, 2011; Cullen, 2012). First, many RNA viruses replicate exclusively in the cytoplasm, thereby avoiding nuclear Drosha cleavage. In addition, if nuclear RNA genomes were processed into miRNAs very little full-length gRNA would be available to be packaged (Umbach and Cullen, 2009; Cullen, 2012). However, some retroviruses have adapted to produce miRNAs by the less frequently used RNA Pol III, addressing the issue of genomic RNA degradation of Pol II transcripts (Klase et al., 2013). Conversely, Drosha and the RNAi pathway can act to inhibit viral replication resulting in potentially beneficial or detrimental outcomes (Barichievy et al., 2015; Umbach and Cullen, 2009). Cellular miRNAs are proposed to regulate viral replication (Houzet and Jeang, 2011; Yeung et al., 2007). To counteract cellular antiviral RNAi responses, retroviruses have adapted by degrading RNAi machinery, diversifying the roles attributed to viral proteins, and producing "decoy" RNA transcripts (Bennasser et al., 2006; Houzet and Jeang, 2011; Yeung et al., 2007).

Cellular MicroRNAs Made in Response to Viral Infections

Mammalian miRNAs are estimated to regulate up to 30% of protein-coding mRNA and the expression patterns of some miRNAs are modulated upon viral infection (Trobaugh and Klimstra, 2017; Houzet and Jeang, 2011). Host miRNAs can impact viral life cycles both directly and indirectly, leading to either enhanced viral production or suppression of the viral life cycle (Trobaugh and Klimstra, 2017; Watanabe et al., 2007). The specificity of the host miRNA response to specific viral infection and whether the RNAi pathway is used as a general antiviral response are active areas of investigation. miRNA pathways identified in specific retroviruses are described below.

Avian leukosis virus subgroup J (ALV-J). ALV can be transmitted vertically and chicken eggs are used in the development of human vaccines (Gao et al., 2010; Hussain et al., 2003). Thus, understanding the mechanism of this oncogenic virus' life cycle has been a major undertaking and seems linked to host miRNAs (Dai et al., 2015; Li et al., 2014, 2015; Wang et al., 2013a). Microarray analysis of chicken (*Gallus gallus*) tumors showed differential expression of several miRNAs known to interact with MAPK and Wnt signaling pathways, both associated with tumorigenesis in exogenous ALV-J infections (Wang et al., 2013a). These miRNAs include gga-miR-125b, gga-miR-193a, gga-miR-193b, and gga-miR-221. gga-miR-1650 was found to bind the 5′ UTR of the ALV-J gRNA resulting in decreased viral replication (Wang et al., 2013b). However, both WT and gga-miR-1650-resistant viruses were able to replicate under endogenous gga-miR-1650 expression indicating a higher level of miRNA expression would be needed to suppress the virus (Wang et al., 2013b). Another miRNA found to be upregulated during ALV-J infection was gga-miR-34b-5p, which enhances viral replication by downregulating interferon regulatory factor-1, preventing host responses to viral infections and promoting tumorigenesis (Li et al., 2015). Another viral sensing target of gga-miR-34b-5p is melanoma differentiation–associated gene 5 (MDA5). MDA5 also controls proliferation, cell cycle progression and migration of ALV-J infected cells, suggesting that gga-miR-34b-5p regulation of host immune responses has a substantial impact on viral replication and disease progression (Li et al., 2017). Like miRNAs, piwi-interacting RNAs (piRNAs) are also able to silence ALV expression (Sun et al., 2017). These RNAs interact with a subclass of Argonaute proteins, known as PIWI proteins, which result in the cleavage of the targeted RNA (Sun et al., 2017). However, the resulting cleavage products can then be used as secondary piRNAs amplifying the effect (Sun et al., 2017). Using RNA-seq, 73 active transposable elements were identified in White Leghorn testis that elicited piRNA-mediated suppression, suggesting piRNAs have evolved to maintain the integrity of germline genomes (Sun et al., 2017).

Prototype foamy virus type 1 (PFV-1). Expression of a silencing suppressor, p19, was found to enhance PFV-1 viral expression in 293T cells leading the authors to speculate that the RNAi pathway was involved in viral regulation (Trobaugh and Klimstra, 2017; Lecellier et al., 2005). This study further showed that human miR-32 shared homology with the 3′UTR of all PFV-1 mRNAs and repressed viral expression in a manner that depended on a complementary sequence (Lecellier et al., 2005). The countermechanism utilized by PVF-1 will be discussed below.

HTLV-1. Human miR-28-3p targets the gRNA of HTLV-1 within the *gag/pol* gene causing a halt in translation of the virus in resting T cells (Trobaugh and Klimstra, 2017; Bai and Nicot, 2015). Interestingly, subtype 1a, the Japanese AKT-1 strain of HTLV-1, contains a single-point mutation within the target sequence creating resistance to miR-28-3p and allows the virus to transmit efficiently (Trobaugh and Klimstra, 2017; Bai and Nicot, 2015). Computational studies predicting host miRNAs that could target HTLV-1 gRNA revealed several potential candidates, including plus-strand miR-653, -648, -596, -644, -496, -431, -326, -125b, -432, -125a, -663, -939, -1538, and -1908 (Houzet and Jeang, 2011; Ruggero et al., 2010; Hakim et al., 2008). Two miRNAs, miRNA-130b and miRNA-93, have been shown to be upregulated in HTLV-1 infected transformed cells, leading to the speculation that miRNAs may play a role in development of ATL as the normally functioning miRNAs regulate tumor suppressor protein TP53INP1 (Houzet and Jeang, 2011; Yeung et al., 2008).

HIV-1. It has been proposed that miRNAs could regulate several steps of the HIV-1 life cycle, either directly by targeting the genomic RNA or indirectly by controlling the cellular environment (Barichievy et al., 2015; Trobaugh and Klimstra, 2017). Expression of 62 miRNAs was reported to be altered between uninfected and HIV-1-infected individuals (Houzet et al., 2008). Interestingly, knockdown of Drosha and Dicer increased HIV-1 expression, suggesting that the RNAi pathway may regulate HIV-1 infection (Houzet and Jeang, 2011; Triboulet et al., 2007). Host miRNAs that have been proposed to target the RNA genome include miR-29a, which is complementary to a region in the *nef* gene (Nathans et al., 2009; Ahluwalia et al., 2008). In addition to simply blocking translation, miR-29a is implicated in the sequestration of HIV-1 RNA in P-bodies and association of the vRNA with RISC components, perhaps functioning in the regulation of viral latency (Nathans et al., 2009; Chable-Bessia et al., 2009). Multiple in silico studies have been carried out to predict additional host miRNAs that could target the HIV-1 genome directly, and the list of hits includes miR-149 (*vpr*), -378 (*env*), and -324-5p (*vif*) (Hariharan et al., 2005). Many studies were performed in specific cell lines or conditions leading to confusion over reproducibility and physiological relevance of miRNAs found to inhibit viral replication; however, the miRNAs uncovered computationally and in some cases experimentally include miR-133b, -138, -149, -326, -92a, -28b, -125b, -150, -223, -382, -29b-3p, -33a-5p, 423, -301a, and -155 (Barichievy et al., 2015; Trobaugh and Klimstra, 2017; Houzet and Jeang, 2011; Ahluwalia et al., 2008; Houzet et al., 2012; Huang et al., 2007; Reynoso et al., 2014; Whisnant et al.,

2013). A cluster of miRNAs, miR-28, miR-125b, miR150, miR-223, and miR-382, are proposed to bind to the 3′UTR of the HIV-1 RNA, suppressing expression of viral proteins and ultimately contributing to viral latency in resting $CD4^+$ T cells (Huang et al., 2007). Certain cells are less susceptible to HIV-1 infection, including peripheral blood monocytes, whereas tissue macrophages are efficiently infected (Kedzierska and Crowe, 2002; Orenstein et al., 1997). As miRNA expression is generally tissue specific, the levels of putative anti-HIV-1 miRNAs (miR28, -150, 223, and 382) were examined in monocytes and macrophages (Wang et al., 2009). Interestingly, monocytes were found to have higher expression levels of these miRNAs than those differentiated into macrophages. Moreover, monocytes in which the targeted miRNAs were knocked down were more permissive to infection, suggesting that the differences in miRNA expression may contribute to viral permissiveness (Wang et al., 2009). Several cellular miRNAs indirectly impacting the HIV-1 life cycle have also been described including miR-20a, -17-5p, -1236, -15a/b, -16, -93, -106b, -198, -27b, -29b, -223, -150, -155, -146a, -888, -132, and -217 (Barichievy et al., 2015; Triboulet et al., 2007; Ma et al., 2014; Shen et al., 2012; Sung and Rice, 2009; Chang et al., 2012; Swaminathan et al., 2012; Chen et al., 2014; Chiang et al., 2013; Zhang et al., 2012). These miRNAs are proposed to have a wide array of functions including suppressing HIV-1 infection by restricting nuclear import of the PIC, cell cycle regulation, and enhancing transcriptional activation of latent viral genomes (Barichievy et al., 2015; Ma et al., 2014; Swaminathan et al., 2012; Chiang et al., 2013). During the progression of HIV-1 infection, many patients develop HIV-associated neurocognitive disorders (HAND) (Xu et al., 2017). Numerous miRNAs were found to have abnormal expression in brain tissue of patients with HAND, including miR-500a-5p, miR-93-3p, miR34c-3p, and miR-381-3p, which were found to be upregulated (Xu et al., 2017). These miRNAs target peroxisome biogenesis factors, resulting in depletion of peroxisomes during HIV-1 infection, which typically plays a role in lipid metabolism and antiviral signaling (Xu et al., 2017). Reduction in peroxisomes over time may contribute to neurological disorders in HIV-1 patients (Xu et al., 2017).

Viral MicroRNAs Encoded Within Genomes

The existence of miRNAs produced from a viral genome was first described for EBV genomic RNA, wherein five miRNAs were initially discovered in 2004 (Pfeffer et al., 2004). Since this early report, computational predictions and experimental evidence support the existence of miRNAs produced from viral genomes, including retroviruses (Tycowski et al., 2015; Groen and Morris, 2013; Ouellet and Provost, 2010; Klase et al., 2013; Houzet and Jeang, 2011). However, many of these miRNAs appear to be expressed at very low levels within the cell and specific functions have yet to be identified for the majority. Coupled with difficulties in reproducibility, the relevance of these transcripts remains a topic of debate (Berkhout and Jeang, 2007; Ouellet and Provost,

2010; Houzet and Jeang, 2011; Lin and Cullen, 2007). A general set of guidelines for the identification of viral miRNAs has been suggested (Tycowski et al., 2015; Klase et al., 2013) including: (1) any ncRNA produced must have importance in the viral life cycle due to space limitations of the genome; (2) the levels of miRNA should be detectable by deep sequencing, indicating that they are expressed at functional concentrations; and (3) the miRNA must have a defined role and should be conserved across clades. Specific examples of miRNAs in retroviral genomes are described below.

Bovine leukemia virus (BLV). One of the early questions regarding miRNA production from retroviruses revolved around how the genome could avoid being processed by the RNAi machinery (Cullen, 2012). Pol II transcripts are the main substrate for the RNAi pathway; however, as mentioned above, Pol III transcripts are occasionally used (Lee et al., 2004; Borchert et al., 2006). Pre-miRNAs transcribed by Pol III are able to bypass Drosha processing, protecting the vRNA from cleavage (Cullen, 2012; Kincaid et al., 2012). As an example, BLV was found to encode a cluster of five miRNAs within SL1 and SL2 stem loops in the 5′UTR named BLV-miR-B1-5 (Kincaid et al., 2012). Interestingly, BLV-miR-B4 has an identical seed sequence to host miR-29, which has previously been implicated in transformation of B cells in mice, and is sufficient to induce tumor formation (Kincaid et al., 2012; Han et al., 2010; Santanam et al., 2010). The natural hosts of BLV are cattle, which develop atypical B-cell ($CD5^+$) tumors that are BLV+ but produce very little viral mRNA, leading to the hypothesis that BLV-miR-B4 may contribute to tumorigenesis (Gillet et al., 2007; Letesson et al., 1990). BLV-miR-B2-5p also shares sequence identity with a host miRNA, miR-943, which has been tied to tumor suppressor p53, hinting at roles for the other BLV-miRs in tumor formation (Kincaid et al., 2012; Chang et al., 2011). In another study, BLV miRNAs were found to be present in leukemic B cells isolated from sheep infected with BLV without detectable levels of Pol II driven viral LTR transcription, supporting the conclusion that miRNAs contribute to a tumorigenic state even in the absence of detectable viral production (Rosewick et al., 2013). The viral miRNAs constituted a high percentage of the host miRNA pool (~40%) and were present in amounts up to 80,000 copies per cell (Rosewick et al., 2013). While the BLV miRNAs may contribute significantly to tumorigenesis, it is also possible that they could lead to viral latency, as suggested for miRNAs produced from other retroviral genomes (Rosewick et al., 2013; Klase et al., 2007). A recent report examined the transcriptomic profile of cells transduced with the BLV miRNAs to identify potential direct and indirect targets and to delineate their impact on the BLV life cycle and B-cell transformation (Gillet et al., 2016). Several mRNAs were found to be targeted by BLV miRNAs, including GZMA, FOS, PPT1, ANXA1, MAP2K1 and PIK3CG, whose gene products are involved in apoptosis, immunity, cell signaling, and oncogenesis pathways (Gillet et al., 2016; Hagn and Jahrsdorfer, 2012; Lieberman, 2010; Grossman et al., 2004). Deletion of miRNAs from the BLV genome leads to decreased viral loads, arguing for a potential role in the viral life cycle; however, the specific role of miRNA is still unknown (Gillet et al., 2016). Intriguingly, BLV is also associated with breast cancer and it has been suggested that a new BLV vaccine in development may exclude attenuated viruses containing miRNAs due to their potential role in tumorigenesis (Gillet et al., 2016; Buehring et al., 2015; Gutierrez et al., 2014).

Bovine foamy Virus (BFV). Three BFV miRNAs produced from the viral genome have been described (Whisnant et al., 2014). BFV miRNAs are transcribed by RNA Pol III as a single transcript from the *bet* gene (Tycowski et al., 2015; Whisnant et al., 2014). However, unlike BLV where the pre-miRNA is exported into the cytoplasm bypassing Drosha, it appears that initial processing of the pri-miRNA transcript may be carried out by Drosha (Whisnant et al., 2014). BFV miR-BF2-5p and miR-BF1-5p are highly expressed in chronically infected cells, accounting for over 50% and 19% of all small RNA within the cell, respectively (Whisnant et al., 2014). These miRNAs are also detected in calves 6 months postinfection, indicating that they are expressed in vivo, although the impact of miRNA expression remains unknown (Whisnant et al., 2014). The seed sequence of miR-BF1-5p is identical to miR-B5, a miRNA expressed from the bovine herpesvirus 1, suggesting they may have a common target (Whisnant et al., 2014; Glazov et al., 2010).

ALV and RSV. A *cis*-acting element within ALV-J genomes, E (XSR), encodes a 148-nt ncRNA that has the ability to fold into a stem–loop structure similar to those processed by Dicer (Yao et al., 2014; Chesters et al., 2006). The XSR element is conserved across most ALV-J isolates and appears to contribute to oncogenesis in some cases (Chesters et al., 2006; Wu et al., 2010). Deep sequencing of an ALV cell line revealed the XSR element produced a miRNA (Yao et al., 2014). Unlike the other retrovirus miRNAs described above, alpha-amanitin blocked expression of the viral miRNA, indicating that transcription is Pol II dependent (Yao et al., 2014). This was the first description of a retrovirus using the canonical RNAi pathway to produce viral miRNAs.

Simian foamy Virus (SFV). Putative viral miRNAs were also predicted to be produced from the SFV genome using a combined computational and synthetic approach (Klase et al., 2013; Kincaid et al., 2014). Two miRNAs, SFVagm-miR-S4 and SFVagm-miR-S6, were shown to be produced in cells transfected with the proviral African green monkey SFV genome and corresponded to sequences in the LTR (Tycowski et al., 2015; Kincaid et al., 2014). While the miRNAs are transcribed by RNA Pol III, knockdown of Drosha resulted in decreased viral miRNA production, indicating Drosha processing is necessary for expression (Kincaid et al., 2014). Like other viral miRNAs, SFVagm-miR-S4 and SFVagm-miR-S6 share seed sequences with host miRNAs, miR-155, and miR-132/212, respectively (Kincaid et al., 2014). Host miR-155 is associated with cell survival and proliferation, and mimics of this miRNA have been found encoded in several herpesvirus genomes (Kincaid et al., 2014; Gottwein et al., 2007; Linnstaedt et al., 2010; Skalsky et al., 2007). Similarly, the miRNA family 132/212 is implicated in innate immunity, and the miRNAs are upregulated during Kaposi's sarcoma–associated herpesvirus infection (Kincaid et al., 2014; Lagos et al., 2010). Although the SFV miRNAs resemble host miRNAs that play roles in herpesvirus infection, the relevance of their expression to the SFV life cycle is unknown.

HIV-1. Although several miRNAs are predicted to originate from the HIV-1 genome, the topic remains controversial due to difficulties in reproducibility (Groen and Morris, 2013; Klase et al., 2013; Umbach and Cullen, 2009; Lin and Cullen, 2007; Pfeffer et al., 2005). Although many retroviral miRNAs appear to mimic host miRNAs that modulate the cellular environment, HIV-1 viral miRNAs appear to target the virus itself to regulate latency (Groen and Morris, 2013; Ouellet and Provost, 2010; Klase et al., 2013; Houzet and Jeang, 2011). In silico methods initially predicted that HIV-1 encoded five miRNAs within the TAR stem loop, R region, *gag*, *gag-pol* frame shift, and *nef* (Houzet and Jeang, 2011; Bennasser et al., 2004). Subsequently, a study was published illustrating the ability of miR-N367, produced from the *nef* region of the viral genome, to negatively regulate HIV-1 expression, potentially contributing to viral latency (Omoto et al., 2004). The 5′UTR of HIV-1 is highly structured and contains many features critical to the viral life cycle including the Psi packaging signal (Balvay et al., 2007) and TAR; the latter forms a stable stem–loop integral to efficient transcription through binding to the Tat protein (Gatignol et al., 1991). In the absence of Tat, for instance, potentially during latency, only short viral transcripts are made that contain TAR RNA (Feinberg et al., 1991; Van Lint et al., 2013). In addition to sequestering TRBP, TAR RNA is capable of being processed by dicer and miRNAs derived from this region were detected by RNase-protection assays in infected T cells (Klase et al., 2007). One TAR-derived miRNA was found to recruit HDAC-1, a histone deacetylase associated with HIV-1 latency, to the viral LTR (Klase et al., 2007). TAR RNA-derived miRNAs, miR-TAR-5p and -3p, have also been identified at low levels (Ouellet et al., 2008). Another conserved double-stranded region of the HIV-1 RNA genome is formed when $tRNA^{Lys3}$ is annealed to the PBS (Hu and Hughes, 2012). This region was found to associate with Ago2 of the RISC complex, potentially activating the RNAi pathway as an antiviral response to infection (Yeung et al., 2009). Intriguingly, the PBS:tRNA hybrid region appears to be processed into miRNAs in ERVs from mice and humans, implicating the host RNAi machinery in viral suppression (Calabrese et al., 2007; Kawaji et al., 2008). Although, the HIV-1 gRNA–tRNA hybrid-derived ncRNA was also found in a later deep sequencing study, the authors concluded it was more likely a degradation product of $tRNA^{Lys3}$ due to its presence in uninfected controls (Schopman et al., 2012). Another in silico alignment of matching regions of the HIV-1 genome and host mRNAs involved in immune responses predicted multiple targets for viral miRNA, though they remain to be validated (Groen and Morris, 2013; Couturier and Root-Bernstein, 2005). An *env* derived miRNA was found to inhibit HIV-1 expression similar to TAR-derived miRNA; however, in the presence of Tat, this inhibitory effect was largely abolished suggesting the virus had found a way to adapt to the host RNAi response (Bennasser et al., 2005).

Viral Suppressors of RNA Silencing

Retroviruses are notorious for their ability to evade host defenses, whether directly through accessory proteins or by modulating translation to favor viral protein production (Malim and Emerman, 2008). Many plant viruses and some insect viruses have adapted to RNAi immune responses by evolving suppressors of RNA silencing (SRS) with the ability to bind and block the function of RISC components or to sequester miRNAs directed against the virus (Voinnet et al., 1999; Umbach and Cullen, 2009; Li et al., 2002). As mentioned above, tomato bushy stunt virus encodes a protein, p19, which functions by binding specifically to short dsRNAs containing the typical 2-nt overhang associated with miRNAs, effectively blocking RISC:miRNA association (Umbach and Cullen, 2009; Qian et al., 2009; Scholthof, 2006). Therefore, it was natural, given the involvement of RNAi in other viral life cycles, to suspect that retroviral genomes might also contain SRS proteins or broaden the functions of existing viral proteins to serve this role (Houzet and Jeang, 2011). Tat has also been implicated as an SRS protein, possibly through hindering Dicer activity (Bennasser et al., 2005; Bennasser and Jeang, 2006). HTLV-1 encodes an accessory protein, Rex, which binds to the Rex-response element and facilitates export of incompletely spliced viral mRNAs (Ballaun et al., 1991; Kusuhara et al., 1999; Younis and Green, 2005). As a potential SRS, Rex is able to block the processing of pri-miRNA by Dicer, although whether the suppression is mediated through dsRNA sequestration or direct binding to Dicer is unknown (Abe et al., 2010). The traditional role of the PFV Tas protein is as a transactivator of PFV transcription (Falcone et al., 2003). As mentioned above, host miR-32 targets PFV to inhibit translation of viral proteins (Lecellier et al., 2005). However, Tas expression caused an accumulation of host miRNA and ameliorated translational repression, supporting Tas as a general RNAi suppressor (Lecellier et al., 2005). Several studies have also been published showing little effect of Tas, Rex, and Tat proteins on the host RNA silencing pathway and argue that the interferon system is much more sensitive and effective than the SRS pathway in mammals (Umbach and Cullen, 2009; Lin and Cullen, 2007).

THERAPEUTICS

Current retroviral treatments for HIV-1 require daily doses of highly active antiretroviral therapy, a combinational regimen targeting multiple steps of the retroviral life cycle (Gulick et al., 1997). Unfortunately, over time the virus is able to adapt, requiring the patient to switch antivirals to avoid viral rebound (Clavel and Hance, 2004; Taylor and Hammer, 2008). Additionally, the side effects caused by treatment are often severe and worsen over time, resulting in nonadherence to the therapy schedule, which can result in selective pressure for viral mutagenesis under suboptimal concentrations of inhibitors (d'Arminio Monforte et al., 2000). With the discovery of RNAi, a potential new approach to viral treatment became available, wherein viral translation could be targeted and suppressed by introduction of engineered small-hairpin RNAs (shRNAs) (Janssen et al., 2013). Applying siRNA strategies to other viral and cancer targets has been employed with varying degrees of success (Trobaugh and Klimstra, 2017; Barnes et al., 2008; van Rooij et al., 2008; Lanford et al., 2010; Dunning et al., 2016; Soifer et al., 2007). Several problems have plagued the development of siRNA-based therapeutics, including efficient delivery to targeted tissues, viral escape through mutagenesis, saturation of RNAi machinery, and stability of introduced siRNAs (Almeida et al., 2011; Trobaugh and Klimstra, 2017; Houzet and Jeang, 2011; van Rooij et al., 2008; Krutzfeldt et al., 2005; Hutvagner et al., 2004). An shRNA vaccine for poliovirus has been tested and found comparable to the Sabin type 1 poliovirus vaccine in mice, although partial deletions of the target viral sequence occur (Barnes et al., 2008). Hepatitis C virus (HCV), a causative agent of liver disease, was the target of another study, as miR-122 is necessary for HCV infection (Jopling et al., 2005, 2008). Weekly injections of a locked nucleic acid–modified oligonucleotide complementary to miR-122 reduced levels of HCV RNA in infected chimpanzees without viral mutation or deleterious side effects (Lanford et al., 2010). Additionally, a more recent study attempted to create a siRNA therapeutic against Ebola virus disease. Unfortunately, the siRNA-loaded lipid nanoparticles did not significantly increase survival rates (Dunning et al., 2016).

Single siRNA species were initially used to target specific regions in the HIV-1 genome (Das et al., 2004; Boden et al., 2003; Leonard et al., 2008; Westerhout et al., 2005). For example, a single shRNA directed against *tat* was able to repress HIV-1 transcription by 95% (Boden et al., 2003). However, when cultured for longer periods of time, HIV-1 was able to adapt by a single-point mutation rendering the shRNA ineffective (Boden et al., 2003). Continuous inhibition of HIV-1 using an siRNA against *nef* was found to induce viral mutation and eventual escape (Das et al., 2004). Interestingly, while viral escape has been linked to point mutations conferring resistance to siRNAs, a follow-up study found that some HIV-1 viral escape mutants resulted in mutations outside of the target sequence resulting in structural changes in the viral RNA (Westerhout et al., 2005). In another example of unusual viral escape, siRNAs against the TAR stem loop resulted in compensatory mutations in the promoter region that enhanced viral transcription (Leonard et al., 2008). As with antiretroviral drugs, siRNA therapy would most likely need to be given in combination for successful inhibition (Das et al., 2004; Boden et al., 2003). In silico predictions suggested that expression of at least four shRNAs would be needed to maintain a high degree of viral repression and decrease the ability of HIV-1 to evade silencing (Applegate et al., 2010). This was corroborated with lentiviral transduction of a combination of four siRNAs targeting Gag, Pol, and R/T-5 regions of HIV-1. Using this approach, HIV-1 replication was inhibited for 40 days without viral escape, although some viral spread was observed, suggesting the RNAi pathway could have been saturated (ter Brake et al., 2008). It also became clear that to sustain expression of siRNAs, lentiviral vectors lacking any sequence repetition would be necessary to avoid recombination and to induce stably expressing cells (Applegate et al., 2010; ter Brake et al., 2008). Intriguingly, HIV-1 mutations tend to follow predictable routes, leading to the concept of coupling second generation siRNA species targeting likely resistance mutations to primary siRNA treatments, thereby enhancing the effectiveness of siRNAs (Schopman et al., 2010). Unfortunately, even with the use of second-generation siRNAs, the virus was still able to escape restriction, adapting with novel mutations (Schopman et al., 2010). An additional concern of shRNA therapy is the reduced expression of Dicer in monocytes (Coley et al., 2010). However, a subset of shorter shRNAs are able to bypass dicer, instead being processed by Ago2 directly and creating a single guide strand, AgoshRNAs (Liu et al., 2013). This approach presents several advantages to traditional shRNAs, including limited off-target toxicity from the passenger strand (Herrera-Carrillo et al., 2017; Herrera-Carrillo and Berkhout, 2017). Successful AgoshRNAs have been designed against the CCR5 coreceptor, Gag, and Pol (Herrera-Carrillo et al., 2017; Herrera-Carrillo and Berkhout, 2017).

Another approach to RNA-based therapeutics is to cleave or delete portions of the integrated or RNA HIV-1 genome through targeted enzymatic activities to disrupt the life cycle (Mitsuyasu et al., 2009; Herrera-Carrillo and Berkhout, 2016; Amado et al., 2004; Rossi, 1999). Transducing CD34$^+$ hematopoietic stem cells with a ribozyme targeted to the *vpr/tat* reading frames and introducing them into HIV+ patients resulted in a slightly decreased viral load (Mitsuyasu et al., 2009). Although the transduced stem cell percentage was likely low, there were no reported negative effects, providing evidence that gene therapy is possible (Mitsuyasu et al., 2009). Zinc-finger nucleases, recombinases and transcription activator–like effector nuclease strategies have been developed for HIV-1 treatment and several have been examined in clinical trials (Sarkar et al., 2007; Tebas et al., 2014; Qu et al., 2013). With the development of the clustered regularly interspaced short palindromic repeat (CRISPR)-CRISPR-associated (Cas) editing system, new approaches to gene therapy have become available (Herrera-Carrillo and Berkhout, 2016). This system relies on guide RNAs that target specific sequences of DNA, which in turn can be cleaved by the endonuclease, Cas9 (Hsu et al., 2014). Although this technique has several advantages, it does require the introduction of the relatively large CRISPR and Cas9 into the genome, which is accomplished by production of stable target cell lines expressing humanized Cas9 and several guide RNA sequences by lentiviral transduction (Wang et al., 2016). However, HIV-1 has been shown to alter nucleotides around the cleavage site to reduce effectiveness of Cas9 (Hsu et al., 2014). Additionally, introduction of CRISPR/Cas can actually accelerate viral escape because once the viral DNA is cleaved, nonhomologous end joining machinery can repair the dsDNA breaks creating resistant viruses within 36 h of infection (Wang et al., 2016; Liang et al., 2016). Alternative nonviral targets have been suggested to limit the spread of HIV-1 infection, including the coreceptors CXCR4 and CCR5. Mutation of CXCR4 and CCR5 has been successful in conferring resistance against viral infection, which could be used in combination therapies (Hou et al., 2015; Ye et al., 2014). Although there are still many aspects of gene therapy that need to be studied and understood in more detail, great strides have been made in the area of RNA-based therapeutics. Many of the lessons learned have been applied to general techniques in the field of virology and cell biology, including the development of stable lentiviral gene–silencing approaches (Hsu et al., 2014; Stewart et al., 2003).

CONCLUDING REMARKS

The discovery that the human genome is predominantly composed of ncRNA has propelled the field of RNA biology into many new areas, such as exogenous siRNA gene regulation, and has led to a deeper understanding of cellular biology. It is now clear that ncRNAs are critical to cellular responses ranging from general homeostasis to the disease states of viral infection and cancer. RNAs that were once considered to be well understood, such as tRNAs, are now thought to possess many alternative functions, including those mediated by tRNA fragments. Many newly discovered ncRNAs, such as vltRNAs, are still being characterized at a basic level. The impact of RNA on retroviral life cycles will undoubtedly grow as novel roles for cellular and viral ncRNA continue to be elucidated. Breakthroughs that have been made in the field to date lay the groundwork for future endeavors that could lead to the development of effective RNA-based therapies.

ACKNOWLEDGMENTS

Support for this work is from NIH grants R01 GM113887 and RO1 GM065056 to KMF and F31 GM119178 to AD. Support from the Ohio State Center for RNA Biology is also gratefully acknowledged.

REFERENCES

Abe, M., Suzuki, H., Nishitsuji, H., Shida, H., Takaku, H., 2010. Interaction of human T-cell lymphotropic virus type I Rex protein with Dicer suppresses RNAi silencing. FEBS Lett. 584, 4313–4318.

Achong, B.G., Trumper, P.A., Giovanella, B.C., 1976. C-type virus particles in human tumours transplanted into nude mice. Br. J. Cancer 34, 203–206.

Adams, J.M., Cory, S., 1992. Oncogene co-operation in leukaemogenesis. Cancer Surv. 15, 119–141.

Agy, M.B., Wambach, M., Foy, K., Katze, M.G., 1990. Expression of cellular genes in CD4 positive lymphoid cells infected by the human immunodeficiency virus, HIV-1: evidence for a host protein synthesis shut-off induced by cellular mRNA degradation. Virology 177, 251–258.

Ahluwalia, J.K., Khan, S.Z., Soni, K., Rawat, P., Gupta, A., Hariharan, M., Scaria, V., Lalwani, M., Pillai, B., Mitra, D., Brahmachari, S.K., 2008. Human cellular microRNA hsa-miR-29a interferes with viral nef protein expression and HIV-1 replication. Retrovirology 5, 117.

Akiyoshi, D.E., Denaro, M., Zhu, H., Greenstein, J.L., Banerjee, P., Fishman, J.A., 1998. Identification of a full-length cDNA for an endogenous retrovirus of miniature swine. J. Virol. 72, 4503–4507.

Alfadhli, A., Still, A., Barklis, E., 2009. Analysis of human immunodeficiency virus type 1 matrix binding to membranes and nucleic acids. J. Virol. 83, 12196–12203.

Ali, L.M., Rizvi, T.A., Mustafa, F., 2016. Cross- and co-packaging of retroviral RNAs and their consequences. Viruses 8.

Almeida, M.I., Reis, R.M., Calin, G.A., 2011. MicroRNA history: discovery, recent applications, and next frontiers. Mutat. Res. 717, 1–8.

Amado, R.G., Mitsuyasu, R.T., Rosenblatt, J.D., Ngok, F.K., Bakker, A., Cole, S., Chorn, N., Lin, L.S., Bristol, G., Boyd, M.P., MacPherson, J.L., Fanning, G.C., Todd, A.V., Ely, J.A., Zack, J.A., Symonds, G.P., 2004. Anti-human immunodeficiency virus hematopoietic progenitor cell-delivered ribozyme in a phase I study: myeloid and lymphoid reconstitution in human immunodeficiency virus type-1-infected patients. Hum. Gene Ther. 15, 251–262.

Anderson, P., Ivanov, P., 2014. tRNA fragments in human health and disease. FEBS Lett. 588, 4297–4304.

Applegate, T.L., Birkett, D.J., McIntyre, G.J., Jaramillo, A.B., Symonds, G., Murray, J.M., 2010. In silico modeling indicates the development of HIV-1 resistance to multiple shRNA gene therapy differs to standard antiretroviral therapy. Retrovirology 7, 83.

Askew, D.S., Bartholomew, C., Buchberg, A.M., Valentine, M.B., Jenkins, N.A., Copeland, N.G., Ihle, J.N., 1991. His-1 and His-2: identification and chromosomal mapping of two commonly rearranged sites of viral integration in a myeloid leukemia. Oncogene 6, 2041–2047.

Askew, D.S., Li, J., Ihle, J.N., 1994. Retroviral insertions in the murine His-1 locus activate the expression of a novel RNA that lacks an extensive open reading frame. Mol. Cell. Biol. 14, 1743–1751.

Bai, X.T., Nicot, C., 2015. miR-28-3p is a cellular restriction factor that inhibits human T cell leukemia virus, type 1 (HTLV-1) replication and virus infection. J. Biol. Chem. 290, 5381–5390.

Ballaun, C., Farrington, G.K., Dobrovnik, M., Rusche, J., Hauber, J., Bohnlein, E., 1991. Functional analysis of human T-cell leukemia virus type I rex-response element: direct RNA binding of Rex protein correlates with in vivo activity. J. Virol. 65, 4408–4413.

Balvay, L., Lopez Lastra, M., Sargueil, B., Darlix, J.L., Ohlmann, T., 2007. Translational control of retroviruses. Nat. Rev. Microbiol. 5, 128–140.

Barbeau, B., Mesnard, J.M., 2011. Making sense out of antisense transcription in human T-cell lymphotropic viruses (HTLVs). Viruses 3, 456–468.

Barichievy, S., Naidoo, J., Mhlanga, M.M., 2015. Non-coding RNAs and HIV: viral manipulation of host dark matter to shape the cellular environment. Front. Genet. 6, 108.

Barnes, D., Kunitomi, M., Vignuzzi, M., Saksela, K., Andino, R., 2008. Harnessing endogenous miRNAs to control virus tissue tropism as a strategy for developing attenuated virus vaccines. Cell Host Microbe 4, 239–248.

Bartosch, B., Weiss, R.A., Takeuchi, Y., 2002. PCR-based cloning and immunocytological titration of infectious porcine endogenous retrovirus subgroup A and B. J. Gen. Virol. 83, 2231–2240.

Bennasser, Y., Jeang, K.T., 2006. HIV-1 tat interaction with dicer: requirement for RNA. Retrovirology 3, 95.

Bennasser, Y., Le, S.Y., Yeung, M.L., Jeang, K.T., 2004. HIV-1 encoded candidate micro-RNAs and their cellular targets. Retrovirology 1, 43.

Bennasser, Y., Le, S.Y., Benkirane, M., Jeang, K.T., 2005. Evidence that HIV-1 encodes an siRNA and a suppressor of RNA silencing. Immunity 22, 607–619.

Bennasser, Y., Yeung, M.L., Jeang, K.T., 2006. HIV-1 TAR RNA subverts RNA interference in transfected cells through sequestration of TAR RNA-binding protein, TRBP. J. Biol. Chem. 281, 27674–27678.

Berkhout, B., Jeang, K.T., 2007. RISCy business: microRNAs, pathogenesis, and viruses. J. Biol. Chem. 282, 26641–26645.

Berkhout, B., 1997. The primer binding site on the RNA genome of human and simian immunodeficiency viruses is flanked by an upstream hairpin structure. Nucleic Acids Res. 25, 4013–4017.

Berkowitz, R., Fisher, J., Goff, S.P., 1996. RNA packaging. Curr. Top. Microbiol. Immunol. 214, 177–218.

Bernstein, E., Caudy, A.A., Hammond, S.M., Hannon, G.J., 2001. Role for a bidentate ribonuclease in the initiation step of RNA interference. Nature 409, 363–366.

Bet, A., Maze, E.A., Bansal, A., Sterrett, S., Gross, A., Graff-Dubois, S., Samri, A., Guihot, A., Katlama, C., Theodorou, I., Mesnard, J.M., Moris, A., Goepfert, P.A., Cardinaud, S., 2015. The HIV-1 antisense protein (ASP) induces CD8 T cell responses during chronic infection. Retrovirology 12, 15.

Birney, E., Stamatoyannopoulos, J.A., Dutta, A., Guigo, R., Gingeras, T.R., Margulies, E.H., Weng, Z., Snyder, M., Dermitzakis, E.T., Thurman, R.E., Kuehn, M.S., Taylor, C.M., Neph, S., Koch, C.M., Asthana, S., Malhotra, A., Adzhubei, I., Greenbaum, J.A., Andrews, R.M., Flicek, P., Boyle, P.J., Cao, H., Carter, N.P., Clelland, G.K., Davis, S., Day, N., Dhami, P., Dillon, S.C., Dorschner, M.O., Fiegler, H., Giresi, P.G., Goldy, J., Hawrylycz, M., Haydock, A., Humbert, R., James, K.D., Johnson, B.E., Johnson, E.M., Frum, T.T., Rosenzweig, E.R., Karnani, N., Lee, K., Lefebvre, G.C., Navas, P.A., Neri, F., Parker, S.C., Sabo, P.J., Sandstrom, R., Shafer, A., Vetrie, D., Weaver, M., Wilcox, S., Yu, M., Collins, F.S., Dekker, J., Lieb, J.D., Tullius, T.D., Crawford, G.E., Sunyaev, S., Noble, W.S., Dunham, I., Denoeud, F., Reymond, A., Kapranov, P., Rozowsky, J., Zheng, D., Castelo, R., Frankish, A., Harrow, J., Ghosh, S., Sandelin, A., Hofacker, I.L., Baertsch, R., Keefe, D., Dike, S., Cheng, J., Hirsch, H.A., Sekinger, E.A., Lagarde, J., Abril, J.F., Shahab, A., Flamm, C., Fried, C., Hackermuller, J., Hertel, J., Lindemeyer, M., Missal, K., Tanzer, A., Washietl, S., Korbel, J., Emanuelsson, O., Pedersen, J.S., Holroyd, N., Taylor, R., Swarbreck, D., Matthews, N., Dickson, M.C., Thomas, D.J., Weirauch, M.T., Gilbert, J., Drenkow, J., Bell, I., Zhao, X., Srinivasan, K.G., Sung, W.K., Ooi, H.S., Chiu, K.P., Foissac, S., Alioto, T., Brent, M., Pachter, L., Tress, M.L., Valencia, A., Choo, S.W., Choo, C.Y., Ucla, C., Manzano, C., Wyss, C., Cheung, E., Clark, T.G., Brown, J.B., Ganesh, M., Patel, S., Tammana, H., Chrast, J., Henrichsen, C.N., Kai, C., Kawai, J., Nagalakshmi, U., Wu, J., Lian, Z., Lian, J., Newburger, P., Zhang, X., Bickel, P., Mattick, J.S., Carninci, P., Hayashizaki, Y., Weissman, S., Hubbard, T., Myers, R.M., Rogers, J., Stadler, P.F., Lowe, T.M., Wei, C.L., Ruan, Y., Struhl, K., Gerstein, M., Antonarakis, S.E., Fu, Y., Green, E.D., Karaoz, U., Siepel, A., Taylor, J., Liefer, L.A., Wetterstrand, K.A., Good, P.J., Feingold, E.A., Guyer, M.S., Cooper, G.M., Asimenos, G., Dewey, C.N., Hou, M., Nikolaev, S., Montoya-Burgos, J.I., Loytynoja, A., Whelan, S., Pardi, F., Massingham, T., Huang, H., Zhang, N.R., Holmes, I., Mullikin, J.C., Ureta-Vidal, A., Paten, B., Seringhaus, M., Church, D., Rosenbloom, K., Kent, W.J., Stone, E.A., Batzoglou, S., Goldman, N., Hardison, R.C., Haussler, D., Miller, W., Sidow, A., Trinklein, N.D., Zhang, Z.D., Barrera, L., Stuart, R., King, D.C., Ameur, A., Enroth, S., Bieda, M.C., Kim, J., Bhinge, A.A., Jiang, N., Liu, J., Yao, F., Vega, V.B., Lee, C.W., Ng, P., Yang, A., Moqtaderi, Z., Zhu, Z., Xu, X., Squazzo, S., Oberley, M.J., Inman, D., Singer, M.A., Richmond, T.A., Munn, K.J., Rada-Iglesias, A., Wallerman, O., Komorowski, J., Fowler, J.C., Couttet, P., Bruce, A.W., Dovey, O.M., Ellis, P.D., Langford, C.F., Nix, D.A., Euskirchen, G., Hartman, S., Urban, A.E., Kraus, P., Van Calcar, S., Heintzman, N., Kim, T.H., Wang, K., Qu, C., Hon, G., Luna, R., Glass, C.K., Rosenfeld,

M.G., Aldred, S.F., Cooper, S.J., Halees, A., Lin, J.M., Shulha, H.P., Xu, M., Haidar, J.N., Yu, Y., Iyer, V.R., Green, R.D., Wadelius, C., Farnham, P.J., Ren, B., Harte, R.A., Hinrichs, A.S., Trumbower, H., Clawson, H., Hillman-Jackson, J., Zweig, A.S., Smith, K., Thakkapallayil, A., Barber, G., Kuhn, R.M., Karolchik, D., Armengol, L., Bird, C.P., de Bakker, P.I., Kern, A.D., Lopez-Bigas, N., Martin, J.D., Stranger, B.E., Woodroffe, A., Davydov, E., Dimas, A., Eyras, E., Hallgrimsdottir, I.B., Huppert, J., Zody, M.C., Abecasis, G.R., Estivill, X., Bouffard, G.G., Guan, X., Hansen, N.F., Idol, J.R., Maduro, V.V., Maskeri, B., McDowell, J.C., Park, M., Thomas, P.J., Young, A.C., Blakesley, R.W., Muzny, D.M., Sodergren, E., Wheeler, D.A., Worley, K.C., Jiang, H., Weinstock, G.M., Gibbs, R.A., Graves, T., Fulton, R., Mardis, E.R., Wilson, R.K., Clamp, M., Cuff, J., Gnerre, S., Jaffe, D.B., Chang, J.L., Lindblad-Toh, K., Lander, E.S., Koriabine, M., Nefedov, M., Osoegawa, K., Yoshinaga, Y., Zhu, B., de Jong, P.J., 2007. Identification and analysis of functional elements in 1% of the human genome by the ENCODE pilot project. Nature 447, 799–816.

Bishop, J.M., Levinson, W.E., Sullivan, D., Fanshier, L., Quintrell, N., Jackson, J., 1970. The low molecular weight RNAs of Rous sarcoma virus. II. The 7 S RNA. Virology 42, 927–937.

Blond, J.L., Beseme, F., Duret, L., Bouton, O., Bedin, F., Perron, H., Mandrand, B., Mallet, F., 1999. Molecular characterization and placental expression of HERV-W, a new human endogenous retrovirus family. J. Virol. 73, 1175–1185.

Boden, D., Pusch, O., Lee, F., Tucker, L., Ramratnam, B., 2003. Human immunodeficiency virus type 1 escape from RNA interference. J. Virol. 77, 11531–11535.

Boeke, J.D., Stoye, J.P., 1997. Retrotransposons, Endogenous Retroviruses, and the Evolution of Retroelements.

Bogerd, H.P., Wiegand, H.L., Doehle, B.P., Lueders, K.K., Cullen, B.R., 2006. APOBEC3A and APOBEC3B are potent inhibitors of LTR-retrotransposon function in human cells. Nucleic Acids Res. 34, 89–95.

Bolinger, C., Yilmaz, A., Hartman, T.R., Kovacic, M.B., Fernandez, S., Ye, J., Forget, M., Green, P.L., Boris-Lawrie, K., 2007. RNA helicase A interacts with divergent lymphotropic retroviruses and promotes translation of human T-cell leukemia virus type 1. Nucleic Acids Res. 35, 2629–2642.

Bolinger, C., Sharma, A., Singh, D., Yu, L., Boris-Lawrie, K., 2010. RNA helicase A modulates translation of HIV-1 and infectivity of progeny virions. Nucleic Acids Res. 38, 1686–1696.

Bonar, R.A., Beard, J.W., 1959. Virus of avian myeloblastosis. XII. Chemical constitution. J. Natl. Cancer Inst. 23, 183–197.

Bonar, R.A., Sverak, L., Bolognesi, D.P., Langlois, A.J., Beard, D., Beard, J.W., 1967. Ribonucleic acid components of BAI strain A (myeloblastosis) avian tumor virus. Cancer Res. 27, 1138–1157.

Borchert, G.M., Lanier, W., Davidson, B.L., 2006. RNA polymerase III transcribes human microRNAs. Nat. Struct. Mol. Biol. 13, 1097–1101.

Brameier, M., Ibing, W., Hofer, K., Montag, J., Stahl-Hennig, C., Motzkus, D., 2013. Mapping the small RNA content of simian immunodeficiency virions (SIV). PLoS One 8, e75063.

Brasey, A., Lopez-Lastra, M., Ohlmann, T., Beerens, N., Berkhout, B., Darlix, J.L., Sonenberg, N., 2003. The leader of human immunodeficiency virus type 1 genomic RNA harbors an internal ribosome entry segment that is active during the G2/M phase of the cell cycle. J. Virol. 77, 3939–3949.

Brian, D.A., Thomason, A.R., Rottman, F.M., Velicer, L.F., 1975. Properties of feline leukemia virus. III. Analysis of the RNA. J. Virol. 16, 535–545.

Brierley, I., Dos Ramos, F.J., 2006. Programmed ribosomal frameshifting in HIV-1 and the SARS-CoV. Virus Res. 119, 29–42.

Broecker, F., Horton, R., Heinrich, J., Franz, A., Schweiger, M.R., Lehrach, H., Moelling, K., 2016. The intron-enriched HERV-K(HML-10) family suppresses apoptosis, an indicator of malignant transformation. Mob. DNA 7, 25.

Buck, C.B., Shen, X., Egan, M.A., Pierson, T.C., Walker, C.M., Siliciano, R.F., 2001. The human immunodeficiency virus type 1 gag gene encodes an internal ribosome entry site. J. Virol. 75, 181–191.

Buehring, G.C., Shen, H.M., Jensen, H.M., Jin, D.L., Hudes, M., Block, G., 2015. Exposure to bovine leukemia virus is associated with breast cancer: a case-control study. PLoS One 10, e0134304.

Cai, X., Hagedorn, C.H., Cullen, B.R., 2004. Human microRNAs are processed from capped, polyadenylated transcripts that can also function as mRNAs. RNA 10, 1957–1966.

Calabrese, J.M., Seila, A.C., Yeo, G.W., Sharp, P.A., 2007. RNA sequence analysis defines Dicer's role in mouse embryonic stem cells. Proc. Natl. Acad. Sci. U.S.A. 104, 18097–18102.

Cann, A.J., 2001. Principles of Molecular Virology (Standard Edition). Elsevier Science.

Cen, S., Khorchid, A., Javanbakht, H., Gabor, J., Stello, T., Shiba, K., Musier-Forsyth, K., Kleiman, L., 2001. Incorporation of lysyl-tRNA synthetase into human immunodeficiency virus type 1. J. Virol. 75, 5043–5048.

Cen, S., Javanbakht, H., Kim, S., Shiba, K., Craven, R., Rein, A., Ewalt, K., Schimmel, P., Musier-Forsyth, K., Kleiman, L., 2002. Retrovirus-specific packaging of aminoacyl-tRNA synthetases with cognate primer tRNAs. J. Virol. 76, 13111–13115.

Cen, S., Javanbakht, H., Niu, M., Kleiman, L., 2004. Ability of wild-type and mutant lysyl-tRNA synthetase to facilitate tRNA(Lys) incorporation into human immunodeficiency virus type 1. J. Virol. 78, 1595–1601.

Chable-Bessia, C., Meziane, O., Latreille, D., Triboulet, R., Zamborlini, A., Wagschal, A., Jacquet, J.M., Reynes, J., Levy, Y., Saib, A., Bennasser, Y., Benkirane, M., 2009. Suppression of HIV-1 replication by microRNA effectors. Retrovirology 6, 26.

Chang, C.J., Chao, C.H., Xia, W., Yang, J.Y., Xiong, Y., Li, C.W., Yu, W.H., Rehman, S.K., Hsu, J.L., Lee, H.H., Liu, M., Chen, C.T., Yu, D., Hung, M.C., 2011. p53 regulates epithelial-mesenchymal transition and stem cell properties through modulating miRNAs. Nat. Cell Biol. 13, 317–323.

Chang, C.F., D'Souza, W.N., Ch'en, I.L., Pages, G., Pouyssegur, J., Hedrick, S.M., 2012. Polar opposites: Erk direction of CD4 T cell subsets. J. Immunol. 189, 721–731.

Cheevers, W.P., Archer, B.G., Crawford, T.B., 1977. Characterization of RNA from equine infectious anemia virus. J. Virol. 24, 489–497.

Chen, L.L., Carmichael, G.G., 2009. Altered nuclear retention of mRNAs containing inverted repeats in human embryonic stem cells: functional role of a nuclear noncoding RNA. Mol. Cell 35, 467–478.

Chen, X., Wolin, S.L., 2004. The Ro 60 kDa autoantigen: insights into cellular function and role in autoimmunity. J. Mol. Med. (Berl.) 82, 232–239.

Chen, X., Smith, J.D., Shi, H., Yang, D.D., Flavell, R.A., Wolin, S.L., 2003. The Ro autoantigen binds misfolded U2 small nuclear RNAs and assists mammalian cell survival after UV irradiation. Curr. Biol. 13, 2206–2211.

Chen, H., Lilley, C.E., Yu, Q., Lee, D.V., Chou, J., Narvaiza, I., Landau, N.R., Weitzman, M.D., 2006. APOBEC3A is a potent inhibitor of adeno-associated virus and retrotransposons. Curr. Biol. 16, 480–485.

Chen, A.K., Sengupta, P., Waki, K., Van Engelenburg, S.B., Ochiya, T., Ablan, S.D., Freed, E.O., Lippincott-Schwartz, J., 2014. MicroRNA binding to the HIV-1 Gag protein inhibits Gag assembly and virus production. Proc. Natl. Acad. Sci. U.S.A. 111, E2676–E2683.

Chendrimada, T.P., Gregory, R.I., Kumaraswamy, E., Norman, J., Cooch, N., Nishikura, K., Shiekhattar, R., 2005. TRBP recruits the Dicer complex to Ago2 for microRNA processing and gene silencing. Nature 436, 740–744.

Chesters, P.M., Smith, L.P., Nair, V., 2006. E (XSR) element contributes to the oncogenicity of Avian leukosis virus (subgroup J). J. Gen. Virol. 87, 2685–2692.

Chiang, K., Liu, H., Rice, A.P., 2013. miR-132 enhances HIV-1 replication. Virology 438, 1–4.

Chiu, I.M., Skuntz, S.F., 1986. Nucleotide sequence analysis of squirrel monkey retrovirus reveals a novel primer-binding site for tRNALys1,2. J. Virol. 58, 983–987.

Chukkapalli, V., Hogue, I.B., Boyko, V., Hu, W.S., Ono, A., 2008. Interaction between the human immunodeficiency virus type 1 Gag matrix domain and phosphatidylinositol-(4,5)-bisphosphate is essential for efficient Gag membrane binding. J. Virol. 82, 2405–2417.

Chukkapalli, V., Oh, S.J., Ono, A., 2010. Opposing mechanisms involving RNA and lipids regulate HIV-1 Gag membrane binding through the highly basic region of the matrix domain. Proc. Natl. Acad. Sci. U.S.A. 107, 1600–1605.

Chukkapalli, V., Inlora, J., Todd, G.C., Ono, A., 2013. Evidence in support of RNA-mediated inhibition of phosphatidylserine-dependent HIV-1 Gag membrane binding in cells. J. Virol. 87, 7155–7159.

Clavel, F., Hance, A.J., 2004. HIV drug resistance. N. Engl. J. Med. 350, 1023–1035.

Clurman, B.E., Hayward, W.S., 1989. Multiple proto-oncogene activations in avian leukosis virus-induced lymphomas: evidence for stage-specific events. Mol. Cell. Biol. 9, 2657–2664.

Coley, W., Van Duyne, R., Carpio, L., Guendel, I., Kehn-Hall, K., Chevalier, S., Narayanan, A., Luu, T., Lee, N., Klase, Z., Kashanchi, F., 2010. Absence of DICER in monocytes and its regulation by HIV-1. J. Biol. Chem. 285, 31930–31943.

Colicelli, J., Goff, S.P., 1986. Isolation of a recombinant murine leukemia virus utilizing a new primer tRNA. J. Virol. 57, 37–45.

Comandur, R., Olson, E.D., Musier-Forsyth, K., 2017. Conservation of tRNA mimicry in the 5′-untranslated region of distinct HIV-1 subtypes. RNA 23, 1850–1859.

Couez, D., Deschamps, J., Kettmann, R., Stephens, R.M., Gilden, R.V., Burny, A., 1984. Nucleotide sequence analysis of the long terminal repeat of integrated bovine leukemia provirus DNA and of adjacent viral and host sequences. J. Virol. 49, 615–620.

Couturier, J.P., Root-Bernstein, R.S., 2005. HIV may produce inhibitory microRNAs (miRNAs) that block production of CD28, CD4 and some interleukins. J. Theor. Biol. 235, 169–184.

Craigie, R., Bushman, F.D., 2012. HIV DNA integration. Cold Spring Harb. Perspect. Med. 2, a006890.

Cross, B.C., Sinning, I., Luirink, J., High, S., 2009. Delivering proteins for export from the cytosol. Nat. Rev. Mol. Cell Biol. 10, 255–264.

Cui, J., Tachedjian, M., Wang, L., Tachedjian, G., Wang, L.F., Zhang, S., 2012. Discovery of retroviral homologs in bats: implications for the origin of mammalian gammaretroviruses. J. Virol. 86, 4288–4293.

Cullen, B.R., 2009. Viral RNAs: lessons from the enemy. Cell 136, 592–597.

Cullen, B.R., 2012. MicroRNA expression by an oncogenic retrovirus. Proc. Natl. Acad. Sci. U.S.A. 109, 2695–2696.

Dai, Z., Ji, J., Yan, Y., Lin, W., Li, H., Chen, F., Liu, Y., Chen, W., Bi, Y., Xie, Q., 2015. Role of gga-miR-221 and gga-miR-222 during tumour formation in chickens infected by subgroup J avian leukosis virus. Viruses 7, 6538–6551.

Dangel, A.W., Mendoza, A.R., Baker, B.J., Daniel, C.M., Carroll, M.C., Wu, L.C., Yu, C.Y., 1994. The dichotomous size variation of human complement C4 genes is mediated by a novel family of endogenous retroviruses, which also establishes species-specific genomic patterns among Old World primates. Immunogenetics 40, 425–436.

d'Arminio Monforte, A., Lepri, A.C., Rezza, G., Pezzotti, P., Antinori, A., Phillips, A.N., Angarano, G., Colangeli, V., De Luca, A., Ippolito, G., Caggese, L., Soscia, F., Filice, G., Gritti, F., Narciso, P., Tirelli, U., Moroni, M., 2000. Insights into the reasons for discontinuation of the first highly active antiretroviral therapy (HAART) regimen in a cohort of antiretroviral naive patients. I.CO.N.A. Study Group. Italian Cohort of Antiretroviral-Naive Patients. AIDS 14, 499–507.

Das, A.T., Klaver, B., Berkhout, B., 1995. Reduced replication of human immunodeficiency virus type 1 mutants that use reverse transcription primers other than the natural tRNA(3Lys). J. Virol. 69, 3090–3097.

Das, A.T., Brummelkamp, T.R., Westerhout, E.M., Vink, M., Madiredjo, M., Bernards, R., Berkhout, B., 2004. Human immunodeficiency virus type 1 escapes from RNA interference-mediated inhibition. J. Virol. 78, 2601–2605.

de Haro, C., Mendez, R., Santoyo, J., 1996. The eIF-2alpha kinases and the control of protein synthesis. FASEB J. 10, 1378–1387.

Deforges, J., de Breyne, S., Ameur, M., Ulryck, N., Chamond, N., Saadi, A., Ponty, Y., Ohlmann, T., Sargueil, B., 2017. Two ribosome recruitment sites direct multiple translation events within HIV1 Gag open reading frame. Nucleic Acids Res. 45 (12), 7382–7400.

Delassus, S., Sonigo, P., Wain-Hobson, S., 1989. Genetic organization of gibbon ape leukemia virus. Virology 173, 205–213.

Desimmie, B.A., Schrijvers, R., Demeulemeester, J., Borrenberghs, D., Weydert, C., Thys, W., Vets, S., Van Remoortel, B., Hofkens, J., De Rijck, J., Hendrix, J., Bannert, N., Gijsbers, R., Christ, F., Debyser, Z., 2013. LEDGINs inhibit late stage HIV-1 replication by modulating integrase multimerization in the virions. Retrovirology 10, 57.

Dewan, V., Wei, M., Kleiman, L., Musier-Forsyth, K., 2012. Dual role for motif 1 residues of human lysyl-tRNA synthetase in dimerization and packaging into HIV-1. J. Biol. Chem. 287, 41955–41962.

Dick, R.A., Kamynina, E., Vogt, V.M., 2013. Effect of multimerization on membrane association of Rous sarcoma virus and HIV-1 matrix domain proteins. J. Virol. 87, 13598–13608.

Diederichs, S., Haber, D.A., 2007. Dual role for argonautes in microRNA processing and posttranscriptional regulation of microRNA expression. Cell 131, 1097–1108.

Dismuke, D.J., Aiken, C., 2006. Evidence for a functional link between uncoating of the human immunodeficiency virus type 1 core and nuclear import of the viral preintegration complex. J. Virol. 80, 3712–3720.

Doria, M., Neri, F., Gallo, A., Farace, M.G., Michienzi, A., 2009. Editing of HIV-1 RNA by the double-stranded RNA deaminase ADAR1 stimulates viral infection. Nucleic Acids Res. 37, 5848–5858.

Duesberg, P.H., Robinson, W.S., 1966. Nucleic acid and proteins isolated from the Rauscher mouse leukemia virus (MLV). Proc. Natl. Acad. Sci. U.S.A. 55, 219–227.

Dunning, J., Sahr, F., Rojek, A., Gannon, F., Carson, G., Idriss, B., Massaquoi, T., Gandi, R., Joseph, S., Osman, H.K., Brooks, T.J., Simpson, A.J., Goodfellow, I., Thorne, L., Arias, A., Merson, L., Castle, L., Howell-Jones, R., Pardinaz-Solis, R., Hope-Gill, B., Ferri, M., Grove, J., Kowalski, M., Stepniewska, K., Lang, T., Whitehead, J., Olliaro, P., Samai, M., Horby, P.W., 2016. Experimental treatment of Ebola virus disease with TKM-130803: a single-arm phase 2 clinical trial. PLoS Med. 13, e1001997.

Eckwahl, M.J., Sim, S., Smith, D., Telesnitsky, A., Wolin, S.L., 2015. A retrovirus packages nascent host noncoding RNAs from a novel surveillance pathway. Genes Dev. 29, 646–657.

Eckwahl, M.J., Telesnitsky, A., Wolin, S.L., 2016a. Host RNA packaging by retroviruses: a newly synthesized story. mBio 7 e02025–02015.

Eckwahl, M.J., Arnion, H., Kharytonchyk, S., Zang, T., Bieniasz, P.D., Telesnitsky, A., Wolin, S.L., 2016b. Analysis of the human immunodeficiency virus-1 RNA packageome. RNA 22, 1228–1238.
Edery, I., Petryshyn, R., Sonenberg, N., 1989. Activation of double-stranded RNA-dependent kinase (dsl) by the TAR region of HIV-1 mRNA: a novel translational control mechanism. Cell 56, 303–312.
Elgar, G., Vavouri, T., 2008. Tuning in to the signals: noncoding sequence conservation in vertebrate genomes. Trends Genet. 24, 344–352.
Elmayan, T., Vaucheret, H., 1996. Single copies of a 35S-transgene can undergo post-transcriptional gene silencing at each generation or can be transcriptionally inactivated in trans by a 35S silencer. Plant J. 9, 787–797.
Emara, M.M., Ivanov, P., Hickman, T., Dawra, N., Tisdale, S., Kedersha, N., Hu, G.F., Anderson, P., 2010. Angiogenin-induced tRNA-derived stress-induced RNAs promote stress-induced stress granule assembly. J. Biol. Chem. 285, 10959–10968.
Erikson, E., Erikson, R.L., Henry, B., Pace, N.R., 1973. Comparison of oligonucleotides produced by RNase T1 digestion of 7 S RNA from avian and murine oncornaviruses and from uninfected cells. Virology 53, 40–46.
Esnault, C., Heidmann, O., Delebecque, F., Dewannieux, M., Ribet, D., Hance, A.J., Heidmann, T., Schwartz, O., 2005. APOBEC3G cytidine deaminase inhibits retrotransposition of endogenous retroviruses. Nature 433, 430–433.
Falcone, V., Schweizer, M., Neumann-Haefelin, D., 2003. Replication of primate foamy viruses in natural and experimental hosts. In: Rethwilm, A. (Ed.), Foamy Viruses. Springer Berlin Heidelberg, Berlin, Heidelberg, pp. 161–180.
Farkasova, H., Hron, T., Paces, J., Hulva, P., Benda, P., Gifford, R.J., Elleder, D., 2017. Discovery of an endogenous Deltaretrovirus in the genome of long-fingered bats (Chiroptera: Miniopteridae). Proc. Natl. Acad. Sci. U.S.A. 114, 3145–3150.
Feinberg, M.B., Baltimore, D., Frankel, A.D., 1991. The role of Tat in the human immunodeficiency virus life cycle indicates a primary effect on transcriptional elongation. Proc. Natl. Acad. Sci. U.S.A. 88, 4045–4049.
Feng, S., Holland, E.C., 1988. HIV-1 tat trans-activation requires the loop sequence within tar. Nature 334, 165–167.
Fisher, A.G., Feinberg, M.B., Josephs, S.F., Harper, M.E., Marselle, L.M., Reyes, G., Gonda, M.A., Aldovini, A., Debouk, C., Gallo, R.C., et al., 1986. The trans-activator gene of HTLV-III is essential for virus replication. Nature 320, 367–371.
Fontana, J., Jurado, K.A., Cheng, N., Ly, N.L., Fuchs, J.R., Gorelick, R.J., Engelman, A.N., Steven, A.C., 2015. Distribution and redistribution of HIV-1 nucleocapsid protein in immature, mature, and integrase-inhibited virions: a role for integrase in maturation. J. Virol. 89, 9765–9780.
Forstemann, K., Tomari, Y., Du, T., Vagin, V.V., Denli, A.M., Bratu, D.P., Klattenhoff, C., Theurkauf, W.E., Zamore, P.D., 2005. Normal microRNA maturation and germ-line stem cell maintenance requires Loquacious, a double-stranded RNA-binding domain protein. PLoS Biol. 3, e236.
Frank, O., Verbeke, C., Schwarz, N., Mayer, J., Fabarius, A., Hehlmann, R., Leib-Mosch, C., Seifarth, W., 2008. Variable transcriptional activity of endogenous retroviruses in human breast cancer. J. Virol. 82, 1808–1818.
Frankel, A.D., Young, J.A., 1998. HIV-1: fifteen proteins and an RNA. Annu. Rev. Biochem. 67, 1–25.
Friant, S., Heyman, T., Poch, O., Wilhelm, M., Wilhelm, F.X., 1997. Sequence comparison of the Ty1 and Ty2 elements of the yeast genome supports the structural model of the tRNAiMet-Ty1 RNA reverse transcription initiation complex. Yeast 13, 639–645.

Frias-Lasserre, D., 2012. Non coding RNAs and viruses in the framework of the phylogeny of the genes, epigenesis and heredity. Int. J. Mol. Sci. 13, 477–490.

Gallo, A., Locatelli, F., 2012. ADARs: allies or enemies? The importance of A-to-I RNA editing in human disease: from cancer to HIV-1. Biol. Rev. Camb. Philos. Soc. 87, 95–110.

Gao, Y.L., Qin, L.T., Pan, W., Wang, Y.Q., Le Qi, X., Gao, H.L., Wang, X.M., 2010. Avian leukosis virus subgroup J in layer chickens, China. Emerg. Infect. Dis. 16, 1637–1638.

Garcia, E.L., Onafuwa-Nuga, A., Sim, S., King, S.R., Wolin, S.L., Telesnitsky, A., 2009. Packaging of host mY RNAs by murine leukemia virus may occur early in Y RNA biogenesis. J. Virol. 83, 12526–12534.

Gatignol, A., Buckler-White, A., Berkhout, B., Jeang, K.T., 1991. Characterization of a human TAR RNA-binding protein that activates the HIV-1 LTR. Science 251, 1597–1600.

Gaudray, G., Gachon, F., Basbous, J., Biard-Piechaczyk, M., Devaux, C., Mesnard, J.M., 2002. The complementary strand of the human T-cell leukemia virus type 1 RNA genome encodes a bZIP transcription factor that down-regulates viral transcription. J. Virol. 76, 12813–12822.

Geisler, S., Coller, J., 2013. RNA in unexpected places: long non-coding RNA functions in diverse cellular contexts. Nat. Rev. Mol. Cell Biol. 14, 699–712.

Gifford, R., Tristem, M., 2003. The evolution, distribution and diversity of endogenous retroviruses. Virus Genes 26, 291–315.

Giles, K.E., Caputi, M., Beemon, K.L., 2004. Packaging and reverse transcription of snRNAs by retroviruses may generate pseudogenes. RNA 10, 299–307.

Gillet, N., Florins, A., Boxus, M., Burteau, C., Nigro, A., Vandermeers, F., Balon, H., Bouzar, A.B., Defoiche, J., Burny, A., Reichert, M., Kettmann, R., Willems, L., 2007. Mechanisms of leukemogenesis induced by bovine leukemia virus: prospects for novel anti-retroviral therapies in human. Retrovirology 4, 18.

Gillet, N.A., Hamaidia, M., de Brogniez, A., Gutierrez, G., Renotte, N., Reichert, M., Trono, K., Willems, L., 2016. Bovine leukemia virus small noncoding RNAs are functional elements that regulate replication and contribute to oncogenesis in vivo. PLoS Pathog. 12, e1005588.

Glazov, E.A., Horwood, P.F., Assavalapsakul, W., Kongsuwan, K., Mitchell, R.W., Mitter, N., Mahony, T.J., 2010. Characterization of microRNAs encoded by the bovine herpesvirus 1 genome. J. Gen. Virol. 91, 32–41.

Gogvadze, E., Stukacheva, E., Buzdin, A., Sverdlov, E., 2009. Human-specific modulation of transcriptional activity provided by endogenous retroviral insertions. J. Virol. 83, 6098–6105.

Gosenca, D., Gabriel, U., Steidler, A., Mayer, J., Diem, O., Erben, P., Fabarius, A., Leib-Mosch, C., Hofmann, W.K., Seifarth, W., 2012. HERV-E-mediated modulation of PLA2G4A transcription in urothelial carcinoma. PLoS One 7, e49341.

Gottwein, E., Mukherjee, N., Sachse, C., Frenzel, C., Majoros, W.H., Chi, J.T., Braich, R., Manoharan, M., Soutschek, J., Ohler, U., Cullen, B.R., 2007. A viral microRNA functions as an orthologue of cellular miR-155. Nature 450, 1096–1099.

Grassmann, R., Jeang, K.T., 2008. The roles of microRNAs in mammalian virus infection. Biochim. Biophys. Acta 1779, 706–711.

Gregory, R.I., Chendrimada, T.P., Cooch, N., Shiekhattar, R., 2005. Human RISC couples microRNA biogenesis and posttranscriptional gene silencing. Cell 123, 631–640.

Griffiths-Jones, S., Grocock, R.J., van Dongen, S., Bateman, A., Enright, A.J., 2006. miRBase: microRNA sequences, targets and gene nomenclature. Nucleic Acids Res. 34, D140–D144.

Griffiths-Jones, S., Saini, H.K., van Dongen, S., Enright, A.J., 2008. miRBase: tools for microRNA genomics. Nucleic Acids Res. 36, D154–D158.

Groen, J.N., Morris, K.V., 2013. Chromatin, non-coding RNAs, and the expression of HIV. Viruses 5, 1633–1645.

Grossman, W.J., Verbsky, J.W., Tollefsen, B.L., Kemper, C., Atkinson, J.P., Ley, T.J., 2004. Differential expression of granzymes A and B in human cytotoxic lymphocyte subsets and T regulatory cells. Blood 104, 2840–2848.

Guilley, H., Richards, K.E., Jonard, G., 1983. Observations concerning the discontinuous DNAs of cauliflower mosaic virus. EMBO J. 2, 277–282.

Gulick, R.M., Mellors, J.W., Havlir, D., Eron, J.J., Gonzalez, C., McMahon, D., Richman, D.D., Valentine, F.T., Jonas, L., Meibohm, A., Emini, E.A., Chodakewitz, J.A., 1997. Treatment with indinavir, zidovudine, and lamivudine in adults with human immunodeficiency virus infection and prior antiretroviral therapy. N. Engl. J. Med. 337, 734–739.

Guo, F., Cen, S., Niu, M., Javanbakht, H., Kleiman, L., 2003. Specific inhibition of the synthesis of human lysyl-tRNA synthetase results in decreases in tRNA(Lys) incorporation, tRNA(3)(Lys) annealing to viral RNA, and viral infectivity in human immunodeficiency virus type 1. J. Virol. 77, 9817–9822.

Gutierrez, G., Rodriguez, S.M., de Brogniez, A., Gillet, N., Golime, R., Burny, A., Jaworski, J.P., Alvarez, I., Vagnoni, L., Trono, K., Willems, L., 2014. Vaccination against delta-retroviruses: the bovine leukemia virus paradigm. Viruses 6, 2416–2427.

Haase, A.D., Jaskiewicz, L., Zhang, H., Laine, S., Sack, R., Gatignol, A., Filipowicz, W., 2005. TRBP, a regulator of cellular PKR and HIV-1 virus expression, interacts with Dicer and functions in RNA silencing. EMBO Rep. 6, 961–967.

Hagn, M., Jahrsdorfer, B., 2012. Why do human B cells secrete granzyme B? Insights into a novel B-cell differentiation pathway. OncoImmunology 1, 1368–1375.

Hakim, S.T., Alsayari, M., McLean, D.C., Saleem, S., Addanki, K.C., Aggarwal, M., Mahalingam, K., Bagasra, O., 2008. A large number of the human microRNAs target lentiviruses, retroviruses, and endogenous retroviruses. Biochem. Biophys. Res. Commun. 369, 357–362.

Halwani, R., Cen, S., Javanbakht, H., Saadatmand, J., Kim, S., Shiba, K., Kleiman, L., 2004. Cellular distribution of Lysyl-tRNA synthetase and its interaction with Gag during human immunodeficiency virus type 1 assembly. J. Virol. 78, 7553–7564.

Han, J., Lee, Y., Yeom, K.H., Kim, Y.K., Jin, H., Kim, V.N., 2004. The Drosha-DGCR8 complex in primary microRNA processing. Genes Dev. 18, 3016–3027.

Han, Y.C., Park, C.Y., Bhagat, G., Zhang, J., Wang, Y., Fan, J.B., Liu, M., Zou, Y., Weissman, I.L., Gu, H., 2010. microRNA-29a induces aberrant self-renewal capacity in hematopoietic progenitors, biased myeloid development, and acute myeloid leukemia. J. Exp. Med. 207, 475–489.

Hanger, J.J., Bromham, L.D., McKee, J.J., O'Brien, T.M., Robinson, W.F., 2000. The nucleotide sequence of koala (Phascolarctos cinereus) retrovirus: a novel type C endogenous virus related to Gibbon ape leukemia virus. J. Virol. 74, 4264–4272.

Harel, J., Huppert, J., Lacour, F., Harel, L., 1965. Demonstration of a ribonucleic acid of very high molecular weight in the avian myeloblastosis virus. C. R. Acad. Sci. Hebd. Seances Acad. Sci. D 261, 2266–2268.

Hariharan, M., Scaria, V., Pillai, B., Brahmachari, S.K., 2005. Targets for human encoded microRNAs in HIV genes. Biochem. Biophys. Res. Commun. 337, 1214–1218.

Harris, R.S., Bishop, K.N., Sheehy, A.M., Craig, H.M., Petersen-Mahrt, S.K., Watt, I.N., Neuberger, M.S., Malim, M.H., 2003. DNA deamination mediates innate immunity to retroviral infection. Cell 113, 803–809.

Hartman, T.R., Qian, S., Bolinger, C., Fernandez, S., Schoenberg, D.R., Boris-Lawrie, K., 2006. RNA helicase A is necessary for translation of selected messenger RNAs. Nat. Struct. Mol. Biol. 13, 509–516.

Hay, J.M., Jones, M.C., Blakebrough, M.L., Dasgupta, I., Davies, J.W., Hull, R., 1991. An analysis of the sequence of an infectious clone of rice tungro bacilliform virus, a plant pararetrovirus. Nucleic Acids Res. 19, 2615–2621.

Hendrick, J.P., Wolin, S.L., Rinke, J., Lerner, M.R., Steitz, J.A., 1981. Ro small cytoplasmic ribonucleoproteins are a subclass of La ribonucleoproteins: further characterization of the Ro and La small ribonucleoproteins from uninfected mammalian cells. Mol. Cell. Biol. 1, 1138–1149.

Herniou, E., Martin, J., Miller, K., Cook, J., Wilkinson, M., Tristem, M., 1998. Retroviral diversity and distribution in vertebrates. J. Virol. 72, 5955–5966.

Herrera-Carrillo, E., Berkhout, B., 2016. Attacking HIV-1 RNA versus DNA by sequence-specific approaches: RNAi versus CRISPR-Cas. Biochem. Soc. Trans. 44, 1355–1365.

Herrera-Carrillo, E., Berkhout, B., 2017. Novel AgoshRNA molecules for silencing of the CCR5 co-receptor for HIV-1 infection. PLoS One 12, e0177935.

Herrera-Carrillo, E., Harwig, A., Berkhout, B., 2017. Silencing of HIV-1 by AgoshRNA molecules. Gene Ther. 24, 453–461.

Holzschu, D.L., Martineau, D., Fodor, S.K., Vogt, V.M., Bowser, P.R., Casey, J.W., 1995. Nucleotide sequence and protein analysis of a complex piscine retrovirus, walleye dermal sarcoma virus. J. Virol. 69, 5320–5331.

Hopper, A.K., 2013. Transfer RNA post-transcriptional processing, turnover, and subcellular dynamics in the yeast *Saccharomyces cerevisiae*. Genetics 194, 43–67.

Hou, P., Chen, S., Wang, S., Yu, X., Chen, Y., Jiang, M., Zhuang, K., Ho, W., Hou, W., Huang, J., Guo, D., 2015. Genome editing of CXCR4 by CRISPR/cas9 confers cells resistant to HIV-1 infection. Sci. Rep. 5, 15577.

Houtgraaf, J.H., Versmissen, J., van der Giessen, W.J., 2006. A concise review of DNA damage checkpoints and repair in mammalian cells. Cardiovasc. Revasc. Med. 7, 165–172.

Houzet, L., Jeang, K.T., 2011. MicroRNAs and human retroviruses. Biochim. Biophys. Acta 1809, 686–693.

Houzet, L., Yeung, M.L., de Lame, V., Desai, D., Smith, S.M., Jeang, K.T., 2008. MicroRNA profile changes in human immunodeficiency virus type 1 (HIV-1) seropositive individuals. Retrovirology 5, 118.

Houzet, L., Klase, Z., Yeung, M.L., Wu, A., Le, S.Y., Quinones, M., Jeang, K.T., 2012. The extent of sequence complementarity correlates with the potency of cellular miRNA-mediated restriction of HIV-1. Nucleic Acids Res. 40, 11684–11696.

Hsu, P.D., Lander, E.S., Zhang, F., 2014. Development and applications of CRISPR-Cas9 for genome engineering. Cell 157, 1262–1278.

Hu, W.S., Hughes, S.H., 2012. HIV-1 reverse transcription. Cold Spring Harb. Perspect. Med. 2.

Huang, Y., Mak, J., Cao, Q., Li, Z., Wainberg, M.A., Kleiman, L., 1994. Incorporation of excess wild-type and mutant tRNA(3Lys) into human immunodeficiency virus type 1. J. Virol. 68, 7676–7683.

Huang, J., Wang, F., Argyris, E., Chen, K., Liang, Z., Tian, H., Huang, W., Squires, K., Verlinghieri, G., Zhang, H., 2007. Cellular microRNAs contribute to HIV-1 latency in resting primary CD4+ T lymphocytes. Nat. Med. 13, 1241–1247.

Huarte, M., Guttman, M., Feldser, D., Garber, M., Koziol, M.J., Kenzelmann-Broz, D., Khalil, A.M., Zuk, O., Amit, I., Rabani, M., Attardi, L.D., Regev, A., Lander, E.S., Jacks, T., Rinn, J.L., 2010. A large intergenic noncoding RNA induced by p53 mediates global gene repression in the p53 response. Cell 142, 409–419.

Hull, S., Boris-Lawrie, K., 2002. RU5 of Mason-Pfizer monkey virus 5′ long terminal repeat enhances cytoplasmic expression of human immunodeficiency virus type 1 gag-pol and nonviral reporter RNA. J. Virol. 76, 10211–10218.

Hussain, A.I., Johnson, J.A., Da Silva Freire, M., Heneine, W., 2003. Identification and characterization of avian retroviruses in chicken embryo-derived yellow fever vaccines: investigation of transmission to vaccine recipients. J. Virol. 77, 1105–1111.

Hutvagner, G., Simard, M.J., Mello, C.C., Zamore, P.D., 2004. Sequence-specific inhibition of small RNA function. PLoS Biol. 2, E98.

Imam, H., Bano, A.S., Patel, P., Holla, P., Jameel, S., 2015. The lncRNA NRON modulates HIV-1 replication in a NFAT-dependent manner and is differentially regulated by early and late viral proteins. Sci. Rep. 5, 8639.

Imamura, K., Imamachi, N., Akizuki, G., Kumakura, M., Kawaguchi, A., Nagata, K., Kato, A., Kawaguchi, Y., Sato, H., Yoneda, M., Kai, C., Yada, T., Suzuki, Y., Yamada, T., Ozawa, T., Kaneki, K., Inoue, T., Kobayashi, M., Kodama, T., Wada, Y., Sekimizu, K., Akimitsu, N., 2014. Long noncoding RNA NEAT1-dependent SFPQ relocation from promoter region to paraspeckle mediates IL8 expression upon immune stimuli. Mol. Cell 53, 393–406.

Inlora, J., Chukkapalli, V., Derse, D., Ono, A., 2011. Gag localization and virus-like particle release mediated by the matrix domain of human T-lymphotropic virus type 1 Gag are less dependent on phosphatidylinositol-(4,5)-bisphosphate than those mediated by the matrix domain of HIV-1 Gag. J. Virol. 85, 3802–3810.

Inlora, J., Collins, D.R., Trubin, M.E., Chung, J.Y., Ono, A., 2014. Membrane binding and subcellular localization of retroviral Gag proteins are differentially regulated by MA interactions with phosphatidylinositol-(4,5)-bisphosphate and RNA. mBio 5, e02202.

Itano, M.S., Arnion, H., Wolin, S.L., Simon, S.M., 2018. Recruitment of 7SL RNA to assembling HIV-1 virus-like particles. Traffic 19, 36–43.

Jackson, R.J., Hellen, C.U., Pestova, T.V., 2010. The mechanism of eukaryotic translation initiation and principles of its regulation. Nat. Rev. Mol. Cell Biol. 11, 113–127.

Jang, S.K., Krausslich, H.G., Nicklin, M.J., Duke, G.M., Palmenberg, A.C., Wimmer, E., 1988. A segment of the 5′ nontranslated region of encephalomyocarditis virus RNA directs internal entry of ribosomes during in vitro translation. J. Virol. 62, 2636–2643.

Janssen, H.L.A., Reesink, H.W., Lawitz, E.J., Zeuzem, S., Rodriguez-Torres, M., Patel, K., van der Meer, A.J., Patick, A.K., Chen, A., Zhou, Y., Persson, R., King, B.D., Kauppinen, S., Levin, A.A., Hodges, M.R., 2013. Treatment of HCV infection by targeting MicroRNA. N. Engl. J. Med. 368, 1685–1694.

Jarmuz, A., Chester, A., Bayliss, J., Gisbourne, J., Dunham, I., Scott, J., Navaratnam, N., 2002. An anthropoid-specific locus of orphan C to U RNA-editing enzymes on chromosome 22. Genomics 79, 285–296.

Javanbakht, H., Halwani, R., Cen, S., Saadatmand, J., Musier-Forsyth, K., Gottlinger, H., Kleiman, L., 2003. The interaction between HIV-1 Gag and human lysyl-tRNA synthetase during viral assembly. J. Biol. Chem. 278, 27644–27651.

Johnson, W.E., Coffin, J.M., 1999. Constructing primate phylogenies from ancient retrovirus sequences. Proc. Natl. Acad. Sci. U.S.A. 96, 10254–10260.

Jones, C.P., Saadatmand, J., Kleiman, L., Musier-Forsyth, K., 2013. Molecular mimicry of human tRNALys anti-codon domain by HIV-1 RNA genome facilitates tRNA primer annealing. RNA 19, 219–229.

Jones, C.P., Cantara, W.A., Olson, E.D., Musier-Forsyth, K., 2014. Small-angle X-ray scattering-derived structure of the HIV-1 5′ UTR reveals 3D tRNA mimicry. Proc. Natl. Acad. Sci. U.S.A. 111, 3395–3400.

Jopling, C.L., Yi, M., Lancaster, A.M., Lemon, S.M., Sarnow, P., 2005. Modulation of hepatitis C virus RNA abundance by a liver-specific MicroRNA. Science 309, 1577–1581.

Jopling, C.L., Schutz, S., Sarnow, P., 2008. Position-dependent function for a tandem microRNA miR-122-binding site located in the hepatitis C virus RNA genome. Cell Host Microbe 4, 77–85.

Judware, R., Li, J., Petryshyn, R., 1993. Inhibition of the dsRNA-dependent protein kinase by a peptide derived from the human immunodeficiency virus type 1 Tat protein. J. Interferon Res. 13, 153–160.

Jurado, K.A., Wang, H., Slaughter, A., Feng, L., Kessl, J.J., Koh, Y., Wang, W., Ballandras-Colas, A., Patel, P.A., Fuchs, J.R., Kvaratskhelia, M., Engelman, A., 2013. Allosteric integrase inhibitor potency is determined through the inhibition of HIV-1 particle maturation. Proc. Natl. Acad. Sci. U.S.A. 110, 8690–8695.

Kao, S.Y., Calman, A.F., Luciw, P.A., Peterlin, B.M., 1987. Anti-termination of transcription within the long terminal repeat of HIV-1 by tat gene product. Nature 330, 489–493.

Karn, J., Stoltzfus, C.M., 2012. Transcriptional and posttranscriptional regulation of HIV-1 gene expression. Cold Spring Harb. Perspect. Med. 2, a006916.

Kawaji, H., Nakamura, M., Takahashi, Y., Sandelin, A., Katayama, S., Fukuda, S., Daub, C.O., Kai, C., Kawai, J., Yasuda, J., Carninci, P., Hayashizaki, Y., 2008. Hidden layers of human small RNAs. BMC Genom. 9, 157.

Kawakami, T., Sherman, L., Dahlberg, J., Gazit, A., Yaniv, A., Tronick, S.R., Aaronson, S.A., 1987. Nucleotide sequence analysis of equine infectious anemia virus proviral DNA. Virology 158, 300–312.

Kedersha, N.L., Rome, L.H., 1986. Isolation and characterization of a novel ribonucleoprotein particle: large structures contain a single species of small RNA. J. Cell Biol. 103, 699–709.

Kedzierska, K., Crowe, S.M., 2002. The role of monocytes and macrophages in the pathogenesis of HIV-1 infection. Curr. Med. Chem. 9, 1893–1903.

Keene, S.E., Telesnitsky, A., 2012. cis-Acting determinants of 7SL RNA packaging by HIV-1. J. Virol. 86, 7934–7942.

Keene, S.E., King, S.R., Telesnitsky, A., 2010. 7SL RNA is retained in HIV-1 minimal virus-like particles as an S-domain fragment. J. Virol. 84, 9070–9077.

Keith, G., Heyman, T., 1990. Heterogeneities in vertebrate tRNAs(Trp) avian retroviruses package only as a primer the tRNA(Trp) lacking modified m2G in position 7. Nucleic Acids Res. 18, 703–710.

Kessl, J.J., Kutluay, S.B., Townsend, D., Rebensburg, S., Slaughter, A., Larue, R.C., Shkriabai, N., Bakouche, N., Fuchs, J.R., Bieniasz, P.D., Kvaratskhelia, M., 2016. HIV-1 integrase binds the viral RNA genome and is essential during virion morphogenesis. Cell 166 1257–1268 e1212.

Khvorova, A., Reynolds, A., Jayasena, S.D., 2003. Functional siRNAs and miRNAs exhibit strand bias. Cell 115, 209–216.

Kickhoefer, V.A., Liu, Y., Kong, L.B., Snow, B.E., Stewart, P.L., Harrington, L., Rome, L.H., 2001. The Telomerase/vault-associated protein TEP1 is required for vault RNA stability and its association with the vault particle. J. Cell Biol. 152, 157–164.

Kikuchi, Y., Ando, Y., Shiba, T., 1986. Unusual priming mechanism of RNA-directed DNA synthesis in copia retrovirus-like particles of Drosophila. Nature 323, 824–826.

Kincaid, R.P., Burke, J.M., Sullivan, C.S., 2012. RNA virus microRNA that mimics a B-cell oncomiR. Proc. Natl. Acad. Sci. U.S.A. 109, 3077–3082.

Kincaid, R.P., Chen, Y., Cox, J.E., Rethwilm, A., Sullivan, C.S., 2014. Noncanonical microRNA (miRNA) biogenesis gives rise to retroviral mimics of lymphoproliferative and immunosuppressive host miRNAs. mBio 5, e00074.

Kinoshita, S., Chen, B.K., Kaneshima, H., Nolan, G.P., 1998. Host control of HIV-1 parasitism in T cells by the nuclear factor of activated T cells. Cell 95, 595–604.

Kiss, T., 2004. Biogenesis of small nuclear RNPs. J. Cell Sci. 117, 5949–5951.

Kissil, J.L., Deiss, L.P., Bayewitch, M., Raveh, T., Khaspekov, G., Kimchi, A., 1995. Isolation of DAP3, a novel mediator of interferon-gamma-induced cell death. J. Biol. Chem. 270, 27932–27936.

Kitazono, M., Sumizawa, T., Takebayashi, Y., Chen, Z.S., Furukawa, T., Nagayama, S., Tani, A., Takao, S., Aikou, T., Akiyama, S., 1999. Multidrug resistance and the lung resistance-related protein in human colon carcinoma SW-620 cells. J. Natl. Cancer Inst. 91, 1647–1653.

Kitazono, M., Okumura, H., Ikeda, R., Sumizawa, T., Furukawa, T., Nagayama, S., Seto, K., Aikou, T., Akiyama, S., 2001. Reversal of LRP-associated drug resistance in colon carcinoma SW-620 cells. Int. J. Cancer 91, 126–131.

Klase, Z., Kale, P., Winograd, R., Gupta, M.V., Heydarian, M., Berro, R., McCaffrey, T., Kashanchi, F., 2007. HIV-1 TAR element is processed by Dicer to yield a viral micro-RNA involved in chromatin remodeling of the viral LTR. BMC Mol. Biol. 8, 63.

Klase, Z.A., Sampey, G.C., Kashanchi, F., 2013. Retrovirus infected cells contain viral microRNAs. Retrovirology 10, 15.

Kleiman, L., Jones, C.P., Musier-Forsyth, K., 2010. Formation of the tRNALys packaging complex in HIV-1. FEBS Lett. 584, 359–365.

Krieg, A.M., Gourley, M.F., Perl, A., 1992. Endogenous retroviruses: potential etiologic agents in autoimmunity. FASEB J. 6, 2537–2544.

Krutzfeldt, J., Rajewsky, N., Braich, R., Rajeev, K.G., Tuschl, T., Manoharan, M., Stoffel, M., 2005. Silencing of microRNAs in vivo with 'antagomirs'. Nature 438, 685–689.

Kurth, R., Bannert, N., 2010. Beneficial and detrimental effects of human endogenous retroviruses. Int. J. Cancer 126, 306–314.

Kusuhara, K., Anderson, M., Pettiford, S.M., Green, P.L., 1999. Human T-cell leukemia virus type 2 Rex protein increases stability and promotes nuclear to cytoplasmic transport of gag/pol and env RNAs. J. Virol. 73, 8112–8119.

Kutluay, S.B., Zang, T., Blanco-Melo, D., Powell, C., Jannain, D., Errando, M., Bieniasz, P.D., 2014. Global changes in the RNA binding specificity of HIV-1 gag regulate virion genesis. Cell 159, 1096–1109.

Lagos, D., Pollara, G., Henderson, S., Gratrix, F., Fabani, M., Milne, R.S., Gotch, F., Boshoff, C., 2010. miR-132 regulates antiviral innate immunity through suppression of the p300 transcriptional co-activator. Nat. Cell Biol. 12, 513–519.

Lanford, R.E., Hildebrandt-Eriksen, E.S., Petri, A., Persson, R., Lindow, M., Munk, M.E., Kauppinen, S., Orum, H., 2010. Therapeutic silencing of microRNA-122 in primates with chronic hepatitis C virus infection. Science 327, 198–201.

Laprevotte, I., Hampe, A., Sherr, C.J., Galibert, F., 1984. Nucleotide sequence of the gag gene and gag-pol junction of feline leukemia virus. J. Virol. 50, 884–894.

Leblanc, J., Weil, J., Beemon, K., 2013. Posttranscriptional regulation of retroviral gene expression: primary RNA transcripts play three roles as pre-mRNA, mRNA, and genomic RNA. Wiley Interdiscip. Rev. RNA 4, 567–580.

Lecellier, C.H., Dunoyer, P., Arar, K., Lehmann-Che, J., Eyquem, S., Himber, C., Saib, A., Voinnet, O., 2005. A cellular microRNA mediates antiviral defense in human cells. Science 308, 557–560.

Lecossier, D., Bouchonnet, F., Clavel, F., Hance, A.J., 2003. Hypermutation of HIV-1 DNA in the absence of the Vif protein. Science 300, 1112.

Lee, Y., Ahn, C., Han, J., Choi, H., Kim, J., Yim, J., Lee, J., Provost, P., Radmark, O., Kim, S., Kim, V.N., 2003. The nuclear RNase III Drosha initiates microRNA processing. Nature 425, 415–419.

Lee, Y., Kim, M., Han, J., Yeom, K.H., Lee, S., Baek, S.H., Kim, V.N., 2004. MicroRNA genes are transcribed by RNA polymerase II. EMBO J. 23, 4051–4060.

Lee, Y., Hur, I., Park, S.Y., Kim, Y.K., Suh, M.R., Kim, V.N., 2006. The role of PACT in the RNA silencing pathway. EMBO J. 25, 522–532.

Lee, K., Ambrose, Z., Martin, T.D., Oztop, I., Mulky, A., Julias, J.G., Vandegraaff, N., Baumann, J.G., Wang, R., Yuen, W., Takemura, T., Shelton, K., Taniuchi, I., Li, Y., Sodroski, J., Littman, D.R., Coffin, J.M., Hughes, S.H., Unutmaz, D., Engelman, A., KewalRamani, V.N., 2010. Flexible use of nuclear import pathways by HIV-1. Cell Host Microbe 7, 221–233.

Lee, J., Mun, S., Kim, D.H., Cho, C.S., Oh, D.Y., Han, K., 2017. Chicken (Gallus gallus) endogenous retrovirus generates genomic variations in the chicken genome. Mob. DNA 8, 2.

Leis, A.J., 1993. Regulation of initation of reverse transcription of retroviruses. In: Goff, S., Skalka, A. (Eds.), Reverse Transcriptase. Cold Spring Harbor laboratory press, Cold Spring Harbor, pp. 33–47.

Leonard, J.N., Shah, P.S., Burnett, J.C., Schaffer, D.V., 2008. HIV evades RNA interference directed at TAR by an indirect compensatory mechanism. Cell Host Microbe 4, 484–494.

Letesson, J.J., Mager, A., Mammerickx, M., Burny, A., Depelchin, A., 1990. B cells from bovine leukemia virus- (BLV) infected sheep with hematological disorders express the CD5 T cell marker. Leukemia 4, 377–379.

Levin, J.G., Grimley, P.M., Ramseur, J.M., Berezesky, I.K., 1974. Deficiency of 60 to 70S RNA in murine leukemia virus particles assembled in cells treated with actinomycin D. J. Virol. 14, 152–161.

Lewis, B.P., Burge, C.B., Bartel, D.P., 2005. Conserved seed pairing, often flanked by adenosines, indicates that thousands of human genes are microRNA targets. Cell 120, 15–20.

Li, H., Li, W.X., Ding, S.W., 2002. Induction and suppression of RNA silencing by an animal virus. Science 296, 1319–1321.

Li, H., Shang, H., Shu, D., Zhang, H., Ji, J., Sun, B., Xie, Q., 2014. gga-miR-375 plays a key role in tumorigenesis post subgroup J avian leukosis virus infection. PLoS One 9, e90878.

Li, Z., Chen, B., Feng, M., Ouyang, H., Zheng, M., Ye, Q., Nie, Q., Zhang, X., 2015. MicroRNA-23b promotes avian leukosis virus subgroup J (ALV-J) replication by targeting IRF1. Sci. Rep. 5, 10294.

Li, J., Chen, C., Ma, X., Geng, G., Liu, B., Zhang, Y., Zhang, S., Zhong, F., Liu, C., Yin, Y., Cai, W., Zhang, H., 2016. Long noncoding RNA NRON contributes to HIV-1 latency by specifically inducing tat protein degradation. Nat. Commun. 7, 11730.

Li, Z., Luo, Q., Xu, H., Zheng, M., Abdalla, B.A., Feng, M., Cai, B., Zhang, X., Nie, Q., 2017. MiR-34b-5p suppresses melanoma differentiation-associated gene 5 (MDA5) signaling pathway to promote avian leukosis virus subgroup J (ALV-J)-infected cells proliferaction and ALV-J replication. Front. Cell Infect. Microbiol. 7, 17. https://doi.org/10.3389/fcimb.2017.00017.

Liang, C., Wainberg, M.A., Das, A.T., Berkhout, B., 2016. CRISPR/Cas9: a double-edged sword when used to combat HIV infection. Retrovirology 13, 37.

Lie, Y.S., Penuel, E.M., Low, M.A., Nguyen, T.P., Mangahas, J.O., Anderson, K.P., Petropoulos, C.J., 1994. Chinese hamster ovary cells contain transcriptionally active full-length type C proviruses. J. Virol. 68, 7840–7849.

Lieberman, J., 2010. Granzyme A activates another way to die. Immunol. Rev. 235, 93–104.

Lin, J., Cullen, B.R., 2007. Analysis of the interaction of primate retroviruses with the human RNA interference machinery. J. Virol. 81, 12218–12226.

Lin, F.H., Thormar, H., 1971. Characterization of ribonucleic acid from visna virus. J. Virol. 7, 582–587.

Linnstaedt, S.D., Gottwein, E., Skalsky, R.L., Luftig, M.A., Cullen, B.R., 2010. Virally induced cellular microRNA miR-155 plays a key role in B-cell immortalization by Epstein-Barr virus. J. Virol. 84, 11670–11678.

Liu, J., Valencia-Sanchez, M.A., Hannon, G.J., Parker, R., 2005. MicroRNA-dependent localization of targeted mRNAs to mammalian P-bodies. Nat. Cell Biol. 7, 719–723.

Liu, Y.P., Schopman, N.C., Berkhout, B., 2013. Dicer-independent processing of short hairpin RNAs. Nucleic Acids Res. 41, 3723–3733.

Liu, S., Comandur, R., Jones, C.P., Tsang, P., Musier-Forsyth, K., 2016. Anticodon-like binding of the HIV-1 tRNA-like element to human lysyl-tRNA synthetase. RNA 22, 1828–1835.

Londin, E., Loher, P., Telonis, A.G., Quann, K., Clark, P., Jing, Y., Hatzimichael, E., Kirino, Y., Honda, S., Lally, M., Ramratnam, B., Comstock, C.E., Knudsen, K.E., Gomella, L., Spaeth, G.L., Hark, L., Katz, L.J., Witkiewicz, A., Rostami, A., Jimenez, S.A., Hollingsworth, M.A., Yeh, J.J., Shaw, C.A., McKenzie, S.E., Bray, P., Nelson, P.T., Zupo, S., Van Roosbroeck, K., Keating, M.J., Calin, G.A., Yeo, C., Jimbo, M., Cozzitorto, J., Brody, J.R., Delgrosso, K., Mattick, J.S., Fortina, P., Rigoutsos, I., 2015. Analysis of 13 cell types reveals evidence for the expression of numerous novel primate- and tissue-specific microRNAs. Proc. Natl. Acad. Sci. U.S.A. 112, E1106–E1115.

Lowe, T.M., Eddy, S.R., 1997. tRNAscan-SE: a program for improved detection of transfer RNA genes in genomic sequence. Nucleic Acids Res. 25, 955–964.

Lower, R., Lower, J., Kurth, R., 1996. The viruses in all of us: characteristics and biological significance of human endogenous retrovirus sequences. Proc. Natl. Acad. Sci. U.S.A. 93, 5177–5184.

Lu, X., Sachs, F., Ramsay, L., Jacques, P.E., Goke, J., Bourque, G., Ng, H.H., 2014. The retrovirus HERVH is a long noncoding RNA required for human embryonic stem cell identity. Nat. Struct. Mol. Biol. 21, 423–425.

Luciano, D.J., Mirsky, H., Vendetti, N.J., Maas, S., 2004. RNA editing of a miRNA precursor. RNA 10, 1174–1177.

Luirink, J., Sinning, I., 2004. SRP-mediated protein targeting: structure and function revisited. Biochim. Biophys. Acta 1694, 17–35.

Lund, A.H., Duch, M., Lovmand, J., Jorgensen, P., Pedersen, F.S., 1993. Mutated primer binding sites interacting with different tRNAs allow efficient murine leukemia virus replication. J. Virol. 67, 7125–7130.

Lund, A.H., Duch, M., Pedersen, F.S., 2000. Selection of functional tRNA primers and primer binding site sequences from a retroviral combinatorial library: identification of new functional tRNA primers in murine leukemia virus replication. Nucleic Acids Res. 28, 791–799.

Ma, L., Shen, C.J., Cohen, E.A., Xiong, S.D., Wang, J.H., 2014. miRNA-1236 inhibits HIV-1 infection of monocytes by repressing translation of cellular factor VprBP. PLoS One 9, e99535.

Ma, G.Y., Yasunaga, J., Matsuoka, M., 2016. Multifaceted functions and roles of HBZ in HTLV-1 pathogenesis. Retrovirology 13, 16.

Mack, M., Bender, K., Schneider, P.M., 2004. Detection of retroviral antisense transcripts and promoter activity of the HERV-K(C4) insertion in the MHC class III region. Immunogenetics 56, 321–332.

Maitra, R.K., McMillan, N.A., Desai, S., McSwiggen, J., Hovanessian, A.G., Sen, G., Williams, B.R., Silverman, R.H., 1994. HIV-1 TAR RNA has an intrinsic ability to activate interferon-inducible enzymes. Virology 204, 823–827.

Majors, J.E., Varmus, H.E., 1983. Nucleotide sequencing of an apparent proviral copy of env mRNA defines determinants of expression of the mouse mammary tumor virus env gene. J. Virol. 47, 495–504.

Mak, J., Kleiman, L., 1997. Primer tRNAs for reverse transcription. J. Virol. 71, 8087–8095.

Mak, J., Jiang, M., Wainberg, M.A., Hammarskjold, M.L., Rekosh, D., Kleiman, L., 1994. Role of Pr160gag-pol in mediating the selective incorporation of tRNA(Lys) into human immunodeficiency virus type 1 particles. J. Virol. 68, 2065–2072.

Malim, M.H., Emerman, M., 2008. HIV-1 accessory proteins–ensuring viral survival in a hostile environment. Cell Host Microbe 3, 388–398.

Malim, M.H., Hauber, J., Le, S.Y., Maizel, J.V., Cullen, B.R., 1989. The HIV-1 rev trans-activator acts through a structured target sequence to activate nuclear export of unspliced viral mRNA. Nature 338, 254–257.

Malmquist, W.A., Van der Maaten, M.J., Boothe, A.D., 1969. Isolation, immunodiffusion, immunofluorescence, and electron microscopy of a syncytial virus of lymphosarcomatous and apparently normal cattle. Cancer Res. 29, 188–200.

Mameli, G., Astone, V., Arru, G., Marconi, S., Lovato, L., Serra, C., Sotgiu, S., Bonetti, B., Dolei, A., 2007. Brains and peripheral blood mononuclear cells of multiple sclerosis (MS) patients hyperexpress MS-associated retrovirus/HERV-W endogenous retrovirus, but not Human herpesvirus 6. J. Gen. Virol. 88, 264–274.

Mangeney, M., Renard, M., Schlecht-Louf, G., Bouallaga, I., Heidmann, O., Letzelter, C., Richaud, A., Ducos, B., Heidmann, T., 2007. Placental syncytins: genetic disjunction between the fusogenic and immunosuppressive activity of retroviral envelope proteins. Proc. Natl. Acad. Sci. U.S.A. 104, 20534–20539.

Mansfield, J.H., Harfe, B.D., Nissen, R., Obenauer, J., Srineel, J., Chaudhuri, A., Farzan-Kashani, R., Zuker, M., Pasquinelli, A.E., Ruvkun, G., Sharp, P.A., Tabin, C.J., McManus, M.T., 2004. MicroRNA-responsive 'sensor' transgenes uncover Hox-like and other developmentally regulated patterns of vertebrate microRNA expression. Nat. Genet. 36, 1079–1083.

Marquet, R., Isel, C., Ehresmann, C., Ehresmann, B., 1995. tRNAs as primer of reverse transcriptases. Biochimie 77, 113–124.

Matera, A.G., Terns, R.M., Terns, M.P., 2007. Non-coding RNAs: lessons from the small nuclear and small nucleolar RNAs. Nat. Rev. Mol. Cell Biol. 8, 209–220.

Matsuoka, M., Green, P.L., 2009. The HBZ gene, a key player in HTLV-1 pathogenesis. Retrovirology 6, 71.

Mattick, J.S., Makunin, I.V., 2006. Non-coding RNA. Hum. Mol. Genet. 15 (Spec No 1), R17–R29.

Mattick, J.S., 2001. Non-coding RNAs: the architects of eukaryotic complexity. EMBO Rep. 2, 986–991.

Maurer, B., Bannert, H., Darai, G., Flugel, R.M., 1988. Analysis of the primary structure of the long terminal repeat and the gag and pol genes of the human spumaretrovirus. J. Virol. 62, 1590–1597.

McAllister, R.M., Nicolson, M., Gardner, M.B., Rongey, R.W., Rasheed, S., Sarma, P.S., Huebner, R.J., Hatanaka, M., Oroszlan, S., Gilden, R.V., Kabigting, A., Vernon, L., 1972. C-type virus released from cultured human rhabdomyosarcoma cells. Nat. N. Biol. 235, 3–6.

Medberry, S.L., Lockhart, B.E., Olszewski, N.E., 1990. Properties of Commelina yellow mottle virus's complete DNA sequence, genomic discontinuities and transcript suggest that it is a pararetrovirus. Nucleic Acids Res. 18, 5505–5513.

Mercer, T.R., Dinger, M.E., Mattick, J.S., 2009. Long non-coding RNAs: insights into functions. Nat. Rev. Genet. 10, 155–159.

Miller, J.T., Ehresmann, B., Hubscher, U., Le Grice, S.F., 2001. A novel interaction of tRNA(Lys,3) with the feline immunodeficiency virus RNA genome governs initiation of minus strand DNA synthesis. J. Biol. Chem. 276, 27721–27730.

Miller, R.H., 1988. Human immunodeficiency virus may encode a novel protein on the genomic DNA plus strand. Science 239, 1420–1422.

Mitobe, Y., Yasunaga, J., Furuta, R., Matsuoka, M., 2015. HTLV-1 bZIP factor RNA and protein impart distinct functions on T-cell proliferation and survival. Cancer Res. 75, 4143–4152.

Mitsuyasu, R.T., Merigan, T.C., Carr, A., Zack, J.A., Winters, M.A., Workman, C., Bloch, M., Lalezari, J., Becker, S., Thornton, L., Akil, B., Khanlou, H., Finlayson, R., McFarlane, R., Smith, D.E., Garsia, R., Ma, D., Law, M., Murray, J.M., von Kalle, C., Ely, J.A., Patino, S.M., Knop, A.E., Wong, P., Todd, A.V., Haughton, M., Fuery, C., Macpherson, J.L., Symonds, G.P., Evans, L.A., Pond, S.M., Cooper, D.A., 2009. Phase 2 gene therapy trial of an anti-HIV ribozyme in autologous CD34+ cells. Nat. Med. 15, 285–292.

Morozov, V.A., Wynyard, S., Matsumoto, S., Abalovich, A., Denner, J., Elliott, R., 2017. No PERV transmission during a clinical trial of pig islet cell transplantation. Virus Res. 227, 34–40.

Muckenfuss, H., Hamdorf, M., Held, U., Perkovic, M., Lower, J., Cichutek, K., Flory, E., Schumann, G.G., Munk, C., 2006. APOBEC3 proteins inhibit human LINE-1 retrotransposition. J. Biol. Chem. 281, 22161–22172.

Nabel, G., Baltimore, D., 1987. An inducible transcription factor activates expression of human immunodeficiency virus in T cells. Nature 326, 711–713.

Nandy, C., Mrazek, J., Stoiber, H., Grasser, F.A., Huttenhofer, A., Polacek, N., 2009. Epstein-barr virus-induced expression of a novel human vault RNA. J. Mol. Biol. 388, 776–784.

Napoli, C., Lemieux, C., Jorgensen, R., 1990. Introduction of a chimeric chalcone synthase gene into petunia results in reversible co-suppression of homologous genes in trans. Plant Cell 2, 279–289.

Nathans, R., Chu, C.Y., Serquina, A.K., Lu, C.C., Cao, H., Rana, T.M., 2009. Cellular microRNA and P bodies modulate host-HIV-1 interactions. Mol. Cell 34, 696–709.

Nikbakht, K.N., Ou, C.Y., Boone, L.R., Glover, P.L., Yang, W.K., 1985. Nucleotide sequence analysis of endogenous murine leukemia virus-related proviral clones reveals primer-binding sites for glutamine tRNA. J. Virol. 54, 889–893.

O'Brien, C.A., Wolin, S.L., 1994. A possible role for the 60-kD Ro autoantigen in a discard pathway for defective 5S rRNA precursors. Genes Dev. 8, 2891–2903.

O'Brien, C.A., Margelot, K., Wolin, S.L., 1993. Xenopus Ro ribonucleoproteins: members of an evolutionarily conserved class of cytoplasmic ribonucleoproteins. Proc. Natl. Acad. Sci. U.S.A. 90, 7250–7254.

Olety, B., Veatch, S.L., Ono, A., 2015. Phosphatidylinositol-(4,5)-bisphosphate acyl chains differentiate membrane binding of HIV-1 gag from that of the phospholipase Cdelta1 Pleckstrin homology domain. J. Virol. 89, 7861–7873.

Omoto, S., Ito, M., Tsutsumi, Y., Ichikawa, Y., Okuyama, H., Brisibe, E.A., Saksena, N.K., Fujii, Y.R., 2004. HIV-1 nef suppression by virally encoded microRNA. Retrovirology 1, 44.

Onafuwa-Nuga, A.A., King, S.R., Telesnitsky, A., 2005. Nonrandom packaging of host RNAs in moloney murine leukemia virus. J. Virol. 79, 13528–13537.

Onafuwa-Nuga, A.A., Telesnitsky, A., King, S.R., 2006. 7SL RNA, but not the 54-kd signal recognition particle protein, is an abundant component of both infectious HIV-1 and minimal virus-like particles. RNA 12, 542–546.

Ono, A., Ablan, S.D., Lockett, S.J., Nagashima, K., Freed, E.O., 2004. Phosphatidylinositol (4,5) bisphosphate regulates HIV-1 gag targeting to the plasma membrane. Mol. Biol. Cell 15, 122a–123a.

Orenstein, J.M., Fox, C., Wahl, S.M., 1997. Macrophages as a source of HIV during opportunistic infections. Science 276, 1857–1861.

Ouellet, D.L., Provost, P., 2010. Current knowledge of microRNAs and noncoding RNAs in virus-infected cells. Methods Mol. Biol. 623, 35–65.

Ouellet, D.L., Plante, I., Landry, P., Barat, C., Janelle, M.E., Flamand, L., Tremblay, M.J., Provost, P., 2008. Identification of functional microRNAs released through asymmetrical processing of HIV-1 TAR element. Nucleic Acids Res. 36, 2353–2365.

Parkin, N.T., Cohen, E.A., Darveau, A., Rosen, C., Haseltine, W., Sonenberg, N., 1988. Mutational analysis of the 5′ non-coding region of human immunodeficiency virus type 1: effects of secondary structure on translation. EMBO J. 7, 2831–2837.

Patience, C., Takeuchi, Y., Weiss, R.A., 1997. Infection of human cells by an endogenous retrovirus of pigs. Nat. Med. 3, 282–286.

Pelletier, J., Sonenberg, N., 1988. Internal initiation of translation of eukaryotic mRNA directed by a sequence derived from poliovirus RNA. Nature 334, 320–325.

Persson, H., Kvist, A., Vallon-Christersson, J., Medstrand, P., Borg, A., Rovira, C., 2009. The non-coding RNA of the multidrug resistance-linked vault particle encodes multiple regulatory small RNAs. Nat. Cell Biol. 11, 1268–1271.

Peters, G., Harada, F., Dahlberg, J.E., Panet, A., Haseltine, W.A., Baltimore, D., 1977. Low-molecular-weight RNAs of moloney murine leukemia virus: identification of the primer for RNA-directed DNA synthesis. J. Virol. 21, 1031–1041.

Pfeffer, S., Zavolan, M., Grasser, F.A., Chien, M., Russo, J.J., Ju, J., John, B., Enright, A.J., Marks, D., Sander, C., Tuschl, T., 2004. Identification of virus-encoded microRNAs. Science 304, 734–736.

Pfeffer, S., Sewer, A., Lagos-Quintana, M., Sheridan, R., Sander, C., Grasser, F.A., van Dyk, L.F., Ho, C.K., Shuman, S., Chien, M., Russo, J.J., Ju, J., Randall, G., Lindenbach, B.D., Rice, C.M., Simon, V., Ho, D.D., Zavolan, M., Tuschl, T., 2005. Identification of microRNAs of the herpesvirus family. Nat. Methods 2, 269–276.

Phizicky, E.M., Hopper, A.K., 2010. tRNA biology charges to the front. Genes Dev. 24, 1832–1860.

Plasterk, R.H., 2002. RNA silencing: the genome's immune system. Science 296, 1263–1265.

Pochart, P., Agoutin, B., Fix, C., Keith, G., Heyman, T., 1993. A very poorly expressed tRNA(Ser) is highly concentrated together with replication primer initiator tRNA(Met) in the yeast Ty1 virus-like particles. Nucleic Acids Res. 21, 1517–1521.

Postler, T.S., Pantry, S.N., Desrosiers, R.C., Ghosh, S., 2017. Identification and characterization of a long non-coding RNA up-regulated during HIV-1 infection. Virology 511, 30–39.

Prats, A.C., Sarih, L., Gabus, C., Litvak, S., Keith, G., Darlix, J.L., 1988. Small finger protein of avian and murine retroviruses has nucleic acid annealing activity and positions the replication primer tRNA onto genomic RNA. EMBO J. 7, 1777–1783.

Pruss, G., Ge, X., Shi, X.M., Carrington, J.C., Bowman Vance, V., 1997. Plant viral synergism: the potyviral genome encodes a broad-range pathogenicity enhancer that transactivates replication of heterologous viruses. Plant Cell 9, 859–868.

Qian, S., Zhong, X., Yu, L., Ding, B., de Haan, P., Boris-Lawrie, K., 2009. HIV-1 Tat RNA silencing suppressor activity is conserved across kingdoms and counteracts translational repression of HIV-1. Proc. Natl. Acad. Sci. U.S.A. 106, 605–610.

Qu, X., Wang, P., Ding, D., Li, L., Wang, H., Ma, L., Zhou, X., Liu, S., Lin, S., Wang, X., Zhang, G., Liu, L., Wang, J., Zhang, F., Lu, D., Zhu, H., 2013. Zinc-finger-nucleases mediate specific and efficient excision of HIV-1 proviral DNA from infected and latently infected human T cells. Nucleic Acids Res. 41, 7771–7782.

Rabin, H., Neubauer, R.H., Woodside, N.J., Cicmanec, J.L., Wallen, W.C., Lapin, B.A., Agrba, V.A., Yakoleva, L.A., Chuvirov, G.N., 1976. Virological studies of baboon (Papio hamadryas) lymphoma: isolation and characterization of foamyviruses. J. Med. Primatol. 5, 13–22.

Ratcliff, F.G., MacFarlane, S.A., Baulcombe, D.C., 1999. Gene silencing without DNA. rna-mediated cross-protection between viruses. Plant Cell 11, 1207–1216.

Reich, P.R., Forget, B.G., Weissman, S.M., 1966. RNA of low molecular weight in KB cells infected with adenovirus type 2. J. Mol. Biol. 17, 428–439.

Reinisch, K.M., Wolin, S.L., 2007. Emerging themes in non-coding RNA quality control. Curr. Opin. Struct. Biol. 17, 209–214.

Rende, F., Cavallari, I., Corradin, A., Silic-Benussi, M., Toulza, F., Toffolo, G.M., Tanaka, Y., Jacobson, S., Taylor, G.P., D'Agostino, D.M., Bangham, C.R., Ciminale, V., 2011. Kinetics and intracellular compartmentalization of HTLV-1 gene expression: nuclear retention of HBZ mRNAs. Blood 117, 4855–4859.

Reynoso, R., Laufer, N., Hackl, M., Skalicky, S., Monteforte, R., Turk, G., Carobene, M., Quarleri, J., Cahn, P., Werner, R., Stoiber, H., Grillari-Voglauer, R., Grillari, J., 2014. MicroRNAs differentially present in the plasma of HIV elite controllers reduce HIV infection in vitro. Sci. Rep. 4, 5915.

Rhim, H., Park, J., Morrow, C.D., 1991. Deletions in the tRNA(Lys) primer-binding site of human immunodeficiency virus type 1 identify essential regions for reverse transcription. J. Virol. 65, 4555–4564.

Riggs, J.L., Oshirls, Taylor, D.O., Lennette, E.H., 1969. Syncytium-forming agent isolated from domestic cats. Nature 222, 1190–1191.

Rinn, J.L., Chang, H.Y., 2012. Genome regulation by long noncoding RNAs. Annu. Rev. Biochem. 81, 145–166.

Roberts, T.M., Boris-Lawrie, K., 2003. Primary sequence and secondary structure motifs in spleen necrosis virus RU5 confer translational utilization of unspliced human immunodeficiency virus type 1 reporter RNA. J. Virol. 77, 11973–11984.

Rosewick, N., Momont, M., Durkin, K., Takeda, H., Caiment, F., Cleuter, Y., Vernin, C., Mortreux, F., Wattel, E., Burny, A., Georges, M., Van den Broeke, A., 2013. Deep sequencing reveals abundant noncanonical retroviral microRNAs in B-cell leukemia/lymphoma. Proc. Natl. Acad. Sci. U.S.A. 110, 2306–2311.

Rossi, J.J., 1999. The application of ribozymes to HIV infection. Curr. Opin. Mol. Ther. 1, 316–322.

Roy, S., Agy, M., Hovanessian, A.G., Sonenberg, N., Katze, M.G., 1991. The integrity of the stem structure of human immunodeficiency virus type 1 Tat-responsive sequence of RNA is required for interaction with the interferon-induced 68,000-Mr protein kinase. J. Virol. 65, 632–640.

Ruby, J.G., Jan, C.H., Bartel, D.P., 2007. Intronic microRNA precursors that bypass drosha processing. Nature 448, 83–86.

Ruggero, K., Corradin, A., Zanovello, P., Amadori, A., Bronte, V., Ciminale, V., D'Agostino, D.M., 2010. Role of microRNAs in HTLV-1 infection and transformation. Mol. Aspects Med. 31, 367–382.

Ruggero, K., Guffanti, A., Corradin, A., Sharma, V.K., De Bellis, G., Corti, G., Grassi, A., Zanovello, P., Bronte, V., Ciminale, V., D'Agostino, D.M., 2014. Small noncoding RNAs in cells transformed by human T-cell leukemia virus type 1: a role for a tRNA fragment as a primer for reverse transcriptase. J. Virol. 88, 3612–3622.

Ruiz, M.T., Voinnet, O., Baulcombe, D.C., 1998. Initiation and maintenance of virus-induced gene silencing. Plant Cell 10, 937–946.

Ruprecht, K., Mayer, J., Sauter, M., Roemer, K., Mueller-Lantzsch, N., 2008. Endogenous retroviruses and cancer. Cell. Mol. Life Sci. 65, 3366–3382.

Russell, R.A., Zeng, Y., Erlwein, O., Cullen, B.R., McClure, M.O., 2001. The R region found in the human foamy virus long terminal repeat is critical for both Gag and Pol protein expression. J. Virol. 75, 6817–6824.

Sandmeyer, S.B., Menees, T.M., 1996. Morphogenesis at the retrotransposon-retrovirus interface: gypsy and copia families in yeast and Drosophila. Curr. Top. Microbiol. Immunol. 214, 261–296.

Santanam, U., Zanesi, N., Efanov, A., Costinean, S., Palamarchuk, A., Hagan, J.P., Volinia, S., Alder, H., Rassenti, L., Kipps, T., Croce, C.M., Pekarsky, Y., 2010. Chronic lymphocytic leukemia modeled in mouse by targeted miR-29 expression. Proc. Natl. Acad. Sci. U.S.A. 107, 12210–12215.

Sarkar, I., Hauber, I., Hauber, J., Buchholz, F., 2007. HIV-1 proviral DNA excision using an evolved recombinase. Science 316, 1912–1915.

Satou, Y., Yasunaga, J.I., Yoshida, M., Matsuoka, M., 2006a. HTLV-1 basic leucine zipper factor gene mRNA supports proliferation of adult T cell leukemia cells. Proc. Natl. Acad. Sci. U.S.A. 103, 8906 vol. 103, p. 720, 2006.

Satou, Y., Yasunaga, J., Yoshida, M., Matsuoka, M., 2006b. HTLV-I basic leucine zipper factor gene mRNA supports proliferation of adult T cell leukemia cells. Proc. Natl. Acad. Sci. U.S.A. 103, 720–725.

Sauter, M., Schommer, S., Kremmer, E., Remberger, K., Dolken, G., Lemm, I., Buck, M., Best, B., Neumann-Haefelin, D., Mueller-Lantzsch, N., 1995. Human endogenous retrovirus K10: expression of Gag protein and detection of antibodies in patients with seminomas. J. Virol. 69, 414–421.

Sawyer, R.C., Dahlberg, J.E., 1973. Small RNAs of Rous sarcoma virus: characterization by two-dimensional polyacrylamide gel electrophoresis and fingerprint analysis. J. Virol. 12, 1226–1237.

Schnitzer, T.J., 1981. Characterization of a simian foamy virus isolated from a spontaneous primate lymphoma. J. Med. Primatol. 10, 312–328.

Scholthof, H.B., 2006. The Tombusvirus-encoded P19: from irrelevance to elegance. Nat. Rev. Microbiol. 4, 405–411.

Schopman, N.C., ter Brake, O., Berkhout, B., 2010. Anticipating and blocking HIV-1 escape by second generation antiviral shRNAs. Retrovirology 7, 1742–4690.

Schopman, N.C., Willemsen, M., Liu, Y.P., Bradley, T., van Kampen, A., Baas, F., Berkhout, B., Haasnoot, J., 2012. Deep sequencing of virus-infected cells reveals HIV-encoded small RNAs. Nucleic Acids Res. 40, 414–427.

Schorn, A.J., Gutbrod, M.J., LeBlanc, C., Martienssen, R., 2017. LTR-retrotransposon control by tRNA-derived small RNAs. Cell 170 61–71.e11.

Schwarz, D.S., Hutvagner, G., Du, T., Xu, Z., Aronin, N., Zamore, P.D., 2003. Asymmetry in the assembly of the RNAi enzyme complex. Cell 115, 199–208.

Seiki, M., Hattori, S., Yoshida, M., 1982. Human adult T-cell leukemia virus: molecular cloning of the provirus DNA and the unique terminal structure. Proc. Natl. Acad. Sci. U.S.A. 79, 6899–6902.

Seiki, M., Hattori, S., Hirayama, Y., Yoshida, M., 1983. Human adult T-cell leukemia virus: complete nucleotide sequence of the provirus genome integrated in leukemia cell DNA. Proc. Natl. Acad. Sci. U.S.A. 80, 3618–3622.

Sen, G.L., Blau, H.M., 2005. Argonaute 2/RISC resides in sites of mammalian mRNA decay known as cytoplasmic bodies. Nat. Cell Biol. 7, 633–636.

SenGupta, D.N., Silverman, R.H., 1989. Activation of interferon-regulated, dsRNA-dependent enzymes by human immunodeficiency virus-1 leader RNA. Nucleic Acids Res. 17, 969–978.

SenGupta, D.N., Berkhout, B., Gatignol, A., Zhou, A.M., Silverman, R.H., 1990. Direct evidence for translational regulation by leader RNA and Tat protein of human immunodeficiency virus type 1. Proc. Natl. Acad. Sci. U.S.A. 87, 7492–7496.

Sheehy, A.M., Gaddis, N.C., Malim, M.H., 2003. The antiretroviral enzyme APOBEC3G is degraded by the proteasome in response to HIV-1 Vif. Nat. Med. 9, 1404–1407.

Shen, C.J., Jia, Y.H., Tian, R.R., Ding, M., Zhang, C., Wang, J.H., 2012. Translation of Pur-alpha is targeted by cellular miRNAs to modulate the differentiation-dependent susceptibility of monocytes to HIV-1 infection. FASEB J. 26, 4755–4764.

Sim, S., Weinberg, D.E., Fuchs, G., Choi, K., Chung, J., Wolin, S.L., 2009. The subcellular distribution of an RNA quality control protein, the Ro autoantigen, is regulated by noncoding Y RNA binding. Mol. Biol. Cell 20, 1555–1564.

Skalsky, R.L., Samols, M.A., Plaisance, K.B., Boss, I.W., Riva, A., Lopez, M.C., Baker, H.V., Renne, R., 2007. Kaposi's sarcoma-associated herpesvirus encodes an ortholog of miR-155. J. Virol. 81, 12836–12845.

Sleeman, J.E., Trinkle-Mulcahy, L., 2014. Nuclear bodies: new insights into assembly/dynamics and disease relevance. Curr. Opin. Cell Biol. 28, 76–83.

Smith, P.R., Venables, P.J., Williams, D., Maini, R.N., 1984. Identification and purification of a 55K polypeptide in Sjogren's syndrome A antigen. Rheumatol. Int. 4, 135–138.

Sobala, A., Hutvagner, G., 2011. Transfer RNA-derived fragments: origins, processing, and functions. Wiley Interdiscip. Rev. RNA 2, 853–862.

Soifer, H.S., Rossi, J.J., Saetrom, P., 2007. MicroRNAs in disease and potential therapeutic applications. Mol. Ther. 15, 2070–2079.

Springer, M.S., Britten, R.J., 1993. Phylogenetic relationships of reverse transcriptase and RNase H sequences and aspects of genome structure in the gypsy group of retrotransposons. Mol. Biol. Evol. 10, 1370–1379.

St Laurent, G., Wahlestedt, C., Kapranov, P., 2015. The Landscape of long noncoding RNA classification. Trends Genet. 31, 239–251.

Stadler, P.F., Chen, J.J., Hackermuller, J., Hoffmann, S., Horn, F., Khaitovich, P., Kretzschmar, A.K., Mosig, A., Prohaska, S.J., Qi, X., Schutt, K., Ullmann, K., 2009. Evolution of vault RNAs. Mol. Biol. Evol. 26, 1975–1991.

Stefanov, Y., Salenko, V., Glukhov, I., 2012. Drosophila errantiviruses. Mob. Genet. Elem. 2, 36–45.

Steiner, E., Holzmann, K., Elbling, L., Micksche, M., Berger, W., 2006. Cellular functions of vaults and their involvement in multidrug resistance. Curr. Drug Targets 7, 923–934.

Stephen, A.G., Raval-Fernandes, S., Huynh, T., Torres, M., Kickhoefer, V.A., Rome, L.H., 2001. Assembly of vault-like particles in insect cells expressing only the major vault protein. J. Biol. Chem. 276, 23217–23220.

Stewart, S.A., Dykxhoorn, D.M., Palliser, D., Mizuno, H., Yu, E.Y., An, D.S., Sabatini, D.M., Chen, I.S., Hahn, W.C., Sharp, P.A., Weinberg, R.A., Novina, C.D., 2003. Lentivirus-delivered stable gene silencing by RNAi in primary cells. RNA 9, 493–501.

Sun, Y.H., Xie, L.H., Zhuo, X., Chen, Q., Ghoneim, D., Zhang, B., Jagne, J., Yang, C., Li, X.Z., 2017. Domestic chickens activate a piRNA defense against avian leukosis virus. Elife 6.

Sung, T.L., Rice, A.P., 2009. miR-198 inhibits HIV-1 gene expression and replication in monocytes and its mechanism of action appears to involve repression of cyclin T1. PLoS Pathog. 5, e1000263.

Swaminathan, G., Rossi, F., Sierra, L.J., Gupta, A., Navas-Martin, S., Martin-Garcia, J., 2012. A role for microRNA-155 modulation in the anti-HIV-1 effects of toll-like receptor 3 stimulation in macrophages. PLoS Pathog. 8, e1002937.

Taft, R.J., Pheasant, M., Mattick, J.S., 2007. The relationship between non-protein-coding DNA and eukaryotic complexity. Bioessays 29, 288–299.

Tam, W., Ben-Yehuda, D., Hayward, W.S., 1997. bic, a novel gene activated by proviral insertions in avian leukosis virus-induced lymphomas, is likely to function through its noncoding RNA. Mol. Cell. Biol. 17, 1490–1502.

Tam, W., Hughes, S.H., Hayward, W.S., Besmer, P., 2002. Avian bic, a gene isolated from a common retroviral site in avian leukosis virus-induced lymphomas that encodes a noncoding RNA, cooperates with c-myc in lymphomagenesis and erythroleukemogenesis. J. Virol. 76, 4275–4286.

Taylor, B.S., Hammer, S.M., 2008. The challenge of HIV-1 subtype diversity. N. Engl. J. Med. 359, 1965–1966.

Taylor, J.M., 1977. An analysis of the role of tRNA species as primers for the transcription into DNA of RNA tumor virus genomes. Biochim. Biophys. Acta 473, 57–71.

Tebas, P., Stein, D., Tang, W.W., Frank, I., Wang, S.Q., Lee, G., Spratt, S.K., Surosky, R.T., Giedlin, M.A., Nichol, G., Holmes, M.C., Gregory, P.D., Ando, D.G., Kalos, M., Collman, R.G., Binder-Scholl, G., Plesa, G., Hwang, W.T., Levine, B.L., June, C.H., 2014. Gene editing of CCR5 in autologous CD4 T cells of persons infected with HIV. N. Engl. J. Med. 370, 901–910.

Telesnitsky, A., Wolin, S.L., 2016. The host RNAs in retroviral particles. Viruses 8.

ter Brake, O., t Hooft, K., Liu, Y.P., Centlivre, M., von Eije, K.J., Berkhout, B., 2008. Lentiviral vector design for multiple shRNA expression and durable HIV-1 inhibition. Mol. Ther. 16, 557–564.

Tian, C., Wang, T., Zhang, W., Yu, X.F., 2007. Virion packaging determinants and reverse transcription of SRP RNA in HIV-1 particles. Nucleic Acids Res. 35, 7288–7302.

Tobaly-Tapiero, J., Bittoun, P., Neves, M., Guillemin, M.C., Lecellier, C.H., Puvion-Dutilleul, F., Gicquel, B., Zientara, S., Giron, M.L., de The, H., Saib, A., 2000. Isolation and characterization of an equine foamy virus. J. Virol. 74, 4064–4073.

Todd, G.C., Duchon, A., Inlora, J., Olson, E.D., Musier-Forsyth, K., Ono, A., 2017. Inhibition of HIV-1 Gag-membrane interactions by specific RNAs. RNA 23, 395–405.

Triboulet, R., Mari, B., Lin, Y.L., Chable-Bessia, C., Bennasser, Y., Lebrigand, K., Cardinaud, B., Maurin, T., Barbry, P., Baillat, V., Reynes, J., Corbeau, P., Jeang, K.T., Benkirane, M., 2007. Suppression of microRNA-silencing pathway by HIV-1 during virus replication. Science 315, 1579–1582.

Tristem, M., 2000. Identification and characterization of novel human endogenous retrovirus families by phylogenetic screening of the human genome mapping project database. J. Virol. 74, 3715–3730.

Trobaugh, D.W., Klimstra, W.B., 2017. MicroRNA regulation of RNA virus replication and pathogenesis. Trends Mol. Med. 23, 80–93.

Tubio, J.M., Costas, J.C., Naveira, H.F., 2004. Evolution of the mdg1 lineage of the Ty3/gypsy group of LTR retrotransposons in *Anopheles gambiae*. Gene 330, 123–131.

Turelli, P., Mangeat, B., Jost, S., Vianin, S., Trono, D., 2004. Inhibition of hepatitis B virus replication by APOBEC3G. Science 303, 1829.

Tycowski, K.T., Guo, Y.E., Lee, N., Moss, W.N., Vallery, T.K., Xie, M., Steitz, J.A., 2015. Viral noncoding RNAs: more surprises. Genes Dev. 29, 567–584.

Ullu, E., Tschudi, C., 1984. Alu sequences are processed 7SL RNA genes. Nature 312, 171–172.

Umbach, J.L., Cullen, B.R., 2009. The role of RNAi and microRNAs in animal virus replication and antiviral immunity. Genes Dev. 23, 1151–1164.

van der Krol, A.R., Mur, L.A., Beld, M., Mol, J.N., Stuitje, A.R., 1990. Flavonoid genes in petunia: addition of a limited number of gene copies may lead to a suppression of gene expression. Plant Cell 2, 291–299.

Van Lint, C., Bouchat, S., Marcello, A., 2013. HIV-1 transcription and latency: an update. Retrovirology 10, 67.

van Meer, G., Voelker, D.R., Feigenson, G.W., 2008. Membrane lipids: where they are and how they behave. Nat. Rev. Mol. Cell Biol. 9, 112–124.

van Rooij, E., Marshall, W.S., Olson, E.N., 2008. Toward microRNA-based therapeutics for heart disease: the sense in antisense. Circ. Res. 103, 919–928.

van 't Wout, A.B., Lehrman, G.K., Mikheeva, S.A., O'Keeffe, G.C., Katze, M.G., Bumgarner, R.E., Geiss, G.K., Mullins, J.I., 2003. Cellular gene expression upon human immunodeficiency virus type 1 infection of CD4(+)-T-cell lines. J. Virol. 77, 1392–1402.

Veneziano, D., Di Bella, S., Nigita, G., Lagana, A., Ferro, A., Croce, C.M., 2016. Noncoding RNA: current deep sequencing data analysis approaches and challenges. Hum. Mutat. 37, 1283–1298.

Voinnet, O., Pinto, Y.M., Baulcombe, D.C., 1999. Suppression of gene silencing: a general strategy used by diverse DNA and RNA viruses of plants. Proc. Natl. Acad. Sci. U.S.A. 96, 14147–14152.

Voinnet, O., 2001. RNA silencing as a plant immune system against viruses. Trends Genet. 17, 449–459.

Voytas, D.F., Boeke, J.D., 1993. Yeast retrotransposons and tRNAs. Trends Genet. 9, 421–427.

Waheed, A.A., Freed, E.O., 2009. Lipids and membrane microdomains in HIV-1 replication. Virus Res. 143, 162–176.

Wain-Hobson, S., Sonigo, P., Danos, O., Cole, S., Alizon, M., 1985. Nucleotide sequence of the AIDS virus, LAV. Cell 40, 9–17.

Wakefield, J.K., Wolf, A.G., Morrow, C.D., 1995. Human immunodeficiency virus type 1 can use different tRNAs as primers for reverse transcription but selectively maintains a primer binding site complementary to tRNA(3Lys). J. Virol. 69, 6021–6029.

Wakefield, J.K., Kang, S.M., Morrow, C.D., 1996. Construction of a type 1 human immunodeficiency virus that maintains a primer binding site complementary to tRNA(His). J. Virol. 70, 966–975.

Walker, T.A., Pace, N.R., Erikson, R.L., Erikson, E., Behr, F., 1974. The 7S RNA common to oncornaviruses and normal cells is associated with polyribosomes. Proc. Natl. Acad. Sci. U.S.A. 71, 3390–3394.

Walter, P., Blobel, G., 1982. Signal recognition particle contains a 7S RNA essential for protein translocation across the endoplasmic reticulum. Nature 299, 691–698.

Wang, T., Tian, C., Zhang, W., Luo, K., Sarkis, P.T., Yu, L., Liu, B., Yu, Y., Yu, X.F., 2007. 7SL RNA mediates virion packaging of the antiviral cytidine deaminase APOBEC3G. J. Virol. 81, 13112–13124.

Wang, T., Tian, C., Zhang, W., Sarkis, P.T., Yu, X.F., 2008. Interaction with 7SL RNA but not with HIV-1 genomic RNA or P bodies is required for APOBEC3F virion packaging. J. Mol. Biol. 375, 1098–1112.

Wang, X., Ye, L., Hou, W., Zhou, Y., Wang, Y.J., Metzger, D.S., Ho, W.Z., 2009. Cellular microRNA expression correlates with susceptibility of monocytes/macrophages to HIV-1 infection. Blood 113, 671–674.

Wang, Q., Gao, Y., Ji, X., Qi, X., Qin, L., Gao, H., Wang, Y., Wang, X., 2013a. Differential expression of microRNAs in avian leukosis virus subgroup J-induced tumors. Vet. Microbiol. 162, 232–238.

Wang, Q., Ji, X., Gao, Y., Qi, X., Wang, X., Wang, Y., Qin, L., Gao, H., 2013b. Overexpression of microRNA gga-miR-1650 decreases the replication of avian leukosis virus subgroup J in infected cells. J. Gen. Virol. 94, 2287–2296.

Wang, J., Xie, G., Singh, M., Ghanbarian, A.T., Rasko, T., Szvetnik, A., Cai, H., Besser, D., Prigione, A., Fuchs, N.V., Schumann, G.G., Chen, W., Lorincz, M.C., Ivics, Z., Hurst, L.D., Izsvak, Z., 2014. Primate-specific endogenous retrovirus-driven transcription defines naive-like stem cells. Nature 516, 405–409.

Wang, Z., Pan, Q., Gendron, P., Zhu, W., Guo, F., Cen, S., Wainberg, M.A., Liang, C., 2016. CRISPR/Cas9-derived mutations both inhibit HIV-1 replication and accelerate viral escape. Cell Rep. 15, 481–489.

Wapinski, O., Chang, H.Y., 2011. Long noncoding RNAs and human disease. Trends Cell Biol. 21, 354–361.

Watanabe, Y., Kishi, A., Yachie, N., Kanai, A., Tomita, M., 2007. Computational analysis of microRNA-mediated antiviral defense in humans. FEBS Lett. 581, 4603–4610.

Waters, L.C., Mullin, B.C., 1977. Transfer RNA into RNA tumor viruses. Prog. Nucleic Acid Res. Mol. Biol. 20, 131–160.

Weiss, R.A., 2006. The discovery of endogenous retroviruses. Retrovirology 3, 67.

Westerhout, E.M., Ooms, M., Vink, M., Das, A.T., Berkhout, B., 2005. HIV-1 can escape from RNA interference by evolving an alternative structure in its RNA genome. Nucleic Acids Res. 33, 796–804.

Whisnant, A.W., Bogerd, H.P., Flores, O., Ho, P., Powers, J.G., Sharova, N., Stevenson, M., Chen, C.H., Cullen, B.R., 2013. In-depth analysis of the interaction of HIV-1 with cellular microRNA biogenesis and effector mechanisms. mBio 4, e000193.

Whisnant, A.W., Kehl, T., Bao, Q., Materniak, M., Kuzmak, J., Lochelt, M., Cullen, B.R., 2014. Identification of novel, highly expressed retroviral microRNAs in cells infected by bovine foamy virus. J. Virol. 88, 4679–4686.

Whitcomb, J.M., Ortiz-Conde, B.A., Hughes, S.H., 1995. Replication of avian leukosis viruses with mutations at the primer binding site: use of alternative tRNAs as primers. J. Virol. 69, 6228–6238.

Wildschutte, J.H., Williams, Z.H., Montesion, M., Subramanian, R.P., Kidd, J.M., Coffin, J.M., 2016. Discovery of unfixed endogenous retrovirus insertions in diverse human populations. Proc. Natl. Acad. Sci. U.S.A. 113, E2326–E2334.

Winter, J., Jung, S., Keller, S., Gregory, R.I., Diederichs, S., 2009. Many roads to maturity: microRNA biogenesis pathways and their regulation. Nat. Cell Biol. 11, 228–234.

Wolin, S.L., Steitz, J.A., 1984. The Ro small cytoplasmic ribonucleoproteins: identification of the antigenic protein and its binding site on the Ro RNAs. Proc. Natl. Acad. Sci. U.S.A. 81, 1996–2000.

Wollmann, R.L., Kirsten, W.H., 1968. Cellular origin of a mouse leukemia viral ribonucleic acid. J. Virol. 2, 1241–1248.

Wu, X., Qian, K., Qin, A., Shen, H., Wang, P., Jin, W., Eltahir, Y.M., 2010. Recombinant avian leukosis viruses of subgroup J isolated from field infected commercial layer chickens with hemangioma and myeloid leukosis possess an insertion in the E element. Vet. Res. Commun. 34, 619–632.

Xu, Z., Asahchop, E.L., Branton, W.G., Gelman, B.B., Power, C., Hobman, T.C., 2017. MicroRNAs upregulated during HIV infection target peroxisome biogenesis factors: implications for virus biology, disease mechanisms and neuropathology. PLoS Pathog. 13, e1006360.

Yamashita, M., Emerman, M., 2004. Capsid is a dominant determinant of retrovirus infectivity in nondividing cells. J. Virol. 78, 5670–5678.

Yao, Y., Smith, L.P., Nair, V., Watson, M., 2014. An avian retrovirus uses canonical expression and processing mechanisms to generate viral microRNA. J. Virol. 88, 2–9.

Ye, L., Wang, J., Beyer, A.I., Teque, F., Cradick, T.J., Qi, Z., Chang, J.C., Bao, G., Muench, M.O., Yu, J., Levy, J.A., Kan, Y.W., 2014. Seamless modification of wild-type induced pluripotent stem cells to the natural CCR5Delta32 mutation confers resistance to HIV infection. Proc. Natl. Acad. Sci. U.S.A. 111, 9591–9596.

Yekta, S., Shih, I.H., Bartel, D.P., 2004. MicroRNA-directed cleavage of HOXB8 mRNA. Science 304, 594–596.

Yeung, M.L., Benkirane, M., Jeang, K.T., 2007. Small non-coding RNAs, mammalian cells, and viruses: regulatory interactions? Retrovirology 4, 74.

Yeung, M.L., Yasunaga, J., Bennasser, Y., Dusetti, N., Harris, D., Ahmad, N., Matsuoka, M., Jeang, K.T., 2008. Roles for microRNAs, miR-93 and miR-130b, and tumor protein 53-induced nuclear protein 1 tumor suppressor in cell growth dysregulation by human T-cell lymphotrophic virus 1. Cancer Res. 68, 8976–8985.

Yeung, M.L., Bennasser, Y., Watashi, K., Le, S.Y., Houzet, L., Jeang, K.T., 2009. Pyrosequencing of small non-coding RNAs in HIV-1 infected cells: evidence for the processing of a viral-cellular double-stranded RNA hybrid. Nucleic Acids Res. 37, 6575–6586.

Yilmaz, A., Bolinger, C., Boris-Lawrie, K., 2006. Retrovirus translation initiation: issues and hypotheses derived from study of HIV-1. Curr. HIV Res. 4, 131–139.

Younis, I., Green, P.L., 2005. The human T-cell leukemia virus Rex protein. Front. Biosci. 10, 431–445.

Zaitseva, L., Myers, R., Fassati, A., 2006. tRNAs promote nuclear import of HIV-1 intracellular reverse transcription complexes. PLoS Biol. 4, e332.

Zapata, J.C., Campilongo, F., Barclay, R.A., DeMarino, C., Iglesias-Ussel, M.D., Kashanchi, F., Romerio, F., 2017. The Human Immunodeficiency Virus 1 ASP RNA promotes viral latency by recruiting the Polycomb Repressor Complex 2 and promoting nucleosome assembly. Virology 506, 34–44.

Zhang, H.S., Wu, T.C., Sang, W.W., Ruan, Z., 2012. MiR-217 is involved in Tat-induced HIV-1 long terminal repeat (LTR) transactivation by down-regulation of SIRT1. Biochim. Biophys. Acta 1823, 1017–1023.

Zhang, Q., Chen, C.Y., Yedavalli, V.S., Jeang, K.T., 2013. NEAT1 long noncoding RNA and paraspeckle bodies modulate HIV-1 posttranscriptional expression. mBio 4 e00596–00512.

Zhao, T.J., Matsuoka, M., 2012. HBZ and its roles in HTLV-1 oncogenesis. Front. Microbiol. 3.

Zhou, L., Sokolskaja, E., Jolly, C., James, W., Cowley, S.A., Fassati, A., 2011. Transportin 3 promotes a nuclear maturation step required for efficient HIV-1 integration. PLoS Pathog. 7, e1002194.

Zhou, W., Parent, L.J., Wills, J.W., Resh, M.D., 1994. Identification of a membrane-binding domain within the amino-terminal region of human immunodeficiency virus type 1 Gag protein which interacts with acidic phospholipids. J Virol. 68, 2556–2569.

Chapter 13

Cellular Control of Endogenous Retroviruses and Retroelements

Suzanne Sandmeyer
University of California, Irvine, CA, United States

Chapter Outline

Introduction 480
Background 480
Long Terminal Repeat Retrotransposons and Endogenous Retroviruses 480
Non-Long Terminal Repeat Retrotransposons 483
Long Interspersed Elements 483
Short Interspersed Elements 485
Chimeric Short Interspersed Elements 485
Pseudogenes 486
Activity 486
Frequencies of Retrotransposition 486
Retroelement Activities and Impacts 486
Control of Retroelement Expression 488
Retroelement Transcription Patterns Favor Self-Preservation 488
Regulation of Retroelements in Somatic Cells by RNA Interference 490
Regulation of Retroelements in Germline and Embryonic Stem Cells 492
Sequence-Specific Suppression of Retroelements 498
Neuronal Activity of Retroelements 499
Roles for Retroelements in Cancer Initiation and Progression 502
Regulation of Retroelements by the Innate Immune System 502
Regulation of Innate Immunity by Endogenous Retrovirus Sequences 502
Sensing of Retroelements by Innate Immunity 504
Direct Control of Retroelements by Innate Immunity 505
Perspectives 507
Acknowledgments 509
References 510

Retrovirus-Cell Interactions. https://doi.org/10.1016/B978-0-12-811185-7.00013-3

INTRODUCTION

Transposable elements (TE) occupy significant fractions of most eukaryotic genomes and our fascination with their dimensionality continues to grow. A subset of these are retroelements. These elements expand through cycles of forward and reverse transcription and genome integration. The several types of retroelements differ significantly in sequence organization, encoded proteins, mechanisms of priming reverse transcription, and genome entry but are united by the common ability to permanently modify the host genome: they can be minimized and mutated, but not easily erased. They vastly outnumber cellular protein–coding genes so that well over half of the human genome can be traced to origins in these sequences. In theory, identified epigenetic mechanisms could completely silence retroelement expression, but this is not the case. One explanation is that animals have evolved using retroelements to enable diversification of species and differentiation within species. This review is focused on results of recent investigations regarding host permissions and restrictions on retroelements and the ways in which these elements have become essential features of animal genomes.

Readily identified retroelement sequences constitute 46.36% of the human genome (International Human Genome Sequencing, 2001) and 38.55% of the mouse genome (Mouse Genome Sequencing et al., 2002). Comparison of post-divergence rates of accumulation of lineage-specific repeats shows that although it occurs at a higher frequency in the mouse genome, both mouse and human genomes continue to undergo retrotransposition (Mouse Genome Sequencing et al., 2002). The ways in which retroelements are an integral part of control of host gene expression, sexual reproduction and differentiation, and recognition of self- and speciation are only beginning to be understood.

BACKGROUND

Retroelements occur in three broad classes: (1) long terminal repeat (LTR) retrotransposons and endogenous retroviruses (ERVs); (2) long interspersed elements (LINEs); and (3) short interspersed elements (SINEs) (reviewed in Levin and Moran, 2011). In this Chapter, LTR retrotransposons and ERVs are collectively referred to as LTR retroelements; LINEs and SINEs are collectively termed non-LTR elements. Non-LTR elements represent the majority of elements in the respective mouse and human genomes: LTR and ERVs (9.87/8.55%); LINEs (19.20/20.99%); and SINEs (8.22/13.64%) (International Human Genome Sequencing, 2001; Mouse Genome Sequencing et al., 2002; Venter et al., 2001).

Long Terminal Repeat Retrotransposons and Endogenous Retroviruses Fig. 13.1

Intact LTR retroelements are 6–9 kb in length with LTRs of 300–1500 bp flanking internal coding sequence. LTRs provide for RNA polymerase 2 (RNAP2) promoter sequences that direct sense transcription of a genomic RNA and potentially antisense transcription of flanking sequences. Genomic transcripts contain

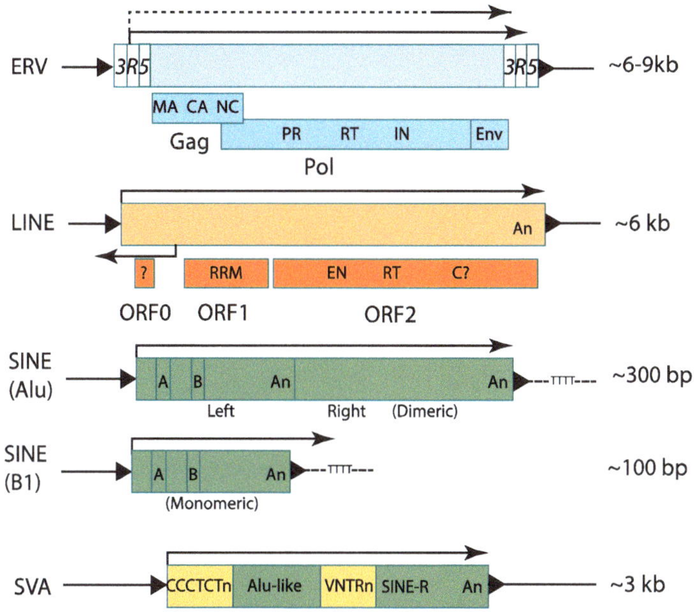

FIGURE 13.1 **Eukaryotic retroelements.** The major classes of mammalian retroelements are endogenous retroviruses (ERVs), long interspersed elements (LINEs), short interspersed elements (SINEs), and chimeric SINEs. Endogenous retroviruses evolved from ancient long terminal repeat (LTR) retrotransposons, not highly represented in mouse and human genomes, by acquiring an envelope gene (*env*). The current major subclasses have lost *env* and/or various amounts of internal coding information. Many are now represented as single LTR remnants of recombination between the LTRs. ERVs are transcribed starting at the upstream end of a repeated region (R) in the LTR and terminating at the downstream end of R. The 5′ unique and 3′ unique sequences are as indicated in *LTR boxes*. Gag is the structural polyprotein that is processed by viral protease (PR) typically into matrix (MA), capsid, and nucleocapsid domains. Pol is the polyprotein precursor encoded in one or two open reading frames (ORFs) that is processed most commonly into major catalytic species including PR, reverse transcriptase (RT), and integrase (IN) by PR. LINEs are non-LTR elements that encode major structural protein ORF1 with nucleic acid binding activity and ORF2 catalytic protein with RT, endonuclease, and a conserved domain of unknown function. In addition, LINEs encode a short antisense ORF0 (0). SINEs are noncoding monomeric or dimeric derivatives of RNAP3-transcripts, most commonly tRNAs or SRP 7SL RNA. They contain RNAP3 promoter elements A and B boxes that are followed by a poly(A) track, which in dimeric elements, separates this sequence from a region variably derived from LINE elements. Human SINEs are Alu elements; in the murine system they are B1, B2 and ID elements. SVA elements are chimeras of a chimeric repeat (CCCTCTn), an Alu-like sequence, copies of a variable nucleotide repeat (VNTRn), and fragments of SINE and ERV sequence (SINE-R). SINEs, SVA, and LINEs terminate in an encoded poly(A) track. Sizes are as shown, *arrows* indicate transcribed region with *dashed lines* indicating intronic information. *Solid arrowheads* flanking elements are short target-site duplications generated during integration.

internal *cis*-acting sequences required for priming plus- and minus-strand reverse transcription to form the double-stranded (ds) complementary (c) DNA and terminate within the downstream LTR. The major classes of LTR retrotransposons are Ty3/Gypsy (Sandmeyer et al., 2015), Ty1/Copia (Curcio et al., 2015) and BEL/Pao (de la Chaux and Wagner, 2011). Of these, only Ty3/Gypsy exists in humans and rodents (Naville et al., 2016). It is also the most similar of the three LTR retrotransposons to human and mouse ERVs in coded proteins and genomic organization. Although LTR retrotransposons exist in fungi, plants, insects, and animals, ERVs are restricted to vertebrates (de Parseval and Heidmann, 2005). LTR retrotransposons and ERVs are endowed with one or two open reading frames (ORFs). The upstream ORF or coding region encodes a major structural polyprotein ("Gag") containing capsid and nucleic acid–binding domains for virus-like particle (VLP) formation, and a downstream ORF encodes a polyprotein ("Pol") containing catalytic domains for replication, including protease, reverse transcriptase (RT)–RNaseH and integrase domains. Retrotransposition initiates with transcription and nuclear export of genomic RNA. In elements with two ORFs, translation of the *gag* reading frame is typically initiated at the upstream start codon, and *pol* is translated via a programmed frameshift from *gag* that generates a Gag-Pol fusion in lower abundance. This mechanism provides for an appropriate excess of Gag to foster VLP assembly. In a minority of LTR retrotransposons, Gag and Pol are translated from a single reading frame. In LTR retrotransposons with one ORF, representation of catalytic proteins is reduced by some alternative to frameshifting, for example, differential proteolytic degradation (Atwood et al., 1996). VLP assembly is followed by proteolytic processing into domain-specific forms, reverse transcription of genomic RNA to yield cDNA, dissociation of the VLP and nuclear access of a preintegration complex, and integration of the cDNA via strand transfer. Integrated elements are flanked by target-site duplications (TSD) of 4–6 bp generated as a consequence of staggered-strand transfer of the cDNA ends of the integrating retroelement into host DNA creating short-single stranded junctions in flanking host DNA and subsequent repair (reviewed in Coffin et al., 1997; Skalka, 2014). Some elements, including the eponymous Gypsy, represent intermediates in the transition from LTR retrotransposons to retroviruses. They retain the ability to be transcribed and replicated in a single cell but have acquired an *env* function and the associated capacity for extracellular transmission (Song et al., 1994). In most bona fide retroviruses, reverse transcription is completed after infection of the new host cell. In a minority of cases, including spumaviruses, similar to LTR retrotransposons, significant reverse transcription occurs in the originating host cell. Most LTR retroelements in the human and mouse genomes are phylogenetically more closely related to retroviruses than to Ty3/Gypsy LTR retrotransposons and thus represent ERVs that invaded genomes through germline infection.

ERV-related sequences occur in essentially four classes: (1) elements that are relatively intact with potentially infectious retrovirus members; (2) elements lacking some coding sequences, typically *env,* but potentially autonomous; (3) non-autonomous elements lacking coding sequences, but retaining essential *cis*-acting sequences for transcription, packaging, and primer binding; and (4) solo LTR products of recombination between LTRs and associated loss of the internal domain and one LTR copy (reviewed in Maksakova et al., 2006). ERVs are designated according to the species in which they were discovered, e.g., human (HERVs) or mouse (MERVs) and the tRNA minus-strand cDNA primer (e.g., HERV-K, indicating primer $tRNA^{Lys}$) (Mayer et al., 2011). In mouse, the first and second classes include potentially infectious ERVs with counterparts among gamma- (e.g., murine leukemia virus [MLV]) and betaretroviruses (e.g., mouse mammary tumor virus [MMTV], MusD, and intracisternal type A particles [IAP]). In humans, HERV-K is the youngest of these classes, with some potentially active members. HERV-I-related forms are ancient and date to species that diverged over 400 million years ago (MYA), whereas younger HERV-L relatives are in common among species that diverged about 100 MYA (de Parseval and Heidmann, 2005). The third class, nonautonomous elements, is represented by mammalian apparent LTR retrotransposons (Smit, 1993) and murine ETns and virus-like 30S elements (Sonigo et al., 1987). Within mice, each of the first three classes has representatives that are active and together they account for ~10% of spontaneous mutations in laboratory strains. IAP and ETn and MusD account for the majority of these de novo germline mutations. The fourth class, LTR-only elements, has accumulated in vast excess over full-length versions through LTR–LTR recombination (Benachenhou et al., 2009; Hughes et al., 2010). About 443,000 LTRs are in the human genome (International Human Genome Sequencing, 2001) and about 631,000 LTRs are in the mouse genome (Mouse Genome Sequencing et al., 2002).

Non-Long Terminal Repeat Retrotransposons

Non-LTR retrotransposons are subdivided into LINEs of about 6 kbp at full length; and nonautonomous monomeric and dimeric SINEs of ~100 bp and ~300–400 bp, respectively (Schmid and Deininger, 1975), and chimeric SINEs. The most abundant human chimeric SINE is SINE-R-VNTR-Alu-like (SVA) of ~2 kbp composed of SINE, variable nucleotide repeat, and HERV sequences (reviewed in Kramerov and Vassetzky, 2011; Richardson et al., 2015). The SVAs are hominid specific and emerged before divergence of gorilla, chimpanzee, and human, 18–25 MYA (Wang et al., 2005), making them the youngest of the non-LTR elements. In mouse and human genomes, autonomous LINEs mediate transposition of SINEs. Despite this relationship, LINEs tend to accumulate in AT-rich regions reflecting insertion bias, but SINEs accumulate preferentially in GC-, gene-rich regions, although younger SINE insertions tend to show a more LINE-like bias for AT-rich regions. This is interpreted to mean that the biased patterns of older SINE insertions reflect their relatively greater ability to be tolerated within coding regions. As discussed more below, these patterns affect the roles that the two types of elements play in genome structure and expression. There are about 868,000 LINEs, 1,448,000 SINEs, and 2700 SVA elements in the human genome, and 660,000 LINEs and 1,498,000 SINEs in the mouse genome (International Human Genome Sequencing, 2001).

Long Interspersed Elements

Full-length LINE sequences have 5′ and 3′ untranslated regions (UTRs) and terminate in an encoded poly(A) sequence (reviewed in Richardson et al., 2015). Sense transcription initiates upstream of a 5′ UTR promoter and terminates downstream of the encoded poly(A) tract or, in the case of about 15% of transcripts, extends downstream into flanking DNA. The LINE genomic RNA is capped and polyadenylated. The UTRs flank two ORFs that are translated into independent proteins (Alisch et al., 2006). *ORF1* encodes a 40-kDa protein with nucleic acid chaperone activity (Hohjoh and Singer, 1996; Martin and Bushman, 2001). *ORF2* encodes a 150-kDa protein containing endonuclease (EN) and RT domains and a cysteine-rich domain of unknown function (Feng et al., 1996; Mathias et al., 1991). Genomic RNA is exported to the cytoplasm and translated with a substantial excess of ORF1 over ORF2 protein. These proteins associate in *cis* with the template RNA, enhancing preferential retrotransposition of 5′-intact RNAs. Some of these complexes associate with stress granules that also contain components of the RNA-induced silencing complex (RISC) (Goodier et al., 2007). LINE proteins bound to genomic RNA can be actively imported into the nucleus as retrotransposition occurs in G1-arrested cells with intact nuclear envelopes (Kubo et al., 2006). However, mitotic nuclear envelope breakdown could also play a major role in promoting nuclear access in dividing cells. Integration is initiated on EN preferentially nicking AT-rich regions with the pattern 5′-(Pyr)n/(Pur)n-3′ (Feng et al., 1996). Reverse transcription is primed from the resulting target DNA 3′ OH and templated starting in the 3′ end of the gRNA. This process is referred to as template-primed reverse transcription (TPRT). Insertions are flanked by 7- to 20-bp TSD created by the variable offset of cDNA joints in the two target strands. EN-independent insertions also occur into breaks in the DNA, particularly in cells deficient in nonhomologous end joining (NHEJ) and at telomeres in cells with deficient telomerase (Morrish et al., 2007) and these characteristically lack TSD (Gilbert et al., 2002; Morrish et al., 2002; Symer et al., 2002). Reverse transcription of LINE genomic RNA is typically incomplete so that the majority of LINEs in the human genome are 5′-truncated and average 900 bp in length (International Human Genome Sequencing, 2001). Completion of reverse transcription is apparently antagonized by components of NHEJ because attenuation of NHEJ components increases the proportion of full-length insertions.

LINEs in rodent and human genomes evolved from a common progenitor. LINE phylogeny follows a pattern in which one or a limited number of element sequences dominates at any one time. LINEs have been active in the human genome from more than 150 MYA (Burton et al., 1986; International Human Genome Sequencing, 2001). Mouse and human LINEs include L1, L2, and L3 subtypes, of which the youngest, L1 derived from L1MA6 is active and most abundant (Mouse Genome Sequencing et al., 2002). Primate-specific L1 subtypes PA1-16 have been identified. Since emergence of great apes ~25 MYA, L1-PA1-5 subtypes have arisen and expanded. Within the human PA1 subtype, there are estimated to be 80–100 active, so-called "hot" elements. These are designated L1Hs. A transcriptionally active subset is further designated Ta (Beck et al., 2010; Brouha et al., 2003; Moran et al., 1996; Sassaman et al., 1997). Ta elements are distinguished by unique single and triplet nucleotide sequences in the 3′ UTR that have been exploited in assays to monitor recent and ongoing retrotransposition (Badge et al., 2003; Boissinot et al., 2000; Ewing and Kazazian, 2010; Huang et al., 2010; Iskow et al., 2010; Sanchez-Luque et al., 2016). Murine L1s have evolved most rapidly in the 5′ noncoding region and are mainly distinguished from primate L1s by the presence of promoter element repeats (Sookdeo et al., 2013). They are differentiated into V, F, A, T_F, and G_F subtypes (Sookdeo et al., 2013). A, T_F, and G_F L1s include most of the ~3000 active elements (Goodier et al., 2001).

Short Interspersed Elements

Monomeric and dimeric SINEs are transcribed by RNAP3. The monomeric segment or the first segment in the dimeric structure contains internal RNAP3 A and B box promoter elements but are lacking the canonical RNAP3 4-6 T-track terminator and have instead a downstream poly(A) tract (Chu et al., 1995). In the dimeric instance, such as represented by the hominid Alu element, the first unit is separated from a second, by an A-rich tract. Intragenic SINE insertions are included in RNAP2 transcripts with consequences for RNA processing, editing, and export (Gong and Maquat, 2011; Ponicsan et al., 2010).

Across eukaryotes, the majority of the RNAP3 class of SINEs have first segments derived from tRNA genes. However, the dominant class of human SINEs, Alu elements, and the murine B1 SINEs are derived from the 7SL signal recognition particle (SRP) RNA (Ullu and Tschudi, 1984). The majority of the 1.448 million Alu elements in the human genome (International Human Genome Sequencing, 2001) fall into J, S, and Y families (Britten et al., 1988; Jurka and Smith, 1988; Slagel et al., 1987). Retrotransposition activity of Alu elements peaked 35–65 MYA. However, AluY elements are polymorphic in the human population and represent the most active class of non-LTR elements (Carroll et al., 2001). Based on sequence conservation, including ability to bind SRP9/14 proteins, mobile Alus fall into levels of mobility with 1836 most mobile and 10,535 and 36,664 in classes of decreasing mobility (Bennett et al., 2008).

The murine genome contains four types of RNAP3-transcribed SINEs: B1, a monomeric 120-bp 7SL RNA derivative (564K copies) (Krayev et al., 1980; Quentin, 1994); B2 a 200-bp dimeric tRNA derivative (348K copies) (Daniels and Deininger, 1985); ID, a tRNA derivative expressed in neuronal tissue (79K copies) (Daniels and Deininger, 1985; DeChiara and Brosius, 1987); and B4, a composite of B1 and ID SINEs (391K copies) (Lee et al., 1998; Serdobova and Kramerov, 1998; Mouse Genome Sequencing et al., 2002). Although there have not been extensive studies of murine SINE mobility, B1 and B2 classes are known to be active in tissue culture (Dewannieux and Heidmann, 2005).

Chimeric Short Interspersed Elements

The SVA class of SINEs is ~2-kbp elements representing composites of GC-rich, variable nucleotide tandem repeats (VNTR) of CCCTCT, inverted Alu-like sequence, SINE-R sequence, resembling portions of ERV HERV-K10, and a cleavage polyadenylation specificity factor (CPSF)–binding site followed by a poly(A) tract (Ostertag et al., 2003; Wang et al., 2005). SVAs appeared in the primate lineage approximately 25 MYA. There are about 2700 SVA elements in the human genome. The polymorphic SVA elements in the human genome are classified as D, E, F, and F1.

Pseudogenes

In addition to mobilizing SINEs, LINEs are implicated in mobilizing mRNAs. Mobilization occurs as a consequence of capture of processed mRNAs through physical association, for example, by copackaging with genomic RNA, or readthrough transcription from LINE elements into flanking genes. This capture or fusion RNA then templates TPRT. Small RNAs including RNAP3-transcribed U6 and some other spliceosomal RNAs are also mobilized by LINE activity (Doucet et al., 2015). As much as 10% of the human genome may be accounted for by this type of retrotransposition (International Human Genome Sequencing, 2001). Although over evolutionary time, these pseudogenes likely do play an important role in expanding gene families, they differ from bona fide retroelements in that they typically are not adapted for efficient proliferation via retrotransposition.

ACTIVITY

Frequencies of Retrotransposition

Although retroelements constitute almost half of the human genome, spontaneous germline mutations attributable to these elements are relatively rare. In humans, L1 germline insertions are estimated to occur in about 1 in 100 individuals and Alu insertions about 1 in 20 individuals (reviewed in Cordaux and Batzer, 2009). However, there is little evidence of ongoing HERV mobilization or contribution to germline mutations (reviewed in Mager and Stoye, 2015; Maldarelli et al., 2014; Stocking and Kozak, 2008). In contrast, TE activity in the laboratory mouse is relatively high and is estimated to account for about 10% of spontaneously arising germline mutations. Mouse retrotransposition frequencies are accounted for by the presence of active ERVs and infectious ERVs generated by recombination (Maksakova et al., 2006).

Retroelement Activities and Impacts

Retroelement-related rearrangements affecting transcription or translation and ectopic recombination of elements resulting in deletions, inversions, translocations, and other rearrangements can have dramatic and deleterious consequences for individual genomes in real time. However, these elements have also contributed to emergence of epigenetic controls, transcription regulatory elements, and gene families that underlie species adaptability and diversification.

One hundred and twenty-four LINE-mediated insertions have been directly implicated in genetic diseases (Hancks and Kazazian, 2016). Furthermore, a significant fraction of L1, Alu, and SVA mobilization is accompanied by potentially impactful rearrangements of flanking sequence (reviewed in Richardson et al., 2015). Somatic L1 mobility is associated with certain types of cancer, although roles in causality and progression have been challenging to establish (reviewed in Burns, 2017; Burns and Boeke, 2012; Iskow et al., 2010; Rodic et al., 2015; Solyom et al., 2012). In addition, there is evidence that retroelements are associated with disease risk: over 18,000 common transposon-related polymorphic variants have been mapped in the human genome and many more are estimated to exist (Sudmant et al., 2015). An early study found that about one-third of AluY family elements retrieved from GenBank were polymorphic (Roy-Engel et al., 2001). A recent genomewide association study in which 809 polymorphic AluYa5/8 and AluYb8/9 elements were mapped relative to 1159 to disease risk loci, concluded that there was a disproportionate association ($P = .013$). In a subset that was further investigated, 44 variants were in linkage disequilibrium with trait-associated SNPs (Payer et al., 2017). These results are consistent with the idea that SINEs are a class of common variants with disease associations and effects on transcription (reviewed in Stocking and Kozak, 2008).

Expanded capabilities of next-generation sequencing (NGS) coupled with sophisticated bioinformatic analysis have further highlighted the potential for retroelement contributions to regulation of gene expression. Estimates of the amounts of the human genome derived from repeated sequences range as high as 69% (de Koning et al., 2011). Furthermore, these contributions are impactful—as up to 30% of mouse and human cap–selected transcripts initiate in TE sequences (Faulkner et al., 2009). In particular, degenerate solo LTRs are a rich source of transcriptional regulatory information. Analysis of 26 orthologous pairs of transcription factors in mouse and human genomes found that fully 20% were within TE sequences, particularly LTRs (Sundaram et al., 2014). For example, 1509 of 319,000 human ERV LTRs had perfect or nearly perfect binding sites for the tumor suppressor p53, accounting for more than one-third of binding sites identified by genomewide ChIP-seq (Wang et al., 2007). Some regulation reflects the sequence bias of retroviruses that evolved to infect cells in which they are under positive regulation and were then repurposed for host gene expression. A dramatic example is HERV sequences that mediate an innate immune transcription network in response to interferon stimulation (Chuong et al., 2016). Other sequences are associated with early pluripotency (Lu et al., 2014; Macfarlan et al., 2012; Muotri et al., 2010; Santoni et al., 2012). Sourcing regulatory sequences from ERVs, LINEs, and SINEs links evolving host control over invading species to an expandable reservoir of regulatory elements as was presciently proposed by McClintock and Britten and Davidson (Davidson and Britten, 1979; McClintock, 1956).

Although co-opting retroelement sequences for gene regulation is a relatively direct extension of functionality to the host, retroelement protein domains have also been exapted for host-related functions. These include resistance to infection by exogenous viruses. The Friend virus susceptibility 1 locus encodes a retrovirus structural protein relic that interferes with replication of incoming MLV subtypes by triggering premature disassembly (Stoye, 1998). Endogenous jaagsiekte sheep retrovirus interferes at a late stage in assembly of wild-type jaagsiekte sheep retrovirus exogenous VLPs (Palmarini et al., 2004). Both of these are forms of host innate immune resistance to infectious retroviruses. Other retroviral-derived proteins include Env, exapted independently in multiple placental lineages to form syncytin proteins that play a key role in placental trophoblast fusion (Blaise et al., 2003; Dewannieux et al., 2005). Recently, Arc/Arg3.1 a protein known for its role in synaptic plasticity and evolutionarily related to Ty3/Gypsy (Campillos et al., 2006; Zhang et al., 2015) was shown to have roles in actin filamentation in dendritic cells (Ufer et al., 2016) and to form VLPs in insect (Ashley et al., 2018) and mammalian (Pastuzyn et al., 2018) neuronal cells that function in packaging and intercellar RNA transmission of Arc/Arg3 mRNA (reviewed in Parrish and Tomonaga, 2018). The Rag1/2 proteins mediate the V(D)J recombination that contributes to immunoglobulin variation evolved from a DNA transposon (Agrawal et al., 1998; Thompson, 1995; van Gent et al., 1996). Interestingly, fewer genes have thus far been identified that represented exapted LINE sequences. One recognized example is striking, however—the LINE ORF1 protein was exapted to yield the pluripotency factor L1TD1 (Wong et al., 2011) (reviewed in Naville et al., 2016). It may act by interfering with endogenous LINE expression during embryonic development.

CONTROL OF RETROELEMENT EXPRESSION

Although we tend to analogize retroelements with invading RNA viruses, it is useful to recall that LTR retrotransposons and LINEs evolutionarily preceded retroviruses and thus have a unique relationship with host genomes. In the last several years, high-throughput genomics technologies, including NGS, have accelerated progress in probing the functions of repetitive sequences. Retroelement expression is regulated by transcription factors, epigenetic factors for transcriptional gene silencing (TGS), posttranscriptional gene silencing (PTGS), and downregulation of replication intermediates by innate immune mechanisms. These avenues of host defense are tuned in diverse ways to detect activation of endogenous and exogenous agents. Intrinsic properties of retroelements that distinguish them for regulation include redundancy, insertion clustering, mimicry of pathogen-associated molecular patterns (PAMPs), highly structured RNA, and generation of cytoplasmic DNA. Countering this is the intrinsic inefficiency of retroelement replication so they may replicate while failing to activate host defenses.

Retroelement Transcription Patterns Favor Self-Preservation

Whereas most host promoter elements are located upstream or coincident with transcription start sites, persistence of retroelements requires that they either target locations in the genome within which they will be expressed from flanking signals or contain promoter elements within the transcribed region. This requirement is resolved in different ways by LTR and non-LTR elements. In the case of LTR elements, transcription of the genomic RNA initiates in the upstream LTR, upstream of the site where it terminates in the downstream LTR. Thus, collectively, the entire LTR sequence is represented in the genomic RNA. Transcription is followed by reverse transcription, during which minus-strand cDNA is primed typically from a tRNA and this short terminal cDNA sequence is transferred to the 3′ end of the genomic RNA, which templates cDNA extension. The plus strand is primed from a RNaseH cleavage site just outside the downstream LTR sequence to generate a second short terminal cDNA, which is transferred to the upstream end and provides the terminal template for minus-strand synthesis and the primer for completing plus-strand synthesis. Thus replication insures that two complete LTRs are reconstituted prior to integration. In the case of both LINE and SINE elements, the promoter is actually positioned downstream of the transcription initiation site thus also ensuring that gRNA replication can regenerate an actively transcribed element. The 3′ ends of EN cleavage sites in genomic DNA prime reverse transcription from the gRNA template (reviewed in Richardson et al., 2015). For L1Hs, the primary promoter activity maps within the first 100 bp of the 5′-UTR and secondary activity maps within the first 670 bp (Swergold, 1990). L1 RNA 3′ sequences contain signals sufficient for efficient polyadenylation and the poly(A) is at least partially copied by reverse transcription so that a poly(A) tract is encoded at the downstream end of LINE insertions (Belancio et al., 2007). Antisense promoter activity has been mapped to two regions, bp 378–431 and bp 480–497 (Yang and Kazazian, 2006). Thus, L1Hs can mobilize production of both single- (ss) and ds transcripts that can extend into flanking sequence. The mouse L1 5′ UTR utilizes a similar mechanism to human L1 but differs by the presence of multiple short regulatory repeats within the promoter (Mouse Genome Sequencing et al., 2002; Sookdeo et al., 2013). SINEs are based on RNAP3-transcribed genes, which have internal promoter A and B boxes (Dieci et al., 2013). TFIIIC interacts at the promoter and loads the TFIIIB initiation factor in a sequence-independent manner, just upstream of the transcription initiation site. The ability of RNAP3-transcribed genes to be self-sufficient transcription units on reinsertion into the genome likely contributed to the multiple instances of their conversion into SINEs.

Mechanisms of retroelement compaction counter ongoing retrotransposition in mammalian genomes. In the case of LINEs, reverse transcription typically fails short of the promoter region so copies exhibit strong bias for the downstream coding sequence and splicing and poly(A) RNA processing signals. Unlike the RNAP3 gene or heterochromatic H3K9Me-targeted distribution of retroelements in model organisms such as fungi and *Drosophila*, mammalian LTR retroelements exhibit bias, if any, for histone modifications that are relatively widespread throughout the genome with significant representation in euchromatic regions (e.g., International Human Genome Sequencing, 2001).

This distribution is likely to have contributed to LTR elements and potentially non-LTR elements having been extensively co-opted as transcription regulatory modules (Wang et al., 2007). Regulators of retroelement expression are reviewed in detail elsewhere (Goodier, 2016; Rowe and Trono, 2011). Some illustrative examples are provided here.

Regulation of Retroelements in Somatic Cells by RNA Interference Fig. 13.2

Regulatory RNAs acting through RNA interference (RNAi) suppress TE and viruses (reviewed in Daugherty and Malik, 2012; Heras et al., 2014; Morris and Mattick, 2014; Sabin et al., 2013). The ability of RNA to act in *trans*, undergo modification and processing, and be derived in sense and antisense strands, suits it for regulation of families of rapidly evolving, RNA-based TE. RNAi in animal cells is mediated by small noncoding (nc)RNAs in three general classes, micro (mi)RNA, short interfering (si)RNA, and piwi (pi)RNA. These are differentiated by processing pathways, protein cofactors, and effector activities (reviewed in Ha and Kim, 2014). These short ncRNAs act in concert with one of the two classes of Argonaute proteins, AGO (*Drosophila* Ago1,2,4) and P-element-induced wimpy testes (PIWI) (*Drosophila* Aubergine [Aub], Ago3, and Piwi). Argonaute proteins are characterized by RNA-binding activity and RNAse III domains. MiRNAs are typically transcribed by RNAP2. These are part of longer capped and polyadenylated primary (pri) RNAs that undergo processing of embedded hairpin structures. PriRNAs are first processed by nuclear Microprocessor, a complex of DiGeorge syndrome RNA-binding protein (DGCR8) and Drosha protein with RNAse III activity generating the pre-miRNA hairpin, with a 5′ P and 3′ 2-nt overhang. This pre-miRNA is exported to the cytoplasm where it is further processed by the RNase III activity of Dicer to ~22 nts by excising the loop connecting the two strands of the hairpin. A strand from this duplex is loaded into the RISC formed with members of the AGO clade. MiRNAs mostly regulate protein-coding genes by imperfect annealing to the 3′ UTR and inhibiting translation and ultimately destabilizing RNA (reviewed in Wilczynska and Bushell, 2015). Endo siRNAs are generated by Dicer cleavage of long duplex RNA structures to generate ~21-nt species generally perfectly complementary to target protein-coding TE, or viral RNAs. They associate with AGO clade proteins in RISC and mediate RNA cleavage. PiRNAs are 24–31 nt and are produced from long priRNAs by an EN distinct from Microprocessor, together with dedicated trimming and modifying enzymes. PiRNAs display signatures of the processing pathway including 5′ U or +10 A bias and 2′ OH methylation at the downstream terminal nucleotide. PiRNAs associate with PIWI proteins in piRISC to target TE and host mRNAs in germline, early embryonic and some differentiating somatic cells for TGS and PTSG (reviewed in Czech and Hannon, 2016; Hirakata and Siomi, 2016; Iwasaki et al., 2015; Ku and Lin, 2014).

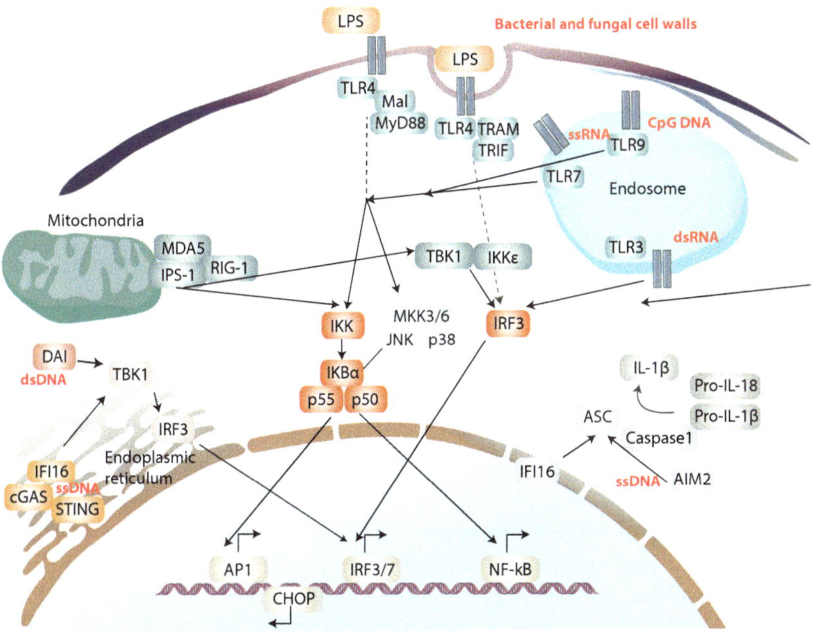

FIGURE 13.2 Retroelement activation of innate immune signaling. TLR1, 2, 4, 5, and 6 sense external bacterial and fungal cell walls via binding of MD2–LPS complex. At the plasma membrane, these TLRs activate a MyD88-dependent ubiquitination and kinase signaling pathway (abbreviated here), culminating in activation of transcription factor NF-κB and JNK and p38 MAP kinases. These kinases activate various transcription factors, regulators of stress response and cytokine expression such as AP-1 and CHOP. Alternatively, internalization of the receptors can activate the TBK1-IKKε pathway and downstream IRF3/7 transcription of Type I IFN genes. TLR7/8, 3, and 9 mediate responses to endosomal ssRNA, dsRNA, and unmethylated CpG DNA, respectively. TLR7/8 and TLR9 act through the MyD88-dependent pathway and TLR3 acts through the MyD88-independent, TRIF-dependent pathway. NF-κB, AP1, and IRF3/7 are transcriptional activators of proinflammatory cytokines and type I IFN responses, respectively. MDA-5 and RIG-1 are sensors of cytosolic long and short dsRNA, respectively. These trigger signaling pathway activation of NF-κB and IRF3/7 transcription factors. DAI and cGAS-IFI16 coupled to endoplasmic reticulum membrane-bound STING sense cytosolic ds and ss DNA, respectively, and activate IRF3. SsDNA is sensed by cytosolic AIM2 and nuclear-exported IFI16 which activate inflammasome cleavage of Pro-IL-1β and Pro-IL-18 to IL-1β. *AIM2*, absent in melanoma 2; *AP1*, activator protein 1; *ASC*, apoptosis-associated speck-like protein containing a caspase recruitment domain; *cGAS*, cyclic GMP-AMP synthase; *CHOP*, CCAAT enhancer-binding protein homologous protein; *DAI*, DNA-dependent activator of IRFs; *IFI16*, IFN-γ-inducible protein 16; *IKK(ε)*, IκB kinase (ε); *IPS-1*, IFN-β promoter stimulator 1; *IRF3*, interferon regulatory protein 3; *IκBα*, NF-κB nuclear import inhibitor; *JNK*, Jun N-terminal kinase; *JNK/p38*, MAP kinases, in part function as activators of survival and stress response; *LINE*, long interspersed element; *Mal*, MyD88 adaptor-like, functions together with MyD88 TLR TIR domain adapters; *MAPK*, mitogen-activated protein kinase; *MD2*, myeloid differentiation factor 2; *MDA5*, melanoma differentiation-associated gene 5; *MKK3/6*, MAPK kinase; *MyD88*, myeloid differentiation response gene 88; *NF-κB/50/65/55*, transcription factor controls cell survival responses to stress including cytokine production; functions together with p50/p55; *RIG-1*, retinoic acid–inducible gene 1; *SINE*, short interspersed element; *STING*, stimulator of IFN genes, TIR (domain), Toll/IL-1R′ resistance domain; *TRAM*, TRIF-related adaptor molecule; *TRIF*, TIR domain-containing adaptor inducing IFN-β.

Although piRISC mediates a significant amount of silencing in the germline and embryonic states, miRNAs also contribute to retroelement suppression in addition to acting in host antiviral defense. This is observed in cells with attenuated DNA methylation, including cancer cells and induced pluripotent stem cells (iPSCs). One example of such a miRNA is MiR128. It antagonizes L1Hs retrotransposition by binding within ORF2 and targeting the RNA for translational suppression and cleavage. Mutations predicted to disrupt MiR128-target association only partially restore reporter LINE L1 retrotransposition, so it is possible that MiR128 affects L1 activity through additional mechanisms (Hamdorf et al., 2015).

LINE and SINE RNAs can also act directly as sources of miRNAs. Classification of RNAs captured in Microprocessor pull-down experiments identified about one-third as repetitive sequences and TE accounted for about two-thirds of these. More specifically, experiments in vitro and in vivo showed that structured regions of the L1Hs 5′ UTR and of the SINE AluYa5 are Microprocessor substrates (Heras et al., 2013; Smalheiser and Torvik, 2006) (reviewed in Heras et al., 2014). Consistent with a miRNA-independent role for Microprocessor, Microprocessor defects were associated with a threefold increase in L1 retrotransposition not observed in a Dicer mutant (Heras et al., 2013). An informatic search for miRNA features in human, mouse and rat sequence databases identified 11 sequences derived from repeated sequences, including LINEs, SINEs, and LTRs. Furthermore, a subset of these miRNAs mapped to the 3′ ends of multiple other mRNAs (Smalheiser and Torvik, 2005). Therefore processing L1 RNA could diminish retrotransposition directly by degradation or mediate downstream targeting of L1 sequence-containing RNAs.

Regulation of Retroelements in Germline and Embryonic Stem Cells

Mammalian epigenetic suppression of TE is relieved in cells undergoing genomewide reprogramming of expression. These contexts include germline cells, preimplantation embryos, and specific somatic tissues. In these instances, PIWI proteins play key roles in sustaining control of TE activity. Despite what are now known to be profound differences between insect and mammalian systems, thinking about the regulatory role of PIWI proteins has been seminally influenced by pioneering studies in *Drosophila* (Aravin et al., 2003, 2004; Brennecke et al., 2007; Vagin et al., 2006) and continues to be influenced by studies in other model systems (reviewed in Ross et al., 2014). These pioneering studies identified small ncRNAs that were enriched in TE sequences and expressed according germline-specific programs. They further established that these RNAs were associated with PIWI proteins leading to their designation as piRNAs. PiRNA roles have recently been reviewed in more detail than is possible here (Crichton et al., 2014; Czech and Hannon, 2016; Hirakata and Siomi, 2016; Iwasaki et al., 2015; Newkirk and An, 2017; Ross et al., 2014). The following discussion summarizes some of the key findings.

In *Drosophila*, piRNAs play a critical role in female germline development and in maternal transmission of TE suppression to the developing embryo via maternal ribonucleoprotein (RNP) particle deposition (reviewed Hirakata and Siomi, 2016). PriRNAs are transcribed by RNAP2 from centromeric and telomeric heterochromatic regions containing multiple, typically fragmented, TE. These are referred to as piRNA clusters. Transcripts originate at flanking or terminal promoters that generate uni- or dual-stranded products or internal bidirectional promoters. Specialized factors stabilize the uncapped RNAs and antagonize transcription termination so that cluster transcripts can be up to hundreds of kb in length (Brennecke et al., 2007; Mohn et al., 2014). Somatic ovarian follicle cells express Piwi and mostly uni-strand transcripts. Pri-piRNAs are exported to the cytoplasm and processed through cleavage, probably by Zucchini (Zuc) EN to create fragments with 5′ U bias. These transcripts are associated with Piwi, trimmed and 2′OH methylated by Hin1 methyltransferase. PiRNA–Piwi RISC complexes are imported into the nucleus. The antisense bias of these piRNAs targets nascent TE transcripts and nucleates association with suppressive factors including H3K9me3 methyltransferase, HDACs, and heterochromatin-promoting factor (HP1). In ovarian germline cells, this primary pathway is joined by a secondary pathway mediated by PIWI clade proteins Aub and Ago3 that generalizes and amplifies the response by slicing sense and antisense RNAs. Transcripts are processed for Aub with a 5′ U bias, similarly to what was described for Piwi. Rather than translocating to the nucleus, Aub (Voronina et al., 2011) remains cytoplasmic and with the piRNA targeting overlapping sense RNAs so that slicing produces a +10 A bias. These are off-loaded to Ago3 so that the active site over the +10 A regenerates 5′ U-bias antisense piRNAs with overlapping complementarity. This cycling of RNA substrates in a so-called ping pong amplification cycle both degrades TE transcripts and regenerates phased piRNAs for nuclear Piwi RISC silencing of TE.

Although aspects of the piRNA pathway remain to be elucidated, key components have been identified. Localization of cytoplasmic RNA granules, also called nuage, differs between somatic and germline contexts. In somatic cells, nuage is associated with the mitochondrial surface and is referred to as Yb bodies. The functional role of this mitochondrial association has not been determined. Although it is obvious to infer ATP availability, most processes in the cell require energy without necessitating this mitochondrial surface association. In germline cells, perinuclear nuage lies opposite the nuclear site of pri-piRNA production. These bodies, also referred to as piP bodies, contain components of RNA processing bodies (Siomi et al., 2011; Voronina et al., 2011). An intriguing parallel occurs in *Saccharomyces cerevisiae*, which lacks RNAi. Induction of Ty3 expression in haploid mating cells sweeps P bodies into large aggregates referred to as retrosomes, associated with which Ty3 assembly occurs (Beliakova-Bethell et al., 2006; Bilanchone et al., 2015). P-body functions are also essential for Ty1 retrotransposition (Checkley et al., 2010; Dutko et al., 2010). PIWI proteins and some other processing factors are targets for dimethyl modification of Arg residues by protein Arg *N*-methyltransferase 5 (Prmt5). Modification of nuage components, including PIWI proteins themselves, enables interactions with the dimethyl-Arg interaction domain of tudor domain (TDRD) proteins that act to scaffold processing reactions. Key proteins with known functions conserved in the murine system are Vasa (murine Maelstrom or MAEL) (Aravin et al., 2009) and DEAD-box helicase Armitage (Armi) (murine MOV10L) but many additional factors are known. Some of these, including a number with unknown contributions, have been identified based on *Drosophila* genetic screens (Frost et al., 2010; Zheng et al., 2010).

MIWI, MILI, and MIWI2 in mouse and PIWIL1, PIWIL2, and PIWIL4 in human take the place of Ago3, Aub, and Piwi, respectively, in *Drosophila*. One major difference between *Drosophila* and *Mus* is that whereas in the *Drosophila* system, oogenesis and male and female fertility depend on the PIWI system, in mice, only male fertility depends on PIWI functions. Although in both *Drosophila* and *Mus*, PIWI proteins are important for suppression of TE, in *Drosophila*, piRNAs recruit H3K9 methyltransferases and HP1, but in mammals, CpG methylation plays an important role (Aravin et al., 2008; Kuramochi-Miyagawa et al., 2008). Numbers of piRNA clusters are relatively similar in *Drosophila* and human—142 and 186, respectively (Brennecke et al., 2007; Girard et al., 2006). However, in the former these are mostly in centromeric and telomeric heterochromatin, whereas mammalian piRNA clusters are scattered throughout unannotated intergenic regions and are mostly euchromatic. These differences in part could reflect patterns in the overall localization of TE. However, in mammals the majority of expressed mouse piRNAs are not TE. Murine, piRNAs are expressed from gene- and repeat-poor regions, with only ~17% representing SINEs, LINEs, and LTR retroelements and differ in sequence from human piRNAs (Girard et al., 2006). In spite of all these, mammalian piRNA clusters are generally syntenic. Mammalian piRNA clusters are generally transcribed from bidirectional promoters marked by the A-MYB transcription factor (Li et al., 2013b). Because TE insertions show little orientation bias, these RNAs reflect both sense and antisense sequences (Aravin et al., 2008). Activities of piRNAs in murine spermatogenesis have been investigated more extensively than in human and represent our best understanding of piRNA activities in a mammalian system. Nonetheless, there may be significant differences in piRNA function between murine and human systems, even in addition to differences between the piRNA sequences themselves. In mice, Dicer, which processes siRNAs, but not piRNAs, is essential for oogenesis and PIWI proteins are not (Murchison et al., 2007). In an additional twist, Dicer expression is driven by a mouse transcript retrotransposon intronic promoter (Flemr et al., 2013; Peaston et al., 2004). Finally, the mouse genome lacks a homolog of the human PIWIL3, the function of which is not yet defined.

DNA methylation is a key mechanism for epigenetic control of mammalian gene expression. Two major periods of development require genomewide erasure of these heritable modifications: primordial germ cells (PGC) undergoing gametogenesis to reset parental imprinting (reviewed in Hayashi and Surani, 2009) and preimplantation embryos to allow for pluripotency and developmental reprogramming. The mechanism of demethylation is not known in detail, but activation-induced deaminases, base excision repair, ten-eleven translocation dioxygenase, and apolipoprotein B mRNA editing enzyme catalytic polypeptide-like (APOBEC1) proteins have been speculated to be involved. Passive demethylation due to low levels of maintenance DNA methyltransferase DNMT1 is also likely to contribute (Hajkova et al., 2010; Popp et al., 2010) (reviewed in Rasmussen and Helin, 2016).

PIWI function is required in the mouse male germline. Migration of primordial germ cells (PGC) and initiation of spermatogenesis occurs at E7.5. After PGC migration, DNA undergoes global demethylation including removal of differential parental epigenetic marks (E7.5-12.5) and later (E15.5-P3) remethylation and parental mark resetting (Guibert et al., 2012; Hajkova et al., 2002; Kuramochi-Miyagawa et al., 2001; Lane et al., 2003; Lees-Murdock et al., 2003) (reviewed in Iwasaki et al., 2015). PIWI clade proteins are expressed differentially during this time. MILI expression occurs during the length of spermatogenesis but is relatively higher during the prepachytene (E12.5–E14) and pachytene phases (P13–P21 after birth). PiRNAs are enriched for TE sequences, particularly promoters of transcriptionally active TE during the prepachytene fetal spermatogenic period, but are less abundant in neonatal and adult male meiosis (Aravin et al., 2006; Molaro et al., 2014). MIWI2 is also active during prepachytene (E15.5–P3). MIWI is expressed from pachytene through formation of elongated spermatids (P14–P27). Mature sperm are produced continuously from P30 onward. Although LINE elements undergo demethylation and are expressed, IAP elements and some other LTR elements in mouse are more resistant to demethylation (Aravin et al., 2009; Trelogan and Martin, 1995). MIWI2 and MILI are both active during remethylation (Aravin et al., 2008; Kuramochi-Miyagawa et al., 2001). Gene disruptions have shed light on the different roles of these PIWI clade proteins.

After cytoplasmic processing generates mature piRNAs, MIWI2 loaded with piRNA enters the nucleus to promote CpG DNA methylation (Aravin et al., 2008; Kuramochi-Miyagawa et al., 2008). This function is likely redundant, potentially with sequence driven methylation, as examination of methylation in mutants defective for piRNA generation showed that methylation, including significant fractions of TE, is actually largely sustained (Manakov et al., 2015; Molaro et al., 2014). Mutations introduced into the EN active sites of MILI and MIWI2 showed infertility and decreased piRNA in the MILI active-site mutants, but little effect of the MIWI2 mutation (De Fazio et al., 2011). Furthermore, in spite of similar reduction in overall methylation, MILI-deficient mice showed increased expression of LINE and IAP elements, whereas MIWI2 mice showed only the former (Kuramochi-Miyagawa et al., 2008). These results together with those of De Fazio et al. suggest a PTGS role for MILI that operates on both LINE and IAP RNAs. Overall, during germline development, deficiencies in MIWI2, MILI, DNMT3L, or TDRD proteins (reviewed in Chen et al., 2011) result in defective DNA methylation, some derepression of retroelements and meiotic catastrophe. An unresolved question is whether meiotic catastrophe stems directly from increased retrotransposition and collateral DNA damage or from some less well-understood function of the piRNA system. In adult spermatogenesis, the repressive histone methyltransferases play redundant roles with DNA methyltransferases and piRNAs. Perhaps not surprisingly, these differ among non-LTR and LTR retroelements. Single deficiencies in G9a H3K9me2 methyltransferase, DNA methyltransferases, or MILI do not relax control of L1, but loss of either G9a or MILI relaxes control of IAPs (Di Giacomo et al., 2013).

The second period of sweeping genomic reduction in methylation occurs in preimplantation mouse blastocysts. At this time parental genomes undergo genomewide demethylation excepting centromeric and imprinting control regions. On implantation, DNA methylation is reestablished to suppress the germline expression program (Borgel et al., 2010; Kobayashi et al., 2012; Lane et al., 2003; Smallwood et al., 2011; Smith et al., 2012). Examination of preimplantation murine L1 and IAP DNA sequences shows about 30% and 60% methylation, respectively. After implantation, methylation increases to 80% and 95%, respectively (Oda et al., 2006) (reviewed in Saitou et al., 2012).

The four mammalian DNA methyltransferases, DNMT1, DNMT3A, and DNMT3B, and catalytically inactive family member, DNMT3L, play distinct roles in retroelement silencing. Maintenance DNMT1 methylates hemi-methylated sequences in newly replicated DNA. DNMT1 is therefore a key embryonic DNA methyltransferase. DNMT1 null mice display an 80% reduction in DNA methylation, elevated TE expression, and die early in development (Gaudet et al., 2004; Lei et al., 1996; Walsh et al., 1998). DNMT3A and DNMT3B function as de novo methyltransferases that reestablish DNA methylation in maturing gametes, postnatal somatic stem cells, and early embryogenesis. Although the DNMT3L homolog of these methyltransferases lacks catalytic activity, it heterodimerizes with DNMT3A or DNMT3B, and links them to nucleosomes lacking the transcription activating modification H3K4me (reviewed in Saitou et al., 2012), thereby perpetuating the repression of these sequences. The heterodimer acts in early embryos and PGC to promote remethylation in male and female germlines. Male mice deficient in DNMT3L undergo meiotic catastrophe and elevated L1 expression (Bourc'his and Bestor, 2004; Smith et al., 2012).

Although interactions between the piRNA system and the other DNA and histone epigenetic systems are not completely understood, evidence supports significant redundancy between modifications in TE control. Important suppressive histone modifications include H3K9me3, H3K9me2, and H3K27me3 (reviewed in Newkirk and An, 2017; Saitou et al., 2012). DNA methylation contributes to recruitment of repressive histone H3K9me3 methyltransferase complexes SUV39H1/2 and SETDB1, and histone deacetylases (reviewed in Cedar and Bergman, 2009). However, murine IAP/MusD, and MERV-L elements that are not upregulated in CpG methylation–deficient *Dnmtl, Dnmt3A Dnmt3B* backgrounds, despite overall decreases in CpG methylation, are upregulated in backgrounds deficient in H3K9me3, H3k9me2, and Lsd1 (reviewed Mager and Stoye, 2015). Several observations also link H3K9me3 to repression of active L1s and ERVs. H3K9me3 is a product of SUV39H1/2 and SETDB1 methyltransferases and is enriched on the promoters of L1 subfamilies (Bulut-Karslioglu et al., 2014; Pezic et al., 2014). In SUV39H1/2-deficient embryonic stem cells (ESC) and MIWI2-deficient mice, expression of these families is elevated (Bulut-Karslioglu et al., 2014; Pezic et al., 2014). However, there may be differential targeting by the two histone methyltransferases because in SETDB1-deficient cells, IAPs are upregulated (Liu et al., 2014). Generation of H3K9me2 is mediated by G9a/GLP. Its role in L1 control is revealed in backgrounds with double deficiencies in DNA methylation and prospermatogonia MILI. In this context, IAPs but not L1s are derepressed, but in G9a-deficient backgrounds, L1s are also derepressed (Di Giacomo et al., 2014).

PRMT5 was mentioned above for its role as an Arg *N*-methyltransferase that modifies targets, including PIWI and related proteins to foster interaction with TDRD scaffolding proteins. It is nuclear between E8.5 and E12.5 spermatogenesis during which period it modifies histones to generate repressive H2A/H4R3me2 (Ancelin et al., 2006). Deficiency in PRMT5 leads to increased expression of L1 and IAP (Kim et al., 2014; Li et al., 2015).

Similar to the mouse model, in humans, SINEs and LINEs are targets of 5mCpG methylation and target methyl-CpG-binding protein 2 (MeCP2). Consensus Alu elements contain ~24 CpG sites for recognition by DNA methylases (Batzer et al., 1996). As they are mostly methylated, they could account for as much as one-third of genomic CpG (Liu et al., 1994; Schmid, 1991). Because Alus are enriched in genic regions, this methylation potentially affects flanking sequence expression (Liu et al., 1994). L1 elements also contain CpG-rich sequence targeted for 5mCpG modification and regulation by MeCP2 (Muotri et al., 2010; Yu et al., 2001). MeCP2 is a member of a family of proteins containing 5mCpG-binding pockets. It associates with SOX2-HDAC to repress LINE expression in pluripotent cells (Muotri et al., 2010; Yu et al., 2001). Because cytosine undergoes a physiologically significant rate of deamination, 5meCpG sequences in Alu and LINE elements are relatively prone to C to T transitions. This inherent mutability potentiates conversion of such sequences to novel functions and drift toward expression. In spite of the requirement for PIWI function during mouse spermatogenesis, the majority of germline retroelement methylation of L1s and LTR elements is not piRNA dependent and is actually established by default de novo methylation (Molaro et al., 2014). PiRNA-dependent methylation may therefore be biased toward younger active families of elements.

Insights into the piRNA system are furthering understanding of the molecular basis of other phenomena as well. One of these is the interstrain mating TE incompatibility first described in *Drosophila* as hybrid dysgenesis (Kidwell et al., 1977).

Oocytes are competent to transmit protective piRNA-PIWI-containing maternal granules to zygotes. Mating between a female lacking specific TE, and therefore corresponding protective piRNA RNPs, and a male having those TE is associated with germline derepression of the male-specific TE and sterility in F1 progeny (Brennecke et al., 2008; Khurana et al., 2011). Knowledge of the mechanics of piRNA activity also sheds light on the phenomenon of paramutation in which epigenetic traits are transmitted. This is explained, at least in some cases, as piRNA-generating alleles acting in *trans* to control the expression of homologous allelic and nonallelic sequences, and potentially even converting them into piRNA producers (de Vanssay et al., 2012). As suggested above by the discovery that piRNAs are not limited to TE, proposed roles of piRNAs have expanded recently. New additions to the catalog of functions include turnover of host mRNAs in embryogenesis. In both *Drosophila* early embryo and murine elongating spermatids, piRNAs cooperate with deadenylation decay enzymes in targeted degradation of maternal RNAs and of certain preexisting mRNA pools, respectively (Barckmann et al., 2015; Gou et al., 2014). Adding to interest in this newly appreciated role, the 3′UTRs of conventional coding mRNAs have also been identified in piRNA pools (Robine et al., 2009). These "genic" piRNAs are presumed to have a regulatory role because piRNAs are also observed in tissues that remain developmentally plastic, such as hippocampal neurons (Lee et al., 2011). Nerve spine density is associated with learning and memory. Based on correlations between nerve spine density and piRNAs in the neuronal model Aplysia, piRNAs have been proposed to have a role in promoting memory by controlling synaptic plasticity (Rajasethupathy et al., 2012).

Retroelements are activated in response to global DNA demethylation in preimplantation embryos. Progress in iPSC technologies raise the possibility that relaxation of epigenetic controls during reprogramming could allow problematic resurgence of TE expression in pluripotent iPSC. Indeed LINE expression is derepressed in iPSC and although expression appears to decline with differentiation, this mechanism would likely be independent of germline piRNAs (Klawitter et al., 2016).

Sequence-Specific Suppression of Retroelements

In addition to global mechanisms of suppression, LINEs, SINEs, and ERVs are subject to differential regulation during development and in response to environmental signals. Alu elements in the human genome contain multiple sites that potentially bind transcription factors responsive to thermal stress with reported bias for metabolic genes and developmental factors (Kim et al., 2001; Polak and Domany, 2006). This topic has been reviewed recently (see Goodier, 2016).

Evolution of sequence-specific repression of both LINEs and ERVs is exemplified by the Kruppel-associated box (KRAB) zinc-finger (KZNF) protein family. Zinc-binding proteins are the largest class of transcription factors, and about half of these contain the KRAB domain required for interaction with the protein KAP1/TRIM28/TIF1β and associated repressive functions. There are over 400 human KRAB-ZNF genes and the number of proteins is expanded by transcript processing. This family of proteins regulates a correspondingly large set of functions (reviewed in Iyengar et al., 2014). The interacting tripartite motif (TRIM) family of proteins, including KAP1/TRIM28/TIF1β, is also relatively large with about 60 members. KAP1/TRIM28/TIF1β has an amino-terminal ring zinc-binding C_3HC_4 (RBCC) domain through which it interacts with KRAB. It mediates repression via a hydrophobic central pentapeptide through which it interacts with HP1, and C-terminal plant homeo- and bromodomains that cooperatively further interact with histone nucleosome remodeling and HDAC complexes. These in turn promote interaction with SETDB1 H3K9me3 histone methyltransferase. Altogether these interactions are self-reinforcing and repressive to targeted sequences. KAP1/TRIM28/TIF1β partners with multiple KZNF DNA-binding proteins to repress ERV and non-LTR elements. The power of associating modular ZNF domains with repressive chromatin effectors to control related sets of retroelements is illustrated by murine KZNF ZFP809 (Wolf and Goff, 2007, 2009). This KZNF family member targets the $tRNA^{Pro}$ primer-binding site of MLV and related ERVs containing that primer tRNA-binding sequence. Other KZNF family members analogously enable repression of other provirus families via KZNF variants (Thomas and Schneider, 2011). Consistent with the role of KZNF as a mediator of repression, L1 expression increases in KAP1-depleted cells. Investigation of primate-specific KZNF variants (Jacobs et al., 2014) showed that ZNF93 targets older L1 families (PA6 to PA3). Interestingly, that binding site is deleted in the 5′-UTR of younger, active L1Hs elements. ZNF91 targets SVA elements. Together these examples underscore the ongoing dynamic between regulatory modules and escape from control by mutational variation in retroelements (Castro-Diaz et al., 2014; Jacobs et al., 2014).

Other ZFP target DNA methylases and HDACs to other L1 sequences. The promyelocytic leukemia zinc finger (PLZF) encoded by Zfp145 is a POZ-ZFP family member that targets DNMTs and HDACs to L1 sequences. It is expressed in male prospermatogonia overlapping the period of demethylation. Zfp145 deficiency is accompanied by hypomethylation of L1. A PLZF-binding site was identified in ORF2 sequence that was speculated to propagate DNA methylation upstream toward the L1 promoter in susceptible elements (Puszyk et al., 2013).

Neuronal Activity of Retroelements

In recent years, there have been intriguing reports of retroelement and piRNA expression in *Drosophila*, mouse, and human brains (reviewed in Erwin et al., 2014; Rajan and Ramasamy, 2014; Richardson et al., 2014). Several regulators of L1 expression are known neurogenesis factors. The human L1 promoter contains binding sites for SOX2 (Tchenio et al., 2000), YY1 (Athanikar et al., 2004), RUNX3 (Yang et al., 2003), T-cell factor (TCF), and lymphoid enhancer factor (LEF)–WNT signaling factor (Kuwabara et al., 2009). The idea that the complexity of the brain and plasticity involved in learning has parallels in retrotransposition-driven somatic mosaicism and piRNA epigenetic inheritance of changes is seductive. Nonetheless, in spite of clear evidence that retrotransposition is elevated in the central nervous system, it has been challenging to quantify events in individual cells and to show causal links to neuronal phenomena.

L1 elements in neural stem cells (NSC) are repressed by methylation and SOX2 and HDAC1 activities (Coufal et al., 2011; Muotri et al., 2010; Nan et al., 1998; Tchenio et al., 2000). The MeCP2 HDAC1–binding partner cooperates in repression of the L1 5′-UTR in rodent and human NSC and defects in MeCP2 increase retrotransposition (Muotri et al., 2010). On initiation of Wnt/β-catenin signaling and NSC differentiation into neural progenitor cells (NPC), inhibitory factors are displaced by TCF/LEF, which can promote expression. Regulation of the LINE promoter parallels that of the neural differentiation factor NeuroD, a driver of neuronal differentiation (Kuwabara et al., 2009). This is consistent with observations of elevated L1 retrotransposition in the cognitive disorder, Rett syndrome, which is associated with defects in MeCP2 (Amir et al., 1999).

Whether transposition in the brain is also regulated by piRNAs is an intriguing question. PiRNA are elevated in the CNS (Lee et al., 2011), and levels of piRNA in Aplysia correlate with targeting serotonin-dependent CpG methylation repression to the promoter of CREB2, an inhibitor of memory (Rajasethupathy et al., 2012) suggesting a positive role for piRNA in memory. In *Drosophila* mushroom bodies, Ago3 and Aub generators of piRNA are reduced in neurons expressing TE and undergoing transposition and elevated in neighboring neurons in which piRNAs are repressed.

Several genetic disorders with neurological ramifications increase our understanding of the interaction of retrotransposons with DNA repair systems and potential links between retrotransposition frequency and neuronal disease. EN-deficient LINE elements can transpose at wt levels in NHEJ repair deficient cells, suggesting that TPRT and repair systems access a shared pool of substrates (Morrish et al., 2002). In wt cells, retrotransposed elements are generally truncated, but this deficiency is ameliorated in cells deficient in NHEJ, also consistent with a role for NHEJ in limiting expansion of active elements (Suzuki et al., 2009). Disruption of the ataxia telangiectasia mutated gene, which is required for checkpoint signaling, results in elevated retrotransposition (Coufal et al., 2011).

L1 retrotransposition in neuronal tissues has been studied in order to understand the role for retrotransposition-based somatic mosaicism in neuronal function. As discussed above, L1 is expressed in neuronal tissue, but efforts to quantify retrotransposition are hampered by the extremely low frequencies of cells with any particular insertion coupled with a high density of preexisting retroelements (Carreira et al., 2014; Goodier, 2014). Nonetheless, studies of neuronal tissues have been informative with respect to roles of host factors in transcription and the potential for L1 retrotransposition in disease states.

Muotri et al. (2005) showed that a tagged human L1 (Moran et al., 1996) transposes in cultured rat hippocampus NSC, and brains of transgenic mice at a higher frequency than in ESC or nonneuronal control cells. Although there are multiple differences between human and rodent model systems, NPC isolated from human fetal brain or derived from human ESC also underwent retrotransposition of the tagged human L1. Furthermore, multiplex PCR assays showed increased copy number in several regions of human postmortem brains compared with liver and heart. This increase was estimated to approach 80 somatic insertions per cell (Coufal et al., 2009). In experiments, designed to molecularly assess the basis of relatively dramatic increases in copy number involving SINE, SVA, and L1 elements, retrotransposon capture sequencing (Sanchez-Luque et al., 2016) was used to first enrich for TE-containing fragments. Samples were then subjected to targeted amplification and sequencing of insertion-genome junctions (Baillie et al., 2011). These experiments showed that all three types of retroelements transposed in the brain and that activity was modestly higher in the hippocampus compared with putamen, temporal gyrus, caudate nucleus, and middle frontal gyrus. These experiments, although valuable reflected differences in only three human donors. Insertions were biased toward protein-coding regions, and several of the targeted genes had neuronal functions suggesting that these were targeted preferentially. Another study took a single-cell approach to surmount the challenge of detecting events that individually are at extremely low-frequency events, although collectively significant. DNA from 300 single neurons from cerebral cortex and caudate nucleus was subjected to multiple displacement amplification and sequenced (Evrony et al., 2012). Although this analysis did not examine hippocampal neurons or compare the brain with other tissues, investigators concluded there were between 0.6 and 0.04 events per neuron, which was significantly lower than some previous estimates. Another heroic single-cell approach used somatic cell nuclear transfer of mouse olfactory bulb neuronal nuclei into enucleated oocytes to amplify sufficient DNA for single-cell sequencing independent of previous more error-prone in vitro methods. This study also found that retrotransposition was rare; although the number of cells that could be examined was limited and olfactory bulb neurons were not previously investigated, complicating direct comparisons (Hazen et al., 2016).

The studies indicating elevated TE mobility in the brain with relatively higher activity in the hippocampus are consistent with results from *Drosophila* brains where expression of TE was found to be severalfold elevated in mushroom body $\alpha\beta$ neurons over $\alpha'\beta'$ and γ neurons. Interestingly, this coincided with lower levels of regulatory Aub and Ago3 RISC proteins in the $\alpha\beta$ neurons, consistent with elevated transposition. DNA sequencing of $\alpha\beta$ neurons led to an estimate of 129 transpositions per genome. An intriguing parallel between hippocampus and mushroom body and the Aplysia study are that both structures are involved in learning and memory.

LINE element retrotransposition in the brain has captured much of the interest in the role of retroelements in the nervous system. However, the recent discovery that the Arc/Arg3.1 protein, which is involved in AMPA receptor recycling, is exapted from an LTR Ty3/Gypsy retrotransposon Gag protein is exciting. Experiments in two laboratories showed that this protein assembles into virus-like particles into which it packages its own RNA and transmits it intercellularly in neuromuscular junctions. Furthermore, this ability was critical for the Arc/Arg3.1 role in neuronal plasticity (Ashley et al., 2018; Pastuzyn et al., 2018).

In summary, retrotransposition in the brain creates somatic mosaicism (reviewed in Erwin et al., 2014; Richardson et al., 2014). The relationships to neurogenesis or neuronal functions remain to be discovered. Intriguing possibilities for this relationship are that retrotransposon expression drives piRNA biogenesis to foster epigenetic changes critical to neuronal plasticity, retrotransposon activation exerts regulatory control of flanking genes, or formation of retroelement RNPs and extracellular particles organizes cytoplasmic RNAs and proteins and promotes intercellular transmission of critical RNAs. None of these possibilities are dependent on high-frequency retrotransposition.

Roles for Retroelements in Cancer Initiation and Progression

Although retrotransposition is elevated in certain cancers, a key question is whether it acts to initiate or drive cancer development. There are few instances in which L1 escape from regulation can be definitively invoked as the cancer-initiating event. Adenomatous polyposis coli gene encodes a regulator of beta-catenin, and deficiencies can play initiating roles in colorectal cancer. Scott et al. (2016) showed that retrotransposition of a hot L1 was the likely initiator of an instance of this disease demonstrating that retrotransposition can cause cancer. In addition, it indicates that hot elements themselves should be evaluated further as risk factors for this and other diseases. Potentially offering more insight into the mechanisms that control retrotransposition itself, two neurological syndromes with implications for innate and autoimmunity are associated with elevated retrotransposition. Rett syndrome, as mentioned above, is a neurodegenerative disorder associated with deficiencies in MeCP2. Affected cells display increased L1 expression (Muotri et al., 2010). Aicardi–Goutieres syndrome can be associated with cerebral atrophy and is caused by defects in innate immune functions APOBEC, MOV10, TREX, RNASEH1, and SAMHD1 that normally antagonize retrotransposition. Schizophrenia is also associated with elevated L1 retrotransposition (Bundo et al., 2014) (reviewed in Guffanti et al., 2014).

REGULATION OF RETROELEMENTS BY THE INNATE IMMUNE SYSTEM

Regulation of Innate Immunity by Endogenous Retrovirus Sequences Fig. 13.3

Retrovirus LTR promoters evolved to be expressed in host target tissues and given the abundance of isolated LTR elements, many are available for

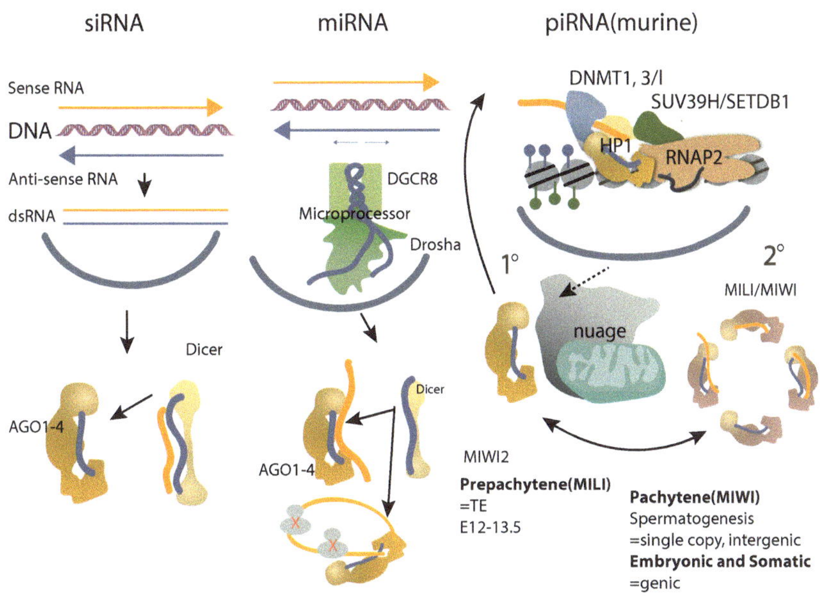

FIGURE 13.3 **Small regulatory RNAs.** Short RNA (si)RNA, micro(mi)RNA, and piwi(pi)RNA participate in transcriptional gene silencing (TGS) and posttranscriptional gene silencing (PTGS). Small regulatory RNAs associated with AGO clade proteins to form RISC complexes. SiRNAs are transcribed as ss, unstructured RNAs that are processed into short RNAs in the cytoplasm by Dicer and off loaded to Ago1-4. They are typically perfectly complementary to target RNAs, which are degraded by Ago slicing activities. MiRNAs are transcribed as structured precursor RNAs which are processed by the microprocessor complex RNAse III activity into hairpin RNAs about 70 nt in length. These precursors are off-loaded to Dicer for processing into miRNAs and then to Ago1-4 to form RISC complexes. These typically anneal to the 3′ ends of target mRNAs with imperfect matches leading to translation inhibition and perfect complementarity resulting in mRNA degradation. PiRNAs are germline-specific RNAs transcribed from long clusters of TE fragments or other intergenic sequences. They are exported from the nucleus and processed into piRNAs in perimitochondrial or perinuclear nuage complexes depending on the cellular context. In mouse and human there are prepachytene and pachytene phases of piRNA production. MIWI2 and MILI are expressed during the former and MIWI and MILI during the latter. During prepachytene TE piwi RNAs are expressed. Mouse Miwi operates on primary antisense RNA reimporting the processed form into the nucleus where it associates with nascent (sense) target RNAs. This leads to association of repressive factors including DNA methyltransferases (DNMT1,3/L), H3K9me3 methyltransferases including SUV39H and SETDB1, HDACs, and heterchromatic factor HP1. Cytoplasmic MILI RISC alone or in concert with MIWI2 can operate in a heterotypic or homotypic ping-pong mode to provoke PTGS in which cytoplasmic target RNA of either strandedness is degraded. Ping-pong cycling also generates substrates for MIWI2 reimport into the nucleus. More details are provided in the text. Sense RNAs are shown in gold and antisense in blue; DNA methylation, blue balls; H3K9me3, green balls. Microprocessor is composed of RNA binding subunit DGCR8 and RNAse III domain RNA cleaving protein Drosha. Prepachytene and pachytene spermatogenesis expression windows are shown for mouse.

regulation of flanking genomic sequence (Faulkner et al., 2009; Kapusta et al., 2013; Sundaram et al., 2014). For example, gammaretroviruses infect immune cells and those ERVs respond to immune signaling. New work has brought into sharper focus how this regulation impacts host immune competency. Interferon-stimulated genes are activated by interferon regulatory factors (IRFs) and signal transducer and activator of transcription (STAT) binding in promoter regions (reviewed in Platanias, 2005). ChIP-seq data were used to identify and map interferon-γ (IFNG)–induced peaks of IRF1 and STAT1 binding to the reference human genome (Chuong et al., 2016). Of 27 TE families enriched, 20 were related to ERV LTRs. More specifically, the ~45 to 60 MY old MER41 gammaretrovirus family represented by ~1000 copies, showed peaks of factor binding and H3K27 acetylation during IFNG-induction. Mutations that decreased expression of pathogen/pattern recognition receptor (PRR) absent in melanoma (AIM2), a key component of the inflammasome, similarly affected MER41-driven reporter expression. These experiments exemplify the path through which an endogenized retrovirus, specialized for a specific tissue, can be repurposed to establish a regulatory network (reviewed in Feschotte, 2008).

Sensing of Retroelements by Innate Immunity

Immune systems interact with retroelements on several levels. Cellular innate immune responses are triggered when PRRs detect conserved PAMPs that may be shared with endogenous retroelements, and these in turn trigger inflammatory responses. The relationships between innate and adaptive immunity and between endogenous retroelements and innate immunity have been recently reviewed (Hurst and Magiorkinis, 2015; Iwasaki and Medzhitov, 2015; Kassiotis and Stoye, 2016; Volkman and Stetson, 2014). First responder components of innate immunity systems operate to limit the levels of pathogen- and endogenous retroelement–associated nucleic acids, but under certain conditions, sensing of these species triggers adaptive responses as well. In the case of expressed ERV proteins, similar to most self-antigens, tolerance can develop. However, for others, expression is either limited, so that tolerance does not fully develop, or is overridden by high levels of retroelement expression coupled with appearance of reactive RNA and cDNA species, which stimulate adaptive or even autoimmune responses.

A number of PRRs that sense incoming pathogens have been identified. Toll-like receptors (TLRs) are among the most extensively studied. These include cell extrinsic plasma membrane receptors (TLR-1, -2, -4, -5, -6, and -11) and intrinsic endosomal membrane receptors (TLR-3, -7, -8, and -9). The plasma membrane TLRs sense protein, lipid, and polysaccharide components of bacterial and fungal cell walls. Endosomal TLR3, 7/8, and 9 receptors sense lumenal dsRNA, ssRNA, and unmethylated CpG DNA, respectively. TLR3, TLR7, and TLR9 are required for control of ERVs. Loss of TLR7 is associated with spontaneous ERV-mediated viremia (Yu et al., 2012). Cytosolic nucleic acid species are monitored as well. The retinoic acid–inducible gene I protein (RIG-I) is representative of RIG-I-like receptors (RLR) that sense nucleic acid and associate with mitochondria. RIG-I responds to short dsRNAs with 5′-triphosphate. In addition, a recent report showed that circular RNAs also stimulate the RIG-1 sensor (Chen et al., 2017). This is made more interesting by the intriguing implication of DBR in retrovirus and retrotransposon retrotransposition (Cheng and Menees, 2004; Galvis et al., 2014). RLR melanoma differentiation-associated protein 5 (MDA5) senses long dsRNAs. These RLRs activate mitochondrial antiviral–signaling and a downstream inflammatory response. Cytoplasmic dsDNA is sensed by the DNA-dependent activator of IFN regulatory factors (DAI). Other cytosolic sensors respond to ssDNA. These include the ER IFI16 and cyclic GMP–AMP synthase (cGAS) in association with the ER adaptor, stimulator of IFN genes (STING). CGMP-AMP acts as a second messenger to activate IFN genes. Nuclear IFI16 and AIM2 receptor cooperate to activate the inflammasome caspase, which processes and activates prointerleukins.

These diverse sensors either act relatively directly, as in the cases of cGAS and IFI16-AIM2, to elevate expression of inflammatory factors, or feed through downstream pathways to activate transcription factors NF-κB, AP1, and IRF3/7 that stimulate expression of proinflammatory cytokines and type I interferons. In the last several years, research to understand the role of innate immunity sensing pathways in ERV control has uncovered fascinating interplay between the regulators and their targets. An early clue was that certain murine ERVs are transcribed in response to bacterial stimulation of the innate immune response (Stoye and Moroni, 1983). In addition, TLR or B-cell receptor activators of NF-κB induce some endogenous retroelements (Zeng et al., 2014) and type I interferon itself induces expression of human SINEs (Hung et al., 2015). However, despite this evidence that ERVs respond to innate immune signaling, defects in TLR7 are associated with elevated levels of ERV intermediates (Young et al., 2012; Yu et al., 2012) and autoimmunity as in murine lupus (Santiago-Raber et al., 2009).

Direct Control of Retroelements by Innate Immunity

Immune signaling responds to retroelement RNA, protein, and DNA intermediates, but intermediates are also directly restricted. Restriction factors for endogenous elements are elucidated based on extensions of studies of retroviral restriction factors and interactomes of retroviruses (e.g., Brass et al., 2008; reviewed in Bushman et al., 2009). Some of these restrictors are upregulated in response to immune signaling triggered by substrate levels. Failures of these restrictors can therefore trigger a positive feedback loop, even culminating in autoimmunity. Identification of contributors to autoimmunity are expanding our understanding of endogenous retroelement restriction (Zeng et al., 2014).

Mechanisms and extent of specificity of discrimination of endogenous retroelement RNAs are varied. A nonexhaustive list illustrates this diversity.

The SKIV2L helicase, a component of the 3′-5′-exosome, acts to limit retroelement RNA levels, and in turn, the immune response (Eckard et al., 2014). Deficiency in this protein has been linked to systemic lupus erythematosus and activation of the RNA sensor RIG-1. L1 retroelement activity is restricted by RNaseL, a member of the 2′, 5′-oligoadenylate synthetase-RNAse L system (Zhang et al., 2014).

A powerful example of host-retroelement conflict is the APOBEC family of cytidine deaminases that edit single-stranded polynucleotides from C to U (reviewed in Koito and Ikeda, 2011; Schumann, 2007). These are found in one or two forms in nonprimate animals, but seven forms in primates, likely expanding in response to retroviruses and retroelements. APOBEC1 edits ApoB and neurofibromin mRNAs resulting in production of tissue-specific forms in primates and other mammals. APOBEC3 includes deaminases A, B, C, D, F, G, H. There are interesting differences between targeting of retroviruses and non-LTR retroelements, which produce cDNA in cytoplasmic particles and nuclear replication complexes, respectively. In tissue culture, anti-ERV activity of APOBEC1 requires deamination activity, whereas anti-L1 activity does not (Ikeda et al., 2011). APOBEC3C and 3G are important against retroviruses and both deaminase-dependent and -independent activities play roles. APOBEC3A and 3B have anti-L1 and SINE activity that is relatively independent of deaminase activity but blocks cDNA production nonetheless. In particular, APOBEC3C and G are well known for antilentiviral activity but also restrict IAP and MusD/HERVK LTR ERVs in mouse and human. Members of the APOBEC3G family restrict Vif-deficient HIV-1 by deaminating cytidine in plus-strand replication intermediates and block replication through a deaminase-independent mechanism (Chiu and Greene, 2008; Malim and Bieniasz, 2012) (reviewed in Salter et al., 2016). Expression of APOBEC3G in yeast restricts retrotransposition of the ERV-like LTR retroelement Ty1 (Dutko et al., 2010; Schumacher et al., 2008). Interestingly, in both yeast and primate cells, APOBEC3G associates with RNA processing bodies referred to as P bodies that contain deadenylation-dependent degradation factors and Argonaute proteins. In *Saccharomyces*, which lacks RNAi, Ty1 and Ty3 depend on P-body components for VLP assembly. However, in animal cells, P-body components such as MOV10 restrict retroelements, but not retroviruses (Arjan-Odedra et al., 2012; Goodier et al., 2012; Li et al., 2013a). Associations of APOBEC and retroelement proteins are driven to differing extents by interactions of each with the respective gRNA and cDNA.

Trex1 is a 3′ to 5′ DNA exonuclease specific for cytosolic ssDNAs. It is upregulated under conditions of TLR dendritic cell activation by NF-κB and type I interferon signaling. Trex was discovered as an HIV-1 restriction factor but also limits levels of ERV cDNA (Stetson et al., 2008). It is part of the ER suppressor of variegation, enhancer of zeste and trithorax (SET) DNA repair complex. In the absence of this activity, HERV cDNA potentially stimulates

IFI16-STING-cGAS signaling. The SAM domain- and HD domain-containing protein 1 (SamHD1) has RNase and nucleotide hydrolase activities. It is a restriction factor for HIV-1 in human cells (Laguette et al., 2011; Ryoo et al., 2014). SamHD1 acts through RNase activity to block HIV-1 replication in noncycling cells. Orthologs in mice also restrict retroviral infection. SAMHD1 limits L1 and Alu/SVA replication in dividing human cell lines (Zhao et al., 2013). However, the understanding of SamHD1 regulation of ERVs is incomplete. In at least some mouse strains, knockouts of *SamHD1* do not show the anticipated elevated MusD, IAP, or Mu-ERV ERVs (Rehwinkel et al., 2013).

Aicardi–Goutieres syndrome has characteristics overlapping those of autoimmune systemic lupus erythematosus including elevated interferon response. It is associated with deficiencies in restriction functions related to control of nucleic acid species, among them endogenous retroelement replication intermediates. A recent study of 374 patients implicated seven genes as contributing to this population (Crow et al., 2015). These included *TREX1* (22%) and *SAMHD1* (11%), *RNASEH2A, B, C* (53%), adenosine deaminase RNA-specific (*ADAR*) (7%), and *MDA5* (3%). ADAR deaminates the adenosine bases to form inosine within dsRNA, potentially affecting nuclease, translation, and splicing factor recognition.

HERV-W Env interacts with TLR4 and CD14 to stimulate proinflammatory cytokines. In vitro evidence suggests that HERV-K dUTPase variants may activate surface membrane TLR2-driven expression of NF-κB related to psoriasis. Expression of MERV gp70 is increased following activation of endosomal TLRs, TLR7, and TLR9, implicated in murine lupus by their role in immune responses to autoantigens. Other studies have linked TLRs to MERV expression, but in contrast, deficiencies in TLR7 and TLR9 were linked to increases in expression of MERV proteins and viral particle formation, development of acute T-cell lymphoblastic leukemia, and inability to generate autoantibodies against MERVs. In humans, NA-TLRs have been linked to activation of autoimmune B cells in pathways likely downstream of responses to HERVs.

PERSPECTIVES

The impact of retroelement function and regulation on the evolving eukaryotic genome represents one of the most exciting new frontiers in biology exposed by two decades of rapid technological and informatic advances. Previous studies were confounded by properties of size, number, diversity, and multitudinous RNAs originating in sense and antisense promoters for complex processing and a dearth of appropriate analytic pipelines. The current explosion in genomics data gathering and analysis tools is empowering researchers with profound insights into retroelement genetic and epigenetic networks and coding and noncoding RNAs that together provide the underpinnings of pluripotency and differentiation, self- and non-self, and genomic structure and expression. A central theme throughout is that these networks are adaptive—able to morph in ways that sample real time and evolutionary space. Current research is unraveling the complexity of relationships of retroelement nucleic acid structure and function and coding and noncoding RNA networks and is being rewarded with insights into the interactions of environment, inheritance, and survival. It is no understatement to say that the impact on understanding of development, learning and memory, and medical diagnosis and treatment will continue to be profound. We are just now addressing the questions posed by McClintock and Britten and Davidson who foresaw how repetitive elements uniquely enable regulatory networks. Retroelements are linked to insertions, deletions, and rearrangements that cause disease in real time but license survival on evolutionary scales. Hence they provide the long sought but controversial link between experience and inheritance and are the basis of who we are as a species.

Retroelement RNAs and DNAs enable the modular building blocks of networks. A key aspect is their intrinsic modular transcriptional independence. The Singer SINE sequences are unmasked as structural elements in DNA and RNA. These genes embed into the genome under multiple guises—as genomic boundary and chromatin structural elements, points of genomic rearrangement and controllers of RNAP2 factor–binding sites responsive to environmental stress and methylation signals. They embed into transcribed regions and provide signals that influence RNA splicing, exonization, termination, polyadenylation, editing, and trafficking. LINE elements mobilize themselves and SINEs. They also serve as transcription factor–binding sites and modulators of RNA expression, processing, and function. However, beyond points of ectopic recombination, LINEs encode genome-modifying capabilities through poorly understood structural protein, EN and RT functions. ERV sequences originating as LTR promoters tissue adapted for infection of immune cells are now uncloaked as networks controlling modulators of innate immune reactions and regulons of pluripotency. More than the other elements, ERV and DNA TE protein-coding sequences have been exapted for a multitude of host functions in DNA rearrangements, RNA organization, RNA transmission and cellular fusion, and control of invading RNA species.

What could the future bring? Eight areas seem poised for high impact enabled by expanding understanding of retroelement regulation: (1) Precision diagnostics and therapies. Not just only the multitude of SINE elements but also the exact array of hot LINE elements have been linked to disease predisposition or initiation. Appreciation of the impact of elements on gene expression will translate into the reality that polymorphisms are not mere markers but have interpretable implications for disease susceptibility and treatment. (2) Autoimmune disease mechanisms and therapies. Genetic defects in multiple controllers of retroelements are now linked to failure or hyperactivation or failure of innate immune systems and some with autoimmune response.

Understanding biochemical mechanisms of these linkages can enable future therapeutics. (3) RNAi is perhaps among the newest of these exciting areas. Retroelements lie at the heart of speciation. Hominids have expanded PIWI genes and functions beyond retroelements. The parallels between rodent models and humans appear incomplete as distinct roles for piRNAs and siRNAs surface for gametogenesis in humans and rodents. What is the basis of sexual dimorphism in piRNA essentiality? What are the roles of piRNAs in epigenetic inheritance, paramutation, and maternal RNA degradation? What secrets lie in the uncharted graveyards of piRNA clusters? Are they the eukaryotic CRISPRs? (4) Neuronal plasticity and retroelement expression. What is the function of expression of LINEs, SINEs, and piRNAs in neuronal tissues? It likely extends beyond the relatively rare retrotransposition events to how retroelement RNAs and proteins influence organization of RNA expression, processing, trafficking, and degradation. (5) Retroelements as the two-edged sword. Advancing biotechnologies have managed to disable immediate undesirable properties and: convert retroviruses into valuable vehicles for gene therapy with considerable promise. However, concerns about interspecies activation of retroelements have plagued organ transplantation therapies. Most recently CRISPRs have been used to purge porcine genomes of ERVs as a route to interspecies organ transplants. The consequences are untested. (6) Retrotransposition in environmental stress, cancer, and aging. One can imagine the links, but biochemical elucidation is required to put insights to therapeutic purpose. (7) Ongoing germline expansions of retroelements. Two instances are especially provocative koalas undergoing endogenization of a gammaretrovirus and expansion of gypsy elements associated with a clam cancer. (8) Stem cell transplantation. The links between retroelement expression and pluripotency are not understood but clearly relevant. Not only is promoter activity directly influenced by retroelement sequences but also sweeps of methylation and demethylation are critical for maintaining suppression of retroelements and imprinting patterns of methylation. Therapies dependent on induced PSC will be more safe if retrotransposition is not uncontrolled.

Retroelement research is enabled by high-throughput genomics and more sensitive proteomic and metabolomics are coupled with powerful analytics. Retroelements are at the core of species and individual identities. Surely we are on the threshold of the golden age of retroelements in the human genome.

ACKNOWLEDGMENTS

This review was substantially contributed to by support from NIH grant GM33281 to study Ty3 for many years and recently by the Grace Beekhuis Bell Chair of Biological Chemistry. I thank the many wonderful colleagues I have had in the field of retrovirus and retrotransposon biology since the beginning. Most especially I thank the many members of my laboratory team over the years and Dr. Virginia Bilanchone who contributed not only to the science but the smooth functioning of the laboratory over that time. I thank Dr. Leslie Parent for her patience and support.

REFERENCES

Agrawal, A., Eastman, Q.M., Schatz, D.G., 1998. Transposition mediated by RAG1 and RAG2 and its implications for the evolution of the immune system. Nature 394, 744–751.

Alisch, R.S., Garcia-Perez, J.L., Muotri, A.R., Gage, F.H., Moran, J.V., 2006. Unconventional translation of mammalian LINE-1 retrotransposons. Genes Dev. 20, 210–224.

Amir, R.E., Van den Veyver, I.B., Wan, M., Tran, C.Q., Francke, U., Zoghbi, H.Y., 1999. Rett syndrome is caused by mutations in X-linked MECP2, encoding methyl-CpG-binding protein 2. Nat. Genet. 23, 185–188.

Ancelin, K., Lange, U.C., Hajkova, P., Schneider, R., Bannister, A.J., Kouzarides, T., Surani, M.A., 2006. Blimp1 associates with Prmt5 and directs histone arginine methylation in mouse germ cells. Nat. Cell Biol. 8, 623–630.

Aravin, A., Gaidatzis, D., Pfeffer, S., Lagos-Quintana, M., Landgraf, P., Iovino, N., Morris, P., Brownstein, M.J., Kuramochi-Miyagawa, S., Nakano, T., Chien, M., Russo, J.J., Ju, J., Sheridan, R., Sander, C., Zavolan, M., Tuschl, T., 2006. A novel class of small RNAs bind to MILI protein in mouse testes. Nature 442, 203–207.

Aravin, A.A., Klenov, M.S., Vagin, V.V., Bantignies, F., Cavalli, G., Gvozdev, V.A., 2004. Dissection of a natural RNA silencing process in the *Drosophila melanogaster* germ line. Mol. Cell. Biol. 24, 6742–6750.

Aravin, A.A., Lagos-Quintana, M., Yalcin, A., Zavolan, M., Marks, D., Snyder, B., Gaasterland, T., Meyer, J., Tuschl, T., 2003. The small RNA profile during *Drosophila melanogaster* development. Dev. Cell 5, 337–350.

Aravin, A.A., Sachidanandam, R., Bourc'his, D., Schaefer, C., Pezic, D., Toth, K.F., Bestor, T., Hannon, G.J., 2008. A piRNA pathway primed by individual transposons is linked to de novo DNA methylation in mice. Mol. Cell 31, 785–799.

Aravin, A.A., van der Heijden, G.W., Castaneda, J., Vagin, V.V., Hannon, G.J., Bortvin, A., 2009. Cytoplasmic compartmentalization of the fetal piRNA pathway in mice. PLoS Genet. 5, e1000764.

Arjan-Odedra, S., Swanson, C.M., Sherer, N.M., Wolinsky, S.M., Malim, M.H., 2012. Endogenous MOV10 inhibits the retrotransposition of endogenous retroelements but not the replication of exogenous retroviruses. Retrovirology 9, 53.

Ashley, J., Cordy, B., Lucia, D., Fradkin, L.G., Budnik, V., Thomson, T., 2018. Retrovirus-like Gag protein Arc1 binds RNA and traffics across synaptic boutons. Cell 172 262–274 e211.

Athanikar, J.N., Badge, R.M., Moran, J.V., 2004. A YY1-binding site is required for accurate human LINE-1 transcription initiation. Nucleic Acids Res. 32, 3846–3855.

Atwood, A., Lin, J.H., Levin, H.L., 1996. The retrotransposon Tf1 assembles virus-like particles that contain excess Gag relative to integrase because of a regulated degradation process. Mol. Cell. Biol. 16, 338–346.

Badge, R.M., Alisch, R.S., Moran, J.V., 2003. ATLAS: a system to selectively identify human-specific L1 insertions. Am. J. Hum. Genet. 72, 823–838.

Baillie, J.K., Barnett, M.W., Upton, K.R., Gerhardt, D.J., Richmond, T.A., De Sapio, F., Brennan, P.M., Rizzu, P., Smith, S., Fell, M., Talbot, R.T., Gustincich, S., Freeman, T.C., Mattick, J.S., Hume, D.A., Heutink, P., Carninci, P., Jeddeloh, J.A., Faulkner, G.J., 2011. Somatic retrotransposition alters the genetic landscape of the human brain. Nature 479, 534–537.

Barckmann, B., Pierson, S., Dufourt, J., Papin, C., Armenise, C., Port, F., Grentzinger, T., Chambeyron, S., Baronian, G., Desvignes, J.P., Curk, T., Simonelig, M., 2015. Aubergine iCLIP reveals piRNA-dependent decay of mRNAs involved in germ cell development in the early embryo. Cell Rep. 12, 1205–1216.

Batzer, M.A., Deininger, P.L., Hellmann-Blumberg, U., Jurka, J., Labuda, D., Rubin, C.M., Schmid, C.W., Zietkiewicz, E., Zuckerkandl, E., 1996. Standardized nomenclature for Alu repeats. J. Mol. Evol. 42, 3–6.

Beck, C.R., Collier, P., Macfarlane, C., Malig, M., Kidd, J.M., Eichler, E.E., Badge, R.M., Moran, J.V., 2010. LINE-1 retrotransposition activity in human genomes. Cell 141, 1159–1170.

Belancio, V.P., Whelton, M., Deininger, P., 2007. Requirements for polyadenylation at the 3′ end of LINE-1 elements. Gene 390, 98–107.

Beliakova-Bethell, N., Beckham, C., Giddings Jr., T.H., Winey, M., Parker, R., Sandmeyer, S., 2006. Virus-like particles of the Ty3 retrotransposon assemble in association with P-body components. RNA 12, 94–101.

Benachenhou, F., Jern, P., Oja, M., Sperber, G., Blikstad, V., Somervuo, P., Kaski, S., Blomberg, J., 2009. Evolutionary conservation of orthoretroviral long terminal repeats (LTRs) and ab initio detection of single LTRs in genomic data. PLoS One 4, e5179.

Bennett, E.A., Keller, H., Mills, R.E., Schmidt, S., Moran, J.V., Weichenrieder, O., Devine, S.E., 2008. Active Alu retrotransposons in the human genome. Genome Res. 18, 1875–1883.

Bilanchone, V., Clemens, K., Kaake, R., Dawson, A.R., Matheos, D., Nagashima, K., Sitlani, P., Patterson, K., Chang, I., Huang, L., Sandmeyer, S., 2015. Ty3 retrotransposon Hijacks mating yeast RNA processing bodies to infect new genomes. PLoS Genet. 11, e1005528.

Blaise, S., de Parseval, N., Benit, L., Heidmann, T., 2003. Genomewide screening for fusogenic human endogenous retrovirus envelopes identifies syncytin 2, a gene conserved on primate evolution. Proc. Natl. Acad. Sci. U.S.A. 100, 13013–13018.

Boissinot, S., Chevret, P., Furano, A.V., 2000. L1 (LINE-1) retrotransposon evolution and amplification in recent human history. Mol. Biol. Evol. 17, 915–928.

Borgel, J., Guibert, S., Li, Y., Chiba, H., Schubeler, D., Sasaki, H., Forne, T., Weber, M., 2010. Targets and dynamics of promoter DNA methylation during early mouse development. Nat. Genet. 42, 1093–1100.

Bourc'his, D., Bestor, T.H., 2004. Meiotic catastrophe and retrotransposon reactivation in male germ cells lacking Dnmt3L. Nature 431, 96–99.

Brass, A.L., Dykxhoorn, D.M., Benita, Y., Yan, N., Engelman, A., Xavier, R.J., Lieberman, J., Elledge, S.J., 2008. Identification of host proteins required for HIV infection through a functional genomic screen. Science 319, 921–926.

Brennecke, J., Aravin, A.A., Stark, A., Dus, M., Kellis, M., Sachidanandam, R., Hannon, G.J., 2007. Discrete small RNA-generating loci as master regulators of transposon activity in Drosophila. Cell 128, 1089–1103.

Brennecke, J., Malone, C.D., Aravin, A.A., Sachidanandam, R., Stark, A., Hannon, G.J., 2008. An epigenetic role for maternally inherited piRNAs in transposon silencing. Science 322, 1387–1392.

Britten, R.J., Baron, W.F., Stout, D.B., Davidson, E.H., 1988. Sources and evolution of human Alu repeated sequences. Proc. Natl. Acad. Sci. U.S.A. 85, 4770–4774.

Brouha, B., Schustak, J., Badge, R.M., Lutz-Prigge, S., Farley, A.H., Moran, J.V., Kazazian Jr., H.H., 2003. Hot L1s account for the bulk of retrotransposition in the human population. Proc. Natl. Acad. Sci. U.S.A. 100, 5280–5285.

Bulut-Karslioglu, A., De La Rosa-Velazquez, I.A., Ramirez, F., Barenboim, M., Onishi-Seebacher, M., Arand, J., Galan, C., Winter, G.E., Engist, B., Gerle, B., O'Sullivan, R.J., Martens, J.H., Walter, J., Manke, T., Lachner, M., Jenuwein, T., 2014. Suv39h-dependent H3K9me3 marks intact retrotransposons and silences LINE elements in mouse embryonic stem cells. Mol. Cell 55, 277–290.

Bundo, M., Toyoshima, M., Okada, Y., Akamatsu, W., Ueda, J., Nemoto-Miyauchi, T., Sunaga, F., Toritsuka, M., Ikawa, D., Kakita, A., Kato, M., Kasai, K., Kishimoto, T., Nawa, H., Okano, H., Yoshikawa, T., Kato, T., Iwamoto, K., 2014. Increased 11 retrotransposition in the neuronal genome in schizophrenia. Neuron 81, 306–313.

Burns, K.H., 2017. Transposable elements in cancer. Nat. Rev. Cancer 17, 415–424.

Burns, K.H., Boeke, J.D., 2012. Human transposon tectonics. Cell 149, 740–752.

Burton, F.H., Loeb, D.D., Voliva, C.F., Martin, S.L., Edgell, M.H., Hutchison, C.A., 1986. Coonservation through-out mammalia and extensive protein-encoding capacity of the highly repeated DNA long interspersed sequence one. J. Mol. Biol. 187, 291–304.

Bushman, F.D., Malani, N., Fernandes, J., D'Orso, I., Cagney, G., Diamond, T.L., Zhou, H., Hazuda, D.J., Espeseth, A.S., Konig, R., Bandyopadhyay, S., Ideker, T., Goff, S.P., Krogan, N.J., Frankel, A.D., Young, J.A., Chanda, S.K., 2009. Host cell factors in HIV replication: meta-analysis of genome-wide studies. PLoS Pathog. 5, e1000437.

Campillos, M., Doerks, T., Shah, P.K., Bork, P., 2006. Computational characterization of multiple Gag-like human proteins. Trends Genet. 22, 585–589.

Carreira, P.E., Richardson, S.R., Faulkner, G.J., 2014. L1 retrotransposons, cancer stem cells and oncogenesis. FEBS J. 281, 63–73.

Carroll, M.L., Roy-Engel, A.M., Nguyen, S.V., Salem, A.H., Vogel, E., Vincent, B., Myers, J., Ahmad, Z., Nguyen, L., Sammarco, M., Watkins, W.S., Henke, J., Makalowski, W., Jorde, L.B., Deininger, P.L., Batzer, M.A., 2001. Large-scale analysis of the Alu Ya5 and Yb8 subfamilies and their contribution to human genomic diversity. J. Mol. Biol. 311, 17–40.

Castro-Diaz, N., Ecco, G., Coluccio, A., Kapopoulou, A., Yazdanpanah, B., Friedli, M., Duc, J., Jang, S.M., Turelli, P., Trono, D., 2014. Evolutionally dynamic L1 regulation in embryonic stem cells. Genes Dev. 28, 1397–1409.

Cedar, H., Bergman, Y., 2009. Linking DNA methylation and histone modification: patterns and paradigms. Nat. Rev. Genet. 10, 295–304.

Checkley, M.A., Nagashima, K., Lockett, S.J., Nyswaner, K.M., Garfinkel, D.J., 2010. P-body components are required for Ty1 retrotransposition during assembly of retrotransposition-competent virus-like particles. Mol. Cell. Biol. 30, 382–398.

Chen, C., Nott, T.J., Jin, J., Pawson, T., 2011. Deciphering arginine methylation: tudor tells the tale. Nat. Rev. Mol. Cell Biol. 12, 629–642.

Chen, Y.G., Kim, M.V., Chen, X., Batista, P.J., Aoyama, S., Wilusz, J.E., Iwasaki, A., Chang, H.Y., 2017. Sensing self and foreign circular RNAs by intron identity. Mol. Cell 67, 228–238.e225.

Cheng, Z., Menees, T.M., 2004. RNA branching and debranching in the yeast retrovirus-like element Ty1. Science 303, 240–243.

Chiu, Y.L., Greene, W.C., 2008. The APOBEC3 cytidine deaminases: an innate defensive network opposing exogenous retroviruses and endogenous retroelements. Annu. Rev. Immunol. 26, 317–353.

Chu, W.M., Liu, W.M., Schmid, C.W., 1995. RNA polymerase III promoter and terminator elements affect Alu RNA expression. Nucleic Acids Res. 23, 1750–1757.

Chuong, E.B., Elde, N.C., Feschotte, C., 2016. Regulatory evolution of innate immunity through co-option of endogenous retroviruses. Science 351, 1083–1087.

Coffin, J.M., Hughes, S.H., Varmus, H.E., 1997. Retroviruses. Cold Spring Harbor Laboratory Press, Plainview, New York.

Cordaux, R., Batzer, M.A., 2009. The impact of retrotransposons on human genome evolution. Nat. Rev. Genet. 10, 691–703.

Coufal, N.G., Garcia-Perez, J.L., Peng, G.E., Marchetto, M.C., Muotri, A.R., Mu, Y., Carson, C.T., Macia, A., Moran, J.V., Gage, F.H., 2011. Ataxia telangiectasia mutated (ATM) modulates long interspersed element-1 (L1) retrotransposition in human neural stem cells. Proc. Natl. Acad. Sci. U.S.A. 108, 20382–20387.

Coufal, N.G., Garcia-Perez, J.L., Peng, G.E., Yeo, G.W., Mu, Y., Lovci, M.T., Morell, M., O'Shea, K.S., Moran, J.V., Gage, F.H., 2009. L1 retrotransposition in human neural progenitor cells. Nature 460, 1127–1131.

Crichton, J.H., Dunican, D.S., Maclennan, M., Meehan, R.R., Adams, I.R., 2014. Defending the genome from the enemy within: mechanisms of retrotransposon suppression in the mouse germline. Cell. Mol. Life Sci. 71, 1581–1605.

Crow, Y.J., Chase, D.S., Lowenstein Schmidt, J., Szynkiewicz, M., Forte, G.M., Gornall, H.L., Oojageer, A., Anderson, B., Pizzino, A., Helman, G., Abdel-Hamid, M.S., Abdel-Salam, G.M., Ackroyd, S., Aeby, A., Agosta, G., Albin, C., Allon-Shalev, S., Arellano, M., Ariaudo, G., Aswani, V., Babul-Hirji, R., Baildam, E.M., Bahi-Buisson, N., Bailey, K.M., Barnerias, C., Barth, M., Battini, R., Beresford, M.W., Bernard, G., Bianchi, M., Billette de Villemeur, T., Blair, E.M., Bloom, M., Burlina, A.B., Carpanelli, M.L., Carvalho, D.R., Castro-Gago, M., Cavallini, A., Cereda, C., Chandler, K.E., Chitayat, D.A., Collins, A.E., Sierra Corcoles, C., Cordeiro, N.J., Crichiutti, G., Dabydeen, L., Dale, R.C., D'Arrigo, S., De Goede, C.G., De Laet, C., De Waele, L.M., Denzler, I., Desguerre, I., Devriendt, K., Di Rocco, M., Fahey, M.C., Fazzi, E., Ferrie, C.D., Figueiredo, A., Gener, B., Goizet, C., Gowrinathan, N.R., Gowrishankar, K., Hanrahan, D., Isidor, B., Kara, B., Khan, N., King, M.D., Kirk, E.P., Kumar, R., Lagae, L., Landrieu, P., Lauffer, H., Laugel, V., La Piana, R., Lim, M.J., Lin, J.P., Linnankivi, T., Mackay, M.T., Marom, D.R., Marques Lourenco, C., McKee, S.A., Moroni, I., Morton, J.E., Moutard, M.L., Murray, K., Nabbout, R., Nampoothiri, S., Nunez-Enamorado, N., Oades, P.J., Olivieri, I., Ostergaard, J.R., Perez-Duenas, B., Prendiville, J.S., Ramesh, V., Rasmussen, M., Regal, L., Ricci, F., Rio, M., Rodriguez, D., Roubertie, A., Salvatici, E., Segers, K.A., Sinha, G.P., Soler, D., Spiegel, R., Stodberg, T.I., Straussberg, R., Swoboda, K.J., Suri, M., Tacke, U., Tan, T.Y., te Water Naude, J., Wee Teik, K., Thomas, M.M., Till, M., Tonduti, D., Valente, E.M., Van Coster, R.N., van der Knaap, M.S., Vassallo, G., Vijzelaar, R., Vogt, J., Wallace, G.B., Wassmer, E., Webb, H.J., Whitehouse, W.P., Whitney, R.N., Zaki, M.S., Zuberi, S.M., Livingston, J.H., Rozenberg, F., Lebon, P., Vanderver, A., Orcesi, S., Rice, G.I., 2015. Characterization of human disease phenotypes associated with mutations in TREX1, RNASEH2A, RNASEH2B, RNASEH2C, SAMHD1, ADAR, and IFIH1. Am. J. Med. Genet. 167A, 296–312.

Curcio, M.J., Lutz, S., Lesage, P., 2015. The Ty1 LTR-retrotransposon of budding yeast, *Saccharomyces cerevisiae*. Microbiol. Spectr. 3 MDNA3-0053-2014.

Czech, B., Hannon, G.J., 2016. One loop to rule them all: the ping-pong cycle and piRNA-guided silencing. Trends Biochem. Sci. 41, 324–337.

Daniels, G.R., Deininger, P.L., 1985. Repeat sequence families derived from mammalian tRNA genes. Nature 317, 819–822.

Daugherty, M.D., Malik, H.S., 2012. Rules of engagement: molecular insights from host-virus arms races. Annu. Rev. Genet. 46, 677–700.

Davidson, E.H., Britten, R.J., 1979. Regulation of gene expression: possible role of repetitive sequences. Science 204, 1052–1059.

De Fazio, S., Bartonicek, N., Di Giacomo, M., Abreu-Goodger, C., Sankar, A., Funaya, C., Antony, C., Moreira, P.N., Enright, A.J., O'Carroll, D., 2011. The endonuclease activity of Mili fuels piRNA amplification that silences LINE1 elements. Nature 480, 259–263.

de Koning, A.P., Gu, W., Castoe, T.A., Batzer, M.A., Pollock, D.D., 2011. Repetitive elements may comprise over two-thirds of the human genome. PLoS Genet. 7, e1002384.

de la Chaux, N., Wagner, A., 2011. BEL/Pao retrotransposons in metazoan genomes. BMC Evol. Biol. 11, 154.

de Parseval, N., Heidmann, T., 2005. Human endogenous retroviruses: from infectious elements to human genes. Cytogenet. Genome Res. 110, 318–332.

de Vanssay, A., Bouge, A.L., Boivin, A., Hermant, C., Teysset, L., Delmarre, V., Antoniewski, C., Ronsseray, S., 2012. Paramutation in Drosophila linked to emergence of a piRNA-producing locus. Nature 490, 112–115.

DeChiara, T.M., Brosius, J., 1987. Neural BC1 RNA: cDNA clones reveal nonrepetitive sequence content. Proc. Natl. Acad. Sci. U.S.A. 84, 2624–2628.

Dewannieux, M., Blaise, S., Heidmann, T., 2005. Identification of a functional envelope protein from the HERV-K family of human endogenous retroviruses. J. Virol. 79, 15573–15577.

Dewannieux, M., Heidmann, T., 2005. L1-mediated retrotransposition of murine B1 and B2 SINEs recapitulated in cultured cells. J. Mol. Biol. 349, 241–247.

Di Giacomo, M., Comazzetto, S., Saini, H., De Fazio, S., Carrieri, C., Morgan, M., Vasiliauskaite, L., Benes, V., Enright, A.J., O'Carroll, D., 2013. Multiple epigenetic mechanisms and the piRNA pathway enforce LINE1 silencing during adult spermatogenesis. Mol. Cell 50, 601–608.

Di Giacomo, M., Comazzetto, S., Sampath, S.C., Sampath, S.C., O'Carroll, D., 2014. G9a co-suppresses LINE1 elements in spermatogonia. Epigenet. Chromatin 7, 24.

Dieci, G., Conti, A., Pagano, A., Carnevali, D., 2013. Identification of RNA polymerase III-transcribed genes in eukaryotic genomes. Biochim. Biophys. Acta 1829, 296–305.

Doucet, A.J., Droc, G., Siol, O., Audoux, J., Gilbert, N., 2015. U6 snRNA pseudogenes: markers of retrotransposition dynamics in mammals. Mol. Biol. Evol. 32, 1815–1832.

Dutko, J.A., Kenny, A.E., Gamache, E.R., Curcio, M.J., 2010. 5′ to 3′ mRNA decay factors colocalize with Ty1 gag and human APOBEC3G and promote Ty1 retrotransposition. J. Virol. 84, 5052–5066.

Eckard, S.C., Rice, G.I., Fabre, A., Badens, C., Gray, E.E., Hartley, J.L., Crow, Y.J., Stetson, D.B., 2014. The SKIV2L RNA exosome limits activation of the RIG-I-like receptors. Nat. Immunol. 15, 839–845.

Erwin, J.A., Marchetto, M.C., Gage, F.H., 2014. Mobile DNA elements in the generation of diversity and complexity in the brain. Nat. Rev. Neurosci. 15, 497–506.

Evrony, G.D., Cai, X., Lee, E., Hills, L.B., Elhosary, P.C., Lehmann, H.S., Parker, J.J., Atabay, K.D., Gilmore, E.C., Poduri, A., Park, P.J., Walsh, C.A., 2012. Single-neuron sequencing analysis of L1 retrotransposition and somatic mutation in the human brain. Cell 151, 483–496.

Ewing, A.D., Kazazian Jr., H.H., 2010. High-throughput sequencing reveals extensive variation in human-specific L1 content in individual human genomes. Genome Res. 20, 1262–1270.

Faulkner, G.J., Kimura, Y., Daub, C.O., Wani, S., Plessy, C., Irvine, K.M., Schroder, K., Cloonan, N., Steptoe, A.L., Lassmann, T., Waki, K., Hornig, N., Arakawa, T., Takahashi, H., Kawai, J., Forrest, A.R., Suzuki, H., Hayashizaki, Y., Hume, D.A., Orlando, V., Grimmond, S.M., Carninci, P., 2009. The regulated retrotransposon transcriptome of mammalian cells. Nat. Genet. 41, 563–571.

Feng, Q., Moran, J.V., Kazazian Jr., H.H., Boeke, J.D., 1996. Human L1 retrotransposon encodes a conserved endonuclease required for retrotransposition. Cell 87, 905–916.

Feschotte, C., 2008. Transposable elements and the evolution of regulatory networks. Nat. Rev. Genet. 9, 397–405.

Flemr, M., Malik, R., Franke, V., Nejepinska, J., Sedlacek, R., Vlahovicek, K., Svoboda, P., 2013. A retrotransposon-driven dicer isoform directs endogenous small interfering RNA production in mouse oocytes. Cell 155, 807–816.

Frost, R.J., Hamra, F.K., Richardson, J.A., Qi, X., Bassel-Duby, R., Olson, E.N., 2010. MOV10L1 is necessary for protection of spermatocytes against retrotransposons by Piwi-interacting RNAs. Proc. Natl. Acad. Sci. U.S.A. 107, 11847–11852.

Galvis, A.E., Fisher, H.E., Nitta, T., Fan, H., Camerini, D., 2014. Impairment of HIV-1 cDNA synthesis by DBR1 knockdown. J. Virol. 88, 7054–7069.

Gaudet, F., Rideout 3rd, W.M., Meissner, A., Dausman, J., Leonhardt, H., Jaenisch, R., 2004. Dnmt1 expression in pre- and postimplantation embryogenesis and the maintenance of IAP silencing. Mol. Cell. Biol. 24, 1640–1648.

Gilbert, N., Lutz-Prigge, S., Moran, J.V., 2002. Genomic deletions created upon LINE-1 retrotransposition. Cell 110, 315–325.

Girard, A., Sachidanandam, R., Hannon, G.J., Carmell, M.A., 2006. A germline-specific class of small RNAs binds mammalian Piwi proteins. Nature 442, 199–202.

Gong, C., Maquat, L.E., 2011. lncRNAs transactivate STAU1-mediated mRNA decay by duplexing with 3′ UTRs via Alu elements. Nature 470, 284–288.

Goodier, J.L., 2014. Retrotransposition in tumors and brains. Mob. DNA 5, 11.

Goodier, J.L., 2016. Restricting retrotransposons: a review. Mob. DNA 7, 16.

Goodier, J.L., Cheung, L.E., Kazazian Jr., H.H., 2012. MOV10 RNA helicase is a potent inhibitor of retrotransposition in cells. PLoS Genet. 8, e1002941.

Goodier, J.L., Ostertag, E.M., Du, K., Kazazian Jr., H.H., 2001. A novel active L1 retrotransposon subfamily in the mouse. Genome Res. 11, 1677–1685.

Goodier, J.L., Zhang, L., Vetter, M.R., Kazazian Jr., H.H., 2007. LINE-1 ORF1 protein localizes in stress granules with other RNA-binding proteins, including components of RNA interference RNA-induced silencing complex. Mol. Cell. Biol. 27, 6469–6483.

Gou, L.T., Dai, P., Yang, J.H., Xue, Y., Hu, Y.P., Zhou, Y., Kang, J.Y., Wang, X., Li, H., Hua, M.M., Zhao, S., Hu, S.D., Wu, L.G., Shi, H.J., Li, Y., Fu, X.D., Qu, L.H., Wang, E.D., Liu, M.F., 2014. Pachytene piRNAs instruct massive mRNA elimination during late spermiogenesis. Cell Res. 24, 680–700.

Guffanti, G., Gaudi, S., Fallon, J.H., Sobell, J., Potkin, S.G., Pato, C., Macciardi, F., 2014. Transposable elements and psychiatric disorders. Am. J. Med. Genet. B Neuropsychiatr. Genet. 165B, 201–216.

Guibert, S., Forne, T., Weber, M., 2012. Global profiling of DNA methylation erasure in mouse primordial germ cells. Genome Res. 22, 633–641.

Ha, M., Kim, V.N., 2014. Regulation of microRNA biogenesis. Nat. Rev. Mol. Cell Biol. 15, 509–524.

Hajkova, P., Erhardt, S., Lane, N., Haaf, T., El-Maarri, O., Reik, W., Walter, J., Surani, M.A., 2002. Epigenetic reprogramming in mouse primordial germ cells. Mech. Dev. 117, 15–23.

Hajkova, P., Jeffries, S.J., Lee, C., Miller, N., Jackson, S.P., Surani, M.A., 2010. Genome-wide reprogramming in the mouse germ line entails the base excision repair pathway. Science 329, 78–82.

Hamdorf, M., Idica, A., Zisoulis, D.G., Gamelin, L., Martin, C., Sanders, K.J., Pedersen, I.M., 2015. miR-128 represses L1 retrotransposition by binding directly to L1 RNA. Nat. Struct. Mol. Biol. 22, 824–831.

Hancks, D.C., Kazazian Jr., H.H., 2016. Roles for retrotransposon insertions in human disease. Mob. DNA 7, 9.

Hayashi, K., Surani, M.A., 2009. Resetting the epigenome beyond pluripotency in the germline. Cell Stem Cell 4, 493–498.

Hazen, J.L., Faust, G.G., Rodriguez, A.R., Ferguson, W.C., Shumilina, S., Clark, R.A., Boland, M.J., Martin, G., Chubukov, P., Tsunemoto, R.K., Torkamani, A., Kupriyanov, S., Hall, I.M., Baldwin, K.K., 2016. The complete genome sequences, unique mutational spectra, and developmental potency of adult neurons revealed by cloning. Neuron 89, 1223–1236.

Heras, S.R., Macias, S., Caceres, J.F., Garcia-Perez, J.L., 2014. Control of mammalian retrotransposons by cellular RNA processing activities. Mob. Genet. Elements 4, e28439.

Heras, S.R., Macias, S., Plass, M., Fernandez, N., Cano, D., Eyras, E., Garcia-Perez, J.L., Caceres, J.F., 2013. The Microprocessor controls the activity of mammalian retrotransposons. Nat. Struct. Mol. Biol. 20, 1173–1181.

Hirakata, S., Siomi, M.C., 2016. piRNA biogenesis in the germline: from transcription of piRNA genomic sources to piRNA maturation. Biochim. Biophys. Acta 1859, 82–92.

Hohjoh, H., Singer, M.F., 1996. Cytoplasmic ribonucleoprotein complexes containing human LINE-1 protein and RNA. EMBO J. 15, 630–639.

Huang, C.R., Schneider, A.M., Lu, Y., Niranjan, T., Shen, P., Robinson, M.A., Steranka, J.P., Valle, D., Civin, C.I., Wang, T., Wheelan, S.J., Ji, H., Boeke, J.D., Burns, K.H., 2010. Mobile interspersed repeats are major structural variants in the human genome. Cell 141, 1171–1182.

Hughes, S., Jenkins, V., Dar, M.J., Engelman, A., Cherepanov, P., 2010. Transcriptional co-activator LEDGF interacts with Cdc7-activator of S-phase kinase (ASK) and stimulates its enzymatic activity. J. Biol. Chem. 285, 541–554.

Hung, T., Pratt, G.A., Sundararaman, B., Townsend, M.J., Chaivorapol, C., Bhangale, T., Graham, R.R., Ortmann, W., Criswell, L.A., Yeo, G.W., Behrens, T.W., 2015. The Ro60 autoantigen binds endogenous retroelements and regulates inflammatory gene expression. Science 350, 455–459.

Hurst, T.P., Magiorkinis, G., 2015. Activation of the innate immune response by endogenous retroviruses. J. Gen. Virol. 96, 1207–1218.

Ikeda, T., Abd El Galil, K.H., Tokunaga, K., Maeda, K., Sata, T., Sakaguchi, N., Heidmann, T., Koito, A., 2011. Intrinsic restriction activity by apolipoprotein B mRNA editing enzyme APOBEC1 against the mobility of autonomous retrotransposons. Nucleic Acids Res. 39, 5538–5554.

International Human Genome Sequencing, C, 2001. Initial sequencing and analysis of the human genome. Nature 409, 860–921.

Iskow, R.C., McCabe, M.T., Mills, R.E., Torene, S., Pittard, W.S., Neuwald, A.F., Van Meir, E.G., Vertino, P.M., Devine, S.E., 2010. Natural mutagenesis of human genomes by endogenous retrotransposons. Cell 141, 1253–1261.

Iwasaki, A., Medzhitov, R., 2015. Control of adaptive immunity by the innate immune system. Nat. Immunol. 16, 343–353.

Iwasaki, Y.W., Siomi, M.C., Siomi, H., 2015. PIWI-interacting RNA: its biogenesis and functions. Annu. Rev. Biochem. 84, 405–433.

Iyengar, B.R., Choudhary, A., Sarangdhar, M.A., Venkatesh, K.V., Gadgil, C.J., Pillai, B., 2014. Non-coding RNA interact to regulate neuronal development and function. Front. Cell. Neurosci. 8, 47.

Jacobs, F.M., Greenberg, D., Nguyen, N., Haeussler, M., Ewing, A.D., Katzman, S., Paten, B., Salama, S.R., Haussler, D., 2014. An evolutionary arms race between KRAB zinc-finger genes ZNF91/93 and SVA/L1 retrotransposons. Nature 516, 242–245.

Jurka, J., Smith, T., 1988. A fundamental division in the Alu family of repeated sequences. Proc. Natl. Acad. Sci. U.S.A. 85, 4775–4778.

Kapusta, A., Kronenberg, Z., Lynch, V.J., Zhuo, X., Ramsay, L., Bourque, G., Yandell, M., Feschotte, C., 2013. Transposable elements are major contributors to the origin, diversification, and regulation of vertebrate long noncoding RNAs. PLoS Genet. 9, e1003470.

Kassiotis, G., Stoye, J.P., 2016. Immune responses to endogenous retroelements: taking the bad with the good. Nat. Rev. Immunol. 16, 207–219.

Khurana, J.S., Wang, J., Xu, J., Koppetsch, B.S., Thomson, T.C., Nowosielska, A., Li, C., Zamore, P.D., Weng, Z., Theurkauf, W.E., 2011. Adaptation to P element transposon invasion in *Drosophila melanogaster*. Cell 147, 1551–1563.

Kidwell, M.G., Kidwell, J.F., Sved, J.A., 1977. Hybrid dysgenesis in *DROSOPHILA MELANOGASTER*: a syndrome of aberrant traits including mutation, sterility and male recombination. Genetics 86, 813–833.

Kim, C., Rubin, C.M., Schmid, C.W., 2001. Genome-wide chromatin remodeling modulates the Alu heat shock response. Gene 276, 127–133.

Kim, S., Gunesdogan, U., Zylicz, J.J., Hackett, J.A., Cougot, D., Bao, S., Lee, C., Dietmann, S., Allen, G.E., Sengupta, R., Surani, M.A., 2014. PRMT5 protects genomic integrity during global DNA demethylation in primordial germ cells and preimplantation embryos. Mol. Cell 56, 564–579.

Klawitter, S., Fuchs, N.V., Upton, K.R., Munoz-Lopez, M., Shukla, R., Wang, J., Garcia-Canadas, M., Lopez-Ruiz, C., Gerhardt, D.J., Sebe, A., Grabundzija, I., Merkert, S., Gerdes, P., Pulgarin, J.A., Bock, A., Held, U., Witthuhn, A., Haase, A., Sarkadi, B., Lower, J., Wolvetang, E.J., Martin, U., Ivics, Z., Izsvak, Z., Garcia-Perez, J.L., Faulkner, G.J., Schumann, G.G., 2016. Reprogramming triggers endogenous L1 and Alu retrotransposition in human induced pluripotent stem cells. Nat. Commun. 7, 10286.

Kobayashi, H., Sakurai, T., Imai, M., Takahashi, N., Fukuda, A., Yayoi, O., Sato, S., Nakabayashi, K., Hata, K., Sotomaru, Y., Suzuki, Y., Kono, T., 2012. Contribution of intragenic DNA methylation in mouse gametic DNA methylomes to establish oocyte-specific heritable marks. PLoS Genet. 8, e1002440.

Koito, A., Ikeda, T., 2011. Intrinsic restriction activity by AID/APOBEC family of enzymes against the mobility of retroelements. Mob. Genet. Elements 1, 197–202.

Kramerov, D.A., Vassetzky, N.S., 2011. Origin and evolution of SINEs in eukaryotic genomes. Heredity (Edinb.) 107, 487–495.

Krayev, A.S., Kramerov, D.A., Skryabin, K.G., Ryskov, A.P., Bayev, A.A., Georgiev, G.P., 1980. The nucleotide sequence of the ubiquitous repetitive DNA sequence B1 complementary to the most abundant class of mouse fold-back RNA. Nucleic Acids Res. 8, 1201–1215.

Ku, H.Y., Lin, H., 2014. PIWI proteins and their interactors in piRNA biogenesis, germline development and gene expression. Natl. Sci. Rev. 1, 205–218.

Kubo, S., Seleme, M.C., Soifer, H.S., Perez, J.L., Moran, J.V., Kazazian Jr., H.H., Kasahara, N., 2006. L1 retrotransposition in nondividing and primary human somatic cells. Proc. Natl. Acad. Sci. U.S.A. 103, 8036–8041.

Kuramochi-Miyagawa, S., Kimura, T., Yomogida, K., Kuroiwa, A., Tadokoro, Y., Fujita, Y., Sato, M., Matsuda, Y., Nakano, T., 2001. Two mouse piwi-related genes: miwi and mili. Mech. Dev. 108, 121–133.

Kuramochi-Miyagawa, S., Watanabe, T., Gotoh, K., Totoki, Y., Toyoda, A., Ikawa, M., Asada, N., Kojima, K., Yamaguchi, Y., Ijiri, T.W., Hata, K., Li, E., Matsuda, Y., Kimura, T., Okabe, M., Sakaki, Y., Sasaki, H., Nakano, T., 2008. DNA methylation of retrotransposon genes is regulated by Piwi family members MILI and MIWI2 in murine fetal testes. Genes Dev. 22, 908–917.

Kuwabara, T., Hsieh, J., Muotri, A., Yeo, G., Warashina, M., Lie, D.C., Moore, L., Nakashima, K., Asashima, M., Gage, F.H., 2009. Wnt-mediated activation of NeuroD1 and retro-elements during adult neurogenesis. Nat. Neurosci. 12, 1097–1105.

Laguette, N., Sobhian, B., Casartelli, N., Ringeard, M., Chable-Bessia, C., Segeral, E., Yatim, A., Emiliani, S., Schwartz, O., Benkirane, M., 2011. SAMHD1 is the dendritic- and myeloid-cell-specific HIV-1 restriction factor counteracted by Vpx. Nature 474, 654–657.

Lane, N., Dean, W., Erhardt, S., Hajkova, P., Surani, A., Walter, J., Reik, W., 2003. Resistance of IAPs to methylation reprogramming may provide a mechanism for epigenetic inheritance in the mouse. Genesis 35, 88–93.

Lee, E.J., Banerjee, S., Zhou, H., Jammalamadaka, A., Arcila, M., Manjunath, B.S., Kosik, K.S., 2011. Identification of piRNAs in the central nervous system. RNA 17, 1090–1099.

Lee, I.Y., Westaway, D., Smit, A.F., Wang, K., Seto, J., Chen, L., Acharya, C., Ankener, M., Baskin, D., Cooper, C., Yao, H., Prusiner, S.B., Hood, L.E., 1998. Complete genomic sequence and analysis of the prion protein gene region from three mammalian species. Genome Res. 8, 1022–1037.

Lees-Murdock, D.J., De Felici, M., Walsh, C.P., 2003. Methylation dynamics of repetitive DNA elements in the mouse germ cell lineage. Genomics 82, 230–237.

Lei, H., Oh, S.P., Okano, M., Juttermann, R., Goss, K.A., Jaenisch, R., Li, E., 1996. De novo DNA cytosine methyltransferase activities in mouse embryonic stem cells. Development 122, 3195–3205.

Levin, H.L., Moran, J.V., 2011. Dynamic interactions between transposable elements and their hosts. Nat. Rev. Genet. 12, 615–627.

Li, X., Zhang, J., Jia, R., Cheng, V., Xu, X., Qiao, W., Guo, F., Liang, C., Cen, S., 2013a. The MOV10 helicase inhibits LINE-1 mobility. J. Biol. Chem. 288, 21148–21160.

Li, X.Z., Roy, C.K., Dong, X., Bolcun-Filas, E., Wang, J., Han, B.W., Xu, J., Moore, M.J., Schimenti, J.C., Weng, Z., Zamore, P.D., 2013b. An ancient transcription factor initiates the burst of piRNA production during early meiosis in mouse testes. Mol. Cell 50, 67–81.

Li, Z., Yu, J., Hosohama, L., Nee, K., Gkountela, S., Chaudhari, S., Cass, A.A., Xiao, X., Clark, A.T., 2015. The Sm protein methyltransferase PRMT5 is not required for primordial germ cell specification in mice. EMBO J. 34, 748–758.

Liu, S., Brind'Amour, J., Karimi, M.M., Shirane, K., Bogutz, A., Lefebvre, L., Sasaki, H., Shinkai, Y., Lorincz, M.C., 2014. Setdb1 is required for germline development and silencing of H3K9me3-marked endogenous retroviruses in primordial germ cells. Genes Dev. 28, 2041–2055.

Liu, W.M., Maraia, R.J., Rubin, C.M., Schmid, C.W., 1994. Alu transcripts: cytoplasmic localisation and regulation by DNA methylation. Nucleic Acids Res. 22, 1087–1095.

Lu, X., Sachs, F., Ramsay, L., Jacques, P.E., Goke, J., Bourque, G., Ng, H.H., 2014. The retrovirus HERVH is a long noncoding RNA required for human embryonic stem cell identity. Nat. Struct. Mol. Biol. 21, 423–425.

Macfarlan, T.S., Gifford, W.D., Driscoll, S., Lettieri, K., Rowe, H.M., Bonanomi, D., Firth, A., Singer, O., Trono, D., Pfaff, S.L., 2012. Embryonic stem cell potency fluctuates with endogenous retrovirus activity. Nature 487, 57–63.

Mager, D.L., Stoye, J.P., 2015. Mammalian endogenous retroviruses. Microbiol. Spectr. 3 MDNA3-0009-2014.

Maksakova, I.A., Romanish, M.T., Gagnier, L., Dunn, C.A., van de Lagemaat, L.N., Mager, D.L., 2006. Retroviral elements and their hosts: insertional mutagenesis in the mouse germ line. PLoS Genet. 2, e2.

Maldarelli, F., Wu, X., Su, L., Simonetti, F.R., Shao, W., Hill, S., Spindler, J., Ferris, A.L., Mellors, J.W., Kearney, M.F., Coffin, J.M., Hughes, S.H., 2014. HIV latency. Specific HIV integration sites are linked to clonal expansion and persistence of infected cells. Science 345, 179–183.

Malim, M.H., Bieniasz, P.D., 2012. HIV restriction factors and mechanisms of evasion. Cold Spring Harb. Perspect. Med. 2, a006940.

Manakov, S.A., Pezic, D., Marinov, G.K., Pastor, W.A., Sachidanandam, R., Aravin, A.A., 2015. MIWI2 and MILI have differential effects on piRNA biogenesis and DNA methylation. Cell Rep. 12, 1234–1243.

Martin, S.L., Bushman, F.D., 2001. Nucleic acid chaperone activity of the ORF1 protein from the mouse LINE-1 retrotransposon. Mol. Cell. Biol. 21, 467–475.

Mathias, S.L., Scott, A.F., Kazazian Jr., H.H., Boeke, J.D., Gabriel, A., 1991. Reverse transcriptase encoded by a human transposable element. Science 254, 1808–1810.

Mayer, J., Blomberg, J., Seal, R.L., 2011. A revised nomenclature for transcribed human endogenous retroviral loci. Mob. DNA 2, 7.

McClintock, B., 1956. Controlling elements and the gene. Cold Spring Harb. Symp. Quant. Biol. 21, 197–216.

Mohn, F., Sienski, G., Handler, D., Brennecke, J., 2014. The rhino-deadlock-cutoff complex licenses noncanonical transcription of dual-strand piRNA clusters in Drosophila. Cell 157, 1364–1379.

Molaro, A., Falciatori, I., Hodges, E., Aravin, A.A., Marran, K., Rafii, S., McCombie, W.R., Smith, A.D., Hannon, G.J., 2014. Two waves of de novo methylation during mouse germ cell development. Genes Dev. 28, 1544–1549.

Moran, J.V., Holmes, S.E., Naas, T.P., DeBerardinis, R.J., Boeke, J.D., Kazazian Jr., H.H., 1996. High frequency retrotransposition in cultured mammalian cells. Cell 87, 917–927.

Morris, K.V., Mattick, J.S., 2014. The rise of regulatory RNA. Nat. Rev. Genet. 15, 423–437.

Morrish, T.A., Garcia-Perez, J.L., Stamato, T.D., Taccioli, G.E., Sekiguchi, J., Moran, J.V., 2007. Endonuclease-independent LINE-1 retrotransposition at mammalian telomeres. Nature 446, 208–212.

Morrish, T.A., Gilbert, N., Myers, J.S., Vincent, B.J., Stamato, T.D., Taccioli, G.E., Batzer, M.A., Moran, J.V., 2002. DNA repair mediated by endonuclease-independent LINE-1 retrotransposition. Nat. Genet. 31, 159–165.

Mouse Genome Sequencing, C., Waterston, R.H., Lindblad-Toh, K., Birney, E., Rogers, J., Abril, J.F., Agarwal, P., Agarwala, R., Ainscough, R., Alexandersson, M., An, P., Antonarakis, S.E., Attwood, J., Baertsch, R., Bailey, J., Barlow, K., Beck, S., Berry, E., Birren, B., Bloom, T., Bork, P., Botcherby, M., Bray, N., Brent, M.R., Brown, D.G., Brown, S.D., Bult, C., Burton, J., Butler, J., Campbell, R.D., Carninci, P., Cawley, S., Chiaromonte, F., Chinwalla, A.T., Church, D.M., Clamp, M., Clee, C., Collins, F.S., Cook, L.L., Copley, R.R., Coulson, A., Couronne, O., Cuff, J., Curwen, V., Cutts, T., Daly, M., David, R., Davies, J., Delehaunty, K.D., Deri, J., Dermitzakis, E.T., Dewey, C., Dickens, N.J., Diekhans, M., Dodge, S., Dubchak, I., Dunn, D.M., Eddy, S.R., Elnitski, L., Emes, R.D., Eswara, P., Eyras, E., Felsenfeld, A., Fewell, G.A., Flicek, P., Foley, K., Frankel, W.N., Fulton, L.A., Fulton, R.S., Furey, T.S., Gage, D., Gibbs, R.A., Glusman, G., Gnerre, S., Goldman, N., Goodstadt, L., Grafham, D., Graves, T.A., Green, E.D., Gregory, S., Guigo, R., Guyer, M., Hardison, R.C., Haussler, D., Hayashizaki, Y., Hillier, L.W., Hinrichs, A., Hlavina, W., Holzer, T., Hsu, F., Hua, A., Hubbard, T., Hunt, A., Jackson, I., Jaffe, D.B., Johnson, L.S., Jones, M., Jones, T.A., Joy, A., Kamal, M., Karlsson, E.K., Karolchik, D., Kasprzyk, A., Kawai, J., Keibler, E., Kells, C., Kent, W.J., Kirby, A., Kolbe, D.L., Korf, I., Kucherlapati, R.S., Kulbokas, E.J., Kulp, D., Landers, T., Leger, J.P., Leonard, S., Letunic, I., Levine, R., Li, J., Li, M., Lloyd, C., Lucas, S., Ma, B., Maglott, D.R., Mardis, E.R., Matthews, L., Mauceli, E., Mayer, J.H., McCarthy, M., McCombie, W.R., McLaren, S., McLay, K., McPherson, J.D., Meldrim, J., Meredith, B., Mesirov, J.P., Miller, W., Miner, T.L., Mongin, E., Montgomery, K.T., Morgan, M., Mott, R., Mullikin, J.C., Muzny, D.M., Nash, W.E., Nelson, J.O., Nhan, M.N., Nicol, R., Ning, Z., Nusbaum, C., O'Connor, M.J., Okazaki, Y., Oliver, K., Overton-Larty, E., Pachter, L., Parra, G., Pepin, K.H., Peterson, J., Pevzner, P., Plumb, R., Pohl, C.S., Poliakov, A., Ponce, T.C., Ponting, C.P., Potter, S., Quail, M., Reymond, A., Roe, B.A., Roskin, K.M., Rubin, E.M., Rust, A.G., Santos, R., Sapojnikov, V., Schultz, B., Schultz, J., Schwartz, M.S., Schwartz, S., Scott, C., Seaman, S., Searle, S., Sharpe, T., Sheridan, A., Shownkeen, R., Sims, S., Singer, J.B., Slater, G., Smit, A., Smith, D.R., Spencer, B., Stabenau, A., Stange-Thomann, N., Sugnet, C., Suyama, M., Tesler, G., Thompson, J., Torrents, D., Trevaskis, E., Tromp, J., Ucla, C., Ureta-Vidal, A., Vinson, J.P., Von Niederhausern, A.C., Wade, C.M., Wall, M., Weber, R.J., Weiss, R.B., Wendl, M.C., West, A.P., Wetterstrand, K., Wheeler, R., Whelan, S., Wierzbowski, J., Willey, D., Williams, S., Wilson, R.K., Winter, E., Worley, K.C., Wyman, D., Yang, S., Yang, S.P., Zdobnov, E.M., Zody, M.C., Lander, E.S., 2002. Initial sequencing and comparative analysis of the mouse genome. Nature 420, 520–562.

Muotri, A.R., Chu, V.T., Marchetto, M.C., Deng, W., Moran, J.V., Gage, F.H., 2005. Somatic mosaicism in neuronal precursor cells mediated by L1 retrotransposition. Nature 435, 903–910.

Muotri, A.R., Marchetto, M.C., Coufal, N.G., Oefner, R., Yeo, G., Nakashima, K., Gage, F.H., 2010. L1 retrotransposition in neurons is modulated by MeCP2. Nature 468, 443–446.

Murchison, E.P., Stein, P., Xuan, Z., Pan, H., Zhang, M.Q., Schultz, R.M., Hannon, G.J., 2007. Critical roles for Dicer in the female germline. Genes Dev. 21, 682–693.

Nan, X., Ng, H.H., Johnson, C.A., Laherty, C.D., Turner, B.M., Eisenman, R.N., Bird, A., 1998. Transcriptional repression by the methyl-CpG-binding protein MeCP2 involves a histone deacetylase complex. Nature 393, 386–389.

Naville, M., Warren, I.A., Haftek-Terreau, Z., Chalopin, D., Brunet, F., Levin, P., Galiana, D., Volff, J.N., 2016. Not so bad after all: retroviruses and long terminal repeat retrotransposons as a source of new genes in vertebrates. Clin. Microbiol. Infect. 22, 312–323.

Newkirk, S.J., An, W., 2017. L1 Regulation in mouse and human germ cells. In: Cristofari, G. (Ed.), Human Retrotransposons in Health and Disease. Springer International Publishing AG, pp. 29–61.

Oda, M., Yamagiwa, A., Yamamoto, S., Nakayama, T., Tsumura, A., Sasaki, H., Nakao, K., Li, E., Okano, M., 2006. DNA methylation regulates long-range gene silencing of an X-linked homeobox gene cluster in a lineage-specific manner. Genes Dev. 20, 3382–3394.

Ostertag, E.M., Goodier, J.L., Zhang, Y., Kazazian Jr., H.H., 2003. SVA elements are nonautonomous retrotransposons that cause disease in humans. Am. J. Hum. Genet. 73, 1444–1451.

Palmarini, M., Mura, M., Spencer, T.E., 2004. Endogenous betaretroviruses of sheep: teaching new lessons in retroviral interference and adaptation. J. Gen. Virol. 85, 1–13.

Parrish, N.F., Tomonaga, K., 2018. A Viral (Arc)hive for Metazoan Memory. Cell 172, 8–10.

Pastuzyn, E.D., Day, C.E., Kearns, R.B., Kyrke-Smith, M., Taibi, A.V., McCormick, J., Yoder, N., Belnap, D.M., Erlendsson, S., Morado, D.R., Briggs, J.A.G., Feschotte, C., Shepherd, J.D., 2018. The Neuronal Gene Arc Encodes a Repurposed Retrotransposon Gag Protein that Mediates Intercellular RNA Transfer. Cell 172, 275–288.e218.

Payer, L.M., Steranka, J.P., Yang, W.R., Kryatova, M., Medabalimi, S., Ardeljan, D., Liu, C., Boeke, J.D., Avramopoulos, D., Burns, K.H., 2017 E3984-E3992. Structural variants caused by Alu insertions are associated with risks for many human diseases. Proc. Natl. Acad. Sci. U.S.A. E3984–E3992.

Peaston, A.E., Evsikov, A.V., Graber, J.H., de Vries, W.N., Holbrook, A.E., Solter, D., Knowles, B.B., 2004. Retrotransposons regulate host genes in mouse oocytes and preimplantation embryos. Dev. Cell 7, 597–606.

Pezic, D., Manakov, S.A., Sachidanandam, R., Aravin, A.A., 2014. piRNA pathway targets active LINE1 elements to establish the repressive H3K9me3 mark in germ cells. Genes Dev. 28, 1410–1428.

Platanias, L.C., 2005. Mechanisms of type-I- and type-II-interferon-mediated signalling. Nat. Rev. Immunol. 5, 375–386.

Polak, P., Domany, E., 2006. Alu elements contain many binding sites for transcription factors and may play a role in regulation of developmental processes. BMC Genom. 7, 133.

Ponicsan, S.L., Kugel, J.F., Goodrich, J.A., 2010. Genomic gems: SINE RNAs regulate mRNA production. Curr. Opin. Genet. Dev. 20, 149–155.

Popp, C., Dean, W., Feng, S., Cokus, S.J., Andrews, S., Pellegrini, M., Jacobsen, S.E., Reik, W., 2010. Genome-wide erasure of DNA methylation in mouse primordial germ cells is affected by AID deficiency. Nature 463, 1101–1105.

Puszyk, W., Down, T., Grimwade, D., Chomienne, C., Oakey, R.J., Solomon, E., Guidez, F., 2013. The epigenetic regulator PLZF represses L1 retrotransposition in germ and progenitor cells. EMBO J. 32, 1941–1952.

Quentin, Y., 1994. A master sequence related to a free left Alu monomer (FLAM) at the origin of the B1 family in rodent genomes. Nucleic Acids Res. 22, 2222–2227.

Rajan, K.S., Ramasamy, S., 2014. Retrotransposons and piRNA: the missing link in central nervous system. Neurochem. Int. 77, 94–102.

Rajasethupathy, P., Antonov, I., Sheridan, R., Frey, S., Sander, C., Tuschl, T., Kandel, E.R., 2012. A role for neuronal piRNAs in the epigenetic control of memory-related synaptic plasticity. Cell 149, 693–707.

Rasmussen, K.D., Helin, K., 2016. Role of TET enzymes in DNA methylation, development, and cancer. Genes Dev. 30, 733–750.

Rehwinkel, J., Maelfait, J., Bridgeman, A., Rigby, R., Hayward, B., Liberatore, R.A., Bieniasz, P.D., Towers, G.J., Moita, L.F., Crow, Y.J., Bonthron, D.T., Reis e Sousa, C., 2013. SAMHD1-dependent retroviral control and escape in mice. EMBO J. 32, 2454–2462.

Richardson, S.R., Doucet, A.J., Kopera, H.C., Moldovan, J.B., Garcia-Perez, J.L., Moran, J.V., 2015. The Influence of LINE-1 and SINE Retrotransposons on Mammalian Genomes. Microbiol. Spectr. 3.

Richardson, S.R., Morell, S., Faulkner, G.J., 2014. L1 retrotransposons and somatic mosaicism in the brain. Annu. Rev. Genet. 48, 1–27.

Robine, N., Lau, N.C., Balla, S., Jin, Z., Okamura, K., Kuramochi-Miyagawa, S., Blower, M.D., Lai, E.C., 2009. A broadly conserved pathway generates 3′UTR-directed primary piRNAs. Curr. Biol. 19, 2066–2076.

Rodic, N., Steranka, J.P., Makohon-Moore, A., Moyer, A., Shen, P., Sharma, R., Kohutek, Z.A., Huang, C.R., Ahn, D., Mita, P., Taylor, M.S., Barker, N.J., Hruban, R.H., Iacobuzio-Donahue, C.A., Boeke, J.D., Burns, K.H., 2015. Retrotransposon insertions in the clonal evolution of pancreatic ductal adenocarcinoma. Nat. Med. 21, 1060–1064.

Ross, R.J., Weiner, M.M., Lin, H., 2014. PIWI proteins and PIWI-interacting RNAs in the soma. Nature 505, 353–359.

Rowe, H.M., Trono, D., 2011. Dynamic control of endogenous retroviruses during development. Virology 411, 273–287.

Roy-Engel, A.M., Carroll, M.L., Vogel, E., Garber, R.K., Nguyen, S.V., Salem, A.H., Batzer, M.A., Deininger, P.L., 2001. Alu insertion polymorphisms for the study of human genomic diversity. Genetics 159, 279–290.

Ryoo, J., Choi, J., Oh, C., Kim, S., Seo, M., Kim, S.Y., Seo, D., Kim, J., White, T.E., Brandariz-Nunez, A., Diaz-Griffero, F., Yun, C.H., Hollenbaugh, J.A., Kim, B., Baek, D., Ahn, K., 2014. The ribonuclease activity of SAMHD1 is required for HIV-1 restriction. Nat. Med. 20, 936–941.

Sabin, L.R., Delas, M.J., Hannon, G.J., 2013. Dogma derailed: the many influences of RNA on the genome. Mol. Cell 49, 783–794.

Saitou, M., Kagiwada, S., Kurimoto, K., 2012. Epigenetic reprogramming in mouse pre-implantation development and primordial germ cells. Development 139, 15–31.

Salter, J.D., Bennett, R.P., Smith, H.C., 2016. The APOBEC protein family: united by structure, divergent in function. Trends Biochem. Sci. 41, 578–594.

Sanchez-Luque, F.J., Richardson, S.R., Faulkner, G.J., 2016. Retrotransposon capture sequencing (RC-Seq): a targeted, high-throughput approach to resolve somatic L1 retrotransposition in humans. Methods Mol. Biol. 1400, 47–77.

Sandmeyer, S., Patterson, K., Bilanchone, V., 2015. Ty3, a position-specific retrotransposon in budding yeast. Microbiol. Spectr. 3.

Santiago-Raber, M.L., Baudino, L., Izui, S., 2009. Emerging roles of TLR7 and TLR9 in murine SLE. J. Autoimmun. 33, 231–238.

Santoni, F.A., Guerra, J., Luban, J., 2012. HERV-H RNA is abundant in human embryonic stem cells and a precise marker for pluripotency. Retrovirology 9, 111.

Sassaman, D.M., Dombroski, B.A., Moran, J.V., Kimberland, M.L., Naas, T.P., DeBerardinis, R.J., Gabriel, A., Swergold, G.D., Kazazian Jr., H.H., 1997. Many human L1 elements are capable of retrotransposition. Nat. Genet. 16, 37–43.

Schmid, C.W., 1991. Human Alu subfamilies and their methylation revealed by blot hybridization. Nucleic Acids Res. 19, 5613–5617.

Schmid, C.W., Deininger, P.L., 1975. Sequence organization of the human genome. Cell 6, 345–358.

Schumacher, A.J., Hache, G., Macduff, D.A., Brown, W.L., Harris, R.S., 2008. The DNA deaminase activity of human APOBEC3G is required for Ty1, MusD, and human immunodeficiency virus type 1 restriction. J. Virol. 82, 2652–2660.

Schumann, G.G., 2007. APOBEC3 proteins: major players in intracellular defence against LINE-1-mediated retrotransposition. Biochem. Soc. Trans. 35, 637–642.

Scott, E.C., Gardner, E.J., Masood, A., Chuang, N.T., Vertino, P.M., Devine, S.E., 2016. A hot L1 retrotransposon evades somatic repression and initiates human colorectal cancer. Genome Res. 26, 745–755.

Serdobova, I.M., Kramerov, D.A., 1998. Short retroposons of the B2 superfamily: evolution and application for the study of rodent phylogeny. J. Mol. Evol. 46, 202–214.

Siomi, M.C., Sato, K., Pezic, D., Aravin, A.A., 2011. PIWI-interacting small RNAs: the vanguard of genome defence. Nat. Rev. Mol. Cell Biol. 12, 246–258.

Skalka, A.M., 2014. Retroviral DNA Transposition: Themes and Variations. Microbiol. Spectr. 2 MDNA300052014.

Slagel, V., Flemington, E., Traina-Dorge, V., Bradshaw, H., Deininger, P., 1987. Clustering and subfamily relationships of the Alu family in the human genome. Mol. Biol. Evol. 4, 19–29.

Smalheiser, N.R., Torvik, V.I., 2005. Mammalian microRNAs derived from genomic repeats. Trends Genet. 21, 322–326.

Smalheiser, N.R., Torvik, V.I., 2006. Alu elements within human mRNAs are probable microRNA targets. Trends Genet. 22, 532–536.

Smallwood, S.A., Tomizawa, S., Krueger, F., Ruf, N., Carli, N., Segonds-Pichon, A., Sato, S., Hata, K., Andrews, S.R., Kelsey, G., 2011. Dynamic CpG island methylation landscape in oocytes and preimplantation embryos. Nat. Genet. 43, 811–814.

Smit, A.F., 1993. Identification of a new, abundant superfamily of mammalian LTR-transposons. Nucleic Acids Res. 21, 1863–1872.

Smith, Z.D., Chan, M.M., Mikkelsen, T.S., Gu, H., Gnirke, A., Regev, A., Meissner, A., 2012. A unique regulatory phase of DNA methylation in the early mammalian embryo. Nature 484, 339–344.

Solyom, S., Ewing, A.D., Rahrmann, E.P., Doucet, T., Nelson, H.H., Burns, M.B., Harris, R.S., Sigmon, D.F., Casella, A., Erlanger, B., Wheelan, S., Upton, K.R., Shukla, R., Faulkner, G.J., Largaespada, D.A., Kazazian Jr., H.H., 2012. Extensive somatic L1 retrotransposition in colorectal tumors. Genome Res. 22, 2328–2338.

Song, S.U., Gerasimova, T., Kurkulos, M., Boeke, J.D., Corces, V.G., 1994. An env-like protein encoded by a Drosophila retroelement: evidence that gypsy is an infectious retrovirus. Genes Dev. 8, 2046–2057.

Sonigo, P., Wain-Hobson, S., Bougueleret, L., Tiollais, P., Jacob, F., Brulet, P., 1987. Nucleotide sequence and evolution of ETn elements. Proc. Natl. Acad. Sci. U.S.A. 84, 3768–3771.

Sookdeo, A., Hepp, C.M., McClure, M.A., Boissinot, S., 2013. Revisiting the evolution of mouse LINE-1 in the genomic era. Mob. DNA 4, 3.

Stetson, D.B., Ko, J.S., Heidmann, T., Medzhitov, R., 2008. Trex1 prevents cell-intrinsic initiation of autoimmunity. Cell 134, 587–598.

Stocking, C., Kozak, C.A., 2008. Murine endogenous retroviruses. Cell. Mol. Life Sci. 65, 3383–3398.

Stoye, J.P., 1998. Fv1, the mouse retrovirus resistance gene. Rev. Sci. Tech. 17, 269–277.

Stoye, J.P., Moroni, C., 1983. Endogenous retrovirus expression in stimulated murine lymphocytes. Identification of a new locus controlling mitogen induction of a defective virus. J. Exp. Med. 157, 1660–1674.

Sudmant, P.H., Rausch, T., Gardner, E.J., Handsaker, R.E., Abyzov, A., Huddleston, J., Zhang, Y., Ye, K., Jun, G., Hsi-Yang Fritz, M., Konkel, M.K., Malhotra, A., Stutz, A.M., Shi, X., Paolo Casale, F., Chen, J., Hormozdiari, F., Dayama, G., Chen, K., Malig, M., Chaisson, M.J., Walter, K., Meiers, S., Kashin, S., Garrison, E., Auton, A., Lam, H.Y., Jasmine Mu, X., Alkan, C., Antaki, D., Bae, T., Cerveira, E., Chines, P., Chong, Z., Clarke, L., Dal, E., Ding, L., Emery, S., Fan, X., Gujral, M., Kahveci, F., Kidd, J.M., Kong, Y., Lameijer, E.W., McCarthy, S., Flicek, P., Gibbs, R.A., Marth, G., Mason, C.E., Menelaou, A., Muzny, D.M., Nelson, B.J., Noor, A., Parrish, N.F., Pendleton, M., Quitadamo, A., Raeder, B., Schadt, E.E., Romanovitch, M., Schlattl, A., Sebra, R., Shabalin, A.A., Untergasser, A., Walker, J.A., Wang, M., Yu, F., Zhang, C., Zhang, J., Zheng-Bradley, X., Zhou, W., Zichner, T., Sebat, J., Batzer, M.A., McCarroll, S.A., Genomes Project, C., Mills, R.E., Gerstein, M.B., Bashir, A., Stegle, O., Devine, S.E., Lee, C., Eichler, E.E., Korbel, J.O., 2015. An integrated map of structural variation in 2,504 human genomes. Nature 526, 75–81.

Sundaram, V., Cheng, Y., Ma, Z., Li, D., Xing, X., Edge, P., Snyder, M.P., Wang, T., 2014. Widespread contribution of transposable elements to the innovation of gene regulatory networks. Genome Res. 24, 1963–1976.

Suzuki, J., Yamaguchi, K., Kajikawa, M., Ichiyanagi, K., Adachi, N., Koyama, H., Takeda, S., Okada, N., 2009. Genetic evidence that the non-homologous end-joining repair pathway is involved in LINE retrotransposition. PLoS Genet. 5, e1000461.

Swergold, G.D., 1990. Identification, characterization, and cell specificity of a human LINE-1 promoter. Mol. Cell. Biol. 10, 6718–6729.

Symer, D.E., Connelly, C., Szak, S.T., Caputo, E.M., Cost, G.J., Parmigiani, G., Boeke, J.D., 2002. Human l1 retrotransposition is associated with genetic instability in vivo. Cell 110, 327–338.

Tchenio, T., Casella, J.F., Heidmann, T., 2000. Members of the SRY family regulate the human LINE retrotransposons. Nucleic Acids Res. 28, 411–415.

Thomas, J.H., Schneider, S., 2011. Coevolution of retroelements and tandem zinc finger genes. Genome Res. 21, 1800–1812.

Thompson, C.B., 1995. New insights into V(D)J recombination and its role in the evolution of the immune system. Immunity 3, 531–539.

Trelogan, S.A., Martin, S.L., 1995. Tightly regulated, developmentally specific expression of the first open reading frame from LINE-1 during mouse embryogenesis. Proc. Natl. Acad. Sci. U.S.A. 92, 1520–1524.

Ufer, F., Vargas, P., Engler, J.B., Tintelnot, J., Schattling, B., Winkler, H., Bauer, S., Kursawe, N., Willing, A., Keminer, O., Ohana, O., Salinas-Riester, G., Pless, O., Kuhl, D., Friese, M.A., 2016. Arc/Arg3.1 governs inflammatory dendritic cell migration from the skin and thereby controls T cell activation. Sci. Immunol. 1, eaaf8665.

Ullu, E., Tschudi, C., 1984. Alu sequences are processed 7SL RNA genes. Nature 312, 171–172.

Vagin, V.V., Sigova, A., Li, C., Seitz, H., Gvozdev, V., Zamore, P.D., 2006. A distinct small RNA pathway silences selfish genetic elements in the germline. Science 313, 320–324.

van Gent, D.C., Mizuuchi, K., Gellert, M., 1996. Similarities between initiation of V(D)J recombination and retroviral integration. Science 271, 1592–1594.

Venter, J.C., Adams, M.D., Myers, E.W., Li, P.W., Mural, R.J., Sutton, G.G., Smith, H.O., Yandell, M., Evans, C.A., Holt, R.A., Gocayne, J.D., Amanatides, P., Ballew, R.M., Huson, D.H., Wortman, J.R., Zhang, Q., Kodira, C.D., Zheng, X.H., Chen, L., Skupski, M., Subramanian, G., Thomas, P.D., Zhang, J., Gabor Miklos, G.L., Nelson, C., Broder, S., Clark, A.G., Nadeau, J., McKusick, V.A., Zinder, N., Levine, A.J., Roberts, R.J., Simon, M., Slayman, C., Hunkapiller, M., Bolanos, R., Delcher, A., Dew, I., Fasulo, D., Flanigan, M., Florea, L., Halpern, A., Hannenhalli, S., Kravitz, S., Levy, S., Mobarry, C., Reinert, K., Remington, K., Abu-Threideh, J., Beasley, E., Biddick, K., Bonazzi, V., Brandon, R., Cargill, M., Chandramouliswaran, I., Charlab, R., Chaturvedi, K., Deng, Z., Di Francesco, V., Dunn, P., Eilbeck, K., Evangelista, C., Gabrielian, A.E., Gan, W., Ge, W., Gong, F., Gu, Z., Guan, P., Heiman, T.J., Higgins, M.E., Ji, R.R., Ke, Z., Ketchum, K.A., Lai, Z., Lei, Y., Li, Z., Li, J., Liang, Y., Lin, X., Lu, F., Merkulov, G.V., Milshina, N., Moore, H.M., Naik, A.K., Narayan, V.A., Neelam, B., Nusskern, D., Rusch, D.B., Salzberg, S., Shao, W., Shue, B., Sun, J., Wang, Z., Wang, A., Wang, X., Wang, J., Wei, M., Wides, R., Xiao, C., Yan, C., Yao, A., Ye, J., Zhan, M., Zhang, W., Zhang, H., Zhao, Q., Zheng, L., Zhong, F., Zhong, W., Zhu, S., Zhao, S., Gilbert, D., Baumhueter, S., Spier, G., Carter, C., Cravchik, A., Woodage, T., Ali, F., An, H., Awe, A., Baldwin, D., Baden, H., Barnstead, M., Barrow, I., Beeson, K., Busam, D., Carver, A., Center, A., Cheng, M.L., Curry, L., Danaher, S., Davenport, L., Desilets, R., Dietz, S., Dodson, K., Doup, L., Ferriera, S., Garg, N., Gluecksmann, A., Hart, B., Haynes, J., Haynes, C., Heiner, C., Hladun, S., Hostin, D., Houck, J., Howland, T., Ibegwam, C., Johnson, J., Kalush, F., Kline, L., Koduru, S., Love, A., Mann, F., May, D., McCawley, S., McIntosh, T., McMullen, I., Moy, M., Moy, L., Murphy, B., Nelson, K., Pfannkoch, C., Pratts, E., Puri, V., Qureshi, H., Reardon, M., Rodriguez, R., Rogers, Y.H., Romblad, D., Ruhfel, B., Scott, R., Sitter, C., Smallwood, M., Stewart, E., Strong, R., Suh, E., Thomas, R., Tint, N.N., Tse, S., Vech, C., Wang, G., Wetter, J., Williams, S., Williams, M., Windsor, S., Winn-Deen, E., Wolfe, K., Zaveri, J., Zaveri, K., Abril, J.F., Guigo, R., Campbell, M.J., Sjolander, K.V., Karlak, B., Kejariwal, A., Mi, H., Lazareva, B., Hatton, T., Narechania, A., Diemer, K., Muruganujan, A., Guo, N., Sato, S., Bafna, V., Istrail, S., Lippert, R., Schwartz, R., Walenz, B., Yooseph, S., Allen, D., Basu, A., Baxendale, J., Blick, L., Caminha, M., Carnes-Stine, J., Caulk, P., Chiang, Y.H., Coyne, M., Dahlke, C., Mays, A., Dombroski, M., Donnelly, M., Ely, D., Esparham, S., Fosler, C., Gire, H., Glanowski, S., Glasser, K., Glodek, A., Gorokhov, M., Graham, K., Gropman, B., Harris, M., Heil, J., Henderson, S., Hoover, J., Jennings, D., Jordan, C., Jordan, J., Kasha, J., Kagan, L., Kraft, C., Levitsky, A., Lewis, M., Liu, X., Lopez, J., Ma, D., Majoros, W., McDaniel, J., Murphy, S., Newman, M., Nguyen, T., Nguyen, N., Nodell, M., Pan, S., Peck, J., Peterson, M., Rowe, W., Sanders, R., Scott, J., Simpson, M., Smith, T., Sprague, A., Stockwell, T., Turner, R., Venter, E., Wang, M., Wen, M., Wu, D., Wu, M., Xia, A., Zandieh, A., Zhu, X., 2001. The sequence of the human genome. Science 291, 1304–1351.

Volkman, H.E., Stetson, D.B., 2014. The enemy within: endogenous retroelements and autoimmune disease. Nat. Immunol. 15, 415–422.

Voronina, E., Seydoux, G., Sassone-Corsi, P., Nagamori, I., 2011. RNA granules in germ cells. Cold Spring Harb. Perspect. Biol. 3.

Walsh, C.P., Chaillet, J.R., Bestor, T.H., 1998. Transcription of IAP endogenous retroviruses is constrained by cytosine methylation. Nat. Genet. 20, 116–117.

Wang, H., Xing, J., Grover, D., Hedges, D.J., Han, K., Walker, J.A., Batzer, M.A., 2005. SVA elements: a hominid-specific retroposon family. J. Mol. Biol. 354, 994–1007.

Wang, T., Zeng, J., Lowe, C.B., Sellers, R.G., Salama, S.R., Yang, M., Burgess, S.M., Brachmann, R.K., Haussler, D., 2007. Species-specific endogenous retroviruses shape the transcriptional network of the human tumor suppressor protein p53. Proc. Natl. Acad. Sci. U.S.A. 104, 18613–18618.

Wilczynska, A., Bushell, M., 2015. The complexity of miRNA-mediated repression. Cell Death Differ. 22, 22–33.

Wolf, D., Goff, S.P., 2007. TRIM28 mediates primer binding site-targeted silencing of murine leukemia virus in embryonic cells. Cell 131, 46–57.

Wolf, D., Goff, S.P., 2009. Embryonic stem cells use ZFP809 to silence retroviral DNAs. Nature 458, 1201–1204.

Wong, R.C., Ibrahim, A., Fong, H., Thompson, N., Lock, L.F., Donovan, P.J., 2011. L1TD1 is a marker for undifferentiated human embryonic stem cells. PLoS One 6, e19355.

Yang, N., Kazazian Jr., H.H., 2006. L1 retrotransposition is suppressed by endogenously encoded small interfering RNAs in human cultured cells. Nat. Struct. Mol. Biol. 13, 763–771.

Yang, N., Zhang, L., Zhang, Y., Kazazian Jr., H.H., 2003. An important role for RUNX3 in human L1 transcription and retrotransposition. Nucleic Acids Res. 31, 4929–4940.

Young, G.R., Eksmond, U., Salcedo, R., Alexopoulou, L., Stoye, J.P., Kassiotis, G., 2012. Resurrection of endogenous retroviruses in antibody-deficient mice. Nature 491, 774–778.

Yu, F., Zingler, N., Schumann, G., Stratling, W.H., 2001. Methyl-CpG-binding protein 2 represses LINE-1 expression and retrotransposition but not Alu transcription. Nucleic Acids Res. 29, 4493–4501.

Yu, P., Lubben, W., Slomka, H., Gebler, J., Konert, M., Cai, C., Neubrandt, L., Prazeres da Costa, O., Paul, S., Dehnert, S., Dohne, K., Thanisch, M., Storsberg, S., Wiegand, L., Kaufmann, A., Nain, M., Quintanilla-Martinez, L., Bettio, S., Schnierle, B., Kolesnikova, L., Becker, S., Schnare, M., Bauer, S., 2012. Nucleic acid-sensing Toll-like receptors are essential for the control of endogenous retrovirus viremia and ERV-induced tumors. Immunity 37, 867–879.

Zeng, M., Hu, Z., Shi, X., Li, X., Zhan, X., Li, X.D., Wang, J., Choi, J.H., Wang, K.W., Purrington, T., Tang, M., Fina, M., DeBerardinis, R.J., Moresco, E.M., Pedersen, G., McInerney, G.M., Karlsson Hedestam, G.B., Chen, Z.J., Beutler, B., 2014. MAVS, cGAS, and endogenous retroviruses in T-independent B cell responses. Science 346, 1486–1492.

Zhang, A., Dong, B., Doucet, A.J., Moldovan, J.B., Moran, J.V., Silverman, R.H., 2014. RNase L restricts the mobility of engineered retrotransposons in cultured human cells. Nucleic Acids Res. 42, 3803–3820.

Zhang, W., Wu, J., Ward, M.D., Yang, S., Chuang, Y.A., Xiao, M., Li, R., Leahy, D.J., Worley, P.F., 2015. Structural basis of arc binding to synaptic proteins: implications for cognitive disease. Neuron 86, 490–500.

Zhao, K., Du, J., Han, X., Goodier, J.L., Li, P., Zhou, X., Wei, W., Evans, S.L., Li, L., Zhang, W., Cheung, L.E., Wang, G., Kazazian Jr., H.H., Yu, X.F., 2013. Modulation of LINE-1 and Alu/SVA retrotransposition by Aicardi-Goutieres syndrome-related SAMHD1. Cell Rep. 4, 1108–1115.

Zheng, K., Xiol, J., Reuter, M., Eckardt, S., Leu, N.A., McLaughlin, K.J., Stark, A., Sachidanandam, R., Pillai, R.S., Wang, P.J., 2010. Mouse MOV10L1 associates with Piwi proteins and is an essential component of the Piwi-interacting RNA (piRNA) pathway. Proc. Natl. Acad. Sci. U.S.A. 107, 11841–11846.

Chapter 14

Strategies to Discover Novel Cellular Factors Involved in Retrovirus Replication

Rebecca J. Kaddis Maldonado, Eunice C. Chen, Leslie J. Parent
Penn State College of Medicine, Hershey, PA, United States

Chapter Outline

Introduction 528
Identifying Protein–Protein Interactions in Retroviruses Using Two-Hybrid Screens 528
- Using Gag Proteins as Bait 528
- Using Integrase as Bait 535
- Additional Retroviral Proteins Used as Bait 535

Mass Spectrometry Approaches to Discovery of Retrovirus–Cell Protein–Protein Interactions 536
- Mass Spectrometry to Identify Proteins Packaged Into Virus Particles 536
- Mass Spectrometry to Identify Host Proteins Interacting With HIV-1 Proteins 537
- Affinity Purification Approaches 537
- Liquid Chromatography Coupled to Tandem Mass Spectrometry 539
- Multidimensional Protein Identification Technology Mass Spectrometry 540

Mass Spectrometry Using Viral DNA as Bait 541
- HIV-1 5′ Untranslated Regions as Bait 541

Mass Spectrometry Using Viral RNA as Bait 542
- HIV-1 Subgenomic RNAs as Bait 542
- Retroviral 5′Untranslated Regions as Bait 543
- Full-Length HIV-1 RNA Used as Bait 543

RNA Interference and Genome-Wide Screens to Assess the Contribution of Host Factors to Virus Replication 544
- Small Interfering RNA Screens 544
- Short Hairpin RNA Screens 546
- Metaanalysis of siRNA Screens in HIV-1 Replication 547
- Targeted RNAi Screens 547

Retrovirus-Cell Interactions. https://doi.org/10.1016/B978-0-12-811185-7.00014-5

Genome-Wide Analysis of Retrotransposition in Yeast 549
Host Genome Editing Using CRISPR/Cas-9 to Find Retrovirus Dependency Genes 549
Gain-of-Function Genetic Approaches Using cDNA Overexpression Screens 550
Identification of Cellular RNAs That Interact With Retroviruses 550
Host RNAs Packaged Into Retrovirus Particles 551
RNA Sequencing Techniques 552
Concluding Remarks 554
Acknowledgments 560
References 560

INTRODUCTION

Retroviruses have limited genomes and depend on host cell factors to facilitate their replication. Multiple host cell proteins and RNAs have been identified that interact with retroviruses to facilitate or interfere with replication, although in many cases, their exact roles during replication remain unclear. Several different experimental approaches have been utilized to identify retrovirus–cell interactions, including mass spectrometry (MS), RNA interference (RNAi), CRISPR, yeast two-hybrid studies, and deep sequencing. This chapter outlines the findings of various studies that sought to use these methods to identify host proteins and RNAs that interact with retroviruses.

IDENTIFYING PROTEIN–PROTEIN INTERACTIONS IN RETROVIRUSES USING TWO-HYBRID SCREENS

To study protein–protein interactions in the context of retrovirus replication, researchers have utilized retrovirus proteins as bait, and a variety of assays and methods were used to identify interacting host protein complexes. As technology advanced, the identification and characterization of proteins and complex protein interactions have increased and become more sophisticated. The results of these experiments have contributed greatly to our knowledge of retrovirus replication and many are described in detail in earlier chapters. Initial studies were performed using two-hybrid screens (Fields and Song, 1989) in yeast or mammalian cells, to identify host protein interactions involved in retrovirus replication. There are numerous examples of important retrovirus–cell interactions that were discovered using two-hybrid approaches, and although not every study can be included in this chapter, several notable examples will be described and included in Table 14.1.

Using Gag Proteins as Bait

Early efforts to identify protein–protein interactions adopted the yeast two-hybrid method, in which a retroviral gene was expressed as bait in yeast cells, and cellular interacting factors were identified (Luban and Goff, 1995). Several

TABLE 14.1 Select Studies Using Two-Hybrid and Mass Spectrometry Screens to Identify Host Proteins that Interact with Retroviruses.

Type of Study	Cell Type	Viral Proteins Used as Bait	Cellular Factors Identified	Citation
Y2H	Undifferentiated human leukemia cell (HL-60) cDNA library in GGY1::171 yeast cells	HIV-1 Gag	CyPA and CyPB	Luban et al. (1993)
Y2H	C57BL/Ka V13 T-lymphoma mouse cDNA library in HF7C yeast cells	BM5def $Pr60^{Gag}$ and MLV Gag	KIF4 (C-terminal region named Y26)	Kim et al. (1998)
Y2H	Hela cDNA library in EGY48 yeast cells	HIV-1 MA with N-term LexA	Translation elongation factor 1-alpha (EF1a)	Cimarelli and Luban (1999)
Y2H	Macrophage/monocyte cells from BALB/c mouse (WEHI-3B) cDNA library in CTY10-5d and GGY::171 yeast cells	MLV Gag with N-term LexADB	Nucleolin, interaction checked with other retroviral Gag proteins	Bacharach et al. (2000)
Y2H	Human B cell cDNA library in Y153 yeast cells	HIV-1 Gag	Tsg101 and cyclophilin B (previously identified)	VerPlank et al. (2001)
Y2H	Mouse T-lymphoma cDNA library	Full-length BM5defGag and MLV Gag with C-term LexA	Endophilin 2	Wang et al. (2003)
Y2H	Hela cDNA library in CG1945 yeast cells	MPMV Gag	hUbc9 and previously identified factors	Weldon et al. (2003)
Y2H	Macrophage/monocyte cells from BALB/c mouse (WEHI-3B) cDNA library in CTY10-5d yeast cells	MLV CA with C-term LexADB	Ubc9 and PIASy	Yueh et al. (2006)

Continued

TABLE 14.1 Select Studies Using Two-Hybrid and Mass Spectrometry Screens to Identify Host Proteins that Interact with Retroviruses—cont'd

Type of Study	Cell Type	Viral Proteins Used as Bait	Cellular Factors Identified	Citation
Y2H	Human testis cDNA library	HIV-1$_{NY5/BRU}$ Gag fragment containing C-term of MA and N-term of CA	PDZD8	Henning et al. (2010)
Y2H	Hela cDNA library	HIV-1 Gag	FLNa and other previously unreported and reported interactors	Cooper et al. (2011)
AP–MS/MS, SILAC/MS	293T (transfected)	HIV-1 Gag-GFP	31 cellular proteins identified across 5 experiments	Engeland et al. (2011)
Y2H	BALB/cJ and C3H/HeN lactating mammary gland cDNA library in CTY10-5d yeast cells	Chimeric MMTV Gag proteins	L9 ribosomal subunit interacted more strongly with more tumorigenic Gag compared to tumor attenuated	Beyer et al. (2013)
AP–MS/MS, SILAC/MS	293T cells	HIV-1 Gag-GFP	1804 proteins identified, 90 were significant in ≥3 screens.	Engeland et al. (2014)
LC–MS/MS	Jurkat cells (transfected)	BirA-tagged HIV-1 Gag	64 unique peptides that represent 47 proteins	Le Sage et al. (2015)
Y2H	HL60 macrophage/monocyte cell cDNA library	HIV-1 IN	INI1	Kalpana et al. (1994)

Edman degradation, MS	293T (stable)	HIV-1 IN	LEDGF/DFS70/p75	Cherepanov et al. (2003)
Y2H	Macrophage/monocyte cells from BALB/c mice (WEHI-3B) cDNA library and T-cell mouse cDNA library in CTY5-10d yeast cells	MLV IN and HIV-1 IN	121 putative interactions identified initially, 27 host proteins characterized with MoMLV IN and further tested with HIV-1 IN	Studamire and Goff (2008)
Y2H	CEMC7 cell cDNA library	HIV-1 IN	LEDGF/p75, Transportin-SR2 (TNPO3), von Hippel–Lindau binding protein 1 (VBP1), and sucrose nonfermenting 5 (SNF5)	Rain et al. (2009)
Y2H	Human T-lymphocyte cDNA library in AH109 yeast cells	Constitutively acetylated HIV-1 IN containing N-term 6xHis, internal TEV, and C-term HA	13 host proteins identified	Allouch and Cereseto (2011)
nano-RP-LC–MS/MS	C8166/45 cells (infected)	HIV-1 RTC and PIC	94 cellular proteins unique to infected samples identified	Schweitzer et al. (2013)
MS-iDIRT	CEM cells and MT-2 cells (infected)	Epitope-tagged HIV-1 Env and Vif	56 host proteins identified (Vif), 189 host proteins identified (Env)	Luo et al. (2016)
Y2H	Human leucocyte cDNA library	HIV-1 NL4-3 strain Vif	SSAT, Triad3, and NVBP	Lake et al. (2003)

Continued

TABLE 14.1 Select Studies Using Two-Hybrid and Mass Spectrometry Screens to Identify Host Proteins that Interact with Retroviruses—cont'd

Type of Study	Cell Type	Viral Proteins Used as Bait	Cellular Factors Identified	Citation
Y2H	Jurkat T-cell cDNA library	Nuclear diffusion inhibitory signal (NIS) of HIV-1 Rev	DDX1	Fang et al. (2004)
2D-DIGE, MALDI—MS/MS	Jurkat cells	HIV-1 Tat	7 proteins identified	Coiras et al. (2006)
TAP, MS/MS	293T cells	TAP-tagged Vpr	2 proteins identified	Schröfelbauer et al. (2007)
1D-GE, LC–MS/MS	Jurkat T-cell nuclear extracts	GST-tagged HIV-1 Tat	183 candidate proteins, 90% novel	Gautier et al. (2009)
AP–MS/MS	HEK293 cells (transient) and Jurkat cells (stable, tet inducible)	2x Strep-3x Flag-Vif	24 Vif-host protein interactions (7 shared between cell types)	Jäger et al. (2011b)
LC–MS/MS	Jurkat cells (transfected)	HIV-1 Nef	10 cellular proteins specifically associated with WT Nef	Mukerji et al. (2012)
MudPIT MS/MS	HeLa cells, nuclear and cytoplasmic fractions	HIV-1 Rev with C-term recombinant maltose binding protein	250 high scoring proteins identified, validate 8 DEAD/H box proteins	Naji et al. (2012)
2D-DIGE, MALDI–TOF MS	Chronically infected T-cell line (CEM)	HIV-1 LAV-1 virions	24 proteins identified within virions	Misumi et al. (2002)

LC–MS/MS, MALDI–TOF	Human monocyte-derived macrophages	HIV-1 virions	253 unique cellular proteins	Chertova et al. (2006)
LC–MS/MS	293T and Jurkat cells (transfected)	HIV-1 virions	10 viral proteins and 15 cellular proteins	Saphire et al. (2006)
1D-GE, LC–MS/MS	293-GPG	Virus particles from VSV-G pseudotyped MMLV retrovirus vector	25 cellular proteins identified	Segura et al. (2008)
LC–MS/MS	Sup-T1 cells and THP1 cells (infected)	Core structures from MLV-pseudotyped HIV-1 NL4-3 virions	202 unique cellular proteins identified, 42 present in all samples	Santos et al. (2012)
LC–MS/MS	293T cells (transfected)	Virus particles from HIV-1 NL4-3 wild-type and Nef-deficient	18 cellular proteins identified	Brégnard et al. (2013)
LC–MS/MS	Nuclear fraction, C8166-45 cells (infected)	Virus particles from VSVg-pseudotyped HIV-1	13 proteins identified in infected nuclei	DeBoer et al. (2014)
LC–MS/MS	HEK293T (transfected)	BirA-tagged FL and MA-deleted HIV-1 Gag, cells, and virions	Over 50 cellular proteins identified	Ritchie et al. (2015)
LC–MS/MS	CD4+ CEMx174	HIV-1 LAI infected cells	687 proteins changed in abundance	Chan et al. (2007)
2D-DIGE, MS/MS	PM1 (T cells)	HIV-1 LAI infected cells	288 showed significant differential expression with 92 unique proteins (7–10 days post), 12 proteins showed significant differential expression with 9 unique proteins (42 h post)	Ringrose et al. (2008)

Continued

TABLE 14.1 Select Studies Using Two-Hybrid and Mass Spectrometry Screens to Identify Host Proteins that Interact with Retroviruses—cont'd

Type of Study	Cell Type	Viral Proteins Used as Bait	Cellular Factors Identified	Citation
AP–MS/MS	HEK293 cells (transient) and Jurkat cells (stable, tetracycline inducible)	All HIV-1 proteins fused to purification tag	497 different HIV-human interactions with 435 individual human proteins	Jäger et al. (2011a)
1D-GE, LC–MS/MS	Chronically HIV-1 infected H9-derived CD4+ cell line (H9/HTLVIIIB)	Biotinylated DNA-PIC complex	19 cellular proteins identified	Raghavendra et al. (2010)
1D-GE, MALDI–TOF	Infected TZM-bl cells	Purinome-capture resin	9 purine-binding proteins identified	Kulkarni et al. (2017)

1D-GE, one-dimensional gel electrophoresis; *2D-DIGE*, two-dimensional differential in-gel electrophoresis; *AP*, affinity purification; *LC*, liquid chromatography; *MALDI–TOF*, matrix-assisted laser desorption ionization–time of flight mass spectrometry; *MS I-DIRT*, mass spectrometric isotopic differentiation of interactions as random or targeted; *MS/MS*, tandem mass spectrometry; *MudPIT*, multidimensional protein identification technology; *PIC*, preintegration complex; *RTC*, reverse transcriptase complex; *SWATH*, sequential window acquisition of all theoretical mass spectra; *Y2H*, yeast two-hybrid.

host factors associated with the HIV-1 (human immunodeficiency virus type 1) Gag protein were identified, including: cyclophilins A and B (Luban et al., 1993), translation elongation factor 1-alpha (Cimarelli and Luban, 1999), Tsg101 (VerPlank et al., 2001), AIP1/ALIX (Strack et al., 2003), PDZD8 (Henning et al., 2010), and Filamin A (Cooper et al., 2011). Using the MLV (Moloney murine leukemia virus) Gag protein as bait, KIF-4, a microtubule-associated protein belonging to the kinesin family, was found as a binding partner (Kim et al., 1998). Subsequent studies demonstrated that KIF-4 played a role in the transport of HIV-1, SIV (simian immunodeficiency virus), MPMV (Mason-Pfizer monkey virus), and RSV (Rous sarcoma virus) Gag proteins (Sekine et al., 1994). Endophilins were also found to interact with MLV Gag, but in this case there was no similar association identified for HIV-1 Gag (Wang et al., 2003). MLV Gag was also noted to bind to a C-terminal fragment of nucleolin, and overexpression of this protein interfered with MLV-inhibited virion release (Bacharach et al., 2000). Two small ubiquitin-like modifier conjugating enzymes, PIASy and Ubc9, bind MLV Gag via the CA domain; rather than affecting virus assembly, they play an early role in infection (Yueh et al., 2006). MMPV Gag was also reported to bind to Ubc9, and when overexpressed in cells, Gag partially relocalized to the nucleus (Weldon et al., 2003). MMTV (mouse mammary tumor virus) Gag protein was noted to associate with the ribosomal L9 protein and overexpression of L9 resulted in accumulation of MMTV Gag in the nucleolus (Beyer et al., 2013).

Using Integrase as Bait

Studamire and Goff (2008) used MLV integrase (IN) as bait and identified 121 interacting proteins across three screens, of which 27 interactions were further characterized and tested for association with MLV or HIV-1 IN proteins. HIV-1 IN was used as bait in another yeast two-hybrid screen, and yeast transcription factor SNF5 was identified (Kalpana et al., 1994). The human homolog of SNF5 was named INI1 (integrase interactor 1) and stimulates DNA joining reaction in human cells. A similar study was performed, in which SNF5/INI1 was also identified, along with human lens epithelium-derived growth factor (LEDGF), transportin-3 (TNPO3), and von Hippel Lindau binding protein 1 (VBP1) (Rain et al., 2009). A later study used an acetylated form of HIV-1 IN as bait, finding 13 binding partners that included transcription factors, chromatin remodeling agents, RNA binding proteins, translation factors, and elements of microtubules (Allouch and Cereseto, 2011).

Additional Retroviral Proteins Used as Bait

Other retroviral factors used in yeast two-hybrid screens included HIV-1 Vif, resulting in identification of 16 initial clones, of which 3 were further characterized: spermine/spermidine N1-acetyltransferase; Traid 3, also known as zinc finger protein-inhibiting NF-kappaB; and a novel protein of unknown function (Feng et al., 2004; Lake et al., 2003). Using HIV-1 Rev as bait, Fang et al. identified the DEAD box helicase DDX1, which plays a role in localization of Rev to the nucleus and nucleolus (Fang et al., 2004). Although there has been great success in the outcomes of these and other studies, there are caveats to two-hybrid studies: protein interactions can fail to be detected if the expression level of the bait protein is too low; the biochemical readout of the assay dictates the protein interaction must take place in the yeast nucleus and either the bait or prey could fail to localize to the nucleus; the viral bait protein might not adopt the proper conformation or function optimally in a nonmammalian intracellular environment; and nonspecific interactions with host factors can result in false positives (Edwards et al., 2002; Luo and Muesing, 2014).

MASS SPECTROMETRY APPROACHES TO DISCOVERY OF RETROVIRUS–CELL PROTEIN–PROTEIN INTERACTIONS

Following the development and optimization of advanced MS techniques, researchers turned to proteomics to study retroviral protein complexes purified from the host cell. Advantages of MS proteomics studies are its high sensitivity and the ability to characterize different protein interactions by changing the conditions used to isolate the protein complexes of interest. The term "proteomics" was originally coined in reference to two-dimensional differential in-gel electrophoresis experiments (2D-DIGE) (Wilkins et al., 1996b). One- and two-dimensional gel electrophoreses are still used to determine differential expression of proteins between two conditions by comparing staining patterns of gels. Differential gel electrophoresis techniques are semiquantitative, as band intensity roughly correlates with protein concentration. The addition of MS allows for identification of differentially expressed proteins and increases quantitative analysis. However, these techniques are limited by their low resolution and the detection of only highly expressed proteins with a bias against membrane proteins. These issues can be overcome using a combination of gel electrophoresis and MS techniques (reviewed in Ong and Mann, 2005; Galeva and Altermann, 2002).

Mass Spectrometry to Identify Proteins Packaged Into Virus Particles

As an example of a successful experiment using this method, Misumi et al. used 2D-DIGE followed by matrix-assisted laser desorption ionization–time of flight MS (MALDI–TOF) to identify host proteins packaged into purified HIV-1 LAV-1 strain particles. Twenty-four proteins were identified by 2D gel electrophoresis and characterized by MALDI–TOF MS, in addition to proteins identified within microvesicles and outside of the virion (Misumi et al., 2002). Segura et al. used a similar technique to study host proteins incorporated into purified VSV-G pseudotyped-MLV particles. Among the 17 bands that were visualized by 1D-DIGE, 12 were also identified by MS (Segura et al., 2008).

Mass Spectrometry to Identify Host Proteins Interacting With HIV-1 Proteins

Coiras et al. and Raghavendra et al. used these techniques to study host proteins that interact with specific HIV-1 viral proteins. Coiras et al. used 2D-DIGE with MALDI–MS to identify seven host proteins interacting with HIV-1 Tat protein. Proteomic analysis was performed in Jurkat cells overexpressing a doxycycline-inducible Tat protein and compared to a Jurkat Tet-Off control cell line (Coiras et al., 2006). Raghavendra et al. used 1D-DIGE with MS to characterize host proteins involved in HIV-1 integration. They used an elegant method: instead of relying on a tagged HIV-1 protein to isolate preintegration complexes (PIC), they utilized a biotinylated target DNA as bait to avoid any steric complications a tagged protein may impose. The biotinylated target DNA–protein complexes were isolated using magnetic streptavidin beads and bound proteins were separated by 1D gel electrophoresis and identified by MS. Noninfected cells were used as a negative control to eliminate nonspecific protein interactions. In total, 19 host proteins were identified to interact with PIC, of which 18 had not been previously characterized (Raghavendra et al., 2010). Lastly Ringrose et al. used 2D-DIGE with MALDI–MS to perform an unbiased proteomic analysis of differential host protein expression in HIV-1 infected PM1 T cells. At the point of peak HIV infection (7–10 days postinoculation), 288 protein spots were differentially expressed and 92 unique viral and host proteins were identified. At 42 h postinoculation, 12 protein spots were differentially expressed and only 9 unique proteins were identified (Ringrose et al., 2008).

A related technique is Edman degradation, which generates an N-terminus tag of 3–4 amino acids, followed by MS, which allows for the "sequencing" of proteins off protein spots on membranes for proteomics studies (Gooley et al., 1997; Wilkins et al., 1996a). This technique was used by Cherepanov et al. to identify LEDGF/p75 as a critical host factor for HIV-1 IN function. Nuclear extracts of 293T cells stably expressing a modified FLAG-tagged HIV-1 IN protein were cross-linked to stabilize protein complexes and IN-interacting complexes were purified by immunoprecipitation and separated by SDS-PAGE. A coprecipitating protein with a molecular mass of approximately 76 kDa was visualized and identified as p76. Subsequent Edman degradation and MS analysis of p76 identified the protein as LEDGF/DFS70/p75 (Cherepanov et al., 2003).

Affinity Purification Approaches

Affinity purification (AP) followed by MS was the first technique used to perform a proteomic analysis to identify protein complexes in cells. Gavin et al.

(2006) identified 491 protein complexes in budding yeast, of which 257 had not been previously characterized. In AP–MS studies, the protein of interest is expressed in cells and any cellular proteins that interact with it are isolated from the cellular environment either by using specific antibodies or by expressing affinity-tagged proteins and pulling down on the tag. This method allows for the unbiased study of the protein of interest in the native cellular environment. Purification techniques using affinity tags are preferable to immunoprecipitation purifications because they produce better MS data with fewer background protein contaminants and because reliable antibodies are not available for all viral proteins (Rigaut et al., 1999; Zeghouf et al., 2004). A variant of AP is tandem affinity purification (TAP), in which the bait protein contains two affinity tags, a calmodulin-binding peptide and *S. Aureus* protein A linked by a TEV protease cleavage site and is purified across two affinity columns to allow efficient purification of protein complexes from a small number of cells (Rigaut et al., 1999). Since the initial characterization of this technique, a variety of TAP tags have been developed and optimized utilizing different tag combinations. However, one caveat of AP–MS is that transient interactions will be missed if they disassociate during the purification process. This can be avoided by the use of proper experimental controls to avoid false positives.

Many studies have used tagged viral proteins and AP–MS techniques to characterize global cellular host factors involved in virus replication. Schröfelbauer et al. (2007) used this technique to identify the damage-specific DNA-binding protein 1 as a cellular cofactor of TAP-tagged HIV-1 Vpr and to characterize the role of Vpr in HIV-1 replication. Similarly, Jäger et al. (2011b) expressed a 2x-Strep 3x-Flag tagged Vif protein in two different mammalian cell types to identify novel Vif-host protein interactions, which included members of the E3 ubiquitin ligase complex CUL5 and factors involved in autophagy. Engeland et al. used a combination of techniques to identify host factors that specifically interact with HIV-1 Gag. Using GFP-tagged Gag, with the tag located at the C-terminus, between MA and CA, and TAP-tagged Gag, the researchers used five independent techniques and identified 31 cellular proteins that were significant in three or more screens. Lyric, a host factor previously found to be induced during HIV-1 infection, was the only protein identified in all of the screens (Kang et al., 2005; Engeland et al., 2014). Assays with the tag alone functioned as experimental controls. Later, Engeland et al. repeated the experiment using the same Gag fusion proteins in six independent AP screens and identified 1804 cellular proteins, with 90 of them considered significant in three or more independent screens. Once again, Lyric was the only protein identified in all screens (Engeland et al., 2011). However, many of the proteins identified had been previously characterized in other proteomic and genome-wide screens. When Engeland et al. classified the interacting proteins, three main categories emerged: SR family proteins, cytoskeleton and microtubule-interacting proteins, and the tRNA synthase complex. The results of these studies allowed further understanding of the biological function of proteins involved in Gag trafficking and function.

In a comprehensive study of host proteins involved in HIV-1 replication, Jäger et al. (2011a) tagged all 18 HIV-1 proteins and polyproteins with 2x Strep-3x Flag affinity tags and performed AP–MS in transiently transfected 293T cells and stably expressed and inducible Jurkat cells. They identified 497 different host proteins, of which approximately 40% (196 interactions) were present in both cell types. The researchers then used coimmunoprecipitation and Western blotting to verify a subset of these interactions and demonstrate the accuracy of their dataset. Comparison of their dataset with previously published studies showed significant overlap, further confirming their findings. By systematically tagging each retroviral protein, Jäger et al. generated a network of all host proteins involved and hypothesized their biological significance in retrovirus replication.

Liquid Chromatography Coupled to Tandem Mass Spectrometry

The previously described MS techniques are considered "top-down" studies, in which mixtures of intact proteins are isolated and analyzed. However, this approach is limited to simple mixtures (Han et al., 2008). In contrast are "bottom-up" approaches to proteomics, which are performed by first extracting and purifying proteins from the sample to remove contaminants, digestion of proteins into peptides, further separation to generate a homogenous mixture of peptides, and lastly analysis by MS (reviewed in Karpievitch et al., 2010). One example of a "bottom-up" approach is liquid chromatography (LC) followed by tandem MS (MS/MS). Saphire et al. performed the first proteomic analysis of host proteins present in the whole virions of enveloped viruses using LC–MS/MS. Jurkat cells and 293T cells were transfected with a HIV-1 provirus and virions were harvested from the supernatant 48 h later. Following tryptic digestion, the peptides were subjected to LC–MS/MS analysis. In addition to identifying virally encoded proteins, Saphire et al. were able to identify 14 host proteins previously characterized to be present in virions as well as a new host protein. To further demonstrate the efficacy of this analysis protein, these investigators introduced a mutation into HIV-1 known to prevent packaging of the host protein cyclophilin A. As expected, virions harvested from the mutant virus did not have detectable cyclophilin A present (Saphire et al., 2006).

A later LC–MS/MS proteomics study expanded the number of host proteins identified in HIV-1 virions (Chertova et al., 2006). In this study, donor monocyte-derived macrophages (MDM) were chosen for study because in these cells, HIV-1 virus particles accumulate within membrane-bound vesicles in the cytoplasmic, including multivesicular bodies and late endosomes. Primary MDM cells were infected with HIV-1 virions harvested from transfected 293T cells. Thin-section electron microscopy was performed to ensure that virus particles were visualized in cytoplasmic vesicles within the infected MDM cells in culture. Supernatants from the infected and uninfected control MDM cells were harvested and subjected to CD45-immunodepletion and ultracentrifugation to remove contaminating microvesicles and cellular proteins. The purified supernatants were analyzed by microcapillary reversed-phase liquid chromatography MS. In addition to virally encoded proteins, 253 unique cellular proteins were identified. Of those proteins, 33 had been identified in prior studies of HIV-1 virions produced from non-MDM cells, including CD63, Alix/AIP1, Tsg101, and eEF1A. In addition, 220 novel proteins not previously found in HIV-1 virus particles were identified in this study, many of which function in MVB and late endosome biology.

Segura et al. identified cellular proteins incorporated into MLV virions produced by using a packaging cell line and purified by ultracentrifugation following subtilisin treatment to remove contaminating microvesicles and proteins. The resulting samples underwent 1D-SDS-PAGE and were visualized by silver stain for in-gel trypsin digestion and analysis by LC–MS/MS. In total, 25 cellular proteins were identified, of which 20 had not been previously characterized as playing a potential role in MLV replication (Segura et al., 2008). Similar to the study by Chertova et al., the identified host proteins were derived from a variety of subcellular compartments and organelles, suggesting that the assembling virus particle had established associations with these host proteins along the way to the plasma membrane for virion budding. These proteomics studies examining host factors packaged into virions can provide clues into the cellular compartments that the Gag-genomic RNA complex passes through on its journey to the plasma membrane and raises questions about whether these cellular pathways contribute directly to the late stages of virus assembly. Moreover, cellular factors packaged into virions are subsequently released into cells on entry of the viral core and therefore could play a role in uncoating or the host response to virus infection.

Multidimensional Protein Identification Technology Mass Spectrometry

Naji et al. (2012) identified host factors that interact with HIV-1 Rev in a unique manner, by comparing the strength of protein–protein interactions in different in vitro intracellular environments. An HIV-1 Rev protein with a mannose binding protein fused to the C-terminus was incubated with nuclear and cytosolic lysate fractions in the presence or absence of the Rev response element (RRE) and the nucleocytoplasmic transport cofactor RanQ69L-GTP. Protein-RNA complexes were isolated and subjected to multidimensional protein identification technology (MudPIT) MS. MudPIT allows for the quick unbiased analysis of an entire proteome by separating peptides by 2-dimensional LC, instead of 2D-DIGE, prior to MS/MS (Washburn et al., 2001). The researchers then used a combination of proteomics and statistical analysis techniques to rank the protein–protein interactions based on the abundance, background significance, and enrichment of proteins identified, resulting in the discovery of 3161 unique proteins, of which 150 proteins strongly bound Rev in the presence of

RRE. From this category, eight DEAD/H helicase family members were found: DDHX36, DDHX24, DDX3X, DDX1, DDX17, DHX9, DDX47, and DDX5. Further small interfering RNAs (siRNA) analysis of DDX3X, DDX5, DDX17, and DDX21 in HIV-1 infected cells confirmed the role of DEAD/H helicase proteins in Rev-dependent HIV-1 replication.

MASS SPECTROMETRY USING VIRAL DNA AS BAIT

In addition to being used to identify protein–protein interactions, MS serves as an informative method to find host factors associated with viral nucleic acids and either proviral DNA or viral RNA can be used as bait. Identification of these host proteins associated with viral nucleic acids sheds light on the mechanisms by which the proviral DNA is integrated and factors that regulate viral RNA expression, processing, nuclear export, and translation. Similar to cellular genes, transcription and processing of the retroviral RNA is regulated by specific transcription factors, mediator proteins, splicing factors, and RNA processing machinery.

HIV-1 5′ Untranslated Regions as Bait

To identify host transcription factors that interact with the HIV-1 5′ long terminal repeat (LTR) to promote viral RNA synthesis, Tacheny et al. (2012) took an unbiased in vitro DNA capture approach using a 5′ desthibiotinylated DNA probe consisting of nucleotides 229–455 of the HIV-1 5′LTR. The DNA probe was prebound to streptavidin magnetic beads before being incubated with either HeLa nuclear extracts that were either unstimulated or stimulated by IL1-β. Following elution and reverse-phase LC, DNA-protein complexes were subjected to MS/MS to identify the bound proteins. Each pull-down was done from both stimulated and unstimulated cells in triplicate. A total of 111 proteins were identified in the unstimulated cells (52 transcription-related protiens, 24 nucleic acid–binding proteins not involved in transcription, 32 proteins with no known nuclear function, and 3 unknown proteins). By contrast, 117 proteins were found in nuclei isolated from IL1-β stimulated cells (58 transcription-related protiens, 24 nucleic acid–binding proteins not involved in transcription, 32 proteins with no known nuclear function, and 3 unknown proteins). The lists of interacting proteins were pooled and all proteins that appeared in only a single experiment were discarded. A total of 125 unique proteins were identified from both the unstimulated and stimulated HeLa cells.

Transcription factors identified include NFκB and YB1, which were previously found to bind the HIV-1 5′ LTR, as well as PBX1 and the homeobox protein Meis. The authors became interested in PBX1 and Meis because these proteins form a heterodimeric complex, and the TRANSFAC database revealed that this region of the HIV-1 5′LTR contains two putative Meis-binding sites. To determine the function role of Meis in 5′ LTR-mediated transcription, a point mutation was created in the first Meis-binding site and fused to a luciferase gene. Interestingly, the luciferase expression from the mutant 5′ LTR was increased compared to the wild-type 5′ LTR, suggesting that Meis serves as a transcriptional repressor of HIV-1.

MASS SPECTROMETRY USING VIRAL RNA AS BAIT

The fate of a cellular RNA is generally determined by the composition of the proteins it associates with to form a ribonucleoprotein complex (RNP) (Giorgi and Moore, 2007; Reed, 2003; Trcek and Singer, 2010). Some proteins promote RNA transcription, 5′ and 3′ end processing, splicing, nuclear export, or RNA degradation of immature transcripts. In an analogous fashion, retroviral RNAs are bound by host factors that facilitate transcriptional levels, splicing, capping, maturation, nuclear export, posttranscriptional control of gene expression, and transport through the cytoplasm. A useful method to determine which host proteins interact with the retroviral RNA is to use the viral RNA as bait in an affinity pull-down assay, followed by MS to identify the bound proteins.

HIV-1 Subgenomic RNAs as Bait

Unlike unspliced cellular RNAs, which are retained in the nucleus and degraded to prevent the synthesis of aberrant proteins, complex retroviruses overcome this gatekeeping mechanism by using viral-encoded accessory proteins such as HIV-1 Rev for nuclear export of incompletely spliced viral transcripts (Malim et al., 1989). To identify cellular binding partners of Rev-dependent HIV-1 RNA, U2OS cells stably express one of two HIV-1 subgenomic constructs containing the 5′ and 3′ LTRs, the *gag* coding region and the Rev response element (RRE) (Kula et al., 2011). Each of these constructs contains 24 tandem repeats of MS2 RNA stem-loops positioned within the exon to label the spliced viral RNA or within the intron to label unspliced viral RNA. These cell lines were cotransfected with Tat-CFP, which is required for efficient expression of the viral subgenomic RNAs, and FLAG-MS2-NLS, which contains a nuclear localization signal and binds to MS2 RNA stem-loops, allowing for high-affinity pull-down of the viral RNA. The cells were fractionated and RNA-protein complexes were purified from nuclear extracts via immunoprecipitation using a FLAG antibody. MS of the eluted complexes yielded 32 proteins, most of which are involved in RNA metabolism. The nuclear matrix protein, matrin3 (MATR3) was identified in MS analysis and utilized for pull-downs conducted from both of the HIV-1 stable cell lines.

To examine the role of MATR3 in HIV-1 replication, the authors used RNAi to knockdown MATR3 in cells expressing an HIV-1 reporter that contained luciferase in place of Nef. MATR3 knockdown led to a decrease in Gag expression and the levels of unspliced viral RNA cytoplasmic accumulation were reduced, even in the presence of Rev. Furthermore, overexpression of MATR3 increased Rev activity, suggesting that MATR3 may serve as an essential cofactor for Rev function. This study presented an innovative in vivo approach in which APs using the MS2 coat protein could be used to identify HIV-1 RNA-binding proteins.

Similarly, Marchand et al. (2011) used purified MS2 coat protein fused to maltose binding protein to isolate RNPs from HeLa extracts containing various truncated HIV-1 RNAs as a method to identify host factors that interact with the *tat/rev* exon to regulate splicing. Following purification of complexes on beads, the proteins were identified via nanocapillary LC (nanoLC)–MS/MS. Thirty-three proteins were identified to bind to the stem-loop 2–splice acceptor site 7 (SLS2-A7) required for the processing of pre-mRNAs that produce the *tat* and *rev* transcripts, including proteins involved in cell division, RNA helicases, nucleolin-binding proteins, translation factors, hnRNP proteins, splicing factors, transcription factors, DNA-binding proteins, and interleukins. Through the use of gel-shift assays, the authors found that hnRNP H and hnRNP K directly bind to SLS2-7 and that overexpression of hnRNP K inhibited splicing of the A7 site in *in vitro* splicing assays.

Retroviral 5′Untranslated Regions as Bait

The 5′ untranslated region (UTR) of unspliced retroviral RNA is highly structured and contains *cis*-acting elements that control several distinct functions of the retroviral replication cycle, including reverse transcription (the primer binding site, PBS), packaging (the psi packaging signal, Ψ), and translation (the *gag* AUG). One of the fundamental questions in retrovirology is how these distinct processes are regulated and whether host factors could influence structure function of the 5′ end of the RNA. To compare the host nuclear proteins that bind to the 5′ UTR of various retroviruses, Stake et al. (2015) performed AP using the 5′UTRs of HIV-1, spleen necrosis virus and RSV as bait and human or chicken nuclear lysates as sources of cellular proteins. RNPs were purified on streptavidin-coated beads and eluted RNPs were subjected to MALDI–TOF tandem MS. A total of 185 proteins were identified, 122 were previously shown to interact with retroviruses, and 63 were novel factors not yet observed to interact with retroviruses. There were 25 factors conserved among all three viruses. The identified proteins were subjected to DAVID (http://abcc.ncifcrf.gov) analysis to characterize protein functions, with the primary roles found to be in transcription/RNA processing, translation/ribosome biogenesis, and nucleocytoplasmic transport.

Full-Length HIV-1 RNA Used as Bait

Several of the above approaches were used to efficiently purify viral RNA-host protein complexes but required that the viral RNA was modified to contain a foreign RNA-binding element. To eliminate the need for modifying the unspliced viral RNA, Knoener et al. (2017) used hybridization purification of RNA-protein complexes followed by MS. HIV-1 infected Jurkat cells were treated with formaldehyde to cross-link RNA-protein complexes, which were purified from whole cell lysates using a biotinylated oligonucleotide hybridized to the unspliced HIV-1 RNA, and MS was performed. This approach resulted in identification of 189 host proteins, 90 of which were previously shown to be important for viral replication, and 24 were identified in earlier studies using HIV-1 RNA as bait (Kula et al., 2011; Marchand et al., 2011). Proteins involved in transcription, splicing, nucleocytoplasmic transport, RNA packaging, budding, translation, cytoskeleton, and early HIV replication were identified. Furthermore, siRNA knockdown of several proteins identified in the study, including MOV10, HNRNPC, HSPD1, and RBBP4 decreased fluorophore expression of HIV-1 reporter constructs, suggesting that some of the proteins identified play functional roles in HIV-1 replication (Knoener et al., 2017).

Even though MS serves as a powerful technique to identify proteins that interact with retroviral proteins and nucleic acids, functional experiments need to be performed to determine whether the identified proteins have a biological role in viral replication. These mechanistic studies are required to validate which factors act to promote or interfere with viral infection. Common methods to answer this question include RNAi and genome editing technologies as well as DNA overexpression studies.

RNA INTERFERENCE AND GENOME-WIDE SCREENS TO ASSESS THE CONTRIBUTION OF HOST FACTORS TO VIRUS REPLICATION

Small Interfering RNA Screens

One method to identify host factors that play a functional role in retroviral replication is RNAi, in which host protein expression is knocked down using either short hairpin RNAs (shRNAs) or siRNAs and the effect on the retroviral lifecycle is assessed. The first RNAi screen to find host proteins that potentially serve as cofactors in HIV-1 replication was conducted using an siRNA library targeting 5000 genes (Nguyen et al., 2006). HeLa CD4 β-gal cells, which contain one integrated copy of the HIV-1 LTR upstream of the reporter β-galactosidase, were cotransfected with a pool of siRNAs, were infected with the HIV-IIIb strain of HIV-1 for 3 days, and the level of β-galactosidase was measured in cell lysates to determine the effect of the siRNA knockdowns at all stages of virus replication. Although several novel factors were identified, the authors focused on the serine/threonine kinase Pak3, which belongs to the group I PAK family. A series of siRNA knockdowns of Pak3, Pak1, and Pak2 in several different cell lines was performed to observe the effect on HIV-1 replication. Pak1 and Pak2 had previously been reported to be associated with Nef (Arora et al., 2000; Krautkramer et al., 2004; Fackler et al., 2000; Frost et al., 1998; Nunn and

Marsh, 1996; Renkema et al., 1999, 2001; Sawai et al., 1994; Wiskerchen and Cheng-Mayer, 1996; Raney et al., 2005; Wolf et al., 2001; Kouwenhoven et al., 2013), and downstream effectors of Pak were previously found to play a role in HIV-1 replication (Fortin et al., 2001; Muthumani et al., 2004; Nabel et al., 1988; Yang et al., 1999). Nguyen et al. found that knockdown of Pak1 and Pak3 inhibited virus replication, whereas overexpression of Pak1 facilitated HIV-1 infection of HeLa CD4 β-gal cells. However, the precise mechanism of action of the Paks proteins in HIV-1 replication remains unclear (Markle et al., 2013).

Although, Nguyen et al. conducted the first study using RNAi to identify factors involved in HIV-1 replication, RNAi screens conducted by Brass et al., Konig et al., and Zhou et al. in 2008 followed suit. Brass et al. (2008) transfected a collection of 21,121 siRNAs into TZM-bl cells, a HeLa-derived cell line that expresses a stably integrated Tat-dependent β-galactosidase reporter, CD4, CXCR4, and CCR5, followed by HIV-IIIb infection. To determine whether the siRNAs affected early events of replication from viral entry through viral protein expression, cells were immunostained for CA/p24 antigen at 48h postinfection. To determine which cellular genes are important for late events, culture supernatants from siRNA treated, HIV-infected cells were used to infect naive cells for 24h, and cells were assayed for Tat-dependent β-galactosidase expression. The authors identified 273 genes that altered HIV-1 dependent gene expression, including 36 that were previously shown to be required for replication. Examples of protein classes identified include members of the nuclear pore complex, the Mediator complex, and endoplasmic reticulum, among others.

Konig et al. (2008) transfected an siRNA library targeting 20,000 genes into 293T cells that were transduced with VSV-G pseudotyped HIV-1 that contained a luciferase reporter to create a single-cycle infection. This method allowed the authors to study the postentry early events of replication. The authors identified 295 genes that were crucial for HIV-1 replication, and only 13 genes overlapped with the siRNA screen reported by Brass et al. (2008). The identified genes encoded for proteins important in nucleocytoplasmic transport (e.g., NUPs and importins such as TNPO3), cytoskeletal proteins, DNA damage response mediators, and nucleic acid–binding proteins.

Zhou et al. (2008) transfected siRNAs targeting 19,709 genes into HeLa P4/P5 cells, which express CD4 on the cell surface and contain an integrated HIV LTR-β-galactosidase reporter. The cells were infected with the HXB2 strain of HIV-1, and Tat-induced transcription of the LTR was measured by β-galactosidase expression 48h postinfection to measure early events up to integration and HIV-dependent gene expression. In a separate experiment, the virus was allowed to spread for 96h before β-galactosidase expression was measured as a means to study late events of replication. In this screen, 311 genes were identified, and only 15 genes overlapped with those identified by Brass et al. (2008), which include members of the Mediator complex.

Genovesio et al. developed a 15-dimensional RNAi screen to identify factors previously found to be involved in HIV-1 replication in HeLa cells (Genovesio et al., 2011). The authors assessed whether HIV-1 infection was established in the cells by combining an RNAi array and a confocal imaging approach. This imaging method measured GFP expression for cells in each spot on the array to determine the effect of a particular siRNA on LTR-driven GFP expression. HeLa-CD4+ LTR-GFP cells were transfected with a collection of siRNAs and subsequently infected with HIV-1. Using this approach, the authors identified 56 genes that were required for HIV-1 replication, including Mediator complex proteins, RNA-binding proteins, mitochondrial proteins, and ribosomal proteins, among others.

Dziuba et al. (2012) took a unique approach to identify host factors essential for replication of HIV-1. Before siRNA knockdown was conducted, gene trap insertional mutagenesis was used to identify candidate genes. This method uses a recombinant MLV vector that contains a promoterless neomycin gene that randomly integrates and disrupts host genes. The cells were then infected with several lytic viruses, and DNA was extracted from the surviving cells and sequenced to determine which genes had been disrupted by the siRNAs. The authors then created siRNA libraries targeting each of the disrupted genes and transfected them into CD4+/CCR5+/CXCR4+ HeLa-derived TZM-bl cells before infecting the cells with HIV-1. siRNAs against *dhx8*, *dnaja1*, *gtf2e1*, *kalrn*, *uba3*, *ube2e3*, and *vmp1* caused a 50% reduction in HIV-1 infection.

Short Hairpin RNA Screens

Yeung et al. (2009) performed an shRNA knockdown in Jurkat CD4 T cells, which are more representative of lymphocytes that are infected by HIV-1 in humans. The authors used lentiviral vectors derived from feline immunodeficiency virus to create Jurkat CD4 T-cell lines, each of which contained a single shRNA targeting one of 54,509 genes. Cells transduced with the shRNA lentiviral vectors underwent puromycin treatment for 3 weeks. Eighty percent of the cells did not survive puromycin selection, thereby reducing the number of genes that could be studied. Next, the selected cells were infected with the NL4-3 strain of HIV-1. Because Jurkat T cells die following infection with HIV-1 in culture, a cell surviving HIV-1 infection was suspected to contain an shRNA that targets a gene essential for HIV-1 replication. Using this method, 252 genes were identified with very little overlap with the other three major siRNA screens previously described (Houzet and Jeang, 2011; Kok et al., 2009; Pache et al., 2011). Using a more targeted approach, Williams et al. expressed a library of siRNAs targeting 59 host RNA helicases. These authors identified five helicases that reliably inhibited the production of HIV-1 virions (DDX5, DDX10, DDX17, DDX28, and DDX52) (Williams et al., 2015).

Landi et al. used an shRNA screen to identify host factors involved in the endocytic events utilized by HIV-1 to downregulate its CD4 receptor, which requires the activities of Env, Vpu, and Nef. The experiment was designed to determine which host factors are involved in Nef-mediated downregulation of CD4 in HeLa CD4+ cells (Landi et al., 2014). A VSV-G pseudotyped FIV lentiviral shRNA library targeting 47,000 human transcripts was transduced into cells, which were subjected to puromycin selection, and then transduced with a HIV construct encoding Nef and a GFP reporter. FACS analysis was used to separate cells having high CD4 surface expression, which is an indicator of shRNA-induced inhibition of CD4 downregulation, from cells demonstrating low CD4 surface expression. A total of 55 genes, including those involved in clathrin-mediated trafficking, were identified.

Metaanalysis of siRNA Screens in HIV-1 Replication

A metaanalysis was performed to compare the results of these three siRNA screens to other large-scale searches for HIV-1 cofactors, such as MS and yeast two-hybrid experiments, revealing that 2410 protein-coding human genes may serve as cofactors for HIV-1 replication (Bushman et al., 2009). A database of overlapping proteins observed from several studies can be viewed at http://hostpathogen.org (Bushman et al., 2009). However, these three siRNA screens had minimal overlap and suggesting that the experimental design of each RNAi screen is a critical determinant of the outcome (Goff, 2008; Bushman et al., 2009). Cell type differences, translational readout (β-galactosidase verse luciferase expression), or differences in the HIV-1 constructs used may account for the differences. Furthermore, HeLa and 293T cells, which can serve as model systems to study HIV-1 replication, are not naturally infected by HIV-1 in vivo. Some researchers have proposed that the inconsistencies among the factors identified in RNAi screens may be due to disruption of normal cellular processes leading to cellular toxicity, rather than a direct effect on host factors that are essential for HIV-1 replication.

Targeted RNAi Screens

The benefit of RNAi screens is that researchers can target specific classes of genes to determine their roles in retroviral replication. Jiang et al. (2014) used shRNAs to specifically target kinases to examine whether any were involved in HIV-1 replication. An shRNA library targeting 474 kinases was introduced into a 293T cell line containing the HIV-1 LTR-hRluc/HSV TK-hLuc dual reporter system. HIV-1 expression was measured by calculating hRLuc intensity divided by hLuc fluorescence. Four kinases were found to specifically inhibit the HIV-1 directed gene expression: AK1, EphB2, PRKACB, and CDK5R2. Further validation experiments revealed that AK1 and PRKACB overexpression stimulated HIV-1 LTR activity (Jiang et al., 2014).

The RNAi screens described up to this point were focused on identifying host factors that promote retroviral replication. Liu et al. (2011), however, sought to use this technique to discover HIV-1 restriction factors present in CD4+ HeLa cells. siRNAs targeting 19,121 human genes were transfected into cells that were subsequently infected with a single-cycle HIV-1 virus pseudotyped with HIV-2 Env and expressing a GFP reporter to study early events of viral replication. Infection was measured by GFP fluorescence, and an increase in fluorescence intensity in siRNA-treated cells indicated that the siRNA targeted an HIV-1 restriction factor. The authors identified 114 genes that potentially restrict HIV-1 infection and will require further study. Kameoka et al. (2007) used an siRNA library targeting 257 genes in human acute monocytic leukemia cells and found six repressive factors. The authors concluded that AP2alpha, ARF6, PARD6alpha, and JAK1 were inhibitory factors that act during early events of infections, whereas Axin1 and DUSO1 had suppressive effects during the last phase of replication.

Investigators interested in discovering cellular proteins that repress HIV-1 gene expression during latency used a pooled shRNA screening regimen (Besnard et al., 2016). Cells latently infected with HIV-1 were transduced with a lentiviral shRNA library and then incubated with CD3/CD28 to induce expression of the latent viral genome. Genes that either promoted or suppressed HIV-1 latency were found (335 and 330 genes, respectively). Factors previously known to be involved in HIV-1 latency were found (e.g., p-TEFb and the PRC2 complex) confirming that the screen was functional. The authors focused more in-depth studies on the mammalian target of rapamycin complex, which had not been discovered as a factor involved in regulating HIV-1 latency.

Although most of the RNAi screens discussed thus far focused on human cellular factors that interact with HIV-1, Wen et al. sought to compare the set of genes involved in the assembly of MPMV with those required for HIV-1 assembly because these viruses follow distinct morphogenetic pathways. MPMV is a type B/D retrovirus that assembles immature capsids in the cytoplasm of an infected cell, whereas HIV-1 is a type C retrovirus, in which particle assembly occurs at the plasma membrane. The goal of this screen was to determine whether the host factors involved in viral assembly differed between these two distinct pathways (Wen et al., 2014). An siRNA library targeting 140 genes involved in membrane trafficking was expressed in Cos-1 cells (MPMV) or HeLa cells (HIV-1), and subsequently, either an MPMV–GFP provirus or an HIV-1 NL4-3-EGFP provirus was transfected. After 48 h, cells were assessed for GFP expression and cell viability, and culture supernatants were subjected to reverse transcription assay to evaluate the efficiency of virus particle release. Seventeen genes were unique to HIV-1, 28 genes were specific to MPMV, and 24 genes were found to be common to both viruses. STRING 9.05 analysis clustered the identified genes into the following three categories: clathrin-mediated vesicle trafficking, actin cytoskeleton regulation, and proteins involved in phosphoinositide metabolism.

Genome-Wide Analysis of Retrotransposition in Yeast

An informative comparative cell biology approach has taken advantage of the powerful, well-established genetic system in yeast to identify candidate cellular genes that influence LTR retrotransposons, which have many replication steps in common with retroviruses (Aye et al., 2004; Irwin et al., 2005; Rai et al., 2017; Griffith et al., 2003). Genome-wide studies of the retrotransposon Ty3 were conducted in strains deleted in 4457 nonessential genes of *Saccharomyces cerevisiae*, revealing dependency of retrotransposition on 130 genes-encoding proteins involved in chromatin regulation and modification, gene transcription, processing of RNAs, translation, posttranslational modification, control of cell cycle, nuclear transport, and vesicular trafficking. The authors identified a subset of genes with overlapping roles in retrovirus replication, several of which were found to affect replication of other retroviruses, including HIV-1, MLV, MPMV, and RSV (Irwin et al., 2005). A follow-up study focusing on the role of cytoplasmic RNA-processing factors, which were identified in a previous genome-wide screen, concluded that Ty3 virus-like particle (VLP) assembly in cytoplasmic foci depends on components of P bodies (Beliakova-Bethell et al., 2006). Using a similar approach, a library of 3, 004 single gene deletions in *Saccharomyces pombe* was probed, and 61 genes were identified that played a role in retrotransposition (Rai et al., 2017). A high degree of overlap was noted among genes identified in Tf1 and Ty3 screens, even though these LTR retrotransposons are not genetically closely related, raising the possibility that these gene products could also be important in the replication of retroviruses.

HOST GENOME EDITING USING CRISPR/Cas-9 TO FIND RETROVIRUS DEPENDENCY GENES

RNAi is the most common genome-wide screen that has been used to date to search for host factors involved in retroviral replication. By contrast, very few screens have been conducted that target the host DNA. In one such experiment, Park et al. (2017) used a CRISPR/Cas-9 screen of human CD4+ T cells to identify host factors required for HIV-1 replication. A lentiviral library of single-guide RNAs (sgRNAs) targeting 18,543 human genes was infected into CD4+ T cells that expressed HIV-1 LTR-GFP, CCR5, and Cas-9. Following puromycin selection, cells that contained the sgRNAs were infected with the CCR5-tropic HIV-1 strain JR-CSF. Viable cells that did not express GFP were isolated by using FACS analysis because these cells contained sgRNAs that targeted a host factor important for viral replication. Five genes were identified: *CD4*, *CCR5*,

TPST2, *SLC35B2*, and *ALCAM*. TPST2 and SLC35B2 were determined to be important for the sulfation of the CCR5 receptor that interacts with gp-120 to facilitate HIV-1 entry, whereas the cell adhesion molecule, ALCAM, is important for T cell aggregation and HIV-1 cell-to-cell spread. This experiment was the first CRISPR screen used to identify host factors involved in retroviral replication. Perhaps this technology will be more commonly used in the future for genome-wide host factor screens that could potentially lead to the discovery of novel host gene that could be developed as drug targets to treat HIV-1 infect patients.

GAIN-OF-FUNCTION GENETIC APPROACHES USING cDNA OVEREXPRESSION SCREENS

Methods that disrupt genes, such as RNAi and CRISPR, rely on a loss-of-function approach. However, gain-of-function approaches have also yielded important discoveries of retrovirus restriction and dependency factors. TRIM5alpha, which restricts the establishment of HIV-1 infection in a species-specific manner, was found by taking advantage of the observation that HIV-1 can infect humans and other primates, but not Old World monkeys (Stremlau et al., 2004). In this experiment, HeLa cells expressing a complementary DNA library from rhesus macaques, an Old World monkey species, were infected with an HIV-1 strain containing GFP as a reporter. Cells that were unable to be infected with HIV-1-GFP were cloned and found to contain rhesus TRIM5alpha.

In a second manuscript using a gain-of-function approach to identify HIV-1 host factors, Nguyen et al. (2007) transfected HeLa CD4 β-galactosidase cells with a cDNA library of 15,000 genes to determine the effect of overexpression of these genes on replication of the HIV-1 HIV-IIIb strain. cDNAs that led to a twofold increase of β-galactosidase expression were considered to be positive hits. There were 315 genes identified in this screen, which were sorted into families including proteins involved in microtubule polymerization, translation elongation, mRNA binding, apoptosis regulation, and membrane proteins.

IDENTIFICATION OF CELLULAR RNAs THAT INTERACT WITH RETROVIRUSES

In addition to host proteins, cellular RNAs have been identified that interact with retroviruses. To identify host RNAs that interact with retroviral components, several techniques have been used, including microarray analysis, SOLiD (sequencing by oligonucleotide ligation and detection) sequencing, cross-linking immunoprecipitation-sequencing (CLIP-seq), photoactivatable ribonucleoside-enhanced cross-linking and immunoprecipitation (PAR-CLIP), and high-throughput sequencing.

Host RNAs Packaged Into Retrovirus Particles

It has been known for decades that host-encoded RNAs are packaged into retrovirus particles, including the U6 small nuclear RNA, the 7SL small noncoding RNA, 5S ribosomal RNA, microRNAs, and transfer RNAs (Giles et al., 2004; Eckwahl et al., 2015, 2016a,b; Berkowitz et al., 1996; Bishop et al., 1970a,b; Onafuwa-Nuga et al., 2005, 2006; Cen et al., 2002; Jiang et al., 1993; Zaitseva et al., 2006; Houzet et al., 2007; Pavon-Eternod et al., 2010; Schopman et al., 2012; Brameier et al., 2013; Tian et al., 2007).

Virus particles assembled in the absence of Ψ-containing viral RNA compensate for the lack of a packageable retroviral genome by incorporating an equivalent amount of RNA derived from the cell. Rulli et al. (2007) used microarray analysis to identify host RNAs encapsidated into VLPs in the presence or absence of Ψ-containing RNA from MLV compared to Rev-dependent HIV-1 VLPs. Cell lysates and VLPs were collected from 293T cells transfected with replication deficient Ψ-plus or Ψ-minus MLV constructs. In parallel, a Rev-independent HIV-1 Gag construct was used (Schneider et al., 1997) to study RNA incorporation into HIV-1 VLPs. RNA was isolated from VLPs, reverse transcribed, and the resulting cDNA was used as a template for synthesis of biotinylated cRNA. The biotinylated cRNA was hybridized to an Affymetrix human U133 array that contained over 47,000 human transcripts, and the signal was amplified using an antibody against biotin. For MLV, the most abundant host mRNAs encapsidated into VLPs in the absence of packageable viral RNA included RNAs that encode ribosomal proteins, TATA element-modulatory factor 1, and a mitochondrial translocase (Rulli et al., 2007). Similarly, the most abundant host mRNAs packaged into HIV-1 VLPs were of ribosomal origin. The ASB-1 and PLEKHB2 RNAs were enriched in the Ψ^- MLV and HIV-1 VLPs. The role of these RNAs, if any, in MLV or HIV-1 replication has not yet been demonstrated.

In a related study, Pavon-Eternod et al. (2010) utilized microarray analysis to identify tRNAs encapsidated into HIV-1 virions and VLPs. Each retrovirus encapsidates a specific host tRNA that binds to the PBS of the 5′ UTR of the genomic RNA and serves as the primer for reverse transcription. tRNA Lysine ($tRNA^{Lys}$) is specifically packaged into HIV-1 virions and VLPs (Jiang et al., 1993; Khorchid et al., 2000; Wei et al., 2005; Gabor et al., 2002). The goal of this experiment was to identify other tRNAs, in addition to $tRNA^{Lys}$, which are packaged into virions. 293T cells were transfected with either a construct expressing the BH10 strain of HIV-1 or with a construct that expressed all HIV-1 proteins except for Gag and Gag-Pol. RNA extracted from the wild-type BH10 virions and VLPs were subjected to microarray analysis using probes for human, bacterial, and yeast tRNAs. As expected, the most abundant tRNAs present in virions were $tRNA^{Lys}$ isoacceptors, comprising 35% of the tRNAs identified. Interestingly, 11% of total tRNAs identified in HIV-1 virions was $tRNA^{Asn(GTT)}$ and 4% was $tRNA^{Ile(UAU)}$. The concentration of tRNA packaged into VLPs was reduced by 84% compared to wild-type virions, suggesting that tRNA encapsidation is dependent on the presence of Gag-Pol, as demonstrated in earlier studies (Cen et al., 2001; Javanbakht et al., 2003; Khorchid et al., 2000). Based on the results of these two microarray studies, this type is useful to identify RNAs packaged into retroviral particles. However, recent technological advances in high-throughput sequencing have allowed researchers to identify a wider range of RNAs packaged into retroviral particles.

RNA Sequencing Techniques

Schopman et al. (2012) utilized SOLiD to identify host miRNAs present in HIV-1 virions. Two hundred and ninety three T cells were transfected with a plasmid containing the LAI strain of HIV-1, and virions produced from these cells were used to infect a variety of cells: primary SupT1 T cells, immature monocyte-derived dendritic cells, CD4 T cells, CD4act (PHA activated in the presence of IL-2), and mature monocyte-derived dendritic cells. Virions were collected from the culture supernatant of the infected primary cells and subjected to SOLiD sequencing. A total of 2,974,136 miRNA sequences were identified from five different cell lines. While the 10 most abundant miRNAs in each cell type differed, the miRNAs common to all cells include the let-7 family, miRNA-21, miRNA-26a, and miRNA-93. The differences in the most abundant miRNAs common to each cell type may contribute to the ability of the virus to establish infection in specific cells. To identify small RNAs in HIV-1 virions, supernatants were collected from the SupT1 cells, which produced the highest amount of virions and RNA was isolated, and multiple species of small RNAs were found, in order of most to least abundant, transposon/repeat RNAs, miRNA, tRNA, rRNA, small cytoplasmic RNA, and snRNA. HIV-1 virions were enriched 10-fold in 38 miRNAs. As an example, miRNA-382, which has been implicated in HIV-1 latency (Huang et al., 2007), was highly enriched in virions compared to the infected cell lysates. These results were validated by analyzing the abundance of $tRNA^{Lys}$ and 7SL RNA in virions as a control. As expected, these RNAs were enriched in the virus particles, providing support for applicability of this method to study the RNA content of virions.

Similarly, Brameier et al. (2013) used ultradeep sequencing to identify small RNAs packaged into SIV $_{mac}$239 virions that were produced from the human T cell line C8166 isolated via centrifugation and PEG precipitation. The virions were subjected to RNase and DNase treatment, and the packaged RNA was extracted and subjected to ultradeep sequencing. More than 428,000 sequences were identified as SIV-derived small RNAs, and 12 of these sequences, which map to regions of the LTR, *pol*, *vif*, *env*, *tat/rev*, and *nef*, were determined to be highly enriched in SIV virions. The authors hypothesized that these virus-derived sequences may have been processed by the host cell RNAi machinery but could find no evidence of a hairpin that would mark them as miRNAs. Similarly, small nonviral RNAs that were present in virions were also analyzed for the presence of human miRNAs. Most miRNAs were present in a low abundance; however, miR-10b-5p was present at 37-fold higher levels compared to supernatant collected from mock-infected cells. Other differentially represented RNAs included U2 and U12 snRNAs, which were decreased in virions, and 7SL, $tRNA^{Lys}$, and snaR-A14, which were enriched. The RNA polymerase III transcript snR-A14 is a member of the small ILF3/NF90-associated RNAs (snaR) that bind to nuclear factor 90, which has been implicated in regulating HIV-1 infection and latency (Parrott and Mathews, 2007; Parrott et al., 2011; Hoque et al., 2011).

Eckwahl et al. utilized high-throughput sequencing to identify RNAs packaged into MLV virions produced from chronically infected NIH3T3 cells (Eckwahl et al., 2015) and HIV-1 virions produced from CEM-SS cells (Eckwahl et al., 2016a). Following ultracentrifugation and micrococcal nuclease digestion, RNA was extracted from MLV virions sequenced (Eckwahl et al., 2015). Whereas 47% of the reads mapped to the MLV genome, 41% mapped to the mouse genome, with most of the reads corresponding to noncoding RNAs (ncRNAs) and the LTR of the VL30 retrotransposons. MLV virions contained abundant levels of host RNAs such as vaults (vt-RNAs), 4.5S RNAs, B2 SINEs, tRNAs, and snoRNA and snRNA precursors, in addition to previously identified host RNAs, 7SL, Y RNAs, U6, and SINEs (Onafuwa-Nuga et al., 2005). Similar methods were used to identify host RNAs encapsidated into HIV-1 virions (Eckwahl et al., 2016a). Approximately 35% of the RNA found in HIV-1 virions contained SINE and LINE RNAs, rRNAs, ncRNAs, intronic RNAs, tRNA, and snRNA precursors and transcripts from ribosomal DNA intergenic spacers. In this study, 7SL was highly enriched in virions, and 5S rRNA, U1 and U6 snRNAs, Y RNAs, Alu element RNAs, and vtRNA1-1 RNA were also identified.

Whereas these two studies aimed to identify host RNAs incorporated into MLV and HIV-1 virions, Kutluay et al. (2014) used PAR-CLIP and HITS-CLIP (high-throughput sequencing of RNA isolated by CLIP-seq) to identify RNA sequences that interact with Gag during virion genesis. Proviral HIV-1_{NL4-3} and HIV-1_{NDK} strains modified to contain three copies of an HA tag in the MA region of Gag were transfected into 293T cells. Cells and virions were UV cross-linked, lysed, Gag-RNA complexes were immunoprecipitated, and the RNA was subjected to high-throughput sequencing. This method allowed for the identification of Gag binding sites at single-nucleotide resolution. In addition to binding the 5′ UTR, where the Ψ packaging signal is located, intracytoplasmic Gag protein was also bound to other regions of the HIV-1 genome with high frequency, including the RRE, the Nef start codon, and UTRs of U3. In immature virions, however, Gag was more distributed across the genome as opposed to being concentrated on Ψ and the RRE. By contrast, the NC domain preferentially bounds to Ψ and the RRE in mature virions, suggesting that the region of the viral RNA to which Gag binds changes during virion formation and maturation. Similarly, the authors used this method to identify the cellular RNAs to which Gag binds. In cells, Gag bounds primarily to tRNAs, with lower amounts of mRNA, 7SL RNA, introns, rRNA, miRNA, lincRNA, and other RNAs also noted. In virions, mRNA was the predominant, representing >50% of the cellular RNA species bound to Gag. Furthermore, the authors concluded that the MA domain of Gag is the primary binding site for tRNAs (Kutluay et al., 2014), confirming prior reports that suggest that MA binding to membranes is facilitated by tRNAs, as reviewed in Alfadhli and Barklis (2014).

CONCLUDING REMARKS

RNAi screening experiments to identify host factors involved in retroviral replication have evolved and became more sophisticated over time. Early RNAi screens aiming to discover novel host factors required for HIV-1 infection spanned the entire human genome, leading to the discovery of hundreds of different factors that may be essential. However, the caveat of these screens was that very few of the factors were found in common among studies performed by different sets of investigators, which could be explained by different experimental conditions or because the siRNAs used to deplete host factors affected genes required for cellular functions that indirectly affected retroviral replication. Later RNAi screens used more rigorous validation studies or focused on specific categories of host factors, allowing the researchers to validate a smaller number of hits. Although there is a benefit to having a more limited list of host factors to probe for their functional relevance to retrovirus replication, this targeted approach is biased and restricts the breadth of cellular genes examined, raising concern that essential genes could be missed. Nonetheless, regardless of the size of an RNAi screen, the identified genes need to be cross-referenced with other large-scale experiments, such as MS screens, and with functional validation experiments to determine the mechanism by which the host genes interfere with or promote virus replication.

Together, the data presented in this chapter highlight the different methods used to identify host factors that interact with retroviral proteins and nucleic acids to facilitate or restrict replication. The accompanying tables (Tables 14.1 and 14.2) provide a comprehensive list of approaches used to identify retrovirus-interacting factors and the references containing the primary data. Most of the data presented are available in searchable online databases, as listed in Table 14.3. The results of these studies have opened the door to further elucidate the mechanisms by which retroviruses propagate by hijacking various host factors at different stages of the replication cycle. Further understanding of the host factors necessary for efficient replication will aid in the identification of other novel drug targets for treatment of retrovirus infections.

TABLE 14.2 **Studies Using Nucleic Acid-Based Methods to Identify Host Proteins and Nucleic Acids that Interact with Retroviruses.**

Type of Study	Cell Type	Viral Construct	Cellular Factors Identified	Citation
Mass spectrometry; viral DNA as bait	HeLa	HIV-1 5′LTR	125 proteins	Tacheny et al. (2012)
Mass spectrometry; viral DNA as bait	HeLa	HIV-1	33 proteins	Marchand et al. (2011)
Mass spectrometry; viral DNA as bait	HEK293T	Subgenomic HIV-1 constructs	32 proteins	Kula et al. (2011)
Mass spectrometry; viral DNA as bait	DF1 (SNV and RSV); HeLa (HIV-1, SNV, RSV)	5′UTR of HIV-1, SNV, RSV	185 proteins	Stake et al. (2015)
Mass spectrometry; viral DNA as bait	Jurkat T cells	HIV-1	189 proteins	Knoener et al. (2017)
Genome-wide Screen	*S. cerevisiae* deletion mutants	Ty1 retrotransposon	101 genes	Griffith et al. (2003)
Genome-wide Screen	*S. cerevisiae* insertional mutants	Ty3 retroelement	53 loci	Aye et al. (2004)
Genome-wide Screen	*S. cerevisiae* deletion mutants	Ty3 retroelement	64 strains- increased transposition, 66 strains decreased transposition	Irwin et al. (2005)
Genome-wide Screen	*S. pombe* deletion mutants	Tf1 LTR-retrotransposon	61 genes promote integration	Rai et al. (2017)
siRNA	J111 (human acute monocytic leukemia)	HIV-1	6 restriction factors	Kameoka et al. (2007)

Continued

TABLE 14.2 **Studies Using Nucleic Acid-Based Methods to Identify Host Proteins and Nucleic Acids that Interact with Retroviruses—cont'd**

Type of Study	Cell Type	Viral Construct	Cellular Factors Identified	Citation
siRNA	HeLa-derived TZM-bl	HIV-1	273 genes	Brass et al. (2008)
siRNA	HEK293T	HIV-1	295 genes	Konig et al. (2008)
siRNA	HeLa P4/P5	HIV-1	311 genes	Zhou et al. (2008)
RNAi meta analysis		HIV-1	Comparison of Brass et al. (2008), Konig et al. (2008) and Zhou et al. (2008)	Bushman et al. (2009). http://hostpathogen.org
shRNA	Jurkat T cells	HIV-1	252 genes	Yeung et al. (2009)
siRNA	LTR GFP HeLa CD4+; GFP-torsin HeLa; wild-type HeLa	HIV-1	56 genes	Genovesio et al. (2011)
siRNA	HeLa-CD4+	HIV-1 restriction factors	114 genes	Liu et al. (2011)
siRNA	HeLa-derived TZM-bl CD4+/CCR5+/CXCR4+	HIV-1	DNAJA1, DHX8, GTF2E1, UBA3, UBE2E3, KALRN, and VMP1	Dziuba et al. (2012)
shRNA	HeLa CD4+	HIV-1 Nef	55 genes	Landi et al. (2014)
shRNA	HEK293T expressing a stable HIV 5′LTR-hRluc/HSV TK-hLuc reporter	HIV-1	AK1, EphB2, PRKACB, and CDK5R2	Jiang et al. (2014)
siRNA	HeLa (HIV-1); Cos-1 (MPMV)	HIV-1 and MPMV	24 genes	Wen et al. (2014)
siRNA	HeLa and HeLa-derived TZM-bl	HIV-1	5 helicases	Williams et al. (2015)

shRNA	J-Lat 5A8-contains an integrated HIV-GFP reporter	HIV-1	335 genes promoted latency, 330 genes inhibited latency	Besnard et al. (2016)
CRISPR	CD4+ T cells	HIV-1	5 genes	Park et al. (2017)
cDNA overexpression	HeLa	HIV-1	TRIM5alpha	Stremlau et al. (2004)
cDNA overexpression	HeLa CD4 β-gal	HIV-1	315 genes	Nguyen et al. (2007)
Microarray of host RNAs	HEK293T	MLV and HIV-1 VLPs	MLV- RNAs that encode ribosomal proteins, mitochondrial translocase, TATA element–modulator factor 1 HIV-1–ribosomal proteins; Both ASB-1 and PLEKHB2 RNAs	Rulli et al. (2007)
Microarray of host tRNAs	HEK293T	HIV-1 virions and VLPs	$tRNA^{Lys}$, $tRNA^{Asn}$, $tRNA^{Ile}$	Pavon-Eternod et al. (2010)
SOLiD sequencing	SupT1 T cell; immature monocyte–derived dendritic cells; CD4+, $CD4_{act}$ (PHA activated in the presence of IL-2); mature monocyte–derived dendritic cells	HIV-1 virions and cells	38 miRNAs in virus particles	Schopman et al. (2012)

Continued

TABLE 14.2 Studies Using Nucleic Acid-Based Methods to Identify Host Proteins and Nucleic Acids that Interact with Retroviruses—cont'd

Type of Study	Cell Type	Viral Construct	Cellular Factors Identified	Citation
Ultradeep sequencing	C8166, human T cells	$SIV_{mac}239$ virions	7SL-RNA, 428,000 sequences-SIV genome, snaR-A14,12 virus-derived small RNAs, $tRNA^{Lys3}$ and its isoacceptors	Brameier et al. (2013)
CLIP-seq of host RNAs bound to Gag	HEK293T	HIV-1 virions and cells	Cells-tRNAs, mRNA, 7SL RNA, introns, rRNA, miRNA, lincRNA. Virions-mRNAs	Kutluay et al. (2014)
High-throughput sequencing	NIH3T3-MLV (chronically infected)	MLV virions	noncoding RNAs, LTR of VL30 retrotransposons, 7SL, Y RNAs, tRNAs, vault RNAs, snoRNAs, snRNA precursors, 4.5S RNAs, B2 SINEs, U6, and SINEs	Eckwahl et al. (2015)
High-throughput sequencing	CEM-SS	HIV-1 virions	SINEs, LINEs, U6, Y RNAs, ribosomal DNA intergenic spacers, 5S rRNA, rRNAs, noncoding RNAs, introns, tRNAs, snRNA precursors, vault RNA1-17SL, U1, Alu elements	Eckwahl et al. (2016a)

TABLE 14.3 Publicly Available Databases of HIV-1 Gene and Protein Network Data From Mass Spectrometry and Genome-Wide Studies.

Name	Database Website	# Interactions Identified	Citations
GPS-Prot	http://gpsprot.org/	7765 nonredundant HIV-1 protein interactions (Sept 2016)	Fahey et al. (2011)
HIV-1 Human Protein Interaction Database (HHPID)	https://www.ncbi.nlm.nih.gov/genome/viruses/retroviruses/hiv-1/interactions/	7762 interactions with 15,665 interaction descriptions (Aug 2016)	Fu et al. (2009) Ptak et al. (2008) Pinney et al. (2009)
IntAct Molecular Interaction Database	http://www.ebi.ac.uk/intact/	99807 interactors in 506367 interactions (IntAct View version 4.2.7)	Hermjakob et al. (2004)
VirusMINT	http://mint.bio.uniroma2.it/virusmint/	557 proteins encoded by 149 viruses. 48 HIV-1 host protein interactions	Chatr-Aryamontri et al. (2009)
Overlap among genome-wide studies of HIV infection	http://www.hostpathogen.org	2410 host protein-coding genes identified to affect HIV-1 replication (Apr, 2009)	Bushman et al. (2009)
HIVed	http://hivlatency.erc.monash.edu/	14,318 genes/proteins and 13,649 HIV-1 interactions (visited Aug 2017)	Li et al. (2017)

ACKNOWLEDGMENTS

We are grateful to have received support from the National Institutes of Health awards R01CA76534 (LJP) and F30 CA214010 (ECC) as well as the Penn State College of Medicine Medical Scientist Training Program and the Department of Medicine.

REFERENCES

Alfadhli, A., Barklis, E., 2014. The roles of lipids and nucleic acids in HIV-1 assembly. Front. Microbiol. 5, 253.

Allouch, A., Cereseto, A., 2011. Identification of cellular factors binding to acetylated HIV-1 integrase. Amino Acids 41, 1137–1145.

Arora, V.K., Molina, R.P., Foster, J.L., Blakemore, J.L., Chernoff, J., Fredericksen, B.L., Garcia, J.V., 2000. Lentivirus Nef specifically activates Pak2. J. Virol. 74, 11081–11087.

Aye, M., Irwin, B., Beliakova-Bethell, N., Chen, E., Garrus, J., Sandmeyer, S., 2004. Host factors that affect Ty3 retrotransposition in *Saccharomyces cerevisiae*. Genetics 168, 1159–1176.

Bacharach, E., Gonsky, J., Alin, K., Orlova, M., Goff, S.P., 2000. The carboxy-terminal fragment of nucleolin interacts with the nucleocapsid domain of retroviral gag proteins and inhibits virion assembly. J. Virol. 74, 11027–11039.

Beliakova-Bethell, N., Beckham, C., Giddings Jr., T.H., Winey, M., Parker, R., Sandmeyer, S., 2006. Virus-like particles of the Ty3 retrotransposon assemble in association with P-body components. RNA 12, 94–101.

Berkowitz, R., Fisher, J., Goff, S.P., 1996. RNA packaging. Curr. Top. Microbiol. Immunol. 214, 177–218.

Besnard, E., Hakre, S., Kampmann, M., Lim, H.W., Hosmane, N.N., Martin, A., Bassik, M.C., Verschueren, E., Battivelli, E., Chan, J., Svensson, J.P., Gramatica, A., Conrad, R.J., Ott, M., Greene, W.C., Krogan, N.J., Siliciano, R.F., Weissman, J.S., Verdin, E., 2016. The mTOR complex controls HIV latency. Cell Host Microbe 20, 785–797.

Beyer, A.R., Bann, D.V., Rice, B., Pultz, I.S., Kane, M., Goff, S.P., Golovkina, T.V., Parent, L.J., 2013. Nucleolar trafficking of the mouse mammary tumor virus gag protein induced by interaction with ribosomal protein L9. J. Virol. 87, 1069–1082.

Bishop, J.M., Levinson, W.E., Quintrell, N., Sullivan, D., Fanshier, L., Jackson, J., 1970a. The low molecular weight RNAs of Rous sarcoma virus. I. The 4 S RNA. Virology 42, 182–195.

Bishop, J.M., Levinson, W.E., Sullivan, D., Fanshier, L., Quintrell, N., Jackson, J., 1970b. The low molecular weight RNAs of Rous sarcoma virus. II. The 7 S RNA. Virology 42, 927–937.

Brameier, M., Ibing, W., Hofer, K., Montag, J., Stahl-Hennig, C., Motzkus, D., 2013. Mapping the small RNA content of simian immunodeficiency virions (SIV). PLoS One 8, e75063.

Brass, A.L., Dykxhoorn, D.M., Benita, Y., Yan, N., Engelman, A., Xavier, R.J., Lieberman, J., Elledge, S.J., 2008. Identification of host proteins required for HIV infection through a functional genomic screen. Science 319, 921–926.

Bregnard, C., Zamborlini, A., Leduc, M., Chafey, P., Camoin, L., Saib, A., Benichou, S., Danos, O., Basmaciogullari, S., 2013. Comparative proteomic analysis of HIV-1 particles reveals a role for Ezrin and EHD4 in the Nef-dependent increase of virus infectivity. J. Virol. 87, 3729–3740.

Bushman, F.D., Malani, N., Fernandes, J., D'orso, I., Cagney, G., Diamond, T.L., Zhou, H., Hazuda, D.J., Espeseth, A.S., Konig, R., Bandyopadhyay, S., Ideker, T., Goff, S.P., Krogan, N.J., Frankel, A.D., Young, J.A., Chanda, S.K., 2009. Host cell factors in HIV replication: meta-analysis of genome-wide studies. PLoS Pathog. 5, e1000437.

Cen, S., Javanbakht, H., Kim, S., Shiba, K., Craven, R., Rein, A., Ewalt, K., Schimmel, P., Musier-Forsyth, K., Kleiman, L., 2002. Retrovirus-specific packaging of aminoacyl-tRNA synthetases with cognate primer tRNAs. J. Virol. 76, 13111–13115.

Cen, S., Khorchid, A., Javanbakht, H., Gabor, J., Stello, T., Shiba, K., Musier-Forsyth, K., Kleiman, L., 2001. Incorporation of lysyl-tRNA synthetase into human immunodeficiency virus type 1. J. Virol. 75, 5043–5048.

Chan, E.Y., Qian, W.J., Diamond, D.L., Liu, T., Gritsenko, M.A., Monroe, M.E., Camp 2nd, D.G., Smith, R.D., Katze, M.G., 2007. Quantitative analysis of human immunodeficiency virus type 1-infected CD4+ cell proteome: dysregulated cell cycle progression and nuclear transport coincide with robust virus production. J. Virol. 81, 7571–7583.

Chatr-Aryamontri, A., Ceol, A., Peluso, D., Nardozza, A., Panni, S., Sacco, F., Tinti, M., Smolyar, A., Castagnoli, L., Vidal, M., Cusick, M.E., Cesareni, G., 2009. VirusMINT: a viral protein interaction database. Nucleic Acids Res. 37, D669–D673.

Cherepanov, P., Maertens, G., Proost, P., Devreese, B., Van Beeumen, J., Engelborghs, Y., De Clercq, E., Debyser, Z., 2003. HIV-1 integrase forms stable tetramers and associates with LEDGF/p75 protein in human cells. J. Biol. Chem. 278, 372–381.

Chertova, E., Chertov, O., Coren, L.V., Roser, J.D., Trubey, C.M., Bess Jr., J.W., Sowder, R.C., Barsov, E., Hood, B.L., Fisher, R.J., Nagashima, K., Conrads, T.P., Veenstra, T.D., Lifson, J.D., Ott, D.E., 2006. Proteomic and biochemical analysis of purified human immunodeficiency virus type 1 produced from infected monocyte-derived macrophages. J. Virol. 80, 9039–9052.

Cimarelli, A., Luban, J., 1999. Translation elongation factor 1-alpha interacts specifically with the human immunodeficiency virus type 1 Gag polyprotein. J. Virol. 73, 5388–5401.

Coiras, M., Camafeita, E., Urena, T., Lopez, J.A., Caballero, F., Fernandez, B., Lopez-Huertas, M.R., Perez-Olmeda, M., Alcami, J., 2006. Modifications in the human T cell proteome induced by intracellular HIV-1 Tat protein expression. Proteomics 6 (Suppl. 1), S63–S73.

Cooper, J., Liu, L., Woodruff, E.A., Taylor, H.E., Goodwin, J.S., D'aquila, R.T., Spearman, P., Hildreth, J.E., Dong, X., 2011. Filamin A protein interacts with human immunodeficiency virus type 1 Gag protein and contributes to productive particle assembly. J. Biol. Chem. 286, 28498–28510.

Deboer, J., Jagadish, T., Haverland, N.A., Madson, C.J., Ciborowski, P., Belshan, M., 2014. Alterations in the nuclear proteome of HIV-1 infected T-cells. Virology 468–470, 409–420.

Dziuba, N., Ferguson, M.R., O'brien, W.A., Sanchez, A., Prussia, A.J., Mcdonald, N.J., Friedrich, B.M., Li, G., Shaw, M.W., Sheng, J., Hodge, T.W., Rubin, D.H., Murray, J.L., 2012. Identification of cellular proteins required for replication of human immunodeficiency virus type 1. AIDS Res. Hum. Retrovir. 28, 1329–1339.

Eckwahl, M.J., Arnion, H., Kharytonchyk, S., Zang, T., Bieniasz, P.D., Telesnitsky, A., Wolin, S.L., 2016a. Analysis of the human immunodeficiency virus-1 RNA packageome. RNA 22, 1228–1238.

Eckwahl, M.J., Sim, S., Smith, D., Telesnitsky, A., Wolin, S.L., 2015. A retrovirus packages nascent host noncoding RNAs from a novel surveillance pathway. Genes Dev. 29, 646–657.

Eckwahl, M.J., Telesnitsky, A., Wolin, S.L., 2016b. Host RNA packaging by retroviruses: a newly synthesized story. mBio 7 e02025–15.

Edwards, A.M., Kus, B., Jansen, R., Greenbaum, D., Greenblatt, J., Gerstein, M., 2002. Bridging structural biology and genomics: assessing protein interaction data with known complexes. Trends Genet. 18, 529–536.

Engeland, C.E., Brown, N.P., Borner, K., Schumann, M., Krause, E., Kaderali, L., Muller, G.A., Krausslich, H.G., 2014. Proteome analysis of the HIV-1 Gag interactome. Virology 460–461, 194–206.

Engeland, C.E., Oberwinkler, H., Schumann, M., Krause, E., Muller, G.A., Krausslich, H.G., 2011. The cellular protein lyric interacts with HIV-1 Gag. J. Virol. 85, 13322–13332.

Fackler, O.T., Lu, X., Frost, J.A., Geyer, M., Jiang, B., Luo, W., Abo, A., Alberts, A.S., Peterlin, B.M., 2000. p21-activated kinase 1 plays a critical role in cellular activation by Nef. Mol. Cell. Biol. 20, 2619–2627.

Fahey, M.E., Bennett, M.J., Mahon, C., Jager, S., Pache, L., Kumar, D., Shapiro, A., Rao, K., Chanda, S.K., Craik, C.S., Frankel, A.D., Krogan, N.J., 2011. GPS-Prot: a web-based visualization platform for integrating host-pathogen interaction data. BMC Bioinf. 12, 298.

Fang, J., Kubota, S., Yang, B., Zhou, N., Zhang, H., Godbout, R., Pomerantz, R.J., 2004. A DEAD box protein facilitates HIV-1 replication as a cellular co-factor of Rev. Virology 330, 471–480.

Feng, F., Davis, A., Lake, J.A., Carr, J., Xia, W., Burrell, C., Li, P., 2004. Ring finger protein ZIN interacts with human immunodeficiency virus type 1 Vif. J. Virol. 78, 10574–10581.

Fields, S., Song, O., 1989. A novel genetic system to detect protein-protein interactions. Nature 340, 245–246.

Fortin, J.F., Barbeau, B., Robichaud, G.A., Pare, M.E., Lemieux, A.M., Tremblay, M.J., 2001. Regulation of nuclear factor of activated T cells by phosphotyrosyl-specific phosphatase activity: a positive effect on HIV-1 long terminal repeat-driven transcription and a possible implication of SHP-1. Blood 97, 2390–2400.

Frost, J.A., Khokhlatchev, A., Stippec, S., White, M.A., Cobb, M.H., 1998. Differential effects of PAK1-activating mutations reveal activity-dependent and -independent effects on cytoskeletal regulation. J. Biol. Chem. 273, 28191–28198.

Fu, W., Sanders-Beer, B.E., Katz, K.S., Maglott, D.R., Pruitt, K.D., Ptak, R.G., 2009. Human immunodeficiency virus type 1, human protein interaction database at NCBI. Nucleic Acids Res. 37, D417–D422.

Gabor, J., Cen, S., Javanbakht, H., Niu, M., Kleiman, L., 2002. Effect of altering the tRNA(Lys)(3) concentration in human immunodeficiency virus type 1 upon its annealing to viral RNA, GagPol incorporation, and viral infectivity. J. Virol. 76, 9096–9102.

Galeva, N., Altermann, M., 2002. Comparison of one-dimensional and two-dimensional gel electrophoresis as a separation tool for proteomic analysis of rat liver microsomes: cytochromes P450 and other membrane proteins. Proteomics 2, 713–722.

Gautier, V.W., Gu, L., O'donoghue, N., Pennington, S., Sheehy, N., Hall, W.W., 2009. In vitro nuclear interactome of the HIV-1 Tat protein. Retrovirology 6, 47.

Gavin, A.C., Aloy, P., Grandi, P., Krause, R., Boesche, M., Marzioch, M., Rau, C., Jensen, L.J., Bastuck, S., Dumpelfeld, B., Edelmann, A., Heurtier, M.A., Hoffman, V., Hoefert, C., Klein, K., Hudak, M., Michon, A.M., Schelder, M., Schirle, M., Remor, M., Rudi, T., Hooper, S., Bauer, A., Bouwmeester, T., Casari, G., Drewes, G., Neubauer, G., Rick, J.M., Kuster, B., Bork, P., Russell, R.B., Superti-Furga, G., 2006. Proteome survey reveals modularity of the yeast cell machinery. Nature 440, 631–636.

Genovesio, A., Kwon, Y.J., Windisch, M.P., Kim, N.Y., Choi, S.Y., Kim, H.C., Jung, S., Mammano, F., Perrin, V., Boese, A.S., Casartelli, N., Schwartz, O., Nehrbass, U., Emans, N., 2011. Automated genome-wide visual profiling of cellular proteins involved in HIV infection. J. Biomol. Screen. 16, 945–958.

Giles, K.E., Caputi, M., Beemon, K.L., 2004. Packaging and reverse transcription of snRNAs by retroviruses may generate pseudogenes. RNA 10, 299–307.

Giorgi, C., Moore, M.J., 2007. The nuclear nurture and cytoplasmic nature of localized mRNPs. Semin. Cell Dev. Biol. 18, 186–193.

Goff, S.P., 2008. Knockdown screens to knockout HIV-1. Cell 135, 417–420.

Gooley, A.A., Ou, K., Russell, J., Wilkins, M.R., Sanchez, J.C., Hochstrasser, D.F., Williams, K.L., 1997. A role for Edman degradation in proteome studies. Electrophoresis 18, 1068–1072.

Griffith, J.L., Coleman, L.E., Raymond, A.S., Goodson, S.G., Pittard, W.S., Tsui, C., Devine, S.E., 2003. Functional genomics reveals relationships between the retrovirus-like Ty1 element and its host *Saccharomyces cerevisiae*. Genetics 164, 867–879.

Han, X., Aslanian, A., Yates, J.R., 2008. Mass spectrometry for proteomics. Curr. Opin. Chem. Biol. 12, 483–490.

Henning, M.S., Morham, S.G., Goff, S.P., Naghavi, M.H., 2010. PDZD8 is a novel Gag-interacting factor that promotes retroviral infection. J. Virol. 84, 8990–8995.

Hermjakob, H., Montecchi-Palazzi, L., Lewington, C., Mudali, S., Kerrien, S., Orchard, S., Vingron, M., Roechert, B., Roepstorff, P., Valencia, A., Margalit, H., Armstrong, J., Bairoch, A., Cesareni, G., Sherman, D., Apweiler, R., 2004. IntAct: an open source molecular interaction database. Nucleic Acids Res. 32, D452–D455.

Hoque, M., Shamanna, R.A., Guan, D., Pe'ery, T., Mathews, M.B., 2011. HIV-1 replication and latency are regulated by translational control of cyclin T1. J. Mol. Biol. 410, 917–932.

Houzet, L., Jeang, K.T., 2011. Genome-wide screening using RNA interference to study host factors in viral replication and pathogenesis. Exp. Biol. Med. (Maywood) 236, 962–967.

Houzet, L., Paillart, J.C., Smagulova, F., Maurel, S., Morichaud, Z., Marquet, R., Mougel, M., 2007. HIV controls the selective packaging of genomic, spliced viral and cellular RNAs into virions through different mechanisms. Nucleic Acids Res. 35, 2695–2704.

Huang, J., Wang, F., Argyris, E., Chen, K., Liang, Z., Tian, H., Huang, W., Squires, K., Verlinghieri, G., Zhang, H., 2007. Cellular microRNAs contribute to HIV-1 latency in resting primary CD4+ T lymphocytes. Nat. Med. 13, 1241–1247.

Irwin, B., Aye, M., Baldi, P., Beliakova-Bethell, N., Cheng, H., Dou, Y., Liou, W., Sandmeyer, S., 2005. Retroviruses and yeast retrotransposons use overlapping sets of host genes. Genome Res. 15, 641–654.

Jäger, S., Cimermancic, P., Gulbahce, N., Johnson, J.R., Mcgovern, K.E., Clarke, S.C., Shales, M., Mercenne, G., Pache, L., Li, K., Hernandez, H., Jang, G.M., Roth, S.L., Akiva, E., Marlett, J., Stephens, M., D'orso, I., Fernandes, J., Fahey, M., Mahon, C., O'donoghue, A.J., Todorovic, A., Morris, J.H., Maltby, D.A., Alber, T., Cagney, G., Bushman, F.D., Young, J.A., Chanda, S.K., Sundquist, W.I., Kortemme, T., Hernandez, R.D., Craik, C.S., Burlingame, A., Sali, A., Frankel, A.D., Krogan, N.J., 2011a. Global landscape of HIV-human protein complexes. Nature 481, 365–370.

Jäger, S., Kim, D.Y., Hultquist, J.F., Shindo, K., Larue, R.S., Kwon, E., Li, M., Anderson, B.D., Yen, L., Stanley, D., Mahon, C., Kane, J., Franks-Skiba, K., Cimermancic, P., Burlingame, A., Sali, A., Craik, C.S., Harris, R.S., Gross, J.D., Krogan, N.J., 2011b. Vif hijacks CBF-beta to degrade APOBEC3G and promote HIV-1 infection. Nature 481, 371–375.

Javanbakht, H., Halwani, R., Cen, S., Saadatmand, J., Musier-Forsyth, K., Gottlinger, H., Kleiman, L., 2003. The interaction between HIV-1 Gag and human lysyl-tRNA synthetase during viral assembly. J. Biol. Chem. 278, 27644–27651.

Jiang, M., Mak, J., Ladha, A., Cohen, E., Klein, M., Rovinski, B., Kleiman, L., 1993. Identification of tRNAs incorporated into wild-type and mutant human immunodeficiency virus type 1. J. Virol. 67, 3246–3253.

Jiang, W.M., Zhang, X.Y., Zhang, Y.Z., Liu, L., Lu, H.Z., 2014. A high throughput RNAi screen reveals determinants of HIV-1 activity in host kinases. Int. J. Clin. Exp. Pathol. 7, 2229–2237.

Kalpana, G.V., Marmon, S., Wang, W., Crabtree, G.R., Goff, S.P., 1994. Binding and stimulation of HIV-1 integrase by a human homolog of yeast transcription factor SNF5. Science 266, 2002–2006.

Kameoka, M., Kitagawa, Y., Utachee, P., Jinnopat, P., Dhepakson, P., Isarangkura-Na-Ayuthaya, P., Tokunaga, K., Sato, H., Komano, J., Yamamoto, N., Oguchi, S., Natori, Y., Ikuta, K., 2007. Identification of the suppressive factors for human immunodeficiency virus type-1 replication using the siRNA mini-library directed against host cellular genes. Biochem. Biophys. Res. Commun. 359, 729–734.

Kang, D.C., Su, Z.Z., Sarkar, D., Emdad, L., Volsky, D.J., Fisher, P.B., 2005. Cloning and characterization of HIV-1-inducible astrocyte elevated gene-1, AEG-1. Gene 353, 8–15.

Karpievitch, Y.V., Polpitiya, A.D., Anderson, G.A., Smith, R.D., Dabney, A.R., 2010. Liquid chromatography mass spectrometry-based proteomics: biological and technological aspects. Ann. Appl. Stat. 4, 1797–1823.

Khorchid, A., Javanbakht, H., Wise, S., Halwani, R., Parniak, M.A., Wainberg, M.A., Kleiman, L., 2000. Sequences within Pr160gag-pol affecting the selective packaging of primer tRNA(Lys3) into HIV-1. J. Mol. Biol. 299, 17–26.

Kim, W., Tang, Y., Okada, Y., Torrey, T.A., Chattopadhyay, S.K., Pfleiderer, M., Falkner, F.G., Dorner, F., Choi, W., Hirokawa, N., Morse, H.C., 1998. Binding of murine leukemia virus Gag polyproteins to KIF4, a microtubule-based motor protein. J. Virol. 72, 6898–6901.

Knoener, R.A., Becker, J.T., Scalf, M., Sherer, N.M., Smith, L.M., 2017. Elucidating the in vivo interactome of HIV-1 RNA by hybridization capture and mass spectrometry. Sci. Rep. 7, 16965.

Kok, K.H., Lei, T., Jin, D.Y., 2009. siRNA and shRNA screens advance key understanding of host factors required for HIV-1 replication. Retrovirology 6, 78.

Konig, R., Zhou, Y., Elleder, D., Diamond, T.L., Bonamy, G.M., Irelan, J.T., Chiang, C.Y., Tu, B.P., De Jesus, P.D., Lilley, C.E., Seidel, S., Opaluch, A.M., Caldwell, J.S., Weitzman, M.D., Kuhen, K.L., Bandyopadhyay, S., Ideker, T., Orth, A.P., Miraglia, L.J., Bushman, F.D., Young, J.A., Chanda, S.K., 2008. Global analysis of host-pathogen interactions that regulate early-stage HIV-1 replication. Cell 135, 49–60.

Kouwenhoven, A., Minassian, V.D., Marsh, J.W., 2013. HIV-1 Nef mediates Pak phosphorylation of Mek1 serine298 and elicits an active phospho-state of Pak2. Curr. HIV Res. 11, 198–209.

Krautkramer, E., Giese, S.I., Gasteier, J.E., Muranyi, W., Fackler, O.T., 2004. Human immunodeficiency virus type 1 Nef activates p21-activated kinase via recruitment into lipid rafts. J. Virol. 78, 4085–4097.

Kula, A., Guerra, J., Knezevich, A., Kleva, D., Myers, M.P., Marcello, A., 2011. Characterization of the HIV-1 RNA associated proteome identifies Matrin 3 as a nuclear cofactor of Rev function. Retrovirology 8, 60.

Kulkarni, M.M., Ratcliff, A.N., Bhat, M., Alwarawrah, Y., Hughes, P., Arcos, J., Loiselle, D., Torrelles, J.B., Funderburg, N.T., Haystead, T.A., Kwiek, J.J., 2017. Cellular fatty acid synthase is required for late stages of HIV-1 replication. Retrovirology 14, 45.

Kutluay, S.B., Zang, T., Blanco-Melo, D., Powell, C., Jannain, D., Errando, M., Bieniasz, P.D., 2014. Global changes in the RNA binding specificity of HIV-1 gag regulate virion genesis. Cell 159, 1096–1109.

Lake, J.A., Carr, J., Feng, F., Mundy, L., Burrell, C., Li, P., 2003. The role of Vif during HIV-1 infection: interaction with novel host cellular factors. J. Clin. Virol. 26, 143–152.

Landi, A., Vermeire, J., Iannucci, V., Vanderstraeten, H., Naessens, E., Bentahir, M., Verhasselt, B., 2014. Genome-wide shRNA screening identifies host factors involved in early endocytic events for HIV-1-induced CD4 down-regulation. Retrovirology 11, 118.

Le Sage, V., Cinti, A., Valiente-Echeverria, F., Mouland, A.J., 2015. Proteomic analysis of HIV-1 Gag interacting partners using proximity-dependent biotinylation. Virol. J. 12, 138.

Li, C., Ramarathinam, S.H., Revote, J., Khoury, G., Song, J., Purcell, A.W., 2017. HIVed, a knowledge base for differentially expressed human genes and proteins during HIV infection, replication and latency. Sci. Rep. 7, 45509.

Liu, L., Oliveira, N.M., Cheney, K.M., Pade, C., Dreja, H., Bergin, A.M., Borgdorff, V., Beach, D.H., Bishop, C.L., Dittmar, M.T., Mcknight, A., 2011. A whole genome screen for HIV restriction factors. Retrovirology 8, 94.

Luban, J., Bossolt, K.L., Franke, E.K., Kalpana, G.V., Goff, S.P., 1993. Human immunodeficiency virus type 1 Gag protein binds to cyclophilins A and B. Cell 73, 1067–1078.

Luban, J., Goff, S.P., 1995. The yeast two-hybrid system for studying protein-protein interactions. Curr. Opin. Biotechnol. 6, 59–64.

Luo, Y., Muesing, M.A., 2014. Mass spectrometry-based proteomic approaches for discovery of HIV-host interactions. Future Virol. 9, 979–992.

Luo, Y., Jacobs, E.Y., Greco, T.M., Mohammed, K.D., Tong, T., Keegan, S., Binley, J.M., Cristea, I.M., Fenyo, D., Rout, M.P., Chait, B.T., Muesing, M.A., 2016. HIV-host interactome revealed directly from infected cells. Nat. Microbiol. 1, 16068.

Malim, M.H., Hauber, J., Le, S.Y., Maizel, J.V., Cullen, B.R., 1989. The HIV-1 rev trans-activator acts through a structured target sequence to activate nuclear export of unspliced viral mRNA. Nature 338, 254–257.

Marchand, V., Santerre, M., Aigueperse, C., Fouillen, L., Saliou, J.M., Van Dorsselaer, A., Sanglier-Cianferani, S., Branlant, C., Motorin, Y., 2011. Identification of protein partners of the human immunodeficiency virus 1 tat/rev exon 3 leads to the discovery of a new HIV-1 splicing regulator, protein hnRNP K. RNA Biol. 8, 325–342.

Markle, T.J., Philip, M., Brockman, M.A., 2013. HIV-1 Nef and T-cell activation: a history of contradictions. Future Virol. 8.

Misumi, S., Fuchigami, T., Takamune, N., Takahashi, I., Takama, M., Shoji, S., 2002. Three isoforms of cyclophilin A associated with human immunodeficiency virus type 1 were found by proteomics by using two-dimensional gel electrophoresis and matrix-assisted laser desorption ionization-time of flight mass spectrometry. J. Virol. 76, 10000–10008.

Mukerji, J., Olivieri, K.C., Misra, V., Agopian, K.A., Gabuzda, D., 2012. Proteomic analysis of HIV-1 Nef cellular binding partners reveals a role for exocyst complex proteins in mediating enhancement of intercellular nanotube formation. Retrovirology 9, 33.

Muthumani, K., Wadsworth, S.A., Dayes, N.S., Hwang, D.S., Choo, A.Y., Abeysinghe, H.R., Siekierka, J.J., Weiner, D.B., 2004. Suppression of HIV-1 viral replication and cellular pathogenesis by a novel p38/JNK kinase inhibitor. AIDS 18, 739–748.

Nabel, G.J., Rice, S.A., Knipe, D.M., Baltimore, D., 1988. Alternative mechanisms for activation of human immunodeficiency virus enhancer in T cells. Science 239, 1299–1302.

Naji, S., Ambrus, G., Cimermancic, P., Reyes, J.R., Johnson, J.R., Filbrandt, R., Huber, M.D., Vesely, P., Krogan, N.J., Yates 3rd, J.R., Saphire, A.C., Gerace, L., 2012. Host cell interactome of HIV-1 Rev includes RNA helicases involved in multiple facets of virus production. Mol. Cell. Proteom. 11 M111 015313.

Nguyen, D.G., Wolff, K.C., Yin, H., Caldwell, J.S., Kuhen, K.L., 2006. "UnPAKing" human immunodeficiency virus (HIV) replication: using small interfering RNA screening to identify novel cofactors and elucidate the role of group I PAKs in HIV infection. J. Virol. 80, 130–137.

Nguyen, D.G., Yin, H., Zhou, Y., Wolff, K.C., Kuhen, K.L., Caldwell, J.S., 2007. Identification of novel therapeutic targets for HIV infection through functional genomic cDNA screening. Virology 362, 16–25.

Nunn, M.F., Marsh, J.W., 1996. Human immunodeficiency virus type 1 Nef associates with a member of the p21-activated kinase family. J. Virol. 70, 6157–6161.

Onafuwa-Nuga, A.A., King, S.R., Telesnitsky, A., 2005. Nonrandom packaging of host RNAs in moloney murine leukemia virus. J. Virol. 79, 13528–13537.

Onafuwa-Nuga, A.A., Telesnitsky, A., King, S.R., 2006. 7SL RNA, but not the 54-kd signal recognition particle protein, is an abundant component of both infectious HIV-1 and minimal virus-like particles. RNA 12, 542–546.

Ong, S.E., Mann, M., 2005. Mass spectrometry-based proteomics turns quantitative. Nat. Chem. Biol. 1, 252–262.

Pache, L., Konig, R., Chanda, S.K., 2011. Identifying HIV-1 host cell factors by genome-scale RNAi screening. Methods 53, 3–12.

Park, R.J., Wang, T., Koundakjian, D., Hultquist, J.F., Lamothe-Molina, P., Monel, B., Schumann, K., Yu, H., Krupzcak, K.M., Garcia-Beltran, W., Piechocka-Trocha, A., Krogan, N.J., Marson, A., Sabatini, D.M., Lander, E.S., Hacohen, N., Walker, B.D., 2017. A genome-wide CRISPR screen identifies a restricted set of HIV host dependency factors. Nat. Genet. 49, 193–203.

Parrott, A.M., Mathews, M.B., 2007. Novel rapidly evolving hominid RNAs bind nuclear factor 90 and display tissue-restricted distribution. Nucleic Acids Res. 35, 6249–6258.

Parrott, A.M., Tsai, M., Batchu, P., Ryan, K., Ozer, H.L., Tian, B., Mathews, M.B., 2011. The evolution and expression of the snaR family of small non-coding RNAs. Nucleic Acids Res. 39, 1485–1500.

Pavon-Eternod, M., Wei, M., Pan, T., Kleiman, L., 2010. Profiling non-lysyl tRNAs in HIV-1. RNA 16, 267–273.

Pinney, J.W., Dickerson, J.E., Fu, W., Sanders-Beer, B.E., Ptak, R.G., Robertson, D.L., 2009. HIV-host interactions: a map of viral perturbation of the host system. AIDS 23, 549–554.

Ptak, R.G., Fu, W., Sanders-Beer, B.E., Dickerson, J.E., Pinney, J.W., Robertson, D.L., Rozanov, M.N., Katz, K.S., Maglott, D.R., Pruitt, K.D., Dieffenbach, C.W., 2008. Cataloguing the HIV type 1 human protein interaction network. AIDS Res Hum Retroviruses 24, 1497–1502.

Raghavendra, N.K., Shkriabai, N., Graham, R., Hess, S., Kvaratskhelia, M., Wu, L., 2010. Identification of host proteins associated with HIV-1 preintegration complexes isolated from infected CD4+ cells. Retrovirology 7, 66.

Rai, S.K., Sangesland, M., Lee Jr., M., Esnault, C., Cui, Y., Chatterjee, A.G., Levin, H.L., 2017. Host factors that promote retrotransposon integration are similar in distantly related eukaryotes. PLoS Genet. 13, e1006775.

Rain, J.C., Cribier, A., Gerard, A., Emiliani, S., Benarous, R., 2009. Yeast two-hybrid detection of integrase-host factor interactions. Methods 47, 291–297.

Raney, A., Kuo, L.S., Baugh, L.L., Foster, J.L., Garcia, J.V., 2005. Reconstitution and molecular analysis of an active human immunodeficiency virus type 1 Nef/p21-activated kinase 2 complex. J. Virol. 79, 12732–12741.

Reed, R., 2003. Coupling transcription, splicing and mRNA export. Curr. Opin. Cell Biol. 15, 326–331.

Renkema, G.H., Manninen, A., Mann, D.A., Harris, M., Saksela, K., 1999. Identification of the Nef-associated kinase as p21-activated kinase 2. Curr. Biol. 9, 1407–1410.

Renkema, G.H., Manninen, A., Saksela, K., 2001. Human immunodeficiency virus type 1 Nef selectively associates with a catalytically active subpopulation of p21-activated kinase 2 (PAK2) independently of PAK2 binding to Nck or beta-PIX. J. Virol. 75, 2154–2160.

Rigaut, G., Shevchenko, A., Rutz, B., Wilm, M., Mann, M., Seraphin, B., 1999. A generic protein purification method for protein complex characterization and proteome exploration. Nat. Biotechnol. 17, 1030–1032.

Ringrose, J.H., Jeeninga, R.E., Berkhout, B., Speijer, D., 2008. Proteomic studies reveal coordinated changes in T-cell expression patterns upon infection with human immunodeficiency virus type 1. J. Virol. 82, 4320–4330.

Ritchie, C., Cylinder, I., Platt, E.J., Barklis, E., 2015. Analysis of HIV-1 Gag protein interactions via biotin ligase tagging. J. Virol. 89, 3988–4001.

Rulli Jr., S.J., Hibbert, C.S., Mirro, J., Pederson, T., Biswal, S., Rein, A., 2007. Selective and nonselective packaging of cellular RNAs in retrovirus particles. J. Virol. 81, 6623–6631.

Santos, S., Obukhov, Y., Nekhai, S., Bukrinsky, M., Iordanskiy, S., 2012. Virus-producing cells determine the host protein profiles of HIV-1 virion cores. Retrovirology 9, 65.

Saphire, A.C., Gallay, P.A., Bark, S.J., 2006. Proteomic analysis of human immunodeficiency virus using liquid chromatography/tandem mass spectrometry effectively distinguishes specific incorporated host proteins. J. Proteome Res. 5, 530–538.

Sawai, E.T., Baur, A., Struble, H., Peterlin, B.M., Levy, J.A., Cheng-Mayer, C., 1994. Human immunodeficiency virus type 1 Nef associates with a cellular serine kinase in T lymphocytes. Proc. Natl. Acad. Sci. U.S.A. 91, 1539–1543.

Schneider, R., Campbell, M., Nasioulas, G., Felber, B.K., Pavlakis, G.N., 1997. Inactivation of the human immunodeficiency virus type 1 inhibitory elements allows Rev-independent expression of Gag and Gag/protease and particle formation. J. Virol. 71, 4892–4903.

Schopman, N.C., Van Montfort, T., Willemsen, M., Knoepfel, S.A., Pollakis, G., Van Kampen, A., Sanders, R.W., Haasnoot, J., Berkhout, B., 2012. Selective packaging of cellular miRNAs in HIV-1 particles. Virus Res. 169, 438–447.

Schröfelbauer, B., Hakata, Y., Landau, N.R., 2007. HIV-1 Vpr function is mediated by interaction with the damage-specific DNA-binding protein DDB1. Proc. Natl. Acad. Sci. U.S.A. 104, 4130–4135.

Schweitzer, C.J., Jagadish, T., Haverland, N., Ciborowski, P., Belshan, M., 2013. Proteomic analysis of early HIV-1 nucleoprotein complexes. J. Proteome Res. 12, 559–572.

Segura, M.M., Garnier, A., Di Falco, M.R., Whissell, G., Meneses-Acosta, A., Arcand, N., Kamen, A., 2008. Identification of host proteins associated with retroviral vector particles by proteomic analysis of highly purified vector preparations. J. Virol. 82, 1107–1117.

Sekine, Y., Okada, Y., Noda, Y., Kondo, S., Aizawa, H., Takemura, R., Hirokawa, N., 1994. A novel microtubule-based motor protein (KIF4) for organelle transports, whose expression is regulated developmentally. J. Cell Biol. 127, 187–201.

Stake, M., Singh, D., Singh, G., Marcela Hernandez, J., Kaddis Maldonado, R., Parent, L.J., Boris-Lawrie, K., 2015. HIV-1 and two avian retroviral 5′ untranslated regions bind orthologous human and chicken RNA binding proteins. Virology 486, 307–320.

Strack, B., Calistri, A., Craig, S., Popova, E., Gottlinger, H.G., 2003. AIP1/ALIX is a binding partner for HIV-1 p6 and EIAV p9 functioning in virus budding. Cell 114, 689–699.

Stremlau, M., Owens, C.M., Perron, M.J., Kiessling, M., Autissier, P., Sodroski, J., 2004. The cytoplasmic body component TRIM5alpha restricts HIV-1 infection in Old World monkeys. Nature 427, 848–853.

Studamire, B., Goff, S.P., 2008. Host proteins interacting with the Moloney murine leukemia virus integrase: multiple transcriptional regulators and chromatin binding factors. Retrovirology 5, 48.

Tacheny, A., Michel, S., Dieu, M., Payen, L., Arnould, T., Renard, P., 2012. Unbiased proteomic analysis of proteins interacting with the HIV-1 5'LTR sequence: role of the transcription factor Meis. Nucleic Acids Res. 40, e168.

Tian, C., Wang, T., Zhang, W., Yu, X.F., 2007. Virion packaging determinants and reverse transcription of SRP RNA in HIV-1 particles. Nucleic Acids Res. 35, 7288–7302.

Trcek, T., Singer, R.H., 2010. The cytoplasmic fate of an mRNP is determined cotranscriptionally: exception or rule? Genes Dev. 24, 1827–1831.

VerPlank, L., Bouamr, F., Lagrassa, T.J., Agresta, B., Kikonyogo, A., Leis, J., Carter, C.A., 2001. Tsg101, a homologue of ubiquitin-conjugating (E2) enzymes, binds the L domain in HIV type 1 Pr55(Gag). Proc. Natl. Acad. Sci. U.S.A. 98, 7724–7729.

Wang, M.Q., Kim, W., Gao, G., Torrey, T.A., Morse, H.C., De Camilli, P., Goff, S.P., 2003. Endophilins interact with Moloney murine leukemia virus Gag and modulate virion production. J. Biol. 3, 4.

Washburn, M.P., Wolters, D., Yates, J.R., 2001. Large-scale analysis of the yeast proteome by multidimensional protein identification technology. Nat. Biotechnol. 19, 242–247.

Wei, M., Cen, S., Niu, M., Guo, F., Kleiman, L., 2005. Defective replication in human immunodeficiency virus type 1 when non-primers are used for reverse transcription. J. Virol. 79, 9081–9087.

Weldon Jr., R.A., Sarkar, P., Brown, S.M., Weldon, S.K., 2003. Mason-Pfizer monkey virus Gag proteins interact with the human sumo conjugating enzyme, hUbc9. Virology 314, 62–73.

Wen, X., Ding, L., Hunter, E., Spearman, P., 2014. An siRNA screen of membrane trafficking genes highlights pathways common to HIV-1 and M-PMV virus assembly and release. PLoS One 9, e106151.

Wilkins, M.R., Ou, K., Appel, R.D., Sanchez, J.C., Yan, J.X., Golaz, O., Farnsworth, V., Cartier, P., Hochstrasser, D.F., Williams, K.L., Gooley, A.A., 1996a. Rapid protein identification using N-terminal "sequence tag" and amino acid analysis. Biochem. Biophys. Res. Commun. 221, 609–613.

Wilkins, M.R., Sanchez, J.C., Gooley, A.A., Appel, R.D., Humphery-Smith, I., Hochstrasser, D.F., Williams, K.L., 1996b. Progress with proteome projects: why all proteins expressed by a genome should be identified and how to do it. Biotechnol. Genet. Eng. Rev. 13, 19–50.

Williams, C.A., Abbink, T.E., Jeang, K.T., Lever, A.M., 2015. Identification of RNA helicases in human immunodeficiency virus 1 (HIV-1) replication - a targeted small interfering RNA library screen using pseudotyped and WT HIV-1. J. Gen. Virol. 96, 1484–1489.

Wiskerchen, M., Cheng-Mayer, C., 1996. HIV-1 Nef association with cellular serine kinase correlates with enhanced virion infectivity and efficient proviral DNA synthesis. Virology 224, 292–301.

Wolf, D., Witte, V., Laffert, B., Blume, K., Stromer, E., Trapp, S., D'aloja, P., Schurmann, A., Baur, A.S., 2001. HIV-1 Nef associated PAK and PI3-kinases stimulate Akt-independent Bad-phosphorylation to induce anti-apoptotic signals. Nat. Med. 7, 1217–1224.

Yang, X., Chen, Y., Gabuzda, D., 1999. ERK MAP kinase links cytokine signals to activation of latent HIV-1 infection by stimulating a cooperative interaction of AP-1 and NF-kappaB. J. Biol. Chem. 274, 27981–27988.

Yeung, M.L., Houzet, L., Yedavalli, V.S., Jeang, K.T., 2009. A genome-wide short hairpin RNA screening of jurkat T-cells for human proteins contributing to productive HIV-1 replication. J. Biol. Chem. 284, 19463–19473.

Yueh, A., Leung, J., Bhattacharyya, S., Perrone, L.A., De Los Santos, K., Pu, S.Y., Goff, S.P., 2006. Interaction of moloney murine leukemia virus capsid with Ubc9 and PIASy mediates SUMO-1 addition required early in infection. J. Virol. 80, 342–352.

Zaitseva, L., Myers, R., Fassati, A., 2006. tRNAs promote nuclear import of HIV-1 intracellular reverse transcription complexes. PLoS Biol. 4, e332.

Zeghouf, M., Li, J., Butland, G., Borkowska, A., Canadien, V., Richards, D., Beattie, B., Emili, A., Greenblatt, J.F., 2004. Sequential Peptide Affinity (SPA) system for the identification of mammalian and bacterial protein complexes. J. Proteome Res. 3, 463–468.

Zhou, H., Xu, M., Huang, Q., Gates, A.T., Zhang, X.D., Castle, J.C., Stec, E., Ferrer, M., Strulovici, B., Hazuda, D.J., Espeseth, A.S., 2008. Genome-scale RNAi screen for host factors required for HIV replication. Cell Host Microbe 4, 495–504.

Index

Note: 'Page numbers followed by "f" indicate figures and "t" indicate tables.'

A

Actin cytoskeleton, 62–64
Affinity purification (AP), 537–539
Aicardi-Goutieres syndrome, 81, 507
AKAP149, PKA regulator, 64
Alix/AIP-1 interactions, 294
Allosteric IN inhibitors (ALLINIs), 440
Alpharetrovirus avian leukosis/sarcoma virus, 200–201
Alpharetroviruses, 22–23
Alu-LTR PCR, 58–59
Amphotericin B methyl ester (AME), 360–361
Annexin 2 (Anx2), 290
Antisense protein RNA, 436
Aotus trivirgatus, 76–77
APOBEC1, 79
APOBEC3G gene, 78–79
APOBECs
  antiviral activity, 80–81
  APOBEC1, 79
  APOBEC3G gene, 78–79
  CEM-SS subclone, 78–79
  deoxyuridines, 79
  glyco-Gag action, 80
  proteasome-independent methods, 79–80
  Skp1–Cullin–Fbox complexes, 79–80
  Vif-minus virus, 78–79
  virus-encoded Vif protein, 79–80
  Zn-containing enzymes, 80–81
Arc/Arg3.1, 487–488
Arg-X-Lys/Arg-Arg (RXK/RR) motifs, 17
ASLV. *See* Avian sarcoma and leukosis virus (ASLV)
Atomic force microscopy, 59–60
Avian leucosis virus (ALV), 204–205
Avian leukosis virus subgroup J (ALV-J), 444
Avian sarcoma and leukosis virus (ASLV), 3, 11–12

B

Basic leucine zipper (bZIP) motif, 208–209
BEL/Pao, 480–482
Betaretroviruses, 2
  glycoproteins, 4
  neoplasia, 25
Bic, 437
Blood–brain barrier, 24–25
bnAbs. *See* Broadly neutralizing antibodies (bnAbs)
Bovine foamy virus (BFV), 448
Bovine leukemia virus (BLV), 173, 447–448
Brain capillary endothelial cell infection, 24–25
Broadly neutralizing antibodies (bnAbs), 26
Bromodomain and extra-terminal domain (BET) proteins, 171–172

C

Calcium, 334–335
Cap–binding proteins
  eIF4E, 266–267
  nonsense RNA–mediated decay, 265–266
Capping proteins, 66
Capsid protein (CA), 53, 55
Capsid stability, 413
Cas9–guide RNA complexes, 168–169
Catalytic core domain (CCD), 165–166
Cathepsin B inhibitors, 22
CD4-bound HIV-2 trimers, 19–20
CD4/feCD134 attachment receptors, 5
CD4 receptor and CXCR4 coreceptor, 62
CDK1/cyclin A2 and CDK2 modification, 82
CDNA overexpression screens, 550
CDR loops. *See* Complementarity-determining region (CDR) loops
Cellular control
  frequencies of retrotransposition, 486
  retroelement activities and impacts, 486–488

Cellular control *(Continued)*
retroelement expression
cancer initiation and progression, 502
germline and embryonic stem cells, 492–498
neuronal activity, 499–502
self-preservation, 488–490
sequence-specific suppression, 498–499
somatic cells by RNA interference, 490–492, 491f
transcription factors, 488
Cellular endocytic pathway, 328–329
Cellular immune responses, 401–402
host pathways
cytosolic sensors, 410
DC-SIGN, 408–409
DDX41, 411
glycoproteins, 408–409
gp55, 407
IFN-stimulated genes (ISG) encoding antiviral factors, 407
macrophages and DCs, 406–407
Myxovirus Resistance 2 (MX2), 411–412
pathogen-associated molecular patterns (PAMPs), 406
pattern recognition receptors (PRRs), 406
retroviral replication, 409
SAM domain and HD domain-containing protein 1 (SAMHD1), 409
TLR2 and TLR4, 407–408
TLR7-mediated immune response, 408
viral proteins, 409, 410t
retrovirus infection pathway, host intrinsic/innate response
chemokine coreceptors, 402–404
incomplete/defective retroviruses, 406
nucleic acid structures, 405–406, 405f
pathogen-associated molecular patterns (PAMPs), 402–404
steps, 402–404, 403f
toll-like receptor 2 (TLR2), 402–404
toll-like receptor 4 (TLR4), 402–404
TRIM5α, 404–405
viral gene products, 404–405, 404f
subversion of, 412–413
Cellular membranes
endoplasmic reticulum (ER), 355–357
glycerophospholipids (GPLs), 354–355
predominant classes, 354–355
sphingolipids, 354–355
sterols, 354–355
Cellular miRNAs, 444–446
Cellular RNAs, identification of, 550–554
CFIm complex, 179
Chemical shift mapping, 325–326
Chimeric short interspersed elements, 485–486
Cholesterol, 354–355
phase behavior regulation, 358
retroviral assembly, 368–370
in retroviral entry
amphotericin B methyl ester (AME), 360–361
CD4 and CCR5, 360–361
HIV-1 infectivity, 360–361
HTLV-1 entry, 361–362
MβCD, 360
2OHpβCD, 360
Cholesterol-binding motifs, 17
Cholesterol-rich lipid domains, 14
Chromatin-binding site (CBS), 285–286
Chromatin-bound protein, 169–170
Chromatin-bound transcription coactivator proteins, 262–263
CLIP-Seq, 433–434
Complementarity-determining region (CDR) loops, 26, 27f
Complex retroviruses, 274
Constitutive transport element (CTE), 277
CPSF6 protein, 88
CRISPR/Cas-9, 549–550
Crm1, 275–277
CRM1-dependent nuclear export signal (NES), 4
Cryo-electron microscopy (cryo-EM), 4
C-terminal chromatin-binding sequence (CBS), 179–180
C-terminal domain (CTD), 165–166
C-terminal PRY/SPRY domain, 76–77
Cyclic guanosine monophosphate–adenosine monophosphate (cGAMP) synthase (cGAS), 86–87
Cyclophilin A (CypA), 60
African green monkeys (AGM), 69–70
cyclosporine efficacy, 68–69
Jurkat cells, 69
nuclear entry, 70
proline residue (P90), 68
structure of, 68
wild-type virus, 69
Cyclophilin A (CypA-DsRed), 61
CypA–Gag interaction, 83
CypA inhibitor cyclosporine A, 177–178
Cysteine residues, 3, 23
Cytidine deaminases. *See* APOBECs

D

DAP3, 438–439
DDX3, 275–277
DDX41, 411
Deltaretroviruses, 172–173
Dendritic cell–specific intercellular adhesion molecule-3-grabbing nonintegrin (DC-SIGN), 12
Dengue virus, 338
Deoxynucleoside triphosphate triphosphohydrolase, 81
Detergent-resistant membranes (DRMs), 358–359
DHX9
  cellular cofactors, 262
  chromatin-bound transcription coactivator proteins, 262–263
  downregulation, 259–261
  gain-of-function experiment, 259–261
  HEK293 cells, 263
  HIV-1 Gag, 261–262
  in-solution assays, 264
  nuclear magnetic resonance spectra, 261–262
  overexpression of, 263–264
  positive transcription elongation factor b (P-TEFb), 263
  primer-binding site (PBS) segment, 261–262
  Tat transactivation responsive element (TAR), 263
  viral reverse transcriptase, 261–262
Diaphanous interacting protein (DIP), 63–64
Disulfide bond isomerization
  receptor- and cellular protease–driven, 20–22
  receptor- and low pH-driven, 22–23
DNA methylation, 494–495
DNA methyltransferases, 496
DNMT1, 496
DNMT3A, 496
DNMT3B, 496
Double-stranded linear DNA, 56–58
Double-stranded RNA-binding domains (dsRBDs), 257
DRMs. *See* Detergent-resistant membranes (DRMs)
Drosophila melanogaster, 421–422
Dynamin-like GTPases, 83–84
Dyneins, 65, 67

E

E2 and E3 small ubiquitin-like modifier (SUMO)-conjugating enzymes, 70–71
E3 ubiquitin ligases, 292
EIAV Gag, 371–373
EIAV p9, 294–295
Encapsidation, 274–275
Endogenized retroviruses sequences, 29
Endogenous retroviruses (ERVs), 421–422, 437–439, 480–483, 481f
  foamy-like ERVs (FLERVs), 201–202
  GATA2 response elements, 202–203
  human endogenous retrovirus subfamily H (HERVH), 202–203
  jawed vertebrates, 201–202
  long-range transcriptional enhancers, 202
  murine endogenous retrovirus with leucine transfer RNA primer (MERVL), 202–203
  pluripotent germ cells, 202–203
  protein-coding genes, 202–203
  stem cells and cancer, 218
  structural protein code, 201–202
Endogenous reverse transcription, 54, 60
Endophilin 2, 295
Endophilins, 535
Endoplasmic reticulum (ER), 355–357
Endosomal sorting complex required for transport (ESCRT), 291–292, 318, 319f
  transport participation
    in cellular processes, 321–322
    in virus budding, 322–323
ENTV. *See* Enzootic nasal tumor virus (ENTV)
Envelope proteins (Env), 2
  class I fusion proteins, 16
  disulfide bond isomerization
    alpharetroviruses, 22–23
    gammaretroviruses, 20–22
  energy barrier, 16
  fusion peptide extension, 16–17
  glycoproteins. *See* Glycoproteins
  host protease cleavage, 16–17
  interferon-induced transmembrane proteins, 28
  low pH, 16–17
  mammalian placenta and male muscle mass, 30–31
  membrane fusion, 16. *See also* Membrane fusion
  motifs, 17
  neoplasia, betaretrovirus infection, 25
  receptor binding, 16–17

Envelope proteins *(Continued)*
six-helix bundle formation, 16–17
structure of, 10–11, 10f
surface subunit (SU) conformation, 18
disulfide bond isomerization, 18–19
lentiviruses, 19–20, 21f
vaccine development
broadly neutralizing antibodies (bnAbs), 26
complementarity-determining region (CDR) loops, 26, 27f
germline antibody genes, 28
mucosal anti-gp41 IgA, 26
Enzootic nasal adenocarcinoma, 25
Enzootic nasal tumor virus (ENTV), 2, 4, 25
Epsilonretrovirus and spumavirus Env, 2
ESCRT. *See* Endosomal sorting complex required for transport (ESCRT)
ESCRT-0, 318–319
ESCRT-I, 320
ESCRT-II, 320
ESCRT-III, 320–321
5-Ethynyl uridine–tagged viral RNA, 56
Eukaryotic cellular metabolism, 114
Eukaryotic elongation factor 1, 64
Eukaryotic retroelements, 480–483, 481f
Exogenous retroviruses, 401–402
Exportin1/XPO1, 275–277
Extracellular deoxynucleosides, 81

F

Fasciculation and elongation protein zeta-1 (FEZ1), 66–67
"Fate of capsid" assay, 59–60
FeLIX protein, 11
FeLV-A entry receptor, 24
FISH, 61
Flaviviruses, 338
Foamy-like ERVs (FLERVs), 201–202
Foamy viruses (FVs), 179–180, 208–209
Foamy virus Gag proteins, 285–286, 294–295
Furin-like protease, 4
Fv1, 72–74

G

Gag-interacting proteins, 177
Gag-membrane interactions, 366–367
Gag-Pol fusion protein, 274–275
GagPol polyproteins, 365
GagPol precursor, 354
Gag proteins, 62, 72–73, 274–275
ASV, 333
as bait, 528–535
EIAV, 333
host factors, 287
localization of, 281
nuclear export signal (NES), 278
restrict virus replication, 296–298
trafficking, 286–287, 365–366
Gain-of-function genetic approaches, 550
Gammaretroviruses, 170–171
disulfide bond isomerization, 20–22
retrovirus entry receptors, 5
Gelsolin, 64
Genome-wide screens
HIV-1 Gene and protein network data, 554, 559t
metaanalysis, 547
retrotransposition in yeast, 549
short hairpin RNA screens, 546–547
small interfering RNA screens, 544–546
targeted RNAi screens, 547–549
GLUT-1, 24
retrovirus entry receptors, 5
Glycerophospholipids (GPLs), 354–355
Glycoproteins
avian sarcoma and leukosis virus (ASLV) subgroups, 3
betaretroviruses, 4
carbohydrate chains, 4
covalent bond, 2, 3f
CRM1-dependent nuclear export signal (NES), 4
cryo-electron microscopy (cryo-EM), 4
cysteine residues, 3
epsilonretrovirus and spumavirus Env, 2
functions, 4
fusion peptide, 3
glycosylation, 4
host cell–specific virus attachment, 2
mRNA, 2
nuclear localization signal (NLS), 4
Rem-like activity, 4
signal peptides, 4
synthesis, 2, 3f
N-terminal receptor-binding domain (RBD), 2
trimer, 2
variable loops, 3
Glycosylphosphatidylinositol (GPI)-anchored proteins, 4–5

H

Heparin-binding sites (HBSs), 12
Heparin sulfate proteoglycans, 24–25
Hepatitis C virus (HCV), 338
Herpesviruses, 171
HERV-H, 438–439
HERV-K, 438–439
HERV-W. *See* Human endogenous retrovirus W (HERV-W)
HERV-W Env, 507
Heterogeneous nuclear ribonucleoprotein (hnRNP), 230–231, 279–280
  cytoplasmic accumulation, 233–234
  Gag (A2RE-1) coding regions, 233–234
  Gag/Env expression, 234–235
  G-rich sequence, 234–235
  mRNA stability, 234
  multiple G-rich sequences, 234–235
  nuclear/cytoplasmic ratio, 234–235
  overexpression and depletion, 234
  putative-binding sites, 234–235
  Rev-dependent reporter, 235–236
  role of, 231, 232t
  splicing enhancer sequences, 231–233
  viral regulatory system, 231–233
  Vpr (A2RE-2) coding regions, 233–234
High-throughput RNA-sequencing, 433–434
Histone acetyltransferases (HATs), 208–209
HIV-1, 445–446, 449
  3′ end processing, 214
  5′ long terminal repeat, 210, 211f
  CDK11, 214
  latent infection
    $CD4^+$ T cells, 214–215
    host cell chromatin, 217–218
    maintenance of, 215–216
    pharmacologic targeting, 217–218
    reactivation of, 216–217
    short-chain fatty acids, 217–218
  proviral transcription, 209–210
  P-TEFb, 213–214
  RNA polymerase II, 210–211
  RNA processing control. *See* RNA processing control
  Tat protein, 212–213, 213f
HIV-1 Gag, 289–290
  Alix/AIP-1 interacts, 294
HIV-1 infectivity, 360–361
HIV-1 Rev, 535–536
HIV-1 Vif, 535–536
HIV-associated neurocognitive disorders (HAND), 445–446
Homologous recombination pathway, 58
Host genome editing, 549–550
Host RNAs
  high-throughput RNA-sequencing, 433–434
  packaged into retrovirus particles, 551–552
  packaged mRNA, 423
  7SL RNA, 430–431
  transfer RNA (tRNA), 423–430, 424t–426t
  U small nuclear RNAs (U snRNAs), 431–432
  vault RNA, 432–433
  Y RNAs, 432
Hepatoma-derived Growth Factor Like 2 (HDGFL2), 170
HTLV-1, 436–437, 445
  entry, 361–362
HTLV-1-associated myelopathy or tropical spastic paraparesis (HAM/TSP), 24
Human endogenous retrovirus subfamily H (HERVH), 202–203
Human endogenous retrovirus W (HERV-W), 30
Human ERVs (HERVs), 437–438
Human immunodeficiency virus 1 (HIV-1), 3, 11–12
Human Staufen (hStau) protein, 280
Human T-lymphotropic virus type I (HTLV-1), 126–127, 172–173, 353–354
Hydrophobic fusion peptides, 14, 16

I

IN-binding domain (IBD), 168
Influenza A viruses, 335–338
Innate immune system
  cyclic guanosine monophosphate–adenosine monophosphate (cGAMP) synthase (cGAS), 86–87
  interferon g-inducible protein 16 (IFI16), 87
  MDA5, 86
  pattern recognition receptors (PRRs), 85
  retroelements
    direct control of, 505–507
    endogenous retrovirus sequences, 502–504, 503f
    sensing of, 504–505
  RIG-I, 86
  TREX1, 88
  TRIM5α, 85–86
Inner nuclear membranes (INMs), 114
Integrase (IN) proteins, 163, 353–354
  protein interaction, 167
  transcriptional control and latency, 201

Interferon g-inducible protein 16 (IFI16), 87
Interferon-induced transmembrane proteins, 28
Interferon-stimulated genes (ISGs), 76
Internal ribosomal entry sites (IRESs), 422
IQGAP1, 288
IQGAPs, 67

J

Jaagsiekte sheep retrovirus (JSRV), 2, 4, 25
Japanese encephalitis virus (JEV), 338
JSRV. *See* Jaagsiekte sheep retrovirus (JSRV)

K

K63-linked polyubiquitin chains, 75–76
KAP1/TRIM28/TIF1β, 498–499
KIF-4, 288, 535
Kinesins, 67–68
Kruppel-associated box (KRAB), 498–499

L

L1 retrotransposition, 500–501
Large intergenic RNA-p21, 435
"Late domains", 363–365
LEDGF/p75, 201
Lens epithelium–derived growth factor (LEDGF), 167
  ALLINI-treated cells, 181–182
  antiviral drug target, 181
  Cas9–guide RNA complexes, 168–169
  chromatin-bound protein, 169–170
  coimmunoprecipitation approach, 167–168
  electronegative and hydrophobic residues, 168
  elements, 168
  HDGFL2, 170
  IN-binding domain (IBD), 168
  intasome complex function, 181
  N-terminal chromatin-binding elements, 170
  polyacrylamide gels, 167–168
  promoter insertion model, 182
  PWWP domain, 168
  quinoline-based compounds, 180
  ribonucleoprotein (RNP) complex, 181–182
  short-hairpin RNA expression, 168–169
  viral protease, 169–170
Lentiviruses, 2, 413
  virus–host interactions, 167
LIM domain kinase, 62
Linear double-stranded DNA, 52
LINE element retrotransposition, 501–502
Lipid microdomains, 377–379
Lipid rafts, 14, 358, 359f
  retrovirus entry receptors, 5, 10
Lipids
  cellular membranes, 354–357, 356f
  membrane microdomains, 358–359, 359f
  retroviral assembly, 363–375
  retroviral cell-cell transfer
    colocalization, 375–377
    HIV-1 Gag and envelope (Env) concentrate, 375, 376f
    lipid microdomains, 377–379
    retroviral accessory proteins, 377–379
    virological synapse, 375
  retroviral entry, 360
    cholesterol, 360–362
    phospholipids, 363
    sphingolipids, 362
  retroviral replication, inhibition of, 379
    cholesterol modifying agents, 379–380
    sphingolipid synthesis, 380–381
  viral membranes, 357–358
Liposomes, 373–374
Long interspersed elements (LINEs), 483–485
Long noncoding RNA 00173, 436
Long noncoding RNA (lncRNAs), 434–439
Long terminal repeat (LTR), 58
  hormone response elements, 205–207, 206f
  retrotransposons, 480–483, 481f
  virus–host interactions, 163
Lv2, 77–78
Lyric, 538
Lys63 Ub modification, 333
LysRS, 428

M

Macropinocytosis, 65
MAP1A and MAP1S, 66
MAP4 and DNAL1, 66
Mass spectrometry (MS), 529t–534t, 536
  affinity purification (AP), 537–539
  HIV-1 Gene and protein network data, 554, 559t
  host proteins with HIV-1 proteins, 537
  liquid chromatography (LC) followed by tandem MS (MS/MS), 539–540
  multidimensional protein identification technology, 540–541
  proteins packaged into virus particles, 536–537
  viral DNA as bait, 541–542
    HIV-1 5′ untranslated regions, 541

viral RNA as bait, 542–544
full-length HIV-1 RNA, 543–544
HIV-1 subgenomic rnas, 542–543
retroviral 5′untranslated regions, 543
Matrin 3 (MATR3), 275–277, 542–543
Matrix-assisted laser desorption ionization-time of flight MS (MALDI-TOF), 536–537
MDA5, 86
Membrane fusion
CCR5 inhibitors, 14
cholesterol-rich lipid domains, 14
class I membrane fusion proteins, 13
and fusion machinery, 14, 15f
host immune recognition, 13
hydration layers, 14
hydrophobic fusion peptides, 14
lentiviral and gammaretroviral membranes, 14–15
lipid rafts, 14
methyl-beta-cyclodextrin, 14–15
peptide–HR1 helix, 14
process, 13
retroviral fusion peptides, 14
transmembrane subunits (TM) proteins, 13
trimers, 16
virus budding, 13
Membrane microdomains, 358–359, 359f
Methyl-beta-cyclodextrin, 14–15
MicroRNA (miRNA), 440–450, 442f
Microtubule affinity–regulating kinases (MARK/PAR-1), 66–67
Microtubule-organizing center (MTOC), 61, 285–286
MLV Gag, 535
MMPV Gag, 535
MMTV. *See* Mouse mammary tumor virus (MMTV)
MMTV Gag, 535
Moloney murine leukemia virus (MoMLV), 20, 166–167
MoMLV, 171–173
Transcriptional control and latency, 207–208
Monoclonal antibodies, 11–12, 67
Mouse mammary tumor virus (MMTV), 2, 4, 277
advantage, 207
DNA-binding sequences, 205
glucocorticoid receptor (GR) gene, 205–207
lactation hormones, 207
“milk factor”, 205
molecular endocrinology and breast cancer, 205–207, 206f
retrovirus entry receptors, 5
steroid-induced retroviral transcription, 205–207
tissue selective expression, 205–207
Mouse transferrin receptor-1 (mTfR1)
retrovirus entry receptors, 5–10
Mov10, 77
MOV10 activity, 262
mTfR1. *See* Mouse transferrin receptor-1 (mTfR1)
Mucosal anti-gp41 IgA, 26
Multidimensional protein identification technology, 540–541
Murine endogenous retrovirus with leucine transfer RNA primer (MERVL), 202–203
Murine leukemia viruses, 72–73
Murine leukemia virus Gag trafficking, 295
Murine leukemia virus (MLV) receptor, 5
Murine retrovirus mouse mammary tumor virus (MMTV), 401–402
MX2, 82–84
MX2, targeting capsid, 82–84
Myristoylation, 291
Myxovirus Resistance 2 (MX2), 411–412

N

Nedd4, 292, 330f, 332
Nef protein, 377–378
Neural stem cells (NSC), 500
Neuropilin type 1 (NRP-1), 12
Neurovirulence, 24–25
Next-generation sequencing (NGS), 487
NGS. *See* Next-generation sequencing (NGS)
Noncoding RNA (ncRNA), 421–422
allosteric IN inhibitors (ALLINIs), 440
endogenous retroviruses, 437–439
host RNAs
high-throughput RNA-sequencing, 433–434
packaged mRNA, 423
7SL RNA, 430–431
transfer RNA (tRNA), 423–430, 424t–426t
U small nuclear RNAs (U snRNAs), 431–432
vault RNA, 432–433
Y RNAs, 432
long noncoding RNA (lncRNAs), 434–439
miRNAs, 440–450

Noncoding RNA *(Continued)*
ribosomal RNA (rRNA), 421–422
RNA interference (RNAi), 440–450
TAR RNA-binding protein (TRBP), 439
therapeutics, 450–453
transfer RNA (tRNA), 421–422
untranslated regions (UTRs), 421–422
Nonhomologous end joining pathway, 58
Non-long terminal repeat retrotransposons
chimeric short interspersed elements, 485–486
long interspersed elements (LINEs), 483–485
short interspersed elements (SINEs), 485
Nonprotein-coding RNA repressor of NFAT (NRON), 434–435
N-terminal domain (NTD), 165–166
Nuage, 493–494
Nuclear envelope (NE), 114
Nuclear export signals (NES), 114–115, 278
Nuclear export, viral RNAs
constitutive transport element, MMTV, 277
Crm1, 275–277
DDX3, 275–277
HIV-1, 275–277
Matrin 3 (MATR3), 275–277
posttranscription processing, 275
Rev, 275–277
RNA trafficking in simple *vs.* complex retroviruses, 275–277, 276f
splicing factors, 275
spumaretrovirus RNA nuclear export, 276f, 278
unspliced RNA, in Rous sarcoma virus (RSV), 278
unspliced viral RNA export, 276f, 277
Nuclear localization signal (NLS), 4, 114–115, 284
Nuclear paraspeckle assembly transcript 1 (NEAT1), 435
Nuclear pore complexes (NPCs), 114
Nuclear pore complex (NPC)
flexible phenylalanine–glycine (FG)-repeat regions
germline mutations, 139–140
HIV-1 replication, 139
host factors, 142
nuclear pore complex (NPC) remodel, 142–143
phylogeny mapping, 139–140, 141f
N-terminal GLFG domain, 139–140
transport receptors, 139
Gag-interacting proteins, 177
Nup62, 132–133
Nup98, 127–128
Nup107–160, 129–131
Nup153, 134–135
Nup214/CAN, 126
Nup358/RanBP2, 124–125
tetratricopeptide repeat (Tpr), 137
Nuclear RNA helicases
ATP-dependant catalytic activity, 253–255
cap–binding proteins
eIF4E, 266–267
nonsense RNA–mediated decay, 265–266
catalytic helicase core, 255–257
and cognate retroviral RNA, 267
cytoplasmic RNA viruses, 254t
DHX9. *See* DHX9
double-stranded RNA-binding domains (dsRBDs), 257
gene expression, 253, 255
MOV10 activity, 262
nonspecific RNA binding, 257
nucleocytoplasmic transport, 264–265
nucleotide triphosphate (NTP) binding, 255
"omics technology", 255, 258–259
retrovirus–RNA helicase interface, 255
reverse transcription and integration, 259
ribonucleoprotein complexes (RNPs), 255
RNA binding domains, 253–255
substrate RNAs and protein cofactors, 255
Nuclear trafficking, 284–285, 285f
Nucleic acid-based methods, 554, 555t–558t
Nucleocapsid protein (NC), 284–285
Nucleoporins (Nups)
eukaryotic cellular metabolism, 114
flexible phenylalanine–glycine (FG)-repeat regions, 114–115
germline mutations, 139–140
HIV-1 replication, 139
host factors, 142
nuclear pore complex (NPC) remodel, 142–143
phylogeny mapping, 139–140, 141f
N-terminal GLFG domain, 139–140
transport receptors, 139
function, 115, 116t–117t
inner nuclear membranes (INMs), 114
multifunctional Nup depletion, 138–139
nuclear envelope (NE), 114

nuclear export signals (NES), 114–115
nuclear localization signal (NLS), 114–115
nuclear pore complexes (NPCs), 114
nucleocytoplasmic transport, 115
retroviral replication
categories, 115–117
HIV-1 interaction, 117–124, 123f
host protein functions, 115–117
Nup62, 132–133
Nup98, 127–129
Nup107–160, 129–132
Nup153, 134–137
Nup155, 134
Nup214/CAN, 126–127
Nup358/RanBP2, 124–125
tetratricopeptide repeat (Tpr), 137–138
tumorigenesis, 117–124
transport receptors/karyopherins, 114–115
viruses replication, 115, 118t–122t
Nucleotidase, 81–82
Nucleotide triphosphate (NTP) binding, 255
Nup62
functions, 133
nuclear pore complex (NPC), 132–133
Nup85 and Nup160 components, 129–131
Nup96, 130
Nup98
functions, 128–129
nuclear pore complex (NPC), 127–128
Nup107–160
functions, 131–132
nuclear pore complex (NPC), 129–131
Nup133, 129–130
Nup153
functions, 135–137
nuclear pore complex (NPC), 134–135
Nup155, 134
Nup214/CAN
functions, 126–127
nuclear pore complex (NPC) protein, 126
Nup358 component, 60
Nup358/RanBP2
functions, 125
nuclear pore complex (NPC), 124–125
NXF1/TAP, 277

O

O-glycosylated proteins, 132
Oligomerization, 75
"Omics technology", 255, 258–259
Open reading frames (ORFs), 480–482
Orthomyxoviruses, 335–338
Ovine pulmonary adenocarcinoma, 25

P

p300 acetyltransferase, 71
Pararetroviruses, 423–427
"Particle-to-pfu" ratio, 56
Pathogen-associated molecular patterns (PAMPs), 86, 402–404, 406
Pattern recognition receptors (PRRs), 85, 406
PDZD8, 60
PF74, 88
PFV Gag, 285–286, 295
Phosphatidylinositol (4,5) bisphosphate ($PI(4,5)P_2$), 333, 429
retroviral assembly
EIAV Gag, 371–373
Gag-leucine zipper constructs (GagLZ), 371–373
HIV-1 assembly, 370–371
HIV-1 Gag association, 371, 372f
liposomes, 373–374
MA domains, 371–373
phosphoinositides, 370
phospholipase C delta 1 (PLCδ 1), 370
WT HIV-1 Gag, 373
Phosphoinositides, 370
Phospholipids, 358
in retroviral entry, 363
Phosphomimetic mutation (T592E), 82
PIWI proteins, 492
Plasma membrane (PM), 354
Polo-like kinases (PLKs), 180, 286
Polyacrylamide gels, 167–168
Polypurine tract (PPT), 56–58
Positive transcription elongation factor b (P-TEFb), 263
Preintegration complex (PIC), 55, 285–286, 429–430
transcriptional control and latency, 201
virus–host interactions, 163–165
Preintegration complex (PICs), 125
Primer-binding site (PBS), 56–58, 261–262
PriRNAs, 493
PRMT5, 497
Proteasome-independent methods, 79–80
Protein kinase R (PKR), 422
Protein phosphatase 2A (B′-PP2A)
BRD2–4 proteins, 175–176
B-type regulatory subunit, 175

Protein phosphatase 2A (B′-PP2A) *(Continued)*
chromatin-associated proteins, 173–175
cytotoxic T-cell response, 176
dephosphorylation, 173–175
holoenzyme, 173–175, 174f
LEDGF/p75, 175–176
phosphorylation and methylation, 175
protein phosphatase 1 (PP1), 173–175
δ-retroviruses, 175–176
Protein-protein interactions
Gag proteins as bait, 528–535
HIV-1 Rev, 535–536
HIV-1 Vif, 535–536
integrase as bait, 535
mass spectrometry (MS) approaches, 529t–534t, 536
affinity purification (AP), 537–539
host proteins with HIV-1 proteins, 537
liquid chromatography (LC) followed by tandem MS (MS/MS), 539–540
multidimensional protein identification technology, 540–541
proteins packaged into virus particles, 536–537
viral DNA as bait, 541–542
viral RNA as bait, 542–544
Traid 3, 535–536
two-hybrid screen, 528, 529t–534t
Proteomics, 536
Prototype foamy virus (PFV), 278
Prototype foamy virus type 1 (PFV-1), 445
Pseudogenes, 486

R

Rag1/2 proteins, 487–488
Restriction factor, 72–73, 76–77
Retinoic acid-inducible gene I protein (RIG-I), 504–505
Retroelements, 507–509
cancer initiation and progression, 502
classes, 480
endogenous retroviruses (ERVs), 480–483, 481f
germline and embryonic stem cells
cytoplasmic RNA granules, 493–494
DNA methylation, 494–495
DNA methyltransferases, 496
MIWI2, 495
nuage, 493–494
piRNAs, 493, 495–496
PIWI proteins, 492
PRMT5, 497
innate immune system
direct control of, 505–507
endogenous retrovirus sequences, 502–504, 503f
sensing of, 504–505
long terminal repeat (LTR) retrotransposons, 480–483, 481f
neuronal activity, 499–502
non-long terminal repeat retrotransposons
chimeric short interspersed elements, 485–486
long interspersed elements (LINEs), 483–485
short interspersed elements (SINEs), 485
self-preservation, 488–490
sequence-specific suppression, 498–499
somatic cells by RNA interference (RNAi)
attenuated DNA methylation, 492
innate immune signaling, 491f
LINE and SINE RNAs, 492
pre-miRNA, 490
small noncoding (nc)RNAs, 490
Retrotransposition, 486
Retrotransposons, 427–428
Retroviral accessory proteins, 377–379
Retroviral assembly
Gag binding to giant unilamellar vesicles, 374–375
Gag-membrane interactions, 366–367
HIV-1 Gag proteins, 363–365, 365f
HIV-1 virions, 363–365, 364f
lipids in
cholesterol, 368–370
$PI(4,5)P_2$, 370–374
trafficking of Gag and sites, 365–366
Retroviral cell-cell transfer
colocalization, 375–377
HIV-1 Gag and envelope (Env) concentrate, 375, 376f
lipid microdomains, 377–379
retroviral accessory proteins, 377–379
virological synapse, 375
Retroviral entry, lipids, 360
cholesterol, 360–362
phospholipids, 363
sphingolipids, 362
Retroviral Env glycoprotein, 354
Retroviral fusion peptides, 14
Retroviral GAG protein trafficking
Alix/AIP-1 interactions, HIV Gag proteins, 294
cytoskeletal proteins, 288–289
E3 ubiquitin ligases, 292

EIAV p9, 294–295
endophilin 2, murine leukemia virus Gag trafficking, 295
foamy virus Gag association, 285–286
host factors, 281, 282t–283t
microtubule-organizing center, 288–289
MLV, 295
multivesicular bodies and endosomal proteins, 285f, 289–290
multivesicular body and endosomal sorting complexes, 291–292
myristoylation and lipid binding, 291
Nedd4, 292
nuclear trafficking of, 284–285, 285f
PFV Gag, 295
transport of, 286–288
tRNA synthetases, 296
Tsg101, 292–294
Retroviral genome trafficking, in cytoplasm, 279
retroviral genomic RNA, dimerization of, 281
unspliced viral RNA transport, 279–280
viral RNA export *vs.* virus assembly, 279
Retrovirus uncoating and reverse transcription
actin cytoskeleton, 62–64
applications, 88–89
capsid protein (CA), 53, 55
CD4 receptor and CXCR4 coreceptor, 62
cell surface plasma membrane, 53–54
cortical F-actin, 64
cyclophilin A (CypA)
African green monkeys (AGM), 69–70
cyclosporine efficacy, 68–69
Jurkat cells, 69
nuclear entry, 70
proline residue (P90), 68
structure of, 68
wild-type virus, 69
cytoskeleton and motors, trafficking
capping proteins, 66
caveola-mediated coated vesicles, 65
dyneins, 65, 67
ecotropic MLVs, 65
fasciculation and elongation protein zeta-1 (FEZ1), 66–67
IQGAPs, 67
kinesins, 67–68
macropinocytosis, 65
MAP1A and MAP1S, 66
MAP4 and DNAL1, 66
microtubule affinity–regulating kinases (MARK/PAR-1), 66–67
microtubule networks, 65–66
monoclonal antibodies, 67
peptides, 65
retrovirus cores, 65
stable MTs (sMTs), 65–66
TRIM5α and TRIM-Cyp restriction factors, 67
deoxynucleotides concentration, 54
dependency factors, 52–53
diaphanous interacting protein (DIP), 63–64
DNA synthesis in vivo, 54
endocytosis, 53–54
eukaryotic elongation factor 1, 64
gelsolin, 64
HIV-1 envelope interaction, 63
HIV-1 Nef protein, 53–54, 63
host factors, 62, 63f
deoxynucleotides concentration, 54
dependency factors, 52–53
diaphanous interacting protein (DIP), 63–64
DNA synthesis in vivo, 54
endocytosis, 53–54
eukaryotic elongation factor 1, 64
gelsolin, 64
HIV-1 envelope interaction, 63
HIV-1 Nef protein, 53–54, 63
host factors, 62, 63f
imaging
bis-arsenical fluorescein derivative, 62
cyclophilin A (CypA-DsRed), 61
FISH, 61
Gag proteins, 62
microtubule-organizing center (MTOC), 61
N-terminal peptide, 61–62
Vpr–GFP, 61
imaging/biochemical detection assays, 56
infection stages
APOBECs, 78–81
factors, 71, 72f
Fv1, 72–74
Lv2, 77–78
Mov10, 77
MX2, 82–84
sterile α motif and histidine-aspartate domain–containing protein 1 (SAMHD1), 81–82
SUN2, 84
TLR7/8 receptors, 84–85
TRIM5α, 74–76
TRIM-Cyp, 76–77
innate immunity. *See* Innate immune system
knockdown/knockout methods, 52–53

Retrovirus uncoating and reverse transcription *(Continued)*
LIM domain kinase, 62
linear double-stranded DNA, 52
nonsense-mediated decay pathway, 64
nuclear entry process, 55–56
nuclear pore component (Nup358), 55–56
overexpression, 54
"particle-to-pfu" ratio, 56
restriction factors, 52–53
simian immunodeficiency virus (SIV), 63–64
tRNA primer, 54
viral genomic RNA, 52
viral nucleic acids
5-ethynyl uridine–tagged viral RNA, 56
Alu-LTR PCR, 58–59
double-stranded linear DNA, 56–58
HIV-1 infection, 59
homologous recombination pathway, 58
long terminal repeat (LTR), 58
minus-strand strong stop DNA, 56–58
nonhomologous end joining pathway, 58
nuclease pretreatment, 58–59
polypurine tract (PPT), 56–58
primer-binding site (PBS), 56–58
process, 56–58, 57f
reverse transcriptase polymerase chain reaction (RT-PCR), 56
viral proteins
CA protein, 71
major homology region (MHR), 70–71
p12 mutants, 71
p300 acetyltransferase, 71
PIASy-binding site, 70–71
replication-competent genome, 70–71
virion membrane, 53
virus life cycle, 52
zinc finger nucleocapsid protein (NC), 53
Rev, 275–277
Reverse Transcriptase Complex (RTC), 55
Reverse transcriptase polymerase chain reaction (RT-PCR), 56
Reverse transcriptase (RT), 353–354
Rev response element (RRE), 231, 542
Rev/RRE-dependent export, 279
Ribonucleoprotein (RNP) complex, 181–182, 274–275
Ribosomal RNA (rRNA), 421–422
RIG-I, 86
RIG-I-like receptors (RLR), 504–505
RING domain, 75–76
RNA-associated early-stage antiviral factor (REAF), 77–78
RNA-binding protein HuR, 64
RNA interference (RNAi), 422, 440–450
and genome-wide screens
metaanalysis, 547
retrotransposition in yeast, 549
short hairpin RNA screens, 546–547
small interfering RNA screens, 544–546
targeted RNAi screens, 547–549
somatic cells, 490–492, 491f
RNA polymerase II, 210–211, 480–482
Tat protein, 212–213, 213f
RNA polymerase II (RNAP II), 200
RNA processing control
and *cis*-elements, 229–230, 230f
CRM1 pathway, 231
heterogenous nuclear ribonucleoproteins (hnRNPs), 230–231. *See also* Heterogenous nuclear ribonucleoproteins (hnRNPs)
Northern blots, 229–230
Rev response element (RRE), 231
secondary structure, 230–231
serine–arginine (SR) proteins, 230–231
Gag protein, 236–238
GAR ESE, 236–240
HeLa/HEK 293 cells, 239
Nef synthesis, 239
overexpression assays, 239
RNA polymerase II, 240
role of, 236, 237t
splicing enhancers, 236
SRSF1 (SF2/ASF), 236–238
SRSF2 (SC35), 238
small molecules
complexity of, 240
HAART treatment, 240–243
multiple indole derivatives, 240–243
regulatory factors, 240
Rev function, 244
splicing factors, 240
SRPK inhibitor, 243–244
structural diversity, 244–245
structure and function, 240, 241t–242t
RNA sequencing techniques, 552–554
RNA silencing (SRS), 450
RNA trafficking, 275–277, 276f
Rous sarcoma virus (RSV) Gag protein, 126–127
RSV Gag protein, 278, 284

S

Saccharomyces cerevisiae, 493–494
SAM domain and HD domain-containing protein 1 (SAMHD1), 409
sCD4-bound trimers, 20
Sequence-specific suppression, 498–499
Serine–arginine (SR) proteins, 230–231
  Gag protein, 236–238
  GAR ESE, 236–240
  HeLa/HEK 293 cells, 239
  Nef synthesis, 239
  overexpression assays, 239
  RNA polymerase II, 240
  role of, 236, 237t
  splicing enhancers, 236
  SRSF1 (SF2/ASF), 236–238
  SRSF2 (SC35), 238
Serine-/threonine-enriched glycoprotein. *See* Nup153
Short interspersed elements (SINEs), 485
Siglec-1, 12
Simian foamy virus (SFV), 448–449
Simian immunodeficiency virus (SIV), 63–64, 402
Simian immunodeficiency virus type 1 (SIV-1), 17
Simple retroviruses, 277
SIV-1. *See* Simian immunodeficiency virus type 1 (SIV-1)
SIV-1 zoonotic transmission, 29
SKIV2L helicase, 505–506
Skp1–Cullin–Fbox complexes, 79–80
7SL RNA, 430–431
Small-hairpin RNAs (shRNAs), 450–451
Sphingolipids, 354–355
  in retroviral entry, 362
Sprouty 2 (Spry2), 322–323
Spumaretrovirus RNA nuclear export, 276f, 278
Sterile α motif and histidine-aspartate domain–containing protein 1 (SAMHD1), 81–82
Sterols, 354–355
SUMOylation, 70–71, 86
SUN2, 84
Suppressor of cytokine signaling 1 (SOCS1), 289–290
Surface subunit (SU) conformation, 18
  disulfide bond isomerization, 18–19
  lentiviruses, 19–20, 21f

T

Tandem affinity purification (TAP), 537–538
Target-site duplications (TSD), 480–482
TAR RNA-binding protein (TRBP), 439
TATA binding protein (TBP), 203–204
Tat protein
  P-TEFb, 213–214
  RNA polymerase II, 212–213, 213f
T cell immunoglobulin and mucin domain-1 (TIM-1), 13
T-cell membrane protein lymphocyte function, 23–24
TE. *See* Transposable elements (TE)
N-Terminal receptor-binding domain (RBD), 2
Tetratricopeptide repeat (Tpr)
  functions, 137–138
  nuclear pore complex (NPC), 137
Thiol-exchange disulfide isomerases, 20–22
Threonine 592, 82
TIM-1. *See* T cell immunoglobulin and mucin domain-1 (TIM-1)
TLR7/8 receptors, 84–85
Toll-like receptor 2 (TLR2), 402–404
Toll-like receptor 4 (TLR4), 402–404
Toll-like receptors (TLRs), 504–505
Traid 3, 535–536
Transcriptional coactivator p75, 167
  ALLINI-treated cells, 181–182
  antiviral drug target, 181
  Cas9–guide RNA complexes, 168–169
  chromatin-bound protein, 169–170
  coimmunoprecipitation approach, 167–168
  electronegative and hydrophobic residues, 168
  elements, 168
  HDGFL2, 170
  IN-binding domain (IBD), 168
  intasome complex function, 181
  N-terminal chromatin-binding elements, 170
  polyacrylamide gels, 167–168
  promoter insertion model, 182
  PWWP domain, 168
  quinoline-based compounds, 180
  ribonucleoprotein (RNP) complex, 181–182
  short-hairpin RNA expression, 168–169
  viral protease, 169–170
Transcriptional control and latency
  alpharetrovirus avian leukosis/sarcoma virus (ASLV), 200–201
  alpharetrovirus genus, 204–205
  avian leucosis virus (ALV), 204–205
  “binding factor A”, 207–208

Transcriptional control and latency (*Continued*)
chromatin-binding properties, 201
complementary DNA (cDNA), 200
deltaretrovirus human T cell leukemia virus (HTLV)-1, 200–201
de novo DNA methyltransferases, 207–208
endogenous retroviruses (ERVs), 200. *See also* Endogenous retroviruses (ERVs)
eukaryotic transcription, 217
gammaretrovirus, 200–201
HIV-1
3′ end processing, 214
5′ long terminal repeat, 210, 211f
CDK11, 214
latent infection, 214–217
proviral transcription, 209–210
P-TEFb, 213–214
RNA polymerase II, 210–211
Tat protein, 212–213, 213f
HIV-1 provirus, 200
host cell genome, 200–201
integrase (IN) proteins, 201
LEDGF/p75, 201
MoMLV provirus, 207–208
mouse mammary tumor virus (MMTV)
advantage, 207
DNA-binding sequences, 205
glucocorticoid receptor (GR) gene, 205–207
lactation hormones, 207
"milk factor", 205
molecular endocrinology and breast cancer, 205–207, 206f
steroid-induced retroviral transcription, 205–207
tissue selective expression, 205–207
multiple histone-modifying enzymes, 207–208
nuclear pore structures, 201
preintegration complex (PIC), 201
retroviral genome complexity, 203–204, 204f
retroviral silencing, 207–208
retroviral transcription, 217–218
RNA polymerase II (RNAP II), 200
TATA binding protein (TBP), 203–204
transactivator proteins, 208–209
tripartite motif-containing 28 (TRIM28), 207–208
Transcriptional start sites (TSSs), 173
Transfer RNA (tRNA), 421–422
HIV-1 Gag, 429
LysRS, 428
maturation, 427
mutagenesis, 428–429
pararetroviruses, 423–427
phosphatidylinositol (4,5) bisphosphate ($PI(4,5)P_2$), 429
preintegration complex (PIC), 429–430
primer-binding site (PBS), 423–427
primer enrichment, 428
primers for retroviral reverse transcription, 423–427, 424t–426t
retrotransposons, 427–428
tRNA-derived fragments (tRFs), 427
$tRNA^{Pro}$ (tRF-3019), 423–427
virus-like particles (VLPs), 427–428
WT PBS-Trp, 429
Transmembrane subunits (TM) proteins, 13
Transposable elements (TE), 480
TREX1, 88
TRIM5α, 74–76, 85–86
CA residues, 74
hexameric array, 75
HIV-1-based vectors, 74
humans and nonhuman primates, 74
interferon-stimulated genes (ISGs), 76
K63-linked polyubiquitin chains, 75–76
mechanism, 75
microtubules and dynein motors, 75–76
murine Fv1 factor, 74
oligomerization, 75
proteasome inhibitor, 76
RING domain, 75–76
TRIM5α and TRIM-Cyp restriction factors, 67
TRIM-Cyp, 76–77
Tripartite motif-containing 28 (TRIM28), 207–208
N-Tropic strains, 72–73
B-tropic viruses, 72–73
Tsg101-UEV domain, 323–325
Tsg101-UEV PTAP-binding pocket, 325
Tsg101-UEV Ub-binding pocket, 325–326
Tumor suppressor gene 101 (Tsg101), 292–294, 320
budding
Co-opt Tsg101, virus replication, 335, 336t–337t
flaviviruses, 338
hepatitis C virus (HCV), 338
orthomyxoviruses, 335–338
cellular factors
Alix, 329–331, 330f
convergence of, 332–335
Nedd4, 330f, 332
Ub, 330f, 331–332

participation
cellular endocytic pathway, 328–329
virus budding, 329
structure
domains, 323, 324f
Tsg101-UEV domain, 323–325
Tsg101-UEV PTAP-binding pocket, 325
Tsg101-UEV Ub-binding pocket, 325–326, 327f
Ub- and PTAP-binding pockets, 326–328, 328f
Tupaia belangeri, 76–77
Two-dimensional differential in-gel electrophoresis experiments (2D-DIGE), 536
Two-hybrid screens, 528–536, 529t–534t
Ty3/Gypsy, 480–482

U

UAP56, 277
UBAP1, 320
Ubc9, 288–289
Unspliced viral mRNA, 274–275
Unspliced viral RNA export, 276f, 277
Untranslated regions (UTRs), 421–422
U small nuclear RNAs (U snRNAs), 431–432
Up-frameshift protein 1 (UPF1), 262

V

Vault RNA, 432–433
VEGF receptor, 12
Viral membranes, 357–358
Viral miRNAs, 446–447
ALV and RSV, 448
bovine foamy virus (BFV), 448
bovine leukemia virus (BLV), 447–448
HIV-1, 449
simian foamy virus (SFV), 448–449
Viral pathogenesis, 2
Viral proteins
atomic force microscopy, 59–60
binding sites, 60
covalent modifications, 70
cyclophilin A, 60
endogenous reverse transcription reactions, 60
“fate of capsid” assay, 59–60
Gag and Pol precursors, 70
nuclear entry, 60
NUP358 component, 60
PDZD8, 60
sedimentation analysis, 59
size and structure, 59
SUMOylation, 70–71
suppressor mutations, 59–60
viral virion infectivity factor (Vif) protein, 59–60
western blot, 59
Viral suppressors, 450
Viral virion infectivity factor (Vif) protein, 59–60
Virological synapse, 375
Virus budding, 322–323
tumor suppressor gene 101 participation, 329
Virus–host interactions
bromodomain and extra-terminal domain (BET) proteins, 171–172
catalytic core domain (CCD), 165–166
C-terminal domain (CTD), 165–166
deltaretroviruses, 172–173
enzymatic activities, 163–165
foamy virus, 179–180
Gag-interacting proteins, 177
gammaretroviruses, 170–171
genomic distributions, 166–167, 166t
HIV-1 CA–interacting proteins
CFIm complex, 179
CPSF6 knockout cells, 178–179
CypA inhibitor cyclosporine A, 177–178
factors, 177–178
gene-tropic integration, 178
LEDGF/p75 knockdown, 178–179
N57A and N74D mutant viruses, 177–178
integrase (IN), 163
protein interaction, 167
lens epithelium-derived growth factor/ transcriptional coactivator p75 (LEDGF/p75)
ALLINI-treated cells, 181–182
antiviral drug target, 181
Cas9–guide RNA complexes, 168–169
chromatin-bound protein, 169–170
coimmunoprecipitation approach, 167–168
electronegative and hydrophobic residues, 168
elements, 168
HDGFL2, 170
IN-binding domain (IBD), 168
intasome complex function, 181
N-terminal chromatin-binding elements, 170
polyacrylamide gels, 167–168

Virus–host interactions *(Continued)*
promoter insertion model, 182
PWWP domain, 168
quinoline-based compounds, 180
ribonucleoprotein (RNP) complex, 181–182
short-hairpin RNA expression, 168–169
viral protease, 169–170
lentiviral integration target, 164f, 165–166
lentiviruses, 167
long terminal repeat (LTR), 163
Moloney murine leukemia virus (MoMLV), 166–167
N-terminal domain (NTD), 165–166
nuclear architecture and chromatin structure, 166–167
preintegration complex (PIC), 163–165
protein phosphatase 2A (B′-PP2A). *See* Protein phosphatase 2A (B′-PP2A)
γ-retroviral integration, 165–166, 165f
stable synaptic complex (SSC), 163–165
Virus-like particles (VLPs), 281, 480–482
Vpr–GFP, 61
Vps22, 320
Vps23, 328–329
Vps25, 320
Vps27, 328–329
Vps36, 320
Vpu protein, 378

X

Xenotropic-like murine retrovirus infection, 402
Xenotropic murine leukemia virus (XMLV), 402

Y

Y-complex. *See* Nup107–160
Y RNAs, 432

Z

Zfp111. *See* Zinc finger protein 111 (Zfp111)
Zinc finger antiviral protein (ZAP), 297
Zinc finger nucleocapsid protein (NC), 53
Zinc finger protein 111 (Zfp111), 25
Zoonosis, 402

FSC
www.fsc.org
MIX
Papier | Fördert gute Waldnutzung
FSC® C083411

Kolibri 360 GmbH
Ferdinand-Jühlke-Straße 7
99095 Erfurt, Deutschland
produktsicherheit@kolibri360.de